Internet and Smartphone Use-Related Addiction Health Problems

Internet and Smartphone Use-Related Addiction Health Problems: Treatment, Education and Research

Editor

Olatz Lopez-Fernandez

MDPI • Basel • Beijing • Wuhan • Barcelona • Belgrade • Manchester • Tokyo • Cluj • Tianjin

Editor
Olatz Lopez-Fernandez
Fundacion Jiménez Díaz University Hospital Health Research Institute
Spain

Editorial Office
MDPI
St. Alban-Anlage 66
4052 Basel, Switzerland

This is a reprint of articles from the Special Issue published online in the open access journal *International Journal of Environmental Research and Public Health* (ISSN 1660-4601) (available at: https://www.mdpi.com/journal/ijerph/special_issues/Internet_Smartphone_Addiction).

For citation purposes, cite each article independently as indicated on the article page online and as indicated below:

LastName, A.A.; LastName, B.B.; LastName, C.C. Article Title. *Journal Name* **Year**, *Volume Number*, Page Range.

ISBN 978-3-0365-1274-7 (Hbk)
ISBN 978-3-0365-1275-4 (PDF)

Contents

About the Editor

Olatz Lopez-Fernandez, Ph.D. She is a Researcher at Fundación Jiménez Díaz University Hospital Health Research Institute (Spain). Her career has spanned positions in Spanish and international universities including University of Barcelona, Catholic University of Louvain, Nottingham Trent University, and Monash University. She is a health and educational sciences researcher with interests in Internet use-related addiction problems, humanistic approach to health sciences, mixed-methods applied research, and educational technology in higher education.

Preface to "Internet and Smartphone Use-Related Addiction Health Problems: Treatment, Education and Research"

A Special Issue on health and educational effects due to problematic Internet and smartphone use has been tackled in the International Journal of Environmental Research and Public Health (IJERPH) titled: "Internet and Mobile Phone Addiction: Health and Educational Effects", developed from 2019–2020. It is the second edition of the previous Special Issue "Internet and Mobile Phone Addiction: Health and Educational Effects", which was developed between 2017 and 2018. The recognition of these Internet use-related addiction problems is now formal, after one of these problems became accredited by the public health organization, World Health Organization (i.e., gaming disorder). However, the debates continue, and the research is quite prolific. For detailed information on the Special Issue referred to, you can visit https://www.mdpi.com/journal/ijerph/special_issues/Internet_Smartphone_Addiction. This Special Issue presents some of the main emerging research on technological topics on health and education approaches to Internet use-related problems, before and during the beginning of coronavirus disease 2019 (COVID-19). The objective is to provide an overview to facilitate a comprehensive and practical approach to these new trends to promote research, interventions, education, and prevention. It contains 40 papers, four reviews and thirty-five empirical papers and an editorial introducing everything in a rapid review format. Overall, the empirical ones are relational, associated with specific behavioral addictive problems with individual factors, and a few with contextual factors, generally in adult populations. Many have adapted scales to measure these problems, and a few cover experiments and mixed methods studies. The reviews tend to be about the concepts and measures of these problems, intervention options, and prevention. The Special Issue covers the following Internet use-related addiction problems: Internet addiction or problematic Internet use, smartphone addiction or problematic mobile phone use, and other specific problems, such as (video) gaming disorder, social media addiction or Facebook addiction, cybersex addiction or problematic usage of pornography, online gambling or eGambling, buying addiction or problematic Internet shopping, among other related behavioral and excessive problems. In summary, it seems that these are a global culture trend impacting health and educational domains. Internet use-related addiction problems have emerged in almost all societies, and strategies to cope with them are under development to offer solutions to these contemporary challenges, especially during the pandemic situation that has highlighted the global health problems that we have and how to holistically tackle them.

Olatz Lopez-Fernandez
Editor

International Journal of
Environmental Research and Public Health

MDPI

Article

Emerging Health and Education Issues Related to Internet Technologies and Addictive Problems

Olatz Lopez-Fernandez [1,2]

1 Turning Point, Eastern Health Clinical School, Monash University, 110 Church Street, Richmond, VIC 3121, Australia; Lopez.Olatz@gmail.com or Olatz.Lopez@quironsalud.es
2 Fundación Jiménez Díaz University Hospital Health Research Institute, Avda. Reyes Católicos 2, 28040 Madrid, Spain

Abstract: This timely editorial paper outlines some of the main emerging research on technological topics on health and education approaches to Internet use-related problems before and during the beginning of coronavirus disease 2019 (COVID-19). Background: The aim is to provide a brief overview to facilitate a rapid comprehensive and practical approach to these new trends to promote research, interventions, education, and prevention. Methods: The rapid review includes an analysis of both health and education technologies studies on Internet use-related addiction problems included in the Special Issue "Internet and Smartphone Use-Related Addiction Health Problems: Treatment, Education and Research" to extract recent findings and a few reflections about the development of the field before and during the first wave of the COVID-19. Results: Main findings highlighted studies which tended to be empirical, with a relational type associating specific addictive problems with individual and a few contextual factors in adult populations. Psychometric studies about scales are prevalent, but predictive and mixed methods ones are starting to emerge, together with reviews about conceptualisation, measure, treatment, and prevention. Conclusions: From the arrival of Internet, our societies have converged in a global culture which has impacted health and educational domains. Internet use-related addiction problems have globally emerged and common knowledge, advances, and strategies exist to overcome challenges which are starting to be tested, and prevention interest has arisen in a pandemic situation with global health problems holistically tackled.

Keywords: Internet addiction; Internet problematic use; problematic mobile phone use; Internet use-related addiction problems; technologies; education; health; treatment; prevention; COVID-19

Citation: Lopez-Fernandez, O. Emerging Health and Education Issues Related to Internet Technologies and Addictive Problems. *Int. J. Environ. Res. Public Health* **2021**, *18*, 321. https://doi.org/10.3390/ijerph18010321

Received: 4 December 2020
Accepted: 31 December 2020
Published: 4 January 2021

1. Introduction

Over the last 25 years there has globally been an increasing use of Internet technologies and electronic devices in our daily lives, especially during the last two decades according to the International Telecommunication Union (ITU [1]). The Internet, usually defined as a set of Information and Communication Technologies (ICTs), is now a global network which has inter-connected more than half of the world population [1], who are informed, and constantly communicated (e.g., via social media, messaging applications (apps)) through domestic and mobile devices (e.g., smartphone). However, negative aspects have commonly emerged producing health, educational, and social problems, such as intercultural or intergenerational gaps (e.g., in African countries there is a dual digital divide [2], and older generations still live the grey digital divide [3]).

Our societies have progressively changed, having become more digital in many ways despite the online inequalities. It has especially impacted our human behaviours because of the ubiquity of the Internet on our everyday life activities [4]. This increase of online uses has had important benefits for health (e.g., eHealth), education (e.g., eLearning), and social life (e.g., social media) sectors, but problems have arisen (e.g., Internet use-related addiction problems, cyberbullying, or cybercrime). Furthermore, the coronavirus disease 2019 (COVID-19) has unexpectedly impacted our societies, strengthening the role of ICT in our new restrictive daily lives to support the rapid public health response worldwide [5].

In the psychological field, the Internet has transformed some of our regular behaviours which have been enriched due to the new contexts we live in (i.e., online, off-line, and both) and opportunities have emerged (e.g., to working remotely via the Internet from home or any other place, to enjoy streaming TV shows online through laptops, smartphones, etc.). Individuals now manage themselves in online and offline domains indistinctively using many devices and apps. The psychology of the Internet (or Cyberpsychology), therefore, is studying the relationship between ICT and the psychological mechanisms, covering both positive and negative issues in a cyberspace which is digitally and humanistic mediated. According to Wallace [6], for instance, the taxonomy of online environments includes the web, the deep (or dark) web, email, forums, chats, blogs, social networks, texting and virtual reality, which are almost all usually managed by individuals connected through the Internet, especially through smartphones.

The COVID-19 pandemic has made these environments part of our daily lives and pushed us to use ICT in great parts of our daily routines, work or study duties, and entertainment, and to become a part of our personal and social lives. These online settings are, therefore, influencing our daily emotions, cognitions, and behaviours, such as the case of some excessive online habits which put a small proportion of the population (e.g., youth) at risk of developing addiction problems. Internet addiction (IA) or problematic Internet use is understood as a set of potential addictive online problems [7] (reason why in this paper it is used the term "Internet use-related addiction problems"); but the only formally recognised problem is (video) gaming disorder (GD [8]). However, a broader spectrum of Internet use-related addiction problems have largely been researched beyond GD [9], mainly through descriptive and relational studies (i.e., studies which look for associations among variables) about social media addiction (SMA; e.g., Facebook addiction), problem mobile phone use (PMPU; e.g., problem smartphone use (PSU)), cybersex addiction (e.g., problematic usage of pornography (PUP)), online gambling (OG), buying addiction (e.g., problem Internet shopping (PIS)), among other related problems (e.g., cyberchondria).

Reasons which drive the generation of this second Special Issue "Internet and Smartphone Use-Related Addiction Health Problems: Treatment, Education and Research" after the successful previous Special Issue "Internet and Mobile Phone Addiction: Health and Educational Effects" [10]. The present introductory article offers a rapid overview of the new advances on Internet use-related addiction problems collected from health and education approaches before and during COVID-19, and to offer a few editorial reflections from environmental and cyberpsychology perspectives.

2. Materials and Methods

This study is a rapid review (i.e., rapid evidence assessment), an evidence-based knowledge synthesis. The systematic review process components are simplified to produce information as a summary of the evidence discovered promptly. The aim was to examine articles included in the cited Special Issue (2019–2020), which were qualitatively reviewed by the author (i.e., invited editor) through a content analysis to provide a rapid outline of the state of the art on these addictive problems associated with using current technologies from a health, educational, or both settings simultaneously. The procedure was the following: (i) to collect all published papers within the Special Issue, (ii) to design a set of variables in a table to rapidly review them one by one (i.e., see Results section and Table 1), (iii) to read article by article extracting all the information previously included in the table (i.e., columns), and (iv) to write the information extracted in the table.

The methodological aim of this type of research study is to collect evidence-based conclusion to rapidly make an informed decision usually by policymakers or to justify the need for further research. In the present editorial paper, the pretension has been to inform within the same year of closing the Special Issue, about how this productive field of Internet-use related addiction problems has grown in pre-COVID times and how it seems the pandemic has also swift the productivity and priorities of this clinical and scientific field.

3. Results

The main content of the articles is presented in Table 1. It includes the following set of variables to analyse the characteristics of the Special Issue:

(a) The authors' names, country where these authors are affiliated, and continent(s);
(b) The setting in which data were collected: if data were collected at an educational institution (e.g., in a secondary school or a university) the setting is labelled as 'Education', if they were collected through a non-educative website (e.g., through a social network) the setting is labelled 'Social', while if data have been collected in a health setting (e.g., a hospital or health centre with patients), the setting is labelled 'Health', and the sector in which the study can be situated (i.e., 'Health' if the main aim is to contribute to health sciences regarding these problems; or 'Education' if the primary aim is to impact educational facets of the problem);
(c) The main method (i.e., empirical-quantitative, qualitative, or mixed method; or theoretical study-review, theoretical model, etc.), type of study (i.e., descriptive study, if the paper only provides detailed definitions and observations of the characteristics of the phenomenon under study; psychometric study—if the aim of the study is testing the psychometric properties of a scale, questionnaire or test assessing Internet use-related addiction problems, such as the reliability and validity to create or adapt a psychometric tool for a specific language or cultural group; relational study, or correlational study, which refers to the aim of associating a set of characteristics of the phenomenon without predictive aims; and predictive study, which implies a statistical or manipulative design to treat a set of variables as predictors or causes of another set of variables, the outcomes or dependent variables), population groups (e.g., children, adolescents, adults), and time of data collection (i.e., by years, if included in the paper);
(d) The main addictive online problem(s) researched: e.g., IA, GD, SMA, PMPU, PSU, PUP, OG, and PIS.
(e) The overall main findings of the study.

Table 1. Articles included in the Special Issue according to authors' names, locations, setting and sector, method, problem(s), and findings.

Authors' Names and Reference	Authors' Location (Continent(s))	Setting (Data Collection) and Sector (Aim)	Method, Type of Study with Population, and Time (Data Collection)	Problem(s) Studied Regarding the ICT and Other Variables	Main Findings
Lutz Wartberg, and Rudolf Kammerl [11]	Germany (Europe)	Education and health	Empirical and Relational study among problem uses in adolescent-parent dyads, in 2019.	Problematic alcohol use, Problematic Internet use (i.e., Internet addiction (IA): i.e., video game addiction (GD) or social media addiction (SMA)), and mental health	Antisocial behaviour was related to all problematic behaviours. Emotional distress, self-esteem, and hyperactivity/inattention were related to substance-unrelated problematic behaviour. Anger control was related to problematic alcohol use and problematic gaming.
Arnold Alejandro Tafur-Mendoza, Julio César Acosta-Prado, Rodrigo Arturo Zárate-Torres, and Duván Emilio Ramírez-Ospina [12]	Perú and Colombia (America)	Education and health	Empirical and Psychometric study about IA in Peruvian university students	Internet addiction test (IAT) adapted to Spanish Peruvian to measure IA.	The Peruvian IAT has been validated with two factors: time/control and stress/compensate. IAT correlated with daily time of use and social skills.
Verónica Marín-Díaz, Juan Manuel Muñoz-González, and Begoña-Esther Sampedro-Requena [13]	Spain (Europe)	Education and health	Empirical and Psychometric-relational study about problematic smartphone use (PSU) in Spanish and Colombian university students in 2017–2018	PSU through the Mobile Phone Problematic Use Scale for Adolescents (MPPUSA)	The MPPUSA was validated with a model of six factors: tolerance, escape route, disconnection, anxiety, negative consequences, social motivations. However, can only be used with Colombian students, young Spanish women, and students in Social Sciences, with those who had problematic behaviour with the devices, as the ones from Health Sciences students did not have the problem.
Wei Hong, Ru-De Liu, Yi Ding, Rui Zhen, Ronghuan Jiang, and Xinchen Fu [14]	China-United States of America (USA) (Asia and America)	Education and health	Empirical and Relational study between autonomy need dissatisfaction and problematic mobile phone use (PMPU) in Chinese high school' students.	PMPU, Mobile gaming, autonomy need dissatisfaction, and boredom proneness	Autonomy need dissatisfaction predicted PMPU. Mediators between both were boredom proneness and mobile phone gaming. Thus, a specific psychological need of the mobile phone was detected.

Table 1. *Cont.*

Authors' Names and Reference	Authors' Location (Continent(s))	Setting (Data Collection) and Sector (Aim)	Method, Type of Study with Population, and Time (Data Collection)	Problem(s) Studied Regarding the ICT and Other Variables	Main Findings
Melina A. Throuvala, Mark D. Griffiths, Mike Rennoldson, and Daria J. Kuss [15]	United Kingdom (UK) (Europe)	Education and Education-Health	Empirical and Predictive study to reduce smartphone uses through a treatment (i.e., an app based on cognitive-behavioural principles to reduce distraction and other psychological outcomes) in undergraduates.	Smartphone use, distraction, nomophobia (NoMO), fear of missing out (FoMO), problem social media use (PSMU), online vigilance, and psychological outcomes	The treatment was effective in reducing smartphone use and distraction, plus other outcomes which were reduced (i.e., impulsivity, stress, anxiety, deficient self-regulation, FoMO, and PSMU) or increased (i.e., self-awareness, mindful attention, self-efficacy), but no effect was observed in PSMU, habitual use or NoMO.
Jui-Kang Tsai, Wei-Hsin Lu, Ray C. Hsiao, Huei-Fan Hu, and Cheng-Fang Yen [16]	Taiwan-USA (Asia and America)	Education and Health	Empirical, and Predictive-relational study about the effect of difficulty in emotion regulation on the occurrence and remission of IA among Taiwanese college students	IA and difficulties in emotion regulation aspects (e.g., impulse control difficulties, lack of emotional awareness, limited access to emotion regulation strategies)	The impulse control difficulties on the difficulties in emotion regulation predicted the incidence of IA during the follow-up period of a year in male participants, but no other aspects predicted the remission of IA. IA did not predict difficulties in emotion regulation.
Seung-Yup Lee, Hae Kook Lee, Jung-Seok Choi, Soo-young Bang, Min-Hyeon Park, Kyu-In Jung, and Yong-Sil Kweon [17]	Korea (Asia)	Health and Health	Empirical, and Longitudinal study (6 months) about the PSU course in South Korean children and adolescents	PSU and mental health indicators	Persistent PSU individuals displayed higher baseline PSU severity and mental health problems at the follow-up. A depressive or anxiety baseline did not influence PSU course. PSU behaved as an addictive disorder. Matthew effect appears in the PSU recovery (i.e., better premorbid psychosocial adjustment prevents it, e.g., more conversations with mothers, perceived happiness, self-esteem)

Table 1. *Cont.*

Authors' Names and Reference	Authors' Location (Continent(s))	Setting (Data Collection) and Sector (Aim)	Method, Type of Study with Population, and Time (Data Collection)	Problem(s) Studied Regarding the ICT and Other Variables	Main Findings
Montserrat Peris, Usue de la Barrera, Konstanze Schoeps, and Inmaculada Montoya-Castilla [18]	Spain (Europe)	Health and Education	Empirical, and Predictive-relational study about how body self-esteem, personality traits, and demographic factors predict adolescents' addictive use of social media and the Internet in high school students	Social networking and IA (i.e., IA symptoms, social media use, geek behaviour, nomophobia), and body self-esteem & personality	Four types of adolescents' IA and as predictors: gender and disinhibition; gender with attractiveness explained social media use; narcissism and neuroticism predict geek behaviour; narcissism best explained nomophobia.
Michelle Colder Carras, Matthew Carras, and Alain B. Labrique [19]	USA (America)	Social and Health	Empirical, and Descriptive mixed-methods study about the techniques to promote healthy play and prevent GD in adult gamers.	GD	Potential targets are specific types of social (e.g., play with others in a group) or self-regulation processes (e.g., set timers or alarms); clear break points and short missions, but loot boxes were not mentioned.
María Luisa Ballestar-Tarín, Conchín Simó-Sanz, Elena Chover-Sierra, Carlos Saus-Ortega, Carmen Casal-Angulo, and Antonio Martínez-Sabater [20]	Spain (Europe)	Health and Education	Empirical and Psychometric-relational study about PSU in Spanish in university students in 2017.	PSU and sociodemographic variables. The Smartphone Addiction Inventory adapted to Spanish (SPAI-Spanish) with a cut-off point.	The SPAI-Spanish has been validated with sociodemographic differences. The cut-off was established for those on perceived risk of PSU.
Siti Rubiaehtul Hassim, Wan Nor Arifin, Yee Cheng Kueh, and Nor Azwany Yaacob [21]	Malasya (Asia)	Education and Health	Empirical and Psychometric-relational study about PSU in Malay medical students in 2017.	PSU, IA, and sociodemographic variables. The Smartphone Addiction Scale adapted to Malay (SAS-M).	The SAS-M has been validated which correlated with IA. The cut-off was established for those on perceived risk of PSU

Authors' Names and Reference	Authors' Location (Continent(s))	Setting (Data Collection) and Sector (Aim)	Method, Type of Study with Population, and Time (Data Collection)	Problem(s) Studied Regarding the ICT and Other Variables	Main Findings
Darryl Mead and Mary Sharpe [22]	UK (Europe)	Health	Theoretical, Descriptive study about the need of recognition of the problematic usage of pornography (PUP) within a European Manifesto.	PUP	The Manifesto for a European research network into Problematic Usage of the Internet in 2018 did not covered the priority of the PUP as other IA problems (e.g., community-focused issues, professional communities affected).
Young-Mi Ko, Sungwon Roh, and Tae Kyung Lee [23]	South Korea (Asia)	Social and Health	Empirical, Predictive-relational study between problematic Internet shopping (PIS) and dissociative experiences in adults.	PIS and sociodemographic characteristics, alcohol use, caffeine intake, online shopping behaviours, and Psychopathological assessments	The risk of PIS was related to an increased tendency toward dissociation and impulsivity. PIS with dissociation have more stress, gambling problems, and impulsivity.
Sarah E. Domoff, Emma Q. Sutherland, Sonja Yokum, and Ashley N. Gearhardt [24]	USA (America)	Education and Health	Empirical, Relational study between PSU and eating behaviours in adolescents between 2015 and 2017.	PSU and emotion regulation difficulties, impulsivity, maladaptive eating behaviours, and adiposity	PSU is associated with these difficulties, dysregulated eating, restrained eating, food addiction, and higher percent body fat; being the difficulties a mediator.
Xinhe Zhang, Xiaoxuan Shi, Shuowei Xu, Jingwen Qiu, Ofir Turel, and Qinghua He [25]	China-USA (Asia and Ameria)	Education and Health	Empirical, Predictive experimental study on IA in Chinese university students in 2018.	To test a treatment for IA (i.e., a 5-week solution-focused group counselling)	The treatment was effective even during a follow up, especially in compulsive-withdrawal and tolerance symptoms.
Sina Ostendorf, Elisa Wegmann, and Matthias Brand [26]	Germany (Europe)	Education and Health	Empirical, Relational study between social-networks-use disorder (SNUD) risks (need to belong, NTB), and protective (online self-regulative competences, OSRC) factors in adolescents during 2017.	SNUD (SMA) and NTB, OSRC with sociodemographic variables	Highest SNUD with high NTB and low OSRC, especially when older. The importance of improving specific competences to prevent SNUD

Table 1. *Cont.*

Authors' Names and Reference	Authors' Location (Continent(s))	Setting (Data Collection) and Sector (Aim)	Method, Type of Study with Population, and Time (Data Collection)	Problem(s) Studied Regarding the ICT and Other Variables	Main Findings
Adnan Veysel Ertemel, and Ela Ari [27]	Turkey (Asia)	Education and Health	Empirical, Predictive study on PSU in adolescent students in 2018	To test a PSU educative program (i.e., consumer behaviour theories and unhook and gamification techniques)	The PSU significantly decreased after the program, depending on gender, mother's education, and class levels.
Jinhee Lee, Joung-Sook Ahn, Seongho Min, and Min-Hyuk Kim [28]	Korea (Asia)	Education and Health	Empirical, Relational study between content type smartphone use and psychological characteristics, addiction propensity, time of its use and PSU in adolescents during 2017.	PSU, Smartphone uses (study, social networking sites (SNS), game, entertainment), Psychology (depressive mood, suicidal ideation),	Psychological characteristics were associated with SNS use, which showed higher addiction (overuse & adverse consequences).
Élodie Verseillié, Stéphanie Laconi, and Henri Chabrol [29]	France (Europe)	Education and Health	Empirical, Relational-predictive study between psychopathology and problematic SNS uses (or SMA) in adults.	SMA (Facebook and Twitter), and psychopathology (personality traits, depression, anxiety symptoms, stress)	Problem SNS uses were predicted by personality traits (depressive and anxiety), only stress predicted problematic Facebook use.
Hiu Yan Wong, Hoi Yi Mo, Marc N. Potenza, Mung Ni Monica Chan, Wai Man Lau, Tsz Kwan Chui, Amir H. Pakpour, and Chung-Ying Lin [30]	Hong Kong, USA, Iran, Sweden (Asia, America and Europe)	Education and Health	Empirical, Relational study between Internet gaming disorder (IGD), SMA, sleep quality, psychological distress among Hong Kong young adults in 2019.	IGD, SMA and distress (depression, anxiety, and stress)	IGD is associated with distress, and a bit with poor sleep quality. SMA similarly associated with distress, but higher with poorer sleep quality.
Anna Faltýnková, Lukas Blinka, Anna Ševčíková, and Daniela Husarova [31]	Czech Republic, Slovakia (Europe)	Education and Health	Empirical, Relational-predictive study between IA, and family environment in Slovakian adolescents.	IA and family type, economic status, parental care, parental control, parental monitoring, communication, and time spent.	Higher parental care and parental monitoring predicted lower IA, while higher parental overprotection and lower socioeconomic status predicted higher IA.

Table 1. *Cont.*

Authors' Names and Reference	Authors' Location (Continent(s))	Setting (Data Collection) and Sector (Aim)	Method, Type of Study with Population, and Time (Data Collection)	Problem(s) Studied Regarding the ICT and Other Variables	Main Findings
Marta Beranuy, Juan M. Machimbarrena, M. Asunción Vega-Osés, Xavier Carbonell, Mark D. Griffiths, Halley M. Pontes, and Joaquín González-Cabrera [32]	Spain (Europe)	Education and Health	Empirical and Psychometric-relational study about IGD short form scale (IGDS9-SF) in Spanish and its relationship with online gambling (OG) and quality life (QoL) in young students in 2019.	To adapt and validate to Spanish the IGDS9-SF, and to relate it to PSU, OG	The Spanish IGDS9-SF was valid and reliable, 1.9% of gamers had IGD. IGD was positively related to PSU and OG. Those with IGD has less QoL.
Vega González-Bueso, Juan José Santamaría, Ignasi Oliveras, Daniel Fernández, Elena Montero, Marta Baño, Susana Jiménez-Murcia, Amparo del Pino-Gutiérrez, and Joan Ribas [33]	Spain (Europe)	Health and Health	Empirical and Descriptive study to classify IGD adolescent patients according to their personality & describe them clinically and demographically.	IGD	There were 2 types: I "higher comorbid symptoms" (introversive, inhibited, doleful, unruly, forceful, oppositional, self-demeaning, borderline) & type II "lower comorbid symptoms" (histrionic, egotistic and conforming)
Xinchen Fu, Jingxuan Liu, Ru-De Liu, Yi Ding, Jia Wang, Rui Zhen, and Fangkai Jin [34]	China and USA (Asia and America)	Education and Health	Empirical, Relational-predictive study between PSU, and parental monitoring in Chinese adolescents	PSU	Parental monitoring predicted PSU, children's escape motivation mediated it, while shyness moderated the path and the impact of monitor PSU
Ningyuan Guo, Tzu Tsun Luk, Sai Yin Ho, Jung Jae Lee, Chen Shen, John Oliffe, Sophia Siu-Chee Chan, Tai Hing Lam, and Man Ping Wang [35]	China and Canada (Asia and America)	Education and Health	Empirical, Relational-predictive study between PSU, and anxiety, depression, and mental well-being in Hong Kong Chinese adults in 2017.	PSU & psychological variables (depression, anxiety, subjective happiness, and mental well-being)	PSU was associated with higher anxiety and depression, and lower happiness and wellbeing)

Table 1. *Cont.*

Authors' Names and Reference	Authors' Location (Continent(s))	Setting (Data Collection) and Sector (Aim)	Method, Type of Study with Population, and Time (Data Collection)	Problem(s) Studied Regarding the ICT and Other Variables	Main Findings
Beifang Fan, Wanxing Wang, Tian Wang, Bo Xie, Huimin Zhang, Yuhua Liao, Ciyong Lu, and Lan Guo [36]	China (Asia)	Education and Health	Empirical, Relational-predictive study between IA, and non-medical use of prescription drugs (NMUPD), and depression in Chinese adults, 2017.	IA and NMUPD (use of opioid, sedative), and depression	IA associated with depression and frequent use of opioid and sedative drugs, elevating the risk of this addiction.
Lijun Chen, and Xiaoliu Jiang [37]	China (Asia)	Education and Health	Empirical, Psychometric mixed methods study about the online PUP or problematic Internet pornography use (IPU) in Chinese adults.	PUP or IPU	The Problematic Pornography Consumption Scale (PPCS) had stronger reliability and validity, including criterion validity, greater sensitivity, and acceptable specificity.
Ki Hyeon Kwak, Hyun Chan Hwang, Sun Mi Kim, and Doug Hyun Han [38]	Korea (Asia)	Health and Health	Empirical, Relational study between IGD, psychology, brain activity between pro-gamers and IGD Korean adolescents between 2016 and 2017.	IGD, psychology (depressed mood, anxiety), and brain activity (resting-state functional magnetic resonance imaging)	Both groups displayed increased brain activity in the parietal lobe (attention), but IGD adolescents showed higher brain activity within the left orbitofrontal cortex (impulsivity and aggression).
Bernadeta Lelonek-Kuleta, Rafał P. Bartczuk, Michał Wiechetek, Joanna Chwaszcz, and Iwona Niewiadomska [39]	Poland (Europe)	Education and Health	Empirical, Relation-predictive study about aspects of gambling among Polish adults.	e-Gambling (OG) and its activities	4.1% did e-gambling (lotteries, sports betting), 26.8% classed as problem gamblers. Men, younger, who earnt less were more often involved in e-gambling.
Sulki Chung, Jaekyoung Lee, and Hae Kook Lee [40]	Korea (Asia)	Education and Health	Empirical, Relation study about personal, family/school, Internet use, and environmental factors among Korean adolescents in 2015.	IA with individual (psychology, family cohesion, attitudes toward academic activities) and environmental factors (Internet characteristics, accessibility to PC cafés, and exposure to Internet game advertising).	6% IA: using the Internet earlier; had higher levels of depression, compulsivity, aggressiveness, lower family cohesion; higher accessibility to PC cafés and exposure to Internet game advertising. Environmental factors had a greater influence than family or school-related factors

Authors' Names and Reference	Authors' Location (Continent(s))	Setting (Data Collection) and Sector (Aim)	Method, Type of Study with Population, and Time (Data Collection)	Problem(s) Studied Regarding the ICT and Other Variables	Main Findings
Amandine Luquiens, Aline Dugravot, Henri Panjo, Amine Benyamina, Stéphane Gaïffas, and Emmanuel Bacry [41]	France (Europe)	Social and Health	Empirical, Predictive study about the effect of self-exclusion in online poker gambling as compared to matched controls, after the end of the self-exclusion period with French adult gamblers in 2016.	OG	Effects of self-exclusion and short-duration self-exclusion were found for money and time spent over a year. Short-duration self-exclusions showed no effect on the heavy gamblers.
Myungsuh Lim [42]	Korea (Asia)	Social and Health	Empirical, Relational study about the effect of Facebook addiction and other demographic and psychological characteristics with USA adults.	SMA (Facebook), and demographics (age and gender), social exclusion, surveillance use, and narcissistic grandiosity	Social exclusion and surveillance were associated with SMA.
Kuan-Ying Hsieh, Ray C. Hsiao, Yi-Hsin Yang, Kun-Hua Lee, and Cheng-Fang Yen [43]	Taiwan and USA (Asia and America)	Education and Health	Empirical, Relational study about self-identity confusion and IA and the mediating effects of psychological inflexibility and experiential avoidance (PI/EA) indicators in Taiwanese college students.	IA, self-identity (Self-Concept and Identity Measure), PI/EA (Acceptance and Action).	The severity of self-identity confusion was associated with both the severity of PI/EA and IA. The severity of PI/EA indicators was associated with IA.
Ju-Yu Yen, Huang-Chi Lin, Wei-Po Chou, Tai-Ling Liu, and Chih-Hung Ko [44]	Taiwan (Asia)	Health and Health	Empirical, Relational-predictive study about resilience, perceived stress, depression, and IGD in Taiwanese young adults in 2013.	IGD and psychological variables.	IGD group had a lower resilience, higher stress, and depression than the control group. Low resilience was associated with a higher IGD and depression. Depression was more associated with IGD than resilience. Depression and stress coping interventions should be provided for those with IGD with low resilience or high stress.

Table 1. *Cont.*

Authors' Names and Reference	Authors' Location (Continent(s))	Setting (Data Collection) and Sector (Aim)	Method, Type of Study with Population, and Time (Data Collection)	Problem(s) Studied Regarding the ICT and Other Variables	Main Findings
Yi-Ping Hsieh, Cheng-Fang Yen, and Wen-Jiun Chou [45]	Taiwan and USA (Asia and America)	Health & Health	Empirical, Psychometric mixed methods study about the Parental Smartphone Use Management Scale (PSUMS) and relationships with PSU and attention deficit hyperactivity disorder (ADHD) in Taiwanese parents of adolescents with ADHD in 2014–2015.	PSU and ADHD	The PSUMS had good factorials validity and high reliabilities. Parents of children with smartphone addiction yielded lower scores on all three PSUMS subscales (i.e., reactive management, proactive management, and monitoring) than parents of children without smartphone addiction
Olatz Lopez-Fernandez, and Daria J. Kuss [46]	Australia and UK (Oceania and Europe)	Health	Theoretical, literature review on Internet use-related addiction problems on a European scale	IA, IGD, OG	Those with problematic uses are educated male adolescents, with comorbid disorders, being gaming and gambling the most severe addictions. Cognitive behavioural therapy was the main treatment. Policy options and preventive strategies are proposed for public health in Europe.
Antonio-José Moreno-Guerrero, Gerardo Gómez-García, Jesús López-Belmonte, and Carmen Rodríguez-Jiménez [47]	Spain (Europe)	Health	Theoretical, literature review on IA in the web of science database	IA	The evolution in the study of the IA is constant, articles in English, in psychiatry, Prof Griffiths, from Nottingham Trent University, publishing in *Computers in Human Behavior* and the *Journal of Behavioral Addictions*, and USA has greater interest in production.
Sheila Yu, and Steve Sussman [48]	USA (America)	Health	Theoretical, systematic review on measures of smartphone addiction (SA) and PSU to check if they are distinct from other addictions, and how may fall on a continuum of addictive behaviors	PSU and SA related to other technological addictions	Most studies neither distinguished SA from other technological addictions nor clarified whether SA was an addiction to the actual smartphone device or its features. No theory to explain the etiologic origins or causal pathways of SA and its associations.

Table 1. *Cont.*

Authors' Names and Reference	Authors' Location (Continent(s))	Setting (Data Collection) and Sector (Aim)	Method, Type of Study with Population, and Time (Data Collection)	Problem(s) Studied Regarding the ICT and Other Variables	Main Findings
Christian Montag, Bernd Lachmann, Marc Herrlich, and Katharina Zweig [49]	Germany (Europe)	Health	Theoretical, review to analyze on several prominent smartphone apps to carve out over-usage smartphone apps	PSU apps (social media apps and Freemium games)	These app-elements are linked to classic psychological/economic theories, e.g., mere-exposure, endowment, and Zeigarnik effects, but also to psychological mechanisms triggering social comparison.

Note: the acronyms used are the following: Internet addiction (IA), (video) gaming disorder (GD), social media addiction (SMA), Internet addiction test (IAT), problematic smartphone use (PSU), Mobile Phone Problematic Use Scale for Adolescents (MPPUSA), problematic mobile phone use (PMPU), nomophobia (NoMO), fear of missing out (FoMO), problem social media use (PSMU), Smartphone Addiction Inventory adapted to Spanish (SPAI-Spanish), Smartphone Addiction Scale adapted to Malay (SAS-M), problematic usage of pornography (PUP), problematic Internet shopping (PIS), social-networks-use disorder (SNUD), need to belong (NTB), online self-regulative competences (OSRC), social networking sites (SNS), Internet gaming disorder (IGD), IGD short form scale (IGDS9-SF), online gambling (OG), quality life (QoL), non-medical use of prescription drugs (NMUPD), Internet pornography use (IPU), psychological inflexibility (PI), experiential avoidance (EA), Parental Smartphone Use Management Scale (PSUMS), attention deficit hyperactivity disorder (ADHD), smartphone addiction (SA).

4. Discussion

The present editorial paper offers a rapid overview of the pre-COVID-19 advances on Internet use-related addiction problems research collected in this second special issue. The main aim was to invite research on these health problems from the areas of treatments, education, and other research advances in the field. After the first wave of the pandemic, the editor would also like to share a few reflections from cyberpsychology and environmental perspectives.

Regarding the first main finding, compared with the previous Special Issue on this theme [10], using the same even stricter peer-review process, the second Special Issue has collected double the number of papers (i.e., 40 versus 20 papers) in approximately a year, which indicates the increasing interest and publication rate in this research field. Related to the authors' affiliations, this second Special Issue includes peer-reviewed papers from international samples (ordered by frequency): Europe (i.e., Spain, Germany, UK, France, Turkey, Sweden, Czech Republic, Slovakia, and Poland), Asia (i.e., China, Korea, Taiwan, Hong Kong, Malaysia, and Iran), America (USA, Canada, Perú, and Colombia), and one from Australia. However, no papers were included from Africa, which is consistent with the existence of the dual digital divide and international data reported [1,2]. This fact does not mean that there is no existence of these type of problems on this continent, but it highlights that no work has been submitted for a review in this case as we were sensitive to it.

On the other hand, the main aim was to cover health and education approaches in the context of these problems, which has partially been achieved, as the majority of papers (74%) included both approaches. However, half of the papers (54%) were studies undertaken in educational settings (i.e., high schools [14,18], universities [12,13]) with a health purpose, such as to adapt a diagnostic tool [12,13,20,21,32,37,45] or to study psychological mechanisms or comorbid problems which affect or mediate these problems to be used in prevention and treatment plans [17,22–25,46]. Only a few studies (20%) were conducted in health settings (i.e., hospitals [44] or health centres [33]) or were health reviews about these problems [46]). In the current literature, the method of extracting health knowledge through educational environments is quite common, but at the same time it shows its maturity is still compromise, as more research is needed in non-educational and clinical sectors with qualitative, mixed methods and manipulative research designs (e.g., quasi-experiments and experiments) to go in-depth on knowing better what the mechanisms behind these problems are, what seems to be the cause and how it affects to the prognostic. Thus, the need for extracting research on treatments was limited in this issue, which is a challenge the field needs to overcome.

The problems studied are more specific IA problems rather than generalised IA (23% of the papers), which is a new trend on the field, and it is conversely opposite to the first two decades of research, in which IA was more prevalent than specific Internet use-related addiction problems. In order of frequency, in this second Special Issue, which was open to research on all addictive Internet and mobile phone use, the most prevalent problems ordered by frequency were: PMPU (38% [27]), IA (26% [16]), GD (23% [19]), SMA (18% [26]), OG (8% [39]), PUP (5% [37]), and PIS (3% [23]). This means the tendency before COVID-19 in the field was to study addictive problems related to the use of mobile technologies (e.g., PSU) and its risks, and this has increased during the pandemic [50]. Reasons for this include those problems being the most researched topics due to the need of increasing and controlling our safety behaviours while travelling or commuting during the pandemic, in addition to other advantages and continuous innovations offered by mobile technologies (e.g., apps, aids, and trackers). Furthermore, as predicted a few years ago [9], other types of problematic addictive use apart from GD are studied and causing concern in population groups (e.g., OG and PUP).

This is consistent with the methodology used in pre-COVID research on Internet use-related addiction problems papers published in the Special Issue (i.e., which collected data between 2013 and 2019). Almost all works were empirical (90%) with quantitative

studies (only two mixed methods and no qualitative ones). Still there are a fifth of the papers in the Special Issue which are psychometric studies, in which usually cultural validations of diagnostic tools have been undertaken, such as the Spanish and Colombian Mobile Phone Problematic Use Scale for Adolescents (MPPUSA [13]), or the Problematic Pornography Consumption Scale (PPCS [37]), which shows the interest to scientifically validate existing measures on problematic use of smartphones, gaming and pornography across cultures [51,52]. This long tradition in psychometrics in the field is probably due to the global phenomena related to the addictive uses of technologies world-wide, and the cross-cultural interest in these common problems in young populations [53]. However, it has also been criticized, as it is quite extensive in comparison with theoretical developments and conceptualisation of these problems and seems not to consolidate enough the field for its recognition as health problems [54].

Regarding the method, over half (59%) of the papers published here were correlational studies looking for associations between these problems and internal or external variables. The novelty is these studies is the inclusion of more contextual or environmental factors related to these technological use problems [31,34] and sometimes were treated as predictive ones (i.e., Relational-predictive studies) using predictive aims through statistical techniques to for example extract the factors which seems to cause an outcome (20% of the papers). Only a few studies have real independent variables manipulated by the researchers to cause an effect in other dependent ones, such as new health and educational interventions studied (e.g., Solution-Focused Group Counselling Intervention [25], an app based on cognitive-behavioural principles to avoid smartphone distractions [15]). This means applied research in the field is starting to look for explanations outside the intra-individual sphere through more controlled research designs, through new clinical and educational interventions to promote preventive strategies to minimize harms. Indeed, it can support provisional solutions and support controlling the incidence, and growth of these problems.

Thus, other external variables, such as inter-individual and environmental factors seem to impact Internet use-related addiction problems as intra-individual factors (e.g., sociodemographic, and psychological characteristics). The external indicators which emerged in this Special Issue are related to the Matthew effect (i.e., good premorbid psychosocial adjustment [17]), parent's education or socio-economic status (i.e., mother´s education [27]), parental care and monitoring (i.e., overprotection promotes these problems [31]), environmental factors (e.g., Internet cafés, advertising [40]), or professional communities (e.g., who manages those affected and their families [22]). Indeed, these contextual factors should also be considered in treatment plans, preventive actions, and policy options [46].

In the first wave of the pandemic in 2020, the lockdowns, restrictions, and isolation have affected individuals' engagement in addictive behaviours [50,55]. Indeed, the problems which depend on the Internet (the medium frequently used in pandemic times) and the need to be interconnected seems to have produced a growth in these problems [50], which means environmental factors, such as a COVID-19, can have a negative effect on these types of addictive problems in some population groups. However, the clinical, academic, and scientific communities have immediately reacted through prevention alerts and strategies [46,56–59]. For instance, according to Montag and Elhai [59], who have discussed technology overuse in young populations during the pandemic through Affective Neuroscience Theory, the neglected indirect media effects should not be overseen (i.e., those brain functions involving positive or negative emotions, such as play or sadness systems have been impacted by COVID-19). During the first lockdowns, children have translated their usual diverse leisure activities to screen time activities at home; it seems the problem has been most parents needed to telework and children, apart from home-schooling online, also spent more time online without a parent or adult guidance, which has caused in some cases lack of care, the increment of sadness, and even anger (i.e., disbalance in primary emotional systems). Thus, again, research on the causality of Internet use-related addiction problems (or Internet use disorders [60]) is needed going beyond descriptive and relational quantitative studies which have occupied almost all the scientific body of this field.

Nevertheless, there are some limitations in the present study. It is based on a limited selection of papers which have been accepted for this Second Special issue based on the editor, the journal academic editorial, and reviewer team through a procedure which usually involved at least two rounds and three reviewers per paper, and a few did not pass the quality check to proceed with the peer-review process. The bias, therefore, in the selection process of the Spanish nationality of the editor an Asiatic journal such as IJERPH, which potentially attracted more proposals from these countries, although an effort has been made to disseminate the invitation world-wide and prioritise locations, topics, and methods not commonly used in this research field to receive a representative pool of a diverse qualified papers. Furthermore, another bias of this editorial rapid review paper is the inherent weakness of conducting a fast review by an author in the same year in which we closed the Special Issue, even with the fact it was extended a few months due to the first wave of COVID-19. However, the IJERPH editorial team and I agreed on this decision and thought it was worth doing this rapid evidence assessment of the papers reviewed, collected, and published in this Special Issue, as it was more successful than initially planned, even with the pandemic factor.

Interestingly, the field is changing and moved to the need to understand the causes of these specific online problems, and the need to intervene and prevent them from individual and environmental perspectives. In the findings, a limited number of theoretical papers have been included, which are those which study the problems in-depth and sustain empirical research, which is a weakness of the field now highlighted during COVID-19 studies [58,59]. Finally, although the first wave of this pandemic and unprecedented situation has also impacted on the delay of finalising the Special Issue, the number of papers finally published are those which started the review process before March 2020, which probably means health care and scientific production even first impacted by the COVID-19 have also continued providing and producing quality outcomes such as this Special Issue.

5. Conclusions

This second Special Issue on "Internet and Smartphone Use-Related Addiction Health Problems: Treatment, Education and Research" has successfully included 40 papers (including this one) from around the globe (except Africa). It includes mainly empirical studies which are relational and a few other types about the adaptation of measures to diagnose these problems, and a few new interventions or preventive strategies to tackle the spectrum of these online usage problems in pre-COVID-19 times. Most of the related issues associated with technological usage problems still seem to be related to individual characteristics of users, who tend to be adults. The problems are about specific online activities and not only focused on gaming disorder. The components which impact their development seem to also be contextual factors, such as the pandemic which seems to have incremented the priorities on researching these health problems from a causal perspective, including environmental factors and intervention and preventive strategies. Thus, although advances have spread around the globe and in diverse Internet use-related addiction problems, maturity of the field still is compromised because there is a need for more qualitative and theoretical research, clinical trials, experimental studies, follow up studies, testing intervention and recovery strategies, policy options, and preventive actions to minimise harms and enhance users' wellbeing and quality of life, even during present pandemic times of mobility restrictions and more online time consumption.

Funding: This research received no external funding.

Institutional Review Board Statement: Not applicable.

Informed Consent Statement: Not applicable.

Data Availability Statement: The data presented in this study are openly available at https://www.mdpi.com/journal/ijerph/special_issues/Internet_Smartphone_Addiction.

Acknowledgments: To Daria J. Kuss for her support in doing a first review.

Conflicts of Interest: The authors declare no conflict of interest.

References

1. International Telecommunication Union (ITU). *Measuring the Information Society Report*; ITU: Geneva, Switzerland, 2018; Volume 1, pp. 1–189. Available online: https://www.itu.int/en/ITU-D/Statistics/Documents/publications/misr2018/MISR-2018-Vol-1-E.pdf (accessed on 1 December 2020).
2. Roycroft, T.R.; Anantho, S. Internet subscription in Africa: Policy for a dual digital divide. *Telecommun. Policy* **2003**, *27*, 61–74. [CrossRef]
3. Millward, P. The 'grey digital divide': Perception, exclusion and barriers of access to the Internet for older people. *First Monday* **2003**, *8*. [CrossRef]
4. Bakardjieva, M. *Internet Society: The Internet in Everyday Life*; Sage: London, UK, 2005; pp. 10–15.
5. Budd, J.; Miller, B.S.; Manning, E.M.; Lampos, V.; Zhuang, M.; Edelstein, M.; Rees, G.; Emery, V.C.; Stevens, M.M.; Keegan, N.; et al. Digital technologies in the public-health response to COVID-19. *Nat. Med.* **2020**, *26*, 1183–1192. [CrossRef] [PubMed]
6. Wallace, P. *The Psychology of the Internet*, 2nd ed.; Cambridge University Press: Cambridge, UK, 2015; pp. 1–21.
7. Cash, H.; Rae, C.D.; Steel, A.H.; Winkler, A. Internet addiction: A brief summary of research and practice. *Curr. Addict. Rep.* **2012**, *8*, 292–298. [CrossRef] [PubMed]
8. World Health Organization [WHO]. ICD-11 Beta Draft—Mortality and Morbidity Statistics. Mental, Behavioural or Neurodevel-Opmental Disorders. 2018. Available online: https://icd.who.int/dev11/l-m/en#/http://id.who.int/icd/entity/1448597234 (accessed on 29 November 2020).
9. Lopez-Fernandez, O. How has internet addiction research evolved since the advent of internet gaming disorder? An overview of cyberaddictions from a psychological perspective. *Curr. Addict. Rep.* **2015**, *2*, 263–271. [CrossRef]
10. Lopez-Fernandez, O. *Internet and Mobile Phone Addiction: Health and Educational Effects*, 1st ed.; MDPI: Basel, Switzerland, 2019; pp. 1–328, ISBN 978-3-03897-604-2 (Pbk); ISBN 978-3-03897-605-9 (PDF).
11. Wartberg, L.; Kammerl, R. Empirical Relationships between Problematic Alcohol Use and a Problematic Use of Video Games, Social Media and the Internet and Their Associations to Mental Health in Adolescence. *Int. J. Environ. Res. Public Health* **2020**, *17*, 6098. [CrossRef] [PubMed]
12. Tafur-Mendoza, A.A.; Acosta-Prado, J.C.; Zárate-Torres, R.A.; Ramírez-Ospina, D.E. Assessing the Psychometric Properties of the Internet Addiction Test in Peruvian University Students. *Int. J. Environ. Res. Public Health* **2020**, *17*, 5782. [CrossRef] [PubMed]
13. Marín-Díaz, V.; Muñoz-González, J.M.; Sampedro-Requena, B.-E. Problematic Relationships with Smartphones of Spanish and Colombian University Students. *Int. J. Environ. Res. Public Health* **2020**, *17*, 5370. [CrossRef]
14. Hong, W.; Liu, R.-D.; Ding, Y.; Zhen, R.; Jiang, R.; Fu, X. Autonomy Need Dissatisfaction in Daily Life and Problematic Mobile Phone Use: The Mediating Roles of Boredom Proneness and Mobile Phone Gaming. *Int. J. Environ. Res. Public Health* **2020**, *17*, 5305. [CrossRef]
15. Throuvala, M.A.; Griffiths, M.D.; Rennoldson, M.; Kuss, D.J. Mind over Matter: Testing the Efficacy of an Online Randomized Controlled Trial to Reduce Distraction from Smartphone Use. *Int. J. Environ. Res. Public Health* **2020**, *17*, 4842. [CrossRef]
16. Tsai, J.-K.; Lu, W.-H.; Hsiao, R.C.; Hu, H.-F.; Yen, C.-F. Relationship between Difficulty in Emotion Regulation and Internet Addiction in College Students: A One-Year Prospective Study. *Int. J. Environ. Res. Public Health* **2020**, *17*, 4766. [CrossRef] [PubMed]
17. Lee, S.-Y.; Lee, H.K.; Choi, J.-S.; Bang, S.-Y.; Park, M.-H.; Jung, K.-I.; Kweon, Y.-S. The Matthew Effect in Recovery from Smartphone Addiction in a 6-Month Longitudinal Study of Children and Adolescents. *Int. J. Environ. Res. Public Health* **2020**, *17*, 4751. [CrossRef] [PubMed]
18. Peris, M.; de la Barrera, U.; Schoeps, K.; Montoya-Castilla, I. Psychological Risk Factors that Predict Social Networking and Internet Addiction in Adolescents. *Int. J. Environ. Res. Public Health* **2020**, *17*, 4598. [CrossRef] [PubMed]
19. Colder Carras, M.; Carras, M.; Labrique, A.B. Stakeholders' Consensus on Strategies for Self- and Other-Regulation of Video Game Play: A Mixed Methods Study. *Int. J. Environ. Res. Public Health* **2020**, *17*, 3846. [CrossRef]
20. Ballestar-Tarín, M.L.; Simó-Sanz, C.; Chover-Sierra, E.; Saus-Ortega, C.; Casal-Angulo, C.; Martínez-Sabater, A. Self-Perception of Dependence as an Indicator of Smartphone Addiction—Establishment of a Cutoff Point in the SPAI–Spain Inventory. *Int. J. Environ. Res. Public Health* **2020**, *17*, 3838. [CrossRef]
21. Hassim, S.R.; Arifin, W.N.; Kueh, Y.C.; Yaacob, N.A. Confirmatory Factor Analysis of the Malay Version of the Smartphone Addiction Scale among Medical Students in Malaysia. *Int. J. Environ. Res. Public Health* **2020**, *17*, 3820. [CrossRef]
22. Mead, D.; Sharpe, M. Aligning the "Manifesto for a European Research Network into Problematic Usage of the Internet" with the Diverse Needs of the Professional and Consumer Communities Affected by Problematic Usage of Pornography. *Int. J. Environ. Res. Public Health* **2020**, *17*, 3462. [CrossRef]
23. Ko, Y.-M.; Roh, S.; Lee, T.K. The Association of Problematic Internet Shopping with Dissociation among South Korean Internet Users. *Int. J. Environ. Res. Public Health* **2020**, *17*, 3235. [CrossRef]
24. Domoff, S.E.; Sutherland, E.Q.; Yokum, S.; Gearhardt, A.N. Adolescents' Addictive Phone Use: Associations with Eating Behaviors and Adiposity. *Int. J. Environ. Res. Public Health* **2020**, *17*, 2861. [CrossRef]
25. Zhang, X.; Shi, X.; Xu, S.; Qiu, J.; Turel, O.; He, Q. The Effect of Solution-Focused Group Counseling Intervention on College Students' Internet Addiction: A Pilot Study. *Int. J. Environ. Res. Public Health* **2020**, *17*, 2519. [CrossRef]

26. Ostendorf, S.; Wegmann, E.; Brand, M. Problematic Social-Networks-Use in German Children and Adolescents—The Interactio of Need to Belong, Online Self-Regulative Competences, and Age. *Int. J. Environ. Res. Public Health* **2020**, *17*, 2518. [CrossRe [PubMed]
27. Ertemel, A.V.; Ari, E. A Marketing Approach to a Psychological Problem: Problematic Smartphone Use on Adolescents. *Int. Environ. Res. Public Health* **2020**, *17*, 2471. [CrossRef] [PubMed]
28. Lee, J.; Ahn, J.-S.; Min, S.; Kim, M.-H. Psychological Characteristics and Addiction Propensity According to Content Type c Smartphone Use. *Int. J. Environ. Res. Public Health* **2020**, *17*, 2292. [CrossRef] [PubMed]
29. Verseillié, É.; Laconi, S.; Chabrol, H. Pathological Traits Associated to Facebook and Twitter among French Users. *Int. J. Enviro Res. Public Health* **2020**, *17*, 2242. [CrossRef]
30. Wong, H.Y.; Mo, H.Y.; Potenza, M.N.; Chan, M.N.M.; Lau, W.M.; Chui, T.K.; Pakpour, A.H.; Lin, C.-Y. Relationships betwee Severity of Internet Gaming Disorder, Severity of Problematic Social Media Use, Sleep Quality and Psychological Distress. *Int. Environ. Res. Public Health* **2020**, *17*, 1879. [CrossRef]
31. Faltýnková, A.; Blinka, L.; Ševčíková, A.; Husarova, D. The Associations between Family-Related Factors and Excessive Interne Use in Adolescents. *Int. J. Environ. Res. Public Health* **2020**, *17*, 1754. [CrossRef]
32. Beranuy, M.; Machimbarrena, J.M.; Vega-Osés, M.A.; Carbonell, X.; Griffiths, M.D.; Pontes, H.M.; González-Cabrera, J. Spar ish Validation of the Internet Gaming Disorder Scale–Short Form (IGDS9-SF): Prevalence and Relationship with Online Gamblin and Quality of Life. *Int. J. Environ. Res. Public Health* **2020**, *17*, 1562. [CrossRef]
33. González-Bueso, V.; Santamaría, J.J.; Oliveras, I.; Fernández, D.; Montero, E.; Baño, M.; Jiménez-Murcia, S.; del Pino-Gutiérre A.; Ribas, J. Internet Gaming Disorder Clustering Based on Personality Traits in Adolescents, and Its Relation with Comorbi Psychological Symptoms. *Int. J. Environ. Res. Public Health* **2020**, *17*, 1516. [CrossRef]
34. Fu, X.; Liu, J.; Liu, R.-D.; Ding, Y.; Wang, J.; Zhen, R.; Jin, F. Parental Monitoring and Adolescent Problematic Mobile Phone Us The Mediating Role of Escape Motivation and the Moderating Role of Shyness. *Int. J. Environ. Res. Public Health* **2020**, *17*, 148 [CrossRef]
35. Guo, N.; Luk, T.T.; Ho, S.Y.; Lee, J.J.; Shen, C.; Oliffe, J.; Chan, S.-C.; Lam, T.H.; Wang, M.P. Problematic Smartphone Use an Mental Health in Chinese Adults: A Population-Based Study. *Int. J. Environ. Res. Public Health* **2020**, *17*, 844. [CrossRef]
36. Fan, B.; Wang, W.; Wang, T.; Xie, B.; Zhang, H.; Liao, Y.; Lu, C.; Guo, L. Problematic Internet Use, Non-Medical Use of Prescriptio Drugs, and Depressive Symptoms among Adolescents: A Large-Scale Study in China. *Int. J. Environ. Res. Public Health* **2020**, *1* 774. [CrossRef] [PubMed]
37. Chen, L.; Jiang, X. The Assessment of Problematic Internet Pornography Use: A Comparison of Three Scales with Mixed Method *Int. J. Environ. Res. Public Health* **2020**, *17*, 488. [CrossRef] [PubMed]
38. Kwak, K.H.; Hwang, H.C.; Kim, S.M.; Han, D.H. Comparison of Behavioral Changes and Brain Activity between Adolescent with Internet Gaming Disorder and Student Pro-Gamers. *Int. J. Environ. Res. Public Health* **2020**, *17*, 441. [CrossRef] [PubMed]
39. Lelonek-Kuleta, B.; Bartczuk, R.P.; Wiechetek, M.; Chwaszcz, J.; Niewiadomska, I. The Prevalence of E-Gambling and of Problen E-Gambling in Poland. *Int. J. Environ. Res. Public Health* **2020**, *17*, 404. [CrossRef] [PubMed]
40. Chung, S.; Lee, J.; Lee, H.K. Personal Factors, Internet Characteristics, and Environmental Factors Contributing to Adolescen Internet Addiction: A Public Health Perspective. *Int. J. Environ. Res. Public Health* **2019**, *16*, 4635. [CrossRef] [PubMed]
41. Luquiens, A.; Dugravot, A.; Panjo, H.; Benyamina, A.; Gaïffas, S.; Bacry, E. Self-Exclusion among Online Poker Gamblers: Effect on Expenditure in Time and Money as Compared to Matched Controls. *Int. J. Environ. Res. Public Health* **2019**, *16*, 4399. [CrossRef
42. Lim, M. Social Exclusion, Surveillance Use, and Facebook Addiction: The Moderating Role of Narcissistic Grandiosity. *Int. Environ. Res. Public Health* **2019**, *16*, 3813. [CrossRef]
43. Hsieh, K.-Y.; Hsiao, R.C.; Yang, Y.-H.; Lee, K.-H.; Yen, C.-F. Relationship between Self-Identity Confusion and Internet Addictior among College Students: The Mediating Effects of Psychological Inflexibility and Experiential Avoidance. *Int. J. Environ. Res Public Health* **2019**, *16*, 3225. [CrossRef]
44. Yen, J.-Y.; Lin, H.-C.; Chou, W.-P.; Liu, T.-L.; Ko, C.-H. Associations among Resilience, Stress, Depression, and Internet Gaming Disorder in Young Adults. *Int. J. Environ. Res. Public Health* **2019**, *16*, 3181. [CrossRef]
45. Hsieh, Y.-P.; Yen, C.-F.; Chou, W.-J. Development and Validation of the Parental Smartphone Use Management Scale (PSUMS) Parents' Perceived Self-Efficacy with Adolescents with Attention Deficit Hyperactivity Disorder. *Int. J. Environ. Res. Public Healt* **2019**, *16*, 1423. [CrossRef]
46. Lopez-Fernandez, O.; Kuss, D.J. Preventing Harmful Internet Use-Related Addiction Problems in Europe: A Literature Review and Policy Options. *Int. J. Environ. Res. Public Health* **2020**, *17*, 3797. [CrossRef] [PubMed]
47. Moreno-Guerrero, A.-J.; Gómez-García, G.; López-Belmonte, J.; Rodríguez-Jiménez, C. Internet Addiction in the Web of Science Database: A Review of the Literature with Scientific Mapping. *Int. J. Environ. Res. Public Health* **2020**, *17*, 2753. [CrossRef] [PubMed]
48. Yu, S.; Sussman, S. Does Smartphone Addiction Fall on a Continuum of Addictive Behaviors? *Int. J. Environ. Res. Public Health* **2020**, *17*, 422. [CrossRef] [PubMed]
49. Montag, C.; Lachmann, B.; Herrlich, M.; Zweig, K. Addictive Features of Social Media/Messenger Platforms and Freemium Games against the Background of Psychological and Economic Theories. *Int. J. Environ. Res. Public Health* **2019**, *16*, 2612 [CrossRef]

Dong, H.; Yang, F.; Lu, X.; Hao, W. Internet addiction and related psychological factors among children and adolescents in china during the coronavirus disease 2019 (COVID-19) epidemic. *Front. Psychiatry* **2020**, *11*, 751. [CrossRef]
Lopez-Fernandez, O.; Honrubia-Serrano, M.L.; Freixa-Blanxart, M.; Gibson, W. Prevalence of problematic mobile phone use in British adolescents. *Cyberpsychol, Behav. Soc. Netw.* **2014**, *17*, 91–98. [CrossRef]
Bőthe, B.; Tóth-Király, I.; Zsila, Á.; Griffiths, M.D.; Demetrovics, Z.; Orosz, G. The development of the problematic pornography consumption scale (PPCS). *J. Sex Res.* **2018**, *55*, 395–406. [CrossRef]
Cheng, C.; Li, A.Y.-L. Internet addiction prevalence and quality of (real) life: A meta-analysis of 31 nations across seven world regions. *Cyberpsychol. Behav. Soc. Netw.* **2014**, *17*, 755–760. [CrossRef]
Van Rooij, A.J.; Prause, N. A critical review of "Internet addiction" criteria with suggestions for the future. *J. Behav. Addict.* **2014**, *3*, 203–213. [CrossRef]
Volkow, N.D. Collision of the COVID-19 and Addiction Epidemics. *Ann. Intern. Med.* **2020**, *173*, 61–62. [CrossRef]
Király, O.; Potenza, M.N.; Stein, D.J.; King, D.L.; Hodgins, D.C.; Saunders, J.B.; Griffiths, M.D.; Gjoneska, B.; Billieux, J.; Brand, M.; et al. Preventing problematic internet use during the COVID-19 pandemic: Consensus guidance. *Compr. Psychiatry* **2020**, *100*, 152180. [CrossRef] [PubMed]
Mestre-Bach, G.; Blycker, G.R.; Potenza, M.N. Pornography use in the setting of the COVID-19 pandemic. *J. Behav. Addict.* **2020**, *9*, 181–183. [CrossRef] [PubMed]
Lopez-Fernandez, O. Considering existing classic and contemporary proposals for preventing online addiction problems: Some old recipes for new problems. *J. Behav. Addict.* **2020**. [CrossRef] [PubMed]
Montag, C.; Elhai, J.D. Discussing digital technology overuse in children and adolescents during the COVID-19 pandemic and beyond: On the importance of considering Affective Neuroscience Theory. *Addict. Behav. Rep.* **2020**, *12*, 100313. [CrossRef] [PubMed]
Elhai, J.D.; Yang, H.; Levine, J.C. Applying fairness in labeling various types of internet use disorders. *J. Behav. Addict.* **2020**. [CrossRef] [PubMed]

International Journal of
Environmental Research and Public Health

MDPI

Article

Development and Validation of the Parental Smartphone Use Management Scale (PSUMS): Parents' Perceived Self-Efficacy with Adolescents with Attention Deficit Hyperactivity Disorder

Yi-Ping Hsieh [1], Cheng-Fang Yen [2,*] and Wen-Jiun Chou [3,*]

1 Department of Social Work, College of Nursing and Professional Disciplines, University of North Dakota, Grand Forks, ND 58202, USA; yiping66@gmail.com
2 Department of Psychiatry, School of Medicine, and Graduate Institute of Medicine, College of Medicine, Kaohsiung Medical University, Kaohsiung 807, Taiwan
3 Department of Child and Adolescent Psychiatry, Chang Gung Memorial Hospital, Kaohsiung Medical Center and College of Medicine, Chang Gung University, Kaohsiung 833, Taiwan
* Correspondence: chfaye@cc.kmu.edu.tw (C.-F.Y.); wjchou@cgmh.org.tw (W.-J.C.)

Received: 7 March 2019; Accepted: 16 April 2019; Published: 21 April 2019

Abstract: The psychometric properties of the Parental Smartphone Use Management Scale (PSUMS) and its prospective relationships with symptoms of smartphone addiction and attention deficit hyperactivity disorder (ADHD) were studies in a sample of parents of adolescents with ADHD. This is a scale to measure parents' perceived self-efficacy on managing their children's smartphone use. Construct validity (exploratory factor analysis and confirmatory factor analysis), criterion-related validity (known-group validity and concurrent validity), and reliability (Cronbach's alpha) were performed for data analyses. The results showed that the PSUMS had good factorials validity and high reliabilities, with Cronbach's alphas ranging between 0.93 and 0.95. The 17-item PSUMS accounted for 78.58% of the total variance and contains three theoretically and statistically appropriate subscales: reactive management, proactive management, and monitoring. Strong relationships were found between parental smartphone use management and symptoms of smartphone addiction and ADHD in expected directions. Moreover, parents of children with smartphone addiction yielded lower scores on all three PSUMS subscales than parents of children without smartphone addiction. The PSUMS is considered a valuable and reliable tool in the study of parental management on their adolescent children's smartphone use, while providing us with important targets for intervention.

Keywords: smartphone use; parental management; scale validation; attention deficit hyperactivity disorder (ADHD); smartphone addiction

1. Introduction

Rapid and continuing growth in the ownership of smartphones means they have now become a central gateway to online services and information. In Taiwan, approximately 73.4% of people own a smartphone, compared with 71.8% in Singapore, 70.4% in South Korea, 63.9% in the United States, and 43.8% in China [1]. In terms of age, smartphone ownership is highest among younger adults aged 18–29 (85%) [1]. The ownership of smartphones among teenagers has also surged. For instance, Adestra [2] found that approximately 87% of teenagers aged 14–18 years have smartphones, compared with 92% of adults aged 19–34 years and 65% of adults aged 56–67 years. In Taiwan, smartphone ownership among adolescents (aged 12–17 years) ranges from 78.8% to 93.3% and is 48.7% among children (aged 6–11 years) [3]. Although half (51.3%) of elementary school children do not own a smartphone, 38.8% of these children often use a family member's smartphone. Among adolescents, smartphones are the most widely used computing devices (87%), followed by laptops (83%), tablets (51%), and desktop computers (43%) [2]. Adolescents often used a smartphone after school (82%), during class recess (74%), while taking transportation (63%), while eating food (57%), and in class (16%) [4].

Compared with adults, adolescents are more vulnerable to smartphone addiction because of the developmental stage of their brains and because they tend to have lower levels of self-control over their impulses to pursue pleasure [5,6]. A 2015 study by the Taiwan Ministry of Education reported that the prevalence of smartphone addiction among adolescents in Taiwan ranges from 13.9% to 25.7% and is even more prevalent than Internet addiction (12.6–19.5%). Smartphone addiction is associated with health problems and negative social, psychological, and behavioral effects such as sleep disturbance, depression, anxiety, low emotional intelligence, poor academic performance, risky behaviors (using a smartphone while driving or walking), and problematic behaviors [7–10]. Adolescent smartphone addiction is therefore considered a serious problem that warrants prevention and intervention.

Aside from smartphone addiction, online safety is also a major parental worry as an increasing number of children and teenagers gain access to the Internet through smartphones and tablets [11,12]. Among the common concerns that parents have about their children's Internet and smartphone use are Internet and smartphone addiction, cyberbullying, online solicitation, sexting, cyberstalking, online friendships, online reputation, and privacy issues [12,13]. Cyberbullying and sexting, for example, are associated with psychological and behavioral health problems among adolescents [14,15] and have even been linked to suicide ideation and attempts [16,17].

Erikson's theory of stages of development suggests that adolescents and young adults (in the fidelity and love stage) are eager to establish friendships and intimate relationships online and to blend their identities with those of their friends; they are also concerned about how they appear to others [18,19]. However, online environments may create a false sense of intimacy and intention. Cyber-stalkers may simulate ideal personas that lead adolescents to misjudge the intentions behind their messages; thereafter, cyber-stalkers may engage in obsessive relational intrusion which can be defined as "the repeated and unwanted pursuit of intimacy through violation of physical and/or symbolic privacy" [20]. An adolescent's stage of cognitive development and their lack of life experience, along with the anonymity of the online environment, leads to greater risk-taking behavior which puts them at risk and compromises their online safety. Given the potential risks of cybercrime and problematic Internet and smartphone use, it is essential for parents to manage, educate, and communicate with their adolescent children and guide their use of digital technology.

Attention deficit hyperactivity disorder (ADHD) is the most common psychiatric disorder among adolescents with an Internet addiction [21]. ADHD has been found to be associated with Internet addiction in both cross-sectional [22] and longitudinal studies [23]. The symptoms and characteristics of ADHD include a tendency to be bored, impaired inhibition, and low achievement, and it is associated with a higher risk of Internet addiction [24] and higher levels of parent–child conflict [25]. Parents should be able to educate their adolescent children and communicate with them about online behavior and safety, as well as executing plans to manage their adolescents' Internet and smartphone use. Therefore, a reliable metric is needed to assess self-perceived parents' self-efficacy regarding management of their ADHD adolescent children's smartphone use. The Parental Smartphone Use Management Scale (PSUMS) was therefore developed and validated in this study. In sum, this study tends to understand parents' self-efficacy and engagement in managing children's smartphone use, explore the useful parenting strategies to prevent children's smartphone addiction and victimization, and develops a reliable and valid scale.

2. Materials and Methods

2.1. Participants and Procedure

Parents of adolescents aged between 11 and 18 years who had received a diagnosis of ADHD according to the diagnostic criteria in the Diagnostic and Statistical Manual of Mental Disorders, Fifth Edition (DSM-5) [26] were recruited for this study between August 2014 and July 2015 from the child and adolescent psychiatric outpatient clinics of two medical centers in Kaohsiung, Taiwan. Two child psychiatrists conducted diagnostic interviews with the parents and made ADHD diagnoses based on the criteria in the DSM-5. Multiple data sources—including clinical observation of each adolescent's behavior and parental ratings of ADHD symptoms on the short version of the Swanson, Nolan, and Pelham, Version IV, Scale—Chinese version (SNAP-IV) [27,28]—were used to support each diagnosis. Adolescents who had an intellectual disability or autism spectrum disorder with difficulties in communication were excluded. Parents who had an intellectual disability, schizophrenia, bipolar disorder, or any cognitive deficits that resulted in significant communication difficulties were also excluded. A total of 237 parents of adolescents with an ADHD diagnosis were invited to participate in the study. Of these, six (2.5%) declined to participate and 20 (8.4%) reported that their children had not used smartphones in the previous month. Thus, 211 (89.0%) of the invited parents participated in the study and were interviewed by research assistants using the research questionnaire—the PSUMS, the Problematic Cellular Phone Use Questionnaire (PCPU-Q), and the short version of the SNAP-IV Chinese version. The Institutional Review Boards (IRBs) of Kaohsiung Medical University and Chang Gung Memorial Hospital, Kaohsiung Medical Center, approved the study (KMUHIRB-20130131). All participants were fully informed about the study and provided informed consent.

2.2. Measures

2.2.1. Parental Smartphone Use Management Scale (PSUMS)

Prior to developing the PSUMS, an item pool was established by conducting a literature review and a focus group. The item pool contained 20 items. A 7-point Likert scale was used to rate the level of agreement with items, ranging from 0 (no efficacy at all) to 6 (very strong efficacy). To ensure the face validity of the scale, psychiatrists, psychologists, and parenting specialists were consulted about the content. In response to their opinions and criticism, the scale was revised to use culture-sensitive wording and remove irrelevant content. Parents were then asked to rate their self-perceived efficacy in managing their children's smartphone use over the previous month. High PSUMS scores indicate high levels of self-efficacy.

2.2.2. Problematic Cellular Phone Use Questionnaire (PCPU-Q)

The PCPU-Q was used to assess the criterion-related validity of the PSUMS. This scale measures adolescents' problematic smartphone use [29] and comprises 12 items, each of which is rated on a 4-point Likert scale. It was developed according to the taxonomies of substance dependence classified in the Diagnostic and Statistical Manual of Mental Disorders, Fourth Edition (DSM-4) (Text Revision) [30] and modified according to the taxonomies of substance use disorder classified in the DSM-5 [26]. The PCPU-Q comprises two subscales—symptoms of problematic cellular phone use (CPU) and functional impairment caused by CPU. The first eight items address symptoms of problematic CPU among adolescents in the preceding month, including tolerance, withdrawal, CPU for longer periods of time or at higher frequencies than intended, persistent desire or unsuccessful attempts to reduce CPU, excessive time spent on CPU, giving up or reducing the number of important social, academic, or recreational activities because of CPU, continued heavy CPU despite knowledge of the physical or psychological problems caused, and an urge to use the cellular phone. The five remaining items gauge the functional impairment exhibited by an adolescent in the preceding month as a result of CPU. Higher overall scores indicate more severe levels of problematic CPU symptoms and CPU-related functional impairments. The internal reliability (Cronbach's α) of the PCPU-Q subscales was 0.92 and 0.92, respectively. According to the scale developers, when participants report four out of eight problematic CPU symptoms and at least one functional impairment, they were categorized into the cellarer-phone-addiction group for screening and analysis purposes.

2.2.3. ADHD Symptoms

The short version of the SNAP-IV (Chinese version) was used to assess the severity of ADHD symptoms exhibited in the preceding month. This version comprises 26-items encompassing the core DSM-4-derived ADHD subscales of inattention, hyperactivity and impulsivity, and the oppositional symptoms of oppositional defiant disorder [27,28]. Each item is rated on a 4-point Likert scale ranging from 0 (not at all) to 3 (very much). The Cronbach's α for inattention, hyperactivity and impulsivity, and oppositional behavior were 0.91, 0.90, and 0.93, respectively, indicating very high internal consistency.

2.3. *Analysis*

The PSUMS was administered to the 211 study participants to determine its psychometric characteristics. Construct validity was determined by using factor analysis. Specifically, 211 participants were randomly divided into two subsamples using the RANDBETWEEN function in EXCEL. The first subsample (n = 103) was then used for exploratory factor analysis (EFA) and the second sample (n = 108) was used for confirmatory factor analysis (CFA). The EFA was conducted using SPSS 24 statistical software (SPSS Inc., Chicago, IL, USA). Based on the assumption that the factors are correlated, principal axis analysis was carried out with Promax rotation, after which the Kaiser–Mayer–Olkin (KMO) measure of sampling adequacy and Bartlett testing were applied. A KMO value of >0.60 and significant statistics from Bartlett testing suggested the data was suitable for factor analysis [31]. The amount of variance accounted for indicates how well a relevant notion or construct can be measured [32].

Since the multivariate skewness (4929.18) and multivariate kurtosis (40.25) indicate the non-normal distribution for the CFA subsample, maximum likelihood with Satorra-Bentler correction was performed to determine whether the model data fit with the item-factor structures obtained from the EFA. The CFA was conducted using LAVAAN package in the R software [33]. In general, the results indicate a good fit when the standardized root mean squared residual (SRMR) below 0.08 [34,35], root mean square error of approximation (RMSEA) below 0.08 [36,37], comparative fit index (CFI) and non-normed fit index (NNFI) above 0.9 [38,39], and a normed chi-square (i.e., chi-square/df) less than 2 [31] or less than 5 [40].

Two approaches for testing criterion-related validity were applied: known-group validity and concurrent validity. Known-group validity was determined by performing independent sample t test between the smartphone-addiction group and non-smartphone-addiction group. The concurrent validity of the PSUMS was determined by examining the correlation between the PSUMS and the PCPU-Q. Internal consistency was tested to assess the reliability of the scale.

3. Results

The preliminary analysis indicated that none of the study variables were skewed or kurtotic. The sociodemographic characteristics of parents and adolescents, adolescents' ADHD symptoms, and smartphone-addiction symptoms are presented in Table 1. Descriptive statistics along with both the EFA and CFA factor loadings of the PSUMS items are presented in Table 2.

Table 1. Sociodemographic characteristics of parents and adolescents and the ADHD symptoms.

Variables	n (%)	Mean (SD)	Range
Sex of parents			
Female	175 (82.9%)		
Male	36 (17.1%)		
Sex of adolescents			
Female	28 (13.3%)		
Male	183 (86.7%)		
Age of parents (years)		43.5 (5.9)	32–64
Age of adolescents (years)		13.7 (1.8)	11–18
Marriage status of parents			
Intact	170 (80.6%)		
Not intact	41 (19.4%)		
Education duration of parents (years)		13.6 (2.8)	6–28
SNAP-IV symptoms of adolescents			
Inattention		12.8 (6.1)	0–27
Hyperactivity/impulsivity		8.9 (6.0)	0–27
Oppositional defiant		9.9 (5.7)	0–24
PCPU-Q			
Smartphone addiction	40 (19%)		
No smartphone addiction	171 (81%)		

Note: SNAP-IV: Short version of the Swanson, Nolan, and Pelham Version IV Scale—Chinese version. PCPU-Q: Problematic Cellular Phone Use Questionnaire. *SD* = standard deviation.

Table 2. Means, standard deviations, Cronbach's α, and factor loadings for the items in the PSUMS.

Items	Mean (SD) (n = 211)	EFA n = 103	CFA n = 108
Reactive Management (α = 0.93)			
I manage my child's smartphone use to prevent it from negatively affecting his/her daily life	4.35 (1.43)	0.73	0.89
I manage how and to what extent my child spends money on his/her smartphone	4.64 (1.44)	0.66	0.81
I don't allow my child to use a smartphone while doing homework	4.42 (1.48)	0.59	0.86
I manage my child's smartphone use outside of the house	3.74 (1.76)	0.59	0.80
I effectively manage when my child can and cannot use a smartphone	4.18 (1.61)	0.57	0.88
I manage my child's activities to prevent him/her from breaking laws	4.30 (1.49)	0.53	0.83
When my child is spending too much time on a smartphone, I manage his/her smartphone use effectively	4.38 (1.40)	0.52	0.90
Proactive Management (α = 0.95)			
I don't distress my child when communicating with him/her about smartphone use	3.72 (1.58)	0.93	0.91
I don't create family tension as a result of enforcing smartphone use guidelines for my child	3.81 (1.55)	0.82	0.93
I discuss and reason with my child	4.11 (1.37)	0.80	0.93
I don't get angry when I manage my child's smartphone use	3.53 (1.61)	0.72	0.83
I communicate with my child effectively and explain why I manage his/her smartphone use	4.22 (1.33)	0.70	0.93
I actively learn new information and skills to manage my child's smartphone use	4.14 (1.44)	0.55	0.82
Monitoring (α = 0.93)			
I know who my child talks with and what they talk about when using a smartphone	3.81 (1.69)	0.91	0.96
I know what my child does on the smartphone	3.86 (1.64)	0.84	0.95
I monitor which apps my child uses	3.79 (1.69)	0.72	0.88
I restrict the type of websites my child is allowed to visit on the smartphone	3.91 (1.69)	0.63	0.85

Note: EFA = exploratory factor analysis; CFA = confirmatory factor analysis; *SD* = standard deviation.

3.1. Construct Validity

3.1.1. Exploratory Factor Analysis (EFA)

An EFA was conducted on a sample of 211 parents of children with ADHD. The KMO coefficient of sampling adequacy was 0.90 which lies within the excellent range. Bartlett's Test of Sphericity, which assesses whether a matrix differs from the identity matrix, yielded significant results, indicating that the matrix did not resemble the identity matrix; this also supported the presence of factors within the data.

Principal axis factor analysis was conducted and the Promax rotation method, which assumes factors are correlated and rotates the factor structure, was used to determine the factor solutions. The initial result confirmed the proposed three-factor solution for all 20 items. After eliminating three items with cross-loadings, a factor analysis was run once more on the remaining 17 items. The results supported a three-factor solution that explained 78.58% of the total variance. The items and factor loadings for the subscales are presented in Table 2. The three-factor solution fits with the theoretical factors used to devise the measurement tool. The first factor, "reactive management" (α = 0.93) includes seven items with factor loadings of 0.52–0.73; the items reflect parents' reactive management of children's smartphone use through rule-setting practices, responding to and controlling this use to avoid negative impacts on children's daily-life functioning. The second subscale, "proactive management" (α = 0.95) includes 6 items with factor loadings of 0.55–0.93; the items reflect parents' perceived efficacy of their proactive management and active mediation of children's smartphone use through positive communication and reasoning. The third factor, "monitoring" (α = 0.93) includes 4 items with factor loadings of 0.63–0.91; the items reflect parents' behavior in monitoring what their children do on their smartphones, whom they talk with, what applications they use, and the websites they visit. The score for each subscale was calculated by its mean. Pearson's correlations among the dimensions of the PSUMS are presented in Table 3.

Table 3. Correlation among the dimensions of the PSUMS.

PSUMS Dimensions	Reactive Management	Proactive Management	Monitoring
Reactive management	-		
Proactive management	0.79 **	-	
Monitoring	0.77 **	0.68 **	-

Note: ** $p < 0.01$.

3.1.2. Confirmatory Factor Analysis

The CFA fit indices were acceptable: *CFI* = 0.934, *NNFI* = 0.923, *RMSEA* = 0.077, and *SRMR* = 0.053, normed chi-square = 1.64; except for the significant chi-square test (chi-square = 189.68, *df* = 116; $p < 0.001$). Moreover, the factor loadings were strong in each factor: loadings = 0.80 to 0.90 in Reactive Management; 0.82 to 0.93 in Proactive Management; and 0.85 to 0.96 in Monitoring. Overall, the current model showed an acceptable fit to the data.

3.2. Criterion Validity

The PCPU-Q was used to determine the criterion-related validity of the PSUMS. After checking the equality of the variance, independent sample *t* test results provided strong evidence for known-groups validity, a subtype of criterion-related validity. We compared the PSUMS results between the parents of children with and without smartphone addiction. The results showed that parents of children with smartphone addiction yielded lower scores on all three PSUMS subscales than parents of children without smartphone addiction (Table 4). The correlation coefficients between three subscales of the PSUMS and two subscales of the PCPU-Q were also calculated. The results of the Pearson's correlation provided strong evidence for concurrent validity (Table 5). The PSUMS was found to be significantly correlated with the PCPU-Q (all $p < 0.001$).

Table 4. Parental smartphone use management among children with and without smartphone addiction.

PSUMS Dimensions	Have Smartphone Addiction (n = 40) Mean (SD)	No Smartphone Addiction (n = 171) Mean (SD)	t	p
Reactive management	3.30 (1.46)	4.52 (1.14)	4.93	<0.001
Proactive management	2.90 (1.39)	4.11 (1.27)	5.34	<0.001
Monitoring	2.94 (1.56)	4.05 (1.47)	4.25	<0.001

Note: SD: Standard Deviation.

Table 5. Correlation between Parental Smartphone Use Management Scale (PSUMS) and PCPU-Q.

PCPU-Q Dimensions	Parental Smartphone Use Management (PSUMS)		
	Reactive Management	Proactive Management	Monitoring
	r	r	r
Smartphone addiction symptoms	−0.35 **	−0.40 **	−0.33 **
Functional impairments	−0.37 **	−0.33 **	−0.26 **

Note: ** $p < 0.01$. PCPU-Q: Problematic Cellular Phone Use Questionnaire.

3.3. Reliability

The internal consistency coefficient (Cronbach's α) of the PSUMS was 0.96. As shown in Table 2, the internal consistency coefficients in the sub-dimensions of the PSUMS were 0.93 for Reactive management, 0.95 for Proactive management, and 0.93 for Monitoring.

4. Discussion

In this study we developed and validated the PSUMS to measure parents' perceived efficacy in managing the smartphone use of adolescents with ADHD. The final version of the 17-item PSUMS accounted for 78.58% of the total variance and contains three theoretically and statistically appropriate subscales: reactive management, proactive management, and monitoring. The three-factor hypothesis was supported by the scale's psychometric properties, including its construct validity (confirmed using EFA and CFA), criterion-related validity (known-group validity and concurrent validity), and reliability (internal consistency measured using Cronbach's alpha). The PSUMS adhered to psychometric standards and was shown to be a concise and promising measure of whether parents of adolescents with ADHD have the knowledge and skills required to successfully manage their children's smartphone use.

The PSUMS consists of three scales. With a rating scale from 0 to 6, an increase in scores corresponds to an increase in parents' efficacy in managing their children's smartphone use. The three smartphone-management items parents feel most confident about were in the sub-dimension "reactive management". These were: "I manage how and to what extent my child spends money on his/her smartphone", "I don't allow my child to use a smartphone while doing homework", and "when my child is spending too much time on a smartphone, I stop his/her smartphone use effectively". The two smartphone-management items they felt least confident about were in the sub-dimension "proactive management", which involved positive communication and reasoning. These were: "I don't get angry when I manage my child's smartphone use", and "I don't distress my child when communicating with him/her about smartphone use". Moreover, parents also produced low scores in regard to managing children's smartphone use outside of the house.

In terms of criterion-related validity, the subscales of the PSUMS were significantly correlated in the expected direction with the PCPU-Q. The first subscale, reactive management, concerns parental intervention in adolescents' smartphone overuse, money spending, cybercrime involvement, poor timing of use, and other smartphone behaviors that may negatively affect their daily lives. Adolescents whose parents use moderate behavioral control reported fewer problem behaviors [41,42]. Similarly, parents who set rules on media time reported that their children (aged 0–6 years) watched less television on average than other children in their age group [43]. As expected, reactive management was negatively correlated with smartphone addiction and functional impairments. Parents who scored lower on reactive management were more likely to have children with a smartphone addiction and functional impairments. The second subscale, proactive management, addresses parents' positive communication, affection toward, and reasoning with adolescents about their smartphone use. Research has shown that adolescents who exhibit the symptoms and characteristics of ADHD are more likely than other children to experience relatively high levels of parent–child conflict [25]. In our study, effective communication, high levels of affection, and reasoning around smartphone use behaviors with adolescents with ADHD was found to be a key element in the successful prevention of problematic smartphone use. Proactive management enables adolescents to discuss their feelings, negotiate rules, and communicate in a safe and caring environment. This type of parenting practice has been linked to prosocial behavior, less externalizing behavior, good mental health, and higher achievement [44–47]. As expected, proactive management was negatively correlated with smartphone addiction and functional impairments. Finally, the third subscale, parental monitoring, was defined as "a set of correlated parenting behaviors involving attention to and tracking of the child's whereabouts, activities, and adaptations" on their smartphone [48]. Parental monitoring has been associated with lower levels of alcohol and substance use [49,50]. Similarly, lower levels of parental monitoring have been found to correlate with a higher percentage of time spent on nonacademic computer use [51]. In this study, parental monitoring was negatively associated with smartphone addiction and functional impairment. The PSUMS was significantly correlated in the expected direction with the PCPU-Q, which confirmed its concurrent validity.

The PSUMS conceptualized and measured parents' self-efficacy regarding their management of the smartphone use of adolescents with ADHD. However, this research had some limitations that need to be addressed. The sample sizes used for EFA and CFA were relatively small (i.e., 103 for EFA and 108 for CFA, total of 211). However, we believed that our contribution adds on the current literature because it is hard to collect more than a hundred adolescents with diagnosed ADHD. Furthermore, studies have shown that our sample sizes were acceptable for both EFA and CFA. For EFA, a rule of thumb to calculate the sample size is using an item-participant ratio of 5 [52]. Given that the original item number of the PSUMS is 20, 100 was a sufficient number for conducting EFA. Regarding CFA, although the consensus is that 200 is a preferable sample size, Iacobucci argue that sample size at 50 is sufficient [53]. Anderson and Gerbing [54] and Kline [55] agree that using a sample size of 100 could be the minimum criterion for doing CFA. Therefore, we tentatively concluded that our sample sizes were acceptable in both EFA and CFA [56]. Nevertheless, future studies are warranted to collect a larger sample size to corroborate our findings. Although construct validity, known-groups validity and concurrent validity were supported by the results, another limitation is that all subscales of PSUMS are significantly correlated with all subscales of PCPU-Q (the criterion scale), which indicated that subscales of PSUMS may not be distinguished. However, when performing the calculation for the test of the differences between two dependent correlations with one variable in common [57], we found a significant difference between two correlations of PSUMS subscales (reactive management and monitoring) with one subscale of PCPU-Q (functional impairment) in common as criterion ($z = 2.49$, $p < 0.01$). This limitation maybe because of the nature of smartphone addiction (PCPU-Q) is related to parental smartphone use management (PSUMS), thus subscales of PSUMS cannot be fully distinguished. Future studies may use monitoring scales as criterion in testing discriminant validity in PSUMS. Despite these limitations, the overall results showed that the PSUMS has good reliability and adequate validity. The reliability and validity of the scale mean it can be used to test different samples. Furthermore, the PSUMS could also be used as a tool for developing smartphone-addiction prevention programs for adolescents.

5. Conclusions

The PSUMS is considered a valuable and reliable tool in the study of parental management on their adolescent children's smartphone use, while providing us with important targets for intervention.

This study provides a foundation for future research in education, family science, and technology science by examining the multidimensional factors associated with parents' efficacy in managing adolescents' smartphone use in the digital age. Because Internet and smartphone are widely used and become an important part of our daily life, it is essential to enhance the traditional parenting and educational programs with consideration of the management on children/adolescents' smartphone use, strength parents' efficacy in doing so, and prepare parents with right tools to face this new challenge in parenting.

Author Contributions: Conceptualization, Y.-P.H., C.-F.Y. and W.-J.C.; methodology, Y.-P.H.; software, Y.-P.H.; validation, Y.-P.H., C.-F.Y. and W.-J.C.; formal analysis, Y.-P.H.; investigation, C.-F.Y. and W.-J.C.; resources, Y.-P.H., C.-F.Y. and W.-J.C.; data curation, Y.-P.H, C.-F.Y. and W.-J.C.; writing—original draft preparation, Y.-P.H.; writing—review and editing, Y.-P.H., C.-F.Y. and W.-J.C.; visualization, Y.-P.H.; supervision, C.-F.Y. and W.-J.C.; project administration, C.-F.Y. and W.-J.C.; funding acquisition, C.-F.Y. and W.-J.C.

Funding: This research was funded by the Ministry of Science and Technology, Taiwan (grand number: MOST 103-2314-B-182A-013 and 104-2314-B-182A-030).

Conflicts of Interest: The authors declare no conflict of interest. The funders had no role in the design of the study; in the collection, analyses, or interpretation of data; in the writing of the manuscript, or in the decision to publish the results.

References

1. Pew Research Center. A Portrait of Smartphone Ownership. Available online: http://www.pewinternet.org/2015/04/01/chapter-one-a-portrait-of-smartphone-ownership/ (accessed on 4 March 2019).
2. Adestra. Consumer Adoption & Usage Study. Available online: http://www.adestra.com/resources/2016-consumer-adoption-usage-study/ (accessed on 4 March 2019).
3. Taiwan Ministry of Education. 2015 Student Network Usage Survey Report. Available online: https://www.edu.tw/News_Content.aspx?n=0217161130F0B192&s=F1AA06D56E8D6B20 (accessed on 4 March 2019).
4. Yang, S.Y.; Chen, M.D.; Huang, Y.C.; Lin, C.Y.; Chang, J.H. Association between smartphone use and musculoskeletal discomfort in adolescent students. *J. Community Health* **2017**, *42*, 423–430. [CrossRef]
5. Chóliz, M. Mobile-phone addiction in adolescence: The Test of Mobile Phone Dependence (TMD). *Prog. Health Sci.* **2012**, *2*, 33–45.
6. Steinbeis, N.; Haushofer, J.; Fehr, E.; Singer, T. Development of behavioral control and associated vmPFC–DLPFC connectivity explains children's increased resistance to temptation in intertemporal choice. *Cereb. Cortex* **2014**, *26*, 32–42. [CrossRef] [PubMed]
7. Choa, K.-S.; Lee, J.-M. Influence of smartphone addiction proneness of young children on problematic behaviors and emotional intelligence: Mediating self-assessment effects of parents using smartphones. *Comput. Hum. Behav.* **2017**, *66*, 303–311. [CrossRef]
8. Hawi, N.S.; Samaha, M. To excel or not to excel: Strong evidence on the adverse effect of smartphone addiction on academic performance. *Comput. Educ.* **2016**, *98*, 81–89. [CrossRef]
9. Thomée, S.; Härenstam, A.; Hagberg, M. Mobile phone use and stress, sleep disturbances, and symptoms of depression among young adults—A prospective cohort study. *BMC Public Health* **2011**, *11*, 1–11. [CrossRef] [PubMed]
10. White, M.P.; Eiser, J.R.; Hagberg, M. Risk perceptions of mobile phone use while driving. *Risk Anal. Int. J.* **2004**, *24*, 323–334. [CrossRef] [PubMed]
11. Boyd, D.; Hargittai, E. Connected and concerned: Variation in parents' online safety concerns. *Policy Internet* **2013**, *5*, 245–269. [CrossRef]
12. Madden, M.; Cortesi, S.; Gasser, U.; Lenhart, A.; Duggan, M. Parents, Teens, and Online Privacy (Pew Internet & American Life Project). Available online: http://pewinternet.org/Reports/2012/Teens-and-Privacy.aspx (accessed on 19 April 2019).
13. George, M.J.; Odgers, C.L. Seven fears and the science of how mobile technologies may be influencing adolescents in the digital age. *Perspect. Psychol. Sci.* **2015**, *10*, 832–851. [CrossRef]
14. Gámez-Guadix, M.; Orue, I.; Smith, P.K.; Calvete, E. Longitudinal and reciprocal relations of cyberbullying with depression, substance use, and problematic internet use among adolescents. *J. Adolesc. Health* **2013**, *53*, 446–452. [CrossRef] [PubMed]

15. Gordon-Messer, D.; Arturo Bauermeister, J.; Grodzinski, A.; Zimmerman, M. Sexting among young adults. *J. Adolesc. Health* **2012**, *52*, 301–306. [CrossRef]
16. Curnutt, H. Flashing your phone: Sexting and the remediation of teen sexuality. *Commun. Q.* **2012**, *60*, 353–369. [CrossRef]
17. Gini, G.; Espelage, D.L. Peer victimization, cyberbullying, and suicide risk in children and adolescents. *JAMA* **2014**, *312*, 545–546. [CrossRef] [PubMed]
18. Subrahmanyam, K.; Greenfield, P.M.; Tynes, B. Constructing sexuality and identity in an online teen chat room. *J. Appl. Dev. Psychol.* **2004**, *25*, 651–666. [CrossRef]
19. Subrahmanyam, K.; Smahel, D.; Greenfield, P. Connecting developmental constructions to the Internet: Identity presentation and sexual exploration in online teen chat rooms. *Dev. Psychol.* **2006**, *42*, 395–406. [CrossRef] [PubMed]
20. Cupach, W.R.; Spitzberg, B.H. *The Dark Side of Relationship Pursuit: From Attraction to Obsession and Stalking*; Lawrence Erlbaum Associates: Mahwah, NJ, USA, 2014.
21. Bozkurt, H.; Coskun, M.; Ayaydin, H.; Adak, I.; Zoroglu, S.S. Prevalence and patterns of psychiatric disorders in referred adolescents with Internet addiction. *Psychiatry Clin. Neurosci.* **2013**, *67*, 352–359. [CrossRef]
22. Yen, J.Y.; Ko, C.H.; Yen, C.F.; Wu, H.Y.; Yang, M.J. The comorbid psychiatric symptoms of Internet addiction: Attention deficit and hyperactivity disorder (ADHD), depression, social phobia, and hostility. *J. Adolesc. Health* **2007**, *41*, 93–98. [CrossRef]
23. Chen, Y.L.; Chen, S.H.; Gau, S.S.F. ADHD and autistic traits, family function, parenting style, and social adjustment for Internet addiction among children and adolescents in Taiwan: A longitudinal study. *Res. Dev. Disabil.* **2015**, *39*, 20–31. [CrossRef]
24. Ko, C.H.; Yen, J.Y.; Yen, C.F.; Chen, C.S.; Chen, C.C. The association between Internet addiction and psychiatric disorder: A review of the literature. *Eur. Psychiatry* **2012**, *27*, 1–8. [CrossRef] [PubMed]
25. Edwards, G.; Barkley, R.A.; Laneri, M.; Fletcher, K.; Metevia, L. Parent–adolescent conflict in teenagers with ADHD and ODD. *J. Abnorm. Child Psychol* **2001**, *29*, 557–572. [CrossRef]
26. American Psychiatric Association. *Diagnostic and Statistical Manual of Mental Disorders (DSM-5®)*; American Psychiatric Association: Washington, DC, USA, 2013.
27. Gau, S.S.F.; Shang, C.Y.; Liu, S.K.; Lin, C.H.; Swanson, J.M.; Liu, Y.C.; Tu, C.L. Psychometric properties of the Chinese version of the Swanson, Nolan, and Pelham, version IV scale–parent form. *Int. J. Methods Psychiatr. Res.* **2008**, *17*, 35–44. [CrossRef]
28. Swanson, J.M.; Kraemer, H.C.; Hinshaw, S.P.; Arnold, L.E.; Conners, C.K.; Abikoff, H.B.; Clevenger, W.; Davies, M.; Elliott, G.R.; Greenhill, L.L.; et al. Clinical relevance of the primary findings of the MTA: Success rates based on severity of ADHD and ODD symptoms at the end of treatment. *J. Am. Acad. Child Adolesc. Psychiatry* **2001**, *40*, 168–179. [CrossRef] [PubMed]
29. Yen, C.F.; Tang, T.C.; Yen, J.Y.; Lin, H.C.; Huang, C.F.; Liu, S.C.; Ko, C.H. Symptoms of problematic cellular phone use, functional impairment and its association with depression among adolescents in Southern Taiwan. *J. Adolesc.* **2009**, *32*, 863–873. [CrossRef]
30. American Psychiatric Association. *Diagnostic and Statistical Manual of Mental Disorders*, 4th ed.; American Psychiatric Association: Washington, DC, USA, 2000.
31. Tabachnick, B.G.; Fidell, L.S. *Using Multivariate Statistics*, 5th ed.; Allyn and Bacon: New York, NY, USA, 2007.
32. Hardy, M.; Bryman, A. *Handbook of Data Analysis*; Sage: Thousand Oaks, CA, USA, 2004.
33. Rosseel, Y. lavaan: An R Package for Structural Equation Modeling. *J. Stat. Softw.* **2012**, *48*, 1–36. Available online: http://www.jstatsoft.org/v48/i02/ (accessed on 6 April 2019). [CrossRef]
34. Hu, L.T.; Bentler, P.M. Fit indices in covariance structure modeling: Sensitivity to under-parameterized model misspecification. *Psychol. Methods* **1998**, *3*, 424–453. [CrossRef]
35. Chang, C.-C.; Su, J.-A.; Chang, K.-C.; Lin, C.-Y.; Koschorke, M.; Thornicroft, G. Perceived stigma of caregivers: Psychometric evaluation for Devaluation of Consumer Families Scale. *Int. J. Clin. Health Psychol* **2018**, *18*, 170–178. [CrossRef] [PubMed]
36. MacCallum, R.C.; Browne, M.W.; Sugawara, H.M. Power analysis and determination of sample size for covariance structure modeling. *Psychol. Methods* **1996**, *1*, 130–149. [CrossRef]
37. Wu, T.-H.; Chang, C.-C.; Chen, C.-Y.; Wang, J.-D.; Lin, C.-Y. Further psychometric evaluation of the Self-Stigma Scale-Short: Measurement invariance across mental illness and gender. *PLoS ONE* **2015**, *10*, e0117592. [CrossRef] [PubMed]

38. Bentler, P.M.; Bonnet, D.C. Significance Tests and Goodness of Fit in the Analysis of Covariance Structures. *Psychol. Bull.* **1980**, *88*, 588–606. [CrossRef]
39. Lin, C.-Y.; Luh, W.-M.; Cheng, C.-P.; Yang, A.-L.; Ma, H.-I. Evaluating the wording effect and psychometric properties of the Kid-KINDL: Using the multitrait-multimethod approach. *Eur. J. Psychol. Assess.* **2014**, *30*, 100–109. [CrossRef]
40. Wheaton, B.; Muthen, B.; Alwin, D.F.; Summers, G. Assessing Reliability and Stability in Panel Models. *Sociol. Methodol.* **1977**, *8*, 84–136. [CrossRef]
41. Furstenberg, F.F.; Cook, T.D.; Eccles, J.; Elder, G.H.; Sameroff, A.; Ardelt, M.; Bartko, W.T.; Chan, W.S.; Geitz, L.M.; Harper, C.; et al. *Managing to Make It: Urban Families and Adolescent Success*; University of Chicago Press: Chicago, IL, USA, 1999.
42. Galambos, N.L.; Barker, E.T.; Almeida, D.M. Parents do matter: Trajectories of change in externalizing and internalizing problems in early adolescence. *Child Dev.* **2003**, *74*, 578–594. [CrossRef]
43. Vandewater, E.A.; Park, S.E.; Huang, X.; Wartella, E.A. "No—you can't watch that" parental rules and young children's media use. *Am. Behav. Sci.* **2005**, *48*, 608–623. [CrossRef]
44. Padilla-Walker, L.M.; Carlo, G.; Christensen, K.J.; Yorgason, J.B. Bidirectional relations between authoritative parenting and adolescents' prosocial behaviors. *J. Res. Adolesc.* **2012**, *22*, 400–408. [CrossRef]
45. Pinquart, M. Associations of parenting dimensions and styles with externalizing problems of children and adolescents: An updated meta-analysis. *Dev. Psychol.* **2017**, *53*, 873–932. [CrossRef]
46. Steinberg, L.; Lamborn, S.D.; Dornbusch, S.M.; Darling, N. Impact of parenting practices on adolescent achievement: Authoritative parenting, school involvement, and encouragement to succeed. *Child Dev.* **1992**, *63*, 1266–1281. [CrossRef] [PubMed]
47. Uji, M.; Sakamoto, A.; Adachi, K.; Kitamura, T. The impact of authoritative, authoritarian, and permissive parenting styles on children's later mental health in Japan: Focusing on parent and child gender. *J. Child Fam. Stud.* **2014**, *23*, 293–302. [CrossRef]
48. Dishion, T.J.; McMahon, R.J. Parental monitoring and the prevention of child and adolescent problem behavior: A conceptual and empirical formulation. *Clin. Child Fam. Psychol. Rev.* **1998**, *1*, 61–75. [CrossRef] [PubMed]
49. Dodge, K.A.; Malone, P.S.; Lansford, J.E.; Miller, S.; Pettit, G.S.; Bates, J.E. A dynamic cascade model of the development of substance-use onset. *Monogr. Soc. Res. Child Dev.* **2009**, *74*, vii-119. [CrossRef]
50. Kopak, A.M.; Ayers, S.; Lopez, V.; Stevenson, P. Parental monitoring, alcohol, and marijuana use among Hispanic and non-Hispanic White adolescents: Findings from the Arizona Youth Survey. *J. Drug Issues* **2011**, *41*, 461–485. [CrossRef]
51. Epstein, J.A. Factors related to adolescent computer use and electronic game use. *ISRN Public Health* **2011**, *2012*, 1–6. [CrossRef]
52. Gorusch, R.L. *Factor Analysis*, 2nd ed.; Lawrence Erlbaum Associates: Hillsdale, NJ, USA, 1983; p. 332.
53. Iacobucci, D. Structural equations modeling: Fit indices, sample size, and advanced topics. *J. Consum. Psychol.* **2010**, *20*, 90–98. [CrossRef]
54. Anderson, J.C.; Gerbing, D.W. The effect of sampling error on convergence, improper solutions, and goodness-of-fit indices for maximum likelihood confirmatory factor analysis. *Psychometrika* **1984**, *49*, 155–173. [CrossRef]
55. Kline, R.B. *Principles and Practice of Structural Equation Modeling*; Guilford: New York, NY, USA, 1998.
56. Su, C.-T.; Ng, H.-S.; Yang, A.-L.; Lin, C.-Y. Psychometric evaluation of the Short Form 36 Health Survey (SF-36) and the World Health Organization Quality of Life Scale Brief Version (WHOQOL-BREF) for patients with schizophrenia. *Psychol. Assess.* **2014**, *26*, 980–989. [CrossRef] [PubMed]
57. Lee, I.A.; Preacher, K.J. Calculation for the Test of the Difference between Two Dependent Correlations with One Variable in Common [Computer Software]. Available online: http://quantpsy.org (accessed on 7 April 2019).

International Journal of
Environmental Research and Public Health

MDPI

Review

Addictive Features of Social Media/Messenger Platforms and Freemium Games against the Background of Psychological and Economic Theories

Christian Montag [1,*], Bernd Lachmann [1], Marc Herrlich [2] and Katharina Zweig [3]

1 Department of Molecular Psychology, Institute of Psychology and Education, Ulm University, 89081 Ulm, Germany
2 Serious Games Engineering, TU Kaiserslautern, 67663 Kaiserslautern, Germany
3 Algorithm Accountability Lab, TU Kaiserslautern, 67663 Kaiserslautern, Germany
* Correspondence: christian.montag@uni-ulm.de

Received: 26 May 2019; Accepted: 18 July 2019; Published: 23 July 2019

Abstract: Currently about 2.71 billion humans use a smartphone worldwide. Although smartphone technology has brought many advances, a growing number of scientists discuss potential detrimental effects due to excessive smartphone use. Of importance, the likely culprit to understand over-usage is not the smartphone itself, but the excessive use of applications installed on smartphones. As the current business model of many app-developers foresees an exchange of personal data for allowance to use an app, it is not surprising that many design elements can be found in social media apps and Freemium games prolonging app usage. It is the aim of the present work to analyze several prominent smartphone apps to carve out such elements. As a result of the analysis, a total of six different mechanisms are highlighted to illustrate the prevailing business model in smartphone app development. First, these app-elements are described and second linked to classic psychological/economic theories such as the mere-exposure effect, endowment effect, and Zeigarnik effect, but also to psychological mechanisms triggering social comparison. It is concluded that many of the here presented app-elements on smartphones are able to prolong usage time, but it is very hard to understand such an effect on the level of a single element. A systematic analysis would require insights into app data usually only being available for the app-designers, but not for independent scientists. Nevertheless, the present work supports the notion that it is time to critically reflect on the prevailing business model of 'user data in exchange for app-use allowance'. Instead of using a service in exchange for data, it ultimately might be better to ban or regulate certain design elements in apps to come up with less addictive products. Instead, users could pay a reasonable fee for an app service.

Keywords: social media/messenger apps; Facebook; WhatsApp; Internet addiction; smartphone addiction; Internet use disorder; smartphone use disorder

1. Introduction

According to current estimates, about 2.71 billion humans use a smartphone worldwide [1]. Without doubt, one of the major driving forces towards a totally connected world with the Internet available in everyone's pocket was the inception of the iPhone in 2007 [2]. The wide distribution of smartphones led to unprecedented opportunities enabling humans to find their way in unknown territory via services such as Google Maps, to reach out to friends and business partners via a myriad of communication channels, and abundant easy access to knowledge earlier only available via visiting a library and the study of lexicons.

1.1. Perils of Smartphone Use

Despite the many advantages of the smartphone, it is heavily discussed among scientists around the globe if problematic usage of the smartphone could have detrimental effects on our mental health [3,4]. Furthermore, associations between anxiety disorder, depression, and problematic use of the smartphone have been observed (underlining the sincerity of this topic), but the causal relationship of effects between the constructs is still unclear [5]. In any case, new research suggests that the many interruptions due to high frequency use and the many daily incoming messages fragmenting everyday life could reduce productivity at work [6–8] and lower a person's well-being [9,10]. Another research group stresses that the availability of smartphones even might lead to "brain drain", when the device is present on the desk while working on a fluid intelligence task or working memory task [11]. In this context, a recent review summarized the literature on cognitive functions [12]. Therefore, this is area of research is not further presented in detail, here.

Beyond the mentioned issues, (excessive use of) smartphones is known to impact on social communication in terms of reducing smiles when interacting with strangers [13], enjoyment of face-to-face interaction [14] or drawing parent's attention away from their children [15]. This might result in a loss of empathy, although so far only correlations between excessive Internet use and lower empathy/social skills could be established [16,17], with only weak associations between excessive smartphone usage and empathy [18]. Again, note that the latter associations base on correlations and so far it is not clear what is cause and what is effect. Moreover, important insights of a study are mentioned which on the one hand observed reduced social skills in excessive Internet users, but on the other hand observed enhanced empathy when the Internet is used in a healthy way [19].

Beyond these observations, a growing number of scientists are focusing on the question of whether excessive smartphone usage could resemble a full-blown addiction [20–22]. Of importance, 'smartphone addiction' is currently not recognized in the official diagnostic manuals such as the International Classification of Diseases-11th Revision (ICD-11) issued by the World Health Organization (WHO) or the Diagnostic and Statistical Manual of Mental Disorders (DSM-5) issued by the American Psychiatric Association (APA). Both ICD-11 and DSM-5 offer practitioners and researchers alike valuable guidance in psycho-diagnostics. With respect to the general debate on the addictive nature of excessive usage of online content, it is noteworthy that Gaming Disorder, as a specific form of 'Internet addiction', has been included in ICD-11 [23]. This means that a first form of 'Internet addiction' is recognized as an official health condition, now.

Given the nomenclature of 'Gaming Disorder' in ICD-11, but also 'Internet Gaming Disorder' in DSM-5 (see Section 3 in the DSM-5's appendix), many researchers currently prefer the term of 'Internet Use Disorder' (IUD) over the classic term 'Internet addiction' [24–26]. With this wording, scientists also stress that a certain kind of online usage and not the Internet per se poses a problem. Therefore, and in line with a prominent model to characterize and understand IUD by Brand et al. [26], the term IUD is also used in the following. Explicitly, it is mentioned that researchers should not over-pathologize everyday life activities [27] and aim at a unification of terms used in the literature in line with the officially suggested nomenclature.

In this context, one currently might prefer the term Smartphone Use Disorder (SUD) over 'smartphone addiction' and understand SUD best as a mobile form of IUD. Of note, many researchers currently use the term problematic smartphone use (PSU), which from our perspective is also 'problematic', because it is not clear if PSU represents a transit zone from healthy towards psychopathological behavior or the end of the spectrum. Correctly, it has been pointed out, that the smartphone does not represent the culprit to understand excessive use of this device [28,29]. Instead one must focus on the manifold installed applications on the smartphones to understand why people are getting 'hooked' to their devices [30]. In this context, it will be of relevance to also distinguish between social and non-social forms of smartphone use to ultimately understand its addictive potential. Interestingly, it has been shown in recent works that higher extraversion (going along with more assertiveness, but also sociality) is not linked to SUD [31,32]. Hence, core social use of social media apps on smartphones (e.g., to stay in touch with friends) might represent the healthy form of smartphone use, reflecting also homo sapiens' urge for social bonding as also outlined in Pankseppian Affective Neuroscience Theory [33]. This all being said, it is understandable that the most prominent categories currently being researched—potentially causing addiction-like symptoms when being overused/or used in the wrong way—are social media/messenger platforms and Freemium games.

As prominent representatives of social media applications, Facebook, Snapchat, Instagram, and Twitter can be named and as a prominent messenger application WhatsApp or Signal. Without doubt, this is the Western view on this topic, whereas in Asia (particularly in China) the application WeChat dominates the market as a hybrid between social media and messenger application plus pay service [34].

The aforementioned term, Freemium, is a hybrid of the English words 'free' and 'premium'. In the realm of Freemium games, this means that the basic version of a game can be downloaded at no monetary cost. However, the person playing a game pays either with his/her data produced while playing, by paying attention to ads, e.g., in exchange for further game-energy, or by paying money to access 'premium' content. Freemium games are very relevant for the present work investigating elements of apps built to prolong usage time, because such games most often run with the business model of 'gaming service in exchange for your data'.

In a work by Montag et al. [35] it has been demonstrated that about 30% of daily smartphone use is accounted for by WhatsApp and Facebook usage in a German sample mainly ranging between ca. 15 and 35 years. Moreover, a recent work by Sha et al. [28] showed that the concepts of SUD and WhatsApp Use Disorder overlap in large parts (correlation of rho = 0.68). A new work underlines the overlap between SUD and excessive social media use, but also links between SUD and (Internet) Gaming Disorder [29]. Once more, this suggests that for the understanding of excessive smartphone use, researchers must also look at the high frequent use of specific applications on the smartphone and not only on the mere use of the total/frequent usage of the smartphone. The focus on social media/messenger applications alone would be a too narrow approach, because according to current numbers also Freemium games play a crucial role when one wants to understand what people are doing on their smartphones. To illustrate this: According to statista.com, "In early 2016 ... U.S. gamers played an average of 3.6 mobile games per month, and 1.3 games on a daily basis" [36].

1.2. Main Research Questions in the Available Literature Investigated so Far

Looking at the growing literature dealing with Internet and Smartphone Use Disorder, it becomes apparent that most studies apply a confirmatory approach to understand IUD/SUD. Since the beginning of this research field with the case study of Kimberly Young [37] scientists tried to test if already existing symptoms in the realm of pathological gambling and/or substance related addictions can be transferred or applied to this potentially new form of a (online) behavioral addiction. It is not surprising that many existing self-report inventories in the field still test if symptoms such as preoccupation with the drug (here the Internet or a certain application), withdrawal symptoms, development of tolerance, loss of control, to name a few, can be associated with excessive Internet usage. Beyond such a confirmatory approach, many studies until this day deal with the important question if and what psychiatric co-morbidities can be observed when dealing with IUD and SUD. Of note, earlier works demonstrated co-morbidities with ADHD and depression [38–40], but also burnout [41] when investigating IUD. Earlier, links between SUD and depression/anxiety disorders have been already mentioned [5]. Of high importance, in the last years the research field has moved forward beyond these questions. Now, researchers also look at the processes explaining how addictive behaviors develop out of habits in the context of IUD and SUD. Up to now, presumably the most promising model to understand this process is the aforementioned I-PACE model by Brand et al. [26]. It argues that a complex interaction between person–affect–cognition and execution variables underlie the development of IUDs. IUDs are explicitly mentioned in plural here because from the introduction it should be clear that distinct forms exist such as Gaming Disorder or Internet Communication Disorder (ICD; or 'social media addiction/use disorder'). This perhaps warrants a more focused research approach—instead of relying too broadly on IUD and SUD (see [31,32] and Montag et al. [42] for overlaps between generalized and specific forms of IUD). Of note, neuroscientific approaches have also been abundantly applied to understand the psycho-biological mechanisms underlying the different forms of IUD/SUD. Due to space restriction, it is referred to a new review study here [43].

Although, the different strains of research presented so far in Section 1 led to valuable insights into the nature of IUDs, only few studies until now aim to understand why specific online applications and their in-built elements are potentially addictive [44,45]. This should be achieved with Section 3 of this work. Of note, in the present work, the term 'application(s)' describes services such as Facebook or Instagram on platform level, whereas the term 'elements/features' refers to certain parts of the applications investigated further in Section 3. Furthermore, it is stated that the terms 'elements' and 'features' are used in an exchangeable way.

2. Methods

In the current work, several features of social media/messenger platforms and Freemium games are presented, which have been developed to grasp a maximum of the online user's attention. The 'classic' app-business model of Silicon Valley asks the user not to pay with money to be able to use a certain application but with his/her personal data. Consequently, app developers have a keen interest in designing their online platforms with the goal to keep users occupied as long as possible [30]. This in turn means that more data per person can be harvested. With an increasing amount of data, revenues tend to rise, due to more effective microtargeting (i.e., sending a person customized ads). This is also shown by Matz et al. [46] demonstrating how clicks and buys can be boosted by the use of personalized ads. In the following, several app mechanisms are presented in alphabetic order. Some of these thoughts have been already covered in a chapter published by the authors in German language [47]. With the present paper, the main points of this earlier work should be made accessible for an international audience. Of importance, the present work is not a systemic review, but rather presents first insights into elements built-in social media and Freemium game apps. These insights have been derived by analysis of prominent and often used apps such as Facebook, WhatsApp, or Candy Crush. What follows is an overview of features built into social media/messenger and Freemium game applications likely prolonging app usage.

3. Results

In this section, we will cover six prominent psychological/economic mechanisms built-in social media apps and/or Freemium games. For an overview, see also Table 1.

3.1. Endless Scrolling/Streaming and the Concept of Flow

App and platform technologies are designed to be immersive, hence producing flow while using a certain app. Flow itself is a positive state of mind, which can promote high productivity. An important prerequisite for flow is that the difficulty of a task matches the ability of a person. If there is no fit, the person is either feeling anxious (the task is too difficult) or bored (the task is too easy). Other flow prerequisites are a clear goal set and a sense of personal control over the task. For a detailed introduction into the concept of flow and additional prerequisites, see the seminal work by Csikszentmihalyi [48].

Flow is needed and warranted in different kinds of work processes, however, it can actually be 'fatal' in the case of applications on the smartphone or other 'ubiquitous' devices, when attention should be focused somewhere else: It is well known that flow goes along with a feeling of time distortion and this is exactly what many developers of social media apps and Freemium games aim to achieve—a person being so immersed that he or she is forgetting about time and space while using a platform or app (cf., [30,49,50]). The distracting aspects of many applications on the smartphone are also a well-known danger for every day traffic [51] and in many countries led to the prohibition of smartphone use while driving a car or riding a bike.

One technique used to prolong usage time in this context is the endless scrolling/streaming feature. For example, on the video platform YouTube one can scroll down endlessly with apparently no end. By endlessly scrolling down, the user is getting more and more immersed (perhaps also on a motor level as perceiving the scrolling as a playful activity) while not coming to a natural stop, where he/she might easily reconsider to quit or change the platform. The behavior of endless scrolling is enhanced, because the user finds from time to time something rewarding (e.g., a funny or interesting video), hence intermittent conditioning principles are observed here (see also research on intermittent conditioning principles in the context of slot machine mechanisms by Harrigan & Dixon [52]). The same phenomenon is at work when 'auto play' is the default: e.g., when watching a video on YouTube or the favorite series on Netflix or Amazon Prime. As soon as one video is at the end, the next video begins with either a similar content (on YouTube) or the second episode of a TV show at Netflix and so forth. By this process, viewers get more and more absorbed, which makes it hard to stop watching.

3.2. Endowment Effect/Mere Exposure Effect

The endowment and the mere exposure effect play important roles in the areas of social media and games. One very prominent game including a simple variant of both effects is called "Hayday" which was released in 2012. This game is still very popular today but already belonged to the most successful Freemium games in 2013 [53]. In this game, the user aims at building a large farmland, where he/she is entitled to feed cattle, harvest crops and follows other agricultural related activities. As with other games, progress without investing money is very slow, which means that after some fast progress at the beginning of the game (to attract people to the game), rewards in terms of the urgently needed relevant game currency are scarce.

Where do the endowment effect and the mere exposure effect come in? The endowment effect describes that buying or owning a product leads to a higher (emotionally felt) value of the product, often way beyond its actual value [54,55]. In real life, this would explain why after buying a cheap souvenir for 2.99€, this thing starts to rise in the subjectively perceived value: reasons for this could be, that carrying the souvenir home feels like an investment and that fond memories of the holiday are also attached to it. This leads to the effect that the value attached to it, if, e.g., asked to sell the souvenir to someone else, often exceeds the original price. Both ownership and loss aversion are discussed as reasons for the endowment effect.

In the context of smartphone games, this plays out as follows: every time players visit the app platform and invest more time in the construction of the virtual world, it will get harder for them to detach from the game or even delete the app. This is in particular the case, if the progress has been slow and it took the users a very long time to construct the online world. It is their own little world now (resulting in the endowment effect). Also, of importance is the mere exposure effect [56]. The mere exposure effect describes that the more often you are exposed to a certain (neutral) thing or application (here a game), the more you like it. This has even been shown for the initials of your name (for an introduction in the initial preference task, e.g., see Sariyska et al. [57]) and seems to be applicable also for many of the gaming or social media apps.

3.3. Social Pressure

A successful way to increase data flow in an application is built into the popular messenger app WhatsApp. WhatsApp represents one of the most successful messenger services used worldwide to exchange text messages and pictures. At the time of writing, WhatsApp has about 1.5 billion users worldwide [58].

Users on WhatsApp are nudged (to encourage someone to do something in a way that is gentle rather than forceful or direct [59]) to communicate fast and often on WhatsApp due to the 'double tick function'. If a user sends a message to a friend, the sender is presented with two gray ticks, which means that the message has successfully arrived at the recipient's phone. If the recipient reads the message, the grey ticks turn blue. As both sides know about these rules, social pressure emerges.

Both parties likely expect a fast answer, above all, if the message apparently has been read. This way, the users of the popular WhatsApp messenger are nudged towards faster communication via social pressure. This process is also known to undermine well-being [60] and might in parts be responsible for rising social concerns when people start to use smartphones [61]. Clearly also a link between features such as the double ticks and the concept of Fear of Missing Out (FoMO) can be established, with FoMO describing the anxiety/fear to miss something in one's own social network [28,62].

Of note, the economic principle of 'nudging' can also be illustrated by the default settings of a fresh WhatsApp installation. The default mode of WhatsApp comes with the 'double tick function' being activated. Most users do not know that they are able to deactivate this app-feature. In general, the 'power of defaults' is a well-known principle applicable to decision making in many different areas and especially in software applications [63,64]. The percentage of users that actually take the time to check or even change the default settings may depend on the specific target group, the importance of the application, and the application design. For instance, there is an anecdotal report by Microsoft that only 5% of Word users change the defaults settings [65].

In sum, system design of apps nudges a person to behave in a certain direction. For person characteristics influencing the perception of these mechanisms, please see Mai et al. [66]. Social pressure plays not only a role on social media platforms to enhance user engagement, but also in computer games. Prominent games such as the massively multiplayer online-role play game (MMORPG) called World of Warcraft (WoW) enable participants to meet in so called 'guilds', hence groups of persons who meet at a certain time online to go on a mission together (in the aforementioned WoW game this could be a 'raid'). As such a guild in particular performs well when being complete, the pressure on the individual of such an online group rises to be at a certain time in the online world to support one's own guild. This goes along with a stronger focus in life on online activities instead on perhaps more pressing issues in the offline world.

3.4. Show Users of an App What They Like

A strong feature for prolonging usage time on social media platforms is the prominent 'Newsfeed' as built-in in the Facebook application. Over the years, Facebook developed machine learning algorithms studying the behavior of their users in detail on the platform. In order to get to know their users better, they do not only record what people 'like' (i.e., give a 'thumbs up') but also how long they hover over a certain post. This could be interpreted as showing a special interest in a certain area. 'Textmining' enables Facebook & Co. to do sentiment analysis and to understand not only what is interesting for their users but also in what mood they are [67–69]. Although, the role of mood for buying decisions/advertising is complex [70,71], such a variable might be of (high) interest to many of the companies behind manifold smartphone/Internet applications. The analysis of the different elements presented in this paragraph technically can all be used to plant the most interesting news in one's own personalized 'Newsfeed'.

The 'Newsfeed' on Facebook is the entry for every user when logging in. Facebook has a great interest in studying the behavior of each person at perfection and in much detail, so that at best only such information is presented in the 'Newsfeed' which is most interesting for the user. Otherwise, people could get bored and close the browser window. For a work dealing with user beliefs considering how the algorithm applied behind the 'Newsfeed' works, please see Rader & Gray [72].

3.5. Social Comparison and Social Reward

Perhaps one of the most prominent features of social reward mechanisms in social media is the iconic 'thumbs up' (giving or getting a 'Like'; (co-)developed by Justin Rosenstein, former worker at Facebook; see also independent.co.uk [73]). Likes demonstrate either positive social feedback on one's own posts or to give another person such a feedback. The power of such feedback has also been proven neuroscientifically, when Instagram users are confronted with their own posted pictures from their account which were manipulated by being presented either with many or few 'Likes' (in this case hearts [74]). Pictures being presented with many Likes elicit stronger activity in the ventral striatum, an area involved of the processing of a rewards [74,75]. It has even been demonstrated that lower gray matter volumes of the nucleus accumbens are associated with longer and higher frequent usage of the Facebook app on smartphones underlining its addictive power [76]. Additionally, several other studies observed that lower gray matter volumes were associated with higher addictive tendencies [77,78].

So far, in this paragraph only social reward via the 'Like' function has been discussed. Beyond that, processes of social comparison also play an important role for revisiting social media platforms because a person gets information on how he/she is perceived by their social network. Without doubt, the social comparison process is not unproblematic, because it might lead to lower self-esteem [79] and the size of one's own real social network to be successfully handled seems to be limited by the so called Dunbar's number (a theoretical limit to the number of people with whom any individual is able to sustain a stable or meaningful social relationship, usually considered to be roughly 150 [80]), likely also valid on social media channels such as Twitter [81]. Social comparison also plays a role in Freemium games, when one thinks of the highscore-boards where you can compare your scores in a game with those from others.

3.6. Zeĭgarnik/Ovsiankina Effect

Classic studies by Zeĭgarnik [82] and Rickers-Ovsiankina [83] made interesting observations concerning memory functions and actions taken after being interrupted while performing a task. In the classic work by Zeĭgarnik, participants of her study got interrupted while solving a puzzle and, in the aftermath, best remembered those tasks where they were interrupted compared to those tasks which they successfully could end. From the same work group, Rickers-Ovsiankina [83] then observed that persons not only seem to be better at remembering the tasks where they got interrupted, but several of the experiment's participants even came back to the unfinished tasks after the experiment ended to ultimately finish the task (and they were not asked to do so).

In sum, these classic works from the field of psychology suggest that individuals involved in the execution of high investment tasks, react with (emotional) strain if interrupted. The final completion of the task will remove this strain. These insights mirror in the design of Freemium games such as Candy Crush Saga. This game has been downloaded an incredible 2.7 billion times in five years [84]. In this game, participants must solve levels in which they follow different missions while playing a Tetris like game (although there are obviously several differences between the games, these cannot be presented in detail here). Of note, when the game has not been played for a while, the gamer is endowed with five lives. Some levels are very hard to solve and in case of Candy Crush Saga it is even mentioned that a "super hard level" is coming up. As some of these levels are "super hard" to solve (rumor has it that it is even impossible at first try), players easily loose several of those free lives ending up with no energy to finish this "super hard level". Being now really attracted by the game, this results in emotional strain which consequently provokes people to spend extra money to buy additional lives/gaming energy, because the next level is only a couple of minutes away.

Table 1. Elements used to prolong usage time of social media apps and/or Freemium games.

Psychological Mechanisms Built-in Social Media/Messenger Apps and/or Freemium Games	Example/Illustration
Endless scrolling/streaming	As soon as one video is at the end on a website such as YouTube, the next video begins with either a similar content or the second episode of a TV show and so forth. By this, viewers get more and more absorbed, which makes it hard to stop watching.
Endowment effect/ mere-exposure effect	Every time players visit the app platform and invest more time in the construction of the virtual world, it will get harder for them to detach from the game or even delete the app. The endowment effect might be both explained by ownership and loss aversion. Also, of importance is the mere exposure effect describing that the more often you are exposed to a certain (neutral) thing or application (here a game), the more you like it.
Social pressure	Illustration from a WhatsApp feature: If a user sends a message to a friend, the sender is presented with two gray ticks, which means that the message has successfully arrived at the recipient's phone. If the recipient reads the message, the grey ticks turn blue. As both sides know about these rules, social pressure emerges. Both parties likely expect a fast answer, above all, if the message apparently has been read.
Show users of an app what they like	Facebook has a great interest in studying the behavior of each person at perfection and in much detail, so that at best only such information is presented in the 'Newsfeed' which is most interesting for the user. Otherwise, people could get bored and close the browser window.
Social comparison and social reward	Perhaps one of the most prominent features of social reward mechanisms in social media is the iconic 'thumbs up'. A 'thumbs up' ('Like') demonstrates either positive social feedback on one's own post or gives another person such a feedback.
Zeïgarnik effect/ Ovsiankina effect	The Zeigarnik effect refers to better remembering of tasks, where a person has been interrupted. Rickers-Ovsiankina then showed that such interrupted tasks are more likely to be finished later on (even if one is not forced to do this). Illustration: Some levels in Freemium games are very hard to solve and in case of Candy Crush Saga it is even mentioned that a "super hard level" is coming up. As some of these levels are "super hard" to solve (rumor has it that it is even impossible at first try), players easily loose several of those free lives ending up with no energy to finish this "super hard level". Being now really attracted by the game, this results in emotional strain which consequently provokes people to spend extra money to buy additional lives/gaming energy, because the next level is only a couple of minutes away.

4. Discussion

The present article provided its readers with an overview on often observed elements/features of smartphone/Internet applications in-built to prolong the time spent using a certain app. The depicted elements/features embedded in psychological and economic theory underline the notion that these elements indeed have been designed to prolong app usage. Implementing such elements in a smartphone-app will also contribute to a more addictive nature of apps. From our perspective, it is of tremendous importance to further study such in-built elements, because SUD has been associated with loneliness [85]. Although it is not clear what is hen and what is egg in this work (hence the direction of causality), a growing number of studies indeed support such associations with further demonstrating robust links between SUD and negative emotionality [5]. According to the I-PACE model [26] (and its recent update [86]), a history of psychopathology might indeed be a vulnerability factor to develop IUD (and its mobile form of SUD). Nevertheless, it is also imaginable that withdrawal from society and escapism to the online world ultimately might cause loneliness and reduce social connectedness, too [87]. Additional relevant areas discussed in the literature—probably being a result of too much online consumption—are links between IUD and BMI [88] and between IUD and body image avoidance [89]. The latter effect is likely driven by social comparison processes, in particular when a person is confronted with thin models or other (thin) attractive persons [90]. These effects might be in particular strong on platforms such as Instagram and for those users with low self-esteem [91] and being female [92,93]. In this context also the role of selfies [3,94] and the smartphone camera needs to be mentioned [95].

Of importance, apps on smartphones are not in general bad and not always cause negative emotionality or dissatisfaction with one's own body image. In contrast, some applications might even foster physical activity and help persons to improve their diet [96]. However, as mentioned, we focused in the present research on the investigation of social media/messenger and Freemium game applications. Such a focus is warranted because several studies already demonstrated that these applications are important drivers of (excessive) smartphone usage (literature on social media/messenger applications and smartphone usage [28,97,98]; literature on gaming and smartphone usage [29,99,100].

Although it needs to be tested further, the here investigated elements of the apps described against psychological and economic theories all likely prolong usage times. Although literature on each of the elements in the study of SUD is scarce, it has been demonstrated that persons have problems in assessing their smartphone consumption, and this is probably due to time distortions [49,101]. Such time distortions are a well-known companion of flow processes as described in 3.1. Beyond the element of flow, a recent study [102] also found empirical support for the endowment effect indeed playing also a role with online products. The authors of this work observed via an online experiment that "consumers become instantaneously attached to and are reluctant to give up digital services once they have obtained them" ([102], p. 311). This all said, the study of elements in-built in smartphone applications present a rather new research endeavor and many questions are still in need to be answered. One of these questions is highlighted in the next section.

The overview so far presented the features separately. In real-world applications, usually several of the explained elements can be found in one application. To our knowledge, studies are largely lacking aiming at an understanding of how strong each of the aforementioned features of the platforms/games impact upon usage time. Clearly, it will be of high interest to know how the different features interact on usage time. For example, when the design of the 'Newsfeed' according to the idea 'show people what they like' would prolong a person's stay on Facebook by 10 min, and implementation of the 'Like'-function by 5 min, will users stay 15 min longer on Facebook or will these effects interact with each other, resulting in 25 min of online app usage? Indeed, this is hard to study for independent scientists because they usually do not get access to data from platforms such as Facebook or Candy Crush Saga. Therefore, one solution will be ultimately to design independent games/social media apps for research purposes where different features are rolled out in different constellations and in different versions of the same game or services. Of note, tech-companies use this approach often and the literature refers here to 'A/B-testing'. By implementing 'A/B-testing', it can easily be tested if version A compared to version B of an online service prolongs usage time. Ethically, it is of great interest, that most of these tests are unsolicited. This also speaks for that ultimately government regulation might be necessary to force the industry to both open up for scientists from different disciplines to be able to study the impact of the aforementioned elements and to build healthier business models.

5. Limitations and Changes of the Current Business Model in Apps?

The present work is of theoretical nature and carved out relevant mechanisms likely being designed to prolong usage time of app-services dominating everyday life. Clearly, our assumption that these mechanisms have been purely designed to prolong usage time of an app is speculative, but please see also a larger serious scientific area dealing with persuasive design/technologies in the computer sciences [103,104]. It cannot be said for sure what the app developers had in mind when constructing a certain feature, but some developers, such as Justin Rosenstein, confess about the addictive nature of some of the presented elements [73]. Aside from this, it is very likely that the main intention of implementing features as introduced above in an application, represents the prolongation of app usage, because companies such as Facebook and Google earn their money with data.

Nevertheless, it is mentioned that in the last months a movement has been seen in the Silicon Valley aiming to foster "Time Well Spent" on digital platforms [105]. Applications such as "Screentime" on Apple's iPhone or "Time Well Spent" on Facebook should help to reduce online time. How good this actually works has not been investigated and please note that some of the authors (together with computer scientists) presented a similar service for Android users already several years ago, hence years before the app industry launched such a laudable initiative (CM and BL as part of a larger development team with their Menthal application [106,107]). However, it seems that even the tech-industry has learned that detrimental effects for mental health can arise due to their developed addictive app mechanisms. Probably implementation of these new control features also reflects a reaction to pressure from regulating bodies. Results from studies investigating the effect of smartphones and/or social media use on well-being came up with diverse results [28,108–110]. Hence, it also needs to be better understood what kind of use-reduction goes along with the healthiest effects.

Although no one can really predict if this is going to happen soon, the future will see if other healthier business models will be built into applications without the addictive potential currently seen in many of the prominent and often used applications. Apart from these new developments, the recent work by Alutaybi et al. [45] needs to be mentioned, because it came up with ideas on how to reduce FoMO and thereby likely also to reduce addictive tendencies towards social media applications. Before going a bit more in detail concerning some of these ideas, it is noteworthy that the authors presented FoMO eliciting features such as 'temporal events' in their article going beyond what is presented in our work, perhaps also resulting in more addictive tendencies. Of note, a 'temporal event' could be defined by an information being available only for a short time period (this was also one of the most prominent Snapchat functions): Either a person is then fast in consuming the content or simply has not seen what was going on. The authors end their interesting article with thoughts on "FoMO-Aware SNSs Design" [45] (p. 4), in which they basically think about designing social media platforms giving users the ability to make a short note on when they are available. This could help in reducing the urge to constantly check social media, because others are aware of a person being absent (and therefore not being able to reply), but also to set up bilateral or even collective protocols providing persons with rules on when they want to respond to each other. The effects of such an intervention clearly still need to be empirically tested. When comparing the work by Alutaybi et al. [45] and the present one, both of them have merits, as the former has more of a computer science design approach to its analysis (also in carving out 'addictive' or FoMO elements), whereas in the present work it is stronger focused on psychological/economic theories as described above.

Finally, the prominent honeycomb model by Kietzmann et al. [111] is mentioned, because it represents one of the first models to systematically analyze the diverse building blocks of social media. Of importance, this early very influential work addressed business persons in order to help them to recognize the most congruent social media platform for their companies in terms of a social media platform's building blocks called identity, conversation, sharing, presence, relationships, reputation, and groups. When focusing solely on addictive tendencies of social media sites, this framework might be fruitful to be brought together with the present psychological theories in order to empirically tackle the question in which area or areas of the honeycomb model, which of our presented features might have what kind of impact on usage time.

Author Contributions: The present work was carried out and designed by all authors. C.M. wrote the first draft of the manuscript, which was revised by all authors (B.L., M.H., K.Z.). Moreover, all authors approved the final version of this work.

Funding: The position of C.M. is funded by a Heisenberg grant awarded to him by the German Research Foundation (MO2363/3-2).

Conflicts of Interest: The authors declare no conflict of interest.

References

1. Statista. Available online: https://www.statista.com/statistics/330695/number-of-smartphone-users-worldwide/ (accessed on 2 April 2019).
2. Maisch, B.; Meckel, M. Innovationskommunikation 2.0—Das Beispiel Apple IPhone. *Mark. Rev. St. Gallen* **2009**, *26*, 42–46. [CrossRef]
3. Montag, C.; Diefenbach, S. Towards Homo Digitalis: Important Research Issues for Psychology and the Neurosciences at the Dawn of the Internet of Things and the Digital Society. *Sustainability* **2018**, *10*, 415. [CrossRef]
4. Scholz, R.; Bartelsman, E.; Diefenbach, S.; Franke, L.; Grunwald, A.; Helbing, D.; Hill, R.; Hilty, L.; Höjer, M.; Klauser, S.; et al. Unintended Side Effects of the Digital Transition: European Scientists' Messages from a Proposition-Based Expert Round Table. *Sustainability* **2018**, *10*, 2001. [CrossRef]
5. Elhai, J.D.; Yang, H.; Montag, C. Cognitive- and Emotion-Related Dysfunctional Coping Processes: Transdiagnostic Mechanisms Explaining Depression and Anxiety's Relations with Problematic Smartphone Use. *Curr. Addict. Rep.* **2019**. [CrossRef]
6. Duke, É.; Montag, C. Smartphone Addiction, Daily Interruptions and Self-Reported Productivity. *Addict. Behav. Rep.* **2017**, *6*, 90–95. [CrossRef] [PubMed]
7. Duke, É.; Montag, C. Smartphone Addiction and Beyond: Initial Insights on an Emerging Research Topic and Its Relationship to Internet Addiction. In *Internet Addiction*; Montag, C., Reuter, M., Eds.; Springer: Cham, Switzerland, 2017; pp. 359–372.
8. Kushlev, K.; Proulx, J.; Dunn, E.W. Silence Your Phones: Smartphone Notifications Increase Inattention and Hyperactivity Symptoms. In Proceedings of the 2016 CHI Conference on Human Factors in Computing Systems, San Jose, CA, USA, 7–12 May 2016; ACM Press: New York, NY, USA, 2016; pp. 1011–1020. [CrossRef]
9. Kushlev, K.; Dunn, E.W. Checking Email Less Frequently Reduces Stress. *Comput. Hum. Behav.* **2015**, *43*, 220–228. [CrossRef]
10. Twenge, J.M.; Martin, G.N.; Campbell, W.K. Decreases in psychological well-being among American adolescents after 2012 and links to screen time during the rise of smartphone technology. *Emotion* **2018**, *18*, 765–780. [CrossRef] [PubMed]
11. Ward, A.F.; Duke, K.; Gneezy, A.; Bos, M.W. Brain drain: The mere presence of one's own smartphone reduces available cognitive capacity. *J. Assoc. Consum. Res.* **2017**, *2*, 140–154. [CrossRef]
12. Wilmer, H.H.; Sherman, L.E.; Chein, J.M. Smartphones and cognition: A review of research exploring the links between mobile technology habits and cognitive functioning. *Front. Psychol.* **2018**, *8*, 605. [CrossRef]
13. Kushlev, K.; Hunter, J.F.; Proulx, J.; Pressman, S.D.; Dunn, E. Smartphones reduce smiles between strangers. *Comput. Hum. Behav.* **2019**, *91*, 12–16. [CrossRef]
14. Dwyer, R.J.; Kushlev, K.; Dunn, E.W. Smartphone use undermines enjoyment of face-to-face social interactions. *J. Exp. Soc. Psychol.* **2018**, *78*, 233–239. [CrossRef]
15. Kushlev, K.; Dunn, E.W. Smartphones distract parents from cultivating feelings of connection when spending time with their children. *J. Soc. Pers. Relatsh.* **2019**, *36*, 1619–1639. [CrossRef]
16. Melchers, M.; Li, M.; Chen, Y.; Zhang, W.; Montag, C. Lower empathy is associated with problematic Internet use: Empirical evidence from Germany and China. *Asian J. Psychiatry* **2015**, *17*, 56–60. [CrossRef] [PubMed]
17. Chou, W.J.; Huang, M.F.; Chang, Y.P.; Chen, Y.M.; Hu, H.F.; Yen, C.F. Social skills deficits and their association with Internet addiction and activities in adolescents with attention-deficit/hyperactivity disorder. *J. Behav. Addict.* **2016**, *6*, 42–50. [CrossRef] [PubMed]
18. Lachmann, B.; Sindermann, C.; Sariyska, R.; Luo, R.; Melchers, M.; Becker, B.; Montag, C. The Role of Empathy and Life Satisfaction in Smartphone Use Disorder. *Front. Psychol.* **2018**, *9*, 398. [CrossRef] [PubMed]
19. Jeon, H.J.; Kim, S.; Chon, W.H.; Ha, J.H. Is Internet overuse associated with impaired empathic ability in Korean college students? *Medicine* **2018**, *97*, e12493. [CrossRef] [PubMed]
20. Lin, Y.-H.; Chang, L.-R.; Lee, Y.-H.; Tseng, H.-W.; Kuo, T.B.J.; Chen, S.-H. Development and Validation of the Smartphone Addiction Inventory (SPAI). *PLoS ONE* **2014**, *9*, e98312. [CrossRef]
21. Kwon, M.; Lee, J.-Y.; Won, W.-Y.; Park, J.-W.; Min, J.-A.; Hahn, C.; Gu, X.; Choi, J.-H.; Kim, D.-J. Development and Validation of a Smartphone Addiction Scale (SAS). *PLoS ONE* **2013**, *8*, e56936. [CrossRef]
22. Samaha, M.; Hawi, N.S. Relationships among smartphone addiction, stress, academic performance, and satisfaction with life. *Comput. Hum. Behav.* **2016**, *57*, 321–325. [CrossRef]

23. Pontes, H.M.; Schivinski, B.; Sindermann, C.; Li, M.; Becker, B.; Zhou, M.; Montag, C. Measurement and conceptualization of Gaming Disorder according to the World Health Organization framework: The development of the Gaming Disorder Test. *Int. J. Ment. Health Addict.* **2019**, 1–21. [CrossRef]
24. Müller, M.; Brand, M.; Mies, J.; Lachmann, B.; Sariyska, R.Y.; Montag, C. The 2D:4D Marker and Different Forms of Internet Use Disorder. *Front. Psychiatry* **2017**, *8*, 213. [CrossRef] [PubMed]
25. Wegmann, E.; Oberst, U.; Stodt, B.; Brand, M. Online-Specific Fear of Missing out and Internet-Use Expectancies Contribute to Symptoms of Internet-Communication Disorder. *Addict. Behav. Rep.* **2017**, *5*, 33–42. [CrossRef] [PubMed]
26. Brand, M.; Young, K.S.; Laier, C.; Wölfling, K.; Potenza, M.N. Integrating Psychological and Neurobiological Considerations Regarding the Development and Maintenance of Specific Internet-Use Disorders: An Interaction of Person-Affect-Cognition-Execution (I-PACE) Model. *Neurosci. Biobehav. Rev.* **2016**, *71*, 252–266. [CrossRef]
27. Billieux, J.; Schimmenti, A.; Khazaal, Y.; Maurage, P.; Heeren, A. Are We Overpathologizing Everyday Life? A Tenable Blueprint for Behavioral Addiction Research. *J. Behav. Addict.* **2015**, *4*, 119–123. [CrossRef] [PubMed]
28. Sha, P.; Sariyska, R.; Riedl, R.; Lachmann, B.; Montag, C. Linking Internet Communication and Smartphone Use Disorder by Taking a Closer Look at the Facebook and WhatsApp Applications. *Addict. Behav. Rep.* **2019**, *9*, 100148. [CrossRef] [PubMed]
29. Leung, H.; Pakpour, A.H.; Strong, C.; Lin, Y.C.; Tsai, M.C.; Griffiths, M.D.; Lin, C.-Y.; Chen, I.H. Measurement invariance across young adults from Hong Kong and Taiwan among three internet-related addiction scales: Bergen Social Media Addiction Scale (BSMAS), Smartphone Application-Based Addiction Scale (SABAS), and Internet Gaming Disorder Scale-Short Form (IGDS-SF9) (Study Part A). *Addict. Behav.* **2019**. [CrossRef]
30. Alter, A. *Irresistible: The Rise of Addictive Technology and the Business of Keeping Us Hooked*; Penguin Press: New York, NY, USA, 2017.
31. Lachmann, B.; Duke, É.; Sariyska, R.; Montag, C. Who's Addicted to the Smartphone and/or the Internet? *Psychol. Pop. Media Cult.* **2019**, *8*, 182–189. [CrossRef]
32. Peterka-Bonetta, J.; Sindermann, C.; Elhai, J.D.; Montag, C. Personality Associations with Smartphone and Internet Use Disorder: A Comparison Study Including Links to Impulsivity and Social Anxiety. *Front. Public Health* **2019**, *7*. [CrossRef]
33. Davis, K.L.; Montag, C. Selected Principles of Pankseppian Affective Neuroscience. *Front. Neurosci.* **2019**, *12*, 1025. [CrossRef]
34. Montag, C.; Becker, B.; Gan, C. The Multipurpose Application WeChat: A Review on Recent Research. *Front. Psychol.* **2018**, *9*, 2247. [CrossRef]
35. Montag, C.; Błaszkiewicz, K.; Sariyska, R.; Lachmann, B.; Andone, I.; Trendafilov, B.; Eibes, M.; Markowetz, A. Smartphone Usage in the 21st Century: Who Is Active on WhatsApp? *BMC Res. Notes* **2015**, *8*, 331. [CrossRef] [PubMed]
36. Statista. Available online: https://www.statista.com/topics/1906/mobile-gaming/ (accessed on 3 April 2019).
37. Young, K.S. Psychology of Computer Use: XL. Addictive Use of the Internet: A Case That Breaks the Stereotype. *Psychol. Rep.* **1996**, *79*, 899–902. [CrossRef] [PubMed]
38. Sariyska, R.; Reuter, M.; Lachmann, B.; Montag, C. Attention Deficit/Hyperactivity Disorder Is a Better Predictor for Problematic Internet Use than Depression: Evidence from Germany. *J. Addict. Res. Ther.* **2015**, *6*, 1–6.
39. Yen, J.-Y.; Ko, C.-H.; Yen, C.-F.; Wu, H.-Y.; Yang, M.-J. The Comorbid Psychiatric Symptoms of Internet Addiction: Attention Deficit and Hyperactivity Disorder (ADHD), Depression, Social Phobia, and Hostility. *J. Adolesc. Health* **2007**, *41*, 93–98. [CrossRef] [PubMed]
40. Yoo, H.J.; Cho, S.C.; Ha, J.; Yune, S.K.; Kim, S.J.; Hwang, J.; Chung, A.; Sung, Y.H.; Lyoo, I.K. Attention Deficit Hyperactivity Symptoms and Internet Addiction. *Psychiatry Clin. Neurosci.* **2004**, *58*, 487–494. [CrossRef] [PubMed]
41. Peterka-Bonetta, J.; Sindermann, C.; Sha, P.; Zhou, M.; Montag, C. The Relationship between Internet Use Disorder, Depression and Burnout among Chinese and German College Students. *Addict. Behav.* **2019**, *89*, 188–199. [CrossRef] [PubMed]

42. Montag, C.; Bey, K.; Sha, P.; Li, M.; Chen, Y.-F.; Liu, W.-Y.; Zhu, Y.-K.; Li, C.-B.; Markett, S.; Keiper, J.; et al. Is It Meaningful to Distinguish between Generalized and Specific Internet Addiction? Evidence from a Cross-Cultural Study from Germany, Sweden, Taiwan and China. *Asia-Pac. Psychiatry* **2015**, *7*, 20–26. [CrossRef] [PubMed]
43. Montag, C.; Becker, M. Psychological and Neuroscientific Advances to Understand Internet Use Disorder. *Neuroforum* **2019**, *25*, 99–107.
44. King, D.L.; Delfabbro, P.H.; Griffiths, M.D. The Role of Structural Characteristics in Problematic Video Game Play: An Empirical Study. *Int. J. Ment. Health Addict.* **2011**, *9*, 320–333. [CrossRef]
45. Alutaybi, A.; Mcalaney, J.; Stefanidis, A.; Phalp, K.; McAlaney, J.; Ali, R. Designing Social Networks to Combat Fear of Missing Out. *Proc. Br. HCI* **2018**, *18*, 1. [CrossRef]
46. Matz, S.C.; Kosinski, M.; Nave, G.; Stillwell, D.J. Psychological Targeting as an Effective Approach to Digital Mass Persuasion. *Proc. Natl. Acad. Sci. USA* **2017**, *114*, 12714–12719. [CrossRef] [PubMed]
47. Zweig, K.A.; Lachmann, B.; Montag, C.; Herrlich, M. Kontinuierliches A/B-Testing Zur Optimierung von Spielerbindung Und Monetarisierung Bei "Freemium"-Spielen. In *Paradoxien des Verbraucherverhaltens*; Springer Fachmedien Wiesbaden: Wiesbaden, Germany, 2019; pp. 43–57.
48. Csikszentmihalyi, M. *Flow: The Classic Work on How to Achieve Happiness*; Random House: London, UK, 2002.
49. Montag, C.; Błaszkiewicz, K.; Lachmann, B.; Sariyska, R.; Andone, I.; Trendafilov, B.; Markowetz, A. Recorded Behavior as a Valuable Resource for Diagnostics in Mobile Phone Addiction: Evidence from Psychoinformatics. *Behav. Sci.* **2015**, *5*, 434–442. [CrossRef] [PubMed]
50. Rau, P.-L.P.; Peng, S.-Y.; Yang, C.-C. Time Distortion for Expert and Novice Online Game Players. *CyberPsychol. Behav.* **2006**, *9*, 396–403. [CrossRef] [PubMed]
51. Kim, H.-J.; Min, J.-Y.; Kim, H.-J.; Min, K.-B. Accident Risk Associated with Smartphone Addiction: A Study on University Students in Korea. *J. Behav. Addict.* **2017**, *6*, 699–707. [CrossRef] [PubMed]
52. Harrigan, K.A.; Dixon, M. PAR Sheets, Probabilities, and Slot Machine Play: Implications for Problem and Non-Problem Gambling. *J. Gambl. Issues* **2009**, 81. [CrossRef]
53. Wikipedia.org. Available online: https://en.wikipedia.org/wiki/Hay_Day (accessed on 2 April 2019).
54. Kahneman, D.; Knetsch, J.L.; Thaler, R.H. Anomalies: The Endowment Effect, Loss Aversion, and Status Quo Bias. *J. Econ. Perspect.* **1991**, *5*, 193–206. [CrossRef]
55. Thaler, R. Toward a Positive Theory of Consumer Choice. *J. Econ. Behav. Organ.* **1980**, *1*, 39–60. [CrossRef]
56. Zajonc, R.B. Mere Exposure: A Gateway to the Subliminal. *Curr. Dir. Psychol. Sci.* **2001**, *10*, 224–228. [CrossRef]
57. Sariyska, R.; Reuter, M.; Bey, K.; Sha, P.; Li, M.; Chen, Y.-F.; Liu, W.-Y.; Zhu, Y.-K.; Li, C.-B.; Suárez-Rivillas, A.; et al. Self-Esteem, Personality and Internet Addiction: A Cross-Cultural Comparison Study. *Personal. Individ. Differ.* **2014**, *61*, 28–33. [CrossRef]
58. Statista. Available online: https://de.statista.com/statistik/daten/studie/285230/umfrage/aktive-nutzer-von-whatsapp-weltweit/ (accessed on 21 July 2019).
59. Thaler, R.; Sunstein, C. *Improving Decisions about Health, Wealth and Happiness*; Penguin Books; Yale University Press: New Haven, CT, USA, 2008.
60. Blabst, N.; Diefenbach, S. WhatsApp and Wellbeing: A Study on WhatsApp Usage, Communication Quality and Stress. In Proceedings of the 31st British Computer Society Human Computer Interaction Conference, Sunderland, UK, 3–6 July 2017; BCS Learning & Development Ltd.: Swindon, UK, 2017; pp. 1–6. [CrossRef]
61. Hadar, A.; Hadas, I.; Lazarovits, A.; Alyagon, U.; Eliraz, D.; Zangen, A. Answering the Missed Call: Initial Exploration of Cognitive and Electrophysiological Changes Associated with Smartphone Use and Abuse. *PLoS ONE* **2017**, *12*. [CrossRef]
62. Przybylski, A.K.; Murayama, K.; DeHaan, C.R.; Gladwell, V. Motivational, Emotional, and Behavioral Correlates of Fear of Missing Out. *Comput. Hum. Behav.* **2013**, *29*, 1841–1848. [CrossRef]
63. Dinner, I.; Johnson, E.J.; Goldstein, D.G.; Liu, K. Partitioning Default Effects: Why People Choose Not to Choose. *J. Exp. Psychol. Appl.* **2011**, *17*, 332–341. [CrossRef] [PubMed]
64. Kesan, J.P.; Shah, R.C. Setting Software Defaults: Perspectives from Law, Computer Science and Behavioral Economics. *Notre Dame Law Rev.* **2006**, *82*, 583–634.
65. Available online: https://archive.uie.com/brainsparks/2011/09/14/do-users-change-their-settings/ (accessed on 17 April 2019).

66. Mai, L.M.; Freudenthaler, R.; Schneider, F.M.; Vorderer, P. "I Know You've Seen It!" Individual and Social Factors for Users' Chatting Behavior on Facebook. *Comput. Hum. Behav.* **2015**, *49*, 296–302. [CrossRef]
67. Eichstaedt, J.C.; Smith, R.J.; Merchant, R.M.; Ungar, L.H.; Crutchley, P.; Preoţiuc-Pietro, D.; Asch, D.A.; Schwartz, H.A. Facebook Language Predicts Depression in Medical Records. *Proc. Natl. Acad. Sci. USA* **2018**, *115*, 11203–11208. [CrossRef] [PubMed]
68. Kramer, A.D.I.; Guillory, J.E.; Hancock, J.T. Experimental Evidence of Massive-Scale Emotional Contagion through Social Networks. *Proc. Natl. Acad. Sci. USA* **2014**, *111*, 8788–8790. [CrossRef] [PubMed]
69. Schwartz, H.A.; Eichstaedt, J.; Kern, M.L.; Park, G.; Sap, M.; Stillwell, D.; Kosinski, M.; Ungar, L. Towards Assessing Changes in Degree of Depression through Facebook. In Proceedings of the Workshop on Computational Linguistics and Clinical Psychology: From Linguistic Signal to Clinical Reality, Baltimore, MD, USA, June 2014; pp. 118–125.
70. Bakamitsos, G.A.; Siomkos, G.J. Context Effects in Marketing Practice: The Case of Mood. *J. Consum. Behav. Int. Res. Rev.* **2004**, *3*, 304–314. [CrossRef]
71. Gardner, M.P. Mood States and Consumer Behavior: A Critical Review. *J. Consum. Res.* **1985**, *12*, 281. [CrossRef]
72. Rader, E.; Gray, R. Understanding User Beliefs about Algorithmic Curation in the Facebook News Feed. In Proceedings of the 33rd Annual ACM Conference on Human Factors in Computing Systems—CHI '15, Seoul, Korea, 8–23 April 2015; ACM Press: New York, NY, USA, 2015; pp. 173–182. [CrossRef]
73. Man Who Invented 'Like' Button Deletes Facebook App over Addiction Fears. Available online: https://www.independent.co.uk/life-style/gadgets-and-tech/facebook-like-inventor-deletes-app-iphone-justin-rosenstein-addiction-fears-a7986566.html (accessed on 3 April 2019).
74. Sherman, L.E.; Payton, A.A.; Hernandez, L.M.; Greenfield, P.M.; Dapretto, M. The Power of the Like in Adolescence: Effects of Peer Influence on Neural and Behavioral Responses to Social Media. *Psychol. Sci.* **2016**, *27*, 1027–1035. [CrossRef]
75. Sherman, L.E.; Greenfield, P.M.; Hernandez, L.M.; Dapretto, M. Peer Influence via Instagram: Effects on Brain and Behavior in Adolescence and Young Adulthood. *Child Dev.* **2018**, *89*, 37–47. [CrossRef]
76. Montag, C.; Markowetz, A.; Blaszkiewicz, K.; Andone, I.; Lachmann, B.; Sariyska, R.; Trendafilov, B.; Eibes, M.; Kolb, J.; Reuter, M.; et al. Facebook Usage on Smartphones and Gray Matter Volume of the Nucleus Accumbens. *Behav. Brain Res.* **2017**, *329*, 221–228. [CrossRef]
77. Makris, N.; Oscar-Berman, M.; Jaffin, S.K.; Hodge, S.M.; Kennedy, D.N.; Caviness, V.S.; Marinkovic, K.; Breiter, H.C.; Gasic, G.P.; Harris, G.J. Decreased Volume of the Brain Reward System in Alcoholism. *Biol. Psychiatry* **2008**, *64*, 192–202. [CrossRef] [PubMed]
78. Seifert, C.L.; Magon, S.; Sprenger, T.; Lang, U.E.; Huber, C.G.; Denier, N.; Vogel, M.; Schmidt, A.; Radue, E.-W.; Borgwardt, S.; et al. Reduced Volume of the Nucleus Accumbens in Heroin Addiction. *Eur. Arch. Psychiatry Clin. Neurosci.* **2015**, *265*, 637–645. [CrossRef] [PubMed]
79. Vogel, E.A.; Rose, J.P.; Roberts, L.R.; Eckles, K. Social Comparison, Social Media, and Self-Esteem. *Psychol. Pop. Media Cult.* **2014**, *3*, 206–222. [CrossRef]
80. Dunbar, R. *How Many Friends Does One Person Need? Dunbar's Number and Other*; Faber & Faber: London, UK, 2010.
81. Gonçalves, B.; Perra, N.; Vespignani, A. Modeling Users' Activity on Twitter Networks: Validation of Dunbar's Number. *PLoS ONE* **2011**, *6*, e22656. [CrossRef] [PubMed]
82. Zeĭgarnik, B.V. *Über das Behalten von Erledigten und Unerledigten Handlungen*; J. Springer: Wiesbaden, Germany, 1927.
83. Rickers-Ovsiankina, M.A. *Die Wiederaufnahme Unterbrochener Handlungen*; Springer: Berlin, Germany, 1928.
84. Candy Crush Saga: 2.73 Billion Downloads in Five Years and still Counting. Available online: https://venturebeat.com/2017/11/17/candy-crush-saga-2-73-billion-downloads-in-five-years-and-still-counting/ (accessed on 2 April 2019).
85. Bian, M.; Leung, L. Linking loneliness, shyness, smartphone addiction symptoms, and patterns of smartphone use to social capital. *Soc. Sci. Comput. Rev.* **2015**, *33*, 61–79. [CrossRef]
86. Brand, M.; Wegmann, E.; Stark, R.; Müller, A.; Wölfling, K.; Robbins, T.W.; Potenza, M.N. The Interaction of Person-Affect-Cognition-Execution (I-PACE) model for addictive behaviors: Update, generalization to addictive behaviors beyond Internet-use disorders, and specification of the process character of addictive behaviors. *Neurosci. Biobehav. Rev.* **2019**, *104*, 1–10. [CrossRef]

87. Savci, M.; Aysan, F. Technological addictions and social connectedness: Predictor effect of internet addiction, social media addiction, digital game addiction and smartphone addiction on social connectedness. *Dusunen Adam J. Psychiatry Neurol. Sci.* **2017**, *30*, 202–216. [CrossRef]
88. Canan, F.; Yildirim, O.; Ustunel, T.Y.; Sinani, G.; Kaleli, A.H.; Gunes, C.; Ataoglu, A. The relationship between internet addiction and body mass index in Turkish adolescents. *Cyberpsychol. Behav. Soc. Netw.* **2014**, *17*, 40–45. [CrossRef]
89. Rodgers, R.F.; Melioli, T.; Laconi, S.; Bui, E.; Chabrol, H. Internet addiction symptoms, disordered eating, and body image avoidance. *Cyberpsychol. Behav. Soc. Netw.* **2013**, *16*, 56–60. [CrossRef]
90. Brown, Z.; Tiggemann, M. Attractive celebrity and peer images on Instagram: Effect on women's mood and body image. *Body Image* **2016**, *19*, 37–43. [CrossRef] [PubMed]
91. Ahadzadeh, A.S.; Sharif, S.P.; Ong, F.S. Self-schema and self-discrepancy mediate the influence of Instagram usage on body image satisfaction among youth. *Comput. Hum. Behav.* **2017**, *68*, 8–16. [CrossRef]
92. Cohen, R.; Newton-John, T.; Slater, A. The relationship between Facebook and Instagram appearance-focused activities and body image concerns in young women. *Body Image* **2017**, *23*, 183–187. [CrossRef] [PubMed]
93. Meier, E.P.; Gray, J. Facebook photo activity associated with body image disturbance in adolescent girls. *Cyberpsychol. Behav. Soc. Netw.* **2014**, *17*, 199–206. [CrossRef] [PubMed]
94. Cohen, R.; Newton-John, T.; Slater, A. 'Selfie'-objectification: The role of selfies in self-objectification and disordered eating in young women. *Comput. Hum. Behav.* **2018**, *79*, 68–74. [CrossRef]
95. Glitsos, L. The camera phone in the concert space: Live music and moving images on the screen. *Music Sound Mov. Image* **2018**, *12*, 33–52, 85. [CrossRef]
96. Schoeppe, S.; Alley, S.; Rebar, A.L.; Hayman, M.; Bray, N.A.; Van Lippevelde, W.; Gnam, J.P.; Bachert, P.; Direito, A.; Vandelanotte, C. Apps to improve diet, physical activity and sedentary behaviour in children and adolescents: A review of quality, features and behaviour change techniques. *Int. J. Behav. Nutr. Phys. Act.* **2017**, *14*, 83. [CrossRef]
97. Haug, S.; Paz Castro, R.; Kwon, M.; Filler, A.; Kowatsch, T.; Schaub, M.P. Smartphone use and smartphone addiction among young people in Switzerland. *J. Behav. Addict.* **2015**, *4*, 299–307. [CrossRef]
98. Zhitomirsky-Geffet, M.; Blau, M. Cross-generational analysis of predictive factors of addictive behavior in smartphone usage. *Comput. Hum. Behav.* **2016**, *64*, 682–693. [CrossRef]
99. Lopez-Fernandez, O.; Männikkö, N.; Kääriäinen, M.; Griffiths, M.D.; Kuss, D.J. Mobile gaming and problematic smartphone use: A comparative study between Belgium and Finland. *J. Behav. Addict.* **2018**, *7*, 88–99. [CrossRef]
100. Lopez-Fernandez, O.; Kuss, D.J.; Romo, L.; Morvan, Y.; Kern, L.; Graziani, P.; Rousseau, A.; Rumpf, H.J.; Bischof, A.; Gässler, A.K.; et al. Self-reported dependence on mobile phones in young adults: A european cross-cultural empirical survey. *J. Behav. Addict.* **2017**, *6*, 168–177. [CrossRef] [PubMed]
101. Lin, Y.H.; Lin, Y.C.; Lee, Y.H.; Lin, P.H.; Lin, S.H.; Chang, L.R.; Tseng, H.W.; Yen, L.Y.; Yang, C.C.; Kuo, T.B. Time distortion associated with smartphone addiction: Identifying smartphone addiction via a mobile application (App). *J. Psychiatr. Res.* **2015**, *65*, 139–145. [CrossRef] [PubMed]
102. Fritze, M.P.; Eisingerich, A.B.; Benkenstein, M. Digital transformation and possession attachment: Examining the endowment effect for consumers' relationships with hedonic and utilitarian digital service technologies. *Electron. Commer. Res.* **2019**, *19*, 311–337. [CrossRef]
103. Hamari, J.; Koivisto, J.; Pakkanen, T. Do persuasive technologies persuade?-a review of empirical studies. In *International Conference on Persuasive Technology*; Springer: Cham, Switzerland, 2014; pp. 118–136.
104. Oinas-Kukkonen, H.; Harjumaa, M. A systematic framework for designing and evaluating persuasive systems. In *International Conference on Persuasive Technology*; Springer: Berlin/Heidelberg, Germany, 2008; pp. 164–176.
105. Facebook. Available online: https://de-de.facebook.com/timewellspent.io/ (accessed on 2 April 2019).
106. Andone, I.; Błaszkiewicz, K.; Eibes, M.; Trendafilov, B.; Montag, C.; Markowetz, A. Menthal: A Framework for Mobile Data Collection and Analysis. In Proceedings of the 2016 ACM International Joint Conference on Pervasive and Ubiquitous Computing Adjunct—UbiComp '16, Heidelberg, Germany, 12–16 September 2016; ACM Press: New York, NY, USA, 2016; pp. 624–629. [CrossRef]
107. Andone, I.; Błaszkiewicz, K.; Trendafilov, B.; Eibes, M.; Montag, C.; Markowetz, A. Menthal—Running a Science Project as a Start-Up. In Proceedings of the Computing in Mental Health, Workshop at CHI 2016, San Jose, CA, USA, 7–12 May 2016.

108. Stieger, S.; Lewetz, D. A week without using social media: Results from an ecological momentary intervention study using smartphones. *Cyberpsychol. Behav. Soc. Netw.* **2018**, *21*, 618–624. [CrossRef] [PubMed]
109. Tromholt, M. The Facebook experiment: Quitting Facebook leads to higher levels of well-being. *Cyberpsychol. Behav. Soc. Netw.* **2016**, *19*, 661–666. [CrossRef] [PubMed]
110. Vanman, E.J.; Baker, R.; Tobin, S.J. The burden of online friends: The effects of giving up Facebook on stress and well-being. *J. Soc. Psychol.* **2018**, *158*, 496–508. [CrossRef] [PubMed]
111. Kietzmann, J.H.; Hermkens, K.; McCarthy, I.P.; Silvestre, B.S. Social Media? Get Serious! Understanding the Functional Building Blocks of Social Media. *Bus. Horiz.* **2011**, *54*, 241–251. [CrossRef]

International Journal of
Environmental Research and Public Health

MDPI

Article

Associations Among Resilience, Stress, Depression, and Internet Gaming Disorder in Young Adults

Ju-Yu Yen [1,2], Huang-Chi Lin [1,3], Wei-Po Chou [3], Tai-Ling Liu [3] and Chih-Hung Ko [3,4,5,*]

1 Department of Psychiatry, Faculty of Medicine, College of Medicine, Kaohsiung Medical University, Kaohsiung City 807, Taiwan
2 Department of Psychiatry, Kaohsiung Municipal Ta-Tung Hospital, Kaohsiung Medical University, Kaohsiung City 812, Taiwan
3 Department of Psychiatry, Kaohsiung Medical University Hospital, Kaohsiung Medical University, Kaohsiung City 807, Taiwan
4 Department of Psychiatry, Kaohsiung Municipal Siaogang Hospital, Kaohsiung Medical University, Kaohsiung 801, Taiwan
5 Substance and Behavior addiction Research Center, Kaohsiung Medical University, Kaohsiung City 807, Taiwan
* Correspondence: chihhungko@gmail.com or cyberko@seed.net.tw

Received: 26 July 2019; Accepted: 29 August 2019; Published: 31 August 2019

Abstract: *Background and Aims*: Using gaming to escape emotional difficulty has been suggested to be a candidate mechanism contributing to Internet gaming disorder (IGD). This study evaluated the associations among resilience, perceived stress, depression, and IGD. *Methods*: A total of 87 participants in an IGD group and 87 participants in a control group were recruited into this study. IGD was diagnosed using the Diagnostic and Statistical Manual of Mental Disorders. Stress levels, resilience, and depression were measured by a self-reported questionnaire. *Results*: The IGD group had a lower resilience, higher perceived stress, and depression than the control group. Hierarchical regression analysis demonstrated that resilience was associated with IGD when perceived stress was controlled. After depression was controlled, resilience and perceived stress were not associated with IGD. Among the IGD group, those with low resilience had higher depression. Furthermore, discipline was the resilience characteristic associated with IGD. *Conclusions*: Low resilience was associated with a higher risk of IGD. IGD individuals with low resilience had higher depression. Depression was more associated with IGD than resilience. Depression assessments and stress coping interventions should be provided for individuals with IGD who exhibit low resilience or high stress.

Keywords: internet gaming disorder; stress; resilience; escape; depression

1. Introduction

Internet gaming disorder (IGD) has been identified as a potential psychiatric disorder because of its negative effects on multiple domains of functioning [1–3]. The international prevalence ranges from 0.3% to 12%, and it is an increasing public health concern, especially in Asian countries. As well as cognitive and behavioral symptoms similar to those of substance-use disorder, IGD also reportedly has associations with psychiatric symptoms, including attention-deficit hyperactivity disorder, depression, anxiety, and psychosomatic symptoms [4–7]. High comorbidity of emotional symptoms suggests that individuals with IGD might use gaming to escape emotional difficulties [8,9].

Addictive behaviors are often initiated as a maladaptive mechanism for coping with stress [10]. Stress may enhance abstinent individuals' memories of addictive behaviors as stress relievers then increase the risk of relapse to addictive behaviors after abstinence [11]. Tao et al. [12] were first to use the escape from stress through gaming as a criterion for Internet addiction. It was subsequently listed as a diagnostic criterion for IGD in the Diagnostic and Statistical Manual of Mental Disorders (DSM-5) [13]. "Negative escapism" describes gaming being negatively reinforced as a means of avoiding stress and was reported among 77.8% of internalizing patients [14]. Kim et al. [6] found escape from negative emotions to be associated with depression in IGD [9]. Furthermore, high levels of escapism were also reported to be associated with more IGD symptoms [15]. Moreover, the motives for escapism mediate the association between psychiatric distress and problematic online gaming [16]. Thus, negative escapism is associated with the symptoms of Internet addiction, psychological distress, and poor life satisfaction among massively multiplayer online role-playing gamers [17], and escapism under stress could play a major role in excessive gaming behavior.

Convincing evidence indicates that stress is a risk factor for addiction and triggers relapses [18]. The negative reinforcement model of addiction is defined as drug-taking or addictive behavior that alleviates a negative emotional state. According to some hypotheses, the negative emotional state that drives such behaviors as negative reinforcement is derived from dysregulation of brain stress systems involved in addiction processes [19]. Thus, investigating the perceived stress among adults with IGD could contribute to understanding its role in developing addictive online gaming behavior.

Not all individuals who face stress develop addictions. Therefore, resilience factors may protect these mentally healthy individuals [20]. For example, psychosocial resilience, such as positive emotions, optimism, humor, cognitive flexibility, and reappraisal could attenuate stress-induced psychopathology [21]. Thus, resilience was reported to be a buffering factor against Internet addiction [22]. Furthermore, resilience was reportedly lower among adolescents with IGD according to questionnaire assessments [15]. Its moderating role in the association between stress and IGD severity was demonstrated by an online questionnaire survey [23]. However, the difference in resilience was not evaluated among adults with IGD based on diagnostic interviewing. Further, resilience included a variety of personal characteristics, such as acceptance, problem-solving skills, capacity to recover, self-regulation, personal competence, and self-efficacy [23,24]. Which characteristics of resilience plays an important role involving development of IGD and should be intervened firstly has not been well-evaluated.

Depression is one of the most reported correlates of IGD [25]. Although the causal relationship had not been confirmed between depression and IGD, a longitudinal study suggested that depression could be an outcome of pathological gaming [26]. Prolong stressful events contribute to depression and resilience plays an important role in stress-related disorder and depression [27]. Resilience has been also reported to decrease the likelihood of stress-induced depression [21]. Both stress and depression were reported to be associated with IGD [23,25]. However, the associations among stress, depression, and resilience have not been well examined through interview studies among adults with IGD. Further, we might hypothesize that resilience could be associated with IGD and contribute to perceived stress and depression.

According to the aforementioned studies, resilience enables people to adapt successfully to stress or emotional difficulty and to avoid stress-related disorders [28]. Thus, we hypothesized that an individual with lower resilience could have higher perceived stress and depression. As IGD could be a maladjustment behavior to depression or stress, we hypothesized that individuals with low resilience were more likely to have IGD. Further, the higher perceived stress and depression could involve the association between low resilience and IGD. Moreover, among individuals of IGD, we hypothesized that those with lower resilience had higher perceived stress and depression. Thus, the aim of the study was to evaluate: (1) the difference in resilience, perceived stress, and depression between IGD and healthy controls; (2) the difference in depression and perceived stress between individuals with lower resilience and those with adequate resilience; (3) the confounding effect of perceived stress and depression in association between resilience and IGD; (4) the difference in depression and perceived stress between individuals with lower resilience and those with adequate resilience among the IGD group; and (5) the most associated resilience characteristics of IGD.

2. Methods

2.1. Participants

Our participants comprised individuals who had IGD at the time of the study (the IGD group) and individuals who had never had IGD (the control group) based on a design of case-control study. All participants were recruited by advertisement from September 2012 to October 2013. The criteria for the IGD group specified that participants be: (1) young adults aged 20 to 30 years and with more than 9 years of education; (2) individuals spending either ≥40 h per week or ≥4 h per day on weekdays and ≥8 h per day on weekends engaged in Internet gaming; and (3) individuals who maintained a pattern of Internet gaming for more than 2 years. Participants who met all of three criteria underwent an additional interview with a psychiatrist using the criteria of the DSM-5 [13] for IGD diagnosis. Those who had IGD at the time were classified into the IGD group.

For each participant in the IGD group, we matched a participant to be included in the control group by gender, education level, and age (within a range of 1 year). The recruitment criterion of participants in the control group was that their daily nonessential Internet use was less than 4 h. These participants were classified into the control group after a diagnostic interview with a psychiatrist.

All participants underwent the interview, which comprised two steps: (1) a diagnostic interview based on the Mini-International Neuropsychiatric Interview (MINI) to assess psychotic disorders, bipolar I disorder, and substance-use disorders; and (2) a history-taking interview to evaluate psychotropic medication use, mental retardation, severe physical disorder, and brain injury. Those who had psychotic disorders, bipolar I disorder, substance-use disorders, mental retardation, severe physical disorder, or brain injury or used psychotropic medication were excluded.

2.2. Measures

The diagnostic criteria of IGD were those of the DSM-5 [13]. The nine criteria comprised preoccupation, withdrawal, tolerance, unsuccessful attempts to control others, loss of interests other than gaming, continued excessive use despite psychosocial problems, deceit regarding online gaming, escape, and functional impairment [13]. We developed a semi-structured interview schedule to assess the DSM-5 criteria for IGD among our participants. Those who met five or more criteria for IGD were classified into the IGD group.

2.2.1. Chinese Version of the MINI

We conducted a diagnostic interview based on the psychotic disorder, bipolar I disorder, and substance-use disorder modules in the Chinese version of the MINI [29] to detect those excluding psychiatric disorders.

2.2.2. 14-Item Resilience Scale (R14)

The R14, which provides reliable internal consistency and external validity, was developed to evaluate the levels of resilience in the general population. Participants' resilience was assessed using this scale [24], in which scores are calculated through summation of the response values for each item, enabling scores to range from 14 to 98. The internal consistency reliability (Cronbach's alpha) of the total scale is 0.93 in the current study. Scores less than 65 indicate low resilience, scores between 65 and 81 indicate moderate resilience, and scores of more than 81 indicate high levels of resilience [24]. In this study, participants scoring more than 64 were classified into an adequate resilience group, and those scoring 64 or less were classified into a low resilience group.

2.2.3. Perceived Stress Scale

The Perceived Stress Scale (PSS) was designed to measure the extent to which situations in one's life are perceived as stressful. The PSS score was correlated with life-event scores, depression, and physical symptomatology [30]. It is suggested that the scale possesses adequate reliability for as outcome measure for experienced stress level. In this study, the 10-item PSS-10 was used to evaluate the level of stress experienced by participants and its internal consistency reliability (Cronbach's alpha) was 0.86.

2.2.4. Center for Epidemiological Studies' Depression Scale

The 20-item Mandarin Chinese version [31] of the Center for Epidemiological Studies' Depression Scale (CES-D) [32] is a self-administered evaluation of the frequency of depressive symptoms during the past week. This was used to evaluate depression. Its internal consistency reliability (Cronbach's alpha) was 0.92 in the current study.

2.2.5. Clinical Global Impression Scale for IGD

The Clinical Global Impression (CGI) scale [33] asks "Considering your total clinical experience with this particular population, how mentally ill is the patient at this time?" Possible responses were 1 for normal, not at all ill; 2 for borderline mentally ill; 3 for mildly ill; 4 for moderately ill; 5 for markedly ill; 6 for severely ill; and 7 for among the most ill patients. We modified the scale for IGD to 1 for normal, not at all ill; 2 for excessive online gaming without fulfilling the IGD criteria; 3 for fulfilling the IGD criteria with mild functional impairment; 4 for moderate functional impairment in health or one field such as academics, socializing, or profession; 5 for moderate functional impairment in multiple dimensions; 6 for severe impairment in one field; and 7 for severe impairment in multiple dimensions of daily life.

2.3. Statistical Analysis

We evaluated the associations of resilience, perceived stress, and depression with IGD using an independent *t*-test. The associations between gender, lower resilience, and IGD were evaluated through Chi-squared analysis. A hierarchical logistic regression model was used to evaluate associations of resilience, perceived stress, and depression with IGD. Finally, we evaluated the associations of resilience, perceived stress, and depression with CGI score among participants with IGD. A logistic regression evaluated the association between items of R14 and IGD with gender, age, and educational level controlled. A value of $p < 0.05$ was considered significant for all analyses, which were performed using the SPSS package (SPSS Inc., Chicago, IL, USA).

A total of 87 participants in the IGD group and 87 in the control group were enrolled in the study after informed consent was obtained. This study was approved by the Institutional Review Board of Kaohsiung Medical University Hospital (KMUH-IRB-990380).

3. Results

A total of 87 participants from the IGD group and 87 participants from the control group were enrolled in this study. No significant differences were exhibited in gender, age, or education level between the IGD and control groups, as revealed in Table 1. The IGD group scored lower on resilience, scored higher on perceived stress, and received a higher CGI score than the control group. When participants were further classified into the adequate and low resilience groups, participants with low resilience were more likely to be diagnosed with IGD. Furthermore, participants with low resilience exhibited higher perceived stress and depression, as shown in Table 2.

Table 1. Associations of age, educational level, gender, perceived stress, depression, Clinical Global Impression score, and resilience with Internet gaming disorder (IGD).

Variables	IGD Diagnosis		χ^2 Test
	Yes (N = 87) N(%)	No (N = 87) N(%)	
Gender			
Female (N = 34)	17(50.0)	17(50.0)	0.000
Male (N = 140)	70(50.0)	70(50.0)	
Resilience			
Lower (N = 53)	37(69.8)	16(30.2)	11.97 **
Adequate (N = 121)	50(41.3)	71(58.7)	
	Mean ± SD	**Mean ± SD**	***t*-Test**
Age	23.29 ± 2.34	23.38 ± 2.40	0.26
Educational level	15.93 ± 1.15	16.14 ± 1.22	1.15
Perceived stress	21.71 ± 5.54	17.76 ± 5.62	−4.67 ***
Resilience	66.57 ± 12.72	73.06 ± 12.66	3.37 **
Depression	20.44 ± 10.00	12.01 ± 7.90	−6.16 ***
CGI score	4.56 ± 0.94	1.09 ± 0.29	−33.02 ***

: <0.01; *: <0.001. Resilience: score on 14-Item Resilience Scale; participants scoring 64 or less were classified into a low resilience group. Perceived stress: score on Perceived Stress Scale-10. Depression: score on Center for Epidemiological Studies' Depression Scale. CGI: Clinical Global Impressions scale.

Table 2. The difference in perceived stress and depression between individuals with low resilience and those with adequate resilience.

Variables	Resilience		*t*-Test
	Low Mean ± SD	Adequate Mean ± SD	
Among all subjects	(N = 53)	(N = 121)	
Perceived stress	22.38 ± 5.87	18.58 ± 5.56	3.99 ***
Depression	21.77 ± 9.19	13.79 ± 9.28	5.24 ***
Among IGD group	(N = 37)	(N = 50)	
Perceived stress	22.97 ± 5.85	20.78 ± 5.16	1.85 [a]
Depression	23.16 ± 9.30	18.42 ± 10.12	2.23 *

*: <0.05; ***: <0.001. Resilience: score on 14-Item Resilience Scale; participants scoring 64 or less were classified into a low resilience group. Perceived stress: score on Perceived Stress Scale-10. Depression: score on Center for Epidemiological Studies' Depression Scale. [a]: $p = 0.07$.

3.1. Associations Among Resilience, Perceived Stress, Depression, and IGD

The hierarchical regression analysis in Table 3 demonstrates that low resilience was positively associated with IGD in Model 1 with gender, age, and educational level controlled. Participants with low resilience had a higher odds ratio (OR, 3.46; 95% confidence interval (CI) = 1.72–6.97) for IGD.

Perceived stress was significantly positively associated with IGD in Model 2. As the Wald χ^2 of perceived stress was higher than that of resilience, perceived stress was more associated with IGD than resilience was (Model 2 in Table 3). This result suggests perceived stress was another important factor to IGD. Further, the resilience was significantly associated with IGD with perceived stress controlled for. It suggested that low resilience had an independent association with IGD from perceived stress.

Depression was significantly positively associated with IGD in Model 3 (OR = 1.1, 95% CI = 1.04–1.17). Resilience and perceived stress exhibited no significant associations with IGD when depression was controlled for. This result demonstrated the mediating effect of depression in the associations of resilience and perceived stress with IGD. It also demonstrated that among resilience, stress, and depression, depression was the factor most associated with IGD.

Table 3. Hierarchical logistic regression model of Internet gaming disorder for resilience, perceived stress, and depression controlling for gender, age, and education level.

Variables	Wald	Exp(β)	95% CI
	Model 1		
Age (year)	0.39	1.05	0.91–1.21
Education level (year)	2.09	0.81	0.60–1.08
Gender	0.00	0.98	0.45–2.14
Resilience (Low)	12.08 **	3.46	1.72–6.97
	Model 2		
Age (year)	0.40	1.05	0.90–1.22
Education level (year)	1.02	0.85	0.63–1.16
Gender	0.05	0.92	0.40–2.08
Resilience (Low)	6.08 *	2.53	1.21–5.30
Perceived stress	11.92 **	1.12	1.05–1.19
	Model 3		
Age (year)	0.52	1.05	0.91–1.23
Education level (year)	1.51	0.82	0.60–1.13
Gender	0.00	1.01	0.43–2.35
Resilience	2.70	1.92	0.88–4.16
Perceived stress	0.03	1.00	0.91–1.09
Depression	9.83 **	1.10	1.04–1.17

*: <0.05; **: <0.01. Resilience: score on 14-Item Resilience Scale; participants scoring 64 or less were classified into a low resilience group. Perceived stress: score on Perceived Stress Scale-10. Depression: score on Center for Epidemiological Studies' Depression Scale.

3.2. *The Difference in Perceived Stress and Depression between Individuals with Low Resilience and Adequate Resilience Among IGD Group*

The t-test in Table 2 demonstrates that individuals with low resilience had higher depression among the IGD group (t = 0.023; p = 0.03). They also had a trend to have higher percieved stress (t = 1.85; p = 0.07); however, this did not reach significance.

3.3. *Association of Resilience Characteristics with IGD*

We firstly evaluated the difference in each item of R14 between IGD and the control group. It demonstrated that the IGD group had a lower score in ability to copy (t = 2.15; p = 0.03), acceptance (t = 2.63; p = 0.01), drive (t = 2.81; p = 0.01), discipline (t = 6.94; p < 0.001), interest/engagement (t = 2.88; p = 0.00), self-efficacy (t = 2.29; p = 0.02), dependable (t = 2.59; p = 0.01), meaning (t = 3.57; p < 0.001), and resourcefulness (t = 2.55; p = 0.01). Then, we regressed IGD on these significant items in R14 with gender, age, and educational level controlled. The logistic regression in Table 4 demonstrates that "I am self-disciplined" was the only resilience characteristic associated with IGD.

Table 4. Logistic regression model of Internet gaming disorder for items of 14-Item Resilience Scale with controlling for gender, age, and education level.

Variables	Wald	Exp(β)	95% CI
	Model 1		
Age (year)	0.00	0.97	0.90–1.25
Education level (year)	0.45	1.06	0.56–1.11
Gender	1.85	0.79	0.40–2.33
Ability to cope	0.05	1.05	0.69–1.59
Acceptance	0.13	0.93	0.63–1.39
Drive	0.36	1.12	0.79–1.60
Discipline	24.44 ***	0.41	0.29–0.58
Interest/engagement	0.30	1.13	0.74–1.72
Self-efficacy	0.86	1.22	0.80–1.84
Dependable	0.02	1.03	0.67–1.58
Meaning	2.46	0.73	0.49–1.08
Resourcefulness	0.00	1.00	0.60–1.65

***: <0.001. Items of 14-Item Resilience Scale: the items were significantly associated with IGD in *t*-test evaluation.

4. Discussion

4.1. Perceived Stress of Individuals with IGD

This study demonstrated that participants with IGD experienced higher perceived stress. This result corresponds with those of other studies [14,23]. Stress is conceptualized as a risk factor contributing to progressive long-term changes in the brain and a subsequent drug-prone state characterized by craving and increased risk of relapse [34]. Accordingly, our results might support the possible role of perceived stress in developing IGD addiction. IGD may also have negative consequences. Ko et al. [35] reported that 46.5%, 25.4%,19.7%, and 4.2% of participants with IGD experienced impaired social interaction, professional failures, impaired family relationships, and failures on examinations related to career opportunities, respectively. Such negative consequences may stress adults with IGD and explain the increased perceived stress levels in our study. Further prospective studies to investigate the addiction process may illuminate the causal relationship between stress and IGD.

4.2. Associations among Resilience, Perceived Stress, Depression, and IGD

As in the studies of Wu et al. [36] and Canale et al. [23], the interview in the present study demonstrated an association between resilience and IGD. Furthermore, we demonstrated that those with low resilience have a 3.46 times odds ratio of being diagnosed with IGD. This suggests that low resilience plays a risk role in online gaming addiction. In the R14, resilience is represented as a stable personal resource and a positive personality trait, such as discipline, that can contribute to personal competence, self-acceptance, and life satisfaction [37]. These characteristics may facilitate stress coping and attenuate stress-induced depression [21]. The lower perceived stress and depression among the adequate resilience group demonstrated in our results might support such a claim. Therefore, the characteristics of resilience can be promoted to prevent IGD and to attenuate perceived stress and depression among the general population.

Two studies have evaluated the buffering effect of resilience on the association between stress and IGD. Wu et al. [36] did not observe significant protective effects of resilience against the effects of stress on IGD. However, Canale et al. [23] identified a moderating role of resilience and demonstrated that higher perceived stress and lower resilience were associated with increased gaming time. The difference in stress measurement methods might have contributed to the inconsistent results. However, the assessment of IGD was performed using questionnaires in the two aforementioned studies. Our study, which used similar measurements to Canale's [23], demonstrated that individuals with IGD had lower resilience and higher perceived stress. Our regression analysis also demonstrated that perceived stress had a higher association with IGD than resilience. Although resilience plays a role in the development of IGD, perceived stress may be another influential factor and should be a target of intervention to prevent IGD.

Individuals with IGD could experience the psychosocial distress results from excessive gaming [35]. However, individuals with IGD usually cope with stress through gaming [15], exhibiting escapist gaming. This might aggravate, not resolve, their psychosocial problems. In the current study, the within-group analysis demonstrated IGD individuals with low reliance had higher, but not significantly, perceived stress than those with adequate resilience. This suggests a limited buffering effect of resilience on stress among individual with IGD relative to that in the general population. This might suggest further intervention aside of promoting resilience should be provided to copy the stress associated with IGD. Therefore, alternative stress coping strategies, such as problem-focus coping, exercise, or mindfulness [38] should be provided to replace escapist gaming for IGD individuals under stress.

Depression is one of the most reported factors associated with IGD [39,40]. With depression controlled, perceived stress and resilience were not significantly associated with IGD. This suggests that depression is the most proximal factor contributing to IGD. According to Baron and Kenny [41], depression may play a mediating role in the association between resilience, perceived stress and IGD. Depression could be the outcome of IGD [26], as IGD leads to psychosocial stress. Without adequate resilience to buffer dysphoric mood and psychosocial stress, individuals of IGD can use gaming as an escape from depression. However, excessive gaming without limitation might lead to further psychosocial problems and result in a vicious cycle. Further, these results also suggest that depression as a mood status has a stronger association with IGD than perceived stress and resilience. This study demonstrated that individuals with IGD had lower resilience and higher depression. Thus, depression could also be the outcome of low resilience under high stress which then contributes to IGD. The causal relationship or bi-direction interaction among resilience, stress, depression, and IGD cannot be confirmed in this cross-sectional study. Nevertheless, comorbid depression should be well assessed and effective intervention—such as antidepressants, cognitive behavior therapy [42], exercise, or promoting resilience—for depression should be provided for individuals with IGD, particularly those with lower resilience or higher perceived stress. Although the depression could be the outcome of IGD [26], gaming could be used to temporarily escape from dysphoric moods for those with limited alternative coping strategies. Thus, depression should be well intervened before aggressive intervention for gaming behavior of individuals with IGD.

4.3. *Resilience Characteristic Most Associated with IGD*

Discipline was the only resilience characteristic associated with IGD when other items were controlled. Individuals with IGD had lower discipline than controls did. Self-regulation was reported to mediate the association between impulsivity and pathological video gaming among youth [43]. Individuals with lower discipline might have had difficulty in controlling their excessive gaming, leading to a risk of IGD. Conversely, a loss of control in gaming with negative consequences might represent a lifestyle with lower discipline. The causal relationship between discipline and IGD cannot be confirmed with a cross-sectional study. However, our study suggests it is important to promote discipline to prevent risk of IGD among the general population and to prevent the deterioration of daily life functioning among individuals with IGD. Future prospective studies regarding the predictive ability of discipline for IGD or the treatment effect of promoting discipline in IGD may clarify this causal relationship.

5. Limitations

This study had limitations that should be considered when interpreting the findings. First, IGD was evaluated only through a diagnostic interview with participants. Additional information from family members or partners could have increased the validity of the diagnosis. Second, the cross-sectional research design could not confirm causal relationships between resilience, perceived stress or escapist online gaming behaviors and IGD. Lastly, we excluded those take psychotropic medication from this study to prevent its effect on depression and perceived stress. However, it could limit the generation of this study's results to populations with psychotropic medication.

6. Conclusions

Individuals with IGD had lower resilience, higher perceived stress, and higher depression. Participants with low resilience exhibited 3.46 times odds ratio for IGD compared with those with adequate resilience. Further, depression was a more proximal factor to IGD than resilience and mediated the association between resilience and IGD. Interventions to promote coping strategies and treat depression should therefore be provided for individuals with IGD and concurrent high stress or depression. Finally, discipline is the resilience characteristic most associated with IGD. Its implications for IGD interventions deserve further study.

Author Contributions: J.-Y.Y. drafted the manuscript; H.-C.L., W.-P.C., and T.-L.L. acquired and analyzed the data; C.-H.K. designed the study, interpreted the data, and gave the final approval of the version to be published.

Funding: This study was founded by the Taiwan Ministry of Science and Technology (MOST105-2314-B-037-027-MY2, MOST107-2314-B-037-101-MY2), Kaohsiung Municipal Hsiao-Kang Hospital (KMHK-104-006), Kaohsiung Medical University Hospital (KMUH105-5R54; KMUH104-4R57), and Research Center for Environmental Medicine of Kaohsiung Medical University in Taiwan from the Featured Areas Research Center Program within the framework of the Higher Education Sprout Project of the Ministry of Education in Taiwan(NHIRD-1060407). These institutions had no role in the design, process, analysis, or production of this study.

Conflicts of Interest: The authors declare no conflict of interest.

References

1. Ferguson, C.J.; Coulson, M.; Barnett, J. A meta-analysis of pathological gaming prevalence and comorbidity with mental health, academic and social problems. *J. Psychiatr. Res.* **2011**, *45*, 1573–1578. [CrossRef] [PubMed]
2. Kuss, D.J.; Griffiths, M.D. Internet and gaming addiction: A systematic literature review of neuroimaging studies. *Brain Sci.* **2012**, *2*, 347–374. [CrossRef] [PubMed]
3. Mentzoni, R.A.; Brunborg, G.S.; Molde, H.; Myrseth, H.; Skouverøe, K.J.M.; Hetland, J.; Pallesen, S. Problematic video game use: Estimated prevalence and associations with mental and physical health. *Cyberpsychol. Behav. Soc. Netw.* **2011**, *14*, 591–596. [CrossRef] [PubMed]
4. Brunborg, G.S.; Mentzoni, R.A.; Frøyland, L.R. Is video gaming, or video game addiction, associated with depression, academic achievement, heavy episodic drinking, or conduct problems? *J. Behav. Addict.* **2014**, *3*, 27–32. [CrossRef] [PubMed]
5. Hellström, C.; Nilsson, K.W.; Leppert, J.; Åslund, C. Effects of adolescent online gaming time and motives on depressive, musculoskeletal, and psychosomatic symptoms. *Upsala J. Med. Sci.* **2015**, *120*, 263–275. [CrossRef]
6. Kim, N.R.; Hwang, S.S.; Choi, J.S.; Kim, D.J.; Demetrovics, Z.; Kiraly, O.; Nagygyorgy, K.; Griffiths, M.D.; Hyun, S.Y.; Choi, S.W. Characteristics and psychiatric symptoms of internet gaming disorder among adults using self-reported DSM-5 criteria. *Psychiatry Investig.* **2016**, *13*, 58–66. [CrossRef] [PubMed]
7. Andreassen, C.S.; Billieux, J.; Griffiths, M.D.; Kuss, D.J.; Demetrovics, Z.; Mazzoni, E.; Pallesen, S. The relationship between addictive use of social media and video games and symptoms of psychiatric disorders: A large-scale cross-sectional study. *Psychol. Addict. Behav.* **2016**, *30*, 252–262. [CrossRef]
8. Kwon, J.H.; Chung, C.S.; Lee, J. The effects of escape from self and interpersonal relationship on the pathological use of Internet games. *Community Ment. Health J.* **2011**, *47*, 113–121. [CrossRef]
9. Park, M.; Kim, Y.J.; Kim, D.J.; Choi, J.S. Differential neurophysiological correlates of information processing in internet gaming disorder and alcohol use disorder measured by event-related potentials. *Sci. Rep.* **2017**, *7*, 9062. [CrossRef]
10. Valentino, R.J.; Lucki, I.; Van Bockstaele, E. Corticotropin-releasing factor in the dorsal raphe nucleus: Linking stress coping and addiction. *Brain Res.* **2010**, *1314*, 29–37. [CrossRef]

11. Zhao, L.Y.; Shi, J.; Zhang, X.L.; Epstein, D.H.; Zhang, X.Y.; Liu, Y.; Kosten, T.R.; Lu, L. Stress enhances retrieval of drug-related memories in abstinent heroin addicts. *Neuropsychopharmacology* **2010**, *35*, 720–726. [CrossRef] [PubMed]
12. Tao, R.; Huang, X.; Wang, J.; Zhang, H.; Zhang, Y.; Li, M. Proposed diagnostic criteria for internet addiction. *Addiction* **2010**, *105*, 556–564. [CrossRef] [PubMed]
13. American Psychiatric Association. *Diagnostic and Statistical Manual of Mental Disorders: DSM-5*, 5th ed.; American Psychiatric Association: Washington, DC, USA, 2013.
14. Martín-Fernández, M.; Matalí, J.L.; García-Sánchez, S.; Pardo, M.; Lleras, M.; Castellano-Tejedor, C. Adolescents with internet gaming disorder (IGD): Profiles and treatment response. *Adicciones* **2016**, *29*, 125–133. [CrossRef] [PubMed]
15. Wu, A.M.; Lai, M.H.; Yu, S.; Lau, J.T.; Lei, M.W. Motives for online gaming questionnaire: Its psychometric properties and correlation with internet gaming disorder symptoms among Chinese people. *J. Behav. Addict.* **2017**, *6*, 11–20. [CrossRef] [PubMed]
16. Király, O.; Tóth, D.; Urbán, R.; Demetrovics, Z.; Maraz, A. Intense video gaming is not essentially problematic. *Psychol. Addict. Behav.* **2017**, *31*, 807–817. [CrossRef] [PubMed]
17. Hagström, D.; Kaldo, V. Escapism among players of MMORPGs—Conceptual clarification, its relation to mental health factors, and development of a new measure. *Cyberpsychol. Behav. Soc. Netw.* **2014**, *17*, 19–25. [CrossRef] [PubMed]
18. Goeders, N.E. The impact of stress on addiction. *European Neuropsychopharmacology* **2003**, *13*, 435–441. [CrossRef] [PubMed]
19. Koob, G.F.; Buck, C.L.; Cohen, A.; Edwards, S.; Park, P.E.; Schlosburg, J.E.; Schmeichel, B.; Vendruscolo, L.F.; Wade, C.L.; Whitfield, T.W.; et al. Addiction as a stress surfeit disorder. *Neuropharmacology* **2014**, *76*, 370–382. [CrossRef]
20. McEwen, B.S. Protective and damaging effects of stress mediators: Central role of the brain. *Dialogues Clin. Neurosci.* **2006**, *8*, 367–381. [CrossRef] [PubMed]
21. Southwick, S.M.; Vythilingam, M.; Charney, D.S. The psychobiology of depression and resilience to stress: Implications for prevention and treatment. *Annu. Rev. Clin. Psychol.* **2005**, *1*, 255–291. [CrossRef]
22. Nam, C.R.; Lee, D.H.; Lee, J.Y.; Choi, A.R.; Chung, S.J.; Kim, D.J.; Bhang, S.Y.; Kwon, J.G.; Kweon, Y.S.; Choi, J.S. The role of resilience in internet addiction among adolescents between sexes: A moderated mediation model. *J. Clin. Med.* **2018**, *7*, 222. [CrossRef] [PubMed]
23. Canale, N.; Marino, C.; Griffiths, M.D.; Scacchi, L.; Monaci, M.G.; Vieno, A. The association between problematic online gaming and perceived stress: The moderating effect of psychological resilience. *J. Behav. Addict.* **2019**, *8*, 1–7. [CrossRef] [PubMed]
24. Wagnild, G. *The Resilience Scale User's Guide for the US English version of the Resilience Scale and the 14-Item Resilience Scale (RS-14)*; Resilience Center: Worden, MT, USA, 2009.
25. Liu, L.; Yao, Y.W.; Li, C.R.; Zhang, J.T.; Xia, C.C.; Lan, J.; Ma, S.S.; Zhou, N.; Fang, X.Y. The Comorbidity between Internet Gaming Disorder and Depression: Interrelationship and Neural Mechanisms. *Front. Psychiatry* **2018**, *9*, 154. [CrossRef] [PubMed]
26. Gentile, D.A.; Choo, H.; Liau, A.; Sim, T.; Li, D.; Fung, D.; Khoo, A. Pathological Video Game Use Among Youths: A Two-Year Longitudinal Study. *Pediatrics* **2011**, *127*, e319–e329. [CrossRef] [PubMed]
27. Strain, J.J. The psychobiology of stress, depression, adjustment disorders and resilience. *World J. Biol. Psychiatry* **2018**, *19*, S14–S20. [CrossRef] [PubMed]
28. Liu, H.; Zhang, C.; Ji, Y.; Yang, L. Biological and psychological perspectives of resilience: Is it possible to improve stress resistance? *Front. Hum. Neurosci.* **2018**, *12*, 326. [CrossRef] [PubMed]
29. Sheehan, D.V.; Lecrubier, Y.; Sheehan, K.H.; Amorim, P.; Janavs, J.; Weiller, E.; Hergueta, T.; Baker, R.; Dunbar, G.C. The mini-international neuropsychiatric interview (M.I.N.I.): The development and validation of a structured diagnostic psychiatric interview for DSM-IV and ICD-10. *J. Clin. Psychiatry* **1998**, *59*, 22–33.
30. Cohen, S.; Kamarck, T.; Mermelstein, R. A global measure of perceived stress. *J. Health Soc. Behav.* **1983**, *24*, 385–396. [CrossRef]
31. Chien, C.P.; Cheng, T.A. Depression in Taiwan: Epidemiological survey utilizing CES-D. *Seishin Shinkeigaku Zasshi* **1985**, *87*, 335–338.
32. Radloff, L.S. The CES-D scale: A self-report depression scale for research in the general population. *Appl. Psychol. Meas.* **1977**, *1*, 385–401. [CrossRef]

33. Busner, J.; Targum, S.D. The clinical global impressions scale: Applying a research tool in clinical practice. *Psychiatry* **2007**, *4*, 28–37. [PubMed]
34. Ruisoto, P.; Contador, I. The role of stress in drug addiction: An integrative review. *Physiol. Behav.* **2019**, *202*, 62–68. [CrossRef] [PubMed]
35. Ko, C.H.; Yen, J.Y.; Chen, S.H.; Wang, P.W.; Chen, C.S.; Yen, C.F. Evaluation of the diagnostic criteria of Internet gaming disorder in the DSM-5 among young adults in Taiwan. *J. Psychiatr. Res.* **2014**, *53*, 103–110. [CrossRef] [PubMed]
36. Wu, A.M.; Chen, J.H.; Tong, K.K.; Yu, S.; Lau, J.T. Prevalence and associated factors of internet gaming disorder among community dwelling adults in Macao, China. *J. Behav. Addict.* **2018**, *7*, 62–69. [CrossRef] [PubMed]
37. Wagnild, G.M.; Young, H.M. Development and psychometric evaluation of the Resilience Scale. *J. Nurs. Meas.* **1993**, *1*, 165–178.
38. Li, W.; Garland, E.L.; Howard, M.O. Therapeutic mechanisms of Mindfulness-Oriented Recovery Enhancement for internet gaming disorder: Reducing craving and addictive behavior by targeting cognitive processes. *J. Addict. Dis.* **2018**, 1–9. [CrossRef]
39. González-Bueso, V.; Santamaría, J.; Fernández, D.; Merino, L.; Montero, E.; Ribas, J. Association between internet gaming disorder or pathological video-game use and comorbid psychopathology: A comprehensive review. *Int. J. Environ. Res. Public Health* **2018**, *15*, 668. [CrossRef]
40. Wang, H.R.; Cho, H.; Kim, D.J. Prevalence and correlates of comorbid depression in a nonclinical online sample with DSM-5 internet gaming disorder. *J. Affect. Disord.* **2018**, *226*, 1–5. [CrossRef]
41. Baron, R.M.; Kenny, D.A. The moderator-mediator variable distinction in social psychological research: Conceptual, strategic, and statistical considerations. *J. Pers. Soc. Psychol.* **1986**, *51*, 1173–1182. [CrossRef]
42. Hickie, I.B. An approach to managing depression in general practice. *Med. J. Aust.* **2000**, *173*, 106–110.
43. Liau, A.K.; Neo, E.C.; Gentile, D.A.; Choo, H.; Sim, T.; Li, D.; Khoo, A. Impulsivity, self-regulation, and pathological video gaming among youth: Testing a mediation model. *Asia Pac. J. Public Health* **2015**, *27*, NP2188–NP2196. [CrossRef] [PubMed]

International Journal of
Environmental Research and Public Health

MDPI

Article

Relationship between Self-Identity Confusion and Internet Addiction among College Students: The Mediating Effects of Psychological Inflexibility and Experiential Avoidance

Kuan-Ying Hsieh [1,2], Ray C. Hsiao [3,4], Yi-Hsin Yang [2,5], Kun-Hua Lee [6,*] and Cheng-Fang Yen [2,7,*]

1 Department of Child and Adolescent Psychiatry, Kaohsiung Municipal Kai-Syuan Psychiatric Hospital, Kaohsiung 80276, Taiwan
2 Graduate Institute of Medicine, College of Medicine, Kaohsiung Medical University, Kaohsiung 80708, Taiwan
3 Department of Psychiatry and Behavioral Sciences, University of Washington School of Medicine, Seattle, WA 98195, USA
4 Department of Psychiatry, Children's Hospital and Regional Medical Center, Seattle, WA 98195, USA
5 School of Pharmacy, Kaohsiung Medical University, Kaohsiung 80708, Taiwan
6 Department of Educational Psychology and Counseling, National Tsing Hua University, Hsin-Chu City 30013, Taiwan
7 Department of Psychiatry, Kaohsiung Medical University Hospital, Kaohsiung 80708, Taiwan
* Correspondence: kunhualee627@gmail.com (K.-H.L.); chfaye@cc.kmu.edu.tw (C.-F.Y.); Tel.: +886-3-5715131 (K.-H.L.); +886-7-3124941 (C.-F.Y.)

Received: 3 July 2019; Accepted: 1 September 2019; Published: 3 September 2019

Abstract: Internet addiction (IA) has become a major public health problem among college students. The aim of this study was to examine the relationship between self-identity confusion and IA and the mediating effects of psychological inflexibility and experiential avoidance (PI/EA) indicators in college students. A total of 500 college students (262 women and 238 men) were recruited. Their levels of self-identity were evaluated using the Self-Concept and Identity Measure. Their levels of PI/EA were examined using the Acceptance and Action Questionnaire-II. The severity of IA was assessed using the Chen Internet Addiction Scale. The relationships among self- identity, PI/EA, and IA were examined using structural equation modeling. The severity of self-identity confusion was positively associated with both the severity of PI/EA and the severity of IA. In addition, the severity of PI/EA indicators was positively associated with the severity of IA. These results demonstrated that the severity of self-identity confusion was related to the severity of IA, either directly or indirectly. The indirect relationship was mediated by the severity of PI/EA. Self-identity confusion and PI/EA should be taken into consideration by the community of professionals working on IA. Early detection and intervention of self-identity confusion and PI/EA should be the objectives for programs aiming to lower the risk of IA.

Keywords: internet addiction; PI; EA; self-identity

1. Introduction

1.1. Internet Addiction in College Students

Internet addiction (IA) has become a major global public health problem. Several studies have proposed diagnostic criteria for IA based on various concepts originally adopted from formal psychiatric disorders such as substance dependence [1] and pathological gambling [2] on the *Diagnostic and Statistical Manual of Mental Disorders, Fourth Edition (DSM-IV)* and impulse control disorder in the

DSM-IV Text Revision [3]. Although IA has not been recognized as a disorder by the DSM-5 or the World Health Organization, the related diagnosis of internet gaming disorder and gaming disorder have been included in the Conditions for Further Study section of the DSM-5 [4] and the International Classification of Diseases, respectively. The common concepts of IA in previous studies include poorly controlled preoccupations, urges, withdrawal, tolerance, or behaviors associated with Internet use which lead to psychosocial distress and significant impairment at home, school, and work, as well as in health or interpersonal relationships [5–7]. The measures developed based on these concepts of IA can be used to evaluate the individuals' severity of IA on the evidence base.

IA is predominant among young people, especially college students. College students normally have free and unlimited access to the Internet, flexible schedules, and are free from their parents' interference. They use the Internet for studying, gaming, social networking, gambling, chatting, shopping, and watching pornographic videos [8,9]. Previous epidemiological studies revealed that the prevalence of IA in college students varied by country. Specifically, IA prevalence ranged from 3.7% in Japan [10] to 13.6% in China [11] and 15.3% in Taiwan [12]. Such deviation has been attributed to differences in study design, sampling, assessment tools, diagnostic criteria, and cultural backgrounds. IA has been associated with sleep disturbance [13], poor physical health [14], low self-esteem [15], poor academic performance [16], and mental health problems [17]. Therefore, IA is a crucial problem among college students and requires investigation.

Investigating the factors related to IA is the first step toward preventing its incidence and providing early treatment to improve mental health. IA is the result of interaction between individuals and their environment [18,19]. Several individual factors have been discovered as being related to IA. Research has demonstrated that depression and anxiety [20,21], a decreased sense of meaning in life [22], personality traits such as psychosis and bizarre mentation [23], low self-esteem [24,25], cognitive distortion [26], high impulsivity [27], and poor emotional regulation [28] are all associated with IA.

1.2. *Self-Identify Confusion and IA*

Self-concept formation, also termed personal identity formation, as a main developmental task in adolescence, includes the acceptance of physical changes and the development of social and emotional competencies and self-efficacy [29]. Previous studies have indicated significant deficits in self-concept formation among individuals with IA compared with those without [30,31]. Deficits in self-concept are characterized by low self-esteem; indecisiveness; inability to concentrate on required or suggested tasks; uncertainty regarding future objectives; an unclear description of self; problems in engaging in intimate relationships; and difficulties in social roles, values, and selections [29]. The Internet provides individuals with deficits in self-concept with a tempting environment to socially interact with others anonymously and to obtain success through Internet gaming or other such activities. These characteristics of the Internet increase the risk of IA in individuals with deficits in self-concept [32,33]. However, a lack of performance-related positive experiences on the Internet might also further deteriorate a self-concept deficit [31]. Therefore, investigating the relationship between self-identify confusion and IA is crucial to understand the mechanism of developing IA.

1.3. *Psychological Inflexibility and Experiential Avoidance and IA*

Psychological inflexibility and experiential avoidance (PI/EA) are defined as rigid strategies guided by psychological reactions and an unwillingness to experience unpleasant events or privations [34]. PI/EA contributes to the development, maintenance, and exacerbation of problematic behaviors [35], problematic substance use [36,37], and addictive behaviors [38]. PI/EA was reported as being positively associated with IA in one cross-sectional study [39].

1.4. Self-Identify Confusion and PI/EA

Research has determined three common types of self-identity confusion: Disturbed identity (a strong tendency to acquire the thoughts, feelings, beliefs, and problems of others in adulthood) [40], unconsolidated identity (failure to take on self-defining roles and demonstrate stable beliefs, attitudes, and values) [41], and lack of identity (sudden and dramatic shifts in self-image with respect to goals, values, vocational aspirations, sexual identity, and types of friends) [4]. People with self-identity confusion may experience difficulties accepting themselves and defusing cognitive stockades, and therefore, tend to develop PI/EA [42]. Some studies have revealed the therapeutic importance of cultivating psychological flexibility in the face of a self-identity crisis resulting from physical illness [43,44]. Throughout the process of accepting one's current self, psychological flexibility is essential for reconstructing one's self-concept, such that behaviors can be changed or redirected according to personal long-term values [44]. Although it is also possible that PI/EA may predate the maturity of self-concept, the present study focused on the influence of self-identity confusion on PI/EA and examined the mediating effects of PI/EA on the relationship between self-identity confusion and IA [45].

1.5. Aims of This Study

No study has examined the relationship between self-identify confusion, PI/EA, and IA. This study investigated the mediating effects of PI/EA on the relationship between self-identity confusion and IA among college students. We hypothesized that self-identity confusion is positively associated with IA and that PI/EA mediate the relationship between self-identity confusion and IA.

2. Materials and Methods

2.1. Participants

Participants were recruited using an advertisement posted for college students aged between 20 and 30 years. Five hundred and three college students responded to the advertisement. Of them, three were excluded due to difficulties in understanding the meaning of research questionnaires. In total, 238 male and 262 female college students coming from 70 colleges located across Taiwan participated in this study. Their mean age was 22.1 years, with a standard deviation of 1.8 years. Written informed consent was obtained from all the participants prior to the assessment. The study was approved by the Institutional Review Board of Kaohsiung Medical University Hospital.

2.2. Measures

Self-Concept and Identity Measure. We used the Self-Concept and Identity Measure (SCIM) to assess the level of self-identity confusion [46]. The 27-item SCIM contains three subscales: Disturbed identity, unconsolidated identity, and lack of identity. Exploratory factor analysis revealed a three-factor structure of the SCIM in the college students. Confirmatory factor analysis validated the three-factor structure in the community sample. The correlations between the three subscales ranged from 0.32 to 0.53 [46]. Each item was rated using a seven-point Likert scale ranging from 1 (strongly disagree) to 7 (strongly agree). High total scores on the three subscales indicated tendencies for disturbed identity, unconsolidated identity, and lack of identity. The Cronbach's α of the subscales in the present study ranged from 0.74 to 0.82.

Chen Internet Addiction Scale. We used the self-administered Chen Internet Addiction Scale (CIAS) to evaluate the participants' severity of IA in the month preceding the study. The CIAS contains 26 items that are rated using a four-point Likert-type scale, with the total score ranging from 26 to 104 [47]. A higher total score indicated a more severe level of IA. The CIAS contains five subscales, including symptoms of compulsive use, withdrawal, tolerance, and problems in interpersonal relationships and health/time management. The Cronbach's α of the five subscales of the CIAS in the present study ranged from 0.78 to 0.90. The 67/68 cutoff point of the CIAS is the optimal diagnostic cutoff point for Internet addiction in college students [48].

The Acceptance and Action Questionnaire-II. The Acceptance and Action Questionnaire-II (AAQ-II) [49] was revised from the original AAQ [50]. The AAQ-II consists of seven statements that represent various aspects of PI (e.g., "My painful experiences and memories make it difficult for me to live a life that I would value") and EA (e.g., "I am afraid of my feelings"). The participants were asked to rate each of these statements on a scale of 1 (never true) to 7 (always true) based on their current experiences. A higher total score indicated a higher level of PI/EA. A study reported that the AAQ-II has adequate internal consistency and convergent and divergent validity [49]. The Cronbach's α of the AAQ-II in the present study was 0.88.

2.3. Procedure and Statistical Analysis

Research assistants explained the procedures and methods for completing the research questionnaires to the participants individually. The participants were allowed to ask questions if they encountered problems while completing the questionnaires, and the research assistants helped them to resolve their problems.

The hypothesized model for the relationships among self-identity confusion, PI/EA, and IA is presented in Figure 1. Structural equation modeling (SEM) was used to estimate the parameters, test the model adequacy, and evaluate the extent of agreement between the observed data and the covariance matrix estimated from the model [51]. Amos version 18.0 software (IBM SPSS) was used to examine the goodness-of-fit index (GFI). The maximum likelihood method was used to analyze the data. In addition, the GFI, non-normed fit index (NFI), incremental fit index (IFI), comparative fit index (CFI), root mean square error of approximation (RMSEA), and standardized root mean square residual (SRMR) were used to evaluate the goodness of fit of the model [52]. On the basis of the goodness-of-fit requirement, the NFI, GFI, IFI, and CFI should be higher than 0.9; values of the RMSEA and SRMR lower than 0.05 are good; and values ranging between 0.05 and 0.09 are acceptable [52]. The Sobel test was applied to examine the mediating effect of PI/EA on the relationship between self-identity confusion and IA. A two-tailed p value of <0.05 was considered statistically significant.

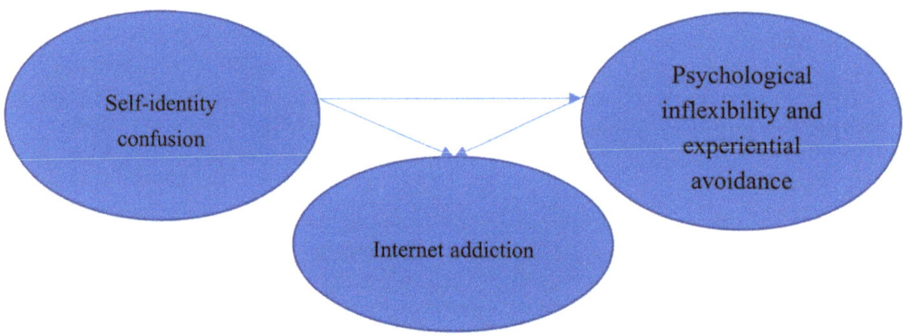

Figure 1. Hypothesized model of the associations among self-identity confusion, Internet addiction, and psychological inflexibility and experiential avoidance (PI/EA).

2.4. Ethics

The study procedures were carried out in accordance with the Declaration of Helsinki. The Institutional Review Boards of Kaohsiung Medical University Hospital approved the study (ethical approval code number: KMUH-IRB-20120249; date of ethical approval: 30 August 2012). All participants were informed about the study and provided written informed consent.

3. Results

In total, 85 (17%) participants were identified as having Internet addiction. The means, standard deviations, and correlation matrices of the measured variables are shown in Table 1. The results revealed significant correlations among the measured variables.

Table 1. The correlation matric of measured variables.

	Mean (SD)	1	2	3	4	5	6	7	8	9	10
1. Tolerance	8.9 (2.5)	1	0.66 *	0.67 *	0.57 *	0.70 *	0.26 *	0.28 *	0.26 *	0.24 *	0.30 *
2. Withdrawal	11.4 (3.3)		1	0.77 *	0.52 *	0.58 *	0.30 *	0.31 *	0.28 *	0.24 *	0.31 *
3. Compulsion	10.6 (3.1)			1	0.58 *	0.69 *	0.28 *	0.32 *	0.30 *	0.27 *	0.37 *
4. Time manage	10.5 (3.4)				1	0.68 *	0.20 *	0.27 *	0.23 *	0.21 *	0.29 *
5. Interpersonal	14.2 (4.2)					1	0.25 *	0.36 *	0.28 *	0.27 *	0.37 *
6. Psychological inflexibility	7.4 (3.3)						1	0.63 *	0.29 *	0.36 *	0.47 *
7. Experiential avoidance	9.8 (3.8)							1	0.35 *	0.39 *	0.52 *
8. Disturbed identity	33.7 (8.6)								1	0.24 *	0.56 *
9. Unconsolidated identity	18.4 (8.7)									1	0.52 *
10. Lack of identity	24.8 (8.4)										1

* $p < 0.001$.

The goodness-of-fit indices of SEM for the hypothesized model on the relationships among self-identity confusion, PI/EA, and IA are listed in Table 2. The estimated coefficients of paths in the hypothesized model are presented in Figure 2. The results found that all NFI, GFI, IFI, and CFI values were higher than 0.9 and the SRMR value was lower than 0.05, which indicated the goodness of fit of the model was good. According to Hu and Bentler [52], the value of the RMSEA (0.085) was at the acceptable level. The result of the Sobel test confirmed the mediating effect of PI/EA on the relationship between self-identity confusion and IA ($Z = 5.135$, $p < 0.05$).

Moreover, all paths in the hypothesized model were significant. The severity of self-identity confusion was positively associated with the severity of IA and PI/EA. In addition, the severity of PI/EA was positively associated with the severity of IA. The severity of self-identity confusion was directly related to the severity of IA and indirectly related to the severity of IA through the increasing severity of PI/EA.

Table 2. The goodness-of-fit index of structural equation modeling for the hypothesized model.

Type	Goodness of Fit Index	The Full Model
Absolute fit indices	χ^2	4.620
	df	32
	RMSEA	0.085 ($p < 0.09$)
	GFI	0.940 ($p > 0.09$)
Relative fit indices	NFI	0.941 ($p > 0.09$)
	IFI	0.953 ($p > 0.09$)
	CFI	0.953 ($p > 0.09$)
	SRMR	0.036 ($p < 0.05$)

χ^2: chi-square; RMSEA: Root Mean Square Error of Approximation; GFI: Goodness-of-Fit Index; NFI: Non-normed-Fit Index; IFI: Incremental Fit Index; CFI: Comparative Fit Index; SRMR: Standardized Root Mean Square Residual.

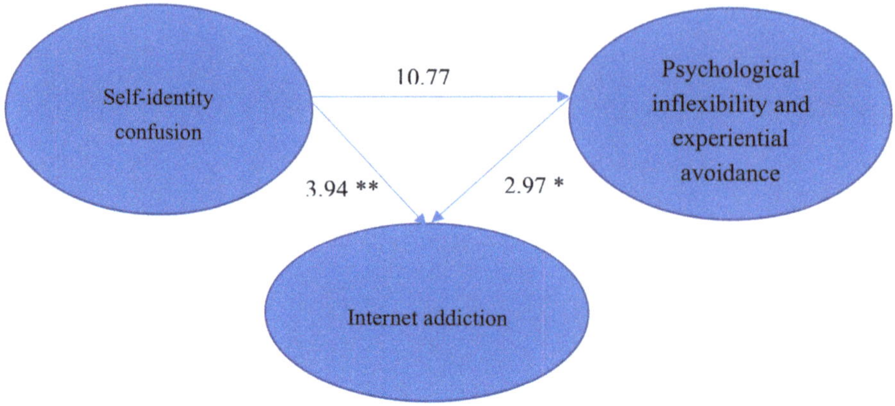

Figure 2. The estimated coefficients of the paths in the hypothesized model for the relationships among self-identity confusion, Internet addiction, and PI/EA. (* $p < 0.05$; ** $p < 0.01$).

4. Discussion

This was the first study to examine the association between self-identity confusion, PI/EA, and IA. The results of this study revealed that self-identity confusion was directly related to IA and indirectly related to IA by the mediation of PI/EA. The results of previous studies on the relationship between self-identity styles and IA have been mixed. Some studies have demonstrated that normative styles of self-identity are protective factors of IA, whereas diffuse-avoidant styles are the risk factors of IA [53,54]. Several possible etiologies may account for the positive association between self-identity confusion and IA. First, the Internet provides college students who have self-identity confusion with an environment to explore their personal values, beliefs, and goals, and college students may use the Internet to escape the daily problems that they encounter which result from self-identity confusion. Second, according to the self-concept fragmentation hypothesis, college students with self-identity confusion may use their own different personalities to interact with others in online environments and feel more at ease within these various online environments than in the real world [55–58]. Third, young adults with self-identity confusion may seek peer advice from others online on issues related to developing self-identity during adolescence [59]. Although college students may use the Internet as a tool to develop their self-identity, IA may limit real-world interaction and therefore slow the development of self-identity [56]. In addition, exposure to different identities in online activities may exacerbate their self-identity confusion [60]. Therefore, college students require help with self-identity confusion to develop adaptive ways to interact with others and develop interpersonal relationships in order to improve their self-identity and consequently reduce their dependence on the Internet. Positive self-identity development could be promoted using Borba's Esteem Builders Curriculum, which is one of the most comprehensive and widely used skills-based curricula [61]. Based on Marcia's identity status theory [62], providing clarity of identity through fostering exploration creates the platform for identity commitment [63] and helps individuals become mature and competent during life transitions [64].

The present study discovered that PI/EA were significantly associated with IA in college students. The relationship between PI and IA has also been supported by the results of an imaging study that suggested that people with IA had difficulties in executive control and attention when switching tasks [65]. People with PI may experience psychosocial maladjustment, and stressful events [66] may cause them to use the Internet to avoid real-world stressors.

In this study, we found that PI/EA mediated the relationship between self-identify confusion and IA. A few possible mechanisms may account for the result. First, an unsuccessful formation of self-concept may lead individuals to experience self-identity diffusion, thereby increasing the risk of mental disorders, such as personality, depressive, and addictive disorders. These disorders usually persist into adulthood if they are not treated appropriately [29,63] and can compromise an individual's self-concept. Persistent self-identity confusion may result in PI/EA [43,44] and result in psychological maladjustment [66]. The Internet may provide individuals with an opportunity to compensate for their self-directed negative feelings by providing opportunities for anonymous social interaction.

Second, people may experience various social-cognitive processes to construct and maintain their self-identity. Beronsky (1992) proposed three processing orientations that contribute to form and maintain a sense of self-identity: A procrastinating, diffuse-avoidant style; an open, informational style; and a conforming, normative style [67]. The individuals with a diffuse-avoidant identity style may tend to use the defense mechanism of avoidance, denial, suppression, and immaturity for masking painful inner conflicts [68,69]. Moreover, the individuals with the normative style may tend to use the defense mechanism of distorting, repressing, or denying reality that may limit self-awareness [70,71]. Therefore, the individuals with diffuse-avoidant or normative identity styles may develop PI/EA and are more likely to use the Internet than those with an informational identity style.

On the basis of our study results, we suggest that school counselors and mental health professionals routinely evaluate whether college students with IA have self-identity confusion and high PI/EA. For self-identity confusion, interventions that provide relational safety and support self-validity and self-exploration may be helpful in reducing the risk of IA [72].

The development of IA should be considered on the basis of ecological systems concepts [73]. Self-identity confusion is an individual factor related to IA. However, self-identity confusion may be influenced by sociocultural factors. Taiwanese culture based on Confucianism is more collectivistic-orientated than Western culture [74], and the difference in value orientation may make college students in Taiwan place greater importance on social relationships than those in Western societies. Whether the collectivistic-orientated culture may influence the development of self-identity and its relationship with IA warrants further study. It is also necessary to replicate the results of the present study in the individualistic-orientated societies.

Our study had a number of limitations. First, the cross-sectional research design limited our ability to draw conclusions regarding the causal relationship among self-identity confusion, PI/EA, and IA. Second, the study data were exclusively self-reported and may have therefore suffered from shared-method variance. Third, the participants in this study were college students who responded to the recruitment advertisement. Therefore, the results of this study might not be generalized to college students who did not participate in this study. Fourth, the present study used the AAQ-II to measure PI/EA. Although research determined that AAQ-II has adequate internal consistency and convergent and divergent validity [49], the items from the AAQ-II were found to be highly correlate with depression, anxiety, and stress, and therefore decreased its discriminant validity [75]. It is suggested that other measures, for example, the CompACT [76], might be needed in combination with the AAQ-II in assessing PI/EA.

Despite these limitations, this study contributes to the literature, as it is the first study to examine the relationship between self-identity confusion and IA and the mediating effects of PI/EA. Further research replicating this SEM model across different samples is suggested.

5. Conclusions

On the basis of our study, we found that self-identity confusion was directly related to IA and indirectly related to IA by the mediation of PI/EA. Therefore, self-identity confusion and PI/EA should be taken into consideration by the professionals working on prevention of IA. Moreover, we proposed that early detection and intervention for self-identity confusion may help college students to develop and consolidate self-identity, reduce PI/EA, and lower the risk of IA.

Author Contributions: Conceptualization, K.-Y.H., R.C.H., K.-H.L. and C.-F.Y.; methodology, C.-F.Y. and K.-H.L.; software, Y.-H.Y.; formal analysis, K.-Y.H., Y.-H.Y., K.-H.L. and C.-F.Y.; writing—original draft preparation, K.-Y.H.;

writing—review and editing, K.-Y.H. and C.-F.Y.; supervision, R.C.H., K.-H.L. and C.-F.Y.; funding acquisition, R.C.H., K.-H.L. and C.-F.Y.

Funding: This research was funded by Chi—Mei Medical Center, Kaohsiung Medical University Research Foundation, grant number 102CM-KMU-02 and Kaohsiung Medical University Hospital, grant number KMUH103-3M38.

Conflicts of Interest: The authors declare no conflict of interest.

References

1. Anderson, K.J. Internet use among college students: An exploratory study. *J. Am. Coll. Health* **2001**, *50*, 21–26. [CrossRef] [PubMed]
2. Young, K.S. Internet addiction: The emergence of a new clinical disorder. *CyberPsychol. Behav.* **1998**, *1*, 237–244. [CrossRef]
3. Shapira, N.A.; Goldsmith, T.D.; Keck, P.E., Jr.; Khosla, U.M.; McElroy, S.L. Psychiatric features of individuals with problematic internet use. *J. Affect. Disord.* **2000**, *57*, 267–272. [CrossRef]
4. American Psychiatric Association. *Diagnostic and Statistical Manual of Mental Disorders*, 5th ed.; American Psychiatric Association: Washington, DC, USA, 2013; ISBN 9780890425558.
5. Weinstein, A.; Lejoyeux, M. Internet addiction or excessive internet use. *Am. J. Drug Alcohol Abuse* **2010**, *36*, 277–283. [CrossRef] [PubMed]
6. Widyanto, L.; Griffiths, M. 'Internet addiction': A critical review. *Int. J. Ment. Health Addict.* **2006**, *4*, 31–51. [CrossRef]
7. Young, K.S. Internet Addiction: A new clinical phenomenon and its consequences. *Am. Behav. Sci.* **2004**, *48*, 402–415. [CrossRef]
8. Kuss, D.J.; Griffiths, M.D. Online social networking and addiction—A review of the psychological literature. *Int. J. Environ. Res. Public Health* **2011**, *8*, 3528–3552. [CrossRef] [PubMed]
9. Kuss, D.J.; Griffiths, M.D.; Binder, J.F. Internet addiction in students: Prevalence and risk factors. *Comput. Hum. Behav.* **2013**, *29*, 959–966. [CrossRef]
10. Tateno, M.; Teo, A.R.; Shirasaka, T.; Tayama, M.; Watabe, M.; Kato, T.A. Internet addiction and self-evaluated attention-deficit hyperactivity disorder traits among Japanese college students. *Psychiatry Clin. Neurosci.* **2016**, *70*, 567–572. [CrossRef]
11. Yang, T.; Yu, L.; Oliffe, J.L.; Jiang, S.; Si, Q. Regional contextual determinants of internet addiction among college students: A representative nationwide study of China. *Eur. J. Public Health* **2017**, *27*, 1032–1037. [CrossRef]
12. Lin, M.P.; Ko, H.C.; Wu, J.Y. Prevalence and psychosocial risk factors associated with internet addiction in a nationally representative sample of college students in Taiwan. *Cyberpsychol. Behav. Soc. Netw.* **2011**, *14*, 741–746. [CrossRef] [PubMed]
13. Abolghasem, P.; Eftekhari, M.; Rezania, S.; Jafarisani, M.; Soleimani, R.; Khalafi, A. Studying the relationship between quality of sleep and addiction to internet among students. *Nova J. Med. Biol. Sci.* **2016**, *5*, 1–7. [CrossRef]
14. Güzel, N.; Kahveci, I.; Solak, N.; Cömert, M.; Turan, F. Internet addiction and its impact on physical health. *Turk. Med. Stud. J.* **2018**, *5*, 32–36. [CrossRef]
15. Budak, E.; Taymur, I.; Askin, R.; Belkiz Gungor, B.; Demirci, H.; Ikra Akgul, A.; Anil Sahin, Z. Relationship between internet addiction, psychopathology and self-esteem among university students. *Eur. Res. J.* **2015**, *1*, 128–135. [CrossRef]
16. Fragkos, C.C.; Fragkos, C.C.; Kiohos, A.P. Internet addiction among Greek university students: Demographic associations with the phenomenon, using the Greek version of young's internet addiction test. *Int. J. Econ. Sci. Appl. Res.* **2010**, *3*, 49–74.
17. Gupta, A.; Khan, A.M.; Rajoura, O.P.; Srivastava, S. Internet addiction and its mental health correlates among undergraduate college students of a university in north India. *J. Family Med. Prim. Care* **2018**, *7*, 721–727. [CrossRef] [PubMed]
18. Li, C.; Dang, J.; Zhang, X.; Zhang, Q.; Guo, J. Internet addiction among Chinese adolescents: The effect of parental behavior and self-control. *Comput. Hum. Behav.* **2014**, *41*, 1–7. [CrossRef]

19. Hyun, G.J.; Han, D.H.; Lee, Y.S.; Kang, K.D.; Yoo, S.K.; Chung, U.-S.; Renshaw, P.F. Risk factors associated with online game addiction: A hierarchical model. *Comput. Hum. Behav.* **2015**, *48*, 706–713. [CrossRef]
20. Bonnaire, C.; Baptista, D. Internet gaming disorder in male and female young adults: The role of alexithymia, depression, anxiety and gaming type. *Psychiatry Res.* **2018**, *272*, 521–530. [CrossRef]
21. Li, G.; Hou, G.; Yang, D.; Jian, H.; Wang, W. Relationship between anxiety, depression, sex, obesity, and internet addiction in Chinese adolescents: A short-term longitudinal study. *Addict. Behav.* **2019**, *90*, 421–427. [CrossRef]
22. Laudet, A.B.; Morgen, K.; White, W.L. The role of social supports, spirituality, religiousness, life meaning and affiliation with 12-step fellowships in quality of life satisfaction among individuals in recovery from alcohol and drug problems. *Alcohol Treat. Q.* **2006**, *24*, 33–73. [CrossRef]
23. Munno, D.; Cappellin, F.; Saroldi, M.; Bechon, E.; Guglielmucci, F.; Passera, R.; Zullo, G. Internet addiction disorder: Personality characteristics and risk of pathological overuse in adolescents. *Psychiatry Res.* **2017**, *248*, 1–5. [CrossRef]
24. Steinfield, C.; Ellison, N.B.; Lampe, C. Social capital, self-esteem, and use of online social network sites: A longitudinal analysis. *J. Appl. Dev. Psychol.* **2008**, *29*, 434–445. [CrossRef]
25. Steinfield, C.; Lampe, C.; Ellison, N.B. The benefits of Facebook "friends": Social capital and college students' use of online social network sites. *J. Comput. Mediat. Commun.* **2007**, *12*, 1143–1168. [CrossRef]
26. Lu, X.; Yeo, K.J. Pathological internet use among Malaysia university students: Risk factors and the role of cognitive distortion. *Comput. Hum. Behav.* **2015**, *45*, 235–242. [CrossRef]
27. Reed, P.A.; Osborne, L.; Romano, M.; Truzoli, R. Higher impulsivity after exposure to the internet for individuals with high but not low levels of self-reported problematic internet behaviours. *Comput. Hum. Behav.* **2015**, *49*, 512–516. [CrossRef]
28. Hormes, J.M.; Kearns, B.; Timko, C.A. Craving Facebook? Behavioral addiction to online social networking and its association with emotion regulation deficits. *Addiction* **2014**, *109*, 2079–2088. [CrossRef]
29. Erikson, E. *Identity: Youth and Crisis*; W. W. Norton & Company: New York, NY, USA, 1968.
30. Lemenager, T.; Dieter, J.; Hill, H.; Hoffmann, S.; Reinhard, I.; Beutel, M.; Vollstadt-Klein, S.; Kiefer, F.; Mann, K. Exploring the neural basis of avatar identification in pathological internet gamers and of self-reflection in pathological social network users. *J. Behav. Addict.* **2016**, *5*, 485–499. [CrossRef]
31. Lemenager, T.; Hoffmann, S.; Dieter, J.; Reinhard, I.; Mann, K.; Kiefer, F. The links between healthy, problematic, and addicted Internet use regarding comorbidities and self-concept-related characteristics. *J. Behav. Addict.* **2018**, *7*, 31–43. [CrossRef]
32. Brand, M.; Young, K.S.; Laier, C.; Wolfling, K.; Potenza, M.N. Integrating psychological and neurobiological considerations regarding the development and maintenance of specific Internet-use disorders: An interaction of person-affect-cognition-execution (I-PACE) model. *Neurosci. Biobehav. Rev.* **2016**, *71*, 252–266. [CrossRef]
33. Tavolacci, M.P.; Ladner, J.; Grigioni, S.; Richard, L.; Villet, H.; Dechelotte, P. Prevalence and association of perceived stress, substance use and behavioral addictions: A cross-sectional study among university students in France, 2009–2011. *BMC Public Health* **2013**, *13*, 724. [CrossRef]
34. Bond, F.W.; Hayes, S.; Barnes-Homes, D. Psychological flexibility, act, and organizational behavior. *J. Organ. Behav. Manag.* **2006**, *26*, 25–54. [CrossRef]
35. Hayes, S.C.; Wilson, K.G.; Gifford, E.V.; Follette, V.M.; Strosahl, K. Experimental avoidance and behavioral disorders: A functional dimensional approach to diagnosis and treatment. *J. Consult. Clin. Psychol.* **1996**, *64*, 1152–1168. [CrossRef]
36. Bricker, J.B.; Schiff, L.; Comstock, B.A. Does avoidant coping influence young adults' smoking? A ten-year longitudinal study. *Nicotine Tob. Res.* **2011**, *13*, 998–1002. [CrossRef]
37. Levin, M.E.; Lillis, J.; Seeley, J.; Hayes, S.C.; Pistorello, J.; Biglan, A. Exploring the relationship between experiential avoidance, alcohol use disorders, and alcohol-related problems among first-year college students. *J. Am. Coll. Health* **2012**, *60*, 443–448. [CrossRef]
38. Kingston, J.; Clarke, S.; Remington, B. Experiential avoidance and problem behavior: A mediational analysis. *Behav. Modif.* **2010**, *34*, 145–163. [CrossRef]
39. Chou, W.P.; Lee, K.H.; Ko, C.H.; Liu, T.L.; Hsiao, R.C.; Lin, H.F.; Yen, C.F. Relationship between PI/EA and internet addiction: Mediating effects of mental health problems. *Psychiatry Res.* **2017**, *257*, 40–44. [CrossRef]

40. Westen, D.; Heim, A.K. Disturbances of self and identity in personality disorders. In *Handbook of Self and Identity*; Leary, M.R., Tangney, J.P., Eds.; Guilford Press: New York, NY, USA, 2003; pp. 643–664. ISBN 978-1593852375.
41. Kroger, J.; Marcia, J.E. The identity statuses: Origins, meanings, and interpretations. In *Handbook of Identity Theory and Research*; Schwartz, S.J., Luyckx, K., Vignoles, V.L., Eds.; Springer: New York, NY, USA, 2011; pp. 31–53. ISBN 978-1-4419-7988-9.
42. Hayes, S.C.; Strosahl, K.; Wilson, K.G. *Acceptance and Commitment Therapy: An Experiential Approach to Behavior Change*; Guilford Press: New York, NY, USA, 1999; ISBN 978-1572304819.
43. Lundgren, T.; Dahl, J.; Hayes, S.C. Evaluation of mediators of change in the treatment of epilepsy with acceptance and commitment therapy. *J. Behav. Med.* **2008**, *31*, 225–235. [CrossRef]
44. Whiting, D.L.; Deane, F.P.; Ciarrochi, J.; McLeod, H.J.; Simpson, G.K. Validating measures of psychological flexibility in a population with acquired brain injury. *Psychol. Assess.* **2015**, *27*, 415–423. [CrossRef]
45. Kwok, S.S.W.; Chan, E.C.C.; Chen, P.P.; Lo, B.C.Y. The "self" in pain: The role of PI in chronic pain adjustment. *J. Behav. Med.* **2016**, *39*, 908–915. [CrossRef]
46. Kaufman, E.A.; Cundiff, J.M.; Crowell, S.E. The development, factor structure, and validation of the self-concept and identity measure (SCIM): A self-report assessment of clinical identity disturbance. *J. Psychopathol. Behav. Assess.* **2015**, *37*, 122–133. [CrossRef]
47. Chen, S.; Weng, L.; Su, Y.; Wu, H.; Yang, P. Development of a Chinese internet addiction scale and its psychometric study. *Chin. J. Psychol.* **2003**, *45*, 279. [CrossRef]
48. Ko, C.H.; Yen, J.Y.; Chen, S.H.; Yang, M.J.; Lin, H.C.; Yen, C.F. Proposed diagnostic criteria and the screening and diagnosing tool of Internet addiction in college students. *Compr. Psychiatry* **2009**, *50*, 378–384. [CrossRef]
49. Bond, F.W.; Hayes, S.C.; Baer, R.A.; Carpenter, K.M.; Guenole, N.; Orcutt, H.K.; Waltz, T.; Zettle, R.D. Preliminary psychometric properties of the acceptance and action questionnaire–II: A revised measure of PI/EA. *Behav. Ther.* **2011**, *42*, 676–688. [CrossRef]
50. McCurry, S.M.; Hayes, S.C.; Strosahl, K.; Wilson, K.G.; Bissett, R.T.; Pistorello, J.; Toarmino, D.; Polusny, M.A.; Dykstra, T.A.; Batten, S.V. Measuring EA: A preliminary test of a working model. *Psychol. Rec.* **2004**, *54*, 553–578. [CrossRef]
51. Jöreskog, K.G.; Sörbom, D. *LISREL 8: Structural Equation Modeling with the SIMPLIS Command Language*; Scientific Software International: Chicago, IL, USA, 1993; ISBN 978-0894980336.
52. Hu, L.-T.; Bentler, P.M. Fit indices in covariance structure modeling: Sensitivity to underparameterized model misspecification. *Psychol. Methods* **1998**, *3*, 424. [CrossRef]
53. Tabaraei, N.; Nikoogoftar, M.; Minoosepehr, S. Determination of problematic internet use: Identity styles and social skills. *IAU Int. J. Soc. Sci.* **2014**, *4*, 27–34.
54. Sinatra, M.; de Palo, V.; Contini, P.; Volpicella, V. Identity styles and internet-related addictive behaviors in adolescents. *Turk. Online J. Educ. Technol.* **2016**, *2016*, 1161–1165.
55. Matsuba, M.K. Searching for self and relationships online. *Cyberpsychol. Behav.* **2006**, *9*, 275–284. [CrossRef]
56. Mazalin, D.; Moore, S. Internet use, identity development and social anxiety among young adults. *Behav. Chang.* **2004**, *21*, 90–102. [CrossRef]
57. Israelashvili, M.; Kim, T.; Bukobza, G. Adolescents' over-use of the cyber world—Internet addiction or identity exploration? *J. Adolesc.* **2012**, *35*, 417–424. [CrossRef]
58. Lee, C.; Aiken, K.D.; Hung, H.C. Effects of college students' video gaming behavior on self-concept clarity and flow. *Soc. Behav. Personal. Int. J.* **2012**, *40*, 673–679. [CrossRef]
59. Allen, K.A.; Ryan, T.; Gray, D.L.; McInerney, D.M.; Waters, L. Social media use and social connectedness in adolescents: The positives and the potential pitfalls. *Aust. Educ. Dev. Psychol.* **2014**, *31*, 18–31. [CrossRef]
60. Sharif, S.P.; Khanekharab, J. Identity confusion and materialism mediate the relationship between excessive social network site usage and online compulsive buying. *Cyberpsychol. Behav. Soc. Netw.* **2017**, *20*, 494–500. [CrossRef]
61. Michele Borba, E.D.; Craig Borba, E.D.; Robert, R. A Research Summary: Effectiveness of Implementing the Esteem Builders Program School-Wide on Elementary Students' Behavior and Academic Self-Concept. 1999. Available online: http://www.micheleborba.com/Pages/PilotStudy.htm (accessed on 30 June 2019).
62. Marcia, J.E. Development and validation of ego-identity status. *J. Personal. Soc. Psychol.* **1966**, *3*, 551–558. [CrossRef]

63. Zaff, J.F.; Hair, E.C. In Well-being: Positive development across the life course. In *Positive Development of the Self: Self-Concept, Self-Esteem, and Identity*; Lawrence Erlbaum Associates Publishers: Mahwah, NJ, USA, 2003; pp. 235–251.
64. Finkenauer, C.; Engels, R.C.M.E.; Meeus, W.H.J.; Oosterwegel, A. Self and identity in early adolescence: The pains and gains of knowing who and what you are. In *Understanding Early Adolescent Self and Identity*; Brinthaupt, T.M., Lipka, R.P., Eds.; State University of New York Press: Albany, NY, USA, 2002; pp. 25–56.
65. Dong, G.; Lin, X.; Zhou, H.; Lu, Q. Cognitive flexibility in internet addicts: fMRI evidence from difficult-to-easy and easy-to-difficult switching situations. *Addict. Behav.* **2014**, *39*, 677–683. [CrossRef]
66. Cheng, C.; Sun, P.; Mak, K.K. Internet addiction and psychosocial maladjustment: Avoidant coping and coping inflexibility as psychological mechanisms. *Cyberpsychol. Behav. Soc. Netw.* **2015**, *18*, 539–546. [CrossRef]
67. Berzonsky, M.D. Identity style and coping strategies. *J. Personal.* **1992**, *60*, 771–788. [CrossRef]
68. Luyckx, K. Identity styles, self-reflective cognition, and identity processes: A study of adaptive and maladaptive dimensions of self-analysis au—Berzonsky, Michael, D. *Identity* **2008**, *8*, 205–219. [CrossRef]
69. Beaumont, S.L. Identity processing styles and defense styles during emerging adulthood: Implications for life distress AU—Seaton, Cherisse, L. *Identity* **2011**, *11*, 1–24. [CrossRef]
70. Berzonsky, M.; Kinney, A. Identity processing style and defense mechanisms. *Pol. Psychol. Bull.* **2008**, *39*, 111–117. [CrossRef]
71. Soenens, B.; Duriez, B.; Goossens, L. Social–psychological profiles of identity styles: Attitudinal and social-cognitive correlates in late adolescence. *J. Adolesc.* **2005**, *28*, 107–125. [CrossRef]
72. Gonzalez-Bueso, V.; Santamaria, J.J.; Fernandez, D.; Merino, L.; Montero, E.; Jimenez-Murcia, S.; Del Pino-Gutierrez, A.; Ribas, J. Internet gaming disorder in adolescents: Personality, psychopathology and evaluation of a psychological intervention combined with parent psychoeducation. *Front. Psychol.* **2018**, *9*, 787. [CrossRef]
73. Bronfenbrenner, U. *The Ecology of Human Development*; Harvard University Press: Cambridge, MA, USA, 1979; ISBN 9780674224575.
74. Triandis, H.C. Individualism-collectivism and personality. *J. Personal.* **2001**, *69*, 907–924. [CrossRef]
75. Tyndall, I.; Waldeck, D.; Pancani, L.; Whelan, R.; Roche, B.; Dawson, D. The Acceptance and Action Questionnaire-II (AAQ-II) as a Measure of EA: Concerns over Discriminant Validity. *J. Contextual Behav. Sci.* **2019**, *12*, 278–284. [CrossRef]
76. Frances, A.W.; Dawson, D.L.; Golijani-Moghaddam, N.G. The development and validation of the comprehensive assessment of Acceptance and Commitment Therapy processes (CompACT). *J. Contextual Behav. Sci.* **2016**, *5*, 134–145. [CrossRef]

International Journal of
Environmental Research and Public Health

MDPI

Article

Social Exclusion, Surveillance Use, and Facebook Addiction: The Moderating Role of Narcissistic Grandiosity

Myungsuh Lim

Department of Business Administration, Sangji University, Wonju 26339, Korea; mslim@sangji.ac.kr

Received: 12 August 2019; Accepted: 3 October 2019; Published: 10 October 2019

Abstract: One hundred and eighty-eight participants completed the online questionnaire with items on demographics (age and gender), social exclusion, surveillance use, Facebook addiction, and narcissistic grandiosity. The findings showed that social exclusion was positively associated with Facebook addiction (B = 0.237, $p < 0.001$) and surveillance use was significantly positively associated with Facebook addiction (B = 0.211, $p < 0.01$). The surveillance use of Facebook was found to be a significant mediator between the risk of social exclusion on Facebook and Facebook addiction (B = 0.054, CI [0.20, 0.113]). Narcissistic grandiosity significantly moderated the associations between social exclusion and Facebook addiction (B = 0.079, $p = 0.012$). These findings suggest that the risk of social exclusion could serve as facilitator of Facebook addiction depending on narcissistic grandiosity.

Keywords: social exclusion; surveillance; Facebook addiction

1. Introduction

Although the use of Facebook has helped increase social capital and build relationships [1], "unfriending" on Facebook leads to negative emotions [2], promotes social media ostracism, and inhibits well-being [3]. A root cause of these negative effects is that Facebook is not always used to maintain or develop relationships. For example, voyeurism, a positive driver of Facebook use, is a form of social surveillance that fulfills the need to belong but attempts to avoid negative feelings and experiences; those who engage in it do not aim to maintain relationships with those they observe [1]. Moreover, Facebook is a platform that makes it easy to self-promote, allowing others to easily surveil the extensive content produced.

On the other hand, cyber-ostracism, which causes social exclusion on the Internet, has been found to be a threat to fundamental human needs, such as the needs of belonging and high self-esteem [4]. The negative effects of feeling excluded are not significantly different whether they arise in the context of a remote communication method or through in-person situations [5]. To cope with their online social exclusion, Facebook users may search for social information to restore their affiliation with other users [6], which facilitates their addiction to Facebook. On the other hand, people with narcissistic tendencies may be more likely to cope with social exclusion in order to maintain their ego-centric aspect [7], although the affiliation need is low [8].

Previous studies have shown that Facebook overuse to fill the need to belong can lead to Facebook addiction [9]. However, there is a lack of research identifying the mediation variables in the process of social exclusion and Facebook addiction and investigating social network attachment in social exclusion situations that do not meet the need to belong [10]. In addition, the self-presentation need [11] and the need for admiration [9] have been cited as the main reasons for using Facebook. Narcissism, in particular, has drawn attention as a major personality characteristic associated with problematic Facebook use [12]. The present research, therefore, investigated the following hypothetical relationships: (1) whether social exclusion on Facebook is positively related to addiction to Facebook;

(2) whether those with a narcissistic personality trait have a greater tendency to become addicted to Facebook as a way to avoid social exclusion; and (3) whether the use of Facebook for surveillance in response to social exclusion leads to Facebook addiction.

2. Hypotheses Development

2.1. Social Exclusion and Facebook Addiction

One of the main reasons that people use the Internet is to reveal themselves and to gain a sense of belonging [13]. A human being is innately motivated to belonging as a fundamental impetus to form and maintain relationships with others [14]. Social exclusion, which undermines the social affiliation that occurs on Facebook, happens when people are restricted from producing content or are unable to receive responses from others [15]. Addiction to social network services (SNS) is classified as Internet-related relationship addiction [16]. While whether SNS addiction should actually be classified as addiction is still debated, it does display relevant symptoms similar to other substance and behavioral addictions, such as mood-repair experience and salience (e.g., full attention to usage) [17]. Facebook is the leading SNS and has been shown to have strong correlations with symptoms classified as Internet addiction [18]. Addiction to Facebook interferes with vital activities in life, and virtual relationships dominate actual relationships [19].

Social exclusion may lead to Facebook addiction for the following reasons. First, social exclusion causes social anxiety. Socially anxious people are prone to becoming immersed in the Internet. According to the social skill model, high levels of social anxiety lead to the increased use of the Internet in order to elevate the sense of self-worth [13]. The fear of social exclusion increases the user's social risk, which leads to the pathological use of the Internet [13].

Second, users who feel socially excluded are motivated to restore their social affiliation, which may lead to Facebook addiction. They want to feel that they belong to a group, which includes affiliation and companionship [20]. Considering that the aim to identify one's social identity within the group to which one belongs and fulfill the need to belong is a significant predictor of the increase in SNS usage [12], social exclusion may make users more obsessed with SNSs. Hence, the lack of relatedness is connected to Facebook addiction [21].

Third, a primary driver of social media use is the fear of being excluded and forgotten, which increases the frequency of Facebook use [10,22]. For example, those snubbed by others on the phone attach to social networks to regain the feeling of social inclusion [10]. Both positive and negative feedback from other users affect Facebook users' well-being, and so does failure to receive feedback because FOMO (Fear of Missing Out) increases the Facebook-related stress stemming from not being popular or involved with peers on Facebook [23].

Hypothesis 1 (H1). *The higher the degree of social exclusion on Facebook is, the greater the Facebook addiction is.*

2.2. Social Exclusion and Surveillance Use

Interpersonal electronic surveillance refers to the use of communication technologies to observe another people's behavior both online and offline. The subjects of surveillance can be close friends, romantic partners, business associates, and family members [24]. Surveillance also allows users to maintain a healthy interpersonal relationship by constantly engaging with others [25]. However, it also compromises privacy settings [24]. Because SNSs, including Facebook, are repositories of messages, pictures, and videos, they provide an environment that facilitates surveillance. In the present study, surveillance concerns the use of surveillance in horizontal relationships between ordinary people [24].

Social exclusion may promote surveillance for the following reasons. Social exclusion promotes the heightened attention to social information and the processing of social information [26]. Social exclusion motivates both the need for belonging and the use of surveillance to cope with the sense of isolation [27] for successful social survival. Social exclusion induces the feeling of not belonging to a group, which generates negative effects, such as anxiety, loneliness, and antisocial behaviors [14]. Users who fear social exclusion continuously monitor their social status to maintain their feeling of social inclusion [28]. This behavior reduces the sense of loneliness caused by social exclusion [29], thus resolving social anxiety. This social monitoring system extends to the social environment and social opportunities [27]. This goal-directed behavior is motivated by the need for social affiliation. The sense of belonging is the innate, evolutionarily adaptive human motive [14]. People seek social information to classify the relationship with another party, to distinguish information that should be strategically stored for the relationship, and to verify the extent to which the social relationship is related to the event, when forming attribution [6]. In addition, the level of information processing at this time can be influenced by the degree of concern about the relationship [6]. Similarly, the social exclusion that users identify on Facebook forces them to face their uncertain social status [30] and to pursue goal-relevant behavior to restore their sense of social affiliation. Users who feel socially excluded are sensitive to social information and actively process it [6]. Based on the above discussion, the following hypothesis is stated:

Hypothesis 2 (H2). *The higher the degree of social exclusion is on Facebook is, the higher the use of Facebook for surveillance is.*

2.3. Surveillance Use and Facebook Addiction

Surveillance use may lead to Facebook addiction for the following reasons. First, remote technologies such as the Internet increase the effect of disinhibition effect on users [31]. Therefore, the effects of social exclusion online to users can be just as powerful as those of social exclusion offline [32]. Because social exclusion online motivates the need to belong [33], Facebook is frequently used for surveillance [34]. People who feel lonely because of social exclusion exhibit higher levels of social monitoring and scanning the environment for social cues [27].

Second, the function of mutual surveillance on Facebook enhances addiction. Facebook's surveillance function keeps people informed about their connections and personal information [35]. Therefore, when social exclusion occurs, people can access social information on Facebook, which increases the number of alternative actions that can compensate for the lack of social relationships. Users pursue surveillance gratification by viewing photographs or newsfeeds on Facebook, which may be associated with the fear of "missing out" [36]. Individuals who have experienced social exclusion may become addicted to Facebook by using Facebook's informative functions to overcome the sense of isolation. Based on the above discussion, the following hypothesis is stated.

Hypothesis 3 (H3). *The higher the use of Facebook for surveillance is, the greater Facebook addiction is.*

2.4. The Mediating Role of Surveillance Use

This study supports recent findings that anxiety about social exclusion (e.g., FOMO) causes surveillance use [37]. As discussed, the FOMO and the need to belong motivate people to overcome uncertainty about their social positions by seeking social information, which encourages surveillance use of Facebook. A lack of a sense of social belonging thus can result from being addicted to Facebook through passive, continuous monitoring of relationships rather than being stimulated to actively form relationships. The Facebook environment also promotes disinhibition in individuals, and the mutual surveillance function of Facebook leads to addiction among those using it for surveillance. Based on this theoretical review, the following hypothesis is proposed:

Hypothesis 4 (H4). *Surveillance use mediates the relationship between social exclusion and Facebook addiction.*

2.5. Moderating Effects of Grandiosity on Social Exclusion and Facebook Addiction

There may be individual differences in the responses to social exclusion [38]. Narcissistic grandiosity is characterized by arrogant, conceited, and domineering attitudes and behaviors based on maladaptive self-enhancement. Unlike others, narcissists have heightened needs for recognition and admiration [39]. In addition, individuals who exhibit the characteristics of narcissistic grandiosity are highly concerned about impression management [40], which frequently occurs on Facebook [41]. If social exclusion occurs on Facebook, the grandiose narcissist will try to find ways to maintain self-enhancement.

Therefore, narcissistic grandiosity may become a variable that has a moderating effect on personal differences in the relation between social exclusion and Facebook addiction. The reasons are as follows. First, the primary reason that narcissists use Facebook is self-exhibition. Therefore, they continuously pursue their intention to use Facebook even in the context of social exclusion. Narcissists have low affiliation needs [8]. Their reason for social networking is not to build social relationships but to present themselves as superior to others [1]. Despite their social exclusion, they can still be addicted to social networking without the surveillance intention and the need to acquire social information about others. Furthermore, they tend to have a high propensity for addiction to SNSs per se [12]. Exhibitionism is a powerful motivator for narcissists to use SNSs [42]. They seek a wide audience rather than social interaction [43]. Their behavior to attract attention becomes stronger when they are socially excluded, which leads to SNS addiction.

Second, a high degree of narcissistic grandiosity promotes unstable self-esteem [44]. Those with unstable self-esteem respond more sensitively to social exclusion than others do [45]. They think that they can stabilize their self-esteem by immersing themselves in online activity [46]. Persons with a high narcissistic tendency and high inconsistency between explicit and implicit self-esteem continually monitor their position because they are insecure and prone to self-doubt [47]. In this case, perceived social anxiety is a major factor in the problematic use of the Internet [13]. In addition, social anxiety is related to low self-esteem [17].

Third, narcissists tend to respond more strongly to social exclusion than others do [48]. They want to avoid the negative result of social exclusion. Narcissists perceive exclusion, such as social rejection, as a serious threat to their ego [48]. Furthermore, users with narcissistic grandiosity are impulsive, which exacerbates their addiction to SNS [49]. Based on the foregoing theoretical review, the following hypothesis is stated.

Hypothesis 5 (H5). *The direct effect of social exclusion on Facebook addiction is contingent on narcissistic grandiosity such that the effect is stronger in those who are high in narcissistic grandiosity.*

2.6. The Research Model

The current study aims to examine the association between social exclusion and Facebook addiction. The research model derived from the findings of the literature review is based on the following assumptions: (1) a positive association exists between social exclusion as predictor variable (X) and Facebook addiction as the dependent variable (Y); (2) this association is mediated by positive surveillance use as mediator variable (M); and (3) grandiosity as moderator variable (W) moderates the direct associations hypothesized, and associations are stronger in individuals with higher grandiosity.

3. Materials and Methods

3.1. Participants and Procedure

Upon their agreement to participate in the study, the participants were informed that they would be asked to discuss their personal experiences regarding Facebook. No conflicts of interest were reported in the current study. The participants recruited from CrowdFlower (Figure eight, San Francisco, USA), a crowdsourcing data collection platform, completed the survey (Mean age = 32.64,

Standard Deviation (SD) = 23.16; females = 60.6%; n = 188). The area of data collection was limited to inside the United States. The average survey completion time was 3.51 min (SD = 3.07). The online introduction explained the policy on privacy protection to all participants before the survey began. The survey introduction described the monetary reward, and after completing the survey, the participants were paid through virtual accounts. All procedures followed were in accordance with the ethical standards of the responsible institutional committee on human experimentation.

3.2. Measures

A 26-item, English-language survey was designed using Qualtrics.com (https://www.qualtrics.com). Before they answered the questions, the survey asked the participants to recall a real interaction with a Facebook friend, such as a romantic partner, co-worker, colleague, virtual friend, etc., whom we named 'X'. While the participants were recalling the relationship with 'X', they indicated the extent of their social exclusion from X, the extent of their use of Facebook for surveillance and the extent of their Facebook addiction. They then checked whether they had the propensity for narcissistic grandiosity (see the Appendix A). Finally, the participants marked gender and age according to their demographic.

Unless otherwise indicated, the measures are based on a 7-point Likert-type scale (1 = strongly disagree, 7 = strongly agree). Social exclusion was measured using four of the six items on the Perceived Risk of Exclusion Scale [50]. Because the two excluded items were assumed to concern social exclusions offline, they were not suitable for the study of social exclusion on Facebook. An example is "I wonder if X might try to avoid me". The participants responded on scales regarding the extent to which the Facebook user perceived the possible risk of exclusion and avoidance by the Facebook friend. The items were summed (Cronbach's $\alpha = 0.921$). Facebook surveillance was measured using four items [51]. These items were selected to assess the extent to which the participants agreed with the behaviors of paying attention to and monitoring the Facebook friend. An example of an item is "I view this person's profile to monitor his/her interactions and watch out for his/her best interests". The participant's responses were averaged (Cronbach's $\alpha = 0.725$). Facebook addiction was measured with six items [19]. These items assess addiction in SNS use. An example is "Since I have been on Facebook my grades/success on work have deteriorated/my performance at work is worse". The participants' responses were averaged (Cronbach's $\alpha = 0.867$). Narcissistic grandiosity was assessed by ten items. These items were used to measure the degree of the propensity for grandiose exhibition according to the Narcissistic Personality Inventory [52], the standard measure of subclinical narcissistic traits. An example is "I really like to be center of attention". The participants responded on a scale ranging from 1 = "Not at all like me" to 7 = "Very much like me". Higher scores indicated higher narcissistic grandiosity. The summed items showed good reliability ($\alpha = 0.921$). Additionally, factor analysis was performed by applying varimax rotation to confirm the validity of the measure. In the results, the set of four factors accounted for 65.348% of the variance. The factor loading of the items ranged from 0.505 to 0.811, and all the items used had acceptable validity.

3.3. Statistical Analyses

The hypotheses were tested using model 5 in PROCESS macro [53], which allows statistical direct, indirect, and conditional direct effects to be assessed simultaneously. In testing the indirect effect of X on Y through M, the conditional direct effect of W was accounted for; furthermore, the conditional direct effect of W on the relationship between X and Y was tested while controlling for M. To test the indirect effects, 5000 bootstrapped resamples were used which generated a bias correction of 95% and adjusted confidence intervals (CI). The CIs that excluded zero demonstrated a statistically significant indirect effect. The two indirect effects were tested simultaneously using the same model that was used to test the conditional direct effect. To test the conditional direct effects, the predictor and moderator variables were mean-centered, and the significant conditional direct effects were decomposed by plotting the slopes at ±1 standard deviation [54].

4. Results

4.1. Preliminary Analyses

First, all the main study variables were tested for non-normality. All measures of skewness (−0.242 to 1.213) and kurtosis (−1.827 to 0.942) were below the cutoff of satisfaction. Second, Mahalanobis distance was used to check for multivariate outliers in the main study variables. Because no participant was found to be a multivariate outlier, the final study sample was 188 participants. Lastly, to decide what, if any, demographic control variables to include in the model testing, the bivariate correlations between the study variables and the demographic characteristics (for age and gender) were examined. The correlations were based on the demographic variable and the study variables. Therefore, age and gender were included as a control variable in all subsequent analyses. The means, standard deviations, and bivariate correlations of the study variables used in the analysis are shown in Table 1. All study variables were intercorrelated in the predicted directions.

Table 1. Means, standard deviations, correlations, and alpha reliabilities [a] for variables.

Variables	M	SD	1	2	3	4
Social Exclusion	3.545	1.612	(0.921)			
Surveillance Use	4.026	1.273	0.328 **	(0.752)		
Facebook Addiction	2.756	1.434	0.515 **	0.455 **	(0.867)	
Grandiosity	3.131	1.333	0.405 **	0.421 **	0.610 **	(0.921)

Note: [a] on diagonal in parentheses, ** $p < 0.01$.

4.2. Testing the Research Model

The results showing the unstandardized regression coefficients are presented in Figure 1. This study first hypothesized that social exclusion is positively associated with Facebook addiction (H1). Figure 1 shows that social exclusion and Facebook addiction were indeed significantly and positively related, indicating that as social exclusion increased, so did Facebook addiction (B = 0.237, t = 4.508, $p < 0.001$, CI [0.133, 0.341]. Thus, H1 was supported. Next, social exclusion was positively associated with surveillance use (B = 0.259, t = 4.524, $p < 0.01$, 95% CI = [0.146, 0.371]). Thus, H2 was supported. In addition, surveillance use was positively associated with Facebook addiction (B = 0.211, t = 3.160, $p < 0.01$, 95% CI = [0.079, 0.342]). Thus, H3 was supported. This study predicted that the mediation model of social exclusion would be related to Facebook addictive tendencies through Facebook surveillance use (H4). The results showed that the indirect effect of social exclusion on Facebook addiction through Facebook surveillance use was significant (B = 0.054, CI [0.020, 0.113]). That is, when Facebook users reported higher levels of social exclusion, they also reported higher levels of Facebook surveillance use. Similarly, as Facebook surveillance use increased, Facebook addiction also increased. Thus, H4 was supported.

Lastly, this study hypothesized the conditional direct effect of narcissistic grandiosity on the relationship between social exclusion and Facebook addiction (H5). The results revealed that narcissistic grandiosity significantly moderated the relationship between social exclusion and Facebook addiction (B = 0.079, t = 2.526, $p = 0.012$, CI [0.017, 0.140]). Figure 2 illustrates this interaction, showing that there were differences between Facebook addiction at high levels of narcissistic grandiosity, at medium levels of narcissistic grandiosity, and at low levels of narcissistic grandiosity, Facebook users reported the highest levels of Facebook addiction when they also reported high levels of social exclusion. Thus, H5 was supported (see Table 2).

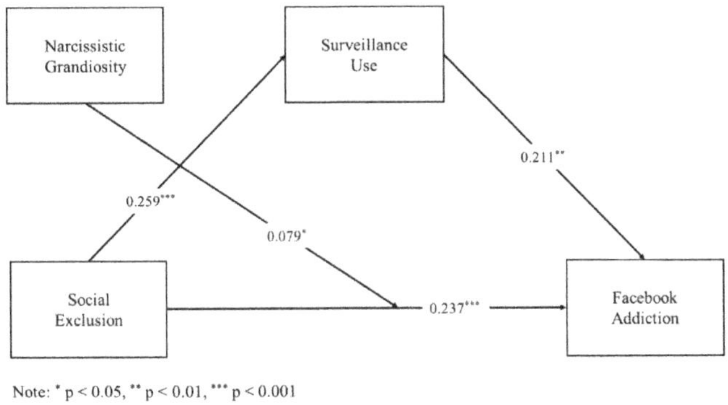

Figure 1. The indirect effect of surveillance use modeling and the conditional direct effect of narcissistic grandiosity on the relationship between social exclusion and Facebook addiction.

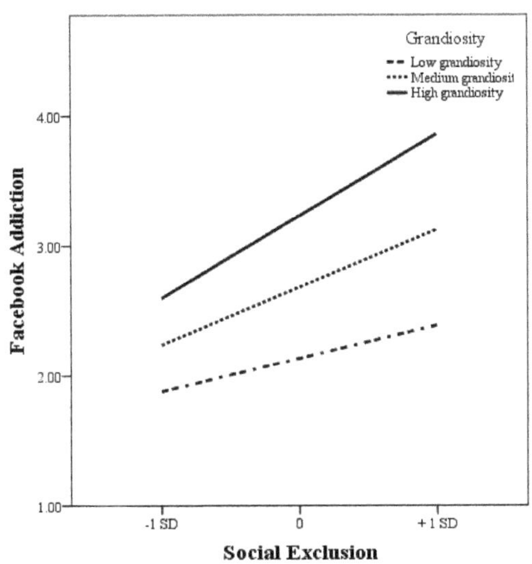

Figure 2. The conditional direct effect of narcissistic grandiosity on social exclusion and Facebook addiction.

Table 2. Conditional direct association of social exclusion on Facebook addiction (N = 188).

Narcissistic Grandiosity	Effect (Slope for Social Exclusion)	SE	*t*	*p*	LLCI	ULCI
Low (1.8)	0.131	0.064	2.073	0.040	0.006	0.257
Medium (3.13)	0.237	0.053	4.508	<0.001	0.133	0.341
High (4.46)	0.342	0.070	4.865	<0.001	0.203	0.481

Note: SE (Standard Error), LLCI & ULCI (Low and Upper Levels for Confidence Interval).

5. Discussion

The theoretical contributions of this research are as follows. This study elaborated an earlier study of Filipkowski and Smyth [5] that confirmed that the effects of online social exclusion on anxiety and emotions were as harmful as social exclusion in personal communication. This result of the present study indicated that the recent phenomenon of social exclusion on SNSs, which is highly common, can lead to surveillance behaviors and result in addiction.

The mediational role of surveillance use can be explained by the social connection theory and the environmental characteristics provided by Facebook. According to the social connection theory [55], social exclusion means that the need of social connection is not met. The loss of an SNS relationship increases the stress regarding social exclusion and further immerses the user in SNS use [56]. From an evolutionary perspective, humans have a basic need to attend to resources that provide social connections and satisfy the affiliation motivation [14,57]. On Facebook, user-generated content is stored through various forms, such as location data, status and mood messages, comments to profiles, videos, and pictures [58]. These resources facilitate surveillance and may also be linked to Facebook addiction behaviors.

The present research refers to a negative aspect of SNS use, but other studies assert that the use of social resources through SNSs can improve user well-being [59,60]. Authentic behavior showing true self-presentation on an SNS can act as a predictor of well-being to enhance the longitudinal health of the user. However, it may not offer appropriate social capital, such as social support, to a person in a low state of well-being due to positivity bias in SNS communication [61]. The narcissistic tendencies this study examines as individual differences are characteristics found in SNS personalities [62], and also narcissistic tendencies are those of a disagreeable extrovert [63]. Therefore, narcissists who do not communicate through their true selves may attempt to pursue relational interests because they do not receive sufficient social support in SNSs where positive bias occurs. In this situation, the previous research on cyber-ostracism related to narcissism focused on aggression as a maladaptive behavior caused by social exclusion [4]. These results of the present study demonstrated that the higher the level of narcissistic grandiosity is, the more severe the Facebook addiction is as another form of maladaptive behavior.

The present study has several limitations that may serve as directions for future research. First, this research did not distinguish among the types of social exclusion and relationships (e.g., romantic partners, co-workers, colleagues, and virtual friends) on Facebook. The typical Facebook friends recalled in the survey comprised heterogeneous groups that differed in the strength of their ties, intimacy, and emotional bonding [64]. Distinguishing these friend groups can produce different results. For example, information processing differs according to whether the social exclusion is caused by collective rejection or dyadic rejection [6]. This distinction could also influence Facebook addiction, so it would be meaningful to distinguish the types of rejection that cause social exclusion. In addition, the intensity of the relationships between romantic partners, coworkers and virtual friends differed. For example, according to the relationship type, social exclusion may result in different anti-social behaviors. Offensive behavior due to social exclusion is more frequent in easily replaced relationships than close relationships [65]. For example, in the case of strong ties (e.g., romantic partners) facing increased relational uncertainty, surveillance can be pursued as negative maintenance of romantic partnerships [66]. Meanwhile, the weak ties among some acquaintances (e.g., work colleagues) may benefit more from direct interactions. Thus, under weak ties, Facebook surveillance in social exclusion can be intensified [64]. Therefore, in the present study, the intensity of the relationship could have affected the causal relationship between the research variables. However, this study did not consider social exclusion and the correlations arising from different type of relationships because it evaluated various relationship types simultaneously in the survey condition. Therefore, in future research, the differences in the indirect effects of surveillance use should be distinguished in the relation with social exclusion and Facebook addiction depending on the type of relationship.

Second, demographic variables including individual differences were not used to confirm the correlations in the model, but future research should introduce them as crucial variables. Individuals have different reactions to social exclusion. For example, individuals with feelings of loneliness or a high need for social belonging are likely to monitor social information [27]. Those in closed romantic relationships might seek information about ex-partners through Facebook surveillance to reduce relational uncertainty, which causes emotional distress [67]. Thus, to deeply understand the specific personal relationship situations (e.g., romantic break-ups) that affect Facebook surveillance use and addiction, the current research model should reflect the characteristics of the users' networks. For example, do other users in Facebook users' networks actively use SNS in terms of the generation? Are there gender differences in responses to relational problems? The results suggested that females respond more negatively than males to interpersonal stressors such as social exclusion [68]. How important are the formation and dissolution of relationships in the life cycle as marital status varies by individual choice? After all, users' personal characteristics can affect their Facebook networks. To increase the validity of this research model, therefore, it is necessary to follow changes in the relationships between the variables by including individual difference variables.

Third, attention should be paid to the interpretation of the narcissistic personal trait, which serves as a moderating variable in this study. This study used narcissistic grandiosity to measure the subclinical narcissistic trait rather than the index to evaluate narcissistic personality disorder. These items used by Gentile et al. [69] reflect the tendency of college students to show more narcissistic personal traits than morbidity than did those of previous decade and also limit the ability of the results to track Facebook usage patterns. Even considering narcissistic tendencies as a minor trait, the amount of time a user takes editing on Facebook before intentionally showing themselves [69] could affect the use of Facebook. In the DSM-5 [70], histrionic personality and narcissistic personality overlap in their pursuit of attention and focus on impression management. In this study, the narcissistic tendencies are framed by emphasis on self-enhancement based on grandiose gestures. On the other hand, research on histrionic disorder tendencies on SNSs explores behavior such as exaggeration of the attractiveness of the selfie [71]. However, there is also a study that defines a focus on selfies as narcissistic behavior [72]. Therefore, it is necessary for future SNS-user research to more clearly discuss differences between the behaviors exhibited by the users with these disorders.

Fourth, to generalize the research results, it is necessary to attempt to reduce future sampling errors. First of all, the number of samples should be increased. Fewer samples were used compared with the items used. Attention, therefore, should be paid to the interpretation of the results. As the sample size decreases, the standard deviation may increase, resulting in less accurate results and a non-response bias as some subjects might not be able to complete the survey. Additionally, if, in a small sample, only those interested in a topic participate in a survey, voluntary response bias occurs, increasing the chances of skewness in the responses [73]. Future studies, therefore, should have more survey participants to be able to generalize the measurement model. Also, this study validated the model limited to Facebook. To generalize the research results, it is necessary to confirm them by collecting variables from various SNS platforms.

Fifth, the constructs of the study were dependent on self-reported data, which are prone to bias [74]. Hence, future researchers are encouraged to employ objective measures to collect data from various sources to strengthen the reliability of the findings.

6. Conclusions

The purpose of this study was to identify the mediation role of surveillance use and the moderating effect of narcissistic grandiosity in the relationship between social exclusion and Facebook addiction. The study yielded several important findings, which have the following implications. First, social exclusion in Facebook is positively associated with Facebook addiction. Second, when social exclusion occurs on Facebook, the higher the degree of narcissistic grandiosity is, the stronger the Facebook addiction is. Third, when social exclusion is a factor that promotes Facebook addiction, surveillance use plays a mediating role to monitor social information between social exclusion and addiction.

Author Contributions: All parts were contributed by M.L.

Funding: This research received no external funding.

Conflicts of Interest: Author declares that he has no conflict of interest.

Informed Consent: All procedures followed were in accordance with the ethical standards of the responsible committee on human experimentation (institutional and national) and with the Helsinki Declaration of 1975, as revised in 2000 (5). Informed consent was obtained from all patients for being included in the study.

Human and Animal Rights: All procedures performed in studies involving human participants were in accordance with the ethical standards of the institutional and/or national research committee and with the 1964 Helsinki declaration and its later amendments or comparable ethical standards.

Appendix A

Social exclusion

I sometimes feel as though X might ignore me.
I wonder if X might try to avoid me.
I suspect I might be shut out of conversation by X.
I think it is possible that X might treat me as if I am not there.

Surveillance use of Facebook

I view X's Facebook profile to monitor his/her interactions and watch out for his/her best interests.
I feel like monitoring X's Facebook page is sufficient to maintain this relationship.
I pay attention to X's Facebook updates as a way to know what he/she is doing w/o actually paying attention to him/her.
I pay attention to X's Facebook updates when I am scrolling through the newsfeed so I feel up-to-date on his/her life.

Facebook addiction

Since I have been on Facebook my grades/success on work have deteriorated/my performance at work is worse.
It happens to me that I sleep much less than usually because I stay longer with Facebook.
Sometimes I have an impression that I live two lives: one private and another virtual.
I would rather spend an afternoon and/or evening on Facebook than devote that time to any other activities.
I have better time with people that I have met over Facebook than with those that I know in person.
I fear that I might meet some of my virtual Facebook friends in real life.

Narcissistic grandiosity

I know that I am good because everybody keeps telling me so.
I like to be the center of attention.
I like to display my body.
I like to look at my body.
I am apt to show off if I get the chance.
I like to be complimented.
I like to start new fads and fashions.
I like to look at myself in the mirror.
I really like to be the center of attention.
I get upset when people don't notice how I look when I go out in public.

References

1. Mäntymäki, M.; Islam, A.N. The Janus face of Facebook: Positive and negative sides of social networking site use. *Comput. Hum. Behav.* **2016**, *61*, 14–26. [CrossRef]
2. Bevan, J.L.; Pfyl, J.; Barclay, B. Negative emotional and cognitive responses to being unfriended on Facebook: An exploratory study. *Comput. Hum. Behav.* **2012**, *28*, 1458–1464. [CrossRef]

3. Schneider, F.M.; Zwillich, B.; Bindl, M.J.; Hopp, F.R.; Reich, S.; Vorderer, P. Social media ostracism: The effects of being excluded online. *Comput. Hum. Behav.* **2017**, *73*, 385–393. [CrossRef]
4. Poon, K.-T.; Chen, Z. Assuring a sense of growth: A cognitive strategy to weaken the effect of cyber-ostracism on aggression. *Comput. Hum. Behav.* **2016**, *57*, 31–37. [CrossRef]
5. Filipkowski, K.B.; Smyth, J.M. Plugged in but not connected: Individuals' views of and responses to online and in-person ostracism. *Comput. Hum. Behav.* **2012**, *28*, 1241–1253. [CrossRef]
6. Gardner, W.L.; Pickett, C.L.; Brewer, M.B. Social exclusion and selective memory: How the need to belong influences memory for social events. *Pers. Soc. Psychol. Bull* **2000**, *26*, 486–496. [CrossRef]
7. Gabriel, M.T.; Critelli, J.W.; Ee, J.S. Narcissistic illusions in self-evaluations of intelligence and attractiveness. *J. Pers.* **1994**, *62*, 143–155. [CrossRef]
8. Elliot, A.J.; Thrash, T.M. Narcissism and motivation. *Psychol. Inq.* **2001**, *12*, 216–219.
9. Casale, S.; Fioravanti, G. Why narcissists are at risk for developing Facebook addiction: The need to be admired and the need to belong. *Addict. Behav.* **2018**, *76*, 312–318. [CrossRef]
10. David, M.E.; Roberts, J.A. Phubbed and alone: Phone snubbing, social exclusion, and attachment to social media. *J. Assoc. Consum. Res.* **2017**, *2*, 155–163. [CrossRef]
11. Nadkarni, A.; Hofmann, S.G. Why do people use Facebook? *Pers. Indiv. Differ.* **2012**, *52*, 243–249. [CrossRef] [PubMed]
12. Kuss, D.J.; Griffiths, M.D. Online social networking and addiction—A review of the psychological literature. *Int. J. Environ. Res. Public Health* **2011**, *8*, 3528–3552. [CrossRef] [PubMed]
13. Lee-Won, R.J.; Herzog, L.; Park, S.G. Hooked on Facebook: The role of social anxiety and need for social assurance in problematic use of Facebook. *Cyberpsych. Beh. Soc. N.* **2015**, *18*, 567–574. [CrossRef] [PubMed]
14. Baumeister, R.F.; Leary, M.R. The need to belong: Desire for interpersonal attachments as a fundamental human motivation. *Pol. Psychol. Bull* **1995**, *117*, 497–529. [CrossRef]
15. Tobin, S.J.; Vanman, E.J.; Verreynne, M.; Saeri, A.K. Threats to belonging on Facebook: Lurking and ostracism. *Soc. Influ.* **2015**, *10*, 31–42. [CrossRef]
16. Young, K.S. Internet addiction: Evaluation and treatment. *BMJ* **1999**, *7*, 351–352. [CrossRef]
17. Hong, F.Y.; Huang, D.H.; Lin, H.Y.; Chiu, S.L. Analysis of the psychological traits, Facebook usage, and Facebook addiction model of Taiwanese university students. *Telemat. Inform.* **2014**, *31*, 597–606. [CrossRef]
18. Lopez-Fernandez, O. Generalised Versus Specific Internet Use-Related Addiction Problems: A Mixed Methods Study on Internet, Gaming, and Social Networking Behaviours. *Int. J. Environ. Res. Public Health* **2018**, *15*, 2913. [CrossRef]
19. Milošević-Đorđević, J.S.; Žeželj, I.L. Psychological predictors of addictive social networking sites use: The case of Serbia. *Comput. Hum. Behav.* **2014**, *32*, 229–234. [CrossRef]
20. Pillow, D.R.; Malone, G.P.; Hale, W.J. The need to belong and its association with fully satisfying relationships: A tale of two measures. *Pers. Indiv. Differ.* **2015**, *74*, 259–264. [CrossRef]
21. Masur, P.K.; Reinecke, L.; Ziegele, M.; Quiring, O. The interplay of intrinsic need satisfaction and Facebook specific motives in explaining addictive behavior on Facebook. *Comput. Hum. Behav.* **2014**, *39*, 376–386. [CrossRef]
22. Przybylski, A.K.; Murayama, K.; DeHaan, C.R.; Gladwell, V. Motivational, emotional, and behavioral correlates of fear of missing out. *Comput. Hum. Behav.* **2013**, *29*, 1841–1848. [CrossRef]
23. Beyens, I.; Frison, E.; Eggermont, S. "I don't want to miss a thing": Adolescents' fear of missing out and its relationship to adolescents' social needs, Facebook use, and Facebook related stress. *Comput. Hum. Behav.* **2016**, *64*, 1–8. [CrossRef]
24. Tokunaga, R.S. Social networking site or social surveillance site? Understanding the use of interpersonal electronic surveillance in romantic relationships. *Comput. Hum. Behav.* **2011**, *27*, 705–713. [CrossRef]
25. Lampe, C.; Ellison, N.; Steinfield, C. A Face (book) in the crowd: Social searching vs. social browsing. In *Paper Presented at the Proceedings of the 2006 20th Anniversary Conference On Computer Supported Cooperative Work*; ACM Publications: New York, NY, USA, 2006; pp. 167–170.
26. Warburton, W.A.; Williams, K.D.; Cairns, D.R. When ostracism leads to aggression: The moderating effects of control deprivation. *J. Exp. Soc. Psychol.* **2006**, *42*, 213–220. [CrossRef]
27. Pickett, C.L.; Gardner, W.L.; Knowles, M. Getting a cue: The need to belong and enhanced sensitivity to social cues. *Pers. Soc. Psychol. Bull* **2004**, *30*, 1095–1107. [CrossRef]
28. Leary, M.R. Making sense of self-esteem. *Curr. Dir. Psychol. Sci.* **1999**, *8*, 32–35. [CrossRef]

29. Peplau, L.A. *Loneliness: A Sourcebook of Current Theory, Research, and Therapy*; John Wiley & Sons Inc.: New York, NY, USA, 1982; Volume 36.
30. Emmers, T.M.; Canary, D.J. The effect of uncertainty reducing strategies on young couples' relational repair and intimacy. *Commun. Q.* **1996**, *44*, 166–182. [CrossRef]
31. Suler, J. The online disinhibition effect. *Cyberpsychol. Behav.* **2004**, *7*, 321–326. [CrossRef]
32. Karlen, C.E.; Daniels, J.R. Cyberostracism and social monitoring: Social anxiety's effects on reactions to exclusion and inclusion online. In *Honors Projects 147*; Digital Commons@IWU: Bloomington, IN, USA, 2011.
33. Mäntymäki, M.; Islam, A.N. Voyeurism and exhibitionism as gratifications from prosuming social networking sites. In *Paper Presented at the 22nd European Conference on Information Systems (ECIS2014)*; Association for Information Systems: Atlanta, GA, USA, 2014.
34. Zhang, Y.; Tang, L.S.-T.; Leung, L. Gratifications, collective self-esteem, online emotional openness, and traitlike communication apprehension as predictors of Facebook uses. *Cyberpsychology Behav. Soc. Netw.* **2011**, *14*, 733–739. [CrossRef]
35. Trottier, D. Interpersonal surveillance on social media. *Can. J. Commun.* **2012**, *37*, 319–332. [CrossRef]
36. Ryan, T.; Chester, A.; Reece, J.; Xenos, S. The uses and abuses of Facebook: A review of Facebook addiction. *J. Behav. Addict.* **2014**, *3*, 133–148. [CrossRef] [PubMed]
37. Blackwell, D.; Leaman, C.; Tramposch, R.; Osborne, C.; Liss, M. Extraversion, neuroticism, attachment style and fear of missing out as predictors of social media use and addiction. *Pers. Individ. Differ.* **2017**, *116*, 69–72. [CrossRef]
38. Williams, K.D. Ostracism. *Annu. Rev. Psychol.* **2007**, *58*, 425–452. [CrossRef] [PubMed]
39. Pincus, A.L.; Cain, N.M.; Wright, A.G. Narcissistic grandiosity and narcissistic vulnerability in psychotherapy. *Pers. Disord. Theory Res. Treat.* **2014**, *5*, 439–443. [CrossRef] [PubMed]
40. Hart, W.; Adams, J.; Burton, K.A.; Tortoriello, G.K. Narcissism and self-presentation: Profiling grandiose and vulnerable Narcissists' self-presentation tactic use. *Pers. Individ. Differ.* **2017**, *104*, 48–57. [CrossRef]
41. Hall, J.A.; Pennington, N.; Lueders, A. Impression management and formation on Facebook: A lens model approach. *New Media Soc.* **2014**, *16*, 958–982. [CrossRef]
42. Panek, E.T.; Nardis, Y.; Konrath, S. Mirror or Megaphone: How relationships between narcissism and social networking site use differ on Facebook and Twitter. *Comput. Hum. Behav.* **2013**, *29*, 2004–2012. [CrossRef]
43. Carpenter, C.J. Narcissism on Facebook: Self-promotional and anti-social behavior. *Pers. Individ. Differ.* **2012**, *52*, 482–486. [CrossRef]
44. Cain, N.M.; Pincus, A.L.; Ansell, E.B. Narcissism at the crossroads: Phenotypic description of pathological narcissism across clinical theory, social/personality psychology, and psychiatric diagnosis. *Clin. Psychol. Rev.* **2008**, *28*, 638–656. [CrossRef] [PubMed]
45. Onoda, K.; Okamoto, Y.; Nakashima, K.I.; Nittono, H.; Yoshimura, S.; Yamawaki, S.; Ura, M. Does low self-esteem enhance social pain? The relationship between trait self-esteem and anterior cingulate cortex activation induced by ostracism. *Soc. Cogn. Affect. Neur.* **2010**, *5*, 385–391. [CrossRef] [PubMed]
46. Krämer, N.C.; Winter, S. Impression management 2.0: The relationship of self-esteem, extraversion, self-efficacy, and self-presentation within social networking sites. *J. Media Psychol.* **2008**, *20*, 106–116. [CrossRef]
47. Zeigler-Hill, V. Discrepancies between implicit and explicit self-esteem: Implications for narcissism and self-esteem instability. *J. Pers.* **2006**, *74*, 119–144. [CrossRef] [PubMed]
48. Twenge, J.M.; Campbell, W.K. "Isn't it fun to get the respect that we're going to deserve?" Narcissism, social rejection, and aggression. *Pers. Soc. Psychol. Bull* **2003**, *29*, 261–272. [CrossRef] [PubMed]
49. Wu, A.M.S.; Cheung, V.I.; Ku, L.; Hung, E.P.W. Psychological risk factors of addiction to social networking sites among Chinese smartphone users. *J. Behav. Addict.* **2013**, *2*, 160–166. [CrossRef] [PubMed]
50. Thau, S.; Derfler-Rozin, R.; Pitesa, M.; Mitchell, M.S.; Pillutla, M.M. Unethical for the sake of the group: Risk of social exclusion and pro-group unethical behavior. *J. Appl. Psychol.* **2015**, *100*, 98–113. [CrossRef] [PubMed]
51. McEwan, B. Sharing, caring, and surveilling: An actor–partner interdependence model examination of Facebook relational maintenance strategies. *Cyberpsychology Behav. Soc. Netw.* **2013**, *16*, 863–869. [CrossRef]
52. Ackerman, R.A.; Witt, E.A.; Donnellan, M.B.; Trzesniewski, K.H.; Robins, R.W.; Kashy, D.A. What does the Narcissistic Personality Inventory really measure? *Assessment* **2010**, *18*, 67–87. [CrossRef]

53. Hayes, A.F. *Introduction to Mediation, Moderation, and Conditional Process Analysis: A Regression-Based Approach*; Guilford Press: New York, NY, USA, 2013.
54. Aiken, L.S.; West, S.G.; Reno, R.R. *Multiple Regression: Testing and Interpreting Interactions*; Sage: New York, NY, USA, 1991.
55. Maner, J.K.; DeWall, C.N.; Baumeister, R.F.; Schaller, M. Does social exclusion motivate interpersonal reconnection? Resolving the "porcupine problem". *J. Pers. Soc. Psychol.* **2007**, *92*, 42–55. [CrossRef]
56. Chiou, W.-B.; Lee, C.-C.; Liao, D.-C. Facebook effects on social distress: Priming with online social networking thoughts can alter the perceived distress due to social exclusion. *Comput. Hum. Behav.* **2015**, *49*, 230–236. [CrossRef]
57. Buss, D.M. Evolutionary personality psychology. *Annu. Rev. Psychol.* **1991**, *42*, 459–491. [CrossRef] [PubMed]
58. Fuchs, C. New media, web 2.0 and surveillance. *Sociol. Compass* **2011**, *5*, 134–147. [CrossRef]
59. Kim, J.; Lee, J.E.R. The Facebook paths to happiness: Effects of the number of Facebook friends and self-presentation on subjective well-being. *Cyberpsychology Behav. Soc. Netw.* **2011**, *14*, 359–364. [CrossRef] [PubMed]
60. Steinfield, C.; Ellison, N.B.; Lampe, C. Social capital, self-esteem, and use of online social network sites: A longitudinal analysis. *J. Appl. Dev. Psychol.* **2008**, *29*, 434–445. [CrossRef]
61. Reinecke, L.; Trepte, S. Authenticity and well-being on social network sites: A two-wave longitudinal study on the effects of online authenticity and the positivity bias in SNS communication. *Comput. Hum. Behav.* **2014**, *30*, 95–102. [CrossRef]
62. Buffardi, L.E.; Campbell, W.K. Narcissism and social networking web sites. *Pers. Soc. Psychol. Bull* **2008**, *34*, 1303–1314. [CrossRef]
63. Paulhus, D.L. Normal narcissism: Two minimalist accounts. *Psychol. Inq.* **2001**, *12*, 228–230.
64. Koban, K.; Krüger, S. Out of sight, (Not yet) out of mind: The impact of tie strength on direct interaction and social surveillance among geographically close and long-distance Facebook friends. *Commun. Res. Report* **2018**, *35*, 74–84. [CrossRef]
65. Elshout, M.; Nelissen, R.; van Beest, I.; Elshout, S.; van Dijk, W. Situational precursors of revenge: Social exclusion, relationship type, and opportunity. *Pers. Relatsh.* **2017**, *24*, 291–305. [CrossRef]
66. Dainton, M.; Goodboy, A.K.; Borzea, D.; Goldman, Z.W. The dyadic effects of relationship uncertainty on negative relational maintenance. *Commun. Rep.* **2017**, *30*, 170–181. [CrossRef]
67. Tong, S.T. Facebook use during relationship termination: Uncertainty reduction and surveillance. *Cyberpsychology Behav. Soc. Netw.* **2013**, *16*, 788–793. [CrossRef] [PubMed]
68. Leary, M.R.; Tambor, E.S.; Terdal, S.K.; Downs, D.L. Self-esteem as an interpersonal monitor: The sociometer hypothesis. *J. Pers. Soc. Psychol.* **1995**, *68*, 518–530. [CrossRef]
69. Gentile, B.; Twenge, J.M.; Freeman, E.C.; Campbell, W.K. The effect of social networking websites on positive self-views: An experimental investigation. *Comput. Hum. Behav.* **2012**, *28*, 1929–1933. [CrossRef]
70. American Psychiatric Association. *Diagnostic and Statistical Manual of Mental Disorders (DSM-5®)*; American Psychiatric Pub.: Washington, DC, USA, 2013.
71. Sorokowski, P.; Sorokowska, A.; Frackowiak, T.; Karwowski, M.; Rusicka, I.; Oleszkiewicz, A. Sex differences in online selfie posting behaviors predict histrionic personality scores among men but not women. *Comput. Hum. Behav.* **2016**, *59*, 368–373. [CrossRef]
72. Arpaci, I.; Yalçın, S.B.; Baloğlu, M.; Kesici, Ş. The moderating effect of gender in the relationship between narcissism and selfie-posting behavior. *Per. Indiv. Differ.* **2018**, *134*, 71–74. [CrossRef]
73. The Disadvantages of a Small Sample Size. Available online: https://sciencing.com/disadvantages-small-sample-size-8448532.html (accessed on 17 September 2019).
74. Straub, D.; Limayem, M.; Karahanna-Evaristo, E. Measuring system usage: Implications for IS theory testing. *Manag. Sci.* **1995**, *41*, 1328–1342. [CrossRef]

International Journal of
Environmental Research and Public Health

MDPI

Article

Self-Exclusion among Online Poker Gamblers: Effects on Expenditure in Time and Money as Compared to Matched Controls

Amandine Luquiens [1,2,3,4,*], Aline Dugravot [3], Henri Panjo [3], Amine Benyamina [1], Stéphane Gaïffas [2,5] and Emmanuel Bacry [2,6]

1 Hôpital Paul Brousse, APHP Villejuif, France. Université Paris-Saclay, 94800 Villejuif, France; amine.benyamina@aphp.fr
2 CMAP, Ecole Polytechnique, 91128 Palaiseau Cedex, France; stephane.gaiffas@polytechnique.edu (S.G.); emmanuel.bacry@polytechnique.edu (E.B.)
3 Université Paris-Sud membre de Université Paris-Saclay, UVSQ, CESP, INSERM, 94800 Villejuif, France; aline.dugravot@inserm.fr (A.D.); henri.panjo@inserm.fr (H.P.)
4 Addictions Department, Nîmes University Hospital, 30000 Nîmes, France
5 LPSM UMR 8001, Université Paris Diderot, 8001 Paris, France
6 CEREMADE UMR 7534 CNRS, Université Paris-Dauphine PSL, 75765 Paris Cedex 16, France
* Correspondence: amandineluquiens@gmail.com; Tel.: +33145594018; Fax: +33140593568

Received: 4 September 2019; Accepted: 6 November 2019; Published: 11 November 2019

Abstract: *Background:* No comparative data is available to report on the effect of online self-exclusion. The aim of this study was to assess the effect of self-exclusion in online poker gambling as compared to matched controls, after the end of the self-exclusion period. *Methods:* We included all gamblers who were first-time self-excluders over a 7-year period ($n = 4887$) on a poker website, and gamblers matched for gender, age and account duration ($n = 4451$). We report the effects over time of self-exclusion after it ended, on money (net losses) and time spent (session duration) using an analysis of variance procedure between mixed models with and without the interaction of time and self-exclusion. Analyzes were performed on the whole sample, on the sub-groups that were the most heavily involved in terms of time or money (higher quartiles) and among short-duration self-excluders (<3 months). *Results:* Significant effects of self-exclusion and short-duration self-exclusion were found for money and time spent over 12 months. Among the gamblers that were the most heavily involved financially, no significant effect on the amount spent was found. Among the gamblers who were the most heavily involved in terms of time, a significant effect was found on time spent. Short-duration self-exclusions showed no significant effect on the most heavily involved gamblers. *Conclusions:* Self-exclusion seems efficient in the long term. However, the effect on money spent of self-exclusions and of short-duration self-exclusions should be further explored among the most heavily involved gamblers.

Keywords: online gambling; self-exclusion; responsible gambling; comparative study; poker

---

1. Introduction

Harmful gambling behaviors are widespread and treatment-seeking is still very low among problem gamblers. Self-exclusion processes could be seen as an accessible tool for problem gamblers who are not ready to seek treatment. In France, gamblers can apply for online self-exclusion per website, for the length of their choice from one week to three years. During this period, they cannot access their gambling account on the website and receive no commercial offer from the gambling service provider. At the end of the period, they can gamble back on the website with no additional procedure. No help is provided nor any counselling during the self-exclusion period.

It has been consistently demonstrated that most self-excluders were indeed heavy gamblers and probably problem gamblers [1,2]. A recent meta-analysis describing gamblers who self-excluded highlighted that this tool, perceived as one of the main "responsible gambling" tools, is still considerably under-used [3]. Main barriers for self-exclusion has been described: complicated enrollment processes, lack of complete exclusion from all venues, little support from venue staff, and lack of adequate information on self-exclusion programs. The proportion of self-excluders could be particularly low among problem internet gamblers [4]. Regulators have reached the conclusion that this tool should be promoted to increase its use. Promoting the use of a potentially therapeutic tool needs to rely on robust efficacy data and not only on empirical data or mere common sense. A recent systematic review of the literature demonstrated that the impact of responsible gambling tools is still poorly supported by scientific evidence [5]. In particular, efficacy data for the effect of self-exclusion on gambling behaviors remains scarce [5]: several studies have shown reduced gambling after a self-exclusion period on both online and offline environments, with variable durations of follow-up, sometimes including the self-exclusion period itself [1,6,7]. Follow-up after online self-exclusion has been reported in only two studies [7,8]. The first one included a limited sample of 20 gamblers, with no control group, and assessed psychosocial outcomes [8]. The other one reported that the majority of online self-excluders returned to gambling after the self-exclusion period expired (n = 1996) [7], and that most of them self-excluded a second time, after another period of more intensive gambling than the first [1]. Only one study has reported follow-up data from matched controls with a comparative research design (n = 86) [5]. One study reported that gambling outcomes did not differ between self-exclusion alone vs self-exclusion combined with counseling or counseling only [9]. No study has reported efficacy data on spontaneous voluntary self-exclusion as compared to no intervention (i.e., no self-exclusion). One experimental study randomized volunteering problem gamblers (but not pathological gamblers, who were excluded) to either a very short 7-day period of self-exclusion or no self-exclusion [10]. The authors reported no significant between-group differences in terms of changes regarding money and time spent gambling at two months.

Skills are important in poker gambling and poker gamblers have been demonstrated to have particular thoughts about their own gambling behavior and to be particularly sensitive to feedbacks on their own practice [11]. Illusion of control could be high in poker gamblers [12] and perception of their own skills could be amplified [13]. The prevalence of problem gambling among online poker gamblers is particularly high, consistently reported between 15% and 20% of active gamblers [14,15]. Several factors predicting excessive gambling in poker gamblers were identified: stress, internal attribution, dissociation, boredom, negative emotions, irrational beliefs, anxiety, and impulsivity [16], lower performance in the emotional intelligence competences (Emotional Quotient inventory Short) and, in particular, those grouped in the Intrapersonal scale (emotional self-awareness, assertiveness, self-regard, independence and self-actualization). Classical financial moderators' relevance have been consistently discussed in poker gambling as financial involvement of problem gamblers can be very low and time involved is critical to take into account [14]. No time moderator is mandatory for online gamblers in France [17]. Self-exclusion could then be one of the most relevant tools currently available for poker online gamblers in France. Poker gamblers are then a particularly interesting population to study to assess the efficacy of self-exclusion.

The aim of the present study was to document the long-term effects of self-exclusion from a poker website as compared to no self-exclusion, using matched controls. Our hypothesis was that self-exclusion would have an effect on time and money spent after the exclusion ended compared to no self-exclusion.

2. Methods

2.1. Population

We included all gamblers who self-excluded for the first time over a 7-year period from June 2010 up to October 2016 (n = 4887) on a poker website, Winamax ®, and1:1 matched gamblers who had never self-excluded up to the time of data collection, matched for gender, age and account duration automatically extracted from the account database following a structured query language (SQL) request. For technical reasons we could not match gamblers for the level of gambling involvement in terms of money/time, which were constructed variables not available from the SQL database. From the matched control group, we removed doubloons where one and the same gambler was matched to several self-excluders (n= 436). In France, self-exclusion is a voluntary process; its duration is fixed by the player from 1 day up to a maximum of 3 years. At the end of the self-exclusion period, the gamblers are notified by email by the provider, and they are then allowed to gamble again on the platform without any additional procedure. At no point during the self-exclusion process is guidance or any kind of help offered. Self-exclusion prevents the gambler from any kind of gambling activity on the website during the chosen period of time.

2.2. Measures

We collected data retrospectively from different prospective databases systematically recorded by the gambling service provider: (a) Gambler data: self-excluders' basic demographics (gender, age, date of opening of the account), characteristics of self-exclusions (date, duration) and detailed gambling variables in the month prior to self-exclusion: cash game winnings, prize amounts for tournaments, buy-in-plus-rake for tournaments, session characteristics (starting date, end date, duration). Sessions were defined as gambling with no period under 10 min without action. This measure was based on our clinical experience and on the information provided by the provider of no systematic disconnection when leaving from the website or the application on wireless devices especially. This measure was built on the experience of difficulty in extracting and interpretation of sessions duration when taking into account connection time only in a previous study [14]. We chose to explore 4-week periods because most employed people in France receive their income once a month. It is therefore important to capture at least 4 weeks per period to avoid any artificially enhanced gambling activity resulting from a possible effect following receipt of income.

Money and time spent in the preceding 4 weeks were the 2 outcomes of interest and were defined as follows: (a): time spent was obtained by summing all session durations in the last 4 weeks. Session duration was obtained by subtracting session end date from session starting date. Money spent in the last 4 weeks was defined as the net losses in the previous 4 weeks, obtained from all cash game and tournament gambling data at table level for players using real money. Table net loss was obtained from the reverse of winnings. Winnings were computed from table data (cash game winnings + prize amount for tournaments-buy-in-plus-rake for tournaments). Account-duration was defined as the time between opening the account on the website and the self-exclusion date, or the self-exclusion date of the matched self-excluders for matched gamblers. Money and time spent in the last 4 weeks on poker on the website were calculated at the self-exclusion date (or self-exclusion date of the matched self-excluders for matched gamblers), and at 3, 4, 6 and 12 months after the end of the self-exclusion period (or after self-exclusion date of the matched self-excluders for matched gamblers). We intentionally took the 12 months period after the end of the self-exclusion period into account to explore any possible changes in patterns over time when gambling was again accessible on the website, the self-exclusion period itself being of no interest for the variables studied, since gamblers were prevented from gambling. There was no missing data.

2.3. Sub-Groups

As we could not match our sample for gambling involvement, we chose to additionally analyze subgroups with similar levels of involvement in terms of money and time. Sub-groups of the gamblers who were the most heavily involved were defined as follows: gamblers from the highest quartile for amounts of money/time spent in the last 4 weeks, respectively >170 €/23 hours. In this sub-group analysis, the matching ratio of 1:1 could not be maintained, and gamblers could no longer be matched on age, sex and account duration. However, the mean age and the proportion of males remained very close across groups: 32.13 years (sd = 9.68) and 86% male among the self-excluders who were the most heavily involved in terms of money (n = 2265) vs 33.08 years (sd = 10.15) and 86% male among the gamblers who were the most heavily involved in money in the matched group (n = 79) and 32.05 years (sd= 9.74) and 87% male among the self-excluders who were the most heavily involved in terms of time (n = 2150) vs 32.73 years (sd = 9.72) and 86% male among the gamblers in the matched group who were the most heavily involved in terms of time (n = 185).

Short-duration self-exclusion was defined as <90 days (n = 1460). In this group, money and time spent in the last 4 weeks were collected at 4, 6 and 12 months after the start of self-exclusion.

2.4. Statistical Analysis

The money and time spent over 12 months after the end of the self-exclusion period were analyzed using a mixed model with the subjects as a random effect. The fixed effects were self-exclusion, time as a categorical variable, and their interaction. This interaction of self-exclusion and time provides a test for the null hypothesis that "the reduction in money/time spent over the 12 months after the end of the self-exclusion would not be different between the two groups": we report here only the p-value of the ANalysis Of Variance (ANOVA) between the mixed models with and without the interaction (i.e., the "null" model), which is in accordance with our hypothesis, testing for an effect of self-exclusion on time/ money spent at any time point over the 12 months. A significant interaction effect means that there are significant differences between groups and over time. In other words, the change in scores over time is different depending on group membership. Analyzes were performed on the whole sample, on the sub-groups with the greatest time or money involvement and on short-duration self-excluders. As sample sizes were smaller in the subgroups of gamblers who were the most heavily involved and led to a lack of power, we completed our analysis with the calculation of effect sizes for self-exclusion at 12 months in these subgroups. We use the Morris d_2 which is a standardized measure of effect size suitable for groups with unequal sample sizes within a pre-post-control design [18]. Additionally, we calculated the effect size for short self-exclusions (< 90 days) at month 12 after self-exclusion among the gamblers who were the most heavily involved (respectively in terms of money / time, self-excluders and matched: n = 683 and 18/n = 665 and 35).The strength of the effect sizes was determined using descriptors of magnitudes of d = 0.01 to 2.0, as initially suggested by Cohen and expanded by Sawilowsky [19].

All tests were 2-sided and performed with R software V3.5.1. (R core Team, free collaborative software).

2.5. Ethics

Gamblers were informed of, and consented to, personal and gambling data collection and analysis in the general conditions of use when opening an account on the website. Data collection and analysis by Winamax were authorized by the "Comité National Informatique et Libertés" (CNIL) and registered with CNIL declaration n° 1430126, which allows the analysis of the routinely recorded data for public health purposes.

The study respected the STrengthening the Reporting of OBservational studies in Epidemiology (STROBE) statement checklist items [20].

3. Results

3.1. Matched Gamblers

The matched gamblers were aged 31.5 (*sd* = 9.5) on average and predominantly male (87%). Money/time spent by the matched gamblers in the preceding 4 weeks amounted to 12.9€ (*sd* = 145.5)/ 3.5h (*sd* = 12.9) on average before the self-exclusion day of their matched self-excluders, and 4.2€ (*sd* = 72.5)/ 1.3h (*sd* = 7.2) at12 months after the end of the self-exclusion period of their matched self-excluders. Account age was 322.33 days on average (*sd* = 445.83).

3.2. Self-Excluders

The characteristics of the first-time self-excluders and short-duration first-time self-excluders are presented in Table 1. The self-excluders were aged 31 on average and predominantly male. The short-duration first-time self-excluders amounted to 30% of all first-time self-excluders over the 7 years. Money/time spent by self-excluders in the last 4 weeks was 398.5€ (*sd* = 1221.4)/ 32.8h (*sd* = 40.1) before self-exclusion, and 32.3€ (*sd*= 386.6)/6.3h (*sd* = 20.0) at12 months after the end of the self-exclusion period. The mean length of self-exclusion was 614 days (*sd* = 499). Short-duration first-time self-excluders were younger and had a greater financial and time involvement in gambling; their account was one month older on average than in the overall sample.

Table 1. Characteristics of first-time self-excluders, and short-duration first-time self-excluders subgroup.

Characteristics	**All 1st Self-Exclusions** *n* **= 4887**	**Short Self-Exclusions** *n* **= 1460**
Account age (days), mean (sd)	272.4 (407.1)	307.59 (415.63)
Age (years), mean (sd)	31.4 (9.6)	30.48 (9.21)
Gender(male), n(%)	4252 (87.0)	1270 (87.99)
Money spent in the last 4 weeks before self-exclusion (€), mean (sd)	398.5 (1221.4)	445.96 (1350.21)
Time spent in the last 4 weeks before self-exclusion (minutes), mean (sd)	1969.8 (2406.0)	2791.3 (2706.4)
Self-exclusion period duration, days, mean (sd)	614.0(499.0)	32.0 (23.0)

3.3. Effects of Self-Exclusion over 12 Months after the End of the Self-Exclusion Period

A significant effect of self-exclusion was found for money and time spent over the 12 months after the end of the self-exclusion period (*p*-value for both models < 2.2e−16) using mixed models with a subject random effect (Figure 1).

3.4. Effect of Self-Exclusion over 12 Months after the End of the Self-Exclusion Period among the Most Heavily Involved Gamblers

The average amount of money spent in the four weeks before and after the self-exclusion period among the gamblers who were the most heavily involved in terms of money, among self-excluders (*n* = 2255) and in the matched group (*n* = 79) are shown in Figure 2. No significant effect of self-exclusions was found on the amounts of money spent (*p* = 0.072) and the effect size was very small (*d* = 0.18).

The average amount of time spent in the 4 weeks before and after the self-exclusion period among the gamblers who were the most heavily involved in terms of time among self-excluders (*n* = 2150) and in the matched group (*n* = 185) are shown in Figure 3. A significant effect of self-exclusion was found for time spent (*p* < 2.2e−16) and the effect size was small (*d* = 0.34).

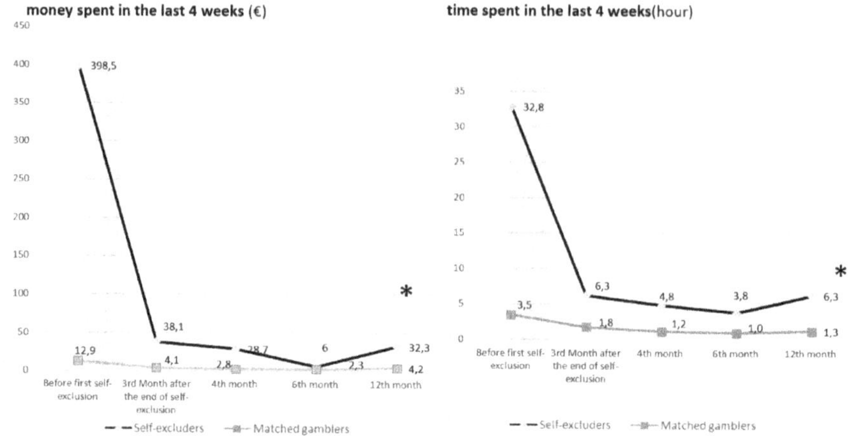

Figure 1. Evolution of money/time spent in the last 4 weeks (€/hours) at baseline and after the end of self-exclusion period (n = 4887 and n = 4451). (* = p-value < 0.05—ANOVA between the mixed model with and the null model without the interaction of self-exclusion X time).

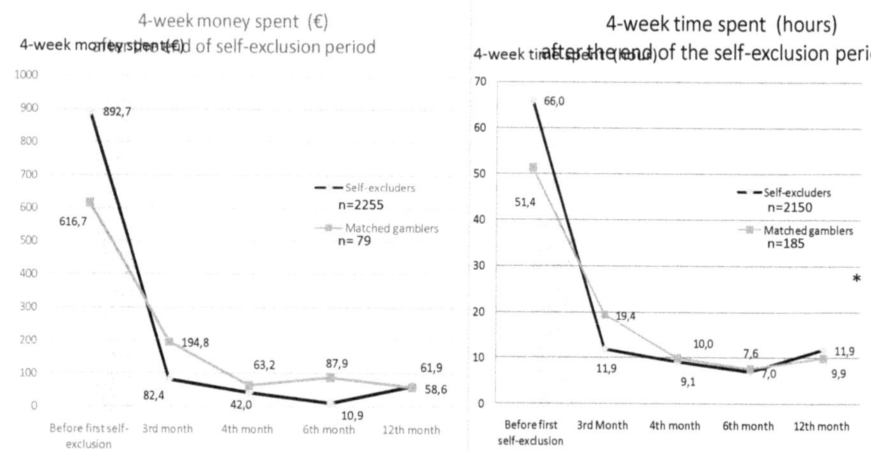

Figure 2. Evolution of money spent (net loss) in the last 4 weeks before and after the self-exclusion period among the gamblers who were the most heavily involved in terms of money (n = 2255 and 79 respectively for the self-excluders and the control group of matched gamblers) and time (n = 2150 and 185 respectively for the self-excluders and the control group of matched gamblers) (* = p-value < 0.05 —ANOVA between the mixed model with and the null model without the interaction of self-exclusion X time). (* = *p-value < 0.05* - ANOVA between the mixed model with and the null model without the interaction of self-exclusion X time).

3.5. Short Self-Exclusions

Significant effect of short self-exclusion was found for money and time spent over the 12 months after a short self-exclusion (p-value in both models <2.2e−16) using mixed models with a subject random effect (Figure 3).

No significant effect of short self-exclusions was found on money/ time spent gambling among the gamblers who were the most heavily involved in terms of money/time (respective p-values = 0.873

and 0.491) (Figure 4) but the sizes of the control groups were very small (respectively $n = 683$ vs 18, and $n = 665$ vs 35). The effect size was very small for money spent ($d = 0.17$), and negative and below the very small level for time spent ($d = -0.09$).

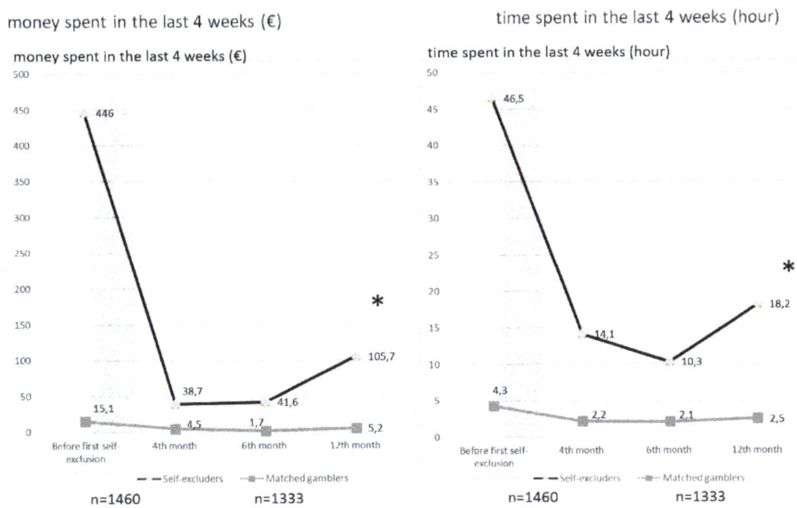

Figure 3. Evolution of money / time spent in the last 4 weeks (€/hours) before and after a short self-exclusion ($n = 1460$ and 1333). (* = p-value < 0.05—ANOVA between the mixed model with and the null model without the interaction of self-exclusion X time).

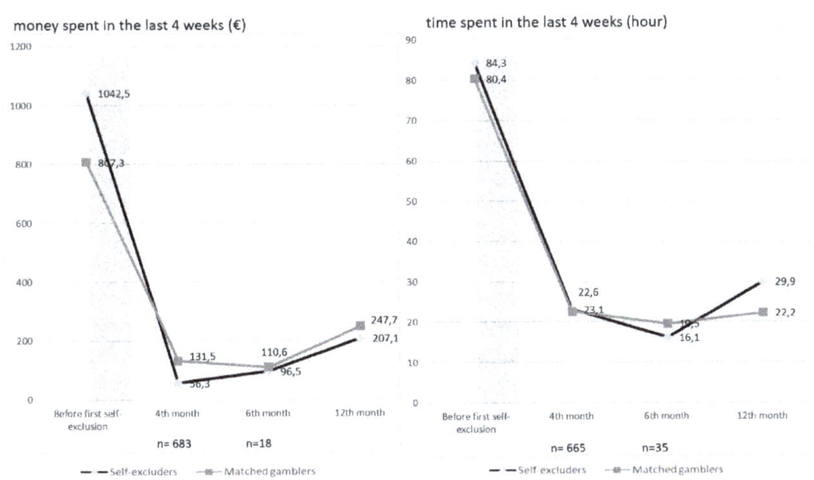

Figure 4. Evolution of money spent (net loss) in the last 4 weeks before and after a short self-exclusion among the gamblers who were the most heavily involved in terms of money ($n = 683$ and 18 respectively for the self-excluders and the control group of matched gamblers) and in terms of time ($n = 665$ and 35 respectively for the self-excluders and the control group of matched gamblers). (* = p-value < 0.05—ANOVA between the mixed model with and the null model without the interaction of self-exclusion X time).

4. Discussion

This is the first real life study, reporting comparative follow-up data on voluntary self-exclusion on the initiative of gamblers and including non-self-selected gamblers. This retrospective study analyzed prospectively registered account -based gambling data. The aim was to assess the efficacy of self-exclusion in the long term in term of time and money involvement.

The analysis of account-based gambling data for all first-time self-excluders on a website over 7 years confirmed the efficacy of self-exclusion on gambling outcomes in the long term. The exhaustiveness of this data is a strength that ensures representativeness and power for the statistical analyzes. However, the effect of self-exclusion among the most heavily involved gamblers was found only for the time spent, and not for the money spent, despite a very high level of expenditure before self-exclusion in this subgroup [14]. One important piece of information here is the spontaneous decrease in gambling involvement among gamblers who were the most heavily involved and who did not self-exclude. This result is congruent with a high rate of spontaneous remissions observed in gambling disorder [21]. This result shows the need to provide comparative data, more informative than a tool that is de facto considered to be efficient and promoted by the regulatory authorities [17]. Another interpretation of this decrease among heavy gamblers who did not self-exclude is that the gamblers were not randomized here, and could have chosen to self-exclude if they lacked confidence in their ability to bring about a change in their gambling without an external constraint such as self-exclusion, the reverse being true for non-self-excluders. The efficacy of short self-exclusions among the most heavily involved gamblers was not supported by our data. This is in line with recent experimental data among problem gamblers suggesting no efficacy of very short self-exclusions on gambling outcomes [10]. Another qualitative study reported a preferred duration to ensure efficacy of 12 months from the perspective of problem gamblers who self-excluded [22].

This study presents some important limitations. First, we included only poker gamblers. As discussed in the introduction, poker gamblers present particular cognitive profiles. Moreover, online poker gamblers are younger than other gamblers [4,12], and their history of gambling and associated damages could differ, as well as their motivation to change. The presented results could reflect some of these particularities and not be true in other gambling activities. No data was available on gambling on other online or offline gambling service providers. Gamblers could have just switched from one website to another during the exclusion period. However, all follow-up data reported here concerns gambling after the end of the exclusion period. Gamblers can gamble back on the website after this period and are commercially encouraged and sometimes offered incentives to do so. Moreover, we have already documented in another study that most gamblers return to the initial website to gamble after a self-exclusion [1]. On the other hand, as gambling is regulated in France, gamblers have to provide their Identity Card when opening an account; this measure theoretically prevents from gambling from an account opened under a false identity. The gambling profiles observed are still informative as such, even if not representative of all gambling activities. The use of account-based gambling data is a strength of the study because it enables objective data to be reported. However, it would be interesting to document the effect of self-exclusion on non-gambling outcomes, such as quality of life. No formal diagnosis of gambling disorder and no information on mental disorders or comorbidities were available. We could not, for technical reasons, match the gamblers for the level of gambling involvement in terms of time and money. Our statistical analysis allows for comparisons between the groups by adjusting the mixed model on the subject, which takes into account all subject characteristics including gambling involvement; however, it does not replace a control group with similar involvement in gambling in term of time and money spent. In addition, self-excluders could be different from non-self-excluders in term of the degree of motivation to change, as self-exclusion is a voluntary process in France. Finally, the naturalistic and ecological design of this study of course prevented any randomization process. We therefore report here results on the effect of self-exclusion rather than on efficacy. Further studies could inform on possible response factors to self-exclusion. No information was available on the health care resources used by the gamblers included.

5. Conclusions

Self-exclusion seems efficient in the long term (i.e., 12 months after the end of the self-exclusion period). However, the effects on money spent as a result of self-exclusion or short self-exclusion should be further explored among the online poker gamblers who are the most heavily involved. A spontaneous, clinically-relevant decrease in gambling activities was demonstrated among most involved gamblers who did not self-exclude. Further study with a randomized design and non-gambling outcomes should be conducted to conclude robustly on the efficacy of short and long self-exclusions in problem gambling, and on response factors.

Author Contributions: Conceptualization, A.L., A.D., H.P., S.G. and E.B.; Methodology, A.L., A.D., H.P., S.G. and E.B.; Software, A.L., A.D., H.P., S.G. and E.B.; Validation, A.L.; Formal Analysis, A.L.,A.D., H.P., S.G. and E.B.; Investigation, A.L.; Resources, A.L.; Data Curation, A.L.; Writing—Original Draft Preparation, A.L.; Writing—Review & Editing, A.L., A.B., A.D., H.P., S.G. and E.B.; Supervision, A.B., S.G. and E.B.; Project Administration, A.L.; Funding Acquisition, A.L.

Funding: This study received a grant from the "Poste d'Accueil" program exchange between the APHP and the Ecole polytechnique.

Acknowledgments: We would like to thank Winamax for allowing access to data through an agreement allowing free analysis and interpretation by our academic team.

Conflicts of Interest: A.L. has received sponsorship to attend scientific meetings, speaker honoraria and consultancy fees from Lundbeck, Indivior, and ARJEL. A.B. has received sponsorship to attend scientific meetings, speaker honoraria and consultancy fees from Lundbeck, Mylan, Gilead, Jansenn Cilag and Indivior. A.D., H.P., S.G. and E.B. have no conflict of interest to report.

References

1. Luquiens, A.; Vendryes, D.; Aubin, H.J.; Benyamina, A.; Gaiffas, S.; Bacry, E. Description and assessment of trustability of motives for self-exclusion reported by online poker gamblers in a cohort using account-based gambling data. *BMJ Open* **2018**, *8*, 022541. [CrossRef] [PubMed]
2. Ladouceur, R.; Sylvain, C.; Gosselin, P. Self-exclusion program: A longitudinal evaluation study. *J. Gambl. Stud.* **2007**, *23*, 85–94. [CrossRef] [PubMed]
3. Motka, F.; Grune, B.; Sleczka, P.; Braun, B.; Ornberg, J.C.; Kraus, L. Who uses self-exclusion to regulate problem gambling? A systematic literature review. *J. Behav. Addict.* **2018**, *7*, 903–916. [CrossRef] [PubMed]
4. Hing, N.; Russell, A.M.; Gainsbury, S.M.; Blaszczynski, A. Characteristics and help-seeking behaviors of Internet gamblers based on most problematic mode of gambling. *J. Med. Internet Res.* **2015**, *17*, 13. [CrossRef] [PubMed]
5. Shaffer, P.; Blaszczynski, A.; Shaffer, H.J. Responsible gambling: A synthesis of the empirical evidence AU-Ladouceur, Robert. *Addict. Res. Theory* **2017**, *25*, 225–235. [CrossRef]
6. Kotter, R.; Kraplin, A.; Buhringer, G. Casino Self-and Forced Excluders' Gambling Behavior Before and After Exclusion. *J. Gambl. Stud.* **2018**, *34*, 597–615. [CrossRef] [PubMed]
7. Drawson, A.S.; Tanner, J.; Mushquash, C.J.; Mushquash, A.; Mazmanian, D. The Use of Protective Behavioural Strategies in Gambling: A Systematic Review. *Int. J. Ment. Health Addict.* **2017**, *15*, 1302–1319. [CrossRef]
8. Hayer, T.; Meyer, G. Internet Self-Exclusion: Characteristics of Self-Excluded Gamblers and Preliminary Evidence for Its Effectiveness. *Int. J. Ment. Health Addict.* **2010**, *9*, 296–307. [CrossRef]
9. Hing, N.; Russell, A.; Tolchard, B. Internet Self-Exclusion: Characteristics of Self-Excluded Gamblers and Preliminary Evidence for Its Effectiveness. *Int. J. Ment. Health Addict.* **2015**, *13*, 296–307. [CrossRef]
10. Caillon, J.; Grall-Bronnec, M.; Perrot, B.; Leboucher, J.; Donnio, Y.; Romo, L.; Challet-Bouju, G. Effectiveness of At-Risk Gamblers' Temporary Self-Exclusion from Internet Gambling Sites. *J. Gambl. Stud.* **2018**, *35*, 601–615. [CrossRef] [PubMed]
11. Gainsbury, S.M.; Abarbanel, B.L.L.; Philander, K.S.; Butler, J.V. Strategies to customize responsible gambling messages: A review and focus group study. *BMC Public Health* **2018**, *18*, 1381. [CrossRef] [PubMed]
12. Barrault, S.; Varescon, I. Cognitive distortions, anxiety, and depression among regular and pathological gambling online poker players. *Cyberpsychol. Behav. Soc. Netw.* **2013**, *16*, 183–188. [CrossRef] [PubMed]
13. MacKay, T.L.; Bard, N.; Bowling, M.; Hodgins, D.C. Do pokers players know how good they are? Accuracy of poker skill estimation in online and offline players. *Comput. Hum. Behav.* **2014**, *31*, 419–424. [CrossRef]

14. Luquiens, A.; Tanguy, M.L.; Benyamina, A.; Lagadec, M.; Aubin, H.J.; Reynaud, M. Tracking online poker problem gamblers with player account-based gambling data only. *Int. J. Methods Psychiatr. Res.* **2016**, *25*, 333–342. [CrossRef] [PubMed]
15. Halme, J. Overseas Internet poker and problem gambling in Finland 2007: A secondary data analysis of a Finnish population survey. *Nord. Stud. Alcohol Drugs* **2011**, *28*, 51–64. [CrossRef]
16. Moreau, A.; Chabrol, H.; Chauchard, E. Psychopathology of Online Poker Players: Review of Literature. *J. Behav. Addict.* **2016**, *5*, 155–168. [CrossRef] [PubMed]
17. *LOI n° 2010-476 du 12 Mai 2010 Relative à L'ouverture à la Concurrence et à la Régulation du Secteur des Jeux D'argent et de Hasard en Ligne-Article 26, 26. Sect. Chapitre Vii: La Lutte Contre Le Jeu Excessif Ou Pathologique*, France, 2010.
18. Morris, S.B. Estimating Effect Sizes From Pretest-Posttest-Control Group Designs. *Organ. Res. Methods* **2008**, *11*, 364–386. [CrossRef]
19. Sawilowsky, S. New effect size rules of thumb. *J. Mod. Appl. Stat. Methods* **2009**, *8*, 467–474. [CrossRef]
20. STROBE Statement. Available online: https://strobe-statement.org/index.php?id=strobe-home (accessed on 6 November 2019).
21. Froberg, F.; Rosendahl, I.K.; Abbott, M.; Romild, U.; Tengstrom, A.; Hallqvist, J. The Incidence of Problem Gambling in a Representative Cohort of Swedish Female and Male 16-24 Year-Olds by Socio-demographic Characteristics, in Comparison with 25-44 Year-Olds. *J. Gambl. Stud.* **2015**, *31*, 621–641. [CrossRef] [PubMed]
22. Ly, C. *Investigating the Use and Effectiveness of the Tasmanian Gambling (Self) Exclusion Programme*; Tasmanian Government: Hobart, Australia, 2010.

International Journal of
Environmental Research and Public Health

MDPI

Article

Personal Factors, Internet Characteristics, and Environmental Factors Contributing to Adolescent Internet Addiction: A Public Health Perspective

Sulki Chung [1,*], Jaekyoung Lee [2] and Hae Kook Lee [3]

1 Department of Social Welfare, Chung-Ang University, Seoul 06974, Korea
2 Korea Center on Gambling Problems, Research and Development Team, Seoul 04554, Korea; jklee@kcgp.or.kr
3 Uijeongbu St. Mary's Hospital, The Catholic University of Korea, Uijeongbu-si 11765, Korea; nplhk@catholic.ac.kr
* Correspondence: chungs@cau.ac.kr; Tel.: +82-2-820-5152

Received: 30 September 2019; Accepted: 19 November 2019; Published: 21 November 2019

Abstract: Individual characteristics, family- and school-related variables, and environmental variables have equal importance in understanding Internet addiction. Most previous studies on Internet addiction have focused on individual factors; those that considered environmental influence typically only examined the proximal environment. Effective prevention and intervention of Internet addiction require a framework that integrates individual- and environmental-level factors. This study examined the relationships between personal factors, family/school factors, perceived Internet characteristics, and environmental variables as they contribute to Internet addiction among adolescents based on the public health model. A representative sample of 1628 junior high school students from 56 regions in Seoul and Gyeonggi-do participated in the study via questionnaires with the cooperation of the Ministry of Health and Welfare and the district office of education. The study analyzed psychological factors, family cohesion, attitudes toward academic activities, Internet characteristics, accessibility to PC cafés, and exposure to Internet game advertising. About 6% of the adolescents were categorized as being in the severely addicted group. Between-group comparisons showed that the addicted group had started using the Internet earlier; had higher levels of depression, compulsivity, and aggressiveness as well as lower family cohesion; and reported higher accessibility to PC cafés and exposure to Internet game advertising. Multiple logistic regression indicated that for adolescents, environmental factors had a greater influence than family or school-related factors. Policy implications for prevention and intervention are discussed.

Keywords: Internet addiction; public health model; Internet game advertising; accessibility; environmental factors

1. Introduction

As Internet usage has rapidly increased, it has become a part of our lives and is both a medium for providing a wealth of information and an important tool for connecting with others around the globe [1]. The advent of smartphones and easy access to the Internet have brought about many advantages. However, many people have developed concerns about such ease of access and social networking, particularly for adolescents; one concern is Internet addiction. Although not a formal diagnosis, Internet addiction is becoming widely accepted as a problem that may require professional treatment [2]. There is a lack of consensus regarding the definition of Internet addiction, and the term has been used in different studies with different connotations. In general, Internet addiction is defined as excessive use of the Internet accompanied by withdrawal, tolerance, and negative repercussions [3]. For consistency reasons, we utilize the term Internet addiction throughout this paper to represent a pathological state that occurs due to Internet overuse. A school-based survey for 9th and 10th grade students in seven European countries (Greece, Spain, Poland, Germany, Romania, the Netherlands, and Iceland) revealed that 0.8% to 1.7% of the students were suspected to have Internet addiction using Young's Internet addiction test [4]. A meta-analysis that reviewed studies from 31 nations reported that the global occurrence of Internet addiction was estimated at 6% [5]. However, higher estimates of the problem have been reported in Korea. Reportedly, almost 100% of South Korean adolescents are Internet users [6], and Internet addiction is a serious public health issue among Korean adolescents [7,8]. According to a national survey, adolescents had the highest proportion of high-risk Internet dependence (13.1%), followed by adults (5.8%), and then children (5.0%) [9]. Survey results from different countries also reported that adolescents as a group are the most vulnerable to Internet addiction [10]. Children and adolescents, having grown up in the digital technology era, have been familiar with digital devices and the Internet from an early age. A typical 15-year-old in 2015 would have been using the Internet since age 10 and the hours spent on the Internet is noticeably increasing among Organization for Economic Co-operation and Development countries [11]. Research studies in addiction have indicated that exposure to substance use at an earlier age increases the risk for addiction and risky behaviors later in life [12,13]. This can also be said for Internet use at an early age, and there is a concern that early access to electronic devices could lead to escalated risks for addiction [14]. Also of importance is the contents of the Internet. Different contents of the Internet affect young people to engage in overuse of the Internet. For instance, Greenfield [15] suggests that the most addictive aspects of the Internet are sexual content and computer gaming. Although these content areas are not limited to the Internet, when these are accessed using the Internet, the addictive potential of the contents are known to be amplified. A recent study also reported that people who frequently engage in web surfing experience the same psychological effects as those who are hooked on gambling [16].

Mental health experts believe that Internet addiction may manifest the same troubling effects as substance abuse or gambling disorders [17]. Internet addiction has been associated with various physical and psychological problems, and Internet addiction mostly manifests in adolescents as social withdrawal, loneliness, low motivation, and low educational performance [18].

Many studies have been conducted to identify the factors associated with Internet addiction. Of these, a number of studies have documented personal risk factors, typically psychological factors. These include depression [18–21], anxiety [20,21], aggression [22], impulsivity [23], and low self-esteem [21,24]. Among many personal risk factors of Internet addiction, depression, aggression, and impulsivity are of particular interest due to their association with other clinical subtypes. Depression is related to emotional vulnerability or internalizing problems, and aggression and impulsivity are related to externalizing problems such as attention deficit hyperactivity disorder, which are both known to increase the risk for Internet addiction [25]. There is strong evidence that individuals with higher levels of depression are more vulnerable to developing Internet addiction [19,26–29]. A recent longitudinal study also confirmed that depression is a positive predictor of adolescent Internet addiction [30]. Some studies have identified aggression as a predictor of Internet addiction. In a cross-sectional study of Lebanese adolescents, higher levels of aggression were associated with a higher level of Internet addiction [27]. A similar pattern was witnessed for young people in Taiwan and Korea [22,31]. Impulsivity is another psychological trait that is positively associated with Internet addiction. Several studies have discussed the correlation between high impulsivity and Internet addiction among adolescents [32,33], and a similar finding was reported among young people with internet gaming disorder [34].

Some studies addressed the extent to which interpersonal variables contribute to Internet addiction. They sought to explain Internet addiction in terms of interpersonal difficulties, and confirmed the role of family- and school-related variables in relation to Internet addiction among adolescents. For instance, conflict with parents, family functioning, and family resilience have been identified to influence Internet addiction [35–37]. School-related variables such as teacher's support and attitude toward school life are also associated with young people's Internet addiction. This is especially true for Korean adolescents, where there is enormous pressure for high academic achievement [38].

Another set of factors that has received relatively less attention is related to the characteristics of the Internet and its influence on addictive behavior. Few studies explicate Internet addiction by understanding what factors of the Internet enhance or reduce a user's excessive form of use. Several researchers have discussed the seductive and gratifying properties specific to the Internet that attract users [39–41]. They claim that certain properties of the Internet are associated with an elevated risk for Internet addiction. For instance, Turkle [42] isolated two dimensions of the Internet that are seductive to the Net generation: the pleasure of control and the perceived fluidity of identity [41]. Several other seductive factors were identified such as self-presentation, diversion, relationship building, and virtual community to name a few [40,41].

Most existing literature on Internet addiction has focused on individual and psychological factors and the proximal environment such as family and peer groups. Extensive meta-analysis or systematic review studies on risk factors of Internet addiction focused mainly on intrapersonal or interpersonal variables [10,31,43], and social environmental factors have received relatively less attention. The studies that have explored environmental factors mostly focused on the proximal environment such as family- and school-related variables [36,38,44]. Only a few studies examined social factors or larger environmental variables (such as exposure to advertisements and accessibility) as they contribute to Internet addiction among young people [45–47]. For instance, in alcohol-related research, factors such as availability of alcohol and exposure to alcohol are known environmental factors that influence young people's drinking behavior [48,49]. It is not difficult to assume that availability and accessibility to the Internet may also affect Internet addiction. However, such factors were infrequently explored in Internet addiction research [46]. Internet addiction is a broader concept that encompasses Internet game addiction. Although this study focuses on Internet addiction in general, we believe that examining the influence of Internet game advertisements as an environmental factor is legitimate given that a large part of Korean adolescents' Internet use is allocated to playing online games [9].

As discussed above, personal variables, family- and school-related variables, characteristics of the Internet, and environmental variables all contribute to the risk of Internet addiction. Effective prevention and intervention of Internet addiction requires a framework that integrates individual- and environmental-level factors. This is well reflected in the public health model or the epidemiologic triangle based on the public health perspective. Originally designed as a model for infectious disease, the public health model considers three areas of causes in understanding and intervening with any health problems [50,51]. One is the external agent that may contain certain destructive potential. Second is the host, who has individual susceptibility. The last factor is the environment that promotes or discourages certain health-related behavior. Applying the model to Internet addiction, the model focuses on the importance of the host (individual characteristics), agent (characteristics of the Internet), and the environment (i.e., social influence and accessibility) in the manifestation of Internet addiction. The public health model posits that any health problem is a result of the interactions among the three factors. Therefore, the development of effective public health measures for prevention and intervention usually requires assessment of all three components and their interactions [51]. To date, few studies have examined Internet addiction from a public health perspective. An awareness of the influence of these three factors may have a significant impact on intervention, prevention, and policy efforts.

The aim of the current study is to fill the gap in addiction literature by examining Internet addiction among adolescents based on a public health framework. Specifically, the study will explore the relationship among individual variables, family/school variables, the Internet characteristics, and environmental factors as they contribute to Internet addiction. The goal is to identify factors that are more influential when all areas are considered within the framework. To achieve this goal, the study utilizes the Internet addiction test (IAT) developed by Young [52,53]. The IAT has been used extensively in many studies to assess Internet addiction in both the clinical and research fields [54]. It has also been used to identify risk factors [31,43]. However, it should be noted that the IAT is not a diagnostic tool and that categorizing addiction based on the IAT does not define a clinical diagnosis.

The following research questions will be addressed using a representative community sample: (1) Are there any differences between the addict and the non-addict group in regard to psychological factors, family- and school-related factors, perceived Internet characteristics, and environmental factors? (2) Which factors that reflect the agent, host, and environment in a public health model are more important than others in predicting Internet addiction?

2. Materials and Methods

2.1. Participants

This study utilized a community-based cross-sectional survey. The survey subjects were junior-high-school students who reported the highest risk of Internet addiction in Korea [9]. A representative sample of 1871 middle (junior-high) school students participated in the study. We randomly selected one school from each of the 56 regions in Seoul and Gyeonggi-do that represented both urban and rural districts. Coeducational schools were selected to ensure equal allocation of genders. All students in one randomly selected classroom from each school were asked to participate in the survey between 1 December and 27 December 2015. Questionnaires were mailed to schools with cooperation from the Ministry of Health and Welfare and the district office of education. The study received approval from the Institutional Review Board. Only students who signed the informed consent participated in the survey. Of the 1871 students who responded to the questionnaire, 1628 were included in the analysis after excluding 243 that had at least one missing value in demographic and Internet use behavior variables. Of the 1628 participants, 52.0% (n = 847) were male and 47.0% (n = 781) were female, and the mean age was 14.9 (SD = 0.34).

2.2. Measures

2.2.1. Internet Addiction

Young's 20-item Internet addiction test was used to measure Internet addiction [52]. Each item was measured on a five-point Likert scale. Responses for each item were added to obtain a final score. Many years after the initial development of the IAT, Young proposed a new cut-off point [53]. People who score more than 50 on the IAT are thought to be experiencing frequent problems because of Internet use. For those who score above 80, Internet use is thought to be causing significant problems with their lives. Because the purpose of this study is to examine risk factors associated with the higher risk of Internet addiction, we used the addiction category based on the IAT rather than the overall score. Young proposed four groups of Internet addiction: normal range: 0–30; mild: 31–49; moderate: 50–79; and severe: 80–100. In order to focus on more severe cases, the normal range and mild addiction (low risk) group were combined into one group. Therefore, respondents were categorized into three groups: normal use: 0–49; moderate addiction: 50–79; and severe addiction: 80–100. The IAT in this study held a good internal consistency with a Cronbach's alpha (∝) of 0.94.

2.2.2. Psychological Variables

The tool used in the study was developed and validated by the National Information Society Agency to identify Internet addiction and coexisting disorders in a national survey on Internet addiction [55]. Each scale consisted of eight questions measured on a four-point Likert scale ranging from one (not at all) to four (very often). Scores were summed for each of the three features with a higher score reflecting a greater degree of impulsivity (e.g., "act on spur of the moment"), depression (e.g., "feeling hopeless"), and aggressiveness (e.g., "have trouble controlling temper"). All three measures possessed good internal consistency (the Cronbach's alpha was 0.86 for impulsivity, 0.91 for depression, and 0.91 for aggressiveness).

2.2.3. Family and School Factors

The Family Cohesion and Adaptability Evaluation Scales (FACES III) was used to measure family relationship. This scale was originally developed by Olson and colleagues [56], and a Korean version of the scale [57] was utilized in the current study. The family cohesion scale in the FACES III consists of ten questions on a five-point Likert scale that measure emotional ties, family support, time spent together, and interest in leisure time with family. Higher scores reflect higher family cohesion, and the Cronbach's alpha (∝) was 0.96.

A school adaptation scale developed for Korean students [58] was employed to examine school-related factors. The scale evaluates the degree of adaptation to the school environment, which includes questions about the relationship with teachers, the relationship with schoolmates, attitude toward academic activities, and school participation. The tool consists of four subscales and 14 questions on a five-point Likert scale. Higher total scores indicate a positive assessment of school life. The Cronbach's alpha (∝) for the subscales were acceptable: 0.94 for relationship with teachers, 0.90 for relationship with schoolmates, 0.62 for attitude toward academic activities, and 0.61 for school participation.

2.2.4. Perceived Internet Characteristics

The agent characteristics in the public health model refer to the traits of the medium in relation to a given problem. In this study, Internet characteristics are defined as the properties of the Internet perceived by individuals in relation to Internet use. Perceived Internet characteristics were measured by the Internet characteristics scale developed by Chung et al. [59] that was designed to identify the characteristics of the Internet perceived by adolescents. It consists of eight factors with 27 items including entertainment, interpersonal relationship, accessibility, profitability, anonymity, immersion, competition, and escape. Based on a previous study that identified three factors to predict young people's Internet addiction [59], our study included three factors as perceived Internet characteristics: namely, interpersonal relationship (four items; e.g., "the Internet provides opportunities to meet with diverse people"), anonymity (four items; e.g., "People can attract attention without revealing themselves"), and entertainment/pleasure (five items; e.g., "the Internet provides entertainment"). Each item was measured using a five-point Likert scale, from strongly disagree to strongly agree. The interpersonal relationship characteristic of the Internet refers to the perception of the Internet as grounds for building relationships; anonymity refers to the perception of the Internet as a site where people can act freely without identifying themselves; entertainment refers to the perception that the Internet provides fun and excitement. The measures for interpersonal relationships, anonymity, and entertainment showed good internal consistency (∝ = 0.85, 0.86, and 0.87, respectively).

2.2.5. Environmental Factors

Social environmental factors consisted of accessibility to PC cafés and exposure to Internet game advertisements. Previous studies have identified access to PC cafés as a risk factor for Internet addiction among Korean adolescents [47]. PC cafés, a unique feature in Korea, are well-equipped with advanced computers and are optimized for playing Internet games, which attracts many young people. We measured the accessibility to PC cafés using the scale reconstructed by Nam and Lee [60] that consists of a three-item (i.e., There is a nearby PC café for Internet use; I have easy access to PC cafés; there are many PC cafés near my home or school) five-point Likert scale. A higher total score indicated higher accessibility to PC cafés. The Cronbach's alpha (∝) was 0.85.

We also included exposure to Internet game advertisements as a major environmental factor. Exposure to Internet game advertisements consists of exposure and acceptability. To our understanding, there is no known instrument with which to measure exposure to Internet game advertisements. We used the tool that examined exposure to alcohol advertisements [61,62], except we restructured it to fit Internet game advertisement exposure. Two questions about respondent's exposure to Internet game advertisements (i.e., How often do you see Internet game advertisements? In the past month, how often did you see Internet game advertisements in the medium that you frequently use?) and the frequency of exposure were measured on a four-point Likert scale. The Cronbach's alpha (∝) was 0.88.

The advertisement acceptability questions used to determine alcohol use among youths were adopted for the current study [62,63]. However, for the purposes of our study, the questions were modified to assess the acceptability of Internet game advertisements. Two questions were asked: the number of game ads they had seen in various media over the past month and the number of memorable ads that they remember.

2.3. Statistical Analysis

To verify the differences between sociodemographic characteristics according to Internet use, a chi-square test (χ^2) was performed. Between-group comparisons of the variables based on the public health model according to the presence of Internet addiction were analyzed using analysis of variance. A logistic regression analysis was utilized to identify the factors associated with Internet addiction. All statistical analyses were performed using SPSS version 20.0 (IBM Corp. Released, Armonk, NY, USA).

3. Results

3.1. Sociodemographic Characteristics

Adolescents' demographic profiles were assessed using four survey items, namely gender, residing region, living with parents, and parents' working status (Table 1). Based on the Internet addiction test cut-off scores, 455 (27.9%) were identified as normal Internet users, 1067 (65.5%) belonged to the moderate-addiction or addiction-risk group, and 106 (6.5%) belonged to the severe addiction group. The group differences according to sociodemographic characteristics showed that 6.6% male and 6.4% female students were in the addiction group, and 66.7% male and 64.3% female students were deemed at risk for addiction (moderate addiction). There were no significant differences between three groups in regard to sociodemographic characteristics.

Table 1. Sociodemographic characteristics (N = 1628).

	n (%)/mean	Internet Addiction			χ^2
		Normal Use (n = 455)	Moderate Addiction (n = 1067)	Severe Addiction (n = 106)	
Gender					1.406
Male	847 (52.0)	226 (26.7)	565 (66.7)	56 (6.6)	(df = 2)
Female	781 (48.0)	229 (29.3)	502 (64.3)	50 (6.4)	p = 0.495
Region					1.552
Seoul	698 (42.9)	187 (26.8)	469 (67.2)	42 (6.0)	(df = 2)
Gyeonggi	930 (57.1)	268 (28.8)	598 (64.3)	64 (6.9)	p = 0.460
Living with parents					
Both	1496 (91.9)	415 (27.7)	987 (66.0)	94 (6.3)	7.019
Father	41 (2.5)	17 (41.5)	21 (51.2)	3 (7.3)	(df = 6)
Mother	82 (5.0)	22 (26.8)	52 (63.4)	8 (9.8)	p = 0.319
Other	9 (0.6)	1 (11.1)	7 (77.8)	1 (11.1)	
Both parents working					0.796
Yes	1075 (66.0)	304 (28.3)	705 (65.6)	66 (6.1)	(df = 2)
No	553 (34.0)	151 (27.3)	362 (65.5)	40 (7.2)	p = 0.672

3.2. Internet Use Behaviors

Table 2 summarizes Internet use behaviors between normal use, moderate addiction, and severe addiction groups. About 74% of students had started using the Internet in grade school, and 19.7%

had started using it while of preschool age. Of those who started using the Internet as early as before preschool, 9.4% were identified as addicted, compared to 10.0% and 5.4% who started using the Internet at preschool age and grade-school age, respectively. These group differences were statistically significant. More adolescents in the normal-use group reported a later age of first Internet use. Of the normal-use group, 81.3% (n = 370) started using the Internet in grade school (vs. 71.8% and 61.3% for the moderate addiction and addiction groups, respectively) (% not shown in Table 2). When students were asked to choose what Internet content they used regularly, the most frequently reported answers were webtoons, movies, or television (91.8%) and Social Networking Service or messenger (91.5%), followed by games (82.1%). Significant group differences were found for games and web surfing, with a higher percentage in the moderate addiction and severe addiction groups compared to the normal-use group.

Table 2. Internet use behavior (N = 1628).

	Totaln (%)	Internet Addiction (n, %)			χ^2	ω
		Normal Use (n = 455)	Moderate Addiction (n = 1067)	Severe Addiction (n = 106)		
Age of first Internet use						
Before preschool	53 (3.2)	12 (22.7)	36 (67.9)	5 (9.4)	36.060 *** (df = 6)	0.15
Preschool age	320 (19.7)	52 (16.2)	236 (73.8)	32 (10.0)		
Grade-school age	1201 (73.8)	370 (30.8)	766 (63.8)	65 (5.4)		
Other	54 (3.3)	21 (38.9)	29 (53.7)	4 (7.4)		
Internet contents						
Game	1337 (82.1)	339 (74.5)	913 (85.6)	85 (80.2)	26.877 *** (df = 2)	0.13
Web toon/movie/TV	1495 (91.8)	409 (89.9)	990 (92.8)	96 (90.6)	3.801 (df = 2)	
Adult materials	168 (10.3)	39 (8.6)	116 (10.9)	13 (12.3)	2.287 (df = 2)	
SNS, messenger	1489 (91.5)	409 (89.9)	985 (92.3)	95 (89.6)	2.893 (df = 2)	
Web surfing	860 (52.8)	214 (47.0)	576 (54.0)	70 (66.0)	14.125 ** (df = 2)	0.09

Note: Multiple answers were allowed for Internet contents. ** $p < 0.01$; *** $p < 0.001$. Effect size (ω) = 0.10 (small), 0.30 (medium), 0.50 (large) [64].

3.3. Relationship between Psychological, Family/School, the Internet, and Environmental Characteristics and Internet Addiction

Table 3 summarizes the group differences in domains based on the public health model, namely psychological features, family/school factors, environmental factors, and Internet characteristics. For each domain, we focused on the variables that previous research had indicated were predictors of Internet addiction. One-way analysis of variance between groups was applied to examine differences between the normal use, moderate addiction, and severe addiction groups, followed by Tukey's post-hoc test. Normality and homogeneity of variance assumptions were assessed. We first tested the normality through Shapiro–Wilk test and examined skewness and kurtosis. Shapiro–Wilk test rejected the null hypothesis and the absolute value for skewness and kurtosis did not exceed 2, which assumes a normal distribution of the variables. Levene's test was performed to assess the homogeneity of variance and the results of the Games–Howell analysis were presented as a post hoc test for variables that do not assume equal variance. In order to confirm the differences among these groups, effect size was suggested. When adolescents in three groups were compared, the severe addiction group consistently had the highest level in all psychological features followed by the moderate addiction group and the normal use group. The differences were significant at the $p < 0.001$ level, although the effect size ranged from 0.01 to 0.24.

Table 3. Relationship between psychological, family/school, Internet, and environmental characteristics, and Internet addiction (N = 1628).

	Min	Max	Mean	SD	Internet Addiction Mean (SD)			F	η^2
					Normal (a)	Moderate Addiction (b)	Severe Addiction (c)		
Psychological factors									
Impulsivity	8	32	15.6	4.6	12.2 (4.2)	16.6 (4.0)	20.2 (4.1)	259.7 *** a < b < c (df = 2)	0.24
Depression	8	32	13.0	4.6	10.5 (3.7)	13.8 (4.3)	16.6 (5.0)	134.4 *** a < b < c (df = 2)	0.14
Aggressiveness	8	32	13.8	5.1	10.8 (4.1)	14.7 (4.7)	18.4 (5.2)	171.4 *** a < b < c (df = 2)	0.18
Family/school factors									
Family cohesion	10	50	37.3	8.2	38.8 (9.8)	36.7 (7.4)	36.2 (8.0)	11.5 *** a > b, c (df = 2)	0.01
Relationship with teachers	3	15	10.9	2.6	11.5 (3.0)	10.6 (2.4)	10.6 (2.5)	20.1 *** a > b, c (df = 2)	0.02
Peer relationships	3	15	11.6	2.6	12.1 (2.8)	11.3 (2.5)	11.5 (2.5)	15.3 *** a > b (df = 2)	0.02
Academic activities	5	25	16.8	3.1	17.6 (3.7)	16.5 (2.8)	17.1 (2.9)	20.2 *** a > b (df = 2)	0.02
School participation	3	15	10.5	2.1	11.0 (2.4)	10.2 (2.0)	10.8 (2.2)	25.8 *** a > b, b>c (df = 2)	0.03
Perceived Internet characteristics									
Relationship	4	20	10.1	4.1	8.0 (4.0)	10.8 (3.7)	13.2 (4.0)	125.8 *** a < b < c (df = 2)	0.14
Anonymity	4	20	9.2	4.0	6.6 (3.5)	10.0 (3.6)	12.5 (4.1)	189.8 *** a < b < c (df = 2)	0.19
Entertainment	5	25	13.2	5.5	10.1 (5.3)	14.3 (4.9)	16.2 (6.0)	122.0 *** a < b < c (df = 2)	0.13
Environmental factors									
Accessibility of PC cafés	3	15	10.2	3.3	9.4 (3.6)	10.4 (3.2)	11.4 (3.1)	21.8 *** a < b < c (df = 2)	0.03
Exposure to Internet game ads	2	8	5.9	1.4	5.6 (1.6)	6.0 (1.3)	6.5 (1.3)	26.5 *** a < b < c (df = 2)	0.03
Acceptability of Internet game ads	1	10	5.6	3.0	5.2 (3.0)	5.7 (3.0)	6.4 (3.1)	8.0 *** a < b < c (df = 2)	0.01

Note: *** $p < 0.001$. Effect size (η^2) = 0.01 (small), 0.06 (medium), 0.14 (large) [64].

A comparison of family and school factors between the three groups showed a similar pattern. Students in the normal use group showed a significantly higher level of family cohesion, relationship with teachers, relationship with peers, attitude toward academic activities, and school participation. All differences were statistically significant.

The characteristics of the Internet content to which adolescents are drawn were compared. Mean scores in interpersonal relationship, anonymity, and entertainment characteristics were highest in the severe addiction group, followed by the moderate addiction group and then the normal use group, with the differences being statistically significant ($p < 0.001$).

Lastly, three environmental factors were compared between normal use, moderate addiction, and severe addiction groups, namely accessibility to PC cafés, exposure to Internet advertisements, and acceptability of Internet game advertisements. The mean score between the three groups was statistically significant ($p < 0.001$). Adolescents with addiction reported higher levels of accessibility to PC cafés, exposure to Internet ads, and acceptability of Internet game ads compared to the two other groups.

3.4. Factors Influencing Internet Addiction

Logistic regression was performed to determine the effect of variables in psychological, parent/school, environmental, and agent domains on the likelihood of having Internet addiction (Table 4). The severe addiction group was tested against the rest of the sample (moderate addiction and normal use groups) in order to determine the risk factors for the higher degree of Internet addiction. The logistic regression model was statistically significant (χ^2 (26) = 231.29, $p < 0.001$), explaining 38.0% of the variance in Internet addiction among adolescents. Controlling for demographic and Internet use characteristics (gender, region, living with parents, age at first exposure to the Internet, etc.), six variables were found to increase the likelihood of Internet addiction. Among psychological factors, impulsiveness (Wald = 36.14, $p < 0.001$, Exp (B) = 1.277) and aggression (Wald = 4.42, $p < 0.05$, Exp (B) = 1.077) were identified as increasing the risk of Internet addiction. Among perceived Internet characteristics, a high perception of the Internet's potential to build relationships (Wald = 12.17, $p < 0.001$, Exp (B) = 1.186) and anonymity (Wald = 8.80, $p < 0.01$, Exp (B) = 1.150) increased the risk of Internet addiction. Family and school factors were not significant predictors of Internet addiction. Two environmental factors, namely accessibility to PC cafés (Wald = 5.58, $p < 0.05$, Exp (B) = 1.116) and exposure to Internet game ads (Wald = 3.91, $p < 0.05$, Exp (B) = 1.26) were significant predictors of Internet addiction. Adolescents with higher accessibility to PC cafés and increased exposure to Internet game ads were respectively 1.12 and 1.26 times more likely to become addicted to the Internet.

Table 4. Logistic regression analysis of factors influencing Internet addiction.

	B	*S. E.*	*Wald*	*OR*	*Exp (B)* 95% *CI*	
					Lower	**Upper**
Psychological factors						
Impulsivity	0.244	0.041	36.140	1.277 ***	1.179	1.382
Depression	0.024	0.036	0.464	1.025	0.956	1.098
Aggressiveness	0.075	0.035	4.417	1.077 *	1.005	1.155
Family/school factors						
Family cohesion	0.026	0.020	1.622	1.026	0.986	1.067
Relationship with teachers	−0.058	0.067	0.746	0.943	0.827	1.077
Peer relationship	0.001	0.060	0.001	1.001	0.891	1.125
Academic activities	0.070	0.060	1.336	1.072	0.953	1.207
School participation	−0.031	0.079	0.155	0.970	0.831	1.131
Internet characteristics						
Relationship	0.170	0.049	12.174	1.186 ***	1.078	1.305
Anonymity	0.140	0.047	8.796	1.150 **	1.049	1.261
Entertainment	−0.014	.032	0.186	0.986	0.926	1.050
Environmental factors						
Accessibility to PC café	0.110	0.047	5.581	1.116 *	1.019	1.223
Exposure to Internet game ads	0.230	0.116	3.914	1.259 *	1.002	1.581
Acceptability of Internet game ads	−0.017	0.045	0.135	0.984	0.900	1.075
Gender (Male = 1)	−0.418	0.309	1.837	0.658	0.360	1.205
Region (Seoul = 1)	−0.507	0.266	3.616	0.603	0.358	1.016
Live with parents	−0.472	0.393	1.445	0.624	0.289	1.347
Both parents working	−0.237	0.267	0.784	0.789	0.468	1.332

Table 4. *Cont.*

	B	*S. E.*	*Wald*	*OR*	*Exp (B)* 95% *CI* Lower	Upper
Age of first Internet use 1	0.259	0.875	0.088	1.296	0.233	7.208
Age of first Internet use 2	−0.285	0.696	0.168	0.752	0.192	2.942
Age of first Internet use 3	−0.651	0.669	0.946	0.522	0.140	1.936
Game	−1.097	0.385	8.105	0.334 **	0.157	0.710
Web toon/movie/TV	−0.388	0.485	0.642	0.678	0.262	1.753
Adult materials	−0.171	0.422	0.165	0.843	0.368	1.927
SNS/messenger	−0.606	0.440	1.899	0.546	0.230	1.292
Web surfing	0.538	0.282	3.653	1.713	0.986	2.976
χ^2			231.287 (*df* = 26) ***			
−2LL			472.806			
Nagelkerke R^2			0.380			

Note: Age of Internet use was dummy coded as the following: Age of first Internet use 1 = before preschool; Age of first Internet use 2 = preschool age; Age of first Internet use 3 = grade school; *B* = coefficient; *SE* = standard error of the coefficient; *OR* = odds ratio; *CI* = confidence interval; * $p < 0.05$; ** $p < 0.01$; *** $p < 0.001$.

4. Discussion

The present study investigated the predictors of Internet addiction based on the public health model using a representative sample of middle-school students in two major regions in Korea. More than half (72%) of Korean Adolescents in the study were categorized as moderate and severe addiction group, indicating that Internet addiction is a public health concern. Applying the public health model, several factors were found to be related to Internet addiction.

In regard to psychological factors, depressive symptoms, impulsivity, and aggressiveness were consistently higher among the addiction and moderate addiction groups compared to the normal use group. This finding is consistent with previous research that discussed the correlation between depression [18,28], impulsivity [23,31,65–68], aggression [22,27], and Internet addiction among young people. However, when other domains of the public health model were introduced, impulsivity and aggressiveness were found to predict Internet addiction. These associations are not surprising and are in line with extensive research on addictive behaviors, as cited in the introduction. Impulsivity is related to craving and dysfunctional coping with affective mood state, thereby increasing the risk of addictive behavior [67,68]. Impulsivity is one of the key symptoms of ADHD. The high comorbidity between ADHD and Internet addiction may be mediated by impulsivity [69]. With respect to aggressiveness, results are also consistent with past research showing that individuals with a high level of aggression also had a higher tendency toward Internet addiction [22,27]. However, since the study design is cross-sectional, it is difficult to determine whether the two traits were significant predictors for Internet addiction or the result of addiction. That is, impulsivity and aggressiveness may result from excessive Internet use and related problems. Future investigations are required to conclude whether impulsivity or aggressiveness result from, or are significant risk factors for, Internet addiction. Interestingly, the influence of depression became non-significant. One possible explanation could be that within the public health framework, depression does not affect one's likelihood for Internet addiction as much as characteristics of the Internet or environmental factors.

This is similar to what we found regarding family and school-related variables. Findings from the bivariate analysis showed that adolescents who used the Internet without problems had a higher level of family cohesion and a more positive attitude toward academic activities compared to the addicted group. Previous studies also indicated that family factors were predictors of Internet addiction. For instance, for Italian adolescents, family members' affective involvement has found to be associated with lower levels of Internet addiction [70]. Another study confirmed that inter-parental conflict predicted the incidence of Internet addiction among Taiwanese adolescents [35]. However, when other domains of the public health model were controlled for, the influence of family and school related variables became non-significant. Again, a possible explanation for this discrepancy may be related to the importance of environmental variables. Although the family and school environment is an important part of their life, larger environmental context related to the Internet may be more crucial in the development of Internet addiction among young people. Adolescents who experience lower family cohesion or find school life stressful may find refuge in nearby PC cafés where they can indulge themselves in games and Internet activities without distraction. This may be especially true for Korean youths, who are constantly under pressure for academic performance both in the family and at school.

Examination of the Internet's characteristics in relation to addiction showed that the addicted group reported a higher level of entertainment (pleasure), anonymity, and interpersonal relationship factors. When other factors of the public health model were controlled, interpersonal relationships and anonymity were found to be associated with Internet addiction. Adolescents who perceived the Internet as a platform for socialization or building relationships were more likely to indulge in Internet use. This may indicate that the Internet is particularly compelling for those who otherwise find it hard to socialize. This result is in line with Chen and Kim's [40] study that found the relationship building property to be positively related to problematic use of social network sites. Those who believe that acting freely without revealing their identity is an attractive aspect of the Internet were at increased risk for addiction. This may be similar to the discussion of having a fluid identity in online life as one of the seductive properties of the Internet [41]. These findings indicate that the characteristics of the Internet can affect adolescents' Internet use behavior. To date, studies examining the influence of the Internet characteristics on Internet addiction within a public health framework are quite limited. More studies are needed to better understand the influence of traits of the Internet on Internet addiction.

Finally, our study results indicated that environmental factors are good predictors of Internet addiction in adolescents. The addicted adolescents reported higher accessibility to PC cafés and greater exposure to Internet game advertising. This held true after controlling for all other domains, namely individual, family and school, and Internet factors. PC café accessibility refers to the physical spaces in which the Internet can be used. Our findings support previous studies that have verified the effect of easy access to PC cafés on Internet addiction among Korean adolescents [47,60]. Availability and accessibility are well documented in substance abuse literature as risk factors for addictive behavior [48,49,71], and the current study's results are consistent with the existing literature [72]. In addition, the influence of exposure to advertising is consistent with previous studies [73,74]. Our study revealed that adolescents who have more exposure to Internet game advertising and those who are more accepting of advertisements are more likely to become addicted. As a major public health policy, the World Health Organization emphasized regulating the availability and marketing of alcohol to prevent drinking-related problems [75]. We believe that the same can be applied to Internet addiction. For example, regulating the density of PC cafés and mandatory staff training to recognize and cope with excessive users can be suggested. Over the past ten years, there has been a tremendous increase in online game marketing in Korea. Despite this rapid growth, there are no regulations regarding the content of advertisements (e.g., violence, adult materials, etc.) or advertising models [76]. Adolescents are frequently exposed to Internet game ads with inappropriate and violent content. In addition, online game advertisements that feature popular actors are increasing significantly, which will likely attract young people. The current study results speak to the need for public policy that considers the effects of these environmental factors.

Most studies on Internet addiction have focused on individual factors, and even those that considered environmental influences were likely to examine the proximal environment such as familial and peer influence. One strength of this study is that it examined the broader environment such as accessibility to PC cafés and the exposure to advertisements that adolescents experienced in their community, which are common in Korea and some Asian countries. The findings confirmed the importance of the environmental influence on adolescents' Internet addiction. At the same time, because PC cafés are unique to only some countries, the findings are applicable to the Korean context and few other cultures. When applying the model, researchers need to explore unique environmental features of the culture. Another limitation of the study is related to the definition of Internet characteristics. We included seductive properties of the Internet as the agent factor in the public health model. However, these properties may reflect the individual's perception rather than the Internet itself. More research and further corroboration is needed to determine the characteristics of the Internet as the agent in relation to addiction. Since this study design was cross-sectional and could not illustrate a causal relationship, its results should be interpreted cautiously. There exists a possibility that some variables found to be influential such as impulsivity, aggressiveness, and the acceptability of online game ads may have been the result of Internet addiction. Further studies should examine whether these are risk factors for Internet addiction, the result of Internet addiction, or both. Despite its limitations, this study fills the gap in addiction literature by attempting to explain adolescents' Internet addiction from a public health perspective.

5. Conclusions

Our study revealed several risk factors in the individual, Internet, and environmental domains. Although misuse of the Internet may be an individual's choice, the current study speaks to the role of other factors such as the characteristics of the Internet and the environment in which individuals make their choice. Since these environmental factors are related to the profit-seeking of related industries, prevention efforts require a social policy approach. That is, strategies should target both changing individuals' behaviors and setting the public health goal of reducing harmful Internet use at the country level. These may include regulations on PC café opening hours for adolescents and restrictions on the promotion and sponsorship of online games in youth activities. In 2019, the World Health Organization announced the inclusion of internet gaming disorder in the International Classification of Diaseses-11, and identified it as a health problem that requires a public health initiative. Further research is needed on intervention policies that specifically target individual and environmental risks based on the public health model. Findings of the current study may help in the planning of suitable strategies geared toward adolescent health.

Author Contributions: S.C. and J.L. conceptualized the study; all authors validated the study; S.C. and J.L. prepared the original draft; all authors contributed to review and editing; S.C. and H.K.L. contributed to funding acquisition.

Funding: This research was funded by Korea Mental Health Technology R & D Project, Ministry of Health and Welfare, Republic of Korea (HM14C2622).

Acknowledgments: Authors would like to thank doctoral students, Suyoung Lee, Soo Bi Lee, Jiseon Kim, for their efforts in administrative and technical support for the original project.

Conflicts of Interest: The authors declare no conflict of interest.

References

1. Young, K.S.; de Abreu, C.N. *Internet Addiction: A Handbook and Guide to Evaluation and Treatment*; John Wiley & Sons Inc.: Hoboken, NJ, USA, 2011.
2. Pies, R. Should DSM-V designate "Internet addiction" a mental disorder? *Psychiatry* **2009**, *6*, 31–37. [PubMed]
3. Young, K.S. Internet Addiction: Diagnosis and treatment considerations. *J. Contemp. Psychother.* **2009**, *39*, 241–246. [CrossRef]
4. Tsitsika, A.; Janikian, M.; Schoenmakers, T.M.; Tzavela, E.C.; Olafsson, K.; Wójcik, S.; Macarie, G.F.; Tzavara, C.; Richardson, C. Internet addictive behavior in adolescence: A cross sectional study in seven European countries. *Cyberpsychol. Behav. Soc. Netw.* **2014**, *17*, 528–535. [CrossRef] [PubMed]
5. Cheng, C.; Li, A.Y.-L. Internet addiction prevalence and quality of (real) life: A meta-analysis of 31 nations across seven world regions. *Cyberpsychol. Behav. Soc. Netw.* **2014**, *17*, 755–760. [CrossRef]
6. Lee, C.S.; McKenzie, K. Socioeconomic and geographic inequalities of internet addiction in Korean adolescents. *Psychiat. Investig.* **2015**, *12*, 559–562. [CrossRef]

7. Ryu, H.; Lee, J.Y.; Choi, A.R.; Chung, S.J.; Park, M.; Bhang, S.; Kwon, J.; Kweon, Y.; Choi, J. Application of Diagnostic Interview for internet addiction (DIA) in clinical practice for Korean adolescents. *J. Clin. Med.* **2019**, *8*, 202. [CrossRef]
8. Ministry of Gender Equality and Family. *2018 Adolescents White Paper*; MOGEF: Seoul, Korea, 2018.
9. National Information Society Agency. *The Survey on Internet Overdependence 2015*; NIA: Seoul, Korea, 2016.
10. Nakayama, H.; Mihara, S.; Higuchi, S. Treatment and risk factors of internet use disorders. *Psychiatry Clin. Neuros.* **2017**, *71*, 492–505. [CrossRef]
11. OECD. *Children & Young People's Mental Health in the Digital Age Shaping the Future*; OECD Publishing: Paris, France, 2018.
12. Odgers, C.L.; Caspi, A.; Nagin, D.S.; Piquero, A.R.; Slutske, W.S.; Milne, B.J.; Dickson, N.; Poulton, R.; Moffitt, T.E. Is It Important to Prevent Early Exposure to Drugs and Alcohol Among Adolescents? *Psychol. Sci.* **2008**, *19*, 1037–1044. [CrossRef]
13. Jordan, C.J.; Andersen, S.L. Sensitive periods of substance abuse: Early risk for the transition to dependence. *Dev. Cogn. Neurosci.* **2017**, *25*, 29–44. [CrossRef]
14. Moon, Y.I.; Koo, H.Y.; Park, H.R. Scope of Internet Addiction and Predictors of Addiction in Korean Children in Early Elementary School. *Korean J. Child Health Nurs.* **2005**, *11*, 263–272.
15. Greenfield, D. The addictive properties of Internet usage. In *Internet Addiction: A Handbook and Guide to Evaluation and Treatment*; Young, K.S., de Abreu, C.N., Eds.; John Wiley & Sons: Hoboken, NJ, USA, 2011; pp. 135–153.
16. Reed, P.; Romano, M.; Re, F.; Roaro, A.; Osborne, L.A.; Viganò, C.; Truzoli, R. Differential physiological changes following internet exposure in higher and lower problematic internet users. *PLoS ONE* **2017**, *12*, 1–11. [CrossRef] [PubMed]
17. Kishore, A.; Pan, T.; Naskar, N.N. Internet addiction and its associated factors: Study among the students of a public health school of Kolkata. *Int. J. Health Allied Sci.* **2019**, *8*, 159–163.
18. Bazrafshan, M.; Jokar, M.; Rahmati, M.; Ahmadi, S.; Kavi, E.; Sookhak, F.; Aliabadi, S.H. The relationship between depression and internet addiction among paramedical students in Larestan, Iran. *J. Clin. Diagn. Res.* **2019**, *13*, 16–19. [CrossRef]
19. Young, K.S.; Rogers, R.C. The relationship between depression and internet addiction. *Cyberpsychol. Behav.* **1998**, *1*, 25–28. [CrossRef]
20. Ho, R.C.; Zhang, M.W.B.; Tsang, T.Y.; Toh, A.H.; Pan, F.; Lu, Y.; Mak, K.-K. The association between internet addiction and psychiatric co-morbidity: A meta-analysis. *BMC Psychiatry* **2014**, *14*, 284–306. [CrossRef] [PubMed]
21. Abdul, A.M.; Wan, I.W.S.; Bahar, N.; Mahadevan, R.; Azhar, S.S. Internet addiction among secondary school students in Klang Valley, Malaysia: What is the Association with depressive symptoms, anxiety symptoms, and self-esteem? *Int. Med. J. Malaysia* **2018**, *17*, 17–25.
22. Ko, C.H.; Yen, J.Y.; Liu, S.C.; Hang, C.F.; Yen, C.F. The associations between aggressive behaviours and internet addiction and online activities in adolescents. *J. Adolescent Health* **2009**, *44*, 598–605. [CrossRef] [PubMed]
23. Lee, H.W.; Choi, J.S.; Shin, Y.; Lee, J.; Jung, H.Y.; Kwon, J.S. Impulsivity in internet addiction: A comparison with pathological gambling. *Cyberpsychol. Behav. Soc. Netw.* **2012**, *15*, 373–377. [CrossRef]
24. Stieger, S.; Burger, C. Implicit and explicit self-esteem in the context of internet addiction. *Cyberpsychol. Behav. Soc. Netw.* **2010**, *13*, 681–688. [CrossRef]
25. Lee, S.-Y.; Lee, H.K.; Choo, H. Typology of Internet gaming disorder and its clinical implications. *Psychiatry Clin. Neurosci.* **2017**, *71*, 479–491. [CrossRef]
26. Yen, J.; Lin, H.; Chou, W.; Liu, T.; Ko, C. Associations among resilience, stress, depression, and Internet gaming disorder in young adults. *Int. J. Environ. Res. Public Health* **2019**, *16*, 3181. [CrossRef] [PubMed]
27. Obeid, S.; Saade, S.; Haddad, C.; Sacre, H.; Khansa, W.; Al Hajj, R.; Kheir, N.; Hallit, S. Internet addiction among Lebanese adolescents: The role of self-esteem, anger, depression, anxiety, social anxiety and fear, impulsivity, and aggression—A cross-sectional study. *J. Nerv. Ment. Dis.* **2019**, *207*, 838–846. [CrossRef] [PubMed]
28. Przepiorka, A.; Blachnio, A.; Cudo, A. The role of depression, personality, and future time perspective in internet addiction in adolescents and emerging adults. *Psychiatry Res.* **2019**, *272*, 340–348. [CrossRef] [PubMed]

29. Kim, Y.J.; Jang, H.M.; Lee, Y.; Lee, D.; Kim, D.J. Effects of internet and smartphone addictions on depression and anxiety based on propensity score matching analysis. *Int. J. Environ. Res. Public Health* **2018**, *15*, 859. [CrossRef]
30. Li, G.; Hou, G.; Yang, D.; Jian, H.; Wang, W. Relationship between anxiety, depression, sex, obesity, and internet addiction in Chinese adolescents: A short-term longitudinal study. *Addict. Behav.* **2019**, *90*, 421–427. [CrossRef]
31. Koo, H.J.; Kwon, J.H. Risk and protective factors of internet addiction: A meta-analysis of empirical studies in Korea. *Yonsei Med. J.* **2014**, *55*, 1691–1711. [CrossRef]
32. Lin, M.-P.; Ko, H.-C.; Wu, J.Y.-W. Prevalence and psychosocial risk factors associated with Internet addiction in a nationally representative sample of college students in Taiwan. *Cyberpsychol. Behav. Soc. Netw.* **2011**, *14*, 741–746. [CrossRef] [PubMed]
33. Liu, S.-J.; Lan, Y.; Wu, L.; Yan, W.-S. Profiles of impulsivity in problematic internet users and cigarette smokers. *Front. Psychol.* **2019**, *10*, 772. [CrossRef]
34. Ryu, H.; Lee, J.Y.; Choi, A.; Park, S.; Kim, D.J.; Choi, J.S. The Relationship between Impulsivity and Internet Gaming Disorder in young adults: Mediating effects of interpersonal relationships and depression. *Int. J. Environ. Res. Public Health* **2018**, *15*, 458. [CrossRef]
35. Ko, C.H.; Wang, P.W.; Liu, T.L.; Yen, C.F.; Chen, C.S.; Yen, J.Y. Bidirectional associations between family factors and internet addiction among adolescents in a prospective investigation. *Psychiatry Clin. Neuros.* **2015**, *69*, 192–200. [CrossRef]
36. Yu, L.; Shek, D.T.-L. Internet addiction in Hong Kong Adolescents: A three-year longitudinal study. *J. Pediatr. Adolesc. Gynec.* **2013**, *26*, S10–S17. [CrossRef] [PubMed]
37. Kim, M.K. Relationship of multicultural adolescents' stress, depression, family resilience and internet game addiction. *J. Korea Converg. Soc.* **2016**, *7*, 205–210. [CrossRef]
38. Jang, S.H.; Park, Y.J. Effects of teenagers' ego-identity, school life adaptation, and stress on internet addiction. *J. Hum.* **2010**, *15*, 195–226.
39. Song, I.; Larose, R.; Eastin, M.S.; Lin, C.A. Internet gratifications and internet addiction: On the uses and abuses of new media. *Cyberpsychol. Behav.* **2004**, *7*, 384–394. [CrossRef] [PubMed]
40. Chen, H.-T.; Kim, Y. Problematic use of social network sites: The interactive relationship between gratifications sought and privacy concerns. *Cyberpsychol. Behav. Soc. Netw.* **2013**, *16*, 806–812. [CrossRef] [PubMed]
41. Leung, L. Net-generation attributes and seductive properties of the internet as predictors of online activities and internet addiction. *Cyberpsychol. Behav.* **2004**, *7*, 333–348. [CrossRef] [PubMed]
42. Turkle, S. *Life on the Screen*; Sion & Schuster: New York, NY, USA, 1995.
43. Lam, L.T. Risk factors of Internet addiction and the health effect of internet addiction on adolescents: A systematic review of longitudinal and prospective studies. *Curr. Psychiatry Rep.* **2014**, *16*, 508. [CrossRef]
44. Yuh, J. Aggression, social environment, and Internet gaming addiction among Korean adolescents. *Soc. Behav. Pers.* **2018**, *46*, 127–138. [CrossRef]
45. Hur, M.H. Demographic, habitual, and socioeconomic determinants of Internet addiction disorder: An empirical study of Korean teenagers. *Cyberpsychol. Behav.* **2006**, *9*, 514–525. [CrossRef]
46. Durkee, T.; Kaess, M.; Carli, V.; Parzer, P.; Wasserman, C.; Floderus, B.; Mandelli, L.; Apter, A.; Balazs, J.; Barzilay, S.; et al. Prevalence of pathological internet use among adolescents in Europe: Demographic and social factors. *Addiction* **2012**, *107*, 2210–2222. [CrossRef]
47. Ju, S.J. A comparative study on risk factors and protective factors for the internet addiction of adolescents from the regional characteristics. *J. Future Oriented Youth Soc.* **2011**, *8*, 121–147.
48. Gruenewald, P.J. Regulating availability: How access to alcohol affects drinking and problems in youth and adults. *Alcohol. Res. Health* **2011**, *34*, 248–256. [PubMed]
49. Paschall, M.J.; Grube, J.W.; Thomas, S.; Cannon, C.; Treffers, R. Relationships between local enforcement, alcohol availability, drinking norms, and adolescent alcohol use in 50 California cities. *J. Stud. Alcohol Drugs* **2012**, *73*, 657–665. [CrossRef] [PubMed]
50. Hester, R.K.; Miller, W.R. *Handbook of Alcoholism Treatment Approaches*, 3rd ed.; Allyn and Bacon: Boston, MA, USA, 2003.
51. Dicker, R.; Coronado, F.; Koo, D.; Parrish, R.G. *Principles of Epidemiology in Public Health Practice*, 3rd ed.; U.S. Department of Health and Human Services: Atlanta, GA, USA, 2006.

52. Young, K.S. Internet addiction: The emergence of a new clinical disorder. *Cyberpsychol. Behav.* **1998**, *1*, 237–244. [CrossRef]
53. Young, K. Clinical assessment of Internet-addicted clients. In *Internet Addiction: A Handbook and Guide to Evaluation and Treatment*; Young, K.S., de Abreu, C.N., Eds.; John Wiley & Sons: Hoboken, NJ, USA, 2011; pp. 19–34.
54. Moon, S.J.; Hwang, J.S.; Kim, J.Y.; Shin, A.L.; Bae, S.M.; Kim, J.W. Psychometric properties of the Internet Addiction Test: A systematic review and meta-analysis. *Cyberpsychol. Behav. Soc. Netw.* **2018**, *21*, 473–484. [CrossRef] [PubMed]
55. National Information Society Agency. *The Survey on Internet Addiction 2011*; NIA: Seoul, Korea, 2012.
56. Olson, D.H.; Portner, J.; Lavee, Y. *FACES III Manual*; University of Minnesota: St. Paul, MN, USA, 1985.
57. Choi, Y.S. A Review of studies of Family Adaptability and Cohesion Evaluation Scales (FACES). *J. Korean Home Manag. Assoc.* **1996**, *14*, 131–145.
58. Son, H.S. A Study on the Adolescents' Needs for School Social Work Services Dependent on their Living Environments. Master's Thesis, Daegu University, Daegu, Korea, 2002.
59. Chung, S.; Kim, Y.S.; Yoo, H.S.; Lee, J.K.; Lee, S.; Lee, S.B. *Public Health Model Development of Internet Game Addiction and Prevalence of Related Harm*; Korea Ministry of Health and Welfare: Seoul, Korea, 2016.
60. Nam, Y.O.; Lee, S.J. Risk and protective factors, and mental health in youth's internet addiction by the addiction types. *Korean J. Soc. Welfare* **2005**, *57*, 195–222.
61. Schooler, C.; Feighery, E.; Flora, J.A. Seventh graders' self-reported exposure to cigarette marketing and its relationship to their smoking behavior. *Am. J. Public Health* **1996**, *86*, 1216–1221. [CrossRef]
62. Unger, J.B.; Schuster, D.; Zogg, J.; Dent, C.W.; Stacy, A.W. Alcohol advertising exposure and adolescent alcohol use: A comparison of exposure measures. *Addict. Res. Theory* **2003**, *11*, 177–193. [CrossRef]
63. Pierce, J.P.; Gilpin, E.A.; Emery, S.L.; Farkas, A.J.; Zhu, S.H.; Choi, W.S.; Berry, C.C.; Distefan, J.M.; White, M.M.; Soroko, S.; et al. *Tobacco Control in California: Who's Winning the War? An Evaluation of the Tobacco Control Program, 1989–1996*; University of California: San Diego, CA, USA, 1998.
64. Cohen, J. *Statistical Power Analysis for the Behavioral Sciences*, 2nd ed.; Lawrence Erlbaum Associates: New York, NY, USA, 1988.
65. Cao, F.; Su, L.; Liu, T.Q.; Gao, X. The Relationship between impulsivity and internet addiction in a sample of Chinese adolescents. *Eur. Psychiat.* **2007**, *22*, 466–471. [CrossRef]
66. Chen, S.; Lo, M.; Lin, S.S.-J. Impulsivity as a Precedent Factor for Problematic Internet Use: How Can We Be Sure? *Int. J. Psychol.* **2017**, *52*, 389–397. [CrossRef] [PubMed]
67. Antons, S.; Mueller, S.M.; Wegmann, E.; Trotzke, P.; Schulte, M.M.; Brand, M. Facets of impulsivity and related aspects differentiate among recreational and unregulated use of Internet pornography. *J. Behav. Addict.* **2019**, *8*, 223–233. [CrossRef] [PubMed]
68. Brand, M.; Young, K.S.; Laier, C.; Wölfling, K.; Potenza, M.N. Integrating psychological and neurobiological considerations regarding the development and maintenance of specific Internet-use disorders: An Interaction of Person-Affect-Cognition-Execution (I-PACE) model. *Neurosci. Biobehav. Rev.* **2016**, *71*, 252–266. [CrossRef] [PubMed]
69. Wang, B.Q.; Yao, N.Q.; Zhou, X.; Liu, J.; Lv, Z.T. The association between attention deficit/hyperactivity disorder and internet addiction: A systematic review and meta-analysis. *BMC Psychiatry* **2017**, *17*, 260. [CrossRef] [PubMed]
70. Cacioppo, M.; Barni, D.; Correale, C.; Mangialavori, S.; Danioni, F.; Gori, A. Do attachment styles and family functioning predict adolescents' problematic internet use? A relative weight analysis. *J. Child Fam. Stud.* **2019**, *28*, 1263–1271. [CrossRef]
71. Broman, C. The availability of substances in adolescence: Influences in emerging adulthood. *J. Child Adoles. Subst.* **2016**, *25*, 487–495. [CrossRef] [PubMed]
72. Dhir, A.; Chen, S.; Nieminen, M. Predicting adolescent internet addiction: The roles of demographics, technology accessibility, unwillingness to communicate and sought internet gratifications. *Comput. Hum. Behav.* **2015**, *51*, 24–33. [CrossRef]
73. Esser, M.B.; Jernigan, D.H. Policy approaches for regulating alcohol marketing in a global context: A public health perspective. *Annu. Rev. Public Health* **2018**, *39*, 385–401. [CrossRef]

74. Jernigan, D.; Noel, J.; Landon, J.; Thornton, N.; Lobstein, T. Alcohol marketing and youth alcohol consumption: A systematic review of longitudinal studies published since 2008. *Addiction* **2016**, *112*, 7–20. [CrossRef]
75. World Health Organization. *Global Status Report on Alcohol and Health 2018*; World Health Organization: Geneva, Switzerland, 2018.
76. Ministry of Science, ICT and Future Planning. *2016 Online Advertising Law System Guide*; KISA: Seoul, Korea, 2016.

International Journal of
Environmental Research and Public Health

MDPI

Article

The Prevalence of E-Gambling and of Problem E-Gambling in Poland

Bernadeta Lelonek-Kuleta *, Rafał P. Bartczuk, Michał Wiechetek, Joanna Chwaszcz and Iwona Niewiadomska

Institute of Psychology, The John Paul II Catholic University of Lublin, 20-950 Lublin, Poland; bartczuk@kul.lublin.pl (R.P.B.); wiechetek@kul.lublin.pl (M.W.); chwaszcz@kul.lublin.pl (J.C.); pfrau@kul.lublin.pl (I.N.)

* Correspondence: bernadetalelonek@kul.lublin.pl; Tel.: +48-81-4453441

Received: 22 November 2019; Accepted: 6 January 2020; Published: 8 January 2020

Abstract: This study estimated the levels of involvement in e-gambling and problem e-gambling in Poland and identified selected sociodemographic variables associated with e-gambling activities. The study was conducted using a representative sample of the adult inhabitants of Poland (n = 2000). The survey contained questions measuring three aspects of gambling (involvement in e-gambling, types of e-gambling activity, and problematic e-gambling). Results suggested that 4.1% of respondents were involved in e-gambling and 26.8% of them could be classified as problem gamblers. The most popular e-gambling games were lotteries and sports betting. Gender, age, size of city of residence, level of education, and income were identified as significant predictors of involvement in e-gambling. The results indicated that men, younger people, and people who earnt less were more often involved in e-gambling. Having children, playing online scratch cards, and online sport betting—but not online lotteries—turned out to be typical for problem online gamblers. The prevalence of problem gambling among Polish e-gamblers suggests that extended research in this area is needed.

Keywords: e-gambling; e-gambling prevalence; forms of e-gambling; problem e-gambling

1. Introduction

The involvement of societies in gambling is a subject that has interested researchers for many years [1,2]. Gambling, as an entertainment form permitted for adults, involves a game in which there is a valuable stake, the result of which partially depends on chance, and it interests the representatives of social sciences mostly due to the potential damages that the activity can cause if specific circumstances arise. In 2013, in the 5th edition of the Diagnostic and Statistical Manual of Mental Disorders (DSM-5), pathological gambling was included in the group of non-substance related disorders (in section: Substance-Related and Addictive Disorders), which made the scientific world admit that pathological gambling and substance addiction have common development mechanisms and analogous symptoms [3]. The World Health Organisation included gambling disorder in the section "Disorders" due to substance use or addictive behaviours in the 11th edition of the International Classification of Diseases (ICD) classification. It is noteworthy that WHO distinguishes two categories of this disorder, including: 6C50.0 gambling disorder, predominantly offline, and 6C50.1 gambling disorder, predominantly online [4]. The development of addiction is most often described in four stages, common for substance and behavioural addictions. Taking the example of exercise addiction, they are as follows: recreational exercise, at-risk exercise, problematic exercise, and exercise addiction [5]. The traditional description of the gambling disorder has included four phases: the reaction to winning, losing, desperation, and hopelessness [6]. The next distinguished phases are accompanied by an increasing intensity of problems resulting from involvement in a given activity, progressive concentration and loss of control. According to researchers, the social consequences of gambling abuse include decreased productivity, social welfare costs resulting from absence from work [7], loss of jobs, early retirement, and even an increase in mortality resulting from suicidal tendencies occurring at advanced stages of gambling addiction [8,9]. Social detriments are also considered in the context of elevated distress and social isolation resulting from gambling [10].

Despite the long tradition of research on the involvement in gambling and problem gambling, a new phenomenon that has very quickly caused concern among the specialists emerged in the recent decades—online gambling [11]. Studies on online gambling have led many researchers to conclude that it has higher addictive potential than any other type of gambling [12–16]. Although results that do not confirm the higher addictive potential of online gambling do exist, most studies confirm this phenomenon [17,18]. Its high addictive potential is further confirmed by the higher rate of gambling addiction among online gamblers than among those who gamble in the traditional form [16,19,20]; this increase may even be three to five times higher [21]. The studies conducted by Effertz et al. on a representative group of 15,023 Germans showed that replacing 10% of offline gambling with online gambling increased the risk of becoming a problem gambler by 8.8–12.6% [22].

As for the increase in problem gambling among online players, the results of international studies conducted by McCormack among 1119 gamblers showed that 14% met the Problem Gambling Severity Index (PGSI) criteria for problem gambling, 29% met the criteria for moderate risk gambling, 32.7% met the criteria for problem gambling at a low level, and 24.3% did not show symptoms indicating problems resulting from gambling [23].

Why online gambling is more addictive remains uncertain; does it result from its higher accessibility, or, perhaps, from the nature of the Internet as a medium via which players gamble? [17,18]. Some studies have reported a lower percentage of addicted gamblers among "pure" online gamblers in comparison with "pure" offline gamblers [18]. Studies have, however, determined certain factors increasing the addictive potential of online gaming, including the games' structure, which consists of, among other things, directness, accessibility, and ease of betting, all of which are particularly dangerous for young gamblers [24]. These factors are said by other gamblers, to have addictive potential. For example, Griffiths et al. also indicated factors related to the games' structure—the directness of reinforcement, the speed of their course, and the frequency of game appearance—but also situational factors, such as accessibility and availability—which other researchers have also confirmed [25–29].

Online gambling seems to be more attractive for various reasons. It offers gamblers additional profits from gambling compared to offline gambling. First and foremost, gamblers can play whenever and wherever they choose (at home or work), which is associated with a high level of comfort and low access costs: players do not need to travel to certain locations, dedicate their time, and so forth [12]. Gamblers also save time because they can play several games simultaneously, which accelerates the course of online games [30,31]. The basic benefit of online gambling is anonymity, which seems to be particularly desirable for certain types of users [32,33].

Recent studies on involvement in online gambling have mainly focused on identifying the risk factors of problem gambling. The studies conducted by Effertz et al. on a representative group of Germans indicated that the risk of problem gambling decreases with higher levels of education and increases in the case of men; people who are unemployed, single, and divorced (respectively); and among migrants [22]. Effertz et al. also noticed that the risk of problem online gambling is the highest among heavy Internet users. Scientists have also drawn attention to the correlation between the type of game and problem gambling. For example, McCormack et al. observed that the risk of problem online gambling was significantly higher among people who gamble regularly and play online betting games, online slot machines, and online roulette, compared to gamblers who do not play regularly. In addition, persons who regularly played two or more types of online games were also at considerably greater risk of developing problem gambling than those who played one game only [23]. Other studies [34] have also highlighted the correlation between the number of the gambling accounts a gamer has, increased involvement in gambling, and increased intensity of problem gaming [35]. In the McCormack studies, people who regularly played only online poker were at lower risk of problem gambling than people who did not play poker regularly, but played other games as well, which is in line with the results of other studies [36]. Although the results suggest that there is a correlation between multi-gambling and problem gambling, researchers have emphasised the shortage of studies in the area [23].

There are few studies on gender differences in online gambling involvement. For example, studies conducted in Ireland have shown that women prefer online games that are more acceptable socially, such as lotteries or scratch cards [37]. Scratch cards are also the game type that women tend to become addicted to, which has also been highlighted by other researchers [32]. Women, on average, play for a shorter period of time (an average of two years for women and seven years for men), and spend less time gambling online than men (1 h per session, compared to 3 h for men).

Legal regulations are another aspect that has also been regarded as significant in recent studies on online gambling, particularly concerning the involvement of citizens in gambling. In his 2016 study, which included 1277 pathological gamblers undergoing addiction treatment, Chóliz discovered significant changes that occurred between 2012 (when online gambling was legalised in Spain) and the turn of 2014/2015. First and foremost, the number of people entering treatment due to pathological gambling in the studied facilities quadrupled. Also, most importantly, patients indicated online gambling as the source of their problems ten times more often than in 2012 (from 2.53% in 2012 to 24.21% in 2014/2015). For a comparison, the number of people indicating slot machines as the main source of their problems decreased (from 80.26% to 65.71%). These results seem particularly important in the context of the tendency to legalise online gambling and liberalise access to it, which has been observed by the specialists [38,39]. Moreover, researchers have also noted that using legal websites for online gambling causes less gambling-related damage [40], and a higher percentage of problem gambling occurs in populations in which legal regulations for online gambling are less restrictive [41].

The gambling market in Poland is regulated by the Act of 19 November 2009 on online gambling, which has so far been amended several times. In light of the Act of 15 December 2016 amending the Act on gambling, online gambling—with the exclusion of pari-mutuel betting and promotional lotteries—is subject to state monopoly [42]. Online games subject to state monopoly are organised by Totalizator Sportowy—a company owned by the State Treasury (Warsaw, Poland). The first online gambling games were introduced by Totalizator Sportowy in December 2018 and included a number of lotteries and games offered by the only legal online casino in Poland—slot machines, roulette, and card games for money. Legal online betting games are currently offered by nine private operators. Until 2017, when Poland tightened its restrictions on the betting market, introducing the possibility of blocking illegal domains, 90% of the industry belonged to the grey market [43].

Due to the relatively recent regulation of online gambling in Poland, there have thus far been no studies on the matter. It is worth noting that the longer tradition of research in Poland on addiction to slot machines resulted in the regulation of the market for these games [44]. The first study to estimate the involvement of Polish people in online gambling, and problem involvement in the activity and its determinants, was therefore, undertaken.

2. Materials and Methods

2.1. Participants

This study was conducted on a nationwide sample of 2000 adult Poles. The sample was representative and randomly selected on the basis of PESEL (Personal Identification Number). The distribution of gender, age, education, the size of place of residence, the region, and the number of people in the household were controlled. The Polish population was layered according to 9 geographical macro-regions, and then, in each of them, the localities were layered into 7 layers, according to size; the municipalities with probabilities proportional to the number of residents older than 18 years of age were randomly selected. Within the framework of each selected municipality, 6 interviews with randomly selected respondents were carried out. The respondents in municipalities were at first selected by means of drawing households proportionally to the number of household members according to the PESEL census records. Upon visiting a selected household, the interviewers selected the respondents using the Kish grid and then conducted computer-assisted personal interviews with them. If conducting the interview with the person from the list was not possible, the interviewer would look for a person of the same age and gender in the same town. The study was carried out by the GfK Polonia, the Polish branch of a well-known international public opinion research institute (GfK SE, Nuremberg, Germany), which also ensured the anonymity and uniformity of testing.

2.2. Measures

2.2.1. E-Gambling

The questionnaire consisted of questions regarding use of online gambling games, frequency, frequency of online gambling, time duration of a single session, and money spent on online gambling. In the first question, the respondent was asked to select the online games on which they have bet money at least once within the last 12 months. The list included 11 categories: online lottery (i.e., receipt lottery), online scratch cards, slot machines and other gambling machines on the Internet, online card games for money, other online casino games (i.e., roulette, dice), Totalizator Sportowy number games via the Internet, other online number games (i.e., bingo), online arcade games for money, online sports betting (including "fantasy sports"), online betting on e-sport or online virtual sports, and online betting on financial markets (i.e., stock exchange, FOREX, binary options). A memory-activating filtering statement was posed: "I am certain that I have not made online cash bets within the last 12 months." The question regarding e-gambling frequency was: How often, within the last 12 months, have you gambled online? (1 = everyday; 2 = several times a week, but not every day; 3 = once a week; 4 = several times a month but more rarely than once a week; 5 = once a month; 6 = several times a year, but more rarely than once a month). The question regarding the time duration of a single session was: How much time did one session of the game usually take? (1 = less than 15 min; 2 = from 15 to 30 min; 3 = from 31 min to an hour; 4 = from over an hour to 2 h; 5 = from over 2 h to 3 h; 6 = more than 3 h). The question regarding spending was open: "How much money exactly have you spent on online gambling within the last 4 months?"

2.2.2. Problem E-Gambling

The Brief Biosocial Gambling Screen (BBGS), adapted to Polish by Niewiadomska et al., was adjusted to the assessment of problem e-gambling [45,46]. It contains three questions about gambling and has been shown to have good screening properties for the criteria of problem gambling compliant with DSM-5 [47]. Each of the questions could be answered with either "Yes" or "No." The risk of problem gambling cut-off at endorsing one symptom is the best indicator of gambling disorder, taking into account that sensitivity and negative predictive value are most important for identifying individuals who potentially need treatment [45,47]. The psychometric properties of the Polish adaptation of BBGS were tested on a representative sample of high school students of the Lublin province. The criterion was fulfilment of at least 5 DSM-IV criteria of pathological hazard, measured by self-report. Sensitivity was 0.82, and specificity was 0.96 [46]. Because not distinguishing by a measure of problem gambling between forms of gambling (online versus offline) is a potentially confounding issue [48], we adjusted the BBGS for the purposes of the current study by rewording the questions in the BBGS to refer to online gambling. The final versions of the questions used in this study were as follows: Within the last 12 months, have you felt powerless, irritated, or anxious when you were trying to quit or limit online gambling? Within the last 12 months, have you tried to keep the fact that you are gambling online from your family and friends? Have you had financial trouble resulting from online gaming, because of which you had to ask your family, friends, or social services for financial support within the last 12 months?

2.2.3. Sociodemographic Variables

Information on the sociodemographic profiles of the respondents was obtained from data collected by the interviewer and from the questionnaire. The sociodemographic questions were answered by 2000 people. Their average age was 45.61 years (SD = 18.456, minimum = 18, maximum = 94). Table 1 presents the sociodemographic descriptive statistics.

Table 1. Descriptive statistics for the sociodemographic variables for the test group (n = 2000).

Variable	Category	n	%
Gender	Male	964	48.2
	Female	1036	51.8
Place of residence	countryside	815	40.8
	town of up to 20,000 residents	263	13.1
	town of 20,000–50,000 residents	213	10.7
	town of 50,000–100,000 residents	163	8.2
	town of 100,000–00,000 residents	151	7.6
	city of 200,000–500,000 residents	182	9.1
	city of more than 500,000 residents	212	10.6
Education	primary	481	24.0
	basic vocational	469	23.4
	secondary	683	34.1
	higher	368	18.4
Frequency of Internet use	nearly every day	1209	60.5
	at least once in a month	276	13.8
	less frequently or does not use at all.	514	25.7
Monthly household income	up to PLN 2000	156	7.8
	PLN 2000–2999	231	11.5
	PLN 3000–4499	487	24.4
	PLN 4500 and above	1125	56.3
Children living in the same household	Yes	268	63.4
	No	732	36.6

Note: PLN—Polish zloty; 1€ ≈ 4.3 PLN.

2.3. *Data Analysis*

All statistical analyses were performed using IBM SPSS version 24 software [49]. We used methods and statistics appropriate to the types of measurement scale and the specific parameters applied. Categorical data are presented as frequencies and percentages, and group comparisons were made using the χ^2-test. We used logistic regression (enter method and simple contras, with the reference category first) to identify the variables that best predict involvement in e-gambling.

3. Results

3.1. *Popularity of Gambling and Types of Games*

The results obtained indicate that 83 (4.1%; 95% CI (3.3%, 5.1%)) of the 2000 respondents surveyed in the last 12 months have made monetary bets using online gambling services. Most respondents played Totalizator Sportowy lotteries, online sports betting, sports betting concerning e-sport or virtual sport, and online card games for money. Online slot and gambling machines and online betting on the financial markets were used least frequently. None of the respondents indicated using online arcade games for money (Table 2).

Table 2. Prevalence of forms of online gambling ($N = 2000$).

Game Type	n	%	95% CILL	95% CIUL
Totalizator Sportowy lotteries	41	2.0	1.5	2.7
Online sports betting (including "fantasy sports")	19	1.0	0.6	1.4
Online betting on e-sports or online virtual sports	11	0.6	0.3	0.9
Online card games for money	10	0.5	0.3	0.9
Online scratch cards	6	0.3	0.1	0.6
Other online casino games (e.g., roulette, dice)	5	0.2	0.1	0.5
Internet lottery (e.g., receipt lottery)	4	0.2	0.1	0.5
Other online lotteries (e.g., bingo)	4	0.2	0.1	0.5
Slot machines or other online gambling machines	1	0.1	0.0	0.2
Online betting on financial markets	1	0.1	0.0	0.2
Online arcade games for money	0	0.0	0.0	0.0

Note: 95% CILL: 95% confidence interval lower limit; 95% CIUL: 95% confidence interval upper limit.

3.2. Sociodemographic Variables Associated with Online Gambling

Logistic regression was used to assess the factors associated with involvement in online gambling (cf. Table 3). The dependent variable had two categories (1—gambling versus 0—not using online gambling services in the last twelve months). Age, gender, population of the place of residence, education, income level, having children, and frequency of Internet use were considered as independent variables. The factors that significantly explained involvement in online gambling were: gender, age, population of the place of residence, education, and monthly family income. Men were more likely to be involved in gambling activities than women. In terms of age, the youngest group (up to age 29) was significantly more likely to be involved in online gambling than older people (over 50). Analysing the size of the place of residence, people living in the countryside were significantly different from those living in towns with 20,000–100,000 residents and those coming from cities of 200,000–500,000 residents. Online gambling activity was much less frequent among people living in towns or cities compared to people living in the countryside. Higher online gambling activity could also be seen among people with primary education compared to those with vocational education. Monthly income was also an important factor explaining involvement in online gambling. People with low monthly incomes were much more likely to devote their time to online gambling than those earning more than PLN 3000. Frequency of Internet use was also an element co-existing with online gambling activity. Individuals using the Internet more frequently were also inclined to become more involved in online gambling.

Table 3. An explanatory model of e-gambling (n = 2000).

Variables	B	SE	Wald	df	p	Exp(B)	95% CI for EXP(B)	
							Lower	Upper
Gender: (ref.: male)	−1.129	0.282	16.055	1	<0.001	0.323	0.186	0.562
Age: (ref.: 15–29 years of age)			21.002	5	0.001			
30–39	−0.527	0.367	2.067	1	0.150	0.590	0.288	1.211
40–49	0.414	0.319	1.685	1	0.194	1.513	0.810	2.829
50–59	−2.904	1.004	8.366	1	0.004	0.055	0.008	0.392
60–69	−1.392	0.681	4.178	1	0.041	0.249	0.065	0.944
70 and above	−2.135	1.166	3.353	1	0.067	0.118	0.012	1.162
Population of the place of residence: (ref.: countryside)			15.507	6	0.017			
town of up to 20,000 residents	0.072	0.377	0.036	1	0.849	1.074	0.513	2.248
town of 20,000–50,000 residents	−1.266	0.559	5.125	1	0.024	0.282	0.094	0.844
town of 50,000–100,000 residents	−1.103	0.580	3.615	1	0.057	0.332	0.106	1.035
town of 100,000–200,000 residents	−0.649	0.509	1.625	1	0.202	0.523	0.193	1.417
city of 200,000–500,000 residents	−4.465	2.268	3.876	1	0.049	0.012	0.000	0.980
city of more than 500,000 residents	0.307	0.379	0.659	1	0.417	1.360	0.647	2.855
Education: (ref.: primary)			14.184	3	0.003			
basic vocational	−1.851	0.547	11.450	1	0.001	0.157	0.054	0.459
secondary	−0.189	0.327	0.335	1	0.563	0.828	0.436	1.570
higher	0.244	0.378	0.415	1	0.519	1.276	0.608	2.675
Children living in the same household: [ref.: no children]	0.260	0.289	0.809	1	0.368	1.297	0.736	2.283
Household net income: (ref.: up to PLN 2000)			24.876	3	0.000			
PLN 2000–2999	−1.105	0.580	3.630	1	0.057	0.331	0.106	1.032
PLN 3000–4499	−1.732	0.532	10.593	1	0.001	0.177	0.062	0.502
PLN 4500 and above	−2.405	0.517	21.649	1	0.000	0.090	0.033	0.249
Internet use: (ref.: nearly every day)			5.077	2	0.079			
at least once in a month	−0.128	0.481	0.071	1	0.789	0.880	0.343	2.256
less frequently or does not use at all	−1.758	0.782	5.054	1	0.025	0.172	0.037	0.798
constantly	−4.682	0.450	108.157	1	0.000	0.009		

Note: Overall model evaluation: Likelihood ratio test: $\chi^2(21) = 167.916$; $p < 0.001$; Cox and Snell $R^2 = 0.081$; Nagelkerke's $R^2 = 0.276$.

3.3. *Prevalence of Problem Gambling among Players and Related Factors*

Out of the 83 people who participated in online gambling, 22 (26.8%; 95% CI (17.9%, 36.7%)) were at risk of becoming problem gamblers. This group consisted of respondents who provided at least one affirmative answer on the BBGS scale. In order to determine the characteristics of problem online gamblers, we compared them to non-problematic gamblers in respect of sociodemographic variables using the chi-square test (cf. Table 4).

A comparison between gamblers at risk of becoming problem gamblers and those who were not at such risk indicated several differentiating variables (cf. Table 4). These included having children and choice of online e-gambling services. Individuals who manifested symptoms of problem use of e-gambling more frequently had children than those with no symptoms of problem gambling. PeGs were also less active in playing Totalizator Sportowy lotteries and more often used online sports betting (including fantasy sports) and online scratch cards.

Table 4. Comparison of sociodemographic variables between non-problem online gamblers (NPeG, $n = 61$) and problem online gamblers (PeG, $n = 22$).

Variables	NPeG		PeG		χ^2	p
	n	%	n	%		
Gender:					0.381	0.537
Male	45	75	15	68.2		
Female	15	25	7	31.8		
Age:					8.998	0.109
15–29	34	55.7	6	28.6		
30–39	10	16.4	3	14.3		
40–49	14	23.0	10	47.6		
50–59	1	1.6	0	0		
60–69	1	1.6	2	9.5		
70 and above	1	1.6	0	0		
Population of the place of residence:					5.320	0.503
countryside	25	41.7	14	60.9		
town of up to 20,000 residents	9	15.0	2	8.7		
town of 20,000–50,000 residents	3	5.0	1	4.3		
town of 50,000–100,000 residents	4	6.7	0	0		
town of 100,000–200,000 residents	5	8.3	0	0		
city of 200,000–500,000 residents	0	0	0	0		
city of more than 500,000 residents	14	23.3	6	26.1		
Education:					4.871	0.181
primary	16	26.2	11	50.0		
basic vocational	4	6.6	1	4.5		
secondary	22	36.1	7	31.8		
higher	19	31.1	3	13.6		
Children living in the same household:	24	40.0	15	68.2	5.126	0.022
Household net income:					3.540	0.316
up to PLN 2000	7	11.7	4	19.0		
PLN 2000–2999	8	13.3	1	4.8		
PLN 3000–4499	20	33.3	4	19.0		
PLN 4500 and above	25	41.7	12	57.1		
Internet use:					1.151	0.562
nearly every day	52	85.2	20	90.9		
at least once in a month	6	9.8	2	9.1		
less frequently or does not use at all.	3	4.9	0	0		
Types of online gambling services:						
Internet lottery	2	3.3	2	9.1	1.150	0.284
Online scratch cards	2	3.3	3	13.6	2.984	0.084
Slot machines or other online gambling machines	1	1.7	0	0	0.371	0.542
Online card games for money	8	13.1	2	9.1	0.247	0.619
Other online casino games (e.g., roulette, dice)	5	8.2	0	0	1.919	0.166
Totalizator Sportowy lotteries	36	59.0	5	21.7	9.289	0.002
Other online lotteries (e.g., bingo)	3	5.0	1	4.5	0.007	0.933
Online arcade games for money	0	0	0	0.0	-	-
Online sports betting (including "fantasy sports")	11	18.3	8	36.4	2.940	0.086
Online betting on e-sports or online virtual sports	10	16.7	1	4.5	2.036	0.154
Online betting on financial markets	1	1.6	0	0	0.365	0.546

Note: PeG—problem online gamblers; NPeG—non-problem online gamblers.

4. Discussion

The results of our study allowed us to determine the extent of the involvement of adult Poles in online gambling. First, it is worth mentioning that the first legitimate online gambling games in Poland were sports betting services, first organised in 2012. The provision of other online gambling services was regulated only by the 2016 Act, under which the provision of other online games is subject to a state monopoly [42]. It should be noted that, in practice, these games were made available on the market only in December 2018, which highlights the specificity of the e-gambling market in Poland and sheds light on the results of this study, which was conducted in December 2018. The lotteries organised by Totalizator Sportowy proved to be the most popular online and offline gambling games, having been the most common type of such games for Poles [50]. It is worth noting here that these games, covered by the state monopoly, may be advertised in public media, which significantly increases their potential accessibility compared to other types of gambling games, the advertising of which is prohibited by law. The second most popular games included sports betting and betting related to e-sports and virtual sports, the popularity of which may be explained by the relatively long history of this type of online gambling in Poland. The popularity of gambling reflects, to a large extent, the cultural specificity or legal regulations of a given country concerning the availability of games. For example, the most popular online gambling game in France is, among others, horse race betting, which illustrates the long-standing tradition of what is considered to be a national sport in France [51]. In Poland, football plays a similar role. The relatively high interest of Poles in e-sports or online virtual sports betting is a new trend. This phenomenon has not yet been assessed, so it is difficult to estimate the extent to which there is an upward trend or whether the behaviour has been long-standing. Taking into account the novelty of the phenomenon of e-sports betting, not only in Poland but also worldwide (e.g., this phenomenon was included in the national survey "e-Games France 2017" for the first time only in 2017), an upward trend may be expected [51]. The interest of Poles in such betting even exceeds the popularity of online card games, which is fourth in terms of popularity. The interest in online gambling as a whole seems to be low in Poland. During the 12-month period prior to the survey, 4.1% of adult Poles made an online bet, which is a very small percentage compared to the 37.1% of Poles engaged in offline gambling during the same period [48]. It is worth mentioning that, in the latest epidemiological study on behavioural addictions in Poland, only 1.2% of Poles declared that they had gambled online in the past year. However, it is also significant that, in this study, the category of online gambling was only one among nine categories of offline gambling.

Due to the fact that this was the first study on e-gambling in Poland, we were interested in determining which individuals choose this form of entertainment most often. The results showed that e-gambling was more popular with men than women, and that interest decreases with age. These data are confirmed by studies conducted in other countries [26,37]. Additionally, online gambling was more popular among those with incomes lower than the national average salary than those with incomes equal to or higher than the national average wage. Involvement in online gambling also decreased with the decline in daily Internet use. This phenomenon was all the more alarming because, in light of the Effertz study, the risk of problem online gambling is largest among highly engaged Internet users [22], which was confirmed by the research of Rémond and Romo [52]. The results obtained are in line with other studies, which show that the popularity of online gambling in the West is attributable, among other things, to low access costs—the gambler does not need to travel to the place where the game is played or devote time to such travel [12]. This makes online gambling more accessible from an economic point of view, and therefore more likely to be selected by people with lower socioeconomic status. Other researchers have also emphasised that accessibility and availability are the factors contributing to increased involvement in online gambling [25–29].

Finally, we were interested in the extent to which problem gambling was exacerbated among online gamblers. The study results showed that 26.8% of gamblers had symptoms indicating a probable gambling addiction on the basis of the BBGS scale. These results reveal that the risk of gambling addiction is higher among Polish online gamblers than among gamblers in general (both online and offline). In light of the latest results of a national survey conducted by the Centre for Public Opinion Research (CBOS: Centrum Badania Opinii Społecznej) on behavioural addictions, 7.7% of all gamblers show a low addiction risk, 0.9% of them a moderate risk level and 0.9% of gamblers are problem gamblers, making a total of 9.5% of gamblers at risk of addiction [53]. It should be mentioned, however, that the CBOS survey was conducted using the Canadian Problem Gambling Index (CPGI) scale.

Relating the results obtained in our study to other studies, a convergence may be observed. International online gambling studies have showed, for instance, that 14% of gamblers met the criteria for problem gambling in accordance with the Problem Gambling Severity Index, 29% met the criteria for risky gambling and 32.7% of gamblers met the criteria for problem gambling at a low level. Other studies also confirm greater exacerbation of problem gambling among online than offline gamblers [16,19,20], which is, according to some, even three to five times greater [21]. In Austrian studies, 31% of online gamblers showed symptoms of problem gambling, while 18% of offline gamblers displayed such symptoms in accordance with Lie-Bet questionnaire results [54].

The last aspect we analysed included sociodemographic variables coexisting with gambling addiction.

The first important factors differentiating gamblers at risk of a gambling disorder from non-problem gamblers were male gender and age from 40 to 49 years old. These results are surprising because Poles addicted to offline gambling are mainly younger people (18–24 years old) [53]. Nevertheless, the results obtained by us are consistent with the Austrian research of Yazdi and Katzian, in the light of which addicted online gamblers most often belong to the age range 30–49 years [54]. It seems that due to the relatively short period of online gambling being available in Poland, these games are mainly used by mature men who have longer experience with offline gambling, and are, therefore, consciously looking for a new offer of already known entertainment.

The level of education also differentiated problem and non-problem gamblers. Problem e-gamblers more often have primary education, which is also the case with offline gamblers in Poland.

The next important factor differentiating gamblers at risk of a gambling disorder from non-problem gamblers was the presence of children in the household. Gamblers at risk constituted a group that more often had children than representatives of the non-addicted group. This may be due to the higher age of problem e-gamblers. At this stage of study, we are still considering how to understand this relationship. It may be argued that people who are interested in gambling—and have children—have been more inclined to opt for online games which are more accessible due to time constraints. As studies confirm the stronger addictive potential of online gambling, this activity is, thus, more likely to turn into addiction. It can also be assumed that gamblers who do not have children gamble offline as well, but also opt for alternative, non-addictive offline entertainment, which is less accessible to those with children. Online gambling is more absorbing, allowing the gambler to play several games at once, as emphasised by both Cotte et al. and Gainsbury et al. [30,31]. The structural characteristics of online games—such as directness, accessibility, ease of betting, and the fact that they pose a particular risk of addiction—are also highlighted by Chóliz [24].

Accessibility factors are also revealed when linking residence to problem gambling. In light of the results, problem e-gamblers come, more often, from the countryside. Despite the fact that rural residents do not gamble online more often than urban residents, they develop problem e-gambling more often. It can be assumed that these people, with limited possibilities of enjoying other entertainment (including offline gambling), engage in e-gambling more intensively, which translates into an increased risk of developing addiction to these games.

Another factor significantly co-existing with the risk of gambling addiction was the type of game being played. In light of the results, gamblers at risk of addiction are more often involved in sports betting (including "fantasy sports"). These results correlate with McCormack's study, which showed that the risk of problem online gambling is significantly higher for, among others, online sports betting gamblers [23]. Given that sports betting is one of the most popular online gambling services in Poland, this is an important discovery. Problem e-gamblers are also more often involved in online scratch cards, which are one the most popular offline gambling game types in Poland [53].

Summarising the results of the study, it is worth once again referring to the Polish legal regulations on gambling. When the study was conducted, legitimate gambling, apart from sports betting, was in its early stages. It would be significant to monitor the development of Poles' involvement in online gambling as it becomes more widespread. Taking into account the relatively low involvement of Poles in online gambling, it may be assumed that the results obtained stem from the fact that these games were not yet very popular at the time the study was conducted. For instance, Chóliz [24] analysed the changes that occurred between 2012 (when online gambling was legalised in Spain) and the turn of 2014/2015, during which period the number of people who started treatment for pathological gambling quadrupled. Additionally, patients indicated online gambling as the main source of their problems nearly ten times more often in 2014/2015 than in 2012. With this in mind, it is extremely important to continue epidemiological studies on participation in online gambling and problem gambling to develop recommendations for legislators based on changes in the behaviour of gamblers resulting from the implementation of legal amendments. Legalisation of online gambling is a very important issue. Researchers note that the use of legitimate gambling sites causes less gambling-related harm than the use of illegal sites [40]. However, the legalisation of online gambling alone cannot be the only preventive factor. Its effects should be monitored and the next steps need to be adapted accordingly. It would be worthwhile to conduct future studies on the relationship between multi-gambling and gambling problems, especially as researchers have pointed out a research deficit in this area [23]. It would also be important to highlight the differences between "pure" online gamblers, "pure" offline gamblers, and "mixed-mode" gamblers. It is also significant to recognise the differences between the genders in terms of online gambling activities. As the online gambling market in Poland is constantly evolving, it is important to use the experience of Western countries when implementing responsible gambling policy. Internet gamblers should be informed by operators about the risks of gambling. In addition, it would be important for operators to monitor online gambler behaviour and identify at risk gamblers and direct messages to them about threats and the possibilities of seeking help. Gamblers should be able to exclude themselves from the site for a certain period of time, and this should also be offered to at risk gamblers. Some activities in this area are already being implemented; however, it is important to conduct research on the effectiveness of preventive measures taken, because cultural factors can modify it.

Limitations

Despite the pioneering character of this study in Poland, this study also has its limitations. The first one is the BBGS, the research tool used, which, despite its psychometric properties, only has a screening character and is used relatively rarely in epidemiological studies. Earlier studies on offline gambling in Poland employed a different scale (CPGI), so it is difficult to compare those results with the results of the present study. The BBGS was used due to the preliminary nature of this study, the continuation of which is being planned. Besides, the adjustment of BBGS to online gambling by rewording the questions, being our attempt to remedy to the lack of distinction between online and offline forms of gambling, is a very unusual technique, and it is uncertain how that step affected the results obtained. The next limitation was the restricted number of questions in the survey and the resulting failure to include more variables, including psychological ones. A more elaborate study is, however, currently underway. There were also no questions about offline gambling in the survey, which makes it impossible to determine whether the outspoken gamblers are "pure" online gamblers. In light of the study by Gainsbury, there are differences between pure online and offline gamblers and "mixed-mode" gamblers [18]. Another limitation of research is the fact that it was conducted in the same month in which the online gambling market was expanded. The result of this may have been that the study only captured gamblers very advanced in using new technologies who were the first to reach games in a new form. This confirms the connection between the use of online gambling games and the intensity of Internet involvement. Another hypothesis may be that the research revealed players looking for new types of gambling. This hypothesis, however, is partly undermined by the results of studies in the light of which most Poles practiced Lottery of Sports Totalizator and online sports betting. Lotteries are the most popular offline games in Poland; they are widely recognized and their publicity is allowed, which translates into their high availability, and therefore—greater involvement of Poles in them. On the other hand, online sports betting has been available in Poland since 2012, which is why it is not a "new" type of game. Research should certainly be continued to learn more about the specifics of Poles' involvement in online gambling.

5. Conclusions

This study provided a characterisation of Poles' involvement in online gambling. This is all the more relevant because in December 2018 new types of online gambling services were introduced. We thus managed to capture the 'initial' state—that is, the very beginning of the new reality of online gambling in Poland. As a result, it is possible to observe changes arising from the introduction of new legal regulations. Studies have shown that 4.1% of Poles made an online bet in the 12-monthnperiod prior to the survey. Lotteries, sports and e-sports betting proved to be the most popular online gambling games. Online gambling was more popular among younger men, with incomes lower than the national average salary, who were highly engaged Internet users. Among all gamblers, 26.8% were reported to be at risk of gambling addiction based on the results of the BBGS screening questionnaire. Addicted online gamblers more often had children and preferred sports betting.

Author Contributions: Conceptualization, B.L.-K. and R.P.B.; methodology, B.L.-K., R.P.B., M.W., and J.C.; validation, B.L.-K. and R.P.B.; formal analysis, R.P.B. and M.W.; investigation, B.L.-K., J.C., M.W., and R.P.B.; resources, B.L.-K., J.C., and R.P.B.; data curation, R.P.B. and M.W.; writing—original draft preparation, B.L.-K., M.W., R.P.B.; writing—review and editing, B.L.-K., R.P.B., J.C., M.W., and I.N.; visualization, M.W.; supervision, I.N.; project administration, B.L.-K.; funding acquisition, J.C. and B.L.-K. All authors have read and agreed to the published version of the manuscript.

Funding: This research was funded by the Ministry of Health—the Republic of Poland Fund for Solving Gambling Problems, grant number 362/HE/2018. The APC was funded by the Ministry of Health—the Republic of Poland Fund for Solving Gambling Problems, grant number 362/HE/2018.

Conflicts of Interest: The authors declare no conflict of interest. The funders had no role in the design of the study; in the collection, analyses, or interpretation of data; in the writing of the manuscript; or in the decision to publish the results.

References

1. Gaboury, A.; Ladouceur, R.; Bussières, O. Structures des loteries et comportements des joueurs. [Structure of lotteries and behaviour of players]. *Rev. Psychol. Appliquée* **1989**, *39*, 197–207.
2. Ladouceur, R.; Mayrand, M. Depressive behaviors and gambling. *Psychol. Rep.* **1987**, *60*, 1019–1022. [CrossRef] [PubMed]
3. American Psychiatric Association. *Diagnostic and Statistical Manual of Mental Disorders [DSM-5]*, 5th ed.; American Psychiatric Association: Washington, DC, USA, 2013.
4. World Health Organization. International Classification of Diseases for Mortality and Morbidity Statistics (11th Revision). Available online: https://icd.who.int/en (accessed on 16 December 2019).
5. Freimuth, M.; Moniz, S.; Kim, S.R. Clarifying exercise addiction: Differential diagnosis, co-occurring disorders, and phases of addiction. *Int. J. Environ. Res. Public Health* **2011**, *8*, 4069–4081. [CrossRef] [PubMed]
6. Lesieur, H.R.; Rosenthal, R.J. Pathological gambling: A review of the literature (prepared for the American Psychiatric Association Task Force on DSM-IV Committee on Disorders of Impulse Control Not Elsewhere Classified). *J. Gambl. Stud.* **1991**, *7*, 5–39. [CrossRef]
7. Thompson, W.N.; Gazel, R.; Rickman, D. Social and legal costs of compulsive gambling. *Gaming Law Rev.* **1997**, *1*, 81–89. [CrossRef]
8. Bischof, A.; Meyer, C.; Bischof, G.; John, U.; Wurst, F.M.; Thon, N.; Lucht, M.; Grabe, H.J.; Rumpf, H.-J. Suicidal events among pathological gamblers: The role of comorbidity of axis I and axis II disorders. *Psychiatry Res.* **2015**, *225*, 413–419. [CrossRef]
9. Bischof, A.; Meyer, C.; Bischof, G.; John, U.; Wurst, F.M.; Thon, N.; Lucht, M.; Grabe, H.-J.; Rumpf, H.-J. Type of gambling as an independent risk factor for suicidal events in pathological gamblers. *Psychol. Addict. Behav.* **2016**, *30*, 263–269. [CrossRef]

10. Walker, D.M.; Barnett, A.H. The social costs of gambling: An economic perspective. *J. Gambl. Stud.* **1999**, *15*, 181–212. [CrossRef]
11. McBride, J.; Derevensky, J. Internet gambling behavior in a sample of online gamblers. *Int. J. Ment. Health Addict.* **2008**, *7*, 149–167. [CrossRef]
12. Griffiths, M.; Parke, A.; Wood, R.; Parke, J. Internet gambling: An overview of psychosocial impacts. *UNLV Gaming Res. Rev. J.* **2005**, *10*, 27–39.
13. LaBrie, R.A.; LaPlante, D.A.; Nelson, S.E.; Schumann, A.; Shaffer, H.J. Assessing the playing field: A prospective longitudinal study of internet sports gambling behavior. *J. Gambl. Stud.* **2007**, *23*, 347–362. [CrossRef] [PubMed]
14. Meyer, G.; Fiebig, M.; Häfeli, J.; Mörsen, C. Development of an assessment tool to evaluate the risk potential of different gambling types. *Int. Gambl. Stud.* **2011**, *11*, 221–236. [CrossRef]
15. Monaghan, S. Internet gambling—Not just a fad. *Int. Gambl. Stud.* **2009**, *9*, 1–4. [CrossRef]
16. Williams, R.J.; Wood, R.T.; Parke, J. *Routledge International Handbook of Internet Gambling*; Routledge: London, UK, 2012.
17. Philander, K.S.; MacKay, T.L. Online gambling participation and problem gambling severity: Is there a causal relationship? *Int. Gambl. Stud.* **2014**, *14*, 214–227. [CrossRef]
18. Gainsbury, S.M. Online gambling addiction: The relationship between internet gambling and disordered gambling. *Curr. Addict. Rep.* **2015**, *2*, 185–193. [CrossRef] [PubMed]
19. Williams, R.J.; Wood, R.T. The proportion of Ontario gambling revenue derived from problem gamblers. *Can. Public Policy* **2007**, *33*, 367–388. [CrossRef]
20. Wood, R.T.; Williams, R.J. *Internet Gambling: Prevalence, Patterns, Problems, and Policy Options*; Ontario Problem Gambling Research Centre: Guelph, ON, Canada, 2009.
21. Canale, N.; Griffiths, M.D.; Vieno, A.; Siciliano, V.; Molinaro, S. Impact of internet gambling on problem gambling among adolescents in Italy: Findings from a large-scale nationally representative survey. *Comput. Hum. Behav.* **2016**, *57*, 99–106. [CrossRef]
22. Effertz, T.; Bischof, A.; Rumpf, H.J.; Meyer, C.; John, U. The effect of online gambling on gambling problems and resulting economic health costs in Germany. *Eur. J. Health Econ.* **2018**, *19*, 967–978. [CrossRef]
23. McCormack, A.; Shorter, G.W.; Griffiths, M.D. An examination of participation in online gambling activities and the relationship with problem gambling. *J. Behav. Addict.* **2013**, *2*, 31–41. [CrossRef]
24. Chóliz, M. The challenge of online gambling: The effect of legalization on the increase in online gambling addiction. *J. Gambl. Stud.* **2016**, *32*, 749–756. [CrossRef]
25. Storer, J.; Abbott, M.; Stubbs, J. Access or adaptation? A meta-analysis of surveys of problem gambling prevalence in Australia and New Zealand with respect to concentration of electronic gaming machines. *Int. Gambl. Stud.* **2009**, *9*, 225–244. [CrossRef]
26. Welte, J.W.; Barnes, G.M.; Wieczorek, W.F.; Tidwell, M.C.O.; Hoffman, J.H. Type of gambling and availability as risk factors for problem gambling: A tobit regression analysis by age and gender. *Int. Gambl. Stud.* **2007**, *7*, 183–198. [CrossRef]
27. Hing, N.; Haw, J. The development of a Multi-dimensional Gambling Accessibility Scale. *J. Gambl. Stud.* **2009**, *25*, 569–581. [CrossRef] [PubMed]
28. Thomas, A.C.; Sullivan, G.B.; Allen, F.C.L. A theoretical model of EGM problem gambling: More than a cognitive escape. *Int. J. Ment. Health Addict.* **2009**, *7*, 97–107. [CrossRef]
29. Chóliz, M. Experimental analysis of the game in pathological gamblers: Effect of the immediacy of the reward in slot machines. *J. Gambl. Stud.* **2010**, *26*, 249–256. [CrossRef]
30. Cotte, J.; Latour, K.A. Blackjack in the kitchen: Understanding online versus casino gambling. *J. Consum. Res.* **2009**, *35*, 742–758. [CrossRef]
31. Gainsbury, S.; Wood, R.; Russell, A.; Hing, N.; Blaszczynski, A. A digital revolution: Comparison of demographic profiles, attitudes and gambling behavior of Internet and non-Internet gamblers. *Comput. Hum. Behav.* **2012**, *28*, 1388–1398. [CrossRef]

32. Hing, N.; Russell, A.M.T.; Gainsbury, S.M.; Nuske, E. The public stigma of problem gambling: Its nature and relative intensity compared to other health conditions. *J. Gambl. Stud.* **2016**, *32*, 847–864. [CrossRef]
33. Barrault, S.; Varescon, I. Online and live regular poker players: Do they differ in impulsive sensation seeking and gambling practice? *J. Behav. Addict.* **2016**, *5*, 41–50. [CrossRef]
34. LaPlante, D.A.; Nelson, S.E.; Gray, H.M. Breadth and depth involvement: Understanding internet gambling involvement and its relationship to gambling problems. *Psychol. Addict. Behav.* **2014**, *28*, 396–403. [CrossRef]
35. Gainsbury, S.M.; Russell, A.; Blaszczynski, A.; Hing, N. Greater involvement and diversity of Internet gambling as a risk factor for problem gambling. *Eur. J. Public Health* **2015**, *25*, 723–728. [CrossRef] [PubMed]
36. Wardle, H.; Moody, A.; Spence, S.; Orford, J.; Volberg, R.; Jotangia, D.; Griffiths, M.; Hussey, D.; Dobbie, F. *British Gambling Prevalence Survey 2010*; National Centre for Social Research: London, UK, 2011.
37. Columb, D.; O'Gara, C. A national survey of online gambling behaviours. *Ir. J. Psychol. Med.* **2017**, *35*, 311–319. [CrossRef] [PubMed]
38. Gainsbury, S.; Wood, R. Internet gambling policy in critical comparative perspective: The effectiveness of existing regulatory frameworks. *Int. Gambl. Stud.* **2011**, *11*, 309–323. [CrossRef]
39. Fiedler, I. Regulation of online gambling. *Econ. Bus. Lett.* **2018**, *7*, 162–168. [CrossRef]
40. Costes, J.M.; Kairouz, S.; Eroukmanoff, V.; Monson, E. Gambling patterns and problems of gamblers on licensed and unlicensed sites in France. *J. Gambl. Stud.* **2016**, *32*, 79–91. [CrossRef]
41. Planzer, S.; Gray, H.M.; Shaffer, H.J. Associations between national gambling policies and disordered gambling prevalence rates within Europe. *Int. J. Law Psychiatry* **2014**, *37*, 217–229. [CrossRef]
42. *Ustawa z Dnia 15 Grudnia 2016 o Zmianie Ustawy o Grach Hazardowych oraz Niektórych Innych Ustaw [Act of 15 December 2016 on the Amendment of the Gambling Activities Act and Certain Other Acts]*; Chancellery of the Sejm of the Republic of Poland: Warsaw, Poland, 2016; pp. 1–22. Available online: http://prawo.sejm.gov.pl/isap.nsf/DocDetails.xsp?id=WDU20170000088 (accessed on 12 November 2019).
43. United Nations Global Compact. *Przeciwdziałanie Szarej Strefie w Polsce [Counteracting the Gray Area in Poland]*; Global Compact Poland: Warsaw, Poland, 2016.
44. Lelonek, B. *Psychospołeczne Korelaty Uzależnień od Gier Hazardowych [Psychosocial Correlates of Gambling Addiction]*; Towarzystwo Naukowe Katolickiego Uniwersytetu Lubelskiego: Lublin, Poland, 2012.
45. Gebauer, L.; LaBrie, R.; Shaffer, H.J. Optimizing DSM-IV-TR classification accuracy: A brief biosocial screen for detecting current gambling disorders among gamblers in the general household population. *Can. J. Psychiatry* **2010**, *55*, 82–90. [CrossRef]
46. Niewiadomska, I.; Augustynowicz, W.; Palacz-Chrisidis, A.; Bartczuk, R.P.; Wiechetek, M.; Chwaszcz, J. *Bateria Metod Służących Do Oceny Ryzyka Zaburzeń Związanych z Hazardem [A Battery of Methods to Assess the Risk of Gambling-Related Disorders]*; Instytut Psychoprofilaktyki i Psychoterapii: Stowarzyszenie Natanaelum: Lublin, Poland, 2014.
47. Brett, E.I.; Weinstock, J.; Burton, S.; Wenzel, K.R.; Weber, S.; Moran, S. Do the DSM-5 diagnostic revisions affect the psychometric properties of the Brief Biosocial Gambling Screen? *Int. Gambl. Stud.* **2014**, *14*, 447–456. [CrossRef]
48. Hing, N.; Russell, A.M.; Browne, M. Risk factors for gambling problems on online electronic gaming machines, race betting and sports betting. *Front. Psychol.* **2017**, *8*, 779. [CrossRef]
49. IBM Corporation. *IBM SPSS Statistics for Windows*; IBM Corporation: Armonk, NY, USA, 2016.
50. Badora, B.; Gwiazda, M.; Herrmann, M.; Kalka, J.; Moskalewicz, J. *Oszacowanie Rozpowszechnienia oraz Identyfikacja Czynników Ryzyka i Czynników Chroniących w Odniesieniu do Hazardu, w Tym Hazardu Problemowego (Patologicznego) oraz Innych Uzależnień Behawioralnych [Estimation of Prevalence and Identification of Risk and Protective Factors in Relation to Gambling, Including Problem (Pathological) Gambling and Other Behavioural Addictions]*; Centrum Badania Opinii Społecznej: Warsaw, Poland, 2012.
51. Costes, J.M.; Eroukmanoff, V. Les pratiques de jeux d'argent sur Internet en France en 2017 [Internet gambling practices in France in 2017]. *Notes L'Observatoire Jeux* **2018**, *9*, 1–8.
52. Rémond, J.J.; Romo, L. Analysis of gambling in the media related to screens: Immersion as a predictor of excessive use? *Int. J. Environ. Res. Public Health* **2018**, *15*, 58. [CrossRef] [PubMed]

53. Moskalewicz, M.; Badora, B.; Feliksiak, M.; Głowacki, A.; Gwiazda, M.; Herrmann, M.; Kawalec, I.; Roguska, B. *Oszacowanie Rozpowszechnienia oraz Identyfikacja Czynników Ryzyka i Czynników Chroniących Hazardu i Innych Uzależnień Behawioralnych—Edycja 2018/2019 [Estimation of Prevalence and Identification of Risk and Protective Factors for Gambling and Other Behavioural Addictions—2018/2019 Edition]*; Centrum Badania Opinii Społecznej: Warsaw, Poland, 2019.
54. Yazdi, K.; Katzian, C. Addictive potential of online-gambling. A prevalence study from Austria. *Psychiatr. Danub.* **2017**, *29*, 376–378. [CrossRef] [PubMed]

International Journal of
Environmental Research and Public Health

MDPI

Review

Does Smartphone Addiction Fall on a Continuum of Addictive Behaviors?

Sheila Yu [1,]* and Steve Sussman [1,2]

[1] Department of Preventive Medicine, University of Southern California, 2001 N Soto St., Los Angeles, CA 90032, USA; ssussma@usc.edu

[2] Department of Psychology, University of Southern California, 3620 South McClintock Ave, Los Angeles, CA 90089, USA

* Correspondence: sheilayu@usc.edu

Received: 7 November 2019; Accepted: 4 January 2020; Published: 8 January 2020

Abstract: Due to the high accessibility and mobility of smartphones, widespread and pervasive smartphone use has become the social norm, exposing users to various health and other risk factors. There is, however, a debate on whether addiction to smartphone use is a valid behavioral addiction that is distinct from similar conditions, such as Internet and gaming addiction. The goal of this review is to gather and integrate up-to-date research on measures of smartphone addiction (SA) and problematic smartphone use (PSU) to better understand (a) if they are distinct from other addictions that merely use the smartphone as a medium, and (b) how the disorder(s) may fall on a continuum of addictive behaviors that at some point could be considered an addiction. A systematic literature search adapted from the Preferred Reporting Items for Systematic Reviews and Meta-Analyses (PRISMA) method was conducted to find all relevant articles on SA and PSU published between 2017 and 2019. A total of 108 articles were included in the current review. Most studies neither distinguished SA from other technological addictions nor clarified whether SA was an addiction to the actual smartphone device or to the features that the device offers. Most studies also did not directly base their research on a theory to explain the etiologic origins or causal pathways of SA and its associations. Suggestions are made regarding how to address SA as an emerging behavioral addiction.

Keywords: smartphone addiction; problematic smartphone use

---

1. Introduction

After the iPhone entered global markets in 2007, significant technological advancements for smartphones have been made [1]. Smartphones offer a myriad of information sources at the touch of one's fingertips and enhance productivity (e.g., synced calendars, email, alerts and alarms, global positioning system (GPS) maps), as well as provide instant access to entertainment and social networking sites and social media channels (e.g., Facebook, Twitter, Instagram) [1,2]. Due to the high accessibility and mobility of smartphones, widespread and pervasive smartphone use has become the social norm [2], which has resulted in an increase in potential distractions that expose users to various health and other risk factors (e.g., distracted driving or walking, resulting in traffic accidents; physical health detriments, such as neck and shoulder issues, blurred vision, and wrist pain; sleep disturbances; highly sedentary habits; poor academic performance; mental health concerns; financial burdens) [1–5].

One other consequence of excessive smartphone use is the possibility of developing an addiction. The concept of addiction historically has focused on recreational drug use, exhibiting classic characteristics related to repetitive use (i.e., tolerance, withdrawal symptoms [4,5], dependence, social problems [5], and loss of control [4]). Addiction now includes behaviors in addition to substances [1–5], such as gambling disorder and Internet gambling disorder [4], as discussed in the latest fifth edition of the Diagnostic and Statistical Manual of Mental Disorders (DSM–5) [4–6] and in the proposed International Classification of Diseases (ICD)-11 for Mortality and Morbidity Statistics [7]. The recognition of these conditions, however, did not occur immediately, as evidence of neurobiological and psychological mechanisms indicating the addictive nature of these behavioral disorders, and accurate assessment of them, took time [8].

The acknowledgement of PSU [3] has led to a debate on whether the concept of SA is a true addiction and, if it is, if it is distinct from other technological addictions engaged in on the smartphone, such as Internet and gaming addiction. Due to the recency of this phenomenon, there has been a paucity of research on its assessment [4,5]. Assessment of SA should involve reliable instruments (e.g., good internal consistency), and demonstrate adequate content validity as a measure of addiction (attempts at appetitive need fulfillment, preoccupation, loss of control, undesired consequences [9]). In addition, such a measure should capture unique consequences of SA (e.g., use while in social situations that interfere with a flow of conversation or upsets a speaker, use while driving). Limitations of knowledge in this new arena of SA/PSU studies include self-report bias from participants, lack of standardized criteria to measure SA/PSU, and different contexts that need to be considered, which may impose variation in external demands for smartphone use (e.g., sociocultural, professional, social, and academic conflicts) [10].

While there are a variety of limitations, the growing empirical research generally seems to support the concept of SA as a genuine addictive disorder, despite the available data being insufficient in establishing a valid and definitive conceptualization of SA [4]. We propose that SA may fall along a continuum of dysregulated behavior, which is problematic and even incapacitating at some point [8,11]. It is important to explore how authors to date have characterized excessive smartphone misuse and understand how various factors gleaned from current research coalesce to discern a clearer picture of this emerging behavioral addiction.

Research Aims

The goal of this review was to examine the most recent studies of SA/PSU to see the extent to which there appears to be unique features of SA/PSU and how studies assess this behavior, to arrive at an understanding regarding at which point there is likely to be consensus on the condition being an addiction along a spectrum of problematic behaviors.

We also explored proneness to SA/PSU as a function of gender, as some researchers have suggested that SA may be more prevalent among females [12–23]. Other research has suggested that subsets of SA (e.g., social media addiction versus gaming addiction) may differentiate males from females in their smartphone use [12,14,15,24–36].

Additionally, we examined variation as a function of age. Some researchers have suggested that SA may be more prevalent among youth and young adults as compared to older adults and elderly who use smartphones less often than youth [12,13,19,26,30,32,34,35,37–55].

2. Materials and Methods

We conducted a systematic review of the literature on and related to SA/PSU utilizing the Preferred Reporting Items for Systematic Reviews and Meta-Analyses (PRISMA) statement [56] as a general guide. Electronic databases utilized included: MEDLINE/PubMed, ProQuest, GALE, Web of Science, Directory of Open Access Journals, Elsevier's collection, Taylor & Francis Online, Wiley Online Library, SpringerLink, and Sage Journal's collection.

2.1. Inclusion Criteria

We utilized the following search terms: "(smartphone OR "mobile phone") AND (addiction OR "problematic use")", published in peer-reviewed journals between 2017 and 2019. Examining Google Scholar, we noticed a steep influx of SA/PSU articles published beginning 2017 (from 22 appearing in 2007; 41 in 2011; 164 in 2013; and 1020 in 2017). Therefore, we began the search from 2017 since this would best capture the current state of research in the area of SA/PSU research. Of the 170 articles returned from the search, 100 were retained for the current review after screening their titles and abstracts. The references of these articles were also searched, of which eight articles were deemed relevant and were included, allowing us to review 108 articles in total.

2.2. Exclusion Criteria

Articles published in a language other than English, duplicate articles (primary articles were retained), studies that did not substantively extend a parent study (i.e., only adapted established SA/PSU scales to a new language with some validation testing), and irrelevant articles (determined after looking through the abstracts or text; examples of irrelevant studies included those assessing the efficacy of an app serving as an intervention, focusing on the risk factors of SA/PSU rather than SA/PSU itself, and measuring invariance among SA-related scales) were excluded from the review pool (see Figure 1).

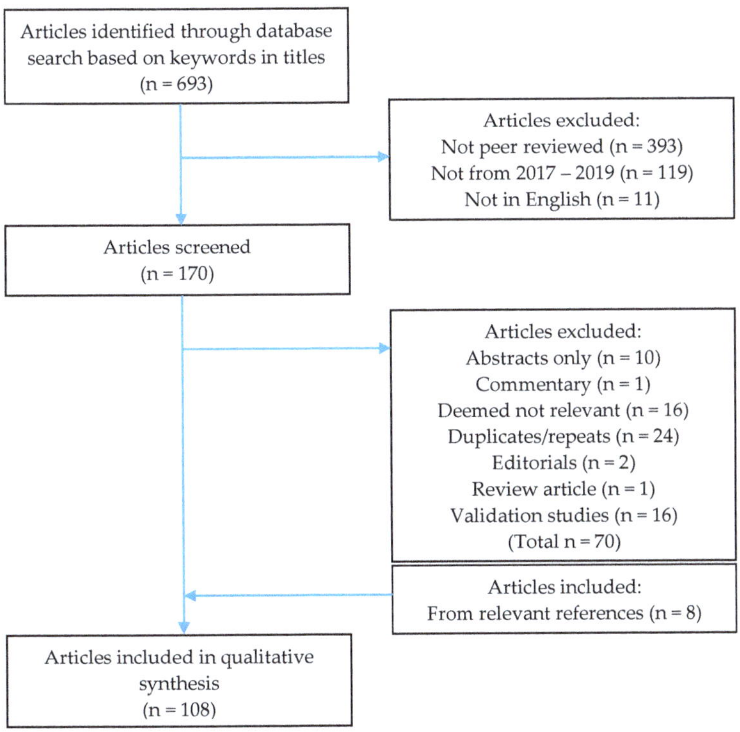

Figure 1. Process of final article selection for the current review.

2.3. Information Synthesis

Since most of the literature on smartphone use were based upon SA and PSU scales, we created a table to compile scales used to measure SA/PSU of smartphones, examining the following (based on Bianchi and Phillips, 2008 [3]): (1) Scale/measurement items, (2) Cronbach's alpha coefficients, and (3) variables and/or factors measured to assess smartphone use. A total of 14 relevant assessments emerged from a review of previous SA/PSU studies (see Table 1).

Table 1. Smartphone addiction/problematic smartphone use assessments in the literature.

Title (Abbreviation): Authors	Year Cat [1]	Items CA [2]	Measures/Factors
Mobile Phone Problem Use Scale (MPPUS): Bianchi and Phillips	2005 PSU	27 0.93	1. Tolerance 2. Escape from other problems 3. Withdrawal 4. Craving 5. Negative life consequences
Mobile Phone Addiction Index (MPAI): Leung	2008 SA	17 0.87	1. Inability to control craving 2. Feeling anxious and lost 3. Withdrawal or escape 4. Productivity loss
Problematic Mobile Phone Use Questionnaire (PMPUQ): Billieux, Van der Linden, and Rochat	2008 PSU	30 0.77	1. Dangerous/prohibited use 2. Dependence 3. Financial problems
Smart Mobile Phone Addiction Scale (MPAS): Hong, Chiu, and Huang	2012 SA	11 0.90	1. Academic problems and effects 2. Time management and problems 3. Substitute satisfaction
Problematic Use of Mobile Phones (PUMP) Scale: Merlo, Stone, and Bibbey	2013 PSU	20 0.94	1. Tolerance 2. Withdrawal 3. Longer time than intended 4. Great deal of time spent 5. Craving 6. Activities given up or reduced 7. Use despite physical/psychological problems 8. Failure to fulfill role obligations 9. Use in physically hazardous situations 10. Use despite social/interpersonal problems
Smartphone Addiction Scale (SAS): Kwon, Lee, Won, Park, Min, Hahn, Gu, Choi, and Kim	2013 SA	48 0.97	1. Daily-life disturbance 2. Positive anticipation 3. Withdrawal 4. Cyberspace-oriented relationship 5. Overuse 6. Tolerance
Short Version for Adolescents (SAS-SV): Kwon, Kim, Cho, Yang	2013 SA	10 0.91	1. Daily-life disturbance 2. Withdrawal 3. Cyberspace-oriented relationship 4. Overuse 5. Tolerance
Smartphone Addiction Scale for College Students (SAS-C): Su, Pan, Liu, Chen, Wang, and Li	2014 SA	22 0.93	1. Withdrawal behavior 2. Salience behavior 3. Social comfort 4. Negative effects 5. Use of application (app) 6. Renewal of app
Smartphone Addiction Management System (SAMS): Lee, Ahn, Choi, and Choi	2014 SA	14 N/A	1. Daily-life disturbance 2. Virtual world orientation 3. Withdrawal 4. Tolerance
Smartphone Addiction Inventory (SPAI): Lin, Chang, Lee, Tseng, Kuo, and Chen	2014 SA	26 0.94	1. Compulsive behavior 2. Functional impairment 3. Withdrawal 4. Tolerance
Korean Smartphone Addiction Proneness Scale (SAPS [3]): Kim, Lee, Lee, Nam, and Chung	2014 SA	29 0.88	1. Disturbance of adaptive functions 2. Withdrawal 3. Tolerance 4. Virtual life orientation
Smartphone Addiction Scale (SPAS): Bian and Leung	2014 SA	19 0.83	1. Disregard of harmful consequences 2. Preoccupation 3. Inability to control craving 4. Productivity loss 5. Feeling anxious and lost
Mobile Phone Problem Use Scale: derivation of a short scale (MPPUS-10): Foerster, Roser, Schoeni, and Röösli	2015 PSU	10 0.85	1. Craving 2. Withdrawal 3. Dependence 4. Loss of control 5. Negative life consequences
Short form of Smartphone Addiction Inventory (SPAI-SF): Lin, Pan, Lin, and Chen	2016 SA	10 0.84	1. Compulsive behavior 2. Functional impairment 3. Withdrawal 4. Tolerance

Notes. [1] Category: Smartphone addiction (SA) or problematic smartphone use (PSU); [2] Cronbach's alpha value; [3] First developed by Korea's National Information Society Agency (2011) "Report on the Development of Korean Smartphone Addiction Proneness Scale for Youth and Adults" (also known as S-Scale) [Korean].

We compared the items across these assessments to see which were common among the scales (see Table 2). Based on Griffiths' (2005) six-component model of addiction, we categorized the measured items based on the components argued to be needed "to be present for a behavior to be operationally defined as addictive": Salience (activity becomes most important, dominating aspect), mood modification (subjective experience of feeling better, whether calming down and/or feeling less destressed), tolerance (more activity needed to feel activity's previous effect), withdrawal (abrupt termination of activity results in unpleasant feeling, psychologically and/or physiologically), conflict (activity causes conflict between addict and others or ones' self-image), and relapse (addictive activity patterns recur) [57].

Table 2. Comparison of major validated scale variables based on Griffiths' component model.

	MPPUS	MPAI	MPAS	SAS	SAS-SV	SPAI	SPAI-SF	SAPS	SPAS	Total
Conflict	√	√	√	√	√	√	√	√	√	9
Withdrawal	√	√	√	√	√	√	√	√	√	9
Tolerance	√	√	√	√	√	√	√	√	√	9
Salience	√	√	√	√	√	√	√	√	√	9
Mood	√	√		√		√		√	√	6
Relapse						√	√	√		3

We also extracted the following information from the studies (see Table 3): (1) First author; (2) year published; (3) whether the study classified smartphone use as an addiction (SA) or problematic smartphone use (PSU); (4) sample size (n); (5) the scale used to assess SA/PSU; (6) if SA/PSU was clearly defined; (7) if studies mentioned the negative attributes and/or risk factors of SA/PSU; (8) if studies distinguished SA/PSU from other technological addictions (e.g., Internet and gaming addiction); (9) if studies differentiated if the addiction is to the actual phone device or to what the phone device offers, such as functions, services, and/or content (e.g., social networking services, apps, etc.); and (10) if studies were structured around a theory or model to explain SA/PSU and how it affects the smartphone user. Under the scale column "adapted", scales refer to scales adapted from SA/PSU assessment scales used in previous studies (e.g., authors compiled x items from scale A and y items from scale B to create their own scale, S) and/or input from field experts.

Table 3. Characteristics of relevant smartphone studies, 2017–2019.

First Author	Year	Cat [1]	Scale	Define [2]	Distinct [3]	D v. F [4]	Theory
Abed	2018	SA	Adapted	No	No	No	n/a
Akodu	2018	SA	SAS-SV	No	No	Yes	n/a
Akturk	2018	SA	SAS-SV	Yes	No	No	n/a
AlAbdulwahab	2017	SA	SAS	No	No	No	n/a
Alavi	2018	SA	MPPUS	No	No	No	Yes [5]
Albursan	2019	SA	Adapted	No	Yes	Yes	n/a
Alhassan	2018	SA	SAS-SV	Yes	No	No	n/a
Alhazmi	2018	SA	SAS-SV	Yes	No	No	n/a
Arefin	2017	SA	Adapted	Yes	Yes	Yes	n/a
Arnavut	2018	SA	SAS-SV	Yes	No	No	n/a
Barnes	2019	SA	Adapted	Yes	Yes	Yes	Yes [6]
Basu	2018	SA	Adapted	Yes	No	No	n/a
Beison	2017	PSU	MPPUS	No	No	No	Yes [7]
Cerit	2018	SA	SAS	Yes	No	No	n/a
Cha	2018	SA	SAPS	Yes	No	No	n/a
Chang	2019	SA	SPAI	Yes	No	No	n/a
Chen	2018	SA	Adapted	No	Yes	Yes	Yes [8]
Chen	2017	SA	Adapted	Yes	No	No	Yes [9]
Chen	2017	SA	SAS-SV	Yes	No	No	n/a
Chiang	2019	SA	SPAI-SF	No	No	No	n/a
Cho	2017	SA	SAPS	Yes	No	No	Yes [10]

Table 3. *Cont.*

First Author	Year	Cat [1]	Scale	Define [2]	Distinct [3]	D v. F [4]	Theory
Cho	2017	SA	AMT [11]	No	No	No	n/a
Choi	2017	SA	SAS	Yes	No	No	n/a
Chou	2019	SA	SPAI-SF	No	No	No	Yes [12]
Chung	2018	SA	SAPS	No	No	No	n/a
Cocoradă	2018	SA	SAS-SV	Yes	No	No	Yes [13]
Contractor	2017	PSU	SAS-SV	Yes	No	No	Yes [14]
De-Sola	2017	SA	MPPUS	Yes	Yes	No	n/a
Ding	2019	PSU	SOCS	No	No	Yes	n/a
Duke	2017	SA	SAS-SV	No	Yes	Yes	n/a
Elhai	2017	PSU	SAS	No	No	No	Yes [15]
Elserty	2018	SA	SAS-SV	Yes	No	No	n/a
Gao	2018	SA	MPAI	Yes	No	No	Yes [16]
Gao	2017	SA	MPAS	No	No	No	n/a
Gezgin	2018	SA	SAS-SV	Yes	No	No	n/a
Gligor	2019	SA	MPDQ [17]	No	No	No	n/a
Gökçearslan	2018	SA	SAS-SV	Yes	No	No	n/a
Habibi	2018	SA	Adapted	Yes	No	No	n/a
Han	2017	SA	MPAI	Yes	Yes	No	Yes [18]
Hao	2019	SA	MPAI	Yes	No	No	n/a
Hawi	2017	SA	SAS-SV	No	No	No	n/a
Heo	2018	SA	SAPS	No	No	No	n/a
Herrero	2019	SA	SPAS	No	No	No	Yes [10]
Herrero	2019	SA	SPAS	Yes	No	No	n/a
Herrero	2019	SA	SPAS	Yes	No	No	n/a
Ihm	2018	SA	Adapted	No	No	No	n/a
Jeong	2019	SA	SAPS	Yes	Yes	No	n/a
Jin	2017	SA	SAS-SV	Yes	No	No	Yes [18,19]
Jo	2018	SA	SAPS	No	No	No	n/a
Khoury	2019	SA	SPAI	No	Yes	No	n/a
Kim	2018	PSU	Adapted	Yes	No	No	Yes [20]
Kim	2017	PSU	Adapted	Yes	No	No	n/a
Kim	2019	SA	SAPS	No	No	No	n/a
Kim	2018	SA	SAPS	Yes	No	No	n/a
Kim	2017	SA	SAPS	No	No	No	Yes [18]
Kim	2017	SA	SAPS	Yes	No	No	Yes [14]
Kim	2017	SA	SAS	No	No	No	n/a
Kim	2018	SA	SAPS	Yes	No	No	Yes [18]
Kita	2018	SA	SAS-SV	Yes	No	No	n/a
Konan	2019	SA	SAS-SV	Yes	No	No	n/a
Konan	2018	SA	SAS-SV	No	No	No	n/a
Kuang-Tsan	2017	SA	MPAS	No	No	No	n/a
Kumcağız	2019	SA	SAS-SV	No	No	No	n/a
Kumcağız	2017	SA	SAS-SV	No	No	No	n/a
Kuss	2018	PSU	PMPUQ	Yes	No	No	Yes [21]
Kwak	2018	SA	Adapted	No	No	No	Yes [22]
Kwan	2017	SA	SAS	Yes	No	Yes	Yes [23,24]
Lan	2018	SA	MPIAS [25]	No	No	No	n/a
Lee	2018	PSU	SAPS	Yes	No	No	Yes [26]
Lee	2017	SA	SAPS	Yes	Yes	No	n/a
Lee	2017	SA	SAPS	No	No	No	n/a
Lee	2017	SA	Adapted	No	No	No	n/a
Lee	2018	SA	Adapted	No	No	No	n/a
Lee	2018	SA	Adapted	No	No	No	n/a
Lee	2018	SA	SAS-SV	No	Yes	No	n/a
Lee	2018	SA	SAS-SV	Yes	Yes	No	Yes [27]
Li	2018	SA	SPAS	No	No	No	Yes [28]
Lian	2018	SA	MPAI	Yes	No	No	Yes [26,29]
Lian	2017	SA	MPAI	Yes	No	No	Yes [26]
Lin	2017	SA	App	Yes	Yes	Yes	n/a
Liu	2018	SA	MPAI	Yes	No	No	Yes [22]

Table 3. *Cont.*

First Author	Year	Cat [1]	Scale	Define [2]	Distinct [3]	D v. F [4]	Theory
Liu	2018	PSU	SAS-SV	Yes	No	Yes	Yes [30]
Liu	2017	SA	MPAI	No	No	No	Yes [31]
Lu	2018	SA	MPAS	No	No	No	n/a
Mahapatra	2018	SA	Adapted	Yes	No	Yes	Yes [32]
Matar B.	2017	SA	SPAI	No	Yes	No	n/a
Megna	2018	SA	SAS-SV	Yes	No	No	n/a
Mei	2018	SA	MPAI	Yes	No	Yes	Yes [33]
Nayak	2018	SA	Adapted	Yes	Yes	No	n/a
Noë	2019	SA	SAS	Yes	Yes	Yes	n/a
Parasuraman	2017	SA	Adapted	No	No	Yes	n/a
Salvi	2018	SA	Adapted	Yes	No	No	n/a
Sekhon	2018	SA	MPPUS	Yes	No	No	n/a
Serin	2019	SA	SAS-SV	Yes	No	No	n/a
Sok	2019	SA	Adapted	No	No	No	n/a
Song	2018	SA	SAS (Lee)	No	No	No	n/a
Sun	2019	SA	SAS (Ku)	Yes	No	No	Yes [34]
Tunc-Aksan	2019	SA	SAS	No	No	Yes	Yes [35]
Wang	2018	SA	SAS-SV	No	No	No	Yes [36,37]
Wang	2017	SA	SAS-SV	No	No	No	Yes [38]
Wolniewicz	2018	PSU	SAS-SV	Yes	Yes	Yes	Yes [15,16]
Xu	2019	PSU	MPAS	Yes	No	No	Yes [22,39]
Yang	2019	SA	MPAI	Yes	No	No	Yes [40]
Yang	2019	PSU	Adapted	Yes	No	No	n/a
Yildiz Durak	2017	SA	SAS	Yes	Yes	No	Yes [41]
Yildiz	2017	SA	SAS-SV	No	No	No	n/a
You	2019	SA	MPAS	No	No	No	Yes [42]
Youn	2018	SA	SAS	No	No	No	n/a

Notes. [1] Category based on study title: Smartphone addiction (SA) or problematic smartphone use (PSU); [2] Define: If SA or PSU was defined; [3] Distinct: If studies distinguished SA from other technological addictions (e.g., Internet, gaming addiction); [4] D v. F: If studies differentiated if the addiction is to the actual phone device or to what the phone device offers, such as functions, services, and/or content (e.g., social networking services, apps, etc.); [5] Psychosocial Theory of Development (Erikson, 1963) and Adolescent Identity Paradigm (Marcia, 1991); [6] Cognitive absorption (Agarwal & Karahanna, 2000); [7] Reward Deficiency Syndrome hypothesis (Blum et al., 1996); [8] Social influence theory (Rashotte, 2007); [9] Four categories of drinking motives (Stewart & Devine, 2000); [10] Big Five personality traits (Norman, 1963); [11] Addiction Measurement Tools of Measuring Smartphone Addiction of Children-Adolescents (Korea Network Information Center); [12] 6-T Internet attitude model (Chou, Wu, & Chen, 2013); [13] Theory of Reasoned Action (Fishbein & Ajzen, 1975); [14] Impulsive pathway perspective (Billieux, 2012); [15] Uses and Gratifications Theory (UGT) (Blumler & Katz, 1974); [16] Compensatory Internet Use Theory (CIUT) (Kardefelt-Winther, 2014); [17] MPDQ: Mobile phone dependence questionnaire; [18] Attachment theory (Bowlby, 1969); [19] Psychoanalytic theory (Kassel et al., 2007); [20] Social enhancement model (Kraut et al., 2002); [21] Biopsychosocial model of addiction (Griffiths, 2005); [22] General strain theory (Agnew, 1992); [23] Theory of parenting style (Baumrind, 1971); [24] First theory of self-regulation (Asgari et al., 2011); [25] MPIAS: Mobile Phone Internet Addiction Scale; [26] Problem behavior theory (PBT) (Jessor, 1977); [27] Power distance belief (Hofstede, 1980); [28] Media system dependency (MSD) theory (Ball-Rokeach & DeFleur, 1976); [29] Social skills deficit theory (Valkenburg & Peter, 2007); [30] Social compensation theory (Zell & Moeller, 2018); [31] Response style theory (Nolen-Hoeksema, 1991); [32] Incentive-Sensitization (Robinson & Berridge, 2003) & Learning Theory (Wallace, 1999); [33] Hierarchy of needs theory (Maslow, 1968); [34] Risky families model (Repetti, 2002); [35] Intrinsic motivation theory (Przybylski, Weinstein, Ryan, & Rigby, 2009); [36] Sensation seeking theory (Zuckerman, 1994); [37] Social support buffering hypothesis (Cohen & Wills, 1985); [38] Cognitive-behavioral model (Davis, 2001); [39] Resilience theory (Fergus & Zimmerman, 2005); [40] Diathesis-stress theories (Monroe & Simons, 1991) and stress-buffering hypothesis (Cohen & Edwards, 1989); [41] Social Cognitive Theory (Bandura, 1986); [42] Sociometer theory of self-esteem (Leary et al., 1995).

To supplement Table 3, we conducted an in-depth analysis on the study samples (see Table 4), specifically (1) the group population, either child/youth (in primary, elementary, middle, or high school), university student (students in college, graduate, and vocational/technical/institute schools were all categorized as university students), or adult; (2) sample size, n; (3) percentage of males in the sample; (4) mean age and age range of participants; (5) participant nationality; (6) sociodemographic characteristics (SDCs) (biometrics, such as body mass index (BMI), height, and weight; number of children; education; income; marital status; region or residence, such as rural or urban; and work/employment status); and (7) reported study biases. SDC data were reported for university students and adults, usually the parents of children.

The point of Table 4 was to determine the generalizability of the studies' results: Can the research findings and conclusions of a particular group of interest be extended to the population at large? If a particular population sample is studied across age and nationality, does that imply that that group is at-risk for SA?

Table 4. Demographic profile of smartphone study samples and reported study biases.

First Author	Pop [1]	n [2]	Males (%)	Age (m (r) [3])	Nationality	SDC [4] (BCEIMRW)	Reported Biases
Abed	Univ	229	35	n/a	Iraqi	MRW	None
Akodu	Univ	77	57	22	Nigerian	B	Sampling
Akturk	Univ	1156	49	n/a	Turkish	CEI	Sampling
AlAbdulwahab	Univ	78	50	21	Saudi	B	None
Alavi	Univ	500	21	28 (18–31+)	Iranian	E M	Sampling
Albursan	Univ	2008	45	22 (17–28)	Arab		Sampling
Alhassan	Adult	935	34	32 (18–55+)	Saudi	EI R	Multiple
Alhazmi	Univ	181	48	24	Saudi	B M	Sampling
Arefin	Univ	247	54	(18–27)	Bangladeshi		Sampling
Arnavut	Adult	714	58	(18–30+)	Turkish		Sampling
Barnes	Univ	140	31	(18–35+)	American	E	Multiple
Basu	Univ	388	60	20	Indian	R	Sampling
Beison	Univ	100	25	20 (18–23)	American [5]	EI	None
Cerit	Univ	214	20	20 (18–26)	Turkish		Recall
Cha	MS	1824	51	16	Korean	I W	Multiple
Chang	ES/Adult	5089	52/31	n/a/43	Taiwanese	EIM	Multiple
Chen	Univ	2000	49	21 (17–23)	Taiwanese		Multiple
Chen	Univ	384	54	n/a	Chinese		Multiple
Chen	Univ	1441	48	20 (17–26)	Chinese	R	Multiple
Chiang	ES/MS	2155	52	n/a	Taiwanese	EIM	Multiple
Cho	Adult	400	52	(20–40+)	Korean	E W	Multiple
Cho	PS/Adult	303	7/51	(20–40+)/(0–6)	Korean	EI W	Sampling
Choi	HS	1020	52	n/a	Korean	R	Sampling
Chou	HS/Adult	1444	43/n/a	n/a/n/a	Taiwanese	EI	Recall
Chung	MS/HS	1796	46	15 (13–18)	Korean		Sampling
Cocoradă	HS/Univ	717	35	20	Romanian		Multiple
Contractor	Adult	346	42	34	American [5]	IM W	Multiple
De-Sola	Adult	1126	48	33 (16–65)	Spanish	E R	Multiple
Ding	Univ	849	56	n/a	Chinese		Multiple
Duke	Adult	262	36	32	German		Recall
Elhai	Adult	308	54	33	American [5]	EIM W	Multiple
Elserty	Univ	420	32	20	Egyptian	B	None
Gao	Univ	1105	29	21 (16–25)	Chinese	I	Multiple
Gao	Univ	722	48	20 (15–24)	Chinese	C I R	Multiple
Gezgin	HS	161	58	16	Turkish		Multiple
Gligor	Univ	150	44	27	Romanian	R	Multiple
Gökçearslan	Univ	885	41	n/a	Turkish	I	None
Habibi	HS	271	n/a	17	Indonesian		None
Han	Univ	543	41	20 (17–22)	Chinese		Sampling
Hao	Univ	847	51	20 (18–24)	Chinese	C R	Sampling
Hawi	Univ	381	59	21 (17–27)	Lebanese		Multiple
Heo	HS	790	23	n/a	Korean		None
Herrero	All	526	52	(15–55+)	Spanish	E R	None
Herrero	All	241	55	(15–55+)	Spanish	E R	Sampling
Herrero	All	416	52	(15–55+)	Spanish	E R	Selection
Ihm	Youth	2000	50	12	Korean		Recall
Jeong	HS	768	58	n/a	Korean	IM	Multiple
Jin	Univ	297	55	20 (17–24)	Chinese	I	Sampling
Jo	All	7003	45	(14–39)	Korean	R	Recall
Khoury	Univ	100	48	(18–25)	Brazilian [5]	IM	Response
Kim	Adult	615	51	30 (19–40)	American [5]		Recall
Kim	All	930	52	26 (13–40)	American [5]	M	Recall
Kim	MS/HS	4512	45	15	Korean	IM	Multiple
Kim	Youth	3380	51	(10–19)	Korean	EIMRW	Information
Kim	Univ	200	37	22 (19–28)	Korean		Sampling
Kim	Univ	608	30	23	Korean	I	Multiple
Kim	HS	1479	48	n/a	Korean	B	Sampling
Kim	Univ	313	42	22 (17–29)	Korean		Sampling
Kita	YA	221	65	19 (17–22)	Israeli		Sampling

Table 4. *Cont.*

First Author	Pop [1]	n [2]	Males (%)	Age (m (r) [3])	Nationality	SDC [4] (BCEIMRW)	Reported Biases
Konan	Univ	496	25	n/a	Turkish		None
Konan	Univ	330	36	(20–24)	Turkish		None
Kuang-Tsan	Univ	332	65	(18–22)	Taiwanese		Sampling
Kumcağız	HS	352	44	16 (14–19)	Turkish	E W	Multiple
Kumcağız	Adult	428	37	40 (21–65)	Turkish	CEIM W	Sampling
Kuss	All	273	26	28 (16–65)	Various		None
Kwak	MS	1170	42	n/a	Korean	I	Sampling
Kwan	Univ	211	35	22	Hong Kong		None
Lan	Univ	1044	48	21	Chinese		Information
Lee	Youth	231	40	16 (13–18)	Korean		Sampling
Lee	MS	370	49	13	Korean		Multiple
Lee	MS/HS	3000	53	n/a	Korean		Sampling
Lee	MS/HS	1125	51	n/a	Korean	EI R	Multiple
Lee	Univ	125	49	n/a	Korean		None
Lee	Univ	324	9	n/a	Korean		Multiple
Lee	MS	490	100	14	Korean	E	Multiple
Lee	Adult	778	63/58 [6]	35/25 [6]	Various	EI	Information
Li	Adult	527	46	27 (18–35)	Chinese	EIM W	Sampling
Lian	Univ	716	54	20 (18–24)	Chinese	R	Sampling
Lian	Univ	682	58	19 (18–24)	Chinese		Recall
Lin	Univ	79	72	22	Taiwanese		Sampling
Liu	HS	899	46	17 (14–19)	Chinese		Sampling
Liu	Univ	465	31	19 (16–24)	Chinese		Sampling
Liu	HS	1196	53	17 (14–20)	Chinese		Multiple
Lu	MS	1311	54	15	Various		Sampling
Mahapatra	HS/Univ	330	58	(15–20)	Indian		Multiple
Matar B.	Univ	688	53	21	Lebanese	W	Recall
Megna	Adult	52	46	27 (18–35)	Italian		None
Mei	Univ	1034	47	20	Chinese	C I R	Multiple
Nayak	Univ	429	35	20 (16–29)	Indian		None
Noë	Adult	64	53	25 (19–46)	British	E W	Sampling
Parasuraman	Adult	409	42	23 (18–55)	Malaysian	E W	Sampling
Salvi	Univ	100	59	21 (18–25)	Indian		None
Sekhon	Univ	80	50	(20–24)	Indian		Sampling
Serin	Univ	287	14	n/a	Turkish	CEIM	Multiple
Sok	Univ	139	16	n/a	Korean		Sampling
Song	PS/Adult	328	n/a/0	(3–5)/n/a	Korean	E W	Multiple
Sun	HS	1041	56	12 (11–15)	Chinese		Multiple
Tunc-Aksan	HS	296	54	n/a	Turkish	E	None
Wang	MS	655	55	17 (15–19)	Chinese		Multiple
Wang	MS	768	44	17 (15–19)	Chinese	E W	Multiple
Wolniewicz	Univ	296	43	20	American [5]	E W	Multiple
Xu	MS	316	47	14 (12–16)	Chinese		Sampling
Yang	HS	1258	53	17 (14–20)	Chinese		Sampling
Yang	Univ	218	58	18 (16–19+)	Taiwanese		Multiple
Yildiz Durak	MS/HS	612	52	13 (10–18)	Turkish	CEI R	Multiple
Yildiz	HS	262	50	17 (14–19)	Turkish		Sampling
You	Univ	653	50	20 (17–25)	Chinese	EI	Sampling
Youn	Youth	158	53	15 (12–19)	Korean	EI	Multiple

Notes. [1] Pop: Population general group (PS: Primary school; ES: Elementary school; MS: Middle school; HS: High school; Univ: University, college, or institute of technology students; YA: Young adult); [2] Sample size n; [3] m (r): mean age (range, if available); [4] SDC: Common sociodemographic characteristics besides gender, age, and ethnicity (B = biometrics: height, weight, BMI; C = number of children; E = education; I = income; M = marital status; R = regional/residence: urban or rural; W: work/employment status); [5] Reports racial status (e.g., African-American/Black, American Indian, Asian, Native Hawaiian/other Pacific Islander, White) and ethnicity (e.g., Hispanic/Latino, Not Hispanic/Latino); [6] Of 778 total, 431 US and 347 Chinese participants: male % and mean age of US and Chinese, respectively.

To supplement Table 4 on the characteristics of the studies' samples (refer to Table 5), we examined the quality of the data collection process by noting if samples were selected as a convenience sample or randomly selected and if a sample size calculation was calculated before participant recruitment. We also categorized the data collected as consequences pertaining to: Academics (grades, academic performance), physiological (headache, eye/neck/hand pain), psychological (experiencing depression, loneliness, cravings), social (peer relationships), or usage; and whether it was objective or subjective (such as an objective clinical diagnosis of depression or a subjective self-report of experiencing depression).

Additionally, we assessed the operationalization of theory by noting if the SA/PSU scale used in the study met Griffiths' criteria for behavioral addictions: Conflict, mood modification, relapse, salience, tolerance, and withdrawal. The 2004 Surgeon General's report, *The Health Consequences of Smoking*, developed a framework for interpreting evidence, specifying a four-level hierarchy for interpreting evidence: (a) Evidence is sufficient to infer a causal relationship (multiple scientifically supported evidence), (b) evidence is suggestive but not sufficient to infer a causal relationship (scientifically supported evidence), (c) evidence is inadequate to infer the presence or absence of a causal relationship (evidence that is not scientifically supported and/or is sparse, of poor quality, or conflicting), or (d) evidence is suggestive of no causal relationship (no evidence) [58]. We also noted if the assumption of addiction, referencing back to the title assumption of whether the authors viewed excessive smartphone use as SA or PSU, influenced the study results.

Table 5. The quality of samples, level of evidence, and theoretical framework of the studies.

First Author	Sel [1]	SSS [2]	Evidence [3]	Obj [4]	Int [5]	Griffiths [6]	Inf [7]
Abed	R	No	PhPs	S	I	n/a	Yes
Akodu	C	No	Ph	O	Sg	C STW	Yes
Akturk	C	Yes	S	S	Sg	C STW	Yes
AlAbdulwahab	C	Yes	Ph	S	I	CM STW	Yes
Alavi	C	Yes	Ps	O	I	CM STW	Yes
Albursan	R	No	Ps	S	I	n/a	Yes
Alhassan	R	No	Ps	S	I	C STW	Yes
Alhazmi	R	Yes	Ph	S	I	C STW	Yes
Arefin	C	No	A	O	Sg	CM STW	Yes
Arnavut	R	No	U	S	I	C STW	Yes
Barnes	C	No	Ps	S	I	C STW	Yes
Basu	R	Yes	U	S	I	CM STW	Yes
Beison	n/a	No	S	S	Sg	CM STW	Yes
Cerit	R	Yes	Ph	S	I	CM STW	Yes
Cha	R	No	PhPsU	S	I	CMRSTW	Yes
Chang	R	No	SU	S	I	CMRSTW	Yes
Chen	C	No	PsS	S	I	n/a	PSU [8]
Chen	C	No	Ps	S	I	CMRSTW	Yes
Chen	R	No	PhPs	S	I	C STW	Yes
Chiang	R	No	Ps	S	I	C TW	Yes
Cho	R	No	Ps	S	I	CMRSTW	Yes
Cho	R	No	Ps	S	I	n/a	Yes
Choi	n/a	No	U	S	I	C TW	Yes
Chou	R	No	U	S	I	C TW	Yes
Chung	R	No	Ph	S	I	CMRSTW	Yes
Cocoradă	C	No	PsU	S	I	C STW	Yes
Contractor	C	No	Ps	S	I	C STW	Yes
De-Sola	C	No	Ps	S	I	CM STW	Yes
Ding	R	No	PsU	S	I	CMRST	Yes
Duke	C	No	U	S	I	C STW	Yes
Elhai	C	No	PsU	S	I	CM STW	Yes
Elserty	C	No	PhU	S	I	C STW	Yes
Gao	C	No	Ps	S	I	CM STW	Yes
Gao	C	No	Ps	S	I	C RSTW	Yes
Gezgin	C	No	PsU	S	I	C STW	Yes
Gligor	n/a	No	Ps	S	I	CM STW	Yes
Gökçearslan	C	No	PsS	S	I	C STW	Yes
Habibi	n/a	No	Ps	S	I	C STW	Yes
Han	R	No	Ps	S	I	CM STW	Yes
Hao	R	No	PsU	S	I	CM STW	Yes
Hawi	R	No	Ps	S	I	C STW	Yes
Heo	C	Yes	Ps	S	I	CMRSTW	Yes
Herrero	R	No	PsSU	S	I	CM STW	Yes
Herrero	R	No	Ps	S	I	CM STW	Yes
Herrero	R	No	Ps	S	I	CM STW	Yes
Ihm	R	No	S	S	I	CMRSTW	Yes
Jeong	n/a	No	PsS	S	I	CMRSTW	Yes
Jin	C	No	Ps	PsS	I	C STW	Yes
Jo	n/a	No	Ps	S	I	CMRSTW	Yes
Khoury	R	No	Ph	O	Sg	CMRSTW	Yes

Table 5. *Cont.*

First Author	Sel [1]	SSS [2]	Evidence [3]	Obj [4]	Int [5]	Griffiths [6]	Inf [7]
Kim	R	No	Ps	S	I	C STW	Yes
Kim	R	No	Ps	S	I	C STW	Yes
Kim	R	No	Ps	S	I	CMRSTW	Yes
Kim	R	No	S	S	I	CMRSTW	Yes
Kim	n/a	Yes	Ps	S	I	CMRSTW	Yes
Kim	C	No	Ph	S	Sg	CMRSTW	Yes
Kim	n/a	No	Ph	S	I	CM STW	Yes
Kim	C	No	Ps	S	I	CMRSTW	Yes
Kita	C	No	S	O	Sg	C STW	Yes
Konan	R	No	PsS	S	I	C STW	Yes
Konan	C	No	Ps	S	I	C STW	Yes
Kuang-Tsan	C	No	S	S	Sg	C RSTW	Yes
Kumcağız	C	No	S	S	I	CM STW	Yes
Kumcağız	C	No	S	S	I	C STW	Yes
Kuss	C	No	Ps	S	I	C RSTW	Yes
Kwak	C	No	PsS	S	Sg	CMRSTW	Yes
Kwan	C	No	PsS	S	I	CM STW	Yes
Lan	R	No	Ps	S	Sg	n/a	Yes
Lee	C	Yes	PsS	S	Sg	CMRSTW	Yes
Lee	C	No	Ps	O	Sg	CMRSTW	PSU [8]
Lee	n/a	No	S	S	Sg	CMRSTW	Yes
Lee	R	No	Ph	S	I	C STW	Yes
Lee	R	No	S	O	Sg	CMRSTW	Yes
Lee	C	No	S	S	I	CM STW	Yes
Lee	n/a	No	Ps	S	I	C STW	Yes
Lee	n/a	No	Ps	S	I	C STW	Yes
Li	C	No	S	S	I	CM STW	Yes
Lian	C	No	S	S	I	CM STW	Yes
Lian	C	No	S	S	I	CM STW	Yes
Lin	n/a	No	U	O	Sg	n/a	Yes
Liu	R	No	Ps	S	I	CM STW	Yes
Liu	C	No	PsS	S	I	C STW	Yes
Liu	C	No	PhPs	S	I	CM STW	Yes
Lu	C	No	S	S	I	C RSTW	Yes
Mahapatra	C	No	APsS	S	I	C STW	Yes
Matar B.	R	No	PsU	S	I	CMRSTW	Yes
Megna	C	No	Ph	O	Sg	C STW	Yes
Mei	C	Yes	Ps	S	I	CM STW	Yes
Nayak	R	No	AU	S	I	C STW	Yes
Noë	C	No	U	O	Sg	CM STW	Yes
Parasuraman	C	No	U	S	I	n/a	Yes
Salvi	n/a	No	Ph	O	Sg	n/a	Yes
Sekhon	n/a	No	Ps	S	I	CM STW	Yes
Serin	C	No	Ps	S	I	C STW	Yes
Sok	C	Yes	PsSU	S	I	n/a	Yes
Song	C	No	S	S	I	CMRSTW	Yes
Sun	n/a	No	PsS	S	I	n/a	Yes
Tunc-Aksan	R	No	APsS	S	I	CM STW	Yes
Wang	n/a	No	PsS	S	I	C STW	Yes
Wang	n/a	No	PsS	S	I	C STW	Yes
Wolniewicz	C	No	PsU	S	I	C STW	Yes
Xu	n/a	No	APs	S	I	C RSTW	Yes
Yang	R	No	Ps	S	I	CM STW	Yes
Yang	C	No	PhPsU	S	I	C S W	Yes
Yildiz Durak	C	No	Ps	S	I	CM STW	Yes
Yildiz	C	No	Ps	S	I	CSTW	Yes
You	C	No	Ps	S	I	C RSTW	Yes
Youn	C	No	PsS	S	I	CM STW	Yes

Notes. [1] Sel: Selection of sample: C = convenience sample, R = randomly selected; [2] SSS: Sufficiency of sample size: Yes or No; [3] Evidence: Type of data presented: A = academic, Ph = physiological, Ps = psychological, S = social, U = usage; [4] Obj: Objectivity: O = objective, S = subjective; [5] IE: Interpretation of the level of evidence: Sf = sufficient, Sg = suggestive but not sufficient, I = inadequate, NR = suggestive of no causal relationship; [6] Griffiths: C = conflict, M = mood modification, R = relapse, S = salience, T = tolerance, W = withdrawal; [7] Inf: Assumption of addiction influencing study results: Yes or No; [8] Was titled an SA study but conclusions use PSU terminology.

3. Results

3.1. The Scales Used to Assess SA/PSU and Their Internal Consistency

The majority of the studies in the current review (96 out of 108) fell under the SA category (89%). Excluding one study that did not use a scale, there were nine major scales used among the studies, in order of frequency: Smartphone Addiction Scale Short Version (SAS-SV) (26%), Smartphone Addiction Proneness Scale (SAPS) (13%), Smartphone Addiction Scale (SAS) (9%), Mobile Phone Addiction Index (MPAI) (8%), Mobile Phone Addiction Scale (MPAS) (5%), Mobile Phone Problem Use Scale (MPPUS) (4%), Smartphone Addiction Scale (SPAS) (4%), Smartphone Addiction Inventory (SPAI) (3%), and SPAI Short Form (SPAI-SF) (2%). Overall, most studies either used SAS or SAS-SV (35%) or some adapted form of an SA/PSU-assessment scale based on scales from previous studies and/or input from field experts (26%). Excluding the adapted scale studies, 8 of the 12 PSU studies (67%) used SA scales to assess problematic use instead of a PSU scale, and 3 of the 95 SA studies (3%) used PSU scales to assess SA instead of SA scales (refer to Table 3).

In terms of meeting Griffiths' criteria for behavioral addictions, all of the major scales met conflict, withdrawal, tolerance, and salience. Six of the nine met mood modification and only three met relapse (refer to Table 2). Excluding nine studies in which specific adapted scale items were unavailable, one study that did not use a scale, and 79 that used one of the nine major scales, the remaining 19 studies used an adapted SA/PSU scale that met at least half of Griffiths' criteria for behavioral addictions (mean of 4.79, SD of 0.92): One study (5% of 19) met three criteria, seven (37%) met four, six (32%) met five, and five (26%) met all six.

All scales demonstrated validity: All scales provided a Cronbach's alpha (CA) value as a measure of internal consistency, and performed factor structure analyses, ensuring that factor fit properly explains correlations among outcomes [59,60]. Other evaluations of validity also included testing the concurrent validity of the scale with other established scales. The CA of the nine major scales ranged from 0.83 to 0.97, with a mean of 0.90 and standard deviation (SD) of 0.05 (refer to Table 1).

3.2. How Studies Defined SA/PSU

In total, 87 of the 108 studies (81%) attempted a working definition of SA/PSU, of which 56 explicitly stated the definition in the paper (64%), while the rest simply characterized SA/PSU by its negative risk factors or listed associated traits. Nearly all of the studies (96%) mentioned the negative risk factors associated with SA/PSU, including physical (insomnia, pain in neck and wrists, eye soreness, etc.) and physiological concerns (depression, anxiety, loneliness, etc.). SA has been defined as a maladaptive dependency on and/or obsessive-compulsive use of the smartphone device [18], a state of being immersed in uncontrollable smartphone usage [41], and the inability to properly regulate smartphone usage to the point of experiencing adverse consequences in one's daily life (Billieux, 2012) [21]. Interestingly, PSU has been similarly defined as "an excessive or uncontrolled use of smartphone" (also Billieux, 2012) [61]. It is interesting to note that the same reference has been used to define both SA and PSU, highlighting the interchangeability of the two terms.

Overall, 64 out of 108 studies (59%) mentioned other technological addictions, such as Internet and gaming addiction, in their paper. Of the 64 studies, only 13 (20%) explicitly argued that SA is separate and distinct from Internet addiction (2 out of 4 PSU studies; 11 out of 47 SA studies). In total, 20 studies out of 108 (19%) (4 out of 13 PSU studies; 16 out of 95 SA studies) mentioned whether SA is an addiction to the actual smartphone device or to the functions/services/content that the smartphone offers users, such as social networking services and various apps that are usually not found on other devices (refer to Table 3).

3.3. Theories Adopted Across These Studies

More than half of the studies (63%) did not explicitly base their study upon a specific established theory (refer to Table 3). Nine of 12 (75%) PSU studies used a theory (five of nine, 42%) or a concept, model, or hypothesis (four of nine, 33%) to guide their research; while 31 of 96 SA studies (29%) used a theory (23 of 96, 24%) or a concept or model (eight of 96, 8%) to guide their study (refer to Table 3).

The following four theories in the current review stood out as guiding points on understanding SA and how its pathways may work. First, social influence theory (Kelman,1974) helps to illustrate how socially influential factors predict a user's intended and actual behavior in virtual settings, which is affected by three social processes: "Compliance (normative influence from others' expectations), internalization (congruence of one's goals with others' goals), and identification (conception of one's self in terms of the group's defining features)" [39]. This theory could help to explain SA among impressionable youths, who are more prone to peer pressure to what their peers may deem cool, which could include engaging in addictive behaviors on the smartphone (e.g., games, social media, streaming videos on Twitch), as well as among adults whose lifestyles are affected by today's heavily connected environment, which makes smartphone use a social norm.

The second theory, the theory of reasoned action (Fishbein & Ajzen, 1975), explains that a person's actual behavior is determined by their intention to behave in a certain way, being influenced both by their own attitudes and the social context. The attitudes towards the smartphone (positive, negative, fear of missing out, and task switching) were considered key predictors of addiction [62]. This perspective focuses on the intrapersonal reasons a person may fall into SA, with positive attitudes potentially bringing about positive reinforcement to keep engaging in the cycle of excessive smartphone use. Other theories helping to illustrate user adoption and acceptance of information technologies include diffusion of innovations (Rogers, 1962), the theory of planned behavior (Ajzen, 1985), and the technology acceptance model (TAM) (Davis, 1989) [25].

However, the previous theories are limited in explaining how attitudes, perceptions, and beliefs are shaped around information technologies. To address this conceptual shortfall, cognitive absorption (CA) (Agarwal & Karahanna, 2000) was proposed as a motivating factor of usage behavior through "cognitive complexity beliefs". CA was defined as "a state of deep involvement with software ... where highly engaging and engrossing experiences result in users' 'deep attention' and complete immersion and engagement with an activity" [25]. The multidimensional construct of CA has five dimensions: Temporal dissociation, focused immersion, heightened enjoyment, control, and curiosity, which follow closely with Griffiths' biopsychosocial model of addiction.

Another relevant theory, the uses and gratifications theory (UGT), "helps understand background characteristics and individual differences motivating people to choose using *particular* types of mass media. UGT can explain how people with certain types of psychological and/or demographic characteristics may be drawn to increasingly use specific types of smartphone features" to achieve gratification [63]. Why are some people more prone to using Instagram and obsessing over how many "likes" they have accrued on their posts? What is it about some people having to constantly check their Twitter to see if their "tweets" are trending or how many "retweets" they have received? This theory could play a role in defining and characterizing social media addiction as a subset of SA.

3.4. Group-Specific Variation in SA

Of the total 108 studies, 84 (78%) focused on adolescents or emerging adults (ranging from elementary, middle, and high school students (32 of 84 studies, 38%) to college/university students (50 of 84, 59%) (the remaining two studies looked at both high school and university students)) and mentioned that smartphone use was highly prevalent among this youth/young adult group as compared to a more elderly group. Fourteen studies (13% of 108) focused on adults, and 10 studies included both child and adult participants, usually family members (10% of 108) (refer to Table 4).

Nineteen studies (18%) reported that gender was a significant predictor of SA, specifically that being female was significantly associated with higher tendencies toward SA or smartphone dependency or that SA/dependency/risk for SA was more prevalent in females than in males. Sixteen studies (15%) suggested that females may be more likely to be addicted to social media while males may be more likely to be addicted to gaming. However, no consistent findings regarding the subtypes of SA by gender were uncovered.

3.5. Generalizability of Study Results and Biases

Across all studies (refer to Table 4), the average percent of male participants was 46% (SD = 13). Twenty-four countries were represented in this review, with 90 studies (83%) categorized as a South, Southeast, or East Asian country (in order of frequency: South Korea, China, Turkey, Taiwan, India, Saudi Arabia, Lebanon, Bangladesh, Hong Kong, Indonesia, Iran, Iraq, and Malaysia).

All studies reported gender and age, and seven reported ethnicity (six studies with American adult participants and one with Brazilian adult participants). We also considered other common sociodemographic characteristics (SDCs) measures, including biometrics (e.g., height, weight, and BMI); number of children; education, income, marital, and work/employment status; and region (northern, southern, western, or eastern) or residence (urban or rural). Fifty of the 108 studies (46%) did not measure any SDCs, 27 (25%) looked at 1 SDC, 18 (17%) looked at 2, 9 (8%) looked at 3, and 4 (4%) looked at 4 SDCs.

About a third of the studies (35%) reported sampling bias (from convenience sampling), which did not allow for generalization of the findings outside of the study group population. For example, the results from studies focusing on university students may not apply to the general public. Recall bias of participants' self-reports on actual smartphone use and/or dependence, information bias, social desirability bias, response bias, and selection bias were also noted. Thirty-nine of the studies (36%) also reported multiple biases while 17 (16%) did not report any biases.

Only two studies used longitudinal data [64,65], and none of the others conducted follow-up assessments on the cross-sectional analyses of smartphone use and behavior. That is, limitations included: The use of cross-sectional data, which limits the ability to draw causal inferences, especially when determining the direction of association between SA/PSU and risk factors of interest; small sample size; not being able to determine whether study characteristics preceded SA/PSU development or were the outcome of smartphone use; use of less-than-optimal instruments tending to be subjective rather than objective; incentive-influenced survey answers; and attrition. The studies that did not report biases or limitations may have overlooked noting the potential ones mentioned above.

Nearly all of the authors recommended one or more of the following: Conducting future studies to further investigate the relationship between SA/PSU and related health risks, thoroughly identifying positive and negative outcomes, conducting longitudinally designed research studies with broader sample profiles; creating public health educational programs to inform the public of the physical and psychological risks associated with SA/PSU, and developing proper evidence-based strategies and interventions to address SA/PSU.

3.6. Quality of Samples and Level of Evidence

Examining data collection methods (refer to Table 5), about a half of the studies (49%) reported using convenience sampling, about a third (34%) reported random sampling, and the rest (17%) did not clearly report a sampling method. Most of the studies (90%) did not calculate an appropriate sample size before recruiting participants. In terms of sample sizes (with a range of 52–7003), 17 studies (16%) had 200 participants or less, 41 (38%) fell in the 201–500 range, 24 (22%) in the 501–1000 range, and 26 (24%) had more than 1001 participants.

In terms of associations with SA/PSU, 1 study focused on academic data, 11 (10%) presented physiological data (e.g., craniovertebral angle, skin conductance), 40 (37%) psychological data (e.g., depression, anxiety), 16 (15%) social data (e.g., social connectedness, alexithymia), 8 (7%) smartphone usage data (e.g., duration, types of functions used), and 28 (26%) reported a mix of those categories.

Most of the studies were subjective reports (90%). Ten of the 11 objective reports (one academic, four physiological, one psychological, two social, and two smartphone usage) were deemed suggestive (but not sufficient or indicative) of a causal association with SA/PSU. The remaining objective report (psychological) was deemed insufficient because the link between SA and national identity was not clear to us. Seven of the 97 subjective reports (six social, one psychological) were deemed suggestive (but not sufficient or indicative) because unlike other psychological reports (within which subjective reports can be verified with objective testing, or physiological measures), certain social aspects (e.g., loneliness, family history of alcohol addiction, need for social assurance, life satisfaction, parental neglect, friends' support, and peer relationships) are hard to measure with objective testing and so subjective data is helpful and could be relatively valid and reliable in those cases.

4. Discussion

4.1. Smartphone Addiction on a Continuum of Addictive Behaviors

4.1.1. Is Smartphone Addiction Really an Entity of Its Own?

With smartphone use engrained as the social norm today, and the pervasive use ever increasing despite an awareness of the health risks and adverse consequences, now the question is if excessive smartphone use is truly an addiction (SA) or just problematic smartphone use (PSU). This review highlights that SA articles already assume that SA is an addiction and frame the research as such, while PSU articles explore reasons why PSU falls short of meeting the necessary criteria to be considered a true addiction. The growing literature on excessive smartphone use has conceptualized the disorder as an addictive behavior. Specifically, advocates of PSU argue that the ethiopathological pathways and processes have not yet been identified in SA research, suggesting that SA interventions are simply targeting the symptoms rather than the underlying causes [66]. We agree that research on SA etiology is necessary in order for the disorder to be properly and accurately diagnosed. Although that will take a considerable amount of time and effort to accrue, hopefully our future capabilities to capture sound psychosocial as well as neurobiological evidence will be established. There has also been criticism that the only support to date of SA being an addiction is limited to "exploratory studies relying on self-report data which is collected via convenience sampl[ing]" [4]. Although the majority of recent reports have been subjective, there have been attempts to collect objective data that are promising, especially if those studies are planning on following up on the data in the future. It seems more studies are also trying to carry out randomized studies with larger sample sizes in order to expand on the empirical data available in the field of SA research.

The current review found that all of the assessment scales measuring SA/PSU met at least half of Griffiths' criteria, which need to be met in order for a disorder to be considered an addiction, according to Griffiths. However, PSU advocates counter that conforming excessive smartphone use within addiction models, such as Griffiths', could oversimplify the disorder and result in clinical irrelevance [66]. Specifically, attempts at conceptualizing tolerance with respect to SA may be insufficient, as inferring tolerance based on the increasing use of the smartphone could vary by several factors, such as age (e.g., teens pressured by peers to participate in social media use), subscription status (subscriptions to apps are paid in full or need to be paid monthly), relationship status (single versus in a relationship), occupation status (student, works desk job), and significant life events (starting or ending a romantic relationship) [4]. Although it could indeed oversimply the condition, Griffiths' model has been widely used as a biopsychosocial framework to operationalize addictive components, and so it is a good starting point to conceptualize the level of addiction to the smartphone. Also, once again, in order to truly verify conceptualizations like tolerance, there is a need for neurobiological evidence (e.g., "alteration/sensitization in specific cerebral circuitries" [4]) to confirm tolerance levels increasing in a smartphone user.

We considered using the "Surgeon General's criteria" (the "Hill criteria") for causality, noting the: (1) Consistency, (2) strength, (3) specificity, (4) temporal relationship, and (5) coherence of the association between SA/PSU and the variables of interest. However, we chose not to include consistency, as none of the studies conducted follow-up studies in different populations under different circumstances, or specificity, which researchers have criticized to be useless or misleading [67]. However, the vast majority of the studies only met the coherence criteria. Additionally, since all but two studies were longitudinal, temporality could not be assessed. Future studies could consider addressing the consistency, strength, and temporal relationship of the association by conducting longitudinal and follow-up studies, with adequate sample sizes ensuring power in the analyses.

4.1.2. Assuming SA Is an Addiction, Is It an Addiction to the Device or on the Device?

What used to be a novelty-type activity about a decade ago has now become more of a normative behavior [68]; excessive smartphone use is one of the more recent forms of human–machine interaction, raising public health alarms. However, the current review notes that despite the growing body of research, many studies still do not clearly define SA. In fact, in some of the literature, SA and PSU are still interchangeable terms. To add to the ambiguity, only about one in five articles make the distinction that SA is an addiction specifically to the mobile features provided by the smartphone (texting; various social media apps, such as Twitter and Instagram) that desktops and even laptops cannot match in terms of ease of portability and handheld capabilities. The vast majority of recent articles did not make a clear distinction of SA from related addictions, such as Internet or gaming addiction, suggesting that researchers may assume that SA is actually a subset of technological addictions. Those articles focused more on the mediating role of associated variables (e.g., emotional intelligence and coping style [48]) on SA and/or the moderating role of variables on the relation between risk factors of SA and SA (e.g., whether perceived social support and depression would moderate the relationship between sensation seeking and adolescent SA [49]) rather than differentiating SA from other technological addictions or specifying that SA is an addiction distinct from other addictions (e.g., gaming, gambling, shopping, socializing, sex) based on what the smartphone offers users.

We believe that the only difference between smartphone and Internet addiction (IA) is that SA is essentially IA presented through a highly portable device. IA is a potential addiction restricted to a stationary type of device. In contrast, excessive smartphone use is more prevalent due to its ease of access and portability, and the smartphone itself has become the most common device of choice for people today to access the Internet. Currently, the smartphone is the medium of problematic overuse of apps, games, and social networking site interactions. In the future, perhaps a new behavioral addiction will be virtual reality addiction through use of high-tech contact lens—the ever-changing and improving technological advancements will make possibilities (and potential problems) endless. However, for today, we argue that SA is an emerging addiction to the smartphone content specifically through the smartphone device.

4.1.3. Generalizability of Results and At-Risk Populations

The emergence of SA as an addiction is highlighted by the fact that people from various countries are affected by SA, which is not localized to just one continent but worldwide. Although several factors (e.g., not all sociodemographic characteristics of samples being measured, biases, cross-sectional limitations of inferencing causality, insufficient sample sizes, inability to determine directionality of SA and its risk factors, attrition) did not allow for generalization of the findings outside of the study group population, the results still suggested that children and young adults are more affected by SA than other age groups and that men and women are affected by SA in different ways.

In terms of SA affecting youth, one study reported that those at risk of developing SA displayed more severe levels of behavioral and emotional problems, lower self-esteem, and poorer quality of communication with their parents compared with those at normal risk [45]. In terms of gender differences, one study observed that girls who frequently use their smartphones may have a greater tendency to use social networking apps (e.g., Facebook) to upload pictures/share their lives online and therefore have a higher degree of smartphone attachment as compared to boys. Girls have also been reported to form and maintain social relationships and be engaged emotionally through constant app connection while boys mostly use smartphones to communicate through texts [36]. However, not all people at risk fall neatly into those observed categories. Implications of these differences require further research among similar group populations (e.g., youths, females versus males, students) around the world to better understand age, gender, and cultural variations in SA that may exist, which could help guide future tailored interventions.

4.2. Limitations

For this review, we narrowed our article pool by searching for only "smartphone"/"mobile phone" and "addiction"/"problematic use" terms. We did not actively search for "android", "iPhone", "cell phone", "cellular device", "compulsive use", "phubbing", "snubbing", and "nomophobia", which could have potentially made us miss articles relevant to this review. However, we were able to locate a few relevant articles after reviewing the reference sections of selected articles. Much more research is needed that investigates differences as a function of demographic variables, such as gender, age, or ethnicity. Another limitation is that our interpretations of categorizing certain variables into their respective groupings may not match others' opinions. For example, while we may propose that a certain subjective report is suggestive of a causal relationship with SA/PSU, another researcher may disagree and say that it is insufficient evidence. Another example would be missing a guiding theory used in articles, making Table 3 incomplete.

5. Conclusions

Most studies to date seem to assume that SA is a valid behavioral addiction, many forming their assumptions based on Griffiths' component model of addiction, and framing their study based on that assumption. The interchangeable use of the terms SA and PSU, inconsistent methodological approaches used to study SA (e.g., varied use of SA/PSU scales among research), lack of standardized diagnostic criteria, and unclear distinctions of SA from other related addictions make it difficult to make a conclusive statement on the status of SA, which could be considered "an ill-defined and heterogeneous construct" [67]. With no unifying theory on SA, all theories mentioned in the current research highlight the complexity of SA: One theory cannot simply explain SA, but rather several theories and models possibly need to be integrated to better explain its distinct addictive traits in this new technologically advanced era. Much more research is needed to confirm the uniqueness of SA, which encompasses the addictive activities engaged in on the smartphone, which includes apps that are not available on other devices. It is most plausible, based on the current studies, to infer that SA falls on a continuum of additive behaviors, from mild PSU to more extreme addictive behavior, where the consequences need to be addressed, prevented, and potentially treated before the adverse health effects debilitate the smartphone user.

Author Contributions: Conceptualization, methodology, analysis, writing—original draft preparation, S.Y.; writing—Review and editing, supervision, S.S. All authors have read and agreed to the published version of the manuscript.

Funding: This research received no external funding.

Acknowledgments: We would like to thank Dmitri Rozgonjuk and Regina van den Eijnden for their support on this paper.

Conflicts of Interest: The authors declare no conflict of interest.

References

1. Elhai, J.D.; Dvorak, R.D.; Levine, J.C.; Hall, B.J. Problematic smartphone use: A conceptual overview and systematic review of relations with anxiety and depression psychopathology. *J. Affect. Disord.* **2017**, *207*, 251–259. [CrossRef] [PubMed]
2. Sapacz, M.; Rockman, G.; Clark, J. Are we addicted to our cell phones? *Comput. Hum. Behav.* **2016**, *57*, 153–159. [CrossRef]

3. Bianchi, A.; Phillips, J.G. Psychological predictors of problem mobile phone use. *Cyberpsychol. Behav.* **2005**, *8*, 39–51. [CrossRef] [PubMed]
4. Billieux, J.; Maurage, P.; Lopez-Fernandez, O.; Kuss, D.J.; Griffiths, M.D. Can Disordered Mobile Phone Use Be Considered a Behavioral Addiction? An Update on Current Evidence and a Comprehensive Model for Future Research. *Curr. Addict. Rep.* **2015**, *2*, 156–162. [CrossRef]
5. Choi, S.W.; Kim, D.J.; Choi, J.S.; Ahn, H.; Choi, E.J.; Song, W.Y.; Kim, S.; Youn, H. Comparison of risk and protective factors associated with smartphone addiction and Internet addiction. *J. Behav. Addict.* **2015**, *4*, 308–314. [CrossRef]
6. American Psychiatric Association. *Diagnostic and Statistical Manual of Mental Disorders*, 5th ed.; American Psychiatric Association: Arlington, VA, USA, 2013.
7. ICD-11 for Mortality and Morbidity Statistics. Available online: https://icd.who.int/browse11/l-m/en (accessed on 1 October 2019).
8. Sussman, S. *Substance and Behavioral Addictions: Concepts, Causes, and Cures*, 1st ed.; Cambridge University Press: Cambridge, UK, 2017.
9. Sussman, S.; Sussman, A.N. Considering the Definition of Addiction. *Int. J. Environ. Res. Public Health* **2011**, *8*, 4025–4038. [CrossRef]
10. Panova, T.; Carbonell, X. Is smartphone addiction really an addiction? *J. Behav. Addict.* **2018**, *7*, 252–259. [CrossRef]
11. Sussman, S.; Rozgonjuk, D.; van den Eijnden, R.J.J.M. Substance and behavioral addictions may share a similar underlying process of dysregulation. *Addiction* **2017**, *112*, 1717–1718. [CrossRef]
12. Alhassan, A.A.; Alqadhib, E.M.; Taha, N.W.; Alahmari, R.A.; Salam, M.; Almutairi, A.F. The relationship between addiction to smartphone usage and depression among adults: A cross sectional study. *BMC Psychiatry* **2018**, *18*, 148. [CrossRef]
13. Chen, C.; Zhang, K.; Gong, X.; Zhao, S.; Lee, M.; Liang, L. Examining the effects of motives and gender differences on smartphone addiction. *Comput. Hum. Behav.* **2017**, *75*, 891–902. [CrossRef]
14. Chun, J.S. Conceptualizing effective interventions for smartphone addiction among Korean female adolescents. *Child. Youth Serv. Rev.* **2018**, *84*, 35–39. [CrossRef]
15. Elserty, N.S.; Helmy, N.A.; Mounir, K.M. Smartphone addiction and its relation to musculoskeletal pain in Egyptian physical therapy students. *Eur. J. Physiother.* **2018**, 1–9. [CrossRef]
16. Hao, Z.; Jin, L.; Li, Y.; Akram, H.; Saeed, M.F.; Ma, J.; Ma, H.; Huang, J. Alexithymia and mobile phone addiction in Chinese undergraduate students: The roles of mobile phone use patterns. *Comput. Hum. Behav.* **2019**, *97*, 51–59. [CrossRef]
17. Jeong, Y.J.; Suh, B.; Gweon, G. Is smartphone addiction different from Internet addiction? comparison of addiction-risk factors among adolescents. *Behav. Inf. Technol.* **2019**, 1–16. [CrossRef]
18. Jin, Y.; Sun, C.; An, J.; Li, J. Attachment Styles and Smartphone Addiction in Chinese College Students: The Mediating Roles of Dysfunctional Attitudes and Self-Esteem. *Int. J. Ment. Health Addict.* **2017**, *15*, 1122–1134. [CrossRef]
19. Jo, H.; Na, E.; Kim, D.J. The relationship between smartphone addiction predisposition and impulsivity among Korean smartphone users. *Addict. Res. Theory* **2017**, *26*, 1–8. [CrossRef]
20. Konan, N.; Çelik, O.T. The Mediator Role of Interaction Anxiety in the Relationship between Social Support Perception and Smartphone Addiction. *J. Educ. Future* **2019**, *15*, 63–75.
21. Kwan, H.; Leung, M. The Structural Model in Parenting Style, Attachment Style, Self-regulation and Self-esteem for Smartphone Addiction. *J. Psychol. Behav. Sci.* **2017**, *3*, 85–103. [CrossRef]
22. Sekhon, A. Gender differences in mobile phone addiction and its association with stress among medical students. *Indian J. Health Wellbeing* **2018**, *9*, 232–234.
23. Yang, Z.; Asbury, K.; Griffiths, M.D. An Exploration of Problematic Smartphone Use among Chinese University Students: Associations with Academic Anxiety, Academic Procrastination, Self-Regulation and Subjective Wellbeing. *Int. J. Ment. Health Addict.* **2019**, *17*, 596–614. [CrossRef]
24. Arnavut, A.; Nuri, C.; Direktör, C. Examination of the relationship between phone usage and smartphone addiction based on certain variables. *An. Psicol.* **2018**, *34*, 446–450. [CrossRef]
25. Barnes, S.; Pressey, A.; Scornavacca, E. Mobile Ubiquity: Understanding the Impact of Cognitive Absorption on Smartphone Addiction. *Comput. Hum. Behav.* **2018**, *90*, 246–258. [CrossRef]

26. Chen, B.; Liu, F.; Ding, S.; Ying, X.; Wang, L.; Wen, Y. Gender differences in factors associated with smartphone addiction: A cross-sectional study among medical college students. *BMC Psychiatry* **2017**, *17*, 1–9. [CrossRef] [PubMed]
27. Chiang, J.T.; Chang, F.C.; Lee, K.W.; Hsu, S.Y. Transitions in smartphone addiction proneness among children: The effect of gender and use patterns. *PLoS ONE* **2019**, *14*, 1–12. [CrossRef]
28. Chung, J.E.; Choi, S.A.; Kim, K.T.; Yee, J.; Kim, J.H.; Seong, J.W.; Seong, J.M.; Kim, J.Y.; Lee, K.E.; Gwak, H.S. Smartphone addiction risk and daytime sleepiness in Korean adolescents: Smartphone addiction and sleep problems. *J. Paediatr. Child Health* **2018**, *54*, 800–806. [CrossRef]
29. Duke, E.; Montag, C. Smartphone addiction, daily interruptions and self-reported productivity. *Addict. Behav. Rep.* **2017**, *19*, 90–95. [CrossRef]
30. Kim, S.G.; Park, J.; Kim, H.T.; Pan, Z.; Lee, Y.; McIntyre, R.S. The relationship between smartphone addiction and symptoms of depression, anxiety, and attention-deficit/hyperactivity in South Korean adolescents. *Ann. Gen. Psychiatry* **2019**, *9*, 1–8. [CrossRef]
31. Kuang-Tsan, C.; Fu-Yuan, H. Study on Relationship Among University Students' Life Stress, Smart Mobile Phone Addiction, and Life Satisfaction. *J. Adult Dev.* **2017**, *24*, 109–118. [CrossRef]
32. Kuss, D.J.; Kanjo, E.; Crook-Rumsey, M.; Kibowski, F.; Wang, G.Y.; Sumich, A. Problematic Mobile Phone Use and Addiction Across Generations: The Roles of Psychopathological Symptoms and Smartphone Use. *J. Technol. Behav. Sci.* **2018**, *3*, 141–149. [CrossRef]
33. Lee, H.; Kim, J.W.; Choi, T.Y. Risk Factors for Smartphone Addiction in Korean Adolescents: Smartphone Use Patterns. *J. Korean Med. Sci.* **2017**, *32*, 1674–1679. [CrossRef]
34. Lee, C.; Lee, S.J. Prevalence and predictors of smartphone addiction proneness among Korean adolescents. *Child. Youth Serv. Rev.* **2017**, *77*, 10–17. [CrossRef]
35. Nayak, J.K. Relationship among smartphone usage, addiction, academic performance and the moderating role of gender: A study of higher education students in India. *Comput. Educ.* **2018**, *123*, 164–173. [CrossRef]
36. Yang, S.Y.; Lin, C.Y.; Huang, Y.C.; Chang, J.H. Gender differences in the association of smartphone use with the vitality and mental health of adolescent students. *J. Am. Coll. Health* **2018**, *66*, 693–701. [CrossRef] [PubMed]
37. Abed, S.N.; Abd, R.K.; Salim, I.D.; Jamal, N.A.R. Health problems of Mobile Phone Addiction for sample of students and their health awareness at institute technical of kut. *J. Pharm. Sci. Res.* **2018**, *10*, 412–415.
38. Alavi, S.S.; Ghanizadeh, M.; Mohammadi, M.R.; Kalhory, S.M.; Jannatifard, F.; Sepahbodi, G. The Survey personal and national identity between individual with cell phone addiction disorder and normal smartphone users. *Iran. J. Psychiatry* **2018**, *13*, 15–21.
39. Chen, C.Y. Smartphone addiction: Psychological and social factors predict the use and abuse of a social mobile application. *Inf. Commun. Soc.* **2018**, 1–14. [CrossRef]
40. Kim, J.H. Smartphone-mediated communication vs. face-to-face interaction: Two routes to social support and problematic use of smartphone. *Comput. Hum. Behav.* **2017**, *67*, 282–291. [CrossRef]
41. Kim, H.J.; Min, J.Y.; Min, K.B.; Lee, T.J.; Yoo, S. Relationship among family environment, self-control, friendship quality, and adolescents' smartphone addiction in South Korea: Findings from nationwide data. *PLoS ONE* **2018**, *13*, 1–13. [CrossRef]
42. Kim, Y.; Lee, N.; Lim, Y. Gender differences in the association of smartphone addiction with food group consumption among Korean adolescents. *Public Health* **2017**, *145*, 132–135. [CrossRef]
43. Kumcagiz, H. Quality of Life as a Predictor of Smartphone Addiction Risk Among Adolescents. *Technol. Knowl. Learn.* **2019**, *24*, 117–127. [CrossRef]
44. Lee, J.E.; Jang, S.I.; Ju, Y.J.; Kim, W.; Lee, H.J.; Park, E.C. Relationship between Mobile Phone Addiction and the Incidence of Poor and Short Sleep among Korean Adolescents: A Longitudinal Study of the Korean Children & Youth Panel Survey. *J. Korean Med. Sci.* **2017**, *32*, 1166–1172. [CrossRef] [PubMed]
45. Lee, J.; Sung, M.J.; Song, S.H.; Lee, Y.M.; Lee, J.J.; Cho, S.M.; Park, M.K.; Shin, Y.M. Psychological Factors Associated with Smartphone Addiction in South Korean Adolescents. *J. Early Adolesc.* **2018**, *38*, 288–302. [CrossRef]
46. Mahapatra, S. Smartphone addiction and associated consequences: Role of loneliness and self-regulation. *Behav. Inf. Technol* **2019**, *38*, 833–844. [CrossRef]

47. Sok, S.R.; Seong, M.H.; Ryu, M.H. Differences of Self-Control, Daily Life Stress, and Communication Skills between Smartphone Addiction Risk Group and General Group in Korean Nursing Students. *Psychiatr. Q* **2019**, *90*, 1–9. [CrossRef] [PubMed]
48. Sun, J.; Liu, Q.; Yu, S. Child neglect, psychological abuse and smartphone addictionamong Chinese adolescents: The roles of emotional intelligence and coping style. *Comput. Hum. Behav.* **2019**, *90*, 74–83. [CrossRef]
49. Wang, P.; Lei, L.; Wang, X.; Nie, J.; Chu, X.; Jin, S. The exacerbating role of perceived social support and the "buffering" role of depression in the relation between sensation seeking and adolescent smartphone addiction. *Personal. Indivd Differ.* **2018**, *130*, 129–134. [CrossRef]
50. Wang, P.; Zhao, M.; Wang, X.; Xie, X.; Wang, Y.; Lei, L. Peer relationship and adolescent smartphone addiction: The mediating role of self-esteem and the moderating role of the need to belong. *J. Behav. Addict.* **2017**, *6*, 708–717. [CrossRef]
51. Xu, T.T.; Wang, H.Z.; Fonseca, W.; Zimmerman, M.A.; Rost, D.H.; Gaskin, J.; Wang, J.L. The relationship between academic stress and adolescents' problematic smartphone usage. *Addict. Res. Theory* **2019**, *27*, 162–169. [CrossRef]
52. Yang, X.; Zhou, Z.; Liu, Q.; Fan, C. Mobile Phone Addiction and Adolescents' Anxiety and Depression: The Moderating Role of Mindfulness. *J. Child Fam. Stud.* **2019**, *28*, 822–830. [CrossRef]
53. Yildiz Durak, H. Investigation of nomophobia and smartphone addiction predictors among adolescents in Turkey: Demographic variables and academic performance. *Soc. Sci. J.* **2018**, in press. [CrossRef]
54. Yıldız, M. Emotion regulation strategies as predictors of internet addiction and smartphone addiction in adolescents. *J Educ. Sci. Psychol.* **2017**, *7*, 66–78.
55. Youn, H.; Lee, S.I.; Lee, S.H.; Kim, J.Y.; Kim, J.H.; Park, E.J.; Park, J.S.; Bhang, S.Y.; Lee, M.S.; Lee, Y.J.; et al. Exploring the Differences between Adolescents' and Parents' Ratings on Adolescents' Smartphone Addiction. *J. Korean Med. Sci.* **2018**, *33*, 1–11. [CrossRef] [PubMed]
56. Moher, D.; Liberati, A.; Tetzlaff, J.; Altman, D.G.; PRISMA Group. Preferred reporting items for systematic reviews and meta-analyses: The PRISMA statement. *PLoS Med.* **2009**, *6*, 1–6. [CrossRef]
57. Griffiths, M. A 'components' model of addiction within a biopsychosocial framework. *J. Subst. Use* **2005**, *10*, 191–197. [CrossRef]
58. Office of the Surgeon General (US); Office on Smoking and Health (US). *The Health Consequences of Smoking: A Report of the Surgeon General*; Centers for Disease Control and Prevention (US): Atlanta, GA, USA, 2004.
59. Kwon, M.; Lee, J.Y.; Won, W.Y.; Park, J.W.; Min, J.A.; Hahn, C.; Gu, X.Y.; Choi, J.H.; Kim, D.J. Development and Validation of a Smartphone Addiction Scale (SAS). *PLoS ONE* **2013**, *8*, 1–7. [CrossRef] [PubMed]
60. Lin, Y.H.; Chang, L.R.; Lee, Y.H.; Tseng, H.W.; Kuo, T.B.; Chen, S.H. Development and validation of the Smartphone Addiction Inventory (SPAI). *PLoS ONE* **2014**, *9*, 1–5. [CrossRef] [PubMed]
61. Kim, J.H. Psychological issues and problematic use of smartphone: ADHD's moderating role in the associations among loneliness, need for social assurance, need for immediate connection, and problematic use of smartphone. *Comput. Hum. Behav.* **2018**, *80*, 390–398. [CrossRef]
62. Cocoradă, E.; Maican, C.I.; Cazan, A.M.; Maican, M.A. Assessing the smartphone addiction risk and its associations with personality traits among adolescents. *Child. Youth Serv. Rev.* **2018**, *93*, 345–354. [CrossRef]
63. Elhai, J.D.; Levine, J.C.; Dvorak, R.D.; Hall, B.J. Non-social features of smartphone use are most related to depression, anxiety and problematic smartphone use. *Comput. Hum. Behav.* **2017**, *69*, 75–82. [CrossRef]
64. Herrero, J.; Torres, A.; Vivas, P.; Urueña, A. Smartphone Addiction and Social Support: A Three-year Longitudinal Study. *Psychosoc. Interv.* **2019**, *28*, 111–118. [CrossRef]
65. Herrero, J.; Urueña, A.; Torres, A.; Hidalgo, A. Socially Connected but Still Isolated: Smartphone Addiction Decreases Social Support Over Time. *Soc. Sci. Comput. Rev.* **2017**, *37*, 73–88. [CrossRef]
66. Billieux, J.; Philippot, P.; Schmid, C.; Maurage, P.; De Mol, J.; Van der Linden, M. Is Dysfunctional Use of the Mobile Phone a Behavioural Addiction? Confronting Symptom-Based Versus Process-Based Approaches. *Clin. Psychol. Psychother.* **2015**, *22*, 460–468. [CrossRef] [PubMed]

67. Rothman, K.J.; Greenland, S. *Hill's Criteria for Causality. Encyclopedia of Biostatistics, 2nd ed*; John Wiley & Sons, Ltd.: Hoboken, NJ, USA, 2005. [CrossRef]
68. Pinna, F.; Dell'Osso, B.; Di Nicola, M.; Janiri, L.; Altamura, A.C.; Carpiniello, B.; Hollander, E. Behavioural addictions and the transition from DSM-IV-TR to DSM-5. *J. Psychopathol.* **2015**, *21*, 380–389.

International Journal of
Environmental Research and Public Health

MDPI

Article

Comparison of Behavioral Changes and Brain Activity between Adolescents with Internet Gaming Disorder and Student Pro-Gamers

Ki Hyeon Kwak, Hyun Chan Hwang, Sun Mi Kim and Doug Hyun Han *

Department of Psychiatry, Chung Ang University Hospital, 102 Heuk seok ro, Dongjak gu, Seoul 06973, Korea; letter2kh@naver.com (K.H.K.); mossflower28@naver.com (H.C.H.); cryslake@naver.com (S.M.K.)
* Correspondence: hduk70@gmail.com; Tel.: +82-2-6299-3132

Received: 23 October 2019; Accepted: 7 January 2020; Published: 9 January 2020

Abstract: While pro-gamers play according to defined living habits and planned schedules, adolescents with internet gaming disorder (IGD) exhibit irregular lifestyles and unregulated impulsive gaming behavior. Fourteen IGD adolescents and 12 pro-gaming students participated in this study. At baseline and after one year, demographic data, the Child Behavior Check List (CBCL), depressed mood, anxiety, and resting-state functional magnetic resonance imaging were assessed. Over the year, IGD adolescents played games as per their usual schedule, while pro-gamer students played according to their school's team schedule. After one year, the pro-gamers' scores had decreased in the CBCL-total (total problematic behaviors), CBCL-externalizing (under-controlled behavior, like impulsivity and aggression), and CBCL-internalizing (over-controlled behavior like depression and anxiety) compared to those of the IGD adolescents. Both groups displayed increased brain activity in the parietal lobe (a component of the attention network) over the years. Compared to pro-gamers, IGD adolescents showed higher brain activity within the left orbitofrontal cortex. Brain activity within the orbitofrontal cortex was associated with CBCL-externalizing scores. These results suggest that gaming had increased the attention network's brain activity, but a well-organized support system could lead to different results, in terms of improved behaviors and suppressing brain activity within the orbitofrontal cortex.

Keywords: internet gaming disorder; pro-gamers; Child Behavior Check List; resting-state functional magnetic resonance imaging

1. Introduction

1.1. Debates on Internet Gaming Disorder

Over the last two decades, many studies have suggested that excessive and problematic online gaming should be defined as a psychiatric disorder along with other addictive behaviors and functional impairments [1–3]. In a national survey of 1178 Americans aged between 8 and 18 years old, Gentile et al. [4] reported that patients with internet gaming disorder (IGD) performed poorly at school and suffered from attentional problems. Stockdale and Coyne [5] asserted that patients with IGD had poor mental health and cognitive functions, including poorer impulse control and attentional problems, compared to healthy controls. Hull et al. [6] reported that mature-rated, risk-glorifying gameplay was associated with substance use (including alcohol and cigarettes), aggression, delinquency, and risky sex. Since 2014, the World Health Organization (WHO) has regarded internet gaming addiction as a significant public health problem. In 2018, the International Classification of Diseases version 11 (ICD-11) defined gaming disorder as a medical disease with a pattern of repetitive or persistent gaming behavior [2,3].

Several studies have suggested that the psychobiological mechanisms underlying behavioral addiction resemble those of chemical addiction, such as that of alcohol and other drugs [7]. Moreover, it is believed that those with IGD share similar characteristics in neural activity and cognitive dysfunctions with those with a gambling disorder [8].

Based on previous correlational studies, the brain neurobiology of IGD can be summarized as increased activity within the orbitofrontal cortex and decreased activity within the dorsolateral prefrontal cortex [9,10]. Kim and Kang [9] found that IGD patients, compared to other gamers, showed a stronger functional connectivity (FC) within the orbitofrontal cortex, which is involved in motivational salience, and a decreased FC within the dorsolateral prefrontal cortex, which is involved in learning and attention. Additionally, Wang et al. [10] reported a decreased cortical thickness within the dorsolateral prefrontal cortex, related to cognitive control, decision making, and reward and loss processing. Motivational salience was thought to be associated with continuous and repetitive behaviors in various addictive diseases [11]. The dorsolateral prefrontal cortex was reported to control the orbitofrontal cortex [12]. In many addictive behaviors, the dorsolateral prefrontal cortex can fail to control the hyper-activated orbitofrontal cortex [12].

However, there is some debate about whether IGD or gaming disorder can be classified as a medical disease [1,13,14]. The American Psychiatric Association (APA) included IGD in Section III of the Diagnostic and Statistical Manual of mental disorders (DSM-5), as it requires further research and data accumulation, due to a lack of IGD cases, high prevalence of non-formal diagnostic criteria, and an unbalanced prevalence rate [1]. Several studies have also suggested that IGD might just be a social phenomenon induced by environmental and social stress [13,14]. Jeong et al. [13] suggested that adolescents excessively played internet games when they were under academic stress or lost self-control. The period of adolescence was known as a particularly at-risk developmental stage for problematic internet game playing [15,16], because individuals were prone to impulsivity and uncontrolled, unplanned internet game play [16,17]. Large, accumulating literature suggests that the prevalence of IGD is dominant in the male population [16]. The initial onset of IGD is not associated with gaming time or the game's genre [13]. Pang et al. [14] reported that gaming motivation in adolescents with IGD was associated with social anxiety and psychological stress.

Several studies expressed opposing opinions regarding the cognitive dysfunction induced by internet gameplay [18–20]. Latham et al. [18] asserted that video gaming could result in extensive improvements in various cognitive functions. Pallavicini et al. [20] suggested that video games could be used as a tool to improve the patient's wellbeing through cognitive and emotional training. Mentiplay et al. [19] also showed that video games could help improve symptoms of patients with developmental coordination disorders. Ballesteros et al. [21] suggested that the working memory and selective attention of older adults could be improved through video games.

1.2. Pro-Gamers vs. Patients with IGD

It's already known that IGD patients not only display excessive internet gameplay, but also have unplanned and irregular lifestyles, and impulsive, unregulated behavior in the absence of support systems [5,22,23]. Kim et al. [24] reported that most IGD patients were unable to control their impulses and displayed delinquent behavior. In a two-year follow up study, Baysak et al. [22] reported that social support systems could be a protective factor for IGD.

There is another population in South Korea that also displays excessive internet gameplay but does not meet the criteria of IGD, called professional gamers (pro-gamers). Pro-gamers belong to the internet game league and have contracts with teams that result in a salary and potential prizes. To perform well, pro-gamers practice for about 10 h a day within a defined schedule [25]. This schedule includes practicing, physical exercise, team strategy conferences, resting, and mealtimes.

For several years, A-hyun High School has recruited students who want to become professional gamers. Before going to A-hyun High School, all adolescents must complete the second grade at a general high school. In their schooling, they have a similar curriculum to third grade general high school students, such as Korean, English, mathematics, and sciences. In addition, they also play internet games for 4–5 h a day, following a schedule the game teacher suggests.

1.3. Resting-State Functional Magnetic Resonance Imaging (MRI) and Fractional Amplitude of Low-Frequency Fluctuation

Resting-state functional MRI (rs-fMRI) is thought to measure spontaneous brain activity, representing brain function [26]. Low-frequency (0.009–0.08 Hz) fluctuations (ALFF) of the blood oxygen level-dependent signal were thought to be related to spontaneous neural activity in rs-fMRI [27]. The fractional amplitude of low-frequency fluctuations (fALFF) is an advanced version of the original ALFF and helps to detect spontaneous brain activity more sensitively [28]. Changes in the fALFF have already been reported in several studies of psychiatric diseases, including schizophrenia [29], autism [30], and attention deficit hyperactivity disorders (ADHD) [31]. Moreover, Kim et al. [32] observed correlations between changes of fALFF within the inferior frontal gyrus and changes in delinquency and externalizing behaviors.

1.4. Hypothesis

We hypothesized that long term internet game play would increase the brain activity within the attentional system in both groups. However, the existence of a support system, including a regular schedule and a supervisor, would lead to different results in terms of behaviors and brain activity. Student pro-gamers with a good support system would show improved behavioral scores compared to IGD patients. In addition, a good support system would prevent the hyperactivity within the orbitofrontal cortex in student pro-gamers, while a poor support system would not prevent hyperactivity within the orbitofrontal cortex in response to impulsive internet game play.

2. Methods

2.1. Participants

The participants in the current study were classified into two groups; pro-gamer students and IGD adolescents. The two groups were both engaged in excessive internet game play. However, the pro-gaming students with a support system had characteristics of a regular lifestyle, while IGD adolescents, without a support system, showed an irregular lifestyle.

The institutional review board of the Chung-Ang University Hospital approved the research protocol for this study. All adolescents were informed about the study's procedures and signed a written informed consent form. Their parents also provided written informed consent. The diagnostic criteria of IGD were based on the DSM-5 [1].

From September 2016 to December 2017, 121 IGD adolescents visited the Department of Psychiatry at the OO University hospital for diagnosis and treatment. In contrast to the systematic caring group (student pro-gamer group), IGD patients who visited for an initial assessment but received no treatment (non-systematic caring group) were regarded as the compared group.

Of the 121 IGD adolescents, we found 27 adolescents who completed psychological and brain-imaging assessments at their first and second visits, but they had not received any treatment or interventions, such as cognitive behavior therapy or psychiatric medications, over the years. Although parents and caretakers had asked that the adolescents receive treatment for IGD, those adolescents refused treatment due to no interest in treatment, no insight of problematic behaviors, and laziness to come to the treatment center. On January 2018, we contacted 27 adolescents via phone to introduce our study, and 14 adolescents and their parents agreed to participate. The other 13 adolescents or their parents were not willing or able to participate in study.

In 2017, 55 adolescents who wanted to be professional gamers applied to A-hyun High School's Pro-Gamer Department. With a ranking from the internet game "League of Legend" competition and a basic academic ability test, 12 adolescents qualified and were accepted to be admitted to the school.

After listening to the purpose of our research, all pro-gaming students and their parents agreed to participate in our study.

The pro-gamer students had school schedules, including regular academic classes (4 h/day), physical sports class, strategy meetings, mealtimes, and game training time (3 h/day in school). Two teachers managed and checked the schedules, while the parents of the IGD adolescents observed the life patterns of the adolescents and reported it. At a baseline and after a year, both groups were asked to give their demographic data, including age, educational year, and internet gameplay time ("How many hours per day do you play internet game?") as well as a number of psychological scales, including the Young Internet Addiction Scale (YIAS), Child Behavior Checklist (CBCL), Child Depressive Inventory (CDI), Beck Anxiety Inventory (BAI), and Korean ADHD Rating Scale (K-ARS). Resting-state functional MRI (rs-fMRI) was also undertaken.

2.2. Clinical Scales

The CBCL is known as a screening tool for assessing problem behaviors in children and adolescents based on the parent's self-report [33,34]. The Korean version of the CBCL (K-CBCL) has standardized reliability and validity [33,34]. Parents assessed their children and adolescents, aged between 4 and 18, using the K-CBCL, in terms of social adaptation and problem behavior. It consisted of 117 questions, with three subscales including a total problem score, as well as externalizing and internalizing scores. Higher scores indicated a greater degree of behavioral and emotional problems [33,34]

The Young Internet Addiction Scale (YIAS), proposed by Young in 1998, is a self-reporting measure for routine internet use [35]. The YIAS consists of 20 self-assessment questions, each graded on a scale of 1 to 5 ("rarely" to "always"). YIAS scores above 50 are considered to reflect problematic internet use. The Korean version of YIAS was verified by Lee et al. The YIAS' internal consistency has been reported to be in the range of 0.90 to 0.91 [36].

The Children's Depression Inventory (CDI), developed in 1977 by Kovacs (1985), is a self-reported measure of depression in children and adolescents aged 7 to 17 years old [37]. The 27 items of the Korean version of CDI, with internal consistency of Cronbach's $\alpha = 0.88$, was verified by Cho and Lee [38].

The Beck Anxiety Inventory (BAI), with 21 questions, is used to measure anxiety severity [39]. The BAI is scored on a scale of 0 to 3 and has a maximum score of 63 points. The Korean version of the BAI, with an internal consistency of Cronbach's $\alpha = 0.93$, was verified by Kwon et al. [40].

The Korean ADHD Rating Scale (K-ARS) is an ADHD symptom severity scale composed of 18 items (9 items for inattentive evaluation and 9 items for hyperactivity evaluation) designed by Dupaul [41]. The Korean version of the ARS, with an internal consistency of 0.77 to 0.89, has been verified by So et al. [42].

2.3. Brain Image Acquisition and Processing

All MRIs were acquired using a 3.0 T Philips Achieva scanner. All participants laid down with their eyes closed and were asked to stay awake. The heads of the participants were stabilized with cushions and taped for severe head movement prevention. Resting-state (Rs-fMRI) images were acquired axially, with an echo-planar imaging sequence, using the following parameters: TR/TE = 3000/40 ms, 40 slices, 64 × 64 matrix, 90° flip angle, 230 mm field of view (FOV), and 3 mm section thickness without a gap. Each scan lasted 720 s, and 230 volumes were obtained. The first 10 volumes were removed for gradient field stabilization.

Data preprocessing and processing were carried out using the Data Processing Assistant for Rs-fMRI (DPARSFA-http://www.restfmri.net), which is a plug-in software that works with Statistical Parametric Mapping (SPM12; http://www.fil.ion.ucl.ac.uk/spm/software/spm12/) and the Rs-fMRI Data Analysis Toolkit (REST; http://resting-fmri.sourceforge.net). Images were corrected for slice acquisition time differences, realigned, normalized, spatially smoothed with a 6 mm full-width half maximum kernel, de-trended, and temporally band-pass filtered to 0.01–0.08 Hz. Based on the results from the realignment processing by SPM, subjects that had a translation or a rotating motion greater than 3 mm or 2°, respectively, in any direction, were excluded from the study. No subject was excluded because of excessive head motion.

To assess brain activity among the groups at baseline, fALFF was performed using the REST software before treatment was administered. During preprocessing, Fisher-transformed correlation coefficients were measured for each pair of the regions of interest (ROIs) in each participant. The fALFF between the ROIs was calculated using the CONN-fMRI FC toolbox (version 15; https://www.nitrc.org/projects/conn). The fALFF method was used to find the regions where the local connectivity was correlated to the clinical scores. As an indicator of fALFF value, Kendall's coefficient of concordance of a given voxel was calculated using the surrounding 26 voxels to evaluate the similarity of the time series. These were then standardized using Z-scores to perform group analyses.

2.4. Statistics

Demographic and psychological data between the pro-gamer group and the IGD adolescent group were analyzed using the Mann–Whitney U test. The changes in the psychological scales were assessed with the Kruskal–Wallis test. The differences in the psychological scales' changes were also evaluated with an analysis of variance (ANOVA).

At baseline, the fALFF between the student pro-gamers and the IGD adolescents was compared using an independent *t*-test with the SPM12 software package. We performed a paired *t*-test, using the SPM12 software, to investigate fALFF changes in both the student pro-gamers and the IGD adolescents. Additionally, the difference in fALFF changes between the student pro-gamers and IGD adolescents were measured with an ANOVA using the SPM12 software package. The correlation was calculated between the fALFF map and the CBCL using SPM12. The resulting maps were set to a threshold using a *p*-value of <0.05, and false discovery rate correction was made for multiple comparisons with an extent of more than 20 contiguous voxels.

3. Results

3.1. Comparison of Demographic and Psychological Data

There were no significant differences in age, intelligence quotient (IQ), internet gaming time, CDI, BAI, K-ARS, CBCL-T, and CBCL-E scores between the student pro-gamer and IGD adolescent groups. However, the IGD adolescents showed increased YIAS and CBCL-I scores compared to the student pro-gamer group (Table 1).

After a year, no difference was seen in the YIAS ($F = 1.12$, $p = 0.30$), internet game playing time ($F = 0.62$, $p = 0.44$), CDI ($F = 3.50$, $p = 0.07$), BAI ($F = 0.02$, $p = 0.89$), and K-ARS ($F = 0.46$, $p = 0.51$) scores between the two groups. However, the CBCL-total scores ($F = 12.76$, $p < 0.01$) and CBCL-externalizing ($F = 19.81$, $p < 0.01$) and CBCL-internalizing ($F = 11.09$, $p < 0.01$) scores decreased in the student pro-gamer group but did not change in the IGD adolescent group (Figure 1).

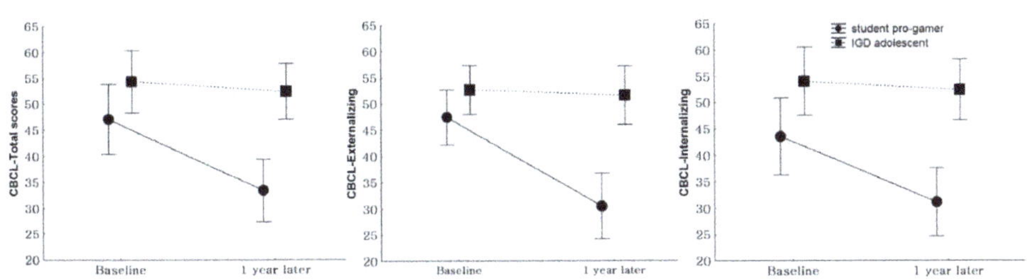

Figure 1. Comparison of the changes in the CBCL between student pro-gamer and IGD adolescent groups.

Table 1. Demographic and psychological data.

		Student Pro-Gamers	IGD Adolescents	Statistics
Age (years)		17.1 ± 0.3	16.5 ± 1.2	z = 1.07, p = 0.28
Education (years)		11.1 ± 0.3	9.7 ± 2.7	z = 1.63, p = 0.10
IQ		86.8 ± 6.3	90.4 ± 10.3	z = −1.29, p = 0.19
YIAS	B	56.3 ± 7.5	65.3 ± 7.5	z = −2.24, p = 0.02 *
	F	55.8 ± 13.0	61.4 ± 11.0	z = −1.46, p = 0.14
Game time (hours/day)	B	6.7 ± 1.4	7.0 ± 1.7	z = −0.24, p = 0.81
	F	7.2 ± 1.2	6.7 ± 2.0	z = 1.19, p = 0.23
CDI	B	7.8 ± 4.7	11.4 ± 5.6	z = −1.73, p = 0.08
	F	5.4 ± 3.9	12.2 ± 6.0	z = −3.10, p < 0.01 *
BAI	B	6.4 ± 2.9	7.9 ± 3.7	z = −1.07, p = 0.28
	F	6.2 ± 2.0	7.8 ± 3.3	z = −1.21, p = 0.81
K-ARS	B	12.5 ± 5.5	12.9 ± 5.8	z = −0.29, p = 0.22
	F	13.1 ± 3.8	12.7 ± 6.5	z = 0.22, p = 0.83
CBCL-T	B	47.1 ± 6.8	54.4 ± 13.9	z = −1.43, p = 0.15
	F	33.4 ± 8.9	52.5 ± 10.9	z = −3.53, p < 0.01 *
CBCL-E	B	47.4 ± 3.9	52.7 ± 11.3	z = −1.82, p = 0.07
	F	30.5 ± 7.8	51.6 ± 12.2	z = −3.81, p < 0.01 *
CBCL-I	B	43.6 ± 8.6	54.1 ± 14.4	z = −2.37, p = 0.02 *
	F	31.2 ± 7.6	52.4 ± 12.9	z = −3.93, p < 0.01 *

Notes: IGD adolescents: adolescents with internet gaming disorder (IGD), B: baseline, F: follow up; Young Internet Addiction Scale (YIAS), Child Behavior Checklist (CBCL), CBCL-T: total, CBCL-E: externalizing, CBCL-I: internalizing; Children's Depressive Inventory (CDI), Beck Anxiety Inventory (BAI), Korean ADHD Rating Scale (K-ARS). * Statistically significant.

The Children Behavior Check List (CBCL) total (F = 12.76, p < 0.01), CBCL-externalizing (F = 19.81, p < 0.01), and CBCL-internalizing (F = 11.09, p < 0.01) scores in student pro-gamers decreased while all CBCL scores in IGD adolescents were unchanged.

3.2. Comparison of the Changes in fALFF between Student Pro-Gamers and IGD Adolescents after a Year

At baseline, there were no regions with different brain activity at resting state between the IGD adolescents and the student pro-gamers.

After a year, both groups showed increased brain activity within the attention networks (parietal lobe). The details were as follows: The fALFF within the parietal lobe (x, y, z, 42, −66, 36, voxels = 26, T = 4.61, uncorrected p < 0.001) in the student pro-gamer group and the fALFF within the parietal lobe gyrus (x, y, z, 36, −24, 45, voxels = 25, T = 4.52, uncorrected p < 0.001) in the IGD adolescents increased.

Only in the IGD adolescents, orbitofrontal cortex activity increased after a year. The details are as follows: IGD adolescents showed an increased fALFF within the left orbitofrontal cortex, including the left subcallosal gyrus (x, y, z, −6, 12, −12, voxels = 121, T = 6.37, uncorrected p < 0.001), left orbital gyrus (x, y, z, −15, 33, −24, voxels = 121, T = 5.99, uncorrected p < 0.001), and left inferior frontal gyrus (x, y, z, −21, 27, −21, voxels = 121, T = 6.37, uncorrected p < 0.001), compared with those of the student pro-gamers (Figure 2).

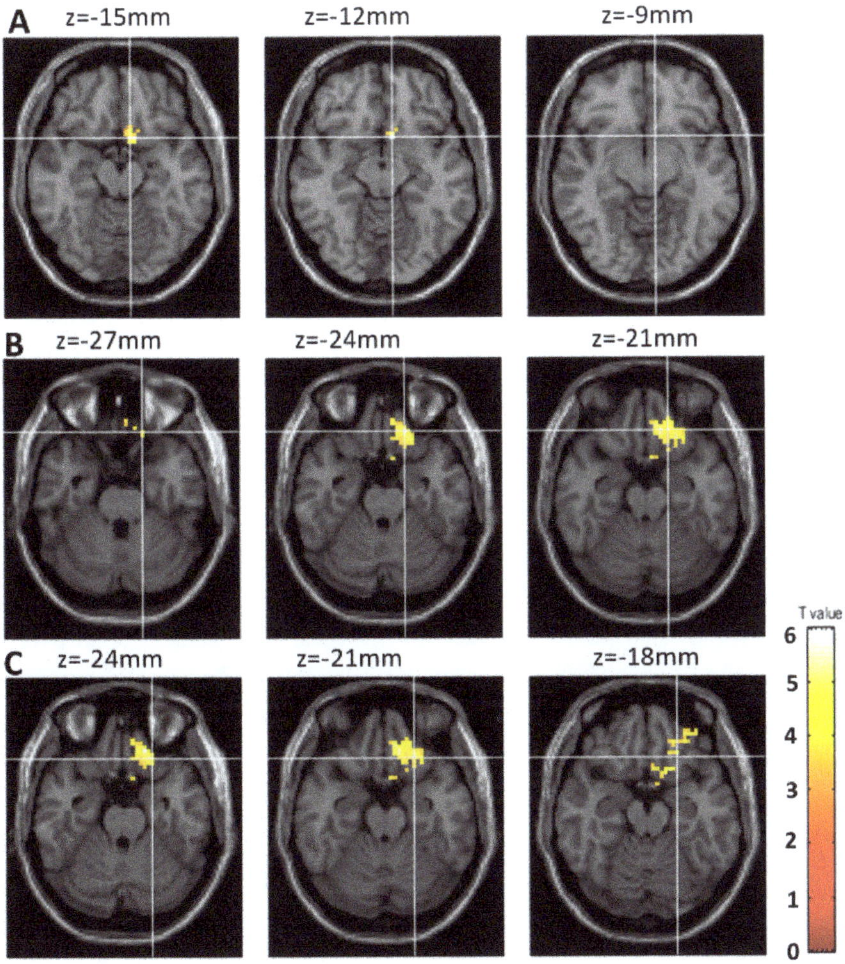

Figure 2. Regions showing differences in the changes of brain activity between the pro-gamer group and IGD adolescent group. (**A**) left subcallosal gyrus (x, y, z, −6, 12, −1), (**B**) left orbital gyrus (x, y, z, −15, 33, −24), (**C**) left inferior frontal gyrus (x, y, z, −21, 27, −21), yellow regions: the IGD adolescent group showed increased brain activity compared to the pro-gamers group.

3.3. Correlation between the fALFF and CBCL Scores in All Adolescents (Student Pro-Gamers and IGD Adolescents)

The fALFF values within the left inferior frontal gyrus were associated with the CBCL-externalizing scores in all adolescents ($r = 0.50$, $p < 0.01$). However, there were no significant correlations between the fALFF values within other areas and the CBCL-total or CBCL-internalizing scores.

4. Discussion

This study compared two groups under the same conditions of excessive internet gaming, one with a support system in place and one with no support system. Even though both groups played internet games for more than seven hours a day, it was assumed that the existence of a support system to keep a regular schedule would lead to different results in terms of behaviors and brain activity. The student pro-gamer group showed improved behavioral scores after a year compared to the baseline. However,

IGD adolescents showed no improvement in behavioral scores and the impulse control network showed dysfunctional brain activity.

4.1. Mproved Problematic Behavioral Scores in Student Pro-Gamers Compared to IGD Adolescents

At baseline, CBCL-internalizing scores in the IGD adolescents were higher than those observed in student pro-gamer group. In addition, the student pro-gamer group showed improved behavioral scores as assessed using the CBCL, compared to IGD adolescent. Moreover, the behavioral score improvements on the CBCL in student pro-gamers included both the CBCL-externalizing and internalizing scores. The CBCL was designed to measure the degree of behavioral and emotional problems in children [43]. The CBCL-externalizing score shows the degree of external problems, including social, thought, and attentional problems, while the CBCL-internalizing score shows the degree of internal problems, including anxiety, depression, and somatic complaints [43]. Additionally, the CBCL-total scores have been suggested as primary screening instruments for ADHD in Korean children [44]. The CBCL-internalizing scores were positively correlated with the Beck Depressive Inventory scores in patients with mood disorders [45]. Altogether, our CBCL results may suggest that the student pro-gamer group showed an improvement in their behavioral and emotional status. With the CBCL results at baseline and follow up, a different interpretation can be suggested. The pro-gamer group was preparing for a pro-gamer career, while the IGD adolescent group had no clear future direction. This situation may have led to the bias of the parental assessment in the CBCL.

Many studies have reported the effects of gaming on adolescent behaviors [46,47]. Greitmeyer et al. [46] suggested that video gaming would affect the social behaviors of gamers, especially when violent and pro-social video games are considered. Shao et al. [47] asserted a positive correlation between the use of violent video games and adolescent aggressiveness. However, our results may indicate that environmental factors and the existence of a support system affects the gamers' behaviors to a greater extent than the game play itself. In a large sample study of adolescents who play violent video games, Przybylski [48] reported that there was no significant correlation between violent video games and aggressive behavior in adolescents. In a survey conducted by the US National Research Council, environmental factors, including family resources and school quality, were seen as crucial factors in preventing mental, emotional, and behavioral disorders in young people [49].

4.2. Increased Brain Activity within the Attention Network (parietal lobe) in Response to One-Year Internet Gameplay in Both Groups

After playing internet games over a year, brain activities within the parietal lobe in both groups increased. The parietal lobe is known as a part of the attention network in the human brain [50]. In previous IGD studies, gaming was seen to affect brain activity within the attention network [25,51,52]. Action video gamers made faster and more precise responses toward targets, using enhanced attention skills [52]. Compared to non-gamers, the frontoparietal network in gamers was used to a greater extent in responding to attention-demanding tasks [51]. Interestingly, repetitive and impulsive internet gaming was thought to be a self-medication for children with ADHD [53,54]. Evren et al. [53] have reported that ADHD was associated with the severity of internet addiction and IGD among university students. Eight weeks of methylphenidate treatment in ADHD children with internet addiction improves ADHD symptoms as well as the severity of internet addiction [55]. Conclusively, increased brain activity within the parietal lobe in both student pro-gamers and the IGD group was associated with internet gaming.

4.3. Increased Brain Activity within the Orbitofrontal Cortex of the IGD Adolescents in Response to a Year of Internet Gaming

Increased brain activity, represented by the fALFF value, within the orbitofrontal cortex was only observed in IGD adolescents in this study. The imbalance between the dorsolateral prefrontal cortex and the orbitofrontal cortex, due to a hyper-activated orbitofrontal cortex, was thought to be related to cognitive control and decision-making dysfunctions, as well as impulsive behaviors in patients with gambling disorders [55]. In the seed base FC analysis, Kim et al. [24] reported the imbalance of FC from the orbitofrontal cortex to the dorsolateral prefrontal cortex in IGD patients. In internet addiction, functional dis-connectivity between the orbitofrontal cortex, dorsolateral prefrontal cortex, and anterior cingulate, due to hyper-activated orbitofrontal cortex, are associated with impulsive and delinquent behaviors [56].

4.4. Association between fALFF Values within the Orbitofrontal and the CBCL-Externalizing Scores

The association between impulsivity, drug seeking behaviors, and dysregulation of the orbitofrontal cortex has been continuously reported in substance use disorder [57,58]. The hyper-activated orbitofrontal cortex might aid to balance the FCs between the dorsolateral prefrontal cortex and the orbitofrontal cortex [11,12].

CBCL-externalizing scores were thought to be associated with impulsive behaviors in children and adolescents [44]. Eisenberg et al. [59] described that problem behaviors in children were related to negative feelings or impulsivity. In other words, difficulty in controlling one's behavior is related to impulsiveness [60].

As the student pro-gamers had a support system in place, this may have helped to promote planning and regular lifestyles. Planning and impulsivity in human behavior have been seen to be closely associated with the functioning of the dorsolateral prefrontal cortex [61,62]. Conclusively, a well-controlled support system that includes planning and a regular life pattern may improve the balance of brain activity between the dorsolateral prefrontal cortex and the orbitofrontal cortex in student pro-gamers.

4.5. Limitations

There were several limitations to our study. First, the number of participants was too small to generalize our results. Because of the small numbers, the brain analysis results have been presented in an uncorrected format. Second, there were few assessment tools used to assess potential environmental factors in our study. Future studies should include a larger number of participants and evaluate the environmental conditions. Third, the assessment tool for excessive internet gaming might be used through objective criteria. Some adolescents may feel that an amount of game playing, that other adolescents feel is not "excessive," can be distressing for them. A subjective perception of "excessive" or "distressing" should be assessed in future studies. Finally, we did not consider different motivations, self-efficacy, and self-worth at the starting point between the pro-gamer students and IGD adolescents. These factors could lead a bias of the results in mental health and brain activity.

5. Conclusions

In response to long term internet game play, increased brain activity within the attention system was found in both groups. However, a well-organized support system, including a regular schedule and a supervisor, would lead to different results in terms of improved behavior scores and suppressing the brain activity within the orbitofrontal cortex.

Author Contributions: Conceptualization, K.H.K., and D.H.H.; methodology, D.H.H.; formal analysis, S.M.K.; investigation, H.C.H.; Writing—Original draft preparation, K.H.K., D.H.H.; Writing—Review and editing, D.H.H.; supervision, D.H.H. All authors have read and agreed to the published version of the manuscript.

Funding: This work was supported by the Ministry of Education of the Korea and the National Research Foundation of Korea (NRF-2018S1A5B6070270).

Acknowledgments: Thanks A-hyun high school for helping research.

Conflicts of Interest: The authors declare no conflict of interest. The funders had no role in the design of the study; in the collection, analyses, or interpretation of data; in the writing of the manuscript, or in the decision to publish the results.

References

1. American Psychiatry Association. *Diagnostic and Statistical Manual of Mental Disorders (DSM-5)*; American Psychiatric Publishing: Arlington, VA, USA, 2013.
2. World Health Organization. *Public Health Implications of Excessive Use of the Internet, Computers, Smartphones and Similar Electronic Devices Meeting Report*; World Health Organization: Tokyo, Japan, 2014.
3. World Health Organization. *The ICD-11 Classification of Mental and Behavioral Disorders: Diagnostic Criteria for Research*; World Health Organization: Geneva, Switzerland, 2018.
4. Gentile, D. Pathological video-game use among youth ages 8 to 18: A national study. *Psychol. Sci.* **2009**, *20*, 594–602. [CrossRef] [PubMed]
5. Stockdale, L.; Coyne, S.M. Video game addiction in emerging adulthood: Cross-sectional evidence of pathology in video game addicts as compared to matched healthy controls. *J. Affect. Disord.* **2018**, *225*, 265–272. [CrossRef] [PubMed]
6. Hull, J.G.; Brunelle, T.J.; Prescott, A.T.; Sargent, J.D. A longitudinal study of risk-glorifying video games and behavioral deviance. *J. Pers. Soc. Psychol.* **2014**, *107*, 300–325. [CrossRef] [PubMed]
7. Weinstein, A.; Lejoyeux, M. New developments on the neurobiological and pharmaco-genetic mechanisms underlying internet and videogame addiction. *Am. J. Addict.* **2015**, *24*, 117–125. [CrossRef] [PubMed]
8. Vaccaro, A.G.; Potenza, M.N. Diagnostic and classification considerations regarding gaming disorder: Neurocognitive and neurobiological features. *Front. Psychiatry* **2019**, *10*, 405. [CrossRef] [PubMed]
9. Kim, J.; Kang, E. Internet game overuse is associated with an alteration of fronto-striatal functional connectivity during reward feedback processing. *Front. Psychiatry* **2018**, *9*, 371. [CrossRef] [PubMed]
10. Wang, Z.; Wu, L.; Yuan, K.; Hu, Y.; Zheng, H.; Du, X.; Dong, G. Cortical thickness and volume abnormalities in internet gaming disorder: Evidence from comparison of recreational internet game users. *Eur. J. Neurosci.* **2018**, *48*, 1654–1666. [CrossRef]
11. Zilverstand, A.; Huang, A.S.; Alia-Klein, N.; Goldstein, R.Z. Neuroimaging Impaired Response Inhibition and Salience Attribution in Human Drug Addiction: A Systematic Review. *Neuron* **2018**, *98*, 886–903. [CrossRef]
12. Koob, G.F.; Volkow, N.D. Neurobiology of addiction: A neurocircuitry analysis. *Lancet Psychiatry* **2016**, *3*, 760–773. [CrossRef]
13. Jeong, E.J.; Ferguson, C.J.; Lee, S.J. Pathological gaming in young adolescents: A longitudinal study focused on academic stress and self-control in South Korea. *J. Youth Adolesc.* **2019**, *48*, 2333–2342. [CrossRef]
14. Pang, S.; Abdin, E.; Chong, S.A.; Chua, B.Y.; Lee, S.P.; Picco, L.; Satghare, P.; Vaingankar, L.; Verma, S.; Subramaniam, M. Correlates of online game play motivations, social anxiety and psychological distress. *Ann. Acad. Med. Singap.* **2017**, *46*, 443–446. [PubMed]
15. Cerniglia, L.; Griffiths, M.D.; Cimino, S.; De Palo, V.; Monacis, L.; Sinatra, M.; Tambelli, R. A latent profile approach for the study of internet gaming disorder, social media addiction, and psychopathology in a normative sample of adolescents. *Psychol. Res. Behav. Manag.* **2019**, *12*, 651–659. [CrossRef] [PubMed]
16. Cerniglia, L.; Zoratto, F.; Cimino, S.; Laviola, G.; Ammaniti, M.; Adriani, W. Internet Addiction in adolescence: Neurobiological, psychosocial and clinical issues. *Neurosci. Biobehav. Rev.* **2017**, *76*, 174–184. [CrossRef] [PubMed]
17. Cerniglia, L.; Guicciardi, M.; Sinatra, M.; Monacis, L.; Simonelli, A.; Cimino, S. The Use of Digital Technologies, Impulsivity and Psychopathological Symptoms in Adolescence. *Behav. Sci.* **2019**, *9*, 82. [CrossRef]
18. Latham, A.J.; Patston, L.L.; Tippett, L.J. The virtual brain: 30 years of video-game play and cognitive abilities. *Front. Psychol.* **2013**, *4*, 629. [CrossRef]
19. Mentiplay, B.F.; FitzGerald, T.L.; Clark, R.A.; Bower, K.J.; Denehy, L.; Spittle, A.J. Do video game interventions improve motor outcomes in children with developmental coordination disorder? A systematic review using the ICF framework. *BMC Pediatr.* **2019**, *19*, 22. [CrossRef]
20. Pallavicini, F.; Ferrari, A.; Mantovani, F. Video games for well-being: A systematic review on the application of computer games for cognitive and emotional training in the adult population. *Front. Psychol.* **2018**, *9*, 2127. [CrossRef]
21. Ballesteros, S.; Mayas, J.; Prieto, A.; Ruiz-Marquez, E.; Toril, P.; Reales, J.M. Effects of video game training on measures of selective attention and working memory in older adults: Results from a randomized controlled trial. *Front. Aging Neurosci.* **2017**, *9*, 354. [CrossRef]

22. Baysak, E.; Yertutanol, F.D.K.; Dalgar, I.; Candansayar, S. How game addiction rates and related psychosocial risk factors change within 2-years: A follow-up study. *Psychiatry Investig.* **2018**, *15*, 984–990. [CrossRef]
23. El-Salhy, M.; Miyazaki, T.; Noda, Y.; Nakajima, S.; Nakayama, H.; Mihara, S.; Kitayuguchi, T.; Higuchi, S.; Muramatsu, T.; Mimura, M. Relationships between internet addiction and clinicodemographic and behavioral factors. *Neuropsychiatr. Dis. Treat.* **2019**, *15*, 739–752. [CrossRef]
24. Kim, J.Y.; Chun, J.W.; Park, C.H.; Cho, H.; Choi, J.; Yang, S.; Ahn, K.J.; Kim, D.J. The correlation between the frontostriatal network and impulsivity in internet gaming disorder. *Sci. Rep.* **2019**, *9*, 1191. [CrossRef] [PubMed]
25. Han, D.H.; Lyoo, I.K.; Renshaw, P.F. Differential regional gray matter volumes in patients with on-line game addiction and professional gamers. *J. Psychiatr. Res.* **2012**, *46*, 507–515. [CrossRef] [PubMed]
26. Wei, X.; Shen, H.; Ren, J.; Li, X.; Xu, X.; Yang, R.; Lai, L.; Chen, L.; Hu, J.; Liu, W.; et al. Altered resting-state connectivity in college students with nonclinical depressive symptoms. *PLoS ONE* **2014**, *9*, e114603. [CrossRef]
27. Lu, H.; Zuo, Y.; Gu, H.; Waltz, J.A.; Zhan, W.; Scholl, C.A.; Rea, W.; Yang, Y.; Stein, E.A. Synchronized delta oscillations correlate with the resting-state functional MRI signal. *Proc. Natl. Acad. Sci. USA* **2007**, *104*, 18265–18269. [CrossRef] [PubMed]
28. Zou, Q.H.; Zhu, C.Z.; Yang, Y.; Zuo, X.N.; Long, X.Y.; Cao, Q.J.; Wang, Y.F.; Zang, Y.F. An improved approach to detection of amplitude of low-frequency fluctuation (ALFF) for resting-state fMRI: Fractional ALFF. *J. Neurosci. Methods* **2008**, *172*, 137–141. [CrossRef] [PubMed]
29. Xu, Y.; Zhuo, C.; Qin, W.; Zhu, J.; Yu, C. Altered spontaneous brain activity in schizophrenia: A meta-analysis and a large-sample study. *Biomed. Res. Int.* **2015**, 204628. [CrossRef] [PubMed]
30. Itahashi, T.; Yamada, T.; Watanabe, H.; Nakamura, M.; Ohta, H.; Kanai, C.; Iwanami, A.; Kato, N.; Hashimoto, R. Alterations of local spontaneous brain activity and connectivity in adults with high-functioning autism spectrum disorder. *Mol. Autism* **2015**, *6*, 30. [CrossRef]
31. Cheng, W.; Ji, X.; Zhang, J.; Feng, J. Individual classification of ADHD patients by integrating multiscale neuroimaging markers and advanced pattern recognition techniques. *Front. Syst. Neurosci.* **2012**, *6*, 58. [CrossRef]
32. Kim, J.I.; Kang, Y.H.; Lee, J.M.; Cha, J.; Park, Y.H.; Kweon, K.J.; Kim, B.N. Resting-state functional magnetic resonance imaging investigation of the neural correlates of cognitive-behavioral therapy for externalizing behavior problems in adolescent bullies. *Prog. Neuro Psychopharmacol. Biol. Psychiatry* **2018**, *86*, 193–202. [CrossRef]
33. Achenbach, T.M. *Manual for the Child Behavior Checklist/4–18 and 1991 Profile*; Burlington: Burlington, NJ, USA, 1991.
34. Lee, S.; Park, J.; Lim, E.; Junh, H. Screening Oppositional Defiant Disorder with the Korean Child Behavior Checklist: The Role of the Subscales of Aggressive and Delinquent Behavior. *J. Korean Acad. Child Adolesc. Psychiatry* **2011**, *22*, 95–102. [CrossRef]
35. Young, K.S. Internet addiction: The emergence of a new clinical disorder. *Cyberpsychol. Behav.* **1998**, *1*, 237–244. [CrossRef]
36. Lee, K.; Lee, H.K.; Gyeong, H.; Yu, B.; Song, Y.M.; Kim, D. Reliability and validity of the Korean version of the internet addiction test among college students. *J. Korean Med. Sci.* **2013**, *28*, 763–768. [CrossRef] [PubMed]
37. Kovacs, M. The Children's Depression, Inventory (CDI). *Psychopharmacol. Bull.* **1985**, *21*, 995–998. [PubMed]
38. Cho, S.C.; Lee, Y.S. Development of the Korean form of the Kovacs' Children's Depression Inventory. *J. Korean Neuropsychiatr. Assoc.* **1990**, *29*, 943–956.
39. Beck, A.T.; Epstein, N.; Brown, G.; Steer, R.A. An inventory for measuring clinical anxiety: Psychometric properties. *J. Consult. Clin. Psychol.* **1998**, *56*, 893. [CrossRef]
40. Kwon, S.M. The assessment of psychopathology in patient with anxiety disorder. *Korean J. Psychopathol.* **1997**, *6*, 37–51.
41. DuPaul, G.J. Parent and teacher ratings of ADHD symptoms: Psychometric properties in a community-based sample. *J. Clin. Child Adolesc. Psychol.* **1991**, *20*, 245–253. [CrossRef]
42. So, Y.K.; Noh, J.S.; Kim, Y.S.; Ko, S.G.; Koh, Y.J. The reliability and validity of Korean parent and teacher ADHD rating scale. *J. Korean Neuropsychiatr. Assoc.* **2002**, *41*, 283–289.
43. Achenbach, T.M.; Ruffle, T.M. The child behavior checklist and related forms for assessing behavioral/emotional problems and competencies. *Pediatr. Rev.* **2000**, *21*, 265–271. [CrossRef]

44. Park, J.I.; Shim, S.H.; Lee, M.; Jung, Y.E.; Park, T.W.; Park, S.H.; Im, Y.J.; Yang, J.C.; Chung, Y.C.; Chung, S.K. The validities and efficiencies of Korean ADHD rating scale and Korean child behavior checklist for screening children with ADHD in the community. *Psychiatry Investig.* **2014**, *11*, 258–265. [CrossRef]
45. Kweon, K.; Lee, H.J.; Park, K.J.; Joo, Y.; Kim, H.W. Child behavior checklist profiles in adolescents with bipolar and depressive disorders. *Compr. Psychiatry* **2016**, *70*, 152–158. [CrossRef] [PubMed]
46. Greitemeyer, T.; Mugge, D.O. Video games do affect social outcomes: A meta-analytic review of the effects of violent and prosocial video game play. *Pers. Soc. Psychol. Bull.* **2014**, *40*, 578–589. [CrossRef] [PubMed]
47. Shao, R.; Wang, Y. The relation of violent video games to adolescent aggression: An examination of moderated mediation effect. *Front. Psychol.* **2019**, *10*, 384. [CrossRef] [PubMed]
48. Przybylski, A.K.; Weinstein, N. Violent video game engagement is not associated with adolescents' aggressive behaviour: Evidence from a registered report. *R. Soc. Open Sci.* **2019**, *6*, 171474. [CrossRef] [PubMed]
49. O'Brien, C. Drug addiction and drug abuse. In *Goodman and Gilman's The Pharmacological Basis of Therapeutics*, 11th ed.; Brunton, L.B., Lazo, J.S., Parker, K.L., Eds.; McGraw-Hill: New York, NY, USA, 2005; pp. 607–629.
50. Ptak, R. The frontoparietal attention network of the human brain: Action, saliency, and a priority map of the environment. *Neuroscientist* **2012**, *18*, 502–515. [CrossRef] [PubMed]
51. Bavelier, D.; Achtman, R.L.; Mani, M.; Focker, J. Neural bases of selective attention in action video game players. *Vis. Res.* **2012**, *61*, 132–143. [CrossRef] [PubMed]
52. Dye, M.W.; Green, C.S.; Bavelier, D. The development of attention skills in action video game players. *Neuropsychologia* **2009**, *47*, 1780–1789. [CrossRef]
53. Evren, C.; Evren, B.; Dalbudak, E.; Topcu, M.; Kutlu, N. Relationships of Internet addiction and Internet gaming disorder symptom severities with probable attention deficit/hyperactivity disorder, aggression and negative affect among university students. *Atten. Defic. Hyperact. Disord.* **2019**, *11*, 413–421. [CrossRef]
54. Han, D.H.; Lee, Y.S.; Na, C.; Ahn, J.Y.; Chung, U.S.; Daniels, M.A.; Haws, C.A.; Renshaw, P.F. The effect of methylphenidate on Internet video game play in children with attention-deficit/hyperactivity disorder. *Compr. Psychiatry* **2009**, *50*, 251–256. [CrossRef]
55. Moccia, L.; Pettorruso, M.; De Crescenzo, F.; De Risio, L.; di Nuzzo, L.; Martinotti, G.; Bifone, A.; Janiri, L.; Di Nicola, M. Neural correlates of cognitive control in gambling disorder: A systematic review of fMRI studies. *Neurosci. Biobehav. Rev.* **2017**, *78*, 104–116. [CrossRef]
56. Park, B.; Han, D.H.; Roh, S. Neurobiological findings related to Internet use disorders. *Psychiatry Clin. Neurosci.* **2017**, *71*, 467–478. [CrossRef] [PubMed]
57. Winstanley, C.A. The orbitofrontal cortex, impulsivity, and addiction: probing orbitofrontal dysfunction at the neural, neurochemical, and molecular level. *Ann. N. Y. Acad. Sci.* **2007**, *1121*, 639–655. [CrossRef] [PubMed]
58. Moorman, D.E. The role of the orbitofrontal cortex in alcohol use, abuse, and dependence. *Prog. Neuropsychopharmacol. Biol. Psychiatry* **2018**, *87*, 85–107. [CrossRef] [PubMed]
59. Eisenberg, N.; Sadovsky, A.; Spinrad, T.L.; Fabes, R.A.; Losoya, S.H.; Valiente, C.; Reiser, M.; Cumberland, A.; Shepard, S.A. The relations of problem behavior status to children's negative emotionality, effortful control, and impulsivity: Concurrent relations and prediction of change. *Dev. Psychol.* **2005**, *41*, 193–211. [CrossRef]
60. Mavrogiorgou, P.; Enzi, B.; Klimm, A.K.; Kohler, E.; Roser, P.; Norra, C.; Juckel, G. Serotonergic modulation of orbitofrontal activity and its relevance for decision making and impulsivity. *Hum. Brain Mapp.* **2017**, *38*, 1507–1517. [CrossRef]
61. Deserno, L.; Wilbertz, T.; Reiter, A.; Horstmann, A.; Neumann, J.; Villringer, A.; Heinze, H.J.; Schlagenhauf, F. Lateral prefrontal model-based signatures are reduced in healthy individuals with high trait impulsivity. *Transl. Psychiatry* **2015**, *5*, e659. [CrossRef]
62. Huang, S.; Zhu, Z.; Zhang, W.; Chen, Y.; Zhen, S. Trait impulsivity components correlate differently with proactive and reactive control. *PLoS ONE* **2017**, *12*, e0176102. [CrossRef]

International Journal of
Environmental Research and Public Health

MDPI

Article

The Assessment of Problematic Internet Pornography Use: A Comparison of Three Scales with Mixed Methods

Lijun Chen * and Xiaoliu Jiang

Department of Psychology, School of Humanities and Social Sciences, Fuzhou University, Fuzhou 350108, China; psyjxl@126.com
* Correspondence: qqkx159@126.com; Tel.: +86-0591-2286-6562

Received: 12 November 2019; Accepted: 10 January 2020; Published: 12 January 2020

Abstract: The primary aim of this study was to compare different screening tools for problematic internet pornography use (IPU) and identify the most accurate measure. The reliability and validity of three scales, namely, the Problematic Pornography Consumption Scale (PPCS), Problematic Pornography Use Scale (PPUS), and Short Internet Addiction Test Adapted to Online Sexual Activities(s-IAT-sex), were examined using three homogeneous groups, respectively. A total of 972 adults (mean age = 24.8) from 28 provinces/regions in China participated in the quantitative part (QUAN). The Brief Pornography Screener served as the reference standard. The PPCS demonstrated stronger reliability and validity, including criterion validity, as well as greater sensitivity and acceptable specificity; therefore, it was considered to be the more accurate screening instrument. In the qualitative part (QUAL), we interviewed 22 volunteers and 11 therapists (who had worked with individuals with problematic IPU) to examine their perspectives on the core features of problematic IPU and dimensions of the PPCS. Almost all the interviewees endorsed the structure of the PPCS. These findings encourage the use of the PPCS in future research studies and underscore its screening applications because of its ability to classify IPU as problematic or nonproblematic.

Keywords: problematic pornography use; internet pornography use; problematic pornography consumption scale; problematic pornography use scale; the short internet addiction test adapted to online sexual activities

---

1. Introduction

Internet pornography use (IPU) is a sexual behavior [1], corresponding to the use of internet to engage in various gratifying sexual activities also known as online pornography use or cybersex [2–4]. It comprises a variety of online sexual activities (OSAs), including watching pornography, online pornography exchange, engaging in sex chats, using sex webcams, searching for sexual partners, or engaging in sexual role playing, among which stands the watching pornography, which is the most popular activity [5]. According to the past findings, engaging in IPU sometimes derives various negative consequences, such as financial, legal, occupational, and relationship trouble or personal problems [6]. Feelings of loss of control and persistent use despite these adverse outcomes constitute compulsive cybersex or problematic IPU. To date, no consensus exists regarding the conceptualization and diagnosis of problematic IPU. For instance, numerous terms have been used to describe the phenomenon (e.g., internet sex addiction [7,8], problematic online sexual activities [9], cybersex addiction [10], and problematic internet pornography use [6]). Although these concepts are slightly different, they all comprise three crucial components: the medium (the internet), the content (sexual behavior), and the problematic use (the compulsive behavior). Regardless of the debate, it is now acknowledged that excessive involvement in IPU or cybersex may become dysfunctional and associated with addiction symptoms (e.g., loss of control, compulsive use). Considering these inconsistent terms sharing crucial components, problematic IPU may be regarded as a subtypes of problematic internet use from a classification perspective, which may help advance clinical and research efforts into its prevalence and impact.

Nevertheless, evidence regarding the problematic IPU is inconsistent, due to the heterogeneity of assessment tool. The fundamental reason is that the definition and diagnostic criteria of problematic IPU is still unclear. In order to address these conceptual ambiguities, researchers have developed several scales that measure different aspects of pornography use [11]. Some briefer scales are more convenient to administer, but they underscore the self-perceived addiction (e.g., Cyber-Pornography Use Inventory-9). Some of these scales have been designed to assess the motivations underlying pornography use among hypersexual men (e.g., Pornography Consumption Inventory) [12]. Some scales fail to capture the different aspects of problematic IPU and focus solely on specific dimensions (e.g., the Pornography Craving Questionnaire, PCQ). Additionally, some globally accessible websites host the Cybersex Addiction Test, Sexaholics Anonymity Test, Sex Addicts Anonymous, and Sexual Addiction Screening Test, which assess difficulties in exercising self-control, its negative consequences, and the social problems that are associated with sexual activities. Furthermore, assessing IPU, using measures of sexual addiction, entails a few challenges. Specifically, these assessments may not be able to capture the characteristics of the activities (e.g., chat-based cybersex, sexual video games that cannot be played offline) and symptoms (e.g., separation from reality due to immersion in the virtual world that are unique to IPU. To address this gap in the literature and conduct further research in this domain, assessments with strong psychometric properties are much needed [5,7].

Several scales of problematic IPU are available to researchers and clinicians. Indeed, a recent meta-analysis identified 22 psychometric instruments that assess problematic pornography use [11]. Otherwise, most of the studies that have been conducted during the past decade had used self-developed items and a few of these measures have been subsequently revalidated [4,5,13]. Therefore, it is difficult to compare the results of different studies because there is a lack of concordance in the assessments that have been used. In order to select suitable tools for comparison from the existing scales, a systematic review was conducted. The following terms and their derivatives were used in multiple combinations: (Cybersex* OR internet porn* OR hypersex*) AND (addict* OR compulsiv* OR problem*) AND (assessment OR scale OR instrument OR measure*), to identify relevant studies in order to address the questions related to assessment and available screening questionnaires. The selection criteria of the literature search were limited to articles focusing specifically on cybersex and/or internet pornography consumption and dysfunctional cybersex, and also describe the development and adaptation of self-reported psychometric instruments that assesses at least one aspect of problematic pornography use. Finally, we found a total of 27 instruments on assess the problematic IPU (cybersex). Through the systematic review process conducted, we decided to retain three scales that were developed to measure problematic pornography use, even if not all of the three scales were specifically designed to measure internet pornography, as a large majority of participants used online pornography, and the developers of these scales suggested that they could be used to measure problematic IPU [14,15], additionally we replaced "pornography" into "internet pornography" in the Chinese version. We selected these three scales for the following reasons: (1) they include fewer items and are thus easily administered measures, (2) all of them cover the core characteristics of IPU, such as loss control, (3) they are grounded in addiction components such as impaired control, conflict, salience [11], (4) they are applicable within the Chinese culture [16–19], and (5) they display strong test-retest (i.e., two weeks) reliability; consequently, these three previously validated scales were identified for further examination. First, the Short Internet Addiction Test Adapted to OSAs (s-IAT-sex), which has demonstrated satisfactory psychometric properties [9]. However, this scale has been validated only among men [5], and a large number of studies have shown that there are substantial gender differences in IPU [18,20,21]. Second, the Problematic Pornography Use Scale (PPUS) [15], which has been validated using a large sample; unfortunately, however, a valid cutoff score has not been specified for this measure. Third, the Problematic Pornography Consumption Scale (PPCS); this scale is founded upon the theoretical framework of Griffiths's components model of addiction [22]. All three scales include strong internal consistency and a valid factorial structure, which has been supported by the results of confirmatory factor analysis (CFA) [9,14,15,19]. Nevertheless, it is difficult to compare the findings of studies that have used these scales because they entail different factor structures. Therefore, it is necessary to select reliable indicators and methods, and identify the most accurate instrument.

In order to effectively compare different scales, a unifying and reliable standard should first be established. The Brief Pornography Screener (BPS), which is a screening tool that measures loss of self-control, overuse of problematic pornography use, may be useful in identifying individuals who are at risk for problematic pornography use or can serve as a proxy measure [23]. Kraus et al., who developed the BPS, have proposed that the diagnostic criteria for compulsive sexual behavior (CSB) should be included in the new International Classification of Diseases (ICD-11) [24], and this proposal has been accepted. According to the upcoming ICD-11's diagnostic criteria for impulse control disorder [25], patterns of failure to control intense sexual impulses or urges and the resultant repetitive sexual behaviors are considered to be the characteristic features of the disorder. The BPS considers compulsive pornography to be the core component of problematic pornography use. Moreover, the BPS has been used with different samples, and it has demonstrated satisfactory psychometric properties among American and Polish pornography users [26]. Many past studies have used the BPS to identify pornography addicts. Furthermore, it has also been used to ascertain the severity of problematic pornography use among men who seek pharmacologic or psychological treatment as a result of their loss of control over their sexual behaviors [27–29]. Therefore, in this study, the BPS scores were used as the reference standard against which the sensitivity and specificity of the three aforementioned scales were ascertained.

Several recent reviews have focused specifically on the conceptualization and assessment of problematic pornography use [4,11,30,31]. Some reviews have briefly summarized and commented on the included instruments [5], whereas others have evaluated their ability to assess the core components of problematic pornography use [11]. However, no past study has compared the different scales and identified the most accurate measure of problematic pornography use using a same standard or indicator. Measures of problematic IPU are heterogeneous, and each scale focuses on a different aspect of problematic IPU. Furthermore, because these scales have not been extensively validated, it is difficult to compare the findings of the studies that have used them. In addition, the sensitivity of the different scales that assess problematic IPU have not been adequately compared. Therefore, in the present study, a QUAN→QUAL mixed-methods design was conducted, including (1) using quantitative methods to identify a scale with a higher sensitivity index from three selected scales (PPCS, PPUS, s-IAT-sex) for assessing problematic IPU. Moreover, the duration of usage, frequency of engagement in OSAs, sexual compulsivity, and pornography cravings were used to examine the criterion validity of the assessments. Subsequently, (2) qualitative interviews were conducted with volunteers and therapists who have serviced the individuals in trouble of problematic IPU to further examine the appropriateness of the "more accurate" scale from the service providers' perspectives, whereby the qualitative part helps to evaluate and interpret the results obtained from the main quantitative study.

2. The Quantitative Part: A Comparison of the Three Retained Scales

2.1. Materials and Methods

2.1.1. Sample

The study sample consisted of 560 men and 412 women, and the mean age of the sample was 24.8 years [*standard deviation (SD)* = 7.2 years; range = 18–48 years]. The group comparisons of the demographic characteristics of the three study samples can be inferred from Table 1.

Table 1. Group comparisons of the demographic characteristics of the three study samples.

The Scales		PPCS [1] (*n* = 317)	PPUS [2] (*n* = 332)	s-IAT-Sex [3] (*n* = 323)	χ^2 (*F*)	*p*
Gender ratio (men/women)		1.39	1.39	1.48	6.92	0.31
Age	Mean ± SD [4]	24.64 ± 7.39	24.47 ± 7.27	25.31 ± 6.93	1.24	0.29
	Range	18–48	18–45	18–45		
Sexual orientation	Homosexual	1.31%	1.76%	0.91%	2.61	0.11
	Heterosexual	91.52%	91.97%	94.69%		
	Bisexual	7.17%	6.37%	4.40%		
Relationship status	Single	46.78%	42.52%	40.64%	12.85	0.23
	Partnered	14.81%	21.41%	22.90%		
	Engaged	0.90%	1.20%	0.96%		
	Married	37.51%	34.91%	35.60%		
Educational level	Primary school or below	0	0	0	2.99	0.08
	Vocational school	1.24%	0.35%	0.34%		
	Middle school	1.55%	0.67%	0.91%		
	University or college	97.21%	99.08%	98.75%		
Work	Full time	46.10%	45.51%	47.43%	0.39	0.53
	Part time	2.51%	2.68%	4.57%		
	Short-term hired labor	1.64%	1.21%	0.56%		
	Unemployed	49.75%	50.60%	47.54%		
Place of residence	Capital	36.01%	47.37%	43.69%	11.70	0.07
	County town	40.38%	31.02%	35.04%		
	Town	12.01%	10.51%	8.36%		
	Village	11.70%	11.10%	13.01%		
Age of first exposure to pornography		16.21 ± 4.27	16.62 ± 4.69	16.62 ± 4.81	75.86	0.08

[1] PPCS = Problematic Pornography Consumption Scale, [2] PPUS = Problematic Pornography Use Scale, [3] s-IAT-sex = Short Internet Addiction Test Adapted to Online Sexual Activities, [4] SD = standard deviation.

2.1.2. Instruments

Three Main IPU Measurements

PPUS. The PPUS is a 12-item self-report scale that assesses four dimensions of IPU [15]: distress and functional problems, excessive use, difficulties in self-control, and IPU to escape or avoid negative emotions. In the Chinese version of the assessment, the term "pornography," which was used in the original scale, was modified as "internet pornography" in all instances (e.g., "I spend too much time being involved in thoughts about internet pornography"). The participants were required to indicate the frequency with which they had engaged in IPU during the past 6 months on a six-point scale that ranged from 0 (never) to 5 (all the time). Higher scores were indicative of a greater severity of engagement in IPU. The Cronbach's alpha of the total scale was 0.95 in this study.

PPCS. The PPCS was used to measure problematic IPU [14]. Responses were recorded on the following 7-point scale: 1 = never, 2 = rarely, 3 = occasionally, 4 = sometimes, 5 = often, 6 = very often, 7 = all the time. PPCS consists of 18 items, and assesses the six core components of addiction: salience, mood modification, conflict, tolerance, relapse, and withdrawal. Each factor is measured by three items (e.g., "I felt that I had to watch more and more internet porn for satisfaction" is an item of measure "tolerance"); the Cronbach's alphas of the aforementioned six factors were 0.77, 0.84, 0.71, 0.78, 0.86, and 0.86, respectively, in the study. The Cronbach's alpha of the total PPCS was 0.96. A cutoff score of 76 was used to ascertain normal and problematic use; specifically, scores that were greater than 76 were indicative of problematic use.

s-IAT-sex. Responses to each of the 12 items of the s-IAT-sex are recorded on a five-point scale that ranges from 1 (never) to 5 (always) [9]. The scale consists of two dimensions. The first factor assesses poor self-control and difficulties in reducing the amount of time that is spent online (six items, e.g., "How often do you find that you stay on Internet sex sites longer than you intended?"), whereas the second factor measures the functional impairments that are associated with engagement in cybersex (six items, e.g., "How often do you feel depressed, moody, or nervous when you are offline, which goes away once you are back on internet sex sites?"). The composite score, which can be computed by summing the individual item scores, can range from 12 to 60; higher scores are indicative of greater problems. The internal consistency (Cronbach's alpha) coefficients of the total scale and first and second factors were 0.89, 0.77, and 0.88, respectively, in this study.

Criterion Validity Questionnaires

PCQ. This 12-item questionnaire is a unidimensional assessment [32,33]. The following are a few sample items: "If the situation permitted, I would watch pornography right now" and "If I were to watch pornography right now, I would have difficulty stopping." The respondents were required to indicate how strongly they agreed with each item using the following seven response options (presented without numerals): "completely disagree," "somewhat disagree," "disagree a little," "neither agree nor disagree," "agree a little," "somewhat agree," and "completely agree." Higher scores are indicative of a greater craving for pornography. The Cronbach's alpha of this scale was 0.92 in the current study. The instructions of the PCQ present a craving-for-pornography vignette, which requires the respondent to imagine that they are alone in their room and seated in front of their computer and that they have a strong urge to watch their favorite type of pornography.

The Sexual Compulsivity Scale (SCS). The extent to which participants exhibit the characteristics of compulsive pornography use was assessed using the 10-item SCS that has been developed by Kalichman et al. [34]. Responses were recorded on a four-point rating scale (1 = not at all like me, 2 = slightly like me, 3 = mainly like me, 4 = very much like me, e.g., "I have to struggle to control my sexual thoughts and behavior"). In this study, the Cronbach's alpha of this scale was 0.86.

Questionnaire of OSAs. Thirteen items were used to measure participants' use of the internet for the following purposes: (1) viewing sexual explicit materials (SEM), (2) seeking sexual partners, (3) cybersex, and (4) flirting and sexual relationship maintenance [35]. Viewing SEM was assessed using five items (e.g., visiting erotic/pornographic websites, viewing and downloading erotic/pornographic videos from the internet, reading erotic/pornographic material online), each of which required responses to be rated on a nine-point scale that ranged from 1 (never) to 9 (at least once a day). The other three subscales assessed frequency using a nine-point scale that ranged from 1 (0 times) to 9 (20 or more times). Two items measured the frequency with which the respondents had sought sexual partners as well as the number of sexual partners that they had sought and found online. The frequency of engagement in cybersex was assessed using four items (e.g., masturbating or viewing strangers masturbating in front of a webcam, describing sexual fantasies either through texts or orally). Internet use for the purposes of flirting and sexual relationship maintenance was measured using two items. The Cronbach's alpha of the entire scale was 0.88 in the study. Higher scores were indicative of more frequent engagement in OSAs.

Additional Questions about IPU. In addition to items that assessed demographic characteristics, a few questions that were related to IPU were also posed to the participants. After providing them with a clear definition of internet pornography, the participants were asked to indicate their age of first exposure to pornography and the duration of time that they typically spent watching internet pornography every week.

The Reference Standard—BPS

The BPS, which has been developed by Kraus et al. [26], was used to assess pornography use during the past 6 months. This five-item assessment uses a three-point rating scale (0 = never, 1 = occasionally, 2 = always, e.g., "You find it difficult to resist strong urges to use sexually explicit material."); a cutoff score of 4 was used to detect problematic pornography use (absolute range = 0–10). Higher scores are indicative of more problematic pornography use. The Cronbach's alpha of the BPS was 0.84.

2.1.3. Procedure

This online study was conducted through a popular Chinese survey website, namely, Wenjuanxing (www.sojump.com). Adult members of the website received an email with a link that redirected them to the survey website and a brief introduction to our survey. This brief introduction informed the recipients that they were eligible for participation if they had engaged in IPU during the past 6 months (e.g., reading online pornographic content, browsing pornographic websites, sharing/watching pornographic videos or pictures, interacting and flirting with others) and were interested in participating in the survey. A total of 972 valid responses were collected from participants from 110 cities in 28 of the 34 provinces/regions in China (i.e., identified using the internet protocol addresses). As expected, all participants obtained scores that were equal to or greater than 14 on the measure of OSAs (the lowest possible score is 13, and it indicates no prior IPU); this indicated that all of them had engaged in at least one OSA during the past 6 months. Three highly homogeneous samples were required to respond to the three measures of problematic IPU, namely, the PPCS, PPUS, and s-IAT-sex, respectively. Each sample also completed the aforementioned mentioned assessments against which their criterion validity was to be examined. This study was conducted in accordance with the Declaration of Helsinki, and the protocol was approved by the Ethics Committee of the Department of Psychology, Fuzhou University (date of approval, 7 April 2019).

2.2. *Analysis*

Statistical analyses were conducted using SPSS 19.0 (IBM, Armonk, NY, USA) and Mplus version 7 [36]. Item-total correlations were computed to identify items that functioned poorly. CFA was used to test the factor structures of the scales of interest. Maximum likelihood estimation with the Satorra-Bentler correction was used to determine the fit between the data and the factor structures. Model fit was tested by inspecting the following indices: root mean square error of approximation (RMSEA; good: $\leq$0.06, acceptable: $\leq$0.08), comparative fit index (CFI; good: $\geq$0.95, acceptable: $\geq$0.90), and Tucker-Lewis index (TLI; good: $\geq$0.95, acceptable: $\geq$0.90). The reliability of the scales was assessed by computing Cronbach's alpha coefficients.

To identify possible groups of at-risk pornography users, latent profile analysis (LPA) was used. LPA was conducted using the original dimensions of each scale as explicit variables, and different groups of individuals with problematic IPU were successively divided into two to four categories for model fitting estimation. Sensitivity was defined as the proportion of persons with positive symptoms (as detected by the BPS) and members of the at-risk group (identified through LPA), whereas specificity was defined as the proportion of persons with negative symptoms and the nonproblematic group [37].

2.3. *Results and Discussion*

2.3.1. Validation of the Three Scales

The results of item analysis, CFA, and tests of reliability and convergent validity are shown in Table 2. Item-sum correlations were computed to examine item functioning. The PPCS and PPUS yielded higher coefficients, and both these scales also yielded good fit indices (i.e., CFA) and stronger reliability coefficients. PPCS, PPUS, and s-IAT-sex significantly positively related with SCS, PCQ, OSAs and usage time severally, and PPCS demonstrated stronger convergent validity.

Table 2. Reliability and validity of the three scales.

Scales	Item Analysis	Confirmatory Factor Analysis				α	External and Convergent Validity			
	rs (Item-Sum Correlation)	χ^2/df	CFI [4]	TLI [5]	RMSEA [6] [90% CI [7]]		SCS [8]	PCQ [9]	OSAs [10]	UT [11]
PPCS [1]	0.62 ***–0.82 ***	210.70/120	0.963	0.952	0.049 [0.038, 0.060]	0.96	0.67 ***	0.70 ***	0.67 ***	0.28 ***
PPUS [2]	0.66 ***–0.85 ***	90.30/48	0.966	0.953	0.052 [0.035, 0.068]	0.95	0.71 ***	0.66 ***	0.56 ***	0.17 ***
s-IAT-sex [3]	0.60 ***–0.79 ***	155.55/52	0.923	0.902	0.079 [0.082, 0.109]	0.89	0.73 ***	0.66 ***	0.46 ***	0.22 ***

[1] PPCS = Problematic Pornography Consumption Scale, [2] PPUS = Problematic Pornography Use Scale, [3] s-IAT-sex = Short Internet Addiction Test Adapted to Online Sexual Activities, [4] CFI = comparative fit index, [5] TLI = Tucker-Lewis index, [6] RMSEA = root mean square error of approximation, [7] CI = confidence interval, [8] SCS = Sexual Compulsivity Scale, [9] PCQ = Pornography Craving Questionnaire, [10] OSAs = online sexual activities, [11] UT = usage time. *** $p < 0.001$.

2.3.2. LPA

The results of LPA are shown in Table 3. For PPCS, the Lo-Mendell-Rubin adjusted likelihood ratio test (LMRT) results were significant when the number of classes was 4, and the entropy value was lower. Thus, the classification accuracy was not as high as that of the three-class solution; accordingly, the three-class solution was selected. For PPUS, when the model consisted of three classes, the LMRT results were significant; furthermore, the entropy value was evidently higher than that of the four-class solution. With regard to the s-IAT-sex, the nonsignificant *p*-value that emerged for the LMRT results suggested that the three- and four-class solutions should be rejected in favor of the two-class solution.

Table 3. Fit indices for latent profile analysis of the three scales assessing problematic internet pornography use.

Scales	Classes [4]	AIC [5]	BIC [6]	SSABIC [7]	Entropy	LMRT [8]
PPCS [1] (n = 317)	2	9298.755	9370.174	9309.910	0.959	1154.76 ***
	3	**8898.213**	**8995.944**	**8913.478**	**0.940**	**404.51** *
	4	8746.574	8870.618	8765.950	0.899	161.63 *
PPUS [2] (n = 332)	2	5718.021	5767.488	5726.251	0.953	799.82 ***
	3	**5424.503**	**5492.995**	**5435.899**	**0.924**	**293.41** **
	4	5348.339	5435.857	5362.900	0.931	83.29
s-IAT-sex [3] (n = 323)	**2**	**3652.433**	**3678.877**	**3656.674**	**0.845**	**205.41** ***
	3	3588.004	3625.780	3594.062	0.771	66.59
	4	3538.775	3587.884	3546.650	0.824	52.22

[1] PPCS = Problematic Pornography Consumption Scale, [2] PPUS = Problematic Pornography Use Scale, [3] s-IAT-sex = Short Internet Addiction Test Adapted to Online Sexual Activities, [4] classes = number of latent classes, [5] AIC = Akaike information criterion, [6] BIC = Bayesian information criterion, [7] SSABIC = sample-size-adjusted Bayesian information criterion, [8] LMRT = Lo-Mendell-Rubin adjusted likelihood ratio test, p = p-value associated with the LMRT results. Bold text is the finally selected models. * $p < 0.05$, ** $p < 0.01$, *** $p < 0.001$.

With regard to the three groups that emerged for the PPCS and PPUS, the first class obtained the lowest averages across all the scale dimensions; thus, this group was referred to as nonproblematic consumption. The second class obtained moderate scores on all the scale dimensions; therefore, these group members were referred to as low-risk pornography users. The third class obtained the highest scores on all the scale dimensions; thus, this group was referred to as at-risk users. As shown in Table 4, with regard to the two classes that emerged for the s-IAT-sex, class 1 obtained lower scores than class 2 on both the scale dimensions; therefore, they were referred to as the nonproblematic and at-risk groups, respectively (group differences in scores on the specific dimensions are shown in Appendix A).

Table 4. Comparisons of the accuracy of the three scales.

Scale	Group	Scores on the BPS					Mean	Range
		<4	$4 \leq x < 6$	≥6	Sensitivity [4]	Specificity [5]		
PPCS [1] (n = 317)	At-risk (n = 29)	3	11	15	89.66%	-	4.60 ± 0.59	1–7
	Low-risk (n = 90)	28	34	28	-	-	2.89 ± 0.46	1–7
	Nonproblematic (n = 198)	170	23	5	-	85.86%	1.41 ± 0.39	1–7
PPUS [2] (n = 332)	At-risk (n = 48)	9	8	31	81.25%	-	2.43 ± 0.48	0–5
	Low-risk (n = 86)	43	25	18	-	-	1.12 ± 0.29	0–5
	Nonproblematic (n = 198)	188	8	2	-	94.95%	0.75 ± 0.84	0–5
s-IAT-sex [3] (n = 323)	At-risk (n = 99)	28	26	45	71.72%	-	2.84 ± 0.44	1–5
	Nonproblematic (n = 224)	195	19	10	-	87.05%	1.54 ± 0.42	1–5

[1] PPCS = Problematic Pornography Consumption Scale, [2] PPUS = Problematic Pornography Use Scale, [3] s-IAT-sex = Short Internet Addiction Test Adapted to Online Sexual Activities, [4] Sensitivity = the proportion of persons with positive symptoms and members of the at-risk group that was identified through LPA, [5] Specificity = the proportion of persons with negative symptoms and the nonproblematic group.

2.3.3. Sensitivity and Specificity Analysis

The results showed that the sensitivity of the PPCS was 89.66%, which is higher than the values that emerged for the PPUS (i.e., 81.25%) and the s-IAT-sex (i.e., 71.72%). There were differences in the specificity of the three scales, and the values ranged from 85.86% to 94.95%. The PPCS demonstrated greater sensitivity (89.66%), and its specificity was 85.86%. This indicates that approximately 10% of problematic users had been classified as nonproblematic users and that approximately 14% of nonproblematic users had not been identified. In general, the PPCS and PPUS performed better than the s-IAT-sex. Since this study aimed to identify the scale with greater sensitivity in detecting problematic IPU, the PPCS was investigated in greater detail.

3. The Qualitative Part: Identification of the Most Accurate Scale

3.1. Methods

3.1.1. Sample

We interviewed 22 (20 men; mean age = 27.2) problematic IPU service volunteers (who provide online services on the following website: http://www.ryeboy.org/; average service time = 3.3 years) and 11 therapists (who have worked with individuals with problematic IPU and had more than 3 years of clinical experience).

3.1.2. The Interview Outline

Since the used scales were easy to administer and consisted of close-ended questions, interviews were conducted to examine participants' perspectives more deeply and comprehensively. The interview guide primarily sought to explore interviewees' understanding of problematic IPU/addiction and their evaluations of the dimensions of the selected scale. The interviewees were required to rate the importance of the dimensions on a scale that ranged from 1 (not at all important) to 7 (very important).

3.1.3. Procedure

In this study, we primarily explored their understanding of the concept of problematic IPU and the dimensions of the recommended scale. Two psychology graduate students served as the interviewers. At the beginning of the interview, the interviewees were informed about the purpose and significance of the interview and assured of the anonymity and strict confidentiality of their interview data; the interviews were recorded with their permission.

3.2. Analysis

The interview recordings were transcribed into verbatim scripts, and participants' identifying information was concealed. Next, we undertook thematic analysis of the text; in other words, we collated different interviewees' responses to the same question to create new text. Tree Nodes were established based on the dimensions of the selected scale, and interviewees' original statements were identified and summarized as a named code. Through this process, NVivo automatically generated statistics for all the references of the texts.

3.3. Results

With regard to the characteristics of problematic IPU, we generated a total of 20 codes by analyzing the interview data. Among these features, preoccupation with IPU (22 mentions), IPU to escape or avoid a negative emotional state (21 mentions), interpersonal conflict (22 mentions), and physiological and psychological symptoms (45 mentions) were most commonly mentioned. Furthermore, the 20 codes were summarized into the six dimensions of the PPCS (see Figure 1).

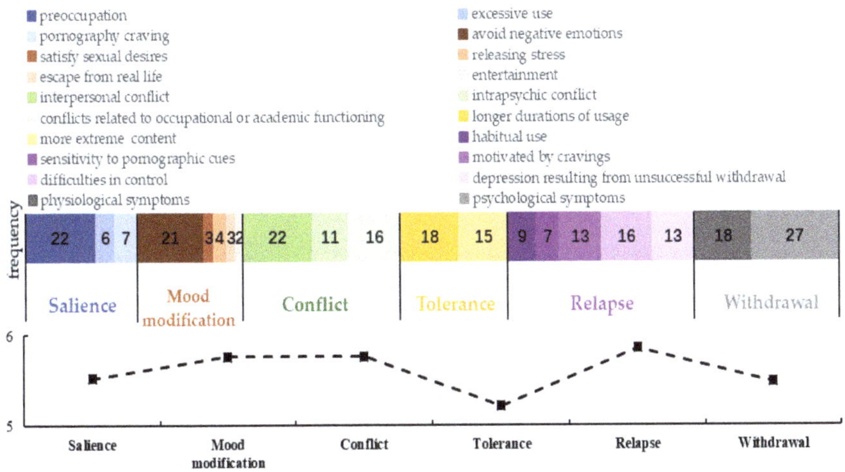

Figure 1. Volunteers and therapists' frequency of mentioning the dimensions of the Problematic Pornography Consumption Scale, features, and importance ratings for the six dimensions (average scores across 33 interviewees). Note: the numbers in the color blocks represent the frequency of mentions, whereas the polyline represents importance ratings for the six dimensions (range = 1–7).

Instance of the interview:

Interviewer: According to your service experience, what do you think is problematic internet pornography use? In other words, what are the expressions/symptoms of problematic internet pornography use?
Interviewee (service volunteer): They (problematic users) show difficulty controlling the craving for internet pornography (code: pornography carving), they are unable to control their own behavior, for instance, browsing pornographic websites, masturbating while watching porn frequently (code: difficulties in control). Their brains are constantly bombarded with sexual materials (code: preoccupation). If they are not exposed to internet pornography, they will feel uncomfortable, or feel that their heart is empty (code: depression resulting from unsuccessful withdrawal).

After presenting interviewees with the definitions of the six components of problematic IPU and further clarifying their meaning using examples, we presented them with questions "Based on your service experience, do you endorse this structure? Which dimension or dimensions do you think are particularly central to IPU?" Most (>95%) participants endorsed the six dimensions. It also can be inferred from Figure 1 that both volunteers and therapists emphasized the centrality of conflict, relapse and withdrawal in IPU (basing the frequency of mentions); at the same time, they weighted the mood modification, relapse and withdrawal as more important features in the problematic use (basing the important rating).

4. General Discussion

Problematic IPU is still a controversial issue; notably, it appears that no real consensus exists regarding the conceptualization and screening tool of problematic IPU. Several scales are available; thus, the assessment of problematic IPU is inconsistent, indicating that findings in this area are not readily comparable. The present study aimed to selected a more sensitive scale to screen problematic IPU, because higher sensitivity implies lower rate of missed diagnosis (i.e., problematic users who have been incorrectly screened as nonproblematic users). Basing on a systematic literature review, three scales were retained. Considering that research with mixed methods combining quantitative and qualitative analyses can enrich and improve our understanding of complicated phenomena [38,39], a quantitative method was used to identify a "more accurate" analysis from the three retained scales. Results of CFA showed that all three scales have good applicability in the wide range of adult groups (age in this case ranged from 18 to 45 years) in three highly homogeneous samples; compared to the other two scales, the PPCS demonstrated greater sensitivity and comparative specificity among samples drawn from the general population (results of the QUAN). Considering that the expression of questionnaire survey is brief and closed, and that the interview can understand the participants' undefined views more deeply and comprehensively, subsequently, results of QUAL showed that symptoms of problematic IPU proposed by the servers (volunteers and therapists) can be grouped into the six dimensions of PPCS and most of the servers supported the six-factor structure of PPCS.

Among the three scales, the PPCS score was most robustly related to the duration of usage, frequency of engagement in OSAs, and pornography cravings. Problematic IPU can appear under the umbrella of hypersexuality similarly to frequent engaging in various forms of cybersex, intense craving for pornography, and compulsive sexual behaviors [40], insofar that the robust relationship not only demonstrated a higher criterion validity, but also implied that co-screening instruments (i.e., pornography craving, frequency and duration of use, compulsive use) are expected to work as auxiliary screening indicators. Recent studies have revealed that for some people, pornographic use gave rise to their feeling of discord and shame contributing to their conflict of actual sexual materials consumption and their belief; in turn, these feelings of distress and shame may drive a morbid self-perception that they are addicted, but this may not be a real behavioral disorder [41,42]. In order to avoid misjudgment due to the self-perceived problematic use, it is more advisable to combine other supporting scales, and the combination diagnosis indexes of the diversity were selected to screen the prevalence of problematic IPU. In this study, with the higher correlation of PPCS with frequency of OSAs, the PCQ showed that combined with other indicators, it can better screen out problematic use and is more likely to avoid the misjudgment caused by subjective self-perceived addiction.

The more robust psychometric properties and higher recognition accuracy of the PPCS may be attributable to the fact that it has been developed in accordance with Griffiths's six-component structural theory of addiction (i.e., in contrast to the PPUS and s-IAT-sex). The PPCS has a very strong theoretical framework, and it assesses more components of addiction [11]. In particular, tolerance and withdrawal are the important dimensions of problematic IPU that are not assessed by the PPUS and s-IAT-sex; PPCS is the only instrument that explicitly assesses the "tolerance" component [11,14]. According to the "two-phased" internet pornography addiction model, in which the first step is characterized by an excessive use to internet pornography, and the second functions as a marker by repeated failures to break free from excessive use, despite negative consequences [43]. Items related to information about salience, carving, and tolerance reflect the engagement in internet pornography, corresponding to the first step, whereas items related to withdrawal, relapse, and conflict measure addiction more, corresponding to the second step. Obviously, components of PPCS includes both engagement in pornography and addiction of IPU, which has an intact theoretical framework of addiction.

The PPCS appears to be a more valid instrument for assessing problematic pornography use, has potential application in detecting prevalence concerning problematic IPU or cybersex addiction, and may be useful in assessing treatment outcomes. Our findings indicate that individuals who score high on the PPCS also report frequent engaging in various forms of online sexual activities, intense craving for pornography, and compulsive sexual behaviors. Thus, it appears important for clinicians to be aware of problematic pornography use and its related associations such as pornography craving, compulsive use. Moreover, it is important to note that the scale PPCS is recommended as a screening instruments to identify problematic users in the public and assess the prevalence rather than a diagnostic tool; future studies should further research its validity and cutoff in clinical sample; we also encourage individuals to visit a clinical therapist after being identified with problematic IPU by the use of PPCS.

This study has several limitations. First, data were collected using self-report measures; therefore, the reliability of the results depends on the respondents' honesty and accuracy of their comprehension of the scale items. Second, the study sample was recruited through an online survey company; therefore, the participants of this study may have been more educated and affluent than the average Chinese person. Furthermore, the study participants primarily lived in the capital/provincial capital, cities, and towns. Third, because the sample consisted of only a small number of non-heterosexual subjects, it was not possible to examine whether the factor structure and meaning of the contents of the PPCS differed across individuals with different sexual orientations.

5. Conclusions

The present study showed that the PPUS, PPCS, and s-IAT-sex are promising measures of problematic IPU. However, when sensitivity and specificity were simultaneously examined, the PPCS emerged as a more suitable measure of problematic IPU. The qualitative findings further confirmed that service providers endorsed the underlying structure of the PPCS.

Author Contributions: Conceptualization, L.C.; Data curation, L.C.; Formal analysis, X.J.; Funding acquisition, L.C.; Investigation, X.J.; Methodology, L.C.; Project administration, L.C.; Resources, L.C.; Supervision, L.C.; Visualization, X.J.; Writing—original draft, L.C.; Writing—review and editing, L.C. and X.J. All authors have read and agreed to the published version of the manuscript.

Funding: This work was supported by the National Social Science Foundation of China (Grant No. CEA150173 and 19BSH117) and the education reform project of Fujian province (FBJG20170038). The funding agencies did not have input into the content of the manuscript and the views described in the manuscript reflect those of the authors and not necessarily those of the funding agencies.

Acknowledgments: We would like to acknowledge Bin Wu and Yan Zhao (the founders of the "*Reyboys*", a non-governmental organization focusing on helping the problematic internet pornography users) for their help to recruit volunteers who serviced the addicts in the qualitative step, and pay tribute to them for their effort in helping problematic users.

Conflicts of Interest: The authors report no conflict of interest with respect to the content of this manuscript.

Appendix A

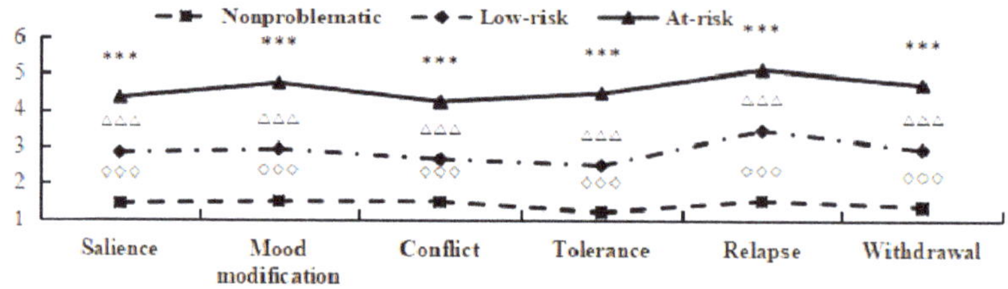

Figure A1. The average scores of the three latent classes based on the dimensions of PPCS. Note: PPCS = Problematic Pornography Consumption Scale, range = 1–7; *** $p < 0.001$ indicate that the score of at-risk group was significantly higher than that of low-risk group; $^{\triangle\triangle\triangle}$ $p < 0.001$ indicate that the score of low-risk group was significantly higher than that of non-problematic group; $^{\diamond\diamond\diamond}$ $p < 0.001$ indicate that the score of at-risk group was significantly higher than that of non-problematic group. The same below.

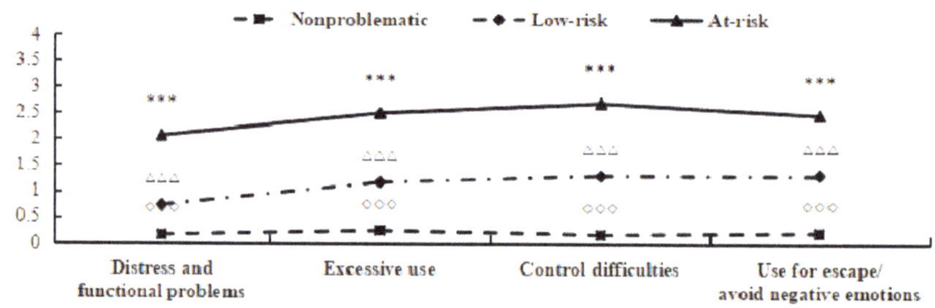

Figure A2. The average scores of the three latent classes based on the dimensions of PPUS. Note: PPUS = Problematic Pornography Use Scale, range = 0–5.

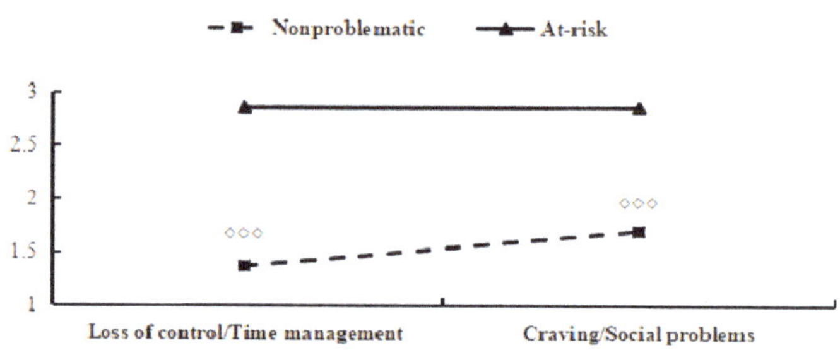

Figure A3. The average scores of the tow latent classes based on the dimensions of s-IAT-sex. Note: s-IAT-sex = short version of the Internet Addiction Test adapted to online sexual activities, range = 1–5.

References

1. Grubbs, J.B.; Wright, P.J.; Braden, A.L.; Wilt, J.A.; Kraus, S.W. Internet pornography use and sexual motivation: A systematic review and integration. *Ann. Int. Commun. Assoc.* **2019**, *43*, 117–155. [CrossRef]
2. Delmonico, D.L. Cybersex: High tech sex addiction. *Sex. Addict. Compulsivity J. Treat. Prev.* **1997**, *4*, 159–167. [CrossRef]
3. Cooper, A.L.; Delmonico, D.L.; Griffin-Shelley, E.; Mathy, R.M. Online sexual activity: An examination of potentially problematic behaviors. *Sex. Addict. Compulsivity* **2004**, *11*, 129–143. [CrossRef]
4. De Alarcón, R.; de la Iglesia, J.I.; Casado, N.M.; Montejo, A.L. Online Porn Addiction: What We Know and What We Don't—A Systematic Review. *Clin. Med.* **2019**, *8*, 91. [CrossRef]

5. Wéry, A.; Billieux, J. Problematic cybersex: Conceptualization, assessment, and treatment. *Addict. Behav.* **2017**, *64*, 238–246. [CrossRef] [PubMed]
6. Grubbs, J.B.; Volk, F.; Exline, J.J.; Pargament, K.I. Internet pornography use: Perceived addiction, psychological distress, and the validation of a brief measure. *Sex. Marital. Ther.* **2015**, *41*, 83–106. [CrossRef]
7. Griffiths, M.D. Internet sex addiction: A review of empirical research. *Addict. Res. Theory* **2012**, *20*, 111–124. [CrossRef]
8. Young, K.S. Internet sex addiction: Risk factors, stages of development, and treatment. *Am. Behav. Sci.* **2008**, *52*, 21–37. [CrossRef]
9. Wéry, A.; Burnay, J.; Karila, L.; Billieux, J. The Short French Internet Addiction Test adapted to online sexual activities: Validation and links with online sexual preferences and addiction symptoms. *Sex Res.* **2015**, *53*, 701–710. [CrossRef]
10. López-Fernández, O. How has internet addiction research evolved since the advent of internet gaming disorder? An overview of cyberaddictions from a psychological perspective. *Curr. Addict. Rep.* **2015**, *2*, 263. [CrossRef]
11. Fernandez, D.P.; Griffiths, M.D. Psychometric Instruments for Problematic Pornography Use: A Systematic Review. *Eval. Health Prof.* **2019**. [CrossRef] [PubMed]
12. Reid, R.C.; Li, D.S.; Gilliland, R.; Stein, J.A.; Fong, T. Reliability, validity, and psychometric development of the Pornography Consumption Inventory in a sample of hypersexual men. *J. Sex Marital Ther.* **2011**, *37*, 359–385. [CrossRef] [PubMed]
13. Short, M.B.; Black, L.; Smith, A.H.; Wetterneck, C.T.; Wells, D.E. A review of Internet pornography use research: Methodology and content from the past 10 years. *Cyberpsychol. Behav. Soc. Netw.* **2012**, *15*, 13–23. [CrossRef] [PubMed]
14. Bőthe, B.; Tóth-Király, I.; Zsila, Á.; Griffiths, M.D.; Demetrovics, Z.; Orosz, G. The development of the problematic pornography consumption scale (PPCS). *Sex Res.* **2018**, *55*, 395–406. [CrossRef] [PubMed]
15. Kor, A.; Zilcha-Mano, S.; Fogel, Y.A.; Mikulincer, M.; Reid, R.C.; Potenza, M.N. Psychometric development of the Problematic Pornography Use Scale. *Addict. Behav.* **2014**, *39*, 861–868. [CrossRef] [PubMed]
16. Chen, L.; Demetrovics, Z.; Potenza, M.N. Does mate preference predict problematic online pornography? Cross-cultural findings. *Behav. Addict.* **2019**, *8*, 63. [CrossRef]
17. Chen, L.; Ding, C.; Jiang, X.; Potenza, M.N. Frequency and duration of use, craving and negative emotions in problematic online sexual activities. *Sex. Addict. Compulsivity* **2018**, *25*, 396–414. [CrossRef]
18. Chen, L.; Yang, Y.; Su, W.; Zheng, L.; Ding, C.; Potenza, M.N. The relationship between sexual sensation seeking and problematic Internet pornography use: A moderated mediation model examining roles of online sexual activities and the third-person effect. *Behav. Addict.* **2018**, *7*, 565–573. [CrossRef]
19. Chen, L.; Wang, X.; Chen, S.M.; Jiang, C.H.; Wang, J.X. Reliability and validity of the Problematic Internet Pornography Use Scale in Chinese college students. *Chin. J. Public Health* **2018**, *34*, 1034–1038. [CrossRef]
20. Ševčíková, A.; Šerek, J.; Barbovschi, M.; Daneback, K. The roles of individual characteristics and liberalism in intentional and unintentional exposure to online sexual material among European youth: A multilevel approach. *Sex. Res. Soc. Policy* **2014**, *11*, 104–115. [CrossRef]
21. Cooper, A.; Delmonico, D.L.; Burg, R. Cybersex users, abusers, and compulsives: New findings and implications. *Sex. Addict. Compulsivity J. Treat. Prev.* **2000**, *7*, 5–29. [CrossRef]
22. Griffiths, M. A 'components' model of addiction within a biopsychosocial framework. *Subst. Use* **2005**, *10*, 191–197. [CrossRef]
23. Sklenarik, S.; Potenza, M.N.; Gola, M.; Kor, A.; Kraus, S.W.; Astur, R.S. Approach bias for erotic stimuli in heterosexual male college students who use pornography. *Behav. Addict.* **2019**, *8*, 234–241. [CrossRef] [PubMed]
24. Kraus, S.W.; Krueger, R.B.; Briken, P.; First, M.B.; Stein, D.J.; Kaplan, M.S.; Voon, V.; Abdo, C.; Grant, J.E.; Atalla, E.; et al. Compulsive sexual behaviour disorder in the ICD-11. *World Psychiatry* **2018**, *17*, 109–110. [CrossRef]
25. Efrati, Y.; Gola, M. Treating Compulsive Sexual Behavior. *Curr. Sex. Health Rep.* **2018**, *10*, 57–64. [CrossRef]
26. Kraus, S.W.; Gola, M.; Kowalewska, E.; Lew-Starowicz, M.; Hoff, R.A.; Porter, E.; Potenza, M.N. Brief Pornography Screener: A comparison of US and Polish pornography users. *J. Behav. Addict.* **2017**, *6*, 27–28. [CrossRef]

27. Kowalewska, E.; Kraus, S.W.; Lew-Starowicz, M.; Gustavsson, K.; Gola, M. Which dimensions of human sexuality are related to compulsive sexual behavior disorder (CSBD)? Study using a multidimensional sexuality questionnaire on a sample of Polish males. *J. Sex. Med.* **2019**, *16*, 1264–1273. [CrossRef]
28. Gola, M.; Wordecha, M.; Sescousse, G.; Lew-Starowicz, M.; Kossowski, B.; Wypych, M.; Makeig, S.; Potenza, M.N.; Marchewka, A. Can pornography be addictive? An fMRI study of men seeking treatment for problematic pornography use. *Neuropsychopharmacology* **2017**, *42*, 2021–2031. [CrossRef]
29. Wordecha, M.; Wilk, M.; Kowalewska, E.; Skorko, M.; Łapiński, A.; Gola, M. "Pornographic binges" as a key characteristic of males seeking treatment for compulsive sexual behaviors: Qualitative and quantitative 10-week-long diary assessment. *Behav. Addict.* **2018**, *7*, 433–444. [CrossRef]
30. Duffy, A.; Dawson, D.L.; Das Nair, R. Pornography addiction in adults: A systematic review of definitions and reported impact. *J. Sex. Med.* **2016**, *13*, 760–777. [CrossRef]
31. Eleuteri, S.; Tripodi, F.; Petruccelli, I.; Rossi, R.; Simonelli, C. Questionnaires and scales for the evaluation of the online sexual activities: A review of 20 years of research. *Cyberpsychol. J. Psychosoc. Res. Cyberspace* **2014**, *8*. [CrossRef]
32. Kraus, S.; Rosenberg, H. The pornography craving questionnaire: Psychometric properties. *Arch. Sex. Behav.* **2014**, *43*, 451–462. [CrossRef] [PubMed]
33. Rosenberg, H.; Kraus, S. The relationship of "passionate attachment" for pornography with sexual compulsivity, frequency of use, and craving for pornography. *Addict. Behav.* **2014**, *39*, 1012–1017. [CrossRef] [PubMed]
34. Kalichman, S.C.; Rompa, D. Sexual sensation seeking and sexual compulsivity scales: Validity, and predicting HIV risk behavior. *J. Personal. Assess* **1995**, *65*, 586–601. [CrossRef]
35. Zheng, L.; Zheng, Y. Online sexual activity in Mainland China: Relationship to sexual sensation seeking and sociosexuality. *Comput. Hum. Behav.* **2014**, *36*, 323–329. [CrossRef]
36. Muthen, L. *Mplus Version 7 User's Guide: Version 7*; Muthen & Muthen: Los Angeles, CA, USA, 2012.
37. Orford, J. *Excessive Appetites: A Psychological View of Addictions*; John Wiley & Sons Ltd.: Hoboken, NJ, USA, 2001.
38. López-Fernández, O.; Molina Azorín, J.F. The use of mixed methods research in the behavioral sciences field. *Qual Quant* **2011**, *45*, 1459–1472. [CrossRef]
39. López-Fernández, O.; Molina-Azorín, J.F. The use of mixed methods research in interdisciplinary educational journals. *Int. J. Mult. Res. Approaches* **2011**, *5*, 269–283. [CrossRef]
40. Kafka, M.P. Hypersexual disorder: A proposed diagnosis for DSM-V. *Arch. Sex. Behav.* **2010**, *39*, 377–400. [CrossRef]
41. Grubbs, J.B.; Perry, S.L.; Wilt, J.A.; Reid, R.C. Pornography problems due to moral incongruence: An integrative model with a systematic review and meta-analysis. *Arch. Sex. Behav.* **2019**, *48*, 397–415. [CrossRef]
42. Grubbs, J.B.; Kraus, S.W.; Perry, S.L. Self-reported addiction to pornography in a nationally representative sample: The roles of use habits, religiousness, and moral incongruence. *Behav. Addict.* **2019**, *8*, 88–93. [CrossRef]
43. Bensimon, P. The role of pornography in sexual offending. *Sex. Addict. Compulsivity* **2007**, *14*, 95–114. [CrossRef]

International Journal of
Environmental Research and Public Health

MDPI

Article

Problematic Internet Use, Non-Medical Use of Prescription Drugs, and Depressive Symptoms among Adolescents: A Large-Scale Study in China

Beifang Fan [1,†], Wanxing Wang [2,3,†], Tian Wang [2,3], Bo Xie [1], Huimin Zhang [1], Yuhua Liao [1], Ciyong Lu [2,3] and Lan Guo [2,3,*]

1 Department of Psychiatry, Shenzhen Nanshan Center for Chronic Disease Control, Shenzhen 518000, China; fanbf@foxmail.com (B.F.); sznsxb@163.com (B.X.); zhangzhanghuimin@163.com (H.Z.); liaoyuhua0011@163.com (Y.L.)
2 Department of Medical Statistics and Epidemiology, School of Public Health, Sun Yat-sen University, Guangzhou 510080, China; wgg0808@163.com (W.W.); wangt97@mail2.sysu.edu.cn (T.W.); luciyong@mail.sysu.edu.cn (C.L.)
3 Guangdong Engineering Technology Research Center of Nutrition Translation, Guangzhou, 510080, China
* Correspondence: guolan3@mail.sysu.edu.cn; Tel.: +86-020-8733-5875
† These authors contributed equally to this work.

Received: 28 November 2019; Accepted: 23 January 2020; Published: 26 January 2020

Abstract: This large-scale study aimed to test, among Chinese adolescents, the association between problematic Internet use (PIU), non-medical use of prescription drugs (NMUPD), and depressive symptoms, as well as the mediating effects of NMUPD on the associations above. This study used the data from the 2017 National School-based Chinese Adolescents Health Survey, and 24,345 students' questionnaires qualified for the analyses. Generalized linear mixed models and path models were performed. In the models without mediation, PIU was associated with depressive symptoms (unstandardized β estimate = 0.26, 95% CI = 0.25–0.27); frequent use of opioid or sedative was also related to depressive symptoms (unstandardized β estimate for opioid = 2.77, 95% CI = 1.90–3.63; unstandardized β estimate for sedative = 4.45, 95% CI = 3.02–5.88). Additionally, the results of the path models indicated that opioid misuse partially mediated the association between PIU and depressive symptoms. PIU and opioid/sedative misuse were related to the increased risk of depressive symptoms, respectively. The association above might be complicated, and PIU may elevate the risk of opioid or sedative misuse and depressive symptoms, which in turn could worsen the situation of PIU and vice versa. Multidisciplinary health intervention programs to prevent adolescents involving in PIU, as well as NMPUD, are recommended to be provided.

Keywords: problematic Internet use; non-medical use of prescription drugs; depressive symptoms; adolescents

1. Introduction

Internet use has proliferated dramatically over the last decades and has become an internal part of contemporary life [1]. Excessive or maladaptive use of the Internet has been termed as "problematic Internet use (PIU)" or "Internet addiction", which has been reported to be related to a series of psychosocial impairments and substance-use problems (e.g., depression and drug use) [2,3]. Recently, Internet gaming disorder has been included in the section on addictive disorders in the 11th Revision of the International Classification of Diseases (ICD-11) as a pattern of behavior characterized by impaired control over gaming, indicating that PIU may develop into a severe social problem [4]. Adolescence represents a vulnerable developmental stage between childhood to adulthood, which is characterized by increased levels of novelty seeking and exploration along with a wide range of risk-taking behaviors or mental health problems [5]. The report of the World Health Organization (WHO) showed that adolescent PIU had been a public health concern worldwide, and China is no exception [6]. Considering the brain undergoes rapid development during adolescence, adolescents involved in PIU may be more vulnerable to psychological disorders, such as depressive symptoms.

Depressive symptoms are one of the most common mental disorders among adolescents. Evidence suggests that adolescent depressive symptoms can not only have long-lasting influences on the development of cognitive abilities and social skills into adulthood, but also may lead to clinical depression, placing a heavy financial burden on individuals, families, and society [7]. Previous studies in Western countries reported that approximately 20–50% of adolescents had suffered from long-term depressive symptoms [8], and our prior study in China also showed that about 6.4% of high school students reported having depressive symptoms with suicidal ideation [9]. Although the association between PIU and depressive symptoms has been documented in prior studies [10,11], the influence path of PIU on depressive symptoms is still unclear.

Over the last two decades, the rapid growth of non-medical use of prescription drugs (NMUPD) among adolescents has drawn increased attention [12]. According to the report of the 2014 National Survey on Drug Use and Health (NSDUH) in the United States, prescription drugs were the second most popular type of drugs among adolescents [13]. A recent study also demonstrated that 2.0% of Chinese high school students reported frequent use of opioids, and 1.8% admitted frequent use of sedatives [14]. Previous studies demonstrated that PIU was reported to be associated with substance use, and PIU and substance use may share similar biological characteristics (e.g., similar vulnerable brain regions) [3,15]. Moreover, substance use may affect an individual's brain function, as well as their ability to self-regulate, leading to the development of depressive symptoms [16,17]. Therefore, NMUPD may play a mediator role between PIU and depressive symptoms. First, Caplan's theory proposed that a preference for online social interaction and the use of the Internet for mood regulation can predict deficient self-regulation of Internet use (i.e., PIU), which was a significant predictor of depressive symptoms [18]. Besides, based on the cognitive–behavioral theory of generalized PIU introduced by Davis, depressive symptoms might predispose individuals to develop maladaptive Internet-related cognitions and behaviors that ultimately result in adverse outcomes (e.g., substance use or PIU) [19,20]. Then, there might be two mediation models (incorporating NMUPD as the mediators) in the association between PIU and depressive symptoms from two orders: (1) PIU—NMUPD—depressive symptoms; and (2) depressive symptoms—NMUPD—PIU.

With the modernization of society, Chinese adolescents are more easily exposed to PIU, NMUPD, and depressive symptoms. However, there is a lack of studies in China considering the effects of NMUPD on the association between PIU and depressive symptoms. Therefore, we conducted this large-scale study aimed to test, among Chinese adolescents, the association between PIU, NMUPD, and depressive symptoms, as well as the mediating effects of NMUPD on the association between PIU and depressive symptoms.

2. Materials and Methods

2.1. Study Design and Participants

This study adopted the data from the 2017 National School-based Chinese Adolescents Health Survey (SCAHS), which is an ongoing study about the health risk behaviors and mental health problems among Chinese adolescents (7th–12th grade) conducted by our group. SCAHS performed a series of large-scale cross-sectional surveys every two years since 2007. The 2017 SCAHS utilized a multi-stage stratified cluster random sampling method to recruit a representative sample of adolescents in Guangdong province, and the procedures for data collection have been described in detail in previous publications [21]. Briefly, Guangdong province was firstly categorized into three stratifications according to geographic locations (Yue Dong, Yue Xi, Yue Bei, and Pearl River Delta) and gross domestic product (GDP) per capita (high-level, middle-level, and low-level). Then, we randomly selected two representative cities from each stratification. Considering adolescence is often described as occurring between 13 and 18 years of age (roughly the period of high school for much of the world) [22], in the present study, four vocational high schools and four general high schools were selected from each chosen city. Finally, two classes were randomly selected from each grade within the chosen schools. All available students were invited to participate in this study voluntarily, and 24,345 students' questionnaires were completed and qualified for analyses, resulting in a response rate of 92.1%. The anonymity of the questionnaires was guaranteed to elevate the validity of self-reports of stigmatized behaviors [23], and our research assistants administered these questionnaires during a regular class period without the presence of teachers (to avoid any potential information bias). Of the total sample, 51.5% (12,526) were boys, and the mean age of the students was 15.2 ± 1.8 years. The study procedures were carried out in accordance with the Declaration of Helsinki. The School of Public Health Institutional Review Board of the Sun Yat-sen University approved the study (the ethic code: L201720). All participants were informed about the study, and all provided written informed consents. If the participant was under 18 years of age, a written informed consent letter was obtained from one of the student's parents (or legal guardian).

2.2. Measures

Depressive symptoms were estimated by the Center for Epidemiologic Studies Depression (CES-D) Scale proposed by Radloff (1977), which has been validated and widely used among Chinese adolescents with satisfactory psychometric properties [24], and the Cronbach's alpha for this CES-D scale was 0.88 in the present study (the Cronbach's alpha ranging from 0.85 to 0.90 in Radloff's study). The total CES-D score ranges from 0 to 60, where higher scores indicate more significant depressive symptoms [25]. A cutoff score of 28 points was utilized to identify students at risk for clinical depression, also calling having depressive symptoms. This cutoff score has been used in previous studies in Chinese adolescents [26,27].

PIU was measured by Young's Internet Addiction Test (IAT) proposed by Young (1998), which has been validated and utilized among Chinese adolescents with satisfactory psychometric properties [28,29], and the Cronbach's alpha for IAT was 0.91 in this study. The IAT includes 20 items rated on a five-point Likert scale (from 1 = not at all to 5 = always) [30], where the total IAT score ranges from 20 to 100 and a higher score representing a greater level of inclination to PIU.

In the present study, NMUPD consists of non-medical use of opioids and sedatives. This list of opioids and sedatives was developed according to the report of the Guangdong Food and Drug Administration, with a focus on the prescription medications that have been widely used by adolescent drug abusers in rehabilitation centers of China. Opioids consist of compounded cough syrup with codeine (codeine), compounded licorice tablets (opium), tramadol hydrochloride, and diphenoxylate. Sedatives included diazepam or triazolam (benzodiazepines), compounded aminopyrine phenacetine tablets (barbiturates), and scopolamine hydrobromide tablets (barbiturates). Non-medical use of opioids or sedatives was measured by asking students how many times have you used the prescription medications as mentioned above for a non-medical purpose in the past year, and response options included "never", "once or twice", and "at least 3 times". Students who selected "never" were considered as abstainers, those who answered "once or twice" were thought as experimenters, and those who selected "at least three times" were treated as frequent users [30].

Information on sociodemographic variables were also collected, including gender (1 = boy, 2 = girl), age, living arrangement (responses were coded as "living in a two-parent family" = 1, "living in a single-parent family" = 2, and "living with others" = 3), household socioeconomic status (HSS, available responses were "above average" = 1, "average" = 2, and "below average" = 3), academic performance (responses were also coded as "above average" = 1, "average" = 2, and "below average" = 3). Family relationships, classmate relations, and relationships with teachers were assessed by asking how students perceived their relationships with family members, classmates, and teachers (available responses included "good" = 1, "average" = 2, and "poor" = 3).

2.3. Statistical Analysis

First, descriptive analyses were used to describe sample characteristics, and the *t*-tests and one-way ANOVA tests were performed to compare the differences in the CES-D scores. Descriptive analyses stratified by depressive symptoms were also conducted, and the chi-square tests or *t*-tests were used to describe the distribution of depressive symptoms in categorical or continuous variables. Second, considering this study used a multi-stage sampling design in which students were clustered into classes, generalized linear mixed models were fitted in which classes were treated as groups, and unstandardized β coefficients were reported. Univariable generalized linear mixed models were first conducted to estimate the potential association of PIU or NMUPD with depressive symptoms. Multivariable generalized linear mixed models, in which variables that were significant at the 0.10 level in the univariate analyses or widely reported in the literature were simultaneously incorporated, were performed to test the independent association of PIU or NMUPD with depressive symptoms. Third, multiplicative interaction items were tested by entering a cross-product term for opioid misuse or sedative misuse and PIU along with the main effect terms for each to the multivariable generalized linear mixed models, and *p*-values for the multiplicative interaction were calculated. Fourth, path models utilizing the maximum likelihood approach were conducted to assess the mediating effects of opioid or sedative misuse on the association between PIU and depressive symptoms. We first ordered the variables as follows: PIU—opioid or sedative misuse—depressive symptoms, and then an alternative model with a different order of the variables were assessed: depressive symptoms—opioid or sedative misuse—PIU. Due to the variables of the CES-D and IAT scores being continuous variables, and the measures of opioid and sedative misuse were categorized, standardized probit coefficients, standardized total effects, and standardized indirect effects were reported, with the bias-corrected 95% confidence intervals (CI) estimated using 1000 bootstrap samples. Path model fit indices were also reported, including comparative fit index (CFI; CFI > 0.90 indicating a good fit), root mean square error of approximation (RMSEA, RMSEA < 0.08 indicating an acceptable fit), and standardized root mean square residual (SRMR; SRMR < 0.08 indicating a good fit) [31,32]. All statistical analyses were conducted using SAS 9.2 (SAS Institute, Inc., Cary, NC, USA) and Mplus version 7.0 (Muthén and Muthén). The percentage of missing data of all relevant variables was less than 0.6, and observations with missing data were eliminated in the generalized linear mixed models and path models. All statistical tests were two-sided, and *p*-values less than 0.05 were considered statistically significant.

3. Results

The sample characteristics are shown in Table 1. A total of 13.5% of students reported living with others, and 13.9% reported below average HSS. The proportion of students who reported poor family relationships, classmate relations, and relationship with teachers, were 4.1%, 1.6%, and 2.4%, respectively. The mean IAT scores among the total students were 35.8 (SD: 12.8). A total of 0.6% and 0.3% of students reported frequent use of opioids and sedatives, respectively. The mean CES-D scores were 13.6 (SD: 8.7), and 6.7% of the students reported having depressive symptoms. There were significant differences in the CES-D scores among the variables of gender, living arrangement, HSS, academic performance, family relationships, classmate relations, relationships with teachers, opioid misuse, and sedative misuse ($p < 0.001$). Additionally, significant differences emerged between the students with and without depressive symptoms in the distribution of gender, living arrangement, HSS, academic performance, family relationships, classmate relations, relationships with teachers, IAT scores, opioid misuse, and sedative misuse ($p < 0.001$).

Table 1. Sample characteristics stratified by depressive symptoms among 24,345 adolescents.

Variable	Total	CES-D Scores, Mean (SD)	p-Value *	Depressive Symptoms		p-Value *
				Yes	No	
Total	24,345 (100)	13.6 (8.7)		1631 (6.7)	22,714 (93.3)	
Gender						
Boys	12,526 (51.5)	12.9 (8.6)	<0.001	731 (45.0)	11,795 (52.1)	<0.001
Girls	11,732 (48.2)	14.4 (8.8)		892 (55.0)	10,840 (47.9)	
Missing data	87 (0.4)					
Living arrangement						
Living in two-parent family	18,094 (74.3)	13.3 (8.6)	<0.001	1125 (69.1)	16,969 (74.9)	<0.001
Living in a single-parent family	2905 (11.9)	15.0 (9.4)		259 (15.9)	2646 (11.7)	
Living with others	3281 (13.5)	14.3 (8.8)		243 (14.9)	3038 (13.4)	
Missing data	65 (0.3)					
HSS						
Above average	6942 (28.5)	12.1 (8.1)	<0.001	315 (19.3)	6627 (29.3)	<0.001
Average	13,944 (57.3)	13.8 (8.6)		928 (57.0)	13,016 (57.5)	
Below average	3388 (13.9)	16.3 (9.8)		385 (23.6)	3003 (13.3)	
Missing data	71 (0.3)					
Academic performance						
Above average	9195 (37.8)	12.3 (8.5))	<0.001	508 (31.2)	8687 (38.5)	<0.001
Average	7576 (31.1)	13.5 (8.2)		434 (26.7)	7142 (31.6)	
Below average	7448 (30.6)	15.5 (9.3)		685 (42.1)	6763 (29.9)	
Missing data	126 (0.5)					
Family relationships						
Good	19,899 (81.7)	12.6 (8.0)	<0.001	946 (58.1)	18,953 (83.8)	<0.001
Average	3362 (13.8)	17.6 (9.8)		429 (26.3)	2933 (13.0)	
Poor	986 (4.1)	21.5 (11.9)		254 (15.6)	732 (3.2)	
Missing data	98 (0.4)					
Classmate relations						
Good	19,561 (80.3)	12.5 (7.9)	<0.001	899 (55.3)	18,662 (82.6)	<0.001
Average	4274 (17.6)	17.9 (9.7)		580 (35.7)	3694 (16.4)	
Poor	381 (1.6)	26.3 (13.7)		146 (9.0)	235 (1.0)	
Missing data	129 (0.5)					
Relationship with teachers						
Good	15,695 (64.5)	12.1 (7.9)	<0.001	706 (43.6)	14,989 (66.6)	<0.001
Average	7844 (32.2)	16.2 (9.2)		790 (48.8)	7054 (31.4)	
Poor	576 (2.4)	21.2 (12.6)		124 (7.7)	452 (2.0)	
Missing data	230 (0.9)					
IAT scores, Mean (SD)	35.8 (12.8)	NA		49.3 (16.5)	34.8 (11.9)	<0.001
Opioid misuse						
Abstainers	23,822 (97.9)	13.6 (8.7)	<0.001	1559 (95.6)	22,263 (98.0)	<0.001
Experimenters	384 (1.6)	17.3 (10.0)		49 (3.0)	335 (1.5)	
Frequent users	139 (0.6)	19.2 (9.8)		23 (1.4)	116 (0.5)	
Sedative misuse						
Abstainers	24,095 (99.0)	13.6 (8.7)	<0.001	1588 (97.4)	22,507 (99.1)	<0.001
Experimenters	188 (0.8)	17.4 (9.1)		23 (1.4)	165 (0.7)	
Frequent users	62 (0.3)	23.8 (14.6)		20 (1.2)	42 (0.2)	

Note: * t-tests or one-way ANOVA tests were performed to test the differences in CES-D score, and Chi-square tests or t-tests were used to compare the differences in adolescents with and without depressive symptoms; HSS—household socioeconomic status; IAT—internet addiction test; SD-standard deviation; CES-D—Center for Epidemiologic Studies Depression; NA—not applicable or no data available.

Without adjusting for other variables, Model 1 demonstrated that PIU, opioid misuse, and sedative misuse were respectively associated with the increase of depressive symptoms ($p < 0.001$). After adjusting for gender, living arrangement, HSS, academic performance, family relationships, classmate relations, and relationships with teachers, Model 2 showed that PIU was significantly associated with depressive symptoms (unstandardized β estimate = 0.26, 95% CI = 0.25–0.27), frequent use of opioids was positively related to depressive symptoms (unstandardized β estimate = 2.77, 95% CI = 1.90–3.63), and frequent use of sedatives was also related to the elevation of depressive symptoms (unstandardized β estimate = 4.45, 95% CI = 3.02–5.88). Additionally, Model 3 and Model 4 did not find any significant multiplicative interaction item between non-medical use of opioids/sedatives and PIU (Table 2).

As shown in Figure 1 and Table 3, the model including the mediator (opioid misuse or sedative misuse) showed that after adjusting for significant covariates, opioid misuse partially mediated the positive association between PIU and depressive symptoms, and the estimate of the standardized indirect effect was 0.003 (95% CI = 0.001–0.005). The obtained indices suggested that the model fit the data well: CFI = 0.92; RMSEA = 0.042, 95% CI = 0.030–0.055; SRMR = 0.061. However, the results also demonstrated that the adjusted standardized indirect effects of PIU on depressive symptoms through sedative misuse was not significant ($p > 0.05$), indicating that the relationship between PIU and depressive symptoms was not mediated by sedative misuse. The model fit indices indicated that the model fit the data satisfactorily: CFI = 0.89; RMSEA = 0.044, 95% CI = 0.032–0.047; SRMR = 0.052.

Table 2. Association between problematic Internet use, opioid misuse, sedative misuse, and depressive symptoms among adolescents.

Variable	CES-D scores							
	Model 1		Model 2		Model 3		Model 4	
	β Estimate # (95% CI)	p-Value	β Estimate # (95% CI)	p-Value	β Estimate # (95% CI)	p-Value	β Estimate # (95% CI)	p-Value
Problematic Internet use (1-score increase)	0.30 (0.29–0.31)	<0.001	0.26 (0.25–0.27)	<0.001	0.26 (0.25–0.27)	<0.001	0.26 (0.25–0.27)	<0.001
Opioid misuse (Ref. = Abstainers)								
Experimenters	3.75 (2.82–4.68)	<0.001	2.77 (1.90–3.63)	<0.001	2.42 (0.19–4.65)	0.034	NA	
Frequent users	5.65 (4.11–7.12)	<0.001	4.45 (3.02–5.88)	<0.001	3.95(−0.32–8.22)	0.071	NA	
Sedative misuse (Ref.= Abstainers)								
Experimenters	3.85 (2.54–5.17)	<0.001	2.86 (1.63–4.09)	<0.001	NA		2.53 (−0.86–5.91)	0.144
Frequent users	10.26 (8.03–12.48)	<0.001	7.18 (5.09–9.26)	<0.001	NA		10.85 (5.15–16.56)	<0.001
Interaction item (opioid misuse)								
Experimenters * Problematic Internet use	NA		NA		−0.02 (−0.07–0.03)	0.502	NA	
Frequent users * Problematic Internet use	NA		NA		−0.04 (−0.13–0.05)	0.387	NA	
Interaction item (sedative misuse)								
Experimenters * Problematic Internet use	NA		NA		NA		−0.01 (−0.09–0.08)	0.884
Frequent users * Problematic Internet use	NA		NA		NA		−0.11 (−0.22–0.01)	0.086

Note: # Unstandardized β coefficient. Model 1: Unadjusted generalized linear mixed models. Model 2: The multivariable generalized linear mixed models were adjusted for gender, living arrangement, HSS, academic performance, family relationships, classmate relations, and relationships with teachers. Model 3: The multivariable generalized linear mixed models simultaneously incorporated the interaction item for opioid misuse and problematic Internet use along with the main effect terms for each, and were adjusted for gender, living arrangement, HSS, academic performance, family relationships, classmate relations, and relationships with teachers. Model 4: The multivariable generalized linear mixed models simultaneously incorporated the interaction item for sedative misuse and problematic Internet use along with the main effect terms for each, and were adjusted for gender, living arrangement, HSS, academic performance, family relationships, classmate relations, and relationships with teachers; Ref.—reference; CES-D—Center for Epidemiologic Studies Depression; NA—not applicable or no data available. * The multiplicative interaction between the two items.

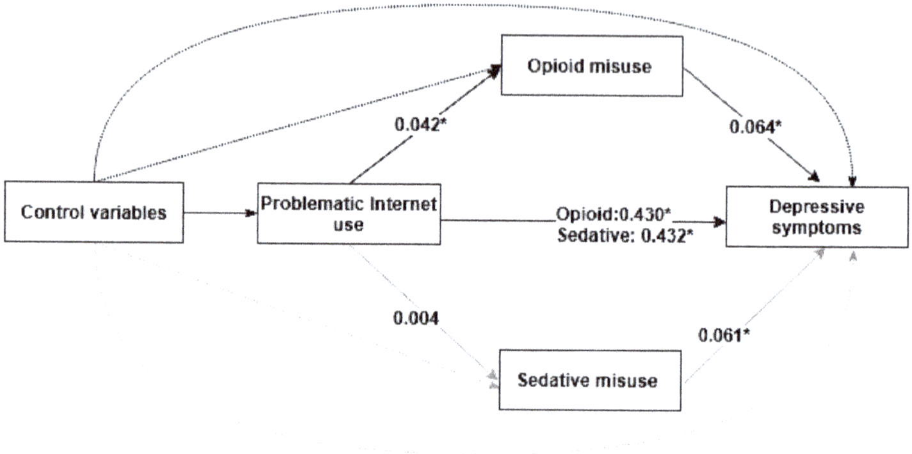

Figure 1. The mediating effects of opioid or sedative misuse on the association between problematic Internet use and depressive symptoms. The solid lines mean the effects between the independent variable, mediator variables, and dependent variable. The dashed lines mean the effects of control variables (gender, living arrangement, HSS, academic performance, family relationships, classmate relations, and relationships with teachers) on the independent variable, mediator variables, and dependent variable.

Table 3. Path analysis showing the effects of problematic Internet use and opioid or sedative misuse on depressive symptoms.

Variable.	Symbol	CES-D Scores	
		Unadjusted Model	Adjusted Model *
		Standardized β Estimate (95% CI)	Standardized β Estimate (95% CI)
Problematic Internet use » Depressive symptoms	Predictor » Outcome	0.437 (0.425–0.449)	0.430 (0.418–0.442)
Opioid misuse » Depressive symptoms	Mediator » Outcome	0.072 (0.058–0.086)	0.064 (0.048–0.080)
Problematic internet use » Opioid misuse	Predictor » Mediator	0.045 (0.033–0.057)	0.042 (0.030–0.054)
Standardized effect			
Indirect		0.003 (0.001–0.005)	0.003 (0.001–0.005)
Total		0.441 (0.429–0.453)	0.430 (0.418–0.442)
Problematic Internet use » Depressive symptoms	Predictor » Outcome	0.4395 (0.4280–0.4518)	0.4318 (0.4202–0.4439)
Sedative misuse » Depressive symptoms	Mediator » Outcome	0.072 (0.058–0.086)	0.061 (0.045–0.077)
Problematic internet use » Sedative misuse	Predictor » Mediator	0.007 (−0.005–0.019)	0.004 (−0.008–0.016)
Standardized effect			
Indirect		0.0005 (−0.002–0.0006)	0.0002 (−0.0018–0.0003)
Total		0.440 (0.428–0.452)	0.432 (0.420–0.444)

Note: * The path models were adjusted for gender, living arrangement, HSS, academic performance, family relationships, classmate relations, and relationships with teachers; CES-D—Center for Epidemiologic Studies Depression.

As shown in Figure 2 and Table 4, the alternative model incorporating the mediator (opioid misuse or sedative misuse) demonstrated that the adjusted indirect effects of depressive symptoms on PIU through opioid misuse were statistically significant (standardized β estimate = 0.002, 95% CI = 0.001–0.002), and the model fit indices were: CFI = 0.91; RMSEA = 0.042, 95% CI = 0.031–0.057; SRMR = 0.063. There were significant standardized indirect effects of depressive symptoms on PIU via sedative misuse (standardized β estimate = 0.001, 95% CI = 0.001–0.001), and the model indices suggested that the model fit the data well: CFI = 0.93; RMSEA = 0.040, 95% CI = 0.029–0.067; SRMR = 0.067. These results suggested that opioid misuse or sedative misuse partially mediated the association of depressive symptoms with PIU, respectively.

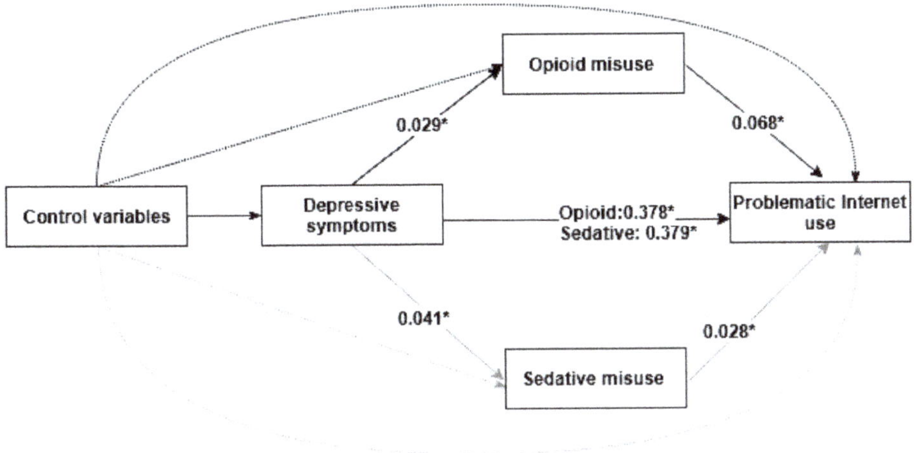

Figure 2. The mediating effects of opioid or sedative misuse on the association between depressive symptoms and problematic Internet use. The solid lines mean the effects between the independent variable, mediator variables, and dependent variable. The dashed lines mean the effects of control variables (gender, living arrangement, HSS, academic performance, family relationships, classmate relations, and relationships with teachers) on the independent variable, mediator variables, and dependent variable.

Table 4. Path analysis showing the effects of depressive symptoms and opioid or sedative misuse on problematic Internet use.

Variable	Symbol	Problematic Internet Use	
		Unadjusted Model	Adjusted Model *
		Standardized β Estimate (95% CI)	Standardized β Estimate (95% CI)
CES-D scores » Problematic Internet use	Predictor » Outcome	0.438 (0.426–0.450)	0.378 (0.366–0.390)
Opioid misuse » Problematic Internet use	Mediator » Outcome	0.076 (0.062–0.090)	0.068 (0.054–0.082)
CES-D scores » Opioid misuse	Predictor » Mediator	0.038 (0.026–0.050)	0.029 (0.017–0.041)
Standardized effect			
Indirect		0.003 (0.001–0.005)	0.002 (0.001–0.002)
Total		0.441 (0.429–0.453)	0.380 (0.368–0.392)
CES-D scores » Problematic Internet use	Predictor » Outcome	0.438 (0.426–0.450)	0.379 (0.367–0.391)
Sedative misuse » Problematic Internet use	Mediator » Outcome	0.039 (0.025–0.053)	0.028 (0.014–0.042)
CES-D scores » Sedative misuse	Predictor » Mediator	0.056 (0.044–0.068)	0.041 (0.029–0.053)
Standardized effect			
Indirect		0.002 (0.001–0.002)	0.001 (0.001–0.001)
Total		0.440 (0.429–0.453)	0.380 (0.368–0.392)

Note: * The path models were adjusted for gender, living arrangement, HSS, academic performance, family relationships, classmate relations, and relationships with teachers; CES-D—Center for Epidemiologic Studies Depression.

4. Discussion

To our knowledge, the present study is the first large-scale study to estimate the direct and indirect association between PIU, opioid or sedative misuse, and depressive symptoms among Chinese adolescents, and to investigate the potential mediating effects of opioid or sedative misuse on the association between PIU and depressive symptoms. This study first found that the mean CES-D scores for the adolescents were13.6 (SD: 8.7), and 6.7% reported having depressive symptoms. These results are parallel with our previous study conducted in 2014 [9], and slightly lower than that described in a prior review revealing a pooled prevalence of depressive symptoms of 24.3% (95% CI, 21.3–27.6%) among adolescents in mainland China [33]. The difference might be related to the variety in the definition of depressive symptoms (i.e., using different measurement scale and cutoff scores), and these results indicate that depressive symptoms among Chinese adolescents are a growing public health problem.

Consistent with previous evidence [9,11], the present study found that there were significant differences in the continuous and categorical variables of depressive symptoms among the groups of gender, living arrangement, HSS, academic performance, family relationships, classmate relations, relationships with teachers, opioid misuse, and sedative misuse. Compared with their corresponding groups, the mean CES-D scores were significantly higher in girls, students living in a single-parent family, students who reported below average HSS or academic performance, and those reporting poor relationships with family members, classmates, or teachers. These results are helpful to identify a profile of adolescents who are vulnerable to depressive symptoms, and particular attention should be paid to the groups with the negative characteristics mentioned above. Moreover, the covariate effects of these variables on the relationship between PIU and depressive symptoms should also be taken into consideration.

In the adjusted generalized linear mixed models without mediation, a positive relationship between depressive symptoms and PIU, opioid misuse, and sedative misuse was found, respectively. First, PIU may increase the risk of depressive symptoms due to adolescents with PIU showing poorer well-being, self-control, and self-esteem, which were reported to be positively associated with psychiatric disorders (including depressive symptoms) [34,35]. Similarly, Park et al. reported that PIU was positively associated with depressive symptoms in Korean adolescents [10]; Dalbudak et al. demonstrated that Internet addiction might increase vulnerability to depressive symptoms in students of Turkey [36]; and Tan et al. found that PIU was significantly related to an increased risk of depressive symptoms among adolescents in Shantou, China [11]. Moreover, the found significant association between opioid or sedative misuse and depressive symptoms might be related to the finding that these drugs may cause negative emotions and urgency, poor concentration, and sleeping problems, which are reported to be associated with an elevated risk of depressive symptoms [17,37]. Adolescents are especially vulnerable to the adverse effects of NMUPD given their still-developing brain [38].

Although there is evidence supporting the role of opioid or sedative misuse in the process through which PIU is related to depressive symptoms, no study before had confirmed the mediating effects of opioid or sedative misuse. Considering mediation analyses require exclusion of an interaction between the exposure and the mediator on the outcome [39], the interaction items between opioid/sedative misuse and PIU on depressive symptoms were first tested, and the results did not find any significant interaction effects. Furthermore, the path models of this study first demonstrated that the association between PIU and depressive symptoms remained significant when opioid or sedative misuse was incorporated as a mediator, and the association above was partially mediated through opioid misuse. These findings were consistent with the theory proposed by Caplan, which showed that PIU might be a reflection of maladaptive cognitions that lead to difficulties with behavioral impulse control and ultimately resulting in adverse outcomes associated with PIU (e.g., NMUPD and depressive symptoms) [18]. Moreover, a possible explanation for the mediating effects of opioid misuse might be that PIU and opioid use may have similarities in biological characteristics [3,15]. PIU may increase the risk of opioid misuse through impaired impulse control [40] and opioid misuse is one of the known factors associated with depressive symptoms [40]. In another aspect, Davis's cognitive–behavioral theory proposed that depressive symptoms may also predispose individuals to develop maladaptive Internet-related cognitions and behaviors that can lead to PIU [19,20]. Moreover, due to the cross-sectional nature of data which may cause bias of the mediating estimates when mediation occurs over time, another path process order was also performed to test the potential the mediating effects opioid or sedative misuse on the association between depressive symptoms and PIU. The present study also demonstrated that opioid and sedative misuse significantly mediated the association of depressive symptoms with PIU, and these results may be related to the finding that adolescents with depressive symptoms may non-medically use opioid or sedatives as a coping strategy to release a negative mood [41], and substance abuse has been reported to be able to predict a high risk of PIU [42]. Taken together, the association between PIU, opioid or sedative misuse, and depressive symptoms might be complicated; for instance, PIU can lead to opioid or sedative misuse and depressive symptoms, which in turn may worsen the problematic Internet use problem among adolescents, and vice versa. The investigation of the mediating effects of opioid or sedative misuse can add evidence for the new understanding of the mechanism of the association between PIU and depressive symptoms.

Based on the findings of the current study, several recommendations for preventing PIU, NMUPD, and depressive symptoms among adolescents are listed below: (1) limiting adolescents' exposure to the Internet for a long time (e.g., playing game); (2) increasing individuals' awareness of the adverse effects of PIU, NMUPD, and depressive symptoms; (3) providing professional health services (e.g., health services provided by clinicians or social workers in the schools or communities) to students who have been involved in PIU, NMUPD, and depressive symptoms. If possible, concomitant treatment for PIU, NMUPD, and depressive symptoms should be considered; and (4) developing a long-term surveillance system to monitor the health-risky behaviors among adolescents in China.

There are several limitations related to this study. First, the use of self-report questionnaires in this study may result in the underestimate of some sensitive data among adolescents (e.g., PIU or NMUPD) for social desirability. Second, due to the cross-sectional nature of the study design, findings should be interpreted with caution, especially regarding the longitudinal indirect and direct effects. Third, because of the school-based study design, the study sample only included students at school, while PIU, NMUPD, or depressive symptoms may be more common among adolescents who were absent from schools. Fourth, PIU and depressive symptoms were measured by the IAT and CES-D scales. Although this measurement of PIU or depressive symptoms has been validated and widely used in previous studies, the answers may still be subjectively biased. Fifth, Axis I comorbidities (such as anxiety or bipolar disorder), which may have an effect on depressive symptoms, were not taken into account in the present study. Despite these limitations, the strength of this study is that it uses a large-scale sample of Chinese adolescents to extend the prior evidence about the association between PIU, opioid or sedative misuse, and depressive symptoms through investigating the mediating effects of opioid or sedative misuse.

5. Conclusions

In conclusion, the present study found that in the models without mediation, PIU and opioid/sedative misuse were positively related to an increased risk of depressive symptoms among adolescents, respectively. Moreover, two path models were conducted to estimate the potential mechanism of the association between PIU, opioid or sedative misuse, and depressive symptoms. The results indicated that the association above might be complex and transactional, and that PIU may elevate the risk of opioid or sedative misuse and depressive symptoms, which in turn could worsen the situation of PIU and vice versa. These study findings have some important implications. Multidisciplinary health intervention programs to prevent adolescents from getting involved in PIU and NMPUD are recommended to be provided, and the concomitant or complex transactional association between PIU, opioid or sedative misuse, and depressive symptoms should be taken into consideration by adolescents, their families and teachers, as well as professional health providers.

Author Contributions: Data curation, L.G., B.F., W.W., T.W., B.X., H.Z. and Y.L.; formal analysis, B.F. and W.W.; funding acquisition, C.L. and L.G.; methodology, L.G.; project administration, C.L. and L.G.; writing—original draft, B.F., W.W. and L.G. All authors have read and agreed to the published version of the manuscript

Funding: This work was supported by National Natural Science Foundation of China (grant No. 81761128030), National Natural Science Foundation of China (grant No. 81903339), Natural Science Foundation of Guangdong Province (grant No. 2018A0303130331; grant No. 2019A1515011091), and the Science Foundation for the Youth Scholars of Sun Yat-sen University (grant No. 18ykpy15). The funders had no role in study design and data analysis.

Acknowledgments: We would like to express our special gratitude and thanks to all the participating schools and students.

Conflicts of Interest: The authors declare no conflict of interest.

References

1. Lin, M.P.; Wu, J.Y.; Chen, C.J.; You, J. Positive outcome expectancy mediates the relationship between social influence and Internet addiction among senior high-school students. *J. Behav. Addict.* **2018**, *7*, 292–300. [CrossRef] [PubMed]
2. An, J.; Sun, Y.; Wan, Y.; Chen, J.; Wang, X.; Tao, F. Associations between problematic internet use and adolescents' physical and psychological symptoms: Possible role of sleep quality. *J. Addict. Med.* **2014**, *8*, 282–287. [CrossRef] [PubMed]
3. Rucker, J.; Akre, C.; Berchtold, A.; Suris, J.C. Problematic Internet use is associated with substance use in young adolescents. *Acta Paediatr.* **2015**, *104*, 504–507. [CrossRef] [PubMed]
4. World Health Organization. Gaming Disoder. Available online: https://www.who.int/features/qa/gaming-disorder/en/ (accessed on 8 December 2019).
5. Tymula, A.; Rosenberg, B.L.; Roy, A.K.; Ruderman, L.; Manson, K.; Glimcher, P.W.; Levy, I. Adolescents' risk-taking behavior is driven by tolerance to ambiguity. *Proc. Natl. Acad. Sci. USA* **2012**, *109*, 17135–17140. [CrossRef] [PubMed]
6. Public Health Implications of Excessive Use of the Internet, Computers, Smartphones and Similar Electronic Devices Meeting report. In *Main Meeting Hall, Foundation for Promotion of Cancer Research National Cancer Research Centre*; World Health Organization: Tokyo, Japan, 2015.
7. Thapar, A.; Collishaw, S.; Pine, D.S.; Thapar, A.K. Depression in adolescence. *Lancet* **2012**, *379*, 1056–1067. [CrossRef]
8. Wesselhoeft, R.; Sorensen, M.J.; Heiervang, E.R.; Bilenberg, N. Subthreshold depression in children and adolescents—A systematic review. *J. Affect. Disord.* **2013**, *151*, 7–22. [CrossRef]
9. Guo, L.; Deng, J.; He, Y.; Deng, X.; Huang, J.; Huang, G.; Gao, X.; Lu, C. Prevalence and correlates of sleep disturbance and depressive symptoms among Chinese adolescents: A cross-sectional survey study. *BMJ Open* **2014**, *4*, e5517. [CrossRef]
10. Park, S.; Hong, K.E.; Park, E.J.; Ha, K.S.; Yoo, H.J. The association between problematic internet use and depression, suicidal ideation and bipolar disorder symptoms in Korean adolescents. *Aust. N. Z. J. Psychiatry* **2013**, *47*, 153–159. [CrossRef]
11. Tan, Y.; Chen, Y.; Lu, Y.; Li, L. Exploring Associations between Problematic Internet Use, Depressive Symptoms and Sleep Disturbance among Southern Chinese Adolescents. *Int. J. Environ. Res. Public Health* **2016**, *13*, 313. [CrossRef]
12. LeClair, A.; Kelly, B.C.; Pawson, M.; Wells, B.E.; Parsons, J.T. Motivations for Prescription Drug Misuse among Young Adults: Considering Social and Developmental Contexts. *Drugs* **2015**, *22*, 208–216. [CrossRef]
13. Center for Behavioral Health Statistics and Quality. *Behavioral Health Trends in the United States: Results from the 2014 National Survey on Drug Use and Health (HHS Publication No. SMA 15-4927, NSDUH Series H-50)*; Substance Abuse and Mental Health Services Administration: Rockville, MD, USA, 2015.
14. Guo, L.; Luo, M.; Wang, W.; Xiao, D.; Xi, C.; Wang, T.; Zhao, M.; Zhang, W.H.; Lu, C. Association between nonmedical use of opioids or sedatives and suicidal behavior among Chinese adolescents: An analysis of sex differences. *Aust. N. Z. J. Psychiatry* **2019**, *53*, 559–569. [CrossRef] [PubMed]
15. Han, D.H.; Hwang, J.W.; Renshaw, P.F. Bupropion sustained release treatment decreases craving for video games and cue-induced brain activity in patients with Internet video game addiction. *Exp. Clin. Psychopharmacol.* **2010**, *18*, 297–304. [CrossRef] [PubMed]
16. Brand, M.; Young, K.S.; Laier, C. Prefrontal control and internet addiction: A theoretical model and review of neuropsychological and neuroimaging findings. *Front. Hum. Neurosci.* **2014**, *8*, 375. [CrossRef] [PubMed]
17. Quello, S.B.; Brady, K.T.; Sonne, S.C. Mood disorders and substance use disorder: A complex comorbidity. *Sci. Pract. Perspect.* **2005**, *3*, 13–21. [CrossRef] [PubMed]
18. Caplan, S.E. Theory and measurement of generalized problematic Internet use: A two-step approach. *Comput. Hum. Behav.* **2010**, *26*, 1089–1097. [CrossRef]
19. Davis, R.A.; Flett, G.L.; Besser, A. Validation of a new scale for measuring problematic internet use: Implications for pre-employment screening. *Cyberpsychol. Behav.* **2002**, *5*, 331–345. [CrossRef]
20. Davis, R.A. A cognitive-behavioral model of pathological Internet use. *Comput. Hum. Behav.* **2001**, *17*, 187–195. [CrossRef]

21. Guo, L.; Luo, M.; Wang, W.X.; Huang, G.L.; Xu, Y.; Gao, X.; Lu, C.Y.; Zhang, W.H. Association between problematic Internet use, sleep disturbance, and suicidal behavior in Chinese adolescents. *J. Behav. Addict.* **2018**, *7*, 965–975. [CrossRef]
22. Crockett, L.J.; Beal, S.J. The life course in the making: Gender and the development of adolescents' expected timing of adult role transitions. *Dev. Psychol.* **2012**, *48*, 1727–1738. [CrossRef]
23. Ramo, D.E.; Liu, H.; Prochaska, J.J. Reliability and validity of young adults' anonymous online reports of marijuana use and thoughts about use. *Psychol. Addict. Behav.* **2012**, *26*, 801–811. [CrossRef]
24. Lee, S.W.; Stewart, S.M.; Byrne, B.M.; Wong, J.P.; Ho, S.Y.; Lee, P.W.; Lam, T.H. Factor structure of the Center for Epidemiological Studies Depression Scale in Hong Kong adolescents. *J. Pers. Assess* **2008**, *90*, 175–184. [CrossRef] [PubMed]
25. Radloff, L.S. The use of the Center for Epidemiologic Studies Depression Scale in adolescents and young adults. *J. Youth Adolesc.* **1991**, *20*, 149–166. [CrossRef] [PubMed]
26. Guo, L.; Xu, Y.; Deng, J.; Huang, J.; Huang, G.; Gao, X.; Wu, H.; Pan, S.; Zhang, W.H.; Lu, C. Association Between Nonmedical Use of Prescription Drugs and Suicidal Behavior Among Adolescents. *JAMA Pediatr.* **2016**, *170*, 971–978. [CrossRef]
27. Yang, H.J.; Soong, W.T.; Kuo, P.H.; Chang, H.L.; Chen, W.J. Using the CES-D in a two-phase survey for depressive disorders among nonreferred adolescents in Taipei: A stratum-specific likelihood ratio analysis. *J. Affect. Disord.* **2004**, *82*, 419–430. [CrossRef] [PubMed]
28. Lai, C.M.; Mak, K.K.; Watanabe, H.; Ang, R.P.; Pang, J.S.; Ho, R.C. Psychometric properties of the internet addiction test in Chinese adolescents. *J. Pediatr. Psychol.* **2013**, *38*, 794–807. [CrossRef] [PubMed]
29. Wang, H.; Zhou, X.; Lu, C.; Wu, J.; Deng, X.; Hong, L. Problematic Internet Use in high school students in Guangdong Province, China. *PLoS ONE* **2011**, *6*, e19660. [CrossRef] [PubMed]
30. Young, K.S. *Caught in the Net: How to Recognize the Signs of Internet Addiction and a Winning Strategy for Recovery*; John Wiley & Sons: New York, NY, USA, 1998.
31. Hu, L.; Bentler, P.M. *Evaluating Model Fit*; Sage Publications, Inc.: Thousand Oaks, CA, USA, 1995.
32. Garver, M.S.; Mentzer, J.T. Logistics research methods: Employing structural equation modeling to test for construct validity. *J. Bus Logist* **1999**, *20*, 33–57.
33. Tang, X.; Tang, S.; Ren, Z.; Wong, D. Prevalence of depressive symptoms among adolescents in secondary school in mainland China: A systematic review and meta-analysis. *J. Affect. Disord.* **2019**, *245*, 498–507. [CrossRef]
34. Mei, S.; Yau, Y.H.; Chai, J.; Guo, J.; Potenza, M.N. Problematic Internet use, well-being, self-esteem and self-control: Data from a high-school survey in China. *Addict. Behav.* **2016**, *61*, 74–79. [CrossRef]
35. Ho, R.C.; Zhang, M.W.; Tsang, T.Y.; Toh, A.H.; Pan, F.; Lu, Y.; Cheng, C.; Yip, P.S.; Lam, L.T.; Lai, C.M.; et al. The association between internet addiction and psychiatric co-morbidity: A meta-analysis. *BMC Psychiatry* **2014**, *14*, 183. [CrossRef]
36. Dalbudak, E.; Evren, C.; Aldemir, S.; Coskun, K.S.; Ugurlu, H.; Yildirim, F.G. Relationship of internet addiction severity with depression, anxiety, and alexithymia, temperament and character in university students. *Cyberpsychol. Behav. Soc. Netw.* **2013**, *16*, 272–278. [CrossRef] [PubMed]
37. Pang, R.D.; Farrahi, L.; Glazier, S.; Sussman, S.; Leventhal, A.M. Depressive symptoms, negative urgency and substance use initiation in adolescents. *Drug Alcohol Depend.* **2014**, *144*, 225–230. [CrossRef] [PubMed]
38. Casey, B.J.; Getz, S.; Galvan, A. The adolescent brain. *Dev. Rev.* **2008**, *28*, 62–77. [CrossRef] [PubMed]
39. Lindley, P.; Walker, S.N. Theoretical and methodological differentiation of moderation and mediation. *Nurs. Res.* **1993**, *42*, 276–279. [CrossRef]
40. Yau, Y.H.; Potenza, M.N.; White, M.A. Problematic Internet Use, Mental Health and Impulse Control in an Online Survey of Adults. *J. Behav. Addict.* **2013**, *2*, 72. [CrossRef]
41. Khantzian, E.J. The self-medication hypothesis of substance use disorders: A reconsideration and recent applications. *Harv. Rev. Psychiatry* **1997**, *4*, 231–244. [CrossRef]
42. Lee, Y.S.; Han, D.H.; Kim, S.M.; Renshaw, P.F. Substance abuse precedes Internet addiction. *Addict. Behav.* **2013**, *38*, 2022–2025. [CrossRef]

International Journal of
Environmental Research and Public Health

MDPI

Article

Problematic Smartphone Use and Mental Health in Chinese Adults: A Population-Based Study

Ningyuan Guo [1], Tzu Tsun Luk [1], Sai Yin Ho [2], Jung Jae Lee [1], Chen Shen [3], John Oliffe [4], Sophia Siu-Chee Chan [1], Tai Hing Lam [2] and Man Ping Wang [1,*]

1 School of Nursing, University of Hong Kong, Hong Kong, China; nyguo@hku.hk (N.G.); luktt@hku.hk (T.T.L.); leejay@hku.hk (J.J.L.); nssophia@hku.hk (S.S.-C.C.)
2 School of Public Health, University of Hong Kong, Hong Kong, China; syho@hku.hk (S.Y.H.); hrmrlth@hku.hk (T.H.L.)
3 Department of Epidemiology and Biostatistics, Imperial College London, London W2 1PG, UK; chen.shen@imperial.ac.uk
4 School of Nursing, University of British Columbia, Vancouver, BC V6T 2B5, Canada; John.Oliffe@ubc.ca
* Correspondence: mpwang@hku.hk; Tel.: +852-3917-6636; Fax: +852-2872-6079

Received: 2 December 2019; Accepted: 24 January 2020; Published: 29 January 2020

Abstract: Problematic smartphone use (PSU) has been associated with anxiety and depression, but few explored its mental well-being correlates that could co-occur with or be independent of mental symptoms. We studied the associations of PSU with anxiety, depression, and mental well-being in Hong Kong Chinese adults in a probability-based survey (N = 4054; 55.0% females; mean age ± SD 48.3 ± 18.3 years). PSU was measured using Smartphone Addiction Scale-Short Version. Anxiety and depression symptoms were evaluated using General Anxiety Disorder screener-2 (GAD-2) and Patient Health Questionnaire-2 (PHQ-2). Mental well-being was measured using Subjective Happiness Scale (SHS) and Short Warwick-Edinburgh Mental Well-Being Scale (SWEMWBS). Multivariable regression analyzed associations adjusting for sociodemographic and lifestyle-related variables. Associations of PSU with mental well-being were stratified by symptom severity of anxiety (GAD-2 cutoff of 3) and depression (PHQ-2 cutoff of 3). We found that PSU was associated with higher odds of anxiety and depression symptom severity and lower scores of SHS and SWEMWBS. Associations of PSU with lower SHS and SWEMWBS scores remained in respondents who screened negative for anxiety or depression symptoms. To conclude, PSU was associated with anxiety, depression, and impaired mental well-being. Associations of PSU with impaired mental well-being could be independent of anxiety or depression symptoms.

Keywords: problematic smartphone use; smartphone addiction; anxiety; depression; mental well-being; population-based study

1. Introduction

The evolving mobile information and communication technologies (ICTs) have raised debates about the potential effects on mental health. Instant messaging (IM) and social networking sites (SNS) can be used for unpleasant moods avoidance, social contacts, and relationship maintenance especially in females [1], whereas a practical and instrumental use such as for social position advancement was found in males [2]. In contrast, many studies have shown the associations between poor mental health outcomes and excessive or intensive use of phone calls, texts, IM, emails, and SNS [3]. These are privileged applications of problematic smartphone use (PSU), an impaired ability to control smartphone use with core symptoms such as loss of control, tolerance, and withdrawal shared with gaming disorder and substance use disorders [4]. PSU has been associated with an array of health outcomes, including self-reported dependence [5], cyberbullying [5], traffic accidence [5], physical symptoms (e.g., eye strain and fatigue) [6], and sleep disturbances [7,8]. Our previous study also showed the lower levels of family communication and well-being associated with PSU [9].

Research in adolescents and young adults have identified psychopathological correlates with PSU, with affective disorders including anxiety and depression were most studied [1]. Small to medium effect size associations were observed between PSU and the severity of anxiety and depression symptoms [10]. Longitudinal studies further supported the predicting effects of PSU on anxiety and depression in college students [7,8]. Although these age groups were deemed to be at higher risk for PSU because of developing self-control and more access to the ICTs, increasing prevalence of both smartphone ownership and PSU in adults of a wider age range warranted general population studies [11–14].

The definition of mental health has been established as not the mere absence of psychopathological symptoms, which leads researchers to broaden the investigation field to positive psychology [15]. Mental well-being investigates hedonic well-being that includes affective emotions and cognitive assessments of life satisfaction and eudemonic well-being that includes psychological functioning and self-realization [16]. Our previous study showed that people with greater mental well-being tended to have lower risks for anxiety and depression [16]. Despite the correlations, the dual continuum model of mental health proposes that mental illness and well-being are on two distinct dimensions [17]. This notion was supported by studies in adults reporting greater well-being despite concurrent affective disorders or impaired mental well-being but without mental illness [18,19]. One study using data from a sample of college students showed that PSU was negatively correlated with mental well-being outcomes [20]. Little is known about whether such associations could co-occur with or operate independently from affective disorders.

The present study took advantage of a representative population-based survey in Chinese adults in Hong Kong, where the smartphone penetration rate (88.6% in people aged 10 years or above in 2017) has been among the highest worldwide due to advanced cyber-infrastructure and low-cost Internet access [21]. We aimed: (a) to confirm the associations of PSU with anxiety and depression symptom severity in the general population; (b) to examine the associations of PSU with mental health outcomes including hedonic and eudemonic well-being; (c) to examine the associations of PSU with mental well-being outcomes with stratification of symptom severity of anxiety and depression.

2. Materials and Methods

2.1. Participants and Procedure

The Hong Kong Family and Health Information Trends Survey (FHInTS) is a periodic territory-wide telephone survey on the general public's behaviors and views regarding information use, individual health, and family well-being, under the project named "FAMILY: A Jockey Club Initiative for a Harmonious Society." The target population was Cantonese-speaking Hong Kong residents aged 18 years old or above. We have conducted five waves of FHInTS since 2009 and reported details of the study design elsewhere [9,11,16]. The present survey was conducted from February to August 2017, as part of the fifth FHInTS. A two-stage probability-based sampling procedure was used. In the first stage, landline telephone numbers were randomly generated using known prefixes assigned to telecommunication services providers under the numbering plan provided by the Office of the Communication Authority. Invalid numbers were then removed according to the computer and manual dialing records. Numbers of successful cases from previous FHInTS were also filtered. In the second stage, once a household was successfully reached, an eligible family member who would have the nearest next birthday was invited to the survey. Interviews were conducted by trained interviewers from the Public Opinion Programme (POP) at the University of Hong Kong. Of 5773 respondents invited, 4054 were successfully interviewed, yielding a response rate of 70.2%.

2.2. Measurements

The ten-item Smartphone Addiction Scale-Short Version (SAS-SV) measured five addiction-like symptoms of PSU, including daily-life disturbance, withdrawal, cyberspace-oriented relationship, overuse, and tolerance [22]. Each item scores on a Likert scale of 1 (strongly disagree) to 6 (strongly agree), with a higher total score (range 10 to 60) indicating a higher PSU level [22]. The Chinese version of SAS-SV was valid and reliable (Cronbach's alpha 0.84) and indicated acceptable fit for the one-factor model (comparative fit index [CFI] 0.98 [> 0.90 acceptable, > 0.95 excellent]; root mean square error of approximation [RMSEA] 0.08 [< 0.08 acceptable, < 0.05 excellent]; non-normed fit index [NNFI] 0.96 [> 0.95 acceptable]) in our previous study using the same sample of the Hong Kong general population [11]. The convergent validity was adequate, with composite reliability of 0.85 (CR > 0.70 acceptable) and average variance extracted of 0.37 (AVE > 0.40 acceptable if CR > 0.60).

The four-item Patient Health Questionnaire (PHQ-4) includes the two-item General Anxiety Disorder screener (GAD-2) and the two-item Patient Health Questionnaire (PHQ-2) [23]. PHQ-2 has two DSM-IV diagnostic core criteria for major depression disorder, and GAD-2 has two core criteria for generalized anxiety disorder that can also screen for panic and social anxiety disorders [23]. Each item scores on a Likert scale of 0 (not at all) to 3 (nearly every day), with total scores of each subscale ranging 0 to 6 [23]. GAD-2 and PHQ-2 Scores of ≥ 3 are recommended to screen positive for anxiety and depression symptoms, respectively [24]. The Chinese version of PHQ-2 has been validated in our previous study in the Hong Kong general population [25]. In the present sample, confirmatory factor analysis (CFA) indicated an excellent fit for the two-factor model of PHQ-4 (Relative Chi-Square 0.24; p = 0.63; incremental fit index [IFI] 1.00 [> 0.95 excellent]; goodness of fit index [GFI] 1.00 [> 0.95 excellent]; adjusted goodness of fit index [AGFI] 1.00 [> 0.95 acceptable]; CFI 1.00; RMSEA 0.01; standardized root mean square residual [SRMR] 0.001 [< 0.08 acceptable, < 0.05 excellent]). GAD-2 had a Cronbach's alpha of 0.74, CR of 0.65, and AVE of 0.48. PHQ-2 had a Cronbach's alpha of 0.73, CR of 0.63, and AVE of 0.46.

The four-item Subjective Happiness Scale (SHS; mean score range 1 to 7) measured the cognitive and affective state characterized by pleasure or satisfaction from hedonic well-being aspect [26]. The Chinese version of SHS has been validated in our previous study in the Hong Kong general population [27]. In the present sample, CFA indicated an excellent fit for the one-factor model (relative Chi-Square 21.68; p < 0.001; IFI 0.993; GFI 0.999; AGFI 0.995; CFI 0.993; RMSEA 0.10; SRMR 0.02). SHS had a Cronbach's alpha of 0.75, CR of 0.77, and AVE of 0.47. The seven-item Short Warwick-Edinburgh Mental Well-Being Scale (SWEMWBS; total score range 7 to 35) was adapted by the WEMWBS that covers both hedonic and eudemonic aspects [28]. A randomly selected subset (n = 1331, 32.8%) were asked the frequency of feeling optimistic about the future, useful, relaxed, close to other people, dealing with problems well, thinking clearly, making up their own mind about things in past two weeks [28]. The Chinese version of SWEMWBS was valid and reliable (Cronbach's alpha 0.85) and indicated excellent fit for the one-factor model (CFI 0.995; RMSEA 0.03; SRMR 0.04; normed-fit index [NFI] 0.991 [> 0.95 acceptable]) in our previous study using the same sample of the Hong Kong general population [16]. The convergent validity was adequate, with CR of 0.85 and AVE of 0.44.

Sociodemographic and lifestyle-related variables included sex, age, marital status, employment status, educational attainment, monthly household income, smoking, and alcohol drinking. This list of potential confounders was selected based on prior associations with PSU and mental health outcomes [11,29,30].

2.3. Statistical Analysis

We checked the distributions of all variables independently, with a skewness value of ≤ |2.0| and a kurtosis value of ≤ |7.0| indicating the normality [31]. All data were weighted by age, sex, and educational attainment distribution of the Hong Kong general population using the random iterative method [32]. Missing data were handled by available case analyses as there were minimal missing values for all variables (< 2.6%). We examined the Spearman correlations of anxiety and depression with mental well-being outcomes, as scores of GAD-2 and PHQ-2 were not normally distributed. Moderate correlation coefficients (r) were observed with scores of SHS (r range −0.35 to −0.38; both $p < 0.001$) and SWEMWBS (both $r = -0.42$; both $p < 0.001$). We examined the associations of SAS-SV score with the odds of severity of anxiety and depression symptoms using bivariate and multivariable logistic regression analyses adjusting for sociodemographic and lifestyle-related variables. Bivariate and multivariable linear regression analyses examined the associations of SAS-SV score with scores of SHS and SWEMWBS. We further stratified the associations of SAS-SV score with scores of SHS and SWEMWBS by symptom severity of anxiety (GAD-2 cutoff of 3) and depression (PHQ-2 cutoff of 3). The interaction effects of anxiety and depression symptoms on the associations of SAS-SV score with mental well-being outcomes were examined using adjusted Wald tests. Values of $p < 0.05$ were considered statistically significant. All analyses were conducted using Stata/MP 15.1 (StataCorp LP, College Station, TX, USA), except for CFA that were conducted using LISREL 9.30 with diagonally weighted least square estimation.

2.4. Ethics

The Institutional Review Board of the University of Hong Kong/Hospital Authority Hong Kong West Cluster approved this study. All respondents provided verbal informed consent. Telephone interviews were tape-recorded for quality checking with respondents' consent. Records were then erased six months after completing the survey.

3. Results

The weighted sample comprised 55.0% females, and the mean age ± standard deviation (SD) was 48.3 ± 18.3 years (Table 1). Smartphone owners (80.0%) reported a mean SAS-SV score ± SD of 28.9 ± 10.1. Using cutoff points ≥ 3 of GAD-2 and PHQ-2, 10.9% ($n = 442$) and 8.1% ($n = 327$) respondents screened positive for anxiety and depression symptoms, respectively. Mean scores of SHS and SWEMWBS were 5.2 ± 1.0 and 23.0 ± 4.2, respectively.

Table 1. Sociodemographic and lifestyle-related characteristics of the sample ($N = 4054$).

Sample Characteristics	Non-Weighted n (%)	Weighted [a] n (%)
Sex		
Male	1535 (37.9)	1826 (45.0)
Female	2519 (62.1)	2228 (55.0)
Age, years		
18–24	417 (10.3)	370 (9.1)
25–44	573 (14.1)	1436 (35.4)
45–64	1437 35.5)	1498 (37.0)
≥ 65	1627 (40.1)	750 (18.5)
Marital status		
Unmarried	852 (21.0)	1164 (28.7)
Cohabitated/married	2577 (63.6)	2478 (61.1)
Divorced/separated/widowed	625 (15.4)	413 (10.2)
Employment status		
Unemployment	128 (3.2)	221 (5.5)
In paid employment	1279 (31.6)	1935 (47.7)
Retired	1632 (40.3)	919 (22.7)
Housekeeper	732 (18.1)	722 (17.8)
Full-time student	283 (7.0)	257 (6.3)
Educational attainment		
Primary or below	959 (23.7)	959 (23.7)
Secondary	1730 (42.7)	1949 (48.1)
Tertiary	1365 (33.7)	1145 (28.3)

Table 1. *Cont.*

Sample Characteristics	Non-Weighted *n* (%)	Weighted [a] *n* (%)
Monthly household income (HK $ [b])		
≤19999	1671 (41.2)	1465 (36.1)
≥20000	2258 (55.7)	2486 (61.3)
Unsteady	125 (3.1)	103 (2.5)
Smoking		
Never	3368 (83.1)	3211 (79.2)
Former	435 (10.7)	467(11.5)
Current	251 (6.2)	376 (9.3)
Alcohol drinking		
Never	2065 (51.0)	1908 (47.1)
Former	256 (6.3)	228 (5.6)
Current	1731 (42.7)	1916 (47.3)
Smartphone ownership		
No	1126 (27.8)	811 (20.0)
Yes	2928 (72.2)	3243 (80.0)
SAS-SV score (range 10–60; mean ± SD)	28.2 ± 10.0	28.9 ± 10.1
Anxiety		
Negative (GAD-2 < 3)	3655 (90.2)	3611 (89.1)
Positive (GAD-2 ≥ 3)	397 (9.8)	442 (10.9)
Depression		
Negative (PHQ-2 < 3)	3777 (93.4)	3718 (91.9)
Positive (PHQ-2 ≥ 3)	269 (6.7)	327 (8.1)
SHS score (range 1–7; mean ± SD)	5.2 ± 1.0	5.2 ± 1.0
SWEMWBS score [c] (range 7–35; mean ± SD)	23.4 ± 4.5	23.0 ± 4.2

SAS-SV, Smartphone Addiction Scale-Short Version; GAD-2, Generalized Anxiety Disorder Questionnaire-2 (range 0–6); PHQ-2, Patient Health Questionnaire-2 (range 0–6); SHS, Subjective Happiness Scale; SWEMWBS, Short Warwick-Edinburgh Mental Well-Being Scale. [a] Weighted by age, sex, and educational attainment distributions of the Hong Kong general population. [b] US $1 = HK $7.8. [c] Subset sample ($n$ = 1331).

Higher SAS-SV scores were observed for respondents who screened positive for symptoms of anxiety (31.9 ± 9.4 vs. 28.5 ± 10.2; $p < 0.001$) and depression (33.2 ± 10.2 vs. 28.5 ± 10.0; $p < 0.001$) than those with negative screening results (Table 2). Multivariable analyses showed that a 1-unit increase in SAS-SV score (range 10–60) was associated with a 3% increase in odds of severity of anxiety symptoms (adjusted odds ratio [AOR] = 1.03; 95% CI: 1.01, 1.04) and a 4% increase in odds of severity of depression symptoms (AOR = 1.04; 95% CI: 1.03, 1.06) after adjusting for sociodemographic and health-related variables.

Table 2. Odds of anxiety and depression associated with SAS-SV score (N = 4054).

Outcomes Associated with SAS-SV Score	SAS-SV Score (Mean ± SD) [a]	Unadjusted Association		Adjusted [b] Association	
		Odds Ratio (95% CI)	*p*	Odds Ratio (95% CI)	*p*
Anxiety					
Negative (GAD-2 < 3)	28.5 ± 10.2	1		1	
Positive (GAD-2 ≥ 3)	31.9 ± 9.4	1.03 (1.02, 1.04)	<0.001	1.03 (1.01, 1.04)	<0.001
Depression					
Negative (PHQ-2 < 3)	28.5 ± 10.0	1		1	
Positive (PHQ-2 ≥ 3)	33.2 ± 10.2	1.04 (1.03, 1.06)	<0.001	1.04 (1.03, 1.06)	<0.001

SAS-SV, Smartphone Addiction Scale-Short Version (range 10-60); CI, Confidence Interval; GAD-2, Generalized Anxiety Disorder Questionnaire-2 (range 0–6); PHQ-2, Patient Health Questionnaire-2 (range 0–6). [a] Weighted by age, sex, and educational attainment distributions of the Hong Kong general population. [b] Adjusted for age, sex, marital status, employment status, educational attainment, monthly household income, smoking, and alcohol drinking.

Multivariable analyses showed that a 1-standard-deviation increase in SAS-SV score was associated with a 0.07-standard-deviation decrease in SHS score (adjusted B = −0.07; SE: 0.002; $p < 0.001$) and a 0.10-standard-deviation decrease in SWEMWBS score (adjusted B = -0.10; SE: 0.01; $p = 0.002$) (Table 3). Reduction in effect size of these associations were observed after stratifications by screening results of anxiety and depression symptoms, except for the stronger association of SAS-SV score with lower SHS score in respondents who screened positive for anxiety symptoms (adjusted B = −0.16; SE: 0.01; $p = 0.013$). The associations of SAS-SV score with lower SHS score remained in respondents who screened negative for anxiety symptoms (adjusted B = −0.04; SE: 0.002; $p = 0.040$) and depression symptoms (adjusted B = −0.05; SE: 0.002; $p = 0.014$). The associations of SAS-SV score with lower SWEMWBS score remained in respondents who screened negative for anxiety symptoms (adjusted B = −0.08; SE: 0.01; $p = 0.022$) and were marginally significant in those who screened negative for depression symptoms (adjusted B = −0.06; SE: 0.01; $p = 0.054$). We observed no interaction effects of anxiety or depression symptoms on the associations of SAS-SV score with lower scores of SHS and SWEMWBS.

Table 3. Standardized beta (B) of SHS and SWEMWBS scores associated with SAS-SV score ($N = 4054$).

Outcomes Associated with SAS-SV Score	Stratification	Unadjusted Association				Adjusted [a] Association				*p* for Interaction [a]
		B (SE)	*p*	F (df = 1)	r^2	B (SE)	*p*	F (df = 21)	R^2	
SHS score (range 1–7)	**Overall**	−0.08 (0.002)	<0.001	16.99	0.01	−0.07 (0.002)	<0.001	9.26	0.06	–
	Anxiety									0.197
	Negative (GAD−2 < 3)	−0.04 (0.002)	0.023	5.15	0.002	−0.04 (0.002)	0.040	7.16	0.06	
	Positive (GAD−2 ≥ 3)	−0.11 (0.01)	0.076	3.17	0.01	−0.16 (0.01)	0.013	0.88	0.07	
	Depression									0.626
	Negative (PHQ−2 < 3)	−0.05 (0.002)	0.008	7.13	0.003	−0.05 (0.002)	0.014	7.89	0.06	
	Positive (PHQ−2 ≥ 3)	−0.01 (0.01)	0.864	0.03	0.0002	−0.03 (0.01)	0.707	0.61	0.07	
SWEMWBS score (range 7–35) [b]	**Overall**	−0.11 (0.01)	0.001	10.84	0.01	−0.10 (0.01)	0.002	5.72	0.12	–
	Anxiety									0.565
	Negative (GAD−2 < 3)	−0.09 (0.01)	0.014	6.10	0.01	−0.08 (0.01)	0.022	4.54	0.11	
	Positive (GAD−2 ≥ 3)	0.01 (0.03)	0.950	0	0	0.02 (0.04)	0.866	1.00	0.19	
	Depression									0.982
	Negative (PHQ−2 < 3)	−0.07 (0.01)	0.028	4.87	0.01	−0.06 (0.01)	0.054	4.85	0.11	
	Positive (PHQ−2 ≥ 3)	−0.03 (0.03)	0.826	0.05	0.001	−0.02 (0.04)	0.895	1.75	0.42	

SAS−SV, Smartphone Addiction Scale−Short Version (range 10–60); SE, Standardized Error; GAD−2, Generalized Anxiety Disorder Questionnaire−2 (range 0–6); PHQ−2, Patient Health Questionnaire−2 (range 0–6); SHS, Subjective Happiness Scale; SWEMWBS, Short Warwick−Edinburgh Mental Well−Being Scale. [a] Adjusted for age, sex, marital status, employment status, educational attainment, monthly household income, smoking, and alcohol drinking. [b] Subset sample ($n = 1331$).

4. Discussion

With a representative sample of Chinese adults in Hong Kong, we confirmed the associations of PSU with anxiety and depression in the general population. Few studies of potential mental health effects of PSU have incorporated both mental illness and mental well−being outcomes. We provided the first evidence of the associations of PSU with impaired hedonic and eudemonic well−being, which remained in respondents who screened negative for anxiety or depression symptoms.

Our study built on young people studies to indicate that the associations of PSU with anxiety and depression could have expanded to adults of all ages. The associations can be explained by the time displacement hypothesis that posits a possible tradeoff between smartphone activities and offline healthier activities such as social interactions [33]. Our previous study supported this explanation by showing that PSU was associated with lower levels of perceived family communication and family well–being [9]. This lack of social support can induce the onset of affective disorders such as anxiety and depression [34]. Other studies showed that PSU symptoms such as overuse and tolerance could risk people to prolong the night–time smartphone usage, which might lead to sleep problems that could mediate the pathway to anxiety and depression [35]. Increasing evidence has suggested that the most problematic application correlated with PSU could be SNS, which could expose people to negative social comparisons with others in perceived more favorable lives and induce affective disorders [36–38]. However, people with symptoms of mental illness might be at higher risk for PSU given the smartphone could be the first and most obvious process to deflect negative cognition and affectivity [39]. Mechanisms in this potential reverse causality can include cognitive– and affective–related maladaptive coping strategies such as repetitive negative thinking and emotion dysregulation [39]. The bidirectional association was hence possible and evident by the reciprocal relations found in prospective cohort studies in young people [7,40].

We observed the association of PSU with lower scores of subjective happiness (i.e., hedonic well–being), which is characterized by affectivity of pleasure and cognition of satisfaction [16]. This finding was consistent with studies of Internet addiction with lower levels of happiness and life satisfaction in young people [41,42]. An intervention restricting night–time smartphone usage also reported the reduced PSU risk and increased levels of subjective happiness at one–week follow–up [43]. PSU was associated with lower scores of SWEMWBS that covers both hedonic and eudemonic aspects of mental well–being in the present study. In contrast, a study showed the improved mental well–being in self–concealers who intentionally withhold personal information in face–to–face settings but engaged more in online communication even driven by PSU [44]. These conflicting findings highlighted the important role of personality traits when evaluating the potential effects of PSU and suggested to balance our findings with potential benefits of ICTs usage such as fostering social inclusion among those who may feel excluded [45].

The magnitude of the association of PSU with impaired subjective happiness increased in respondents who screened positive for anxiety symptoms, which suggested the co–occurrence of lower levels of hedonic well–being with anxiety disorder. This finding can be supported by the cognitive–and affective–related coping processes in the pathway from anxiety symptoms to PSU [39]. Another explanation can be the moderate correlation between scores of SHS and GAD–2 found in the present sample. Previous studies also showed that people with symptoms of affective disorders had lower levels of mental well–being than those without [46]. Despite the correlated relations, the independence of mental well–being from mental illness was supported by the remained associations of PSU with mental well–being in respondents who screened negative for anxiety or depression. This finding provided insights that the absence of psychopathological symptoms might have non–buffering effects on the impaired mental well–being outcomes associated with PSU.

Our findings need to be interpreted with caution. Consistent with a systematic review that reported the small effect size associations of PSU with symptom severity of anxiety and depression (adjusted B range 0.12 to 0.18) [10], a 1–unit increase in SAS–SV score was associated with 3%–4% increase in the odds of positive screening results of anxiety and depression in the present study. The small effect size was also observed for the association of PSU with mental well–being outcomes (adjusted B range 0.07 to 0.10). A study across three large–scale datasets (total N = 355358) showed a much smaller association (median adjusted B = −0.07) of adolescents' digital technologies use with combined mental illness and well–being outcomes [47]. However, unaccounted factors might affect both PSU and mental health in such cross–sectional associations. Longitudinal and experimental studies are warranted to distill causal and predictive models.

One of the study's limitations is that the cross–sectional data restricts the causal inference of the findings. Residual confounding by unmeasured or unknown confounders might exist even after adjusting for many sociodemographic and lifestyle–related variables. We used the landline telephone survey. Sampling bias might exist due to the lack of data on mobile phone–only households that may have different smartphone use patterns. To increase the sample's representativeness, we weighted data according to the age, sex, and educational attainment distributions of the Hong Kong general population. We used self−reported data, which are subject to recall bias and social desirability bias. Future studies of PSU could include behavioral methods for collecting data on smartphone use, such as objectively examining participants' screen time and usage of individual apps. We used screening instruments rather than diagnostic instruments to measure PSU and mental health outcomes. However, more accurate diagnoses by face−to−face assessments in clinical settings would limit the generalizability of findings compared with the population−based study.

5. Conclusions

Our population−based study indicated that PSU was associated with higher odds of severity of anxiety and depression symptoms and lower levels of mental well−being. The associations of PSU with impaired mental well−being outcomes could remain in people who screened negative for anxiety or depression symptoms. Such findings highlighted the importance to investigate other psychological constructs with PSU, particularly from aspects of positive psychology. Longitudinal and experimental studies are warranted to explain causality and possible mechanisms of the associations between PSU and mental health.

Author Contributions: Conceptualization, M.P.W., S.Y.H., C.S., S.S.-C.C., and T.H.L.; Formal analysis, N.G.; Funding acquisition, T.H.L.; Methodology, M.P.W., S.Y.H., C.S., and T.H.L.; Supervision, M.P.W. and S.Y.H.; Writing—original draft, N.G.; Writing—review & editing, N.G., T.T.L., S.Y.H., J.J.L., C.S., J.O., T.H.L., and M.P.W. All authors have read and agreed to the published version of the manuscript.

Funding: This work was supported by the Hong Kong Jockey Club Charities Trust as part of the project: "FAMILY: A Jockey Club Initiative for a Harmonious Society" (https://www.family.org.hk/en).

Acknowledgments: This project was funded by the Hong Kong Jockey Club Charities Trust. We would like to thank the respondents who completed the telephone surveys and the Public Opinion Programme (HKU) for conducting the interviews.

Conflicts of Interest: The authors declare no conflict of interest. The funders had no role in the design of the study; in the collection, analyses, or interpretation of data; in the writing of the manuscript, or in the decision to publish the results.

References

1. De-Sola Gutiérrez, J.; Rodríguez de Fonseca, F.; Rubio, G. Cell−phone addiction: A review. *Front. Psychiatry* **2016**, *7*, 175. [CrossRef] [PubMed]
2. Ling, R. Adolescent girls and young adult men: Two sub−cultures of the mobile telephone. *Revista de Estudios de juventud* **2002**, *52*, 33–46.
3. Thomée, S. Mobile phone use and mental health. A review of the research that takes a psychological perspective on exposure. *Int. J. Environ. Res. Public Health* **2018**, *15*, 2692. [CrossRef] [PubMed]
4. Chóliz, M. Mobile phone addiction: A point of issue. *Addiction* **2010**, *105*, 373–374. [PubMed]
5. Billieux, J.; Maurage, P.; Lopez-Fernandez, O.; Kuss, D.J.; Griffiths, M.D. Can disordered mobile phone use be considered a behavioral addiction? An update on current evidence and a comprehensive model for future research. *Curr. Addict. Rep.* **2015**, *2*, 156–162. [CrossRef]
6. Xie, X.; Dong, Y.; Wang, J. Sleep quality as a mediator of problematic smartphone use and clinical health symptoms. *J. Behav. Addict.* **2018**, *7*, 466–472. [CrossRef]
7. Liu, S.; Wing, Y.K.; Hao, Y.; Li, W.; Zhang, J.; Zhang, B. The associations of long−time mobile phone use with sleep disturbances and mental distress in technical college students: A prospective cohort study. *Sleep* **2019**, *42*, 1–10. [CrossRef]
8. Thomée, S.; Eklöf, M.; Gustafsson, E.; Nilsson, R.; Hagberg, M. Prevalence of perceived stress, symptoms of depression and sleep disturbances in relation to information and communication technology (ICT) use among young adults—an explorative prospective study. *Comput. Hum. Behav.* **2007**, *23*, 1300–1321. [CrossRef]

9. Guo, N.; Wang, M.P.; Luk, T.T.; Ho, S.Y.; Fong, D.Y.T.; Chan, S.S.; Lam, T.H. The association of problematic smartphone use with family well–being mediated by family communication in Chinese adults: A population–based study. *J. Behav. Addict.* **2019**, *8*, 412–419. [CrossRef] [PubMed]
10. Elhai, J.D.; Dvorak, R.D.; Levine, J.C.; Hall, B.J. Problematic smartphone use: A conceptual overview and systematic review of relations with anxiety and depression psychopathology. *J. Affect. Disord.* **2017**, *207*, 251–259. [CrossRef] [PubMed]
11. Luk, T.T.; Wang, M.P.; Shen, C.; Wan, A.; Chau, P.H.; Oliffe, J.; Viswanath, K.; Chan, S.S.; Lam, T.H. Short version of the Smartphone Addiction Scale in Chinese adults: Psychometric properties, sociodemographic, and health behavioral correlates. *J. Behav. Addict.* **2018**, *7*, 1157–1165. [CrossRef]
12. De-Sola, J.; Talledo, H.; de Fonseca, F.R.; Rubio, G. Prevalence of problematic cell phone use in an adult population in Spain as assessed by the Mobile Phone Problem Use Scale (MPPUS). *PLoS ONE* **2017**, *12*, e0181184. [CrossRef] [PubMed]
13. Lopez-Fernandez, O. Short version of the Smartphone Addiction Scale adapted to Spanish and French: Towards a cross–cultural research in problematic mobile phone use. *Addict. Behav.* **2017**, *64*, 275–280. [CrossRef] [PubMed]
14. Nahas, M.; Hlais, S.; Saberian, C.; Antoun, J. Problematic smartphone use among Lebanese adults aged 18–65 years using MPPUS–10. *Comput. Hum. Behav.* **2018**, *87*, 348–353. [CrossRef]
15. World Health Organization. Promoting mental health: Concepts, emerging evidence, practice: Summary report. Available online: https://apps.who.int/iris/bitstream/handle/10665/42940/9241591595.pdf (accessed on 11 November 2019).
16. Sun, Y.; Luk, T.T.; Wang, M.P.; Shen, C.; Ho, S.Y.; Viswanath, K.; Chan, S.S.C.; Lam, T.H. The reliability and validity of the Chinese short Warwick–Edinburgh Mental Well–being Scale in the general population of Hong Kong. *Qual. Life Res.* **2019**, *28*, 2813–2820. [CrossRef] [PubMed]
17. Westerhof, G.J.; Keyes, C.L.M. Mental illness and mental health: The two continua model across the lifespan. *J. Adult. Dev.* **2010**, *17*, 110–119. [CrossRef]
18. Seow, L.S.E.; Vaingankar, J.A.; Abdin, E.; Sambasivam, R.; Jeyagurunathan, A.; Pang, S.; Chong, S.A.; Subramaniam, M. Positive mental health in outpatients with affective disorders: Associations with life satisfaction and general functioning. *J. Affect. Disord.* **2016**, *190*, 499–507. [CrossRef]
19. Keyes, C.L.M. Mental illness and/or mental health? Investigating axioms of the complete state model of health. *J. Consult. Clin. Psychol.* **2005**, *73*, 539–548. [CrossRef]
20. Horwood, S.; Anglim, J. Problematic smartphone usage and subjective and psychological well–being. *Comput. Hum. Behav.* **2019**, *97*, 44–50. [CrossRef]
21. Census and Statistics Department. Thematic household survey report No. 64. Available online: https://www.statistics.gov.hk/pub/B11302642018XXXXB0100.pdf (accessed on 11 November 2019).
22. Kwon, M.; Kim, D.-J.; Cho, H.; Yang, S. The Smartphone Addiction Scale: Development and validation of a short version for adolescents. *PLoS ONE* **2013**, *8*, e83558. [CrossRef]
23. Kroenke, K.; Spitzer, R.L.; Williams, J.B.W.; Löwe, B. An ultra–brief screening scale for anxiety and depression: The PHQ–4. *Psychosomatics* **2009**, *50*, 613–621. [PubMed]
24. Löwe, B.; Wahl, I.; Rose, M.; Spitzer, C.; Glaesmer, H.; Wingenfeld, K.; Schneider, A.; Brähler, E. A 4–item measure of depression and anxiety: Validation and standardization of the Patient Health Questionnaire–4 (PHQ–4) In the general population. *J. Affect. Disord.* **2010**, *122*, 86–95. [CrossRef] [PubMed]
25. Yu, X.; Stewart, S.M.; Wong, P.T.K.; Lam, T.H. Screening for depression with the Patient Health Questionnaire–2 (PHQ–2) among the general population in Hong Kong. *J. Affect. Disord.* **2011**, *134*, 444–447. [CrossRef] [PubMed]
26. Lyubomirsky, S.; Lepper, H.S. A measure of subjective happiness: Preliminary reliability and construct validation. *Soc. Indic. Res.* **1999**, *46*, 137–155. [CrossRef]
27. Nan, H.; Ni, M.Y.; Lee, P.H.; Tam, W.W.S.; Lam, T.H.; Leung, G.M.; McDowell, I. Psychometric evaluation of the Chinese version of the Subjective Happiness Scale: Evidence from the Hong Kong FAMILY Cohort. *Int. J. Behav. Med.* **2014**, *21*, 646–652. [CrossRef] [PubMed]
28. Stewart-Brown, S.; Tennant, A.; Tennant, R.; Platt, S.; Parkinson, J.; Weich, S. Internal construct validity of the Warwick-Edinburgh Mental Well–being Scale (WEMWBS): A rasch analysis using data from the Scottish health education population survey. *Health Qual. Life Outcomes* **2009**, *7*, 15. [CrossRef]

29. Haug, S.; Castro, R.P.; Kwon, M.; Filler, A.; Kowatsch, T.; Schaub, M.P. Smartphone use and smartphone addiction among young people in Switzerland. *J. Behav. Addict.* **2015**, *4*, 299–307. [CrossRef]
30. Lee, K.E.; Kim, S.-H.; Ha, T.-Y.; Yoo, Y.-M.; Han, J.-J.; Jung, J.-H.; Jang, J.-Y. Dependency on smartphone use and its association with anxiety in Korea. *Public Health Rep.* **2016**, *131*, 411–419. [CrossRef]
31. West, S.G.; Finch, J.F.; Curran, P.J. *Structural Equation Models with Non-Normal Variables: Problems and Remedies*; SAGE Publishing: Thousand Oaks, CA, USA, 1995.
32. Izrael, D.; Hoaglin, D.C.; Battaglia, M.P. To rake or not to rake is not the question anymore with the enhanced raking macro. In Proceedings of the Twenty–Ninth Annual SAS Users Group International Conference, Montreal, QC, Canada, 9–12 May 2004.
33. Kraut, R.; Patterson, M.; Lundmark, V.; Kiesler, S.; Mukophadhyay, T.; Scherlis, W. Internet paradox: A social technology that reduces social involvement and psychological well-being? *Am. Psychol.* **1998**, *53*, 1017. [CrossRef]
34. Priel, B.; Shamai, D. Attachment style and perceived social support: Effects on affect regulation. *Pers. Individ. Dif.* **1995**, *19*, 235–241. [CrossRef]
35. Adams, S.K.; Kisler, T.S. Sleep quality as a mediator between technology-related sleep quality, depression, and anxiety. *Cyberpsychol. Behav. Soc. Netw.* **2013**, *16*, 25–30. [CrossRef]
36. Shensa, A.; Escobar-Viera, C.G.; Sidani, J.E.; Bowman, N.D.; Marshal, M.P.; Primack, B.A. Problematic social media use and depressive symptoms among U.S. young adults: A nationally–representative study. *Soc. Sci. Med.* **2017**, *182*, 150–157. [PubMed]
37. Vannucci, A.; Flannery, K.M.; Ohannessian, C.M. Social media use and anxiety in emerging adults. *J. Affect. Disord.* **2017**, *207*, 163–166. [CrossRef] [PubMed]
38. Boers, E.; Afzali, M.H.; Newton, N.; Conrod, P. Association of screen time and depression in adolescence. *JAMA Pediatr.* **2019**, *173*, 853–859. [CrossRef] [PubMed]
39. Elhai, J.D.; Yang, H.; Montag, C. Cognitive– and emotion–related dysfunctional coping processes: Transdiagnostic mechanisms explaining depression and anxiety's relations with problematic smartphone use. *Curr. Addict. Rep.* **2019**, *6*, 410–417. [CrossRef]
40. Li, J.-B.; Mo, P.K.H.; Lau, J.T.F.; Su, X.-F.; Zhang, X.; Wu, A.M.S.; Mai, J.-C.; Chen, Y.-X. Online social networking addiction and depression: The results from a large–scale prospective cohort study in Chinese adolescents. *J. Behav. Addict.* **2018**, *7*, 686–696. [CrossRef]
41. Akın, A. The relationships between internet addiction, subjective vitality, and subjective happiness. *Cyberpsychol. Behav. Soc. Netw.* **2012**, *15*, 404–410. [CrossRef]
42. Lachmann, B.; Sindermann, C.; Sariyska, R.Y.; Luo, R.; Melchers, M.C.; Becker, B.; Cooper, A.J.; Montag, C. The role of empathy and life satisfaction in internet and smartphone use disorder. *Front. Psychol.* **2018**, *9*, 398. [CrossRef]
43. Hughes, N.; Burke, J. Sleeping with the frenemy: How restricting 'bedroom use' of smartphones impacts happiness and wellbeing. *Comput. Hum. Behav.* **2018**, *85*, 236–244. [CrossRef]
44. Magsamen-Conrad, K.; Billotte-Verhoff, C.; Greene, K. Technology addiction's contribution to mental wellbeing: The positive effect of online social capital. *Comput. Hum. Behav.* **2014**, *40*, 23–30. [CrossRef]
45. Council on communications and media. Media use in school–aged children and adolescents. *Pediatrics* **2016**, *138*, e20162592. [CrossRef] [PubMed]
46. Sambasivam, R.; Vaingankar, J.A.; Chong, S.A.; Abdin, E.; Jeyagurunathan, A.; Seow, L.S.E.; Pang, S.; Subramaniam, M. Positive mental health in outpatients: Comparison within diagnostic groups. *BMC Psychiatry* **2016**, *16*, 412. [CrossRef] [PubMed]
47. Orben, A.; Przybylski, A.K. The association between adolescent well–being and digital technology use. *Nat. Hum. Behav.* **2019**, *3*, 173–182. [CrossRef] [PubMed]

International Journal of
Environmental Research and Public Health

MDPI

Article

Parental Monitoring and Adolescent Problematic Mobile Phone Use: The Mediating Role of Escape Motivation and the Moderating Role of Shyness

Xinchen Fu [1], Jingxuan Liu [2], Ru-De Liu [1,*], Yi Ding [3], Jia Wang [4], Rui Zhen [5] and Fangkai Jin [1]

1 Beijing Key Laboratory of Applied Experimental Psychology, National Demonstration Center for Experimental Psychology Education (Beijing Normal University), Faculty of Psychology, Beijing Normal University, Beijing 100875, China; fxc_psy@163.com (X.F.); jfk2016@163.com (F.J.)
2 Department of Psychology and Neuroscience, Duke University, Durham, NC 27708, USA; liujingxuan1503@gmail.com
3 Graduate School of Education, Fordham University, New York, NY 10023, USA; yding4@fordham.edu
4 Teachers' College, Beijing Union University, Beijing 100011, China; wangjia@mail.bnu.edu.cn
5 Institute of Psychological Sciences, College of Education, Hangzhou Normal University, No. 2318 Yuhangtang Road, Hangzhou 311121, China; zhenrui1206@126.com
* Correspondence: rdliu@bnu.edu.cn; Tel.: +86-10-58806324

Received: 5 February 2020; Accepted: 24 February 2020; Published: 25 February 2020

Abstract: In an attempt to reduce the negative consequences of adolescent media use, parents often monitor their children's online activities. However, research suggests that parental monitoring often does not reduce children's problematic mobile phone use as expected. Based on the results of a survey of 584 Chinese adolescents, we found that parental monitoring positively predicted children's problematic mobile phone use (PMPU) within a Chinese cultural context. The results also showed that children's escape motivation partially mediated this relationship, while their level of shyness moderated both the mediated path and the direct impact of parental monitoring on children's PMPU. The findings suggested that a higher level of shyness increased the likelihood that parental monitoring would increase the child's escape motivation and PMPU. The study results provide guidelines for parents and educators regarding interventions for adolescents' problematic phone use.

Keywords: parental monitoring; problematic mobile phone use; escape motivation; shyness

1. Introduction

Mobile phones are the most common platform used by adolescents to access the internet. In Europe, adolescents use mobile phones more frequently than they do computers to access online information [1]. In China, 98.6% of internet users, or 817 million people, of whom 18% are adolescents, connect to the internet through mobile phones [2]. Some common internet activities adolescents engage in via mobile phones are socializing [3,4], accessing entertainment [5,6], and obtaining study resources [7].

Because mobile phones offer such a wide range of attractions, some adolescents overuse their mobile phones, which can lead to problematic mobile phone use (PMPU) [8,9]. PMPU, also known as mobile phone dependency [10] and mobile phone addiction [11], can elicit unpleasant withdrawal symptoms when an individual's mobile phone access is restricted. Studies have found that adolescents are more prone to mobile phone addiction than adults [12,13].

Considering the wide range of negative health and psychological consequences of PMPU for adolescents [14–16], parents often mediate their child's mobile phone use in an attempt to reduce PMPU. One type of mediation strategy used frequently by parents, parental monitoring, has drawn wide social attention [17]. Parental monitoring refers to parents' tracking of children's online activities [18] and is often thought by parents to be an effective way to prevent their child's PMPU. However, research has shown that parental monitoring may not reduce adolescent PMPU behaviors as expected [8,19,20]. Therefore, the current study aimed to further explore the impact of parental monitoring on children's problematic mobile phone use and to investigate the mechanism underlying such an impact in the Chinese cultural context.

1.1. Parental Monitoring and Problematic Mobile Phone Use (PMPU)

PMPU is a type of behavioral addiction [21,22]. It can produce a series of negative health and psychological consequences, such as sleep problems and physical harm [16,23]; decreased level of physical activity [24,25]; social problems [16]; decreased life satisfaction [26]; academic problems [15,27–29]; anxiety when separated from mobile phones [30,31]; negative emotions including stress, anxiety, and depression [14,32]; alexithymia [33,34]; and decreased empathy [26].

In an attempt to reduce the negative effects of children's media use, parents often adopt different mediation strategies to influence their child's media-use behaviors [35]. Parental mediation is a multi-dimensional concept. It encompasses all types of parental strategies, including mediation, controlling, and providing instruction and interpretation regarding media content, to reduce the negative effect of media use on children [36]. Parental mediation was originally divided into three dimensions: restrictive mediation, active mediation, and co-use [37]. These three dimensions were found to be present in parental mediation of children's TV viewing [38,39], gaming [40], and internet use behaviors [35]. However, as internet use became more prevalent among adolescents, researchers suggested that the traditional three dimensions needed to be expanded to further address parental strategies regarding children's internet use behaviors [17,41,42]. Nikken and Jansz [18] proposed a five-dimensional model, in which the restrictive mediation dimension was divided into time restriction and special content restriction, and supervision (i.e., monitoring) was added as a new dimension. Parental monitoring refers to parents' tracking of their child's online activities and history, such as email interactions and website access [17,18,20].

Parental monitoring is the most common mediation strategy for adolescent internet use [43] and can predict adolescent internet-use behaviors [44]. Yet surprisingly, many studies have found that parental monitoring is ineffective in reducing those problematic behaviors. Livingstone and Helsper [17] found that parental monitoring failed to reduce risky online behaviors (such as porn, violence, and privacy viewing) in children aged 9–16 years; a follow-up study showed that parental monitoring was positively correlated with increased internet-related risk [45]. Studies also found that parental monitoring positively predicted children's internet-use behaviors [44] and adolescent internet addiction [46]. In addition, cross-cultural research conducted by Bayraktar [47] found that whereas parental monitoring was negatively correlated with adolescent risky internet-use behaviors (such as porn, viewing violence and excessive internet use) in Europe, it positively predicted adolescent involvement in such risky behaviors in Turkey. With increased adolescent access to the internet via mobile phones, parental monitoring of internet use has evolved to monitoring of internet-related mobile phone-use behaviors [48]. A study conducted in Germany [19] found the correlation between parental monitoring and adolescent mobile phone dependency to be insignificant: Parental monitoring was inefficient in reducing the negative effects of children's PMPU. Similar results were obtained in Taiwan [8]. These study results suggested that parental monitoring may not reduce children's PMPU. The first goal of our study was to examine whether parental monitoring was related to or positively predicted adolescent PMPU.

1.2. Mediating Role of Escape Motivation

Parental monitoring refers to parental supervision of children's internet use behaviors with the aim to reduce the negative impacts of media use [48]. Why would parental monitoring perpetuate children's PMPU behaviors instead of reducing them? The underlying mechanism warrants further investigation. Based on the existing theories and study results, we hypothesized that escape motivation might mediate the path between parental monitoring and children's PMPU.

Within the context of adolescent PMPU, escape motivation refers to the motivation that drives adolescents to escape negative emotions via mobile phone use, leading to PMPU [49]. Escaping reality is one of two major functions of PMPU [50]. In many cases, the underlying motivation of addictive behaviors is to escape reality in an effort to reduce painful and negative emotions [51]. Among all types of motivation, escape motivation is often thought to be the most important predictor of internet addiction; it serves as a strong predictor for a wide range of internet and mobile phone-related addictions, including internet addiction [52–54], video game addiction [55,56], online game addiction [57,58], online video apps overuse (i.e., YouTube) [59], and PMPU [60–62].

Parental monitoring can also induce negative emotions in adolescents. According to the self-determination theory, all humans strive for freedom, and therefore their motivation is optimal when they are void of external influence and interference [63]. Thus, when children's freedom is restricted, they may experience reactance [64]. Some studies have already examined the mediating roles of escape motivation in the relations between negative emotions, addictive behaviors, and PMPU. For example, escape motivation was found to mediate the relation between psychological distress (such as emotional imbalance, depression, and anxiety) and video game addiction [65,66] and the relation between psychiatric disorders (such as somatization and OCD) and video game addiction [67]. Escape motivation was also found to mediate the path from loneliness to PMPU [67]. Accordingly, parental monitoring may result in conflicts between parents and child, and thus cause the child to experience stress and negative emotions. As a result, through escape motivation, the child's mobile phone-use behaviors might increase [68,69], leading to PMPU [62].

1.3. Moderating Role of Shyness

When choosing mediation strategies to reduce children's PMPU, parents should take their child's temperament into consideration [70]. Shyness, one of the most stable temperament types, should be given special attention [71].

Shyness is a common social experience that involves timidity, discomfort, embarrassment, and fear of being evaluated. It is often accompanied by a desire to minimize social interactions [72–74]. Research suggests that shyness can induce a series of negative consequences, including loneliness [75], anxiety, and depression [76,77]. To reduce or avoid those negative consequences, shy individuals often view online communication as a means to avoid face-to-face interaction [77]. An individual's degree of shyness can positively predict online social activities [78] and internet addictive behaviors [79,80]. Such a predictive effect was shown to be consistent over time [81]. Similarly, many studies suggest that shy individuals increase their mobile phone use to avoid face-to-face contact, which leads to PMPU [82–84].

Research concerning parenting styles suggests that parents should adopt adaptive parenting that is suitable to their child's temperament [85]. Studies have found that inhibitive temperament (similar to shyness) mediates the relation between maternal authoritative parenting and girls' prosocial behaviors [86]. Zarra-Nezhad et al. [87] discovered that parental emotional support positively predicted prosocial behavior only among shy children, while parental control positively predicted prosocial behaviors only among children who were not shy. Because parental mediation strategies are similar to parenting styles to some degree [88], parental monitoring of children's mobile phone use should be similar to parental control. Thus, it could be inferred that parental monitoring exerts different effects based on the shyness level of a child. Furthermore, a study [89] showed that shyness also predicts an individual's drinking motivation; shy individuals are more inclined to reduce their negative emotions and navigate social contexts via drinking, which can eventually lead to alcohol addiction. This suggests that shyness is related to behavioral motivation. Therefore, we hypothesized that parental monitoring would positively predict more escape motivation for shy adolescents than for those who are not shy.

1.4. The Present Study

Based on the above, the present study aimed to investigate the mechanism and impact of parental monitoring on adolescent PMPU through a hypothesized moderated mediation model (Figure 1). We anticipated that (a) parental monitoring would positively predict adolescent PMPU (H1) via the mediation of escape motivation (H2), and (b) adolescent level of shyness would moderate the impact of parental monitoring on adolescent escape motivation (H3) and PMPU (H4).

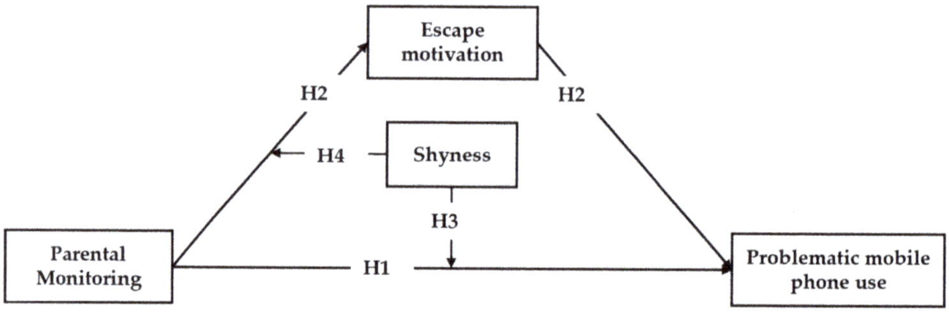

Figure 1. The hypothesized model.

2. Materials and Methods

2.1. Participants

In the present study, we contacted an urban middle school in Beijing, China, and informed them of the purpose of our study. All students recruited in grades 7, 8, 10, and 11 were voluntary to participate. A total of 584 students were recruited. The mean age of these participants was 16.13 years (standard deviation (SD) = 2.80) with a range from 13 to 18; 267 (45.7%) were boys, 281 (49.0%) were girls, and 31 (5.3%) did not report gender. Among these students, 23.3% of fathers and 25.8% of mothers received a high-school education or below and 76.7% of fathers and 74.2% of mothers received an undergraduate level of education or above.

2.2. Procedures

We obtained the approval to conduct the study from the Research Ethics Committee of a major research university in Beijing and the principals of the participating schools. The students were informed of the voluntary nature of this study and their right to opt out at any time during the course of the study. Then, they were asked to complete a paper-pencil questionnaire that included demographic information, Parental Monitoring, Escape Motivation, Shyness and Problematic mobile phone-use measures.

2.3. Measures

2.3.1. Parental Monitoring

Parental monitoring was assessed by the Parental Mediation of Children's Internet Use Scale [18], which was validated in Chinese and exhibited satisfactory reliability and validity [90]. In this study, the word "internet use" was changed to "mobile phone use" to assess parental monitoring of children's mobile phone-use behaviors. The scale included four items (e.g., "My parents check my mobile phone use behaviors; My parents check my mobile phone chatting records; My parents keep an eye on me when I use mobile phone") on a 5-point Likert scale (1 = *never*, 5 = *always*). Higher scores reflected higher levels of parent monitoring of children's mobile phone use. Cronbach's α for the present study was 0.892.

2.3.2. Escape Motivation

Escape motivation was assessed by the Mobile Phone-Use Motivations Scale [49]. It consisted of six items (e.g., I use/play with my smartphone to feel less lonely; to fill uncomfortable silence; to make myself feel better when I feel down) using a 5-point scale ranging from 1 (*strongly disagree*) to 5 (*strongly agree*); higher scores indicated a higher degree of escape motivation. This measure yielded a Cronbach's α of 0.865 in the present study.

2.3.3. Shyness

Shyness was assessed by the Shyness Scale [72], which was validated in Chinese and showed satisfactory reliability and validity [91]. This scale included 13 items (e.g., I have trouble looking someone right in the eye; I feel tense when I'm with people I don't know well; I feel inhibited in social situations.) on a 5-point Likert scale, ranging from 1 (*strongly disagree*) to 5 (*strongly agree*), with higher scores indicating a higher degree of shyness. Cronbach's α for the scale was 0.803 in the present study.

2.3.4. Problematic Mobile Phone Use (PMPU)

Problematic mobile phone use (PMPU) was assessed by the Problematic Mobile Phone-Use Scale [92], which was translated into Chinese and proved to be valid [93]. It consisted of 10 items (e.g., I find it difficult to switch off my mobile phone; I feel anxious if I have not checked for messages or switched on my mobile phone for some time; I find myself engaged on the mobile phone for longer periods of time than intended) on a 5-point scale that ranged from 1 (*strongly disagree*) to 5 (*strongly agree*), with higher scores indicating a higher degree of problematic mobile phone use. Cronbach's α for the present study was 0.821.

2.3.5. Statistical Analysis

We conducted descriptive analyses and Pearson correlations with SPSS 22.0 (IBM, New York, NY, USA). The pattern of missing data was first evaluated. The results showed that 1.25% of the data was missing, and the missing rates on all variables were less than 10%. Therefore, we used the listwise method to handle the missing data in the following structural equation model [94]. Among the 583 participants, 518 provided complete data on all the variables. Next, the moderated mediation model was tested using the SPSS macro PROCESS 3.0 (model 8) (http://www.afhayes.com) recommended by Hayes [95]. We generated 5000 bootstrapped samples to estimate the confidence interval of the model effect. A 95% confidence interval without zero indicates statistical significance.

3. Results

3.1. Preliminary Analyses

The descriptive statistics and correlation matrix are presented in Table 1. Parental monitoring was positively correlated with children's escape motivation, shyness, and PMPU. Children's escape motivation was positively correlated with their shyness level and PMPU. Children's shyness level was positively correlated with PMPU.

Table 1. Means, standard deviations, and correlations between variables.

Variables	M	SD	1	2	3	4	5
1. Age	16.13	2.80	_				
2. Parental monitoring	2.06	1.14	0.114 **	_			
3. Escape motivation	2.86	1.05	−0.023	0.122 **	_		
4. Shyness	2.75	0.70	−0.031	0.135 **	0.147 ***	_	
5. PMPU	2.45	0.76	−0.092 *	0.163 ***	0.607 ***	0.213 ***	_

Note. PMPU = Problematic Mobile Phone Use. * $p < 0.05$. ** $p < 0.01$. *** $p < 0.001$.

3.2. Testing for the Proposed Model

The analysis results of SPSS macro PROCESS are presented in Table 2, which consists of four parts: Model 1, Model 2, conditional indirect effect analysis of Model 1, and conditional direct effect analysis of Model 2. Model 1 was used to test the effects of parental monitoring on children's escape motivation (part of H2), and the interaction between parental monitoring and children's shyness on children's escape motivation (H4), after controlling for age and gender. Model 2 examined the effects of parental monitoring on children's PMPU (H1), children's escape motivation on children's PMPU (part of H2), and the interaction between parental monitoring and children's shyness on children's PMPU (H3).

Table 2. Bootstrap test on moderated mediation effect.

Conditional Process Analysis	β	*SE*	*t*	*p*	*LLCI–ULCI*
Model 1					
Outcome: Escape motivation					
Predictors:					
Age	−0.009	0.014	−0.668	0.505	−0.037–0.018
Gender	0.150	0.079	1.902	0.058	−0.005–0.305
Parental monitoring	0.101 *	0.042	2.392	0.017	0.018–0.183
Shyness	0.129 **	0.041	3.164	0.002	0.050–0.209
Monitoring × Shyness	0.102 *	0.041	2.474	0.014	0.021–0.183
Model 2					
Outcome: PMPU					
Predictors:					
Age	−0.031 **	0.012	−2.672	0.008	−0.054–0.008
Gender	−0.156 *	0.066	−2.352	0.019	−0.287–0.026
Parental monitoring	0.078 *	0.035	2.198	0.028	0.008–0.148
Escape motivation	0.612 ***	0.035	17.571	0.000	0.544–0.681
Shyness	0.124 ***	0.035	3.582	0.000	0.056–0.191
Monitoring × Shyness	0.063 +	0.035	1.806	0.071	−0.005–0.131
Conditional indirect effect analysis of model 1			β	*Boot SE*	*BootLLCI-BootULCI*
M − 1 SD			0.007	0.060	−0.111–0.125
M			0.103 *	0.042	0.021–0.186
M + 1 SD			0.188 ***	0.051	0.087–0.289
Conditional direct effect analysis of model 2			β	*Boot SE*	*BootLLCI-BootULCI*
M − 1 SD			0.020	0.050	−0.079–0.119
M			0.080 *	0.035	0.010–0.149
M + 1 SD			0.132 **	0.044	0.046–0.217

Note. Bootstrap sample size = 5000. SE = standard error, LL = low limit, CI = confidence interval, UL = upper limit. + $p < 0.10$. * $p < 0.05$. ** $p < 0.01$. *** $p < 0.001$.

The conditional indirect effect analysis of Model 1 tested the effects of parental monitoring at its mean, plus one, and minus one standard deviation on children's escape motivation at the mean of shyness. The conditional direct effect analysis of Model 2 tested the effects of parental monitoring at its mean, plus one standard deviation, and minus one standard deviation on children's PMPU at the mean of the shyness. According to Model 1 ($F = 5.96$, $R^2 = 0.05$, $p < 0.001$) and Model 2 ($F = 65.74.45$, $R^2 = 0.41$, $p < 0.001$), after controlling for gender and age, parental monitoring positively predicted children's PMPU ($\beta = 0.078$, $p < 0.05$), supporting H1 (see Figure 2). Parental monitoring positively predicted children's escape motivation ($\beta = 0.101$, $p < 0.05$), and children's escape motivation positively predicted children's PMPU ($\beta = 0.612$, $p < 0.001$), supporting H2.

The interaction of parental monitoring and shyness showed significant effects on children's escape motivation ($\beta = 0.102$, $p < 0.05$). Thus, H4 was supported. This finding suggests that the relation between parental monitoring and children's escape motivation was moderated by children's level of shyness (see Figure 3). In addition, two of the three conditional indirect effects (based on the moderator values at the mean and at plus and minus one standard deviation) were positive and significantly different from zero (see conditional indirect effect analysis of Model 1). That is, according to the interaction of parental monitoring and children's shyness, the indirect effects of parental monitoring on children's escape motivation were stronger when children's shyness level was moderate to high, but lower when children's shyness level was low.

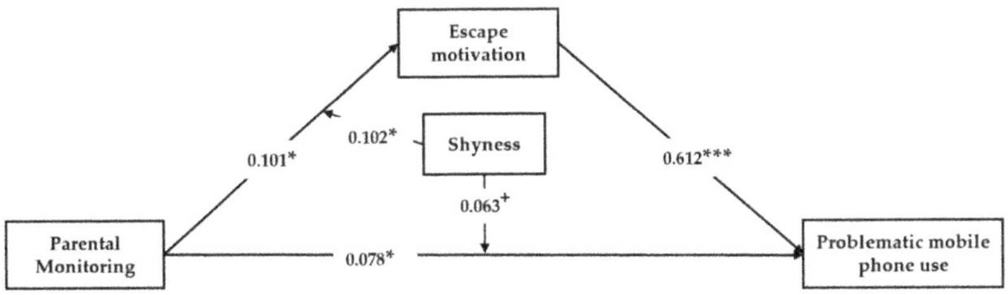

Figure 2. Moderation effects of shyness in the mediation model. All values shown are standardized coefficients. $^{+}$ $p < 0.10$. * $p < 0.05$. *** $p < 0.001$.

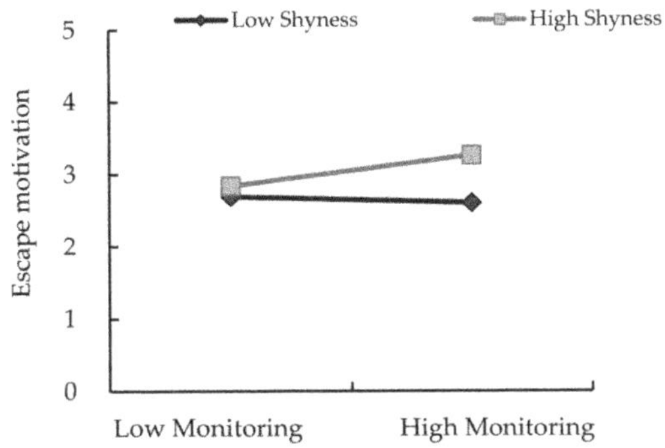

Figure 3. Moderation effect of shyness in the relation between parental monitoring and children's escape motivation.

Furthermore, the interaction of parental monitoring and shyness showed marginally significant effects on children's PMPU ($\beta = 0.063$, $p < 0.10$). Therefore, H3 was supported. Two of the three conditional direct effects (based on the moderator values at the mean and at plus and minus one standard deviation) were positive and significantly different from zero (see conditional direct effect analysis of Model 2). These findings suggest that the relation between parental monitoring and children's PMPU was moderated by children's level of shyness (see Figure 4). The direct effects of parental monitoring on children's PMPU were higher when children's shyness was moderate to high, but lower when children's shyness was low. In conclusion, the above results indicated that parental monitoring affects children's PMPU through a moderated mediation path, with children's escape motivation as the mediator and children's shyness level as the moderator.

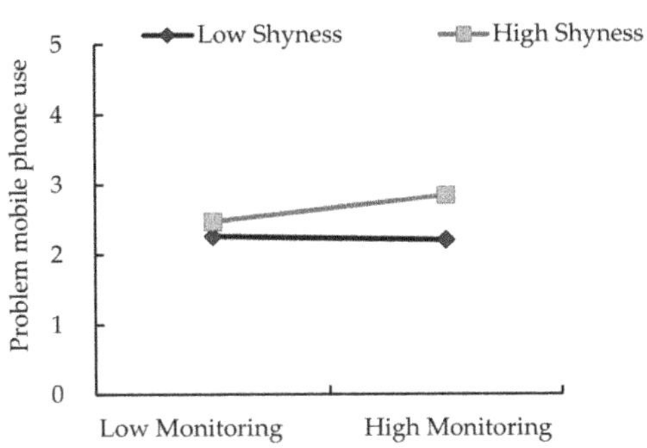

Figure 4. Moderation effect of shyness the relation between parental monitoring and children's PMPU.

4. Discussion

To reduce the negative impact of children's mobile phone use, parental monitoring has not achieved consistent results [45,48]; its underlying mechanism remains unclear. The present study examined a moderated mediation model and found that children's escape motivation partially mediated the association of parental monitoring and their PMPU. In addition, children's degree of shyness moderated the path from parental monitoring to their escape motivation and to PMPU.

4.1. Escape Motivation Partially Mediates the Relation between Parental Monitoring and Adolescent PMPU

Parental monitoring positively correlated with adolescent PMPU, which supports H1. This is consistent with the findings of previous studies regarding internet addiction [46]. Research has shown that adolescents are especially vulnerable to the negative impacts of increased mobile phone use, and parents hope to reduce these negative influences via mediation. Considering that teenage years are a critical period of learning and social development, most parents employ monitoring strategies to mediate their children's mobile phone use. However, research findings suggest that parental monitoring often leads to unexpected results. Because of adolescents' desire for freedom and psychological reactance induced by restriction, they may be subjected to the "Pandora effect" [63,68,96]; as the strength of parental monitoring of mobile phone use increases, adolescents' mobile phone-use behaviors also increase, which can eventually lead to PMPU. Similar results were found in other areas of addiction, including adolescent internet addiction [97], internet dependency [98], and sexual behaviors [99]. From the perspective of the sociology of emotions, emotions serve as an important precursor to effective parental mediation. However, parental monitoring often results in the conflicts between parents and child, causing negative emotions. The model of compensatory internet use suggests that internet use is viewed as a compensatory means to escape the reality. An individual with relatively low overall happiness tends to relieve negative emotions and escape from real-life problems via mobile phone use [100–102]. Moreover, in the family environment, whether the parents' supervision is effective or not partially depends on their own behaviors. According to Bandura's social cognitive theory, children observe the behaviors of others around them and are especially prone to observing, and imitating their parents' behaviors [103]. It has been found that parents' looking down at their own mobile phone (parent phubbing) in the process of communication with their children will not only aggravate children's addiction [104], but also affect children's attitude towards self-control mobile phone use [105]. Therefore, if parents want to achieve efficient monitoring results, they should control their own mobile phone-use behaviors. In general, although parental monitoring aims to improve children's mobile phone-use behaviors, it is often ineffective or leads to undesirable outcomes.

In addition, as we hypothesized, escape motivation mediated the relation between parental monitoring and adolescent PMPU, which supports H2. This result is consistent with similar studies. One study found that escape motivation mediated the predictive effect of negative emotions on PMPU [67]. When individuals face real-life struggles, escape motivation can prompt them to compensate through video games and/or the internet, which can eventually lead to video game or internet addiction [54,106,107]. Furthermore, Erikson [108] proposed that there is a developmental issue in every psychosocial stage of human development; adolescence is a crucial period for identity development, as well as a time when psychological reactance peaks. To adolescents, parental monitoring acts as a restriction of freedom, which could induce their escape motivation. In addition, parental monitoring often creates parent-child conflict, which could cause adolescents to develop negative emotions, thus leading to escape motivation.

4.2. Shyness Moderates Both the Relations of Parental Monitoring with Adolescent Escape Motivation and Adolescent PMPU

The impact of parental monitoring on adolescent PMPU differs across individuals. The present study found that adolescent shyness level moderated the relation between parental monitoring and adolescent escape motivation, and between parental monitoring and adolescent PMPU, which supports

H4 and H3. As children's shyness level increased, increases in parental monitoring strengthened their escape motivation and PMPU. This could be explained by the fact that shy individuals rely more on mobile phones to socialize, and parental monitoring can cause them to generate more negative reactions, such as psychological reactance and feelings of insecurity; this in turn perpetuates their escape motivation and exacerbates PMPU. There are three possible reasons why shy individuals rely more on mobile-phone socialization. First, shyness can be related to social anxiety; to avoid the embarrassment and discomfort elicited by face-to-face interactions, shy individuals tend to socialize via the internet, thus increasing internet use behaviors that can lead to internet and mobile-phone addiction [83,109]. Second, shy individuals often avoid occasions that expose them to evaluation [110]. Internet and mobile phone interactions can disguise their identity and thus protect them from others' evaluations. Third, shyness that originates from social anxiety can expose individuals to psychological challenges [111]. To mitigate and overcome these challenges, shy individuals tend to avoid face-to-face interactions, devote themselves to internet use, and achieve satisfaction through online interactions. Overall, when their parents employ monitoring strategies, shy adolescents might feel insecure and avoid being evaluated, which prompts them to escape reality via mobile phone use.

4.3. Limitation and Implication

The present study has a few limitations. First, our data are cross-sectional, and thus cannot infer strong causational relationships. Experimental or longitudinal designs could be used to further prove the relations between these variables. Second, all of our data came from adolescents' subjective responses. Although our measures have relatively high reliability/validity, the addition of responses from other sources (such as parents) would make our results more persuasive. Data of parental monitoring were obtained from children's self-report, but not from their parents' reports due to the limitation of our research conditions. In future research, obtaining data of parental monitoring from parents will help to reduce potential bias. Third, the findings in this study are in the context of Chinese culture, and its generalization to other culture should be made with caution. In order to avoid the negative effects of mobile phones, many parents in China prohibit children from bringing mobile phones to school. Using mobile phones at home is also often restricted. In comparison to children in other cultural contexts, Chinese children might have less freedom to use mobile phones. Moreover, studies on parental monitoring and children's PMPU have yielded different results under different cultural and national contexts (such as Turkey and Germany) [19,47]. Therefore, future studies can explore the effects of cultural differences and cultural contexts on the effect of parental monitoring on children's PMPU. Finally, future research should involve more schools and students to explore the differences between children in different developmental stages, such as early adolescence (12–13 years) and late adolescence (16–18).

Despite the above limitations, the contributions of the present study are relevant to educators, parents, and adolescents. First, our study results found that parental monitoring positively predicts adolescents' PMPU. However, as the forms of online activities increase and the access to the internet becomes easier, parents are more likely to use monitoring strategies [20]. Moreover, parental monitoring is the most direct mediation method and, therefore, the most convenient for parents [18]. One study points out parents' greater tendency to monitor their child's mobile phone-use behaviors when their use increases [46]. When parental monitoring increases children's mobile phone use, parental monitoring in turn increases, leading to a maladaptive cycle. Parents and educators should be cautious and optimize the use of monitoring strategies to control adolescent mobile phone use. Second, the present study found that parental monitoring positively predicted adolescents' escape motivation, which predicted their PMPU. Third, the effect of parental monitoring on adolescents differed across temperaments. When parents strengthen their monitoring, shy adolescents are more likely to display escape motivation and develop problematic mobile phone-use behaviors. Parents should adopt appropriate mediation strategies according to their child's temperament.

5. Conclusions

Parental monitoring positively predicts children's PMPU, and this predictive effect is partially mediated by children's escape motivation. Children's level of shyness moderates the relation between parental monitoring and children's escape motivation, and the relation between parental monitoring and children's PMPU, Thus, increasing the strength of parental monitoring can lead to an increase in escape motivation and PMPU among shy adolescents.

Author Contributions: Conceptualization, X.F. and J.L.; Formal analysis, X.F. and F.J.; Funding acquisition, R.-D.L.; Investigation, R.Z., F.J. and J.W.; Project administration, R.-D.L.; Supervision, R.-D.L.; Validation, R.-D.L.; Writing—review and editing, X.F., J.L., R.-D.L. and Y.D. All authors have read and agreed to the published version of the manuscript.

Funding: This research was funded by National Social Science Fund of China: 17BSH102.

Conflicts of Interest: The authors declare no conflict of interest.

References

1. Mascheroni, G.; Ólafsson, K. *Net Children Go Mobile: Risks and Opportunities*, 2nd ed.; Educatt: Milan, Italy, 2014.
2. CNNIC. The 43st China Statistical Report on Internet Development. 2018. Available online: http://www.cnnic.net.cn/hlwfzyj/hlwxzbg/hlwtjbg/201902/P020190318523029756345.pdf (accessed on 20 December 2018).
3. Bolin, G.; Westlund, O. Mobile generations: The role of mobile technology in the shaping of Swedish media generations. *Int. J. Commun.* **2009**, *3*, 108–124.
4. Villanti, A.C.; Johnson, A.L.; Ilakkuvan, V.; Jacobs, M.A.; Graham, A.L.; Rath, J.M. Social media use and access to digital technology in US young adults in 2016. *J. Med. Internet Res.* **2017**, *19*, e196. [CrossRef] [PubMed]
5. Chóliz, M. Mobile-phone addiction in adolescence: The test of mobile phone dependence (TMD). *Prog. Health Sci.* **2012**, *2*, 33–44.
6. Westlund, O. New(s) functions for the mobile: A cross-cultural study. *New Media Soc.* **2010**, *12*, 91–108. [CrossRef]
7. Ott, T.; Magnusson, A.G.; Weilenmann, A.; af Segerstad, Y.H. It must not disturb, it's as simple as that: Students' voices on mobile phones in the infrastructure for learning in Swedish upper secondary school. *Educ. Inf. Technol.* **2018**, *23*, 517–536. [CrossRef]
8. Chang, F.C.; Chiu, C.H.; Chen, P.H.; Chiang, J.T.; Miao, N.F.; Chuang, H.Y.; Liu, S. Children's use of mobile devices, smartphone addiction and parental mediation in Taiwan. *Comput. Hum. Behav.* **2019**, *93*, 25–32. [CrossRef]
9. Nikhita, C.S.; Jadhav, P.R.; Ajinkya, S.A. Prevalence of mobile phone dependence in secondary school adolescents. *J. Clin. Diagn. Res. JCDR* **2015**, *9*, 6–9. [CrossRef]
10. Seo, D.G.; Park, Y.; Kim, M.K.; Park, J. Mobile phone dependency and its impacts on adolescents' social and academic behaviors. *Comput. Hum. Behav.* **2016**, *63*, 282–292. [CrossRef]
11. Kwon, M.; Kim, D.J.; Cho, H.; Yang, S. The smartphone addiction scale: Development and validation of a short version for adolescents. *PLoS ONE* **2013**, *8*, e83558. [CrossRef]
12. Bhise, A.T.; Ghatule, A.A.; Ghatule, A.P. Study of mobile addiction among students wRT Gender and education. *Indian J. Res. Manag. Bus. Soc. Sci.* **2014**, *2*, 17–21.
13. Bianchi, A.; Phillips, J.G. Psychological predictors of problem mobile phone use. *Cyberpsychol. Behav.* **2005**, *8*, 39–51. [CrossRef] [PubMed]
14. Hawi, N.S.; Samaha, M. Relationships among smartphone addiction, anxiety, and family relations. *Behav. Inf. Technol.* **2017**, *36*, 1046–1052. [CrossRef]
15. Kates, A.W.; Wu, H.; Coryn, C.L. The effects of mobile phone use on academic performance: A meta-analysis. *Comput. Educ.* **2018**, *127*, 107–112. [CrossRef]
16. Thomée, S.; Härenstam, A.; Hagberg, M. Mobile phone use and stress, sleep disturbances, and symptoms of depression among young adults-a prospective cohort study. *BMC Public Health* **2011**, *11*, 66. [CrossRef]
17. Livingstone, S.; Helsper, E.J. Parental mediation of children's internet use. *J. Broadcast. Electron. Media* **2008**, *52*, 581–599. [CrossRef]

18. Nikken, P.; Jansz, J. Developing scales to measure parental mediation of young children's internet use. *Learn. Media Technol.* **2014**, *39*, 250–266. [CrossRef]
19. Hefner, D.; Knop, K.; Schmitt, S.; Vorderer, P. Rules? Role model? Relationship? The impact of parents on their children's problematic mobile phone involvement. *Media Psychol.* **2019**, *22*, 82–108. [CrossRef]
20. Sonck, N.; Nikken, P.; de Haan, J. Determinants of Internet mediation: A comparison of the reports by Dutch parents and children. *J. Child. Media* **2013**, *7*, 96–113. [CrossRef]
21. Panova, T.; Carbonell, X. Is smartphone addiction really an addiction? *J. Behav. Addict.* **2018**, *7*, 252–259. [CrossRef]
22. Sunthlia, A.; Ahmad, S.; Singh, S.P. Menace of mobile phone overuse: An emerging public health concern. *Int. J. Commun. Med. Public Health* **2017**, *3*, 153–156. [CrossRef]
23. Randler, C.; Wolfgang, L.; Matt, K.; Demirhan, E.; Horzum, M.B.; Beşoluk, Ş. Smartphone addiction proneness in relation to sleep and morningness–eveningness in German adolescents. *J. Behav. Addict.* **2016**, *5*, 465–473. [CrossRef] [PubMed]
24. Haug, S.; Castro, R.P.; Kwon, M.; Filler, A.; Kowatsch, T.; Schaub, M.P. Smartphone use and smartphone addiction among young people in Switzerland. *J. Behav. Addict.* **2015**, *4*, 299–307. [CrossRef] [PubMed]
25. Rebold, M.J.; Sheehan, T.; Dirlam, M.; Maldonado, T.; O'Donnell, D. The impact of cell phone texting on the amount of time spent exercising at different intensities. *Comput. Hum. Behav.* **2016**, *55*, 167–171. [CrossRef]
26. Lachmann, B.; Sindermann, C.; Sariyska, R.Y.; Luo, R.; Melchers, M.C.; Becker, B.; Montag, C. The role of empathy and life satisfaction in internet and smartphone use disorder. *Front. Psychol.* **2018**, *9*, 398. [CrossRef] [PubMed]
27. Alosaimi, F.D.; Alyahya, H.; Alshahwan, H.; Al Mahyijari, N.; Shaik, S.A. Smartphone addiction among university students in Riyadh, Saudi Arabia. *Saudi Med. J.* **2016**, *37*, 675–683. [CrossRef]
28. Lepp, A.; Barkley, J.E.; Karpinski, A.C. The relationship between cell phone use, academic performance, anxiety, and satisfaction with life in college students. *Comput. Hum. Behav.* **2014**, *31*, 343–350. [CrossRef]
29. Samaha, M.; Hawi, N.S. Relationships among smartphone addiction, stress, academic performance, and satisfaction with life. *Comput. Hum. Behav.* **2016**, *57*, 321–325. [CrossRef]
30. Cheever, N.A.; Rosen, L.D.; Carrier, L.M.; Chavez, A. Out of sight is not out of mind: The impact of restricting wireless mobile device use on anxiety levels among low, moderate and high users. *Comput. Hum. Behav.* **2014**, *37*, 290–297. [CrossRef]
31. Tams, S.; Legoux, R.; Leger, P.M. Smartphone withdrawal creates stress: A moderated mediation model of nomophobia, social threat, and phone withdrawal context. *Comput. Hum. Behav.* **2018**, *81*, 1–9. [CrossRef]
32. Elhai, J.D.; Dvorak, R.D.; Levine, J.C.; Hall, B.J. Problematic smartphone use: A conceptual overview and systematic review of relations with anxiety and depression psychopathology. *J. Affect. Disord.* **2017**, *207*, 251–259. [CrossRef]
33. Hao, Z.; Jin, L.; Li, Y.; Akram, H.R.; Saeed, M.F.; Ma, J.; Huang, J. Alexithymia and mobile phone addiction in Chinese undergraduate students: The roles of mobile phone use patterns. *Comput. Hum. Behav.* **2019**, *97*, 51–59. [CrossRef]
34. Volkmer, S.A.; Lermer, E. Unhappy and addicted to your phone? Higher mobile phone use is associated with lower well-being. *Comput. Hum. Behav.* **2019**, *93*, 210–218. [CrossRef]
35. Warren, R.; Bluma, A. Parental mediation of children's Internet use: The influence of established media. *Commun. Res. Rep.* **2002**, *19*, 8–17. [CrossRef]
36. Nathanson, A.I. Parent and child perspectives on the presence and meaning of parental television mediation. *J. Broadcast. Electron. Media* **2001**, *45*, 201–220. [CrossRef]
37. Bybee, C.R.; Robinson, D.; Turow, J. Determinants of parental guidance of children's television viewing for a special subgroup: Mass media scholars. *J. Broadcast. Electron. Media* **1982**, *26*, 697–710. [CrossRef]
38. Valkenburg, P.M.; Krcmar, M.; Peeters, A.L.; Marseille, N.M. Developing a scale to assess three styles of television mediation: "Instructive mediation," "restrictive mediation," and "social coviewing". *J. Broadcast. Electron. Media* **1999**, *43*, 52–66. [CrossRef]
39. Warren, R.; Gerke, P.; Kelly, M.A. Is there enough time on the clock? Parental involvement and mediation of children's television viewing. *J. Broadcast. Electron. Media* **2002**, *46*, 87–111. [CrossRef]
40. Nikken, P.; Jansz, J. Parental mediation of children's videogame playing: A comparison of the reports by parents and children. *Learn. Media Technol.* **2006**, *31*, 181–202. [CrossRef]

41. Livingstone, S.; Haddon, L.; Görzig, A.; Ólafsson, K. *Risks and Safety on the Internet: The Perspective of European Children: Full Findings and Policy Implications from the EU Kids Online Survey of 9–16 Year Olds and Their Parents in 25 Countries*; EU Kids Online: London, UK, 2011.
42. Mendoza, K. Surveying parental mediation: Connections, challenges and questions for media literacy. *J. Media Lit. Educ.* **2009**, *1*, 28–41.
43. Shin, W.; Li, B. Parental mediation of children's digital technology use in Singapore. *J. Child. Media* **2017**, *11*, 1–19. [CrossRef]
44. Vaala, S.E.; Bleakley, A. Monitoring, mediating, and modeling: Parental influence on adolescent computer and Internet use in the United States. *J. Child. Media* **2015**, *9*, 40–57. [CrossRef]
45. Duerager, A.; Livingstone, S. *How Can Parents Support Children's Internet Safety*; LSE, EU Kids Online: London, UK, 2012.
46. Kalmus, V.; Blinka, L.; Olafsson, K. Does it matter what mama says: Evaluating the role of parental mediation in European adolescents' excessive Internet use. *Child. Soc.* **2015**, *29*, 122–133. [CrossRef]
47. Bayraktar, F. Online Risks and Parental Mediation Strategies Comparison of Turkish Children/Adolescents Who Live in Turkey and Europe. *Educ. Sci. Egit. Ve Bilim* **2017**, *42*, 25–37. [CrossRef]
48. Livingstone, S.; Haddon, L.; Vincent, J.; Mascheroni, G.; Ólafsson, K. *Net Children Go Mobile: The UK Report*; London School of Economics and Political Science: London, UK, 2014.
49. Kim, J.H. Smartphone-mediated communication vs. face-to-face interaction: Two routes to social support and problematic use of smartphone. *Comput. Hum. Behav.* **2017**, *67*, 282–291. [CrossRef]
50. Lowry, P.B.; Gaskin, J.; Moody, G.D. Proposing the multi-motive information systems continuance model (MISC) to better explain end-user system evaluations and continuance intentions. *J. Assoc. Inf. Syst.* **2015**, *16*, 515–579.
51. Peele, S. *The Meaning of Addiction: Compulsive Experience and Its Interpretation*; Lexington Books DC Heath and Com: Lexington, MA, USA, 1985.
52. Parker, B.J.; Plank, R.E. A uses and gratifications perspective on the Internet as a new information source. *Am. Bus. Rev.* **2000**, *18*, 43–49.
53. Song, I.; Larose, R.; Eastin, M.S.; Lin, C.A. Internet gratifications and Internet addiction: On the uses and abuses of new media. *Cyberpsychol. Behav.* **2004**, *7*, 384–394. [CrossRef]
54. Yee, N. Motivations for play in online games. *Cyberpsychol. Behav.* **2006**, *9*, 772–775. [CrossRef]
55. Billieux, J.; Chanal, J.; Khazaal, Y.; Rochat, L.; Gay, P.; Zullino, D.; Van der Linden, M. Psychological predictors of problematic involvement in massively multiplayer online role-playing games: Illustration in a sample of male cybercafé players. *Psychopathology* **2011**, *44*, 165–171. [CrossRef]
56. Dauriat, F.Z.; Zermatten, A.; Billieux, J.; Thorens, G.; Bondolfi, G.; Zullino, D.; Khazaal, Y. Motivations to play specifically predict excessive involvement in massively multiplayer online role-playing games: Evidence from an online survey. *Eur. Addict. Res.* **2011**, *17*, 185–189. [CrossRef]
57. Király, O.; Urbán, R.; Griffiths, M.D.; Ágoston, C.; Nagygyörgy, K.; Kökönyei, G.; Demetrovics, Z. The mediating effect of gaming motivation between psychiatric symptoms and problematic online gaming: An online survey. *J. Med. Internet Res.* **2015**, *17*, e88. [CrossRef] [PubMed]
58. Kuss, D.J.; Griffiths, M.D.; Pontes, H.M. Chaos and confusion in DSM-5 diagnosis of Internet Gaming Disorder: Issues, concerns, and recommendations for clarity in the field. *J. Behav. Addict.* **2017**, *6*, 103–109. [CrossRef] [PubMed]
59. Klobas, J.E.; McGill, T.J.; Moghavvemi, S.; Paramanathan, T. Compulsive YouTube usage: A comparison of use motivation and personality effects. *Comput. Hum. Behav.* **2018**, *87*, 129–139. [CrossRef]
60. Bae, S.M. The relationship between the type of smartphone use and smartphone dependence of Korean adolescents: National survey study. *Child. Youth Serv. Rev.* **2017**, *81*, 207–211. [CrossRef]
61. Khang, H.; Kim, J.K.; Kim, Y. Self-traits and motivations as antecedents of digital media flow and addiction: The Internet, mobile phones, and video games. *Comput. Hum. Behav.* **2013**, *29*, 2416–2424. [CrossRef]
62. Wang, J.L.; Wang, H.Z.; Gaskin, J.; Wang, L.H. The role of stress and motivation in problematic smartphone use among college students. *Comput. Hum. Behav.* **2015**, *53*, 181–188. [CrossRef]
63. Grolnick, W.S.; Deci, E.L.; Ryan, R.M. Internalization within the family: The self-determination theory perspective. In *Parenting and Children's Internalization of Values: A Handbook of Contemporary Theory*; Grusec, J.E., Kuczynski, L., Eds.; John Wiley & Sons Inc.: Hoboken, NJ, USA, 1997; pp. 135–161.

64. White, S.R.; Rasmussen, E.E.; King, A.J. Restrictive mediation and unintended effects: Serial multiple mediation analysis explaining the role of reactance in US adolescents. *J. Child. Media* **2015**, *9*, 510–527. [CrossRef]
65. Maroney, N.; Williams, B.J.; Thomas, A.; Skues, J.; Moulding, R. A stress-coping model of problem online video game use. *Int. J. Ment. Health Addict.* **2019**, *17*, 845–858. [CrossRef]
66. Ballabio, M.; Griffiths, M.D.; Urbán, R.; Quartiroli, A.; Demetrovics, Z.; Király, O. Do gaming motives mediate between psychiatric symptoms and problematic gaming? An empirical survey study. *Addict. Res. Theory* **2017**, *25*, 397–408. [CrossRef]
67. Shen, X.; Wang, J.L. Loneliness and excessive smartphone use among Chinese college students: Moderated mediation effect of perceived stressed and motivation. *Comput. Hum. Behav.* **2019**, *95*, 31–36. [CrossRef]
68. Fikkers, K.M.; Piotrowski, J.T.; Valkenburg, P.M. A matter of style? Exploring the effects of parental mediation styles on early adolescents' media violence exposure and aggression. *Comput. Hum. Behav.* **2017**, *70*, 407–415. [CrossRef]
69. Valkenburg, P.M.; Piotrowski, J.T.; Hermanns, J.; De Leeuw, R. Developing and validating the perceived parental media mediation scale: A self-determination perspective. *Hum. Commun. Res.* **2013**, *39*, 445–469. [CrossRef]
70. Demirhan, E.; Randler, C.; Horzum, M.B. Is problematic mobile phone use explained by chronotype and personality? *Chronobiol. Int.* **2016**, *33*, 821–831. [CrossRef]
71. Rapee, R.M.; Coplan, R.J. Conceptual relations between anxiety disorder and fearful temperament. *New Dir. Child Adolesc. Dev.* **2010**, *127*, 17–31. [CrossRef]
72. Cheek, J.M.; Buss, A.H. Shyness and sociability. *J. Personal. Soc. Psychol.* **1981**, *41*, 330–339. [CrossRef]
73. Henderson, L.; Zimbardo, P.G. Shyness as a clinical condition: The Stanford model. In *International Handbook of Social Anxiety: Concepts, Research and Interventions Relating to the Self and Shyness*; Crozier, W.R., Alden, L.E., Eds.; John Wiley & Sons Ltd.: New York, NY, USA, 2001; pp. 431–447.
74. Rubin, K.H.; Coplan, R.J.; Bowker, J.C. Social withdrawal in childhood. *Annu. Rev. Psychol.* **2009**, *60*, 141–171. [CrossRef] [PubMed]
75. Tan, J.; Ai, Y.; Wen, X.; Wu, Y.; Wang, W. Relationship between shyness and loneliness among Chinese adolescents: Social support as mediator. *Soc. Behav. Personal. Int. J.* **2016**, *44*, 201–208. [CrossRef]
76. Barstead, M.G.; Smith, K.A.; Laursen, B.; Booth-LaForce, C.; King, S.; Rubin, K.H. Shyness, preference for solitude, and adolescent internalizing: The roles of maternal, paternal, and best-friend support. *J. Res. Adolesc.* **2018**, *28*, 488–504. [CrossRef] [PubMed]
77. Laghi, F.; Schneider, B.H.; Vitoroulis, I.; Coplan, R.J.; Baiocco, R.; Amichai-Hamburger, Y.; Flament, M. Knowing when not to use the Internet: Shyness and adolescents' on-line and off-line interactions with friends. *Comput. Hum. Behav.* **2013**, *29*, 51–57. [CrossRef]
78. Appel, M.; Gnambs, T. Shyness and social media use: A meta-analytic summary of moderating and mediating effects. *Comput. Hum. Behav.* **2019**, *98*, 294–301. [CrossRef]
79. Ang, C.S.; Chan, N.N.; Lee, C.S. Shyness, Loneliness Avoidance, and Internet Addiction: What are the Relationships? *J. Psychol.* **2018**, *152*, 25–35. [CrossRef] [PubMed]
80. Chak, K.; Leung, L. Shyness and locus of control as predictors of internet addiction and internet use. *Cyberpsychol. Behav.* **2004**, *7*, 559–570. [CrossRef] [PubMed]
81. Tian, Y.; Bian, Y.; Han, P.; Gao, F.; Wang, P. Associations between psychosocial factors and generalized pathological internet use in Chinese university students: A longitudinal cross-lagged analysis. *Comput. Hum. Behav.* **2017**, *72*, 178–188. [CrossRef]
82. Bian, M.; Leung, L. Linking loneliness, shyness, smartphone addiction symptoms, and patterns of smartphone use to social capital. *Soc. Sci. Comput. Rev.* **2015**, *33*, 61–79. [CrossRef]
83. Han, L.; Geng, J.; Jou, M.; Gao, F.; Yang, H. Relationship between shyness and mobile phone addiction in Chinese young adults: Mediating roles of self-control and attachment anxiety. *Comput. Hum. Behav.* **2017**, *76*, 363–371. [CrossRef]
84. Hong, W.; Liu, R.D.; Oei, T.P.; Zhen, R.; Jiang, S.; Sheng, X. The mediating and moderating roles of social anxiety and relatedness need satisfaction on the relationship between shyness and problematic mobile phone use among adolescents. *Comput. Hum. Behav.* **2019**, *93*, 301–308. [CrossRef]
85. Luke, F.; Chan, C.C.; Au, A.; Lai, S.M. Adaptive parenting for alleviating young children's shyness: A randomized controlled trial of an early intervention program. *Infant Child Dev.* **2017**, *26*, 1–19. [CrossRef]

86. Hastings, P.D.; Rubin, K.H.; DeRose, L. Links among gender, inhibition, and parental socialization in the development of prosocial behavior. *Merrill-Palmer Q.* **2005**, *51*, 467–493. [CrossRef]
87. Zarra-Nezhad, M.; Moazami-Goodarzi, A.; Nurmi, J.E.; Eklund, K.; Ahonen, T.; Aunola, K. Children's Shyness Moderates the Associations between Parenting Behavior and the Development of Children's Pro-Social Behaviors. *J. Child Fam. Stud.* **2018**, *27*, 3008–3018. [CrossRef]
88. Eastin, M.S.; Greenberg, B.S.; Hofschire, L. Parenting the internet. *J. Commun.* **2006**, *56*, 486–504. [CrossRef]
89. Young, C.M.; DiBello, A.M.; Traylor, Z.K.; Zvolensky, M.J.; Neighbors, C. A Longitudinal Examination of the Associations between Shyness, Drinking Motives, Alcohol Use, and Alcohol-Related Problems. *Alcohol. Clin. Exp. Res.* **2015**, *39*, 1749–1755. [CrossRef] [PubMed]
90. Wu, N. The Relationship between Children Internet use with Academic Performance: The Role of Parental Mediation. Ph.D. Thesis, Central China Normal University, Wuhan, China, 2016. (In Chinese).
91. Li, F.F.; Luo, Q.; Zhou, Z.K.; Sun, X.J.; Wei, H. Mediating effect of interpersonal communication disturbances between college students' shyness and dependence of online communication. *J. Clin. Psychol.* **2012**, *20*, 182–184. (In Chinese)
92. Foerster, M.; Roser, K.; Schoeni, A.; Röösli, M. Problematic mobile phone use in adolescents: Derivation of a short scale MPPUS-10. *Int. J. Public Health* **2015**, *60*, 277–286. [CrossRef] [PubMed]
93. Zhuang, H.; Liu, R.-D.; Liu, Y. The impact of peer attachment on problematic mobile phone use among adolescents: Moderated mediation effect of loneliness and self-construa. *Psychol. Sci.* **2017**, *40*, 89–95. (In Chinese)
94. Little, R.J.; Rubin, D.B. *Statistical Analysis with Missing Data*; John Wiley & Sons: Hoboken, NJ, USA, 2019; Volume 793.
95. Hayes, A.F. *Introduction to Mediation, Moderation, and Conditional Process Analysis: A Regression-Based Approach*; The Guilford Press: New York, NY, USA, 2013.
96. Brehm, S.S.; Brehm, J.W. *Psychological Reactance: A Theory of Freedom and Control*; Academic Press: Cambridge, MA, USA, 2013.
97. Yen, J.Y.; Yen, C.F.; Chen, C.C.; Chen, S.H.; Ko, C.H. Family factors of internet addiction and substance use experience in Taiwanese adolescents. *Cyberpsychol. Behav.* **2007**, *10*, 323–329. [CrossRef] [PubMed]
98. Xiuqin, H.; Huimin, Z.; Mengchen, L.; Jinan, W.; Ying, Z.; Ran, T. Mental health, personality, and parental rearing styles of adolescents with Internet addiction disorder. *Cyberpsychol. Behav. Soc. Netw.* **2010**, *13*, 401–406. [CrossRef]
99. Nikken, P.; de Graaf, H. Reciprocal relationships between friends' and parental mediation of adolescents' media use and their sexual attitudes and behavior. *J. Youth Adolesc.* **2013**, *42*, 1696–1707. [CrossRef]
100. Hasan, M.R.; Jha, A.K.; Liu, Y. Excessive use of online video streaming services: Impact of recommender system use, psychological factors, and motives. *Comput. Hum. Behav.* **2018**, *80*, 220–228. [CrossRef]
101. Kardefelt-Winther, D. A conceptual and methodological critique of internet addiction research: Towards a model of compensatory internet use. *Comput. Hum. Behav.* **2014**, *31*, 351–354. [CrossRef]
102. Kardefelt-Winther, D. Conceptualizing Internet use disorders: Addiction or coping process? *Psychiatry Clin. Neurosci.* **2017**, *71*, 459–466. [CrossRef]
103. Bandura, A. Social learning theory of aggression. *J. Commun.* **1978**, *28*, 12–29. [CrossRef] [PubMed]
104. Xie, X.; Chen, W.; Zhu, X.; He, D. Parents' phubbing increases Adolescents' Mobile phone addiction: Roles of parent-child attachment, deviant peers, and gender. *Child. Youth Serv. Rev.* **2019**, *105*, 104426. [CrossRef]
105. Fu, X.; Liu, J.; Liu, R.D.; Ding, Y.; Hong, W.; Jiang, S. The Impact of Parental Active Mediation on Adolescent Mobile Phone Dependency: A Moderated Mediation Model. *Comput. Hum. Behav.* **2020**, *107*, 106280. [CrossRef]
106. Caplan, S.; Williams, D.; Yee, N. Problematic Internet use and psychosocial well-being among MMO players. *Comput. Hum. Behav.* **2009**, *25*, 1312–1319. [CrossRef]
107. Billieux, J.; Van der Linden, M.; Achab, S.; Khazaal, Y.; Paraskevopoulos, L.; Zullino, D.; Thorens, G. Why do you play World of Warcraft? An in-depth exploration of self-reported motivations to play online and in-game behaviours in the virtual world of Azeroth. *Comput. Hum. Behav.* **2013**, *29*, 103–109. [CrossRef]
108. Erikson, E.H. *Identity and the Life Cycle: Selected Papers*; International Universities Press: Oxford, UK, 1959.
109. Odacı, H.; Çelik, Ç.B. Who are problematic internet users? An investigation of the correlations between problematic internet use and shyness, loneliness, narcissism, aggression and self-perception. *Comput. Hum. Behav.* **2013**, *29*, 2382–2387. [CrossRef]

110. Cheek, J.M.; Melchior, L.M. Shyness and anxious self-preoccupation during a social interaction. *J. Soc. Behav. Personal.* **1990**, *5*, 117–130.
111. Jackson, T.; Fritch, A.; Nagasaka, T.; Gunderson, J. Towards explaining the association between shyness and loneliness: A path analysis with American college students. *Soc. Behav. Personal. Int. J.* **2002**, *30*, 263–270. [CrossRef]

International Journal of
Environmental Research and Public Health

MDPI

Article

Internet Gaming Disorder Clustering Based on Personality Traits in Adolescents, and Its Relation with Comorbid Psychological Symptoms

Vega González-Bueso [1,*,†], Juan José Santamaría [1,*,†], Ignasi Oliveras [2], Daniel Fernández [3,4,5], Elena Montero [2], Marta Baño [1], Susana Jiménez-Murcia [6,7,8], Amparo del Pino-Gutiérrez [6,9] and Joan Ribas [1]

1. Atención e Investigación en Socioadicciones (AIS), Mental Health and Addictions Network, Generalitat de Catalunya (XHUB), 08014 Barcelona, Spain; mbano@ais-info.org (M.B.); 38039jrs@comb.cat (J.R.)
2. Department of Psychiatry & Forensic Medicine, Institute of Neurosciences, Universitat Autònoma de Barcelona, Bellaterra, 08193 Barcelona, Spain; ignasi.oliveras@uab.cat (I.O.); emontero@ais-info.org (E.M.)
3. Research and Development Unit, Parc Sanitari Sant Joan de Déu, Fundació Sant Joan de Déu, Sant Boi de Llobregat, 08830 Barcelona, Spain; df.martinez@pssjd.org
4. Department of Statistics and Operations Research, Polytechnic University of Catalonia-BarcelonaTech, 08028 Barcelona, Spain
5. Centro de Investigación Biomédica en Red de Salud Mental (CIBERSAM), Instituto de Salud Carlos III, 28029 Madrid, Spain
6. Pathological Gambling Unit, Department of Psychiatry, Bellvitge University Hospital-IDIBELL, 08907 Barcelona, Spain; sjimenez@bellvitgehospital.cat (S.J.-M.); adelpino@ub.edu (A.d.P.-G.)
7. Ciber Fisiopatología Obesidad y Nutrición (CIBERObn), Instituto de Salud Carlos III, 28029 Madrid, Spain
8. Department of Clinical Sciences, School of Medicine and Health Sciences, University of Barcelona, 08907 Barcelona, Spain
9. Nursing Department of Mental Health, Public Health, Maternal and Child Health, Nursing School of the University of Barcelona, 08907 Barcelona, Spain

* Correspondence: vgonzalez@ais-info.org (V.G.-B.); jsantamaria@ais-info.org (J.J.S.)
† These authors contributed equally to this work and share first authorship.

Received: 20 December 2019; Accepted: 23 February 2020; Published: 26 February 2020

Abstract: In recent years, the evidence regarding Internet Gaming Disorder (IGD) suggests that some personality traits are important risk factors for developing this problem. The heterogeneity involved in problematic online gaming and differences found in the literature regarding the comorbid psychopathology associated with the problem could be explained through different types of gamers. Clustering analysis can allow organization of a collection of personality traits into clusters based on similarity. The objectives of this study were: (1) to obtain an empirical classification of IGD patients according to personality variables and (2) to describe the resultant groups in terms of clinical and sociodemographic variables. The sample included 66 IGD adolescent patients who were consecutive referrals at a mental health center in Barcelona, Spain. A Gaussian mixture model cluster analysis was used in order to classify the subjects based on their personality. Two clusters based on personality traits were detected: type I "higher comorbid symptoms" (n = 24), and type II "lower comorbid symptoms" (n = 42). The type I included higher scores in introversive, inhibited, doleful, unruly, forceful, oppositional, self-demeaning and borderline tendency traits, and lower scores in histrionic, egotistic and conforming traits. The type I obtained higher scores on all the Symptom Check List-90 items-Revised, all the State-Trait Anxiety Index scales, and on the DSM-5 IGD criteria. Differences in personality can be useful in determining clusters with different types of dysfunctionality.

Keywords: internet gaming disorder; cluster analysis; video game; video game addiction; personality; comorbidity

1. Introduction

According to the ICD-11 (International Classification of Diseases) [1], Internet Gaming Disorder (IGD) is defined as "a pattern of persistent or recurrent gaming behavior ('digital gaming' or 'video-gaming'), which may be online (i.e., over the Internet) or offline, manifested by: (1) impaired control over gaming (e.g., onset, frequency, intensity, duration, termination, context); (2) increasing priority given to gaming to the extent that gaming takes precedence over other life interests and daily activities; and (3) continuation or escalation of gaming despite the occurrence of negative consequences. The behavior pattern is of sufficient severity to result in significant impairment in personal, family, social, educational, occupational or other important areas of functioning. The pattern of gaming behavior may be either continuous or, on the other hand, episodic and recurrent. The gaming behavior and other features are normally evident over a period of at least 12 months for a diagnosis to be assigned, although the required duration may be shortened if all diagnostic requirements are met and symptoms are severe".

Nowadays, behavioral addictions, including IGD, are increasingly being documented worldwide [2]. The current versions of the official diagnostic classification manuals have included addictions without substances in the behavioral addictions category. At the moment, only Gambling Disorder has been included in this category, and although IGD seems to share many factors with this disorder such as the negative reinforcement as a maintaining variable in the long-term maintenance of the behavior, or the use of positive reinforcement as a developing mechanism at the beginning of the problem [3,4], the DSM-5 work group decided to include IGD in Section III of the diagnostic manual DSM-5 [5] as a condition that requires further study.

In recent years, numerous factors involved in the etiology of problematic online gaming and of IGD have been identified. Although more research is needed, the evidence suggests that some personality traits are important risk factors for developing this problem [6,7]. Personality traits are different across individuals, influence their behavior, reflect people's characteristic patterns of thoughts, attitudes, emotions, and behaviors [8], and they have relatively high stability over a lifetime, even though some authors have found changes after cognitive behavioral therapies [9]. Specifically, researchers have linked patients with IGD to personality traits common in other addictive disorders.

Regarding the Big Five Model of personality, some of the factors of the model seem to have relevance in the addiction process [10]. Among them, high neuroticism is the most commonly present in IGD, and has been interpreted as a way to use online gaming to overcome negative life effects [11] or to modify negative emotions [12,13]. The results regarding the other traits of the model and male gamers are various, with low agreeableness [14], low conscientiousness and low extraversion [15] being the most consistent findings. These domains reflect an impaired offline social cooperation with their peers, low self-discipline, and low motivation to maintain positive interpersonal relations.

Taking other personality models as a reference, some authors have found that low sociability [16], low openness to experience [17], and the combination of low self-directedness and cooperativeness [18] were correlated with problematic gaming and IGD. Finally, low self-directness [14] has been found as a predictor of IGD. This trait reflects the lack of regulation and adaptation of one's own behavior in order to achieve personally chosen goals and values, and the tendency to be laid back.

All these traits combined can predispose people to avoid social interactions and new activities. This diminished social participation in the "real world" and the need to establish some kind of social interactions with others could lead gamers to find online relationships, which are commonly more distant and superficial, with people with similar interests. Thus, they expand their social network [19] and find themselves in a safe environment when they are in the online world.

Clustering can be defined as the statistical methods that allow us to organize a collection of data points into clusters based on similarity [20]. These methods are part of what we call unsupervised learning and they are used in a wide range of research fields such as psychology, biology and market research [21–23]. Clustering methods have been used to classify individuals based upon behavioral patterns [24], personality traits [25] and severity of mental disorder [26].

Several studies have focused on classifying the personality traits of patients with behavioral addictions using clusters, and their relationships with psychiatric comorbidity and sociodemographic characteristics. The study of personality profiles for gambling disorder [25,27,28] and compulsive buying [29] shows that there is a heterogeneity in the personality patterns of the affected people, and these different profiles are associated with differences in the number and severity of other psychological comorbid symptoms. These results suggest that the experience with problematic behavior varies between patients, and that the processing of the negative consequences derived from it may affect them in different ways.

Focusing on the use of videogames, Billieux et al. [30] classified a group of 1057 general population online gamers, playing a massive multiplayer online roleplaying-game, into five reliable clusters (three problematic and two non-problematic clusters) according to several psychological risk factors (impulsivity, motives to play, self-esteem) and potential consequences of playing (addiction symptoms, positive and negative affect). Members of the two non-problematic clusters were defined by low impulsivity and high and low levels of self-esteem, respectively. The motivations to play of these two clusters vary between non-fundamental, to motivations related to social exchange and roleplaying.

With respect to the three problematic clusters, the first one was composed of gamers with poor self-esteem and high impulsivity (but low sensation seeking), and by low achievement and high escapism motives. The second problematic cluster included high self-esteem, high impulsivity, and motivations regarding achievements in the game. The last cluster comprise gamers with high self-esteem, high impulsivity, and motives related to roleplaying, achievements and escapism. This study demonstrated the existence of distinct subtypes of problematic online gamers, emphasizing the high heterogeneity and the wide range of psychological factors involved in the problem.

Therefore, it could be possible that, similar to what happens in other behavioral addictions, the different personality types of the gamers have a role not only in the development or perpetuation of the problem, but also in the comorbid psychopathology associated to the disorder. In that sense, several authors have analyzed the different personality and psychopathological features among IGD patients [31,32], finding fewer functional personality traits and higher psychopathological scores compared with a normative population.

Nevertheless, the relationship between personality and psychopathology in IGD remains unclear, and the associations between comorbidity and IGD in adolescents and young adults have shown inconsistent results with depression, anxiety, ADHD or hyperactivity, social phobia/anxiety, and obsessive-compulsive disorder, finding among different authors full, partial and no associations with these symptoms [33]. It is possible that these inconsistencies could be related to the analysis as a whole of different personality types among video game players.

In other words, cluster analyses can help in the conceptualization of patients consulting for IGD, and these profiles of similarities and differences among individuals can contribute to clarifying some of the results found in previous research regarding clinical profiles, and can help to improve the clinical treatments. Considering this, the objectives of this study are as follows: (1) to obtain an empirical classification of IGD patients according to personality variables and (2) to describe the resultant groups in terms of clinical and sociodemographic variables.

2. Materials and Methods

2.1. Participants

The sample included 66 IGD patients who were consecutive referrals for assessment and outpatient treatment at the Behavioral Addiction Unit in the mental health center AIS-PRO JUVENTUD (Care and Research in Behavioral Addiction) (AIS), located in Barcelona, Spain.

The required sample size was calculated based on the standard deviations of the questionnaire Millon Adolescent Personality Inventory (MACI). Thus, by setting an alpha risk of 0.05 and a beta risk of 0.20 in a two-sided test with a 10% estimated dropout rate, we required a sample size of 59 individuals to detect a minimum expected difference between groups of 6 units. We therefore decided to recruit 66 patients.

The exclusion criteria for being included in the analyses were: (1) had neurological disorders or primary psychiatric conditions that could affect cognitive function (assessed through semi-structured, face-to-face, clinical interview, (2) had a head injury with loss of consciousness for more than 2 min or a learning disorder, (3) used psychostimulants or drugs that could interfere with the assessment, (4) were older than 21 years or younger than 12. No potential participants were excluded based on exclusion criteria 1, 2, or 3.

This study was carried out according to the latest version of the Declaration of Helsinki. The Ethics Committee of CEIC Fundació Unió Catalana d'Hospitals (CEIC14/71) approved the study, and informed consent (signed document) was obtained from parents of adolescents under the age of 18 years and adolescents over the age of 18 years (and assent in adolescents under the age of 18 years).

The characteristics of the sample were as follows: all patients were Caucasian and male. All the adolescents included have as their main and problematic videogame an online videogame, with a mean age of 15.80 (SD 2.18) years. Most patients had elementary education (92.4%). Regarding their main problem, all the patients included played online videogames, and the mean duration of the problem was 2.2 (SD 1.7) years. There was no consumption of alcohol or drugs and only a 3.0% of the sample were smokers.

2.2. Instruments

2.2.1. Millon Adolescent Personality Inventory (MACI)

This personality test [34] has 160 items and is self-administered. It measures thirty-one scales: twelve Personality Patterns scales (Axis II), eight Expressed Concerns Scales, seven Clinical Syndrome Scales, three Modifying Indices (particular response styles), and a Validity scale. The instrument has been translated and validated into a Spanish population with a good internal consistency of 0.82 (mean Cronbach's alpha) [35]. The MACI is one of the most widely used personality assessment tests for adolescents [36–40]. The MACI is constructed using an underlying theory of personality and psychopathology, and can identify and assess a wide range of psychological difficulties in adolescents. Studies have examined the potential utility of the MACI for assessing substance use disorders [41], reporting support for MACI as a screening instrument. We used this instrument in order to facilitate current and future comparisons in other studies regarding IGD or other psychological problems.

2.2.2. Symptom CheckList-90 Items-Revised (SCL-90-R)

This questionnaire [42] evaluates psychological problems and psychopathological symptoms. It contains ninety items and measures nine primary symptom dimensions: somatization, obsession-compulsion, interpersonal sensitivity, depression, anxiety, hostility, phobic anxiety, paranoid ideation, and psychoticism. It also includes three global indices, such as a global severity index (GSI), that measures overall psychological distress, a positive symptom distress index (PSDI), to measure the intensity of symptoms; and a positive symptom total (PST). This scale has been translated to Spanish and validated in a Spanish population [43], and presents a good internal consistency (mean Cronbach's alpha = 0.75).

2.2.3. State-Trait Anxiety Index (STAI)

This questionnaire [44] includes forty items on a 4-point rating scale and is self-reported, measuring state anxiety (twenty items) and trait anxiety (twenty items). The minimum score is 20 and the maximum is 80 points. The state anxiety uses items that measure subjective feelings of apprehension, tension, nervousness, worry, and activation/arousal of the autonomic nervous system and evaluates the current state of anxiety. The trait anxiety scale includes general states of calmness, confidence, and security and evaluates relatively stable aspects of "anxiety proneness". The STAI has been translated to Spanish and validated in the Spanish population with a mean Cronbach's alpha coefficient of 0.92 [45].

2.2.4. DSM-5 IGD Criteria

This instrument is a questionnaire evaluating the criteria for IGD proposed in the DSM-5 [5]. The diagnostic criteria of IGD are composed of 9 items: preoccupation, withdrawal, tolerance, unsuccessful attempts to control, loss of other interests, continued excessive use despite psychosocial problems, deceiving regarding online gaming, escape, and functional impairment. Five or more DSM-5 criteria of IGD indicates Internet gaming problems. The criteria were asked in a questionnaire form using a "yes" or "no" response.

2.2.5. Sociodemographical Variables

Additional demographic, clinical, and social/family variables related to internet gaming were measured using a semi-structured face-to-face clinical interview, including age, sex, duration of the problem, and education level.

2.3. Procedure

First, at intake, a face-to-face specific clinical interview and a functional analysis of IGD was carried by experienced psychologists (more than 5 years of clinical experience in behavioral addictions) using the semi-structured clinical interview SCID-I [46]. The questions included in this interview were about tolerance, preoccupation, withdrawal, loss of control, playing for long periods, escaping from adverse mental states, risking or losing relationships or opportunities because of the behavior, deception/covering up, giving up other activities, persistence of the behavior despite problems, and functional impairment (e.g., functional impairment in familial relationships, other social relationships, and academic achievement), and questions regarding demographic data.

During a second session (with an average duration of 90 min) before starting the treatment, were administered the above-mentioned questionnaires.

With regard to meeting the diagnostic criteria for IGD, the results obtained through the DSM-5 diagnostic criteria questionnaire were compared post hoc with the results obtained through the face-to-face clinical interview, and only patients who met the DSM-5 criteria for IGD were included in our analysis.

2.4. Statistical Analysis

Statistical analysis was carried out using the statistical software R version 3.5.3 (R Core Team, Vienna, Austria) and, in particular, its R packages mclust [47] and factoextra [48].

We obtained clusters based on the scores of the 12 sub-scales of the personality patterns scale of the MACI (Introversive, Inhibited, Doleful, Submissive, Histrionic, Egotistic, Unruly, Forceful, Conforming, Oppositional, Self-Demeaning, and Borderline Tendency) conducting a Gaussian Mixture Model (GMM) cluster analysis. Model-fitting is performed using the expectation-maximization (EM) algorithm [49,50], which later is used to initialize a hierarchical model-based agglomerative clustering [51,52]. The optimal number of clusters was selected based on the Bayesian Information Criterion (BIC), where the lowest values indicate better fit. Each individual was allocated to one class only according to their highest probability of membership.

Chi-square tests (χ^2) for categorical variables and *t*-test (2 groups) for quantitative measures were computed to assess differences between clusters. When the normality assumption was not accomplished according to Shapiro–Wilk test, the equivalent non-parametric Mann–Whitney U test was performed. A two-sided *p*-value < 0.05 was considered statistically significant. Cohen's *d* was used to measure the effect size for power analysis purposes. The effect size was classified as high ($d = 0.8$), medium ($d = 0.5$) or low ($d = 0.2$) according to Cohen [53].

3. Results

3.1. Cluster Composition: Description for the Cluster Indicators

In the MACI scale (n = 66 men, age: 15.80 ± 2.18) there are 12 sub-scales: Introversive, Inhibited, Doleful, Submissive, Histrionic, Egotistic, Unruly, Forceful, Conforming, Oppositional, Self-Demeaning, and Borderline Tendency. The descriptive statistics of the sample are presented in Table 1.

Table 1. Descriptive statistics of the personality traits of the Millon Adolescent Personality Inventory (MACI) scale (n = 66).

Personality Traits	Mean	SD	Median	IQR	Min	Max
Introversive	25.20	11.61	23	17.80	9	52
Inhibited	20.86	10.82	18.50	25.50	4	48
Doleful	11.47	9.92	8	14	0	41
Submissive	43.85	10.00	45.50	13	21	69
Histrionic	36.55	9.97	37	14.30	12	54
Egotistic	33.47	10.85	35	15	3	51
Unruly	30.23	9.39	29.50	11.80	10	52
Forceful	10.59	6.81	9	7.6	0	34
Conforming	45.41	9.05	46	13.30	18	62
Oppositional	20.30	9.86	20	12.80	4	44
Self-Demeaning	19.42	13.84	16	20	0	55
Borderline Tendency	11.70	7.72	11	12.50	0	30

SD = Standard Deviation, IQR = Interquartile range.

We conducted a Gaussian finite mixture model cluster analysis. The results, for the personality patterns scale, showed that a 2 cluster (VEE: ellipsoidal, equal shape and orientation) was the optimal solution. The cluster classification for each subject is shown in Figure 1. The first cluster has 24 subjects and the second cluster has 42 subjects.

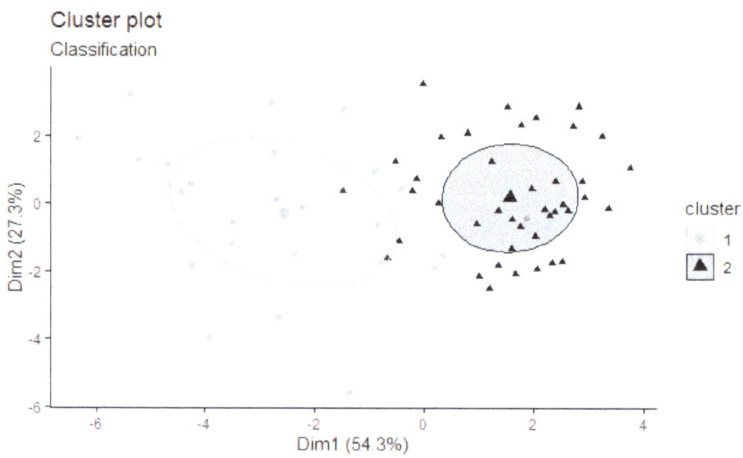

Figure 1. Classification of subjects for each cluster.

The results of the t-test and the Mann–Whitney U test comparing the two clusters created by the model-based clustering algorithm show that there are significant differences in all the sub-scales, except in the Submissive. The first cluster has higher values in the following sub-scales: introversive, inhibited, doleful, unruly, forceful, oppositional, self-demeaning and borderline tendency, while in the histrionic, egotistic and conforming sub-scales the first cluster has lower values (see Tables 2 and 3).

The results of the Cohen's d show that the effect size was high in all scales apart from the Submissive and Unruly scales (*d*-values < 0.8; Tables 2 and 3).

Table 2. T-test comparison between clusters of the personality traits of the MACI and the State-Trait Anxiety Index (STAI) scales (cluster 1 $n = 24$, cluster 2 $n = 42$).

Independent Samples Test						
	CLUSTER 1	**CLUSTER 2**				
	Mean (SD)	**Mean (SD)**	**t**	**df**	**Sig. (2 tailed)**	**Cohen's d**
MACI						
Submissive	46.48 (11.86)	42.24 (8.42)	1.69	64	0.126	0.43
Egotistic	27.80 (3.03)	36.93 (7.33)	−3.16	33.15	0.003	**−0.92**
Unruly	34.12 (9.70)	27.85 (8.45)	2.76	64	0.010	0.70
Conforming	39.44 (8.19)	49.04 (7.49)	−4.86	64	0.000	**−1.23**
Oppositional	28.76 (8.58)	15.14 (6.46)	7.32	64	0.000	**1.86**
STAI						
Anxiety State	18.72 (8.35)	11.38 (7.52)	3.67	63	0.000	**0.94**
Anxiety Trait	23.48 (8.11)	12.64 (7.00)	5.75	64	0.000	**1.46**

SD = Standard Deviation; df = degrees of freedom (Cohen's d > 0.80 in bold).

Table 3. Mann–Whitney U-test for the MACI, Symptom CheckList-90 items-Revised (SCL-90-R), and DSM 5 criteria mean scores (cluster 1 $n = 24$, cluster 2 $n = 42$).

Mann–Whitney U Test					
	CLUSTER 1	**CLUSTER 2**			
	Mean (SD)	**Mean (SD)**	**U**	**Sig. (2 tailed)**	**Cohen's d**
MACI					
Introversive	35.76 (9.27)	18.76 (7.44)	948.50	0.000	**2.08**
Inhibited	29.88 (11.04)	15.37 (5.94)	883	0.000	**1.76**
Doleful	20.44 (9.42)	6.00 (5.07)	941	0.000	**2.05**
Histrionic	29.88 (10.95)	40.61 (6.70)	223.50	0.000	**−1.26**
Forceful	14.56 (7.49)	8.17 (5.07)	781	0.000	**1.05**
Self-Demeaning	32.96 (11.51)	11.17 (6.97)	974	0.000	**2.45**
Borderline Tend.	19.44 (5.34)	6.97 (4.39)	998.5	0.000	**2.61**
SCL-90-R					
Somatization	0.55 (0.54)	0.26 (0.27)	718	0.006	0.75
Obsessive-comp	1.03 (0.60)	0.61 (0.51)	719	0.006	0.76
Interp. sens.	1.00 (0.81)	0.38 (0.39)	781	0.000	**1.04**
Depression	0.93 (0.72)	0.28 (0.30)	840.50	0.000	**1.28**
Anxiety	0.69 (0.81)	0.21 (0.22)	700	0.012	**0.90**
Hostility	1.10 (0.91)	0.56 (0.45)	708	0.009	**0.81**
Phobia	0.40 (0.57)	0.10 (0.19)	782.50	0.000	0.77
Paranoid ideation	1.03 (0.82)	0.38 (0.43)	773	0.000	**1.06**
Psychoticism	0.56 (0.50)	0.14 (0.20)	845	0.000	**1.18**
Global severity	0.80(0.56)	0.33(0.25)	833	0.000	**1.20**
DSM 5 criteria	5.84 (1.82)	4.95 (1.55)	684	0.019	0.54

SD = Standard Deviation; df = degrees of freedom (Cohen's d > 0.80 in bold).

3.2. *Comparison between the Clusters in Sociodemographic and Clinical Variables*

After that, we compared the two clusters regarding a series of sociodemographic variables, i.e., age, age of onset of the disorder, disorder duration, and education level. The results of the *t*-tests showed that there were no significant differences between clusters regarding age, age of onset, and years with IGD. Besides, the Fisher exact test, to elucidate whether the education level was related to the cluster classification, was also non-significant.

In order to investigate if there were other significant psychopathological differences between clusters, we conducted Student's t-tests and Mann–Whitney U-tests to compare the clusters in the Symptom Check List-90 items-Revised scores, in the DSM 5 diagnostic criteria scores and the STAI. The SCL-R has nine dimensions: somatization, obsessive-compulsive, interpersonal sensitivity, depression, anxiety, hostility, phobic anxiety, paranoid ideation and psychoticism. The results of the analyses showed that the first cluster had significantly increased scores in all dimensions of the SCL-R, except for the somatization. Regarding the diagnostic criteria, the analysis comparing the cluster regarding the DSM scores yielded a significant result ($U = 684$, $p < 0.05$) indicating that the subjects in the first cluster had higher scores. Regarding Cohen's d values, we observed high effect sizes except for the three dimensions of the SCL-90-R (Somatization, Obsessive-Compulsive and Phobia), and for the DSM-5 scores, the effect size was moderate.

In a similar manner, we conducted the same analyses for the anxiety-trait and anxiety-state of the STAI and all the comparisons were statistically significant (all $t = 5.54$; $p < 0.001$ for anxiety trait and $t = 3.58$; $p < 0.001$ for anxiety state), indicating that the first cluster of each scale had higher anxiety levels than the members of the second cluster.

4. Discussion

This work explores the existence of empirical clusters for IGD starting from the patients' personality trait scores, using a person-centered methodology.

Two clusters were detected: type I "higher comorbid symptoms" and type II "lower comorbid symptoms". The main differences in the clustering variables between both clusters were the mean scores in the personality traits introversive, inhibited, doleful, unruly, forceful, oppositional, self-demeaning and borderline tendency (higher scores for the type I), histrionic, egotistic and conforming (lower scores for the type I). It should be noted the similarity between this cluster I and the IGD dimensional personality profile found in other studies.

In line with our results, literature about personality traits and IGD suggests that high scores in neuroticism [54], impulsivity [55] and introversion [15,18], and lower levels in agreeableness [14,15], cooperativeness [18] and self-regulation [56], are risk factors for this disorder. Furthermore, some authors have found that low levels of self- directedness uniquely predicted video game abuse in an adult population with Gambling Disorder [14], and others have identified lower responsibility as a risk factor associated with this IGD [57]. Summarizing the conclusions of this previous research, this combination of personality traits could lead the affected people to use online gaming as a maladaptive mood modifier and/or a strategy to overcome negative life events, to having higher tendency toward competition, lower life satisfaction, lower expectations of self-efficacy, less face-to-face social support, and increased feelings of anger.

Analyzing the comorbidity level of both clusters, the type I has the higher comorbidity, obtaining higher scores on all SCL-90R scales (i.e., somatization, obsessive-compulsive, sensitivity, depression, anxiety, hostility, phobia, paranoid ideation and psychoticism) all STAI scales (anxiety trait and state), and on the IGD diagnostic criteria, being the obsessive-compulsive, sensitivity, depression, anxiety, hostility, paranoid ideation, psychoticism and anxiety trait above general population scoring. Despite the literature regarding comorbid psychopathology in adolescents and young adults with IGD having found full associations with anxiety [55], depression [58], ADHD or hyperactivity symptoms [59], and social phobia/anxiety [60], other authors have found partial associations [61,62] and no associations [63] with these same symptoms. These contradictory results and a possible publication bias [33] makes it difficult to detect the directionality and the relationship of these associations and shows the existing complexity of the relationship between IGD, personality and psychopathology.

Some authors have found relationships between personality factors and psychopathology in adolescents. According to the results found by Castellanos-Ryan [64], general psychopathology was related to high disinhibition/impulsivity, low agreeableness, high neuroticism and hopelessness, high delay-discounting and poor response inhibition. Of all of this, high neuroticism is the Big Five trait most strongly associated with several psychopathological symptoms, especially with anxiety and mood disorders and their comorbidity [65]. It has been hypothesized that the negative emotionality related with this trait could mediate this association [66]. Other personality trait predictors of psychopathology in adolescents are low conscientiousness, low extraversion [67], hopelessness and high impulsivity [68]. Regarding the Egotistic personality scale, where cluster 2 scored higher, adolescents who score high in the Egotistic Personality subscale have a passive-independent pattern, are perceived as conceited, have strong self-esteem, may take others for granted, and may fantasize about future success and power [69]. Some authors showed that certain narcissistic features are adaptive when paired with high levels of self-esteem [70]. Taking these results into account, the differences between the Type I and Type II personality traits could suggest that the higher negative affect, the lower positive affect, the difficulties in establishing social support, and the poor general behavioral control involved in the Type I personality profiles may be characterizing the chances of developing comorbid psychopathology, and could partially explain the inconsistent results regarding IGD and psychopathology.

When interpreting the results of this study, the effects of maturity and personality development in adolescence must be taken into account. Recent research has shown a spontaneous recovery of video game addiction [71,72] and it must be noted that personality traits may change during individual development [73]. Therefore, it is possible that the dysfunctional personality traits found will change or disappear in the future, and with them the influence they are having on the development or maintenance of IGD and the associated comorbidity.

In addition, our results show that there is a group of patients with IGD whose comorbid mental health symptoms are non-existent, or similar to the non-psychiatric population. Since there are no differences in the age at which the disorder begins or in the duration of the disorder between both cluster groups, it seems that in at least some individuals, IGD presents itself as a solitary condition or as an originating condition.

Regardless of the debate about the diagnosis and comorbidity [74], across the years the prevalence of the disordered gaming and the incidence of patients seeking treatment for IGD has remained stable [75]. This study is focused on the heterogeneity of the disorder and in the existence of different subgroups of IGD patients, based on personality, with differences in the seriousness of their psychological comorbidity. Our results suggest that these variables appear to be useful in determining clusters which represent different clinical subtypes with different degrees of severity. Such differences among online gamers imply that the experience of playing may vary in patients and the general population. Therefore, in order to succeed in developing instruments, planning efficient prevention programs for general population and targeting at-risk gamers, these dependent effects should be considered. Furthermore, the data regarding standard therapy for IGD are limited, with cognitive-behavioral therapy, family therapy and pharmacological intervention showing some significant results [76,77]. In order to develop more specific and accurate treatment interventions, the existence of IGD clusters should be considered.

Concerning limitations, the results of this study are based on a small sample, but considering that the sample is formed by a clinical sample, evaluated in a controlled setting, there may be a reason to anticipate that our findings may be of value for similar profile patients. Second, the measures used are based on self-administered questionnaires, although a trained psychologist supervised the entire process. Third, the cross-sectional design does not allow us to determine causality of the variables assessed. Fourth, the severity of IGD has not been reflected in this paper, as the objective of this work was to describe the different types of IGD-diagnosed adolescents with respect to personality traits and to analyze the comorbidity of each cluster groups.

Future research could complement these results using longitudinal designs addressing the potential mediating role of personality in the etiological factors and clinical course of IGD, in order to try to develop a predictive model of problematic video games use and its comorbidity.

5. Conclusions

Problematic use of online videogames is a growing clinical issue in developed countries. There is ongoing debate about the suitability of the current proposed definition and diagnostic criteria and about its related dysfunctional consequences and clinical correlates. The current literature shows that IGD is a complex problem where neurobiological systems, personality, and socioeconomic and environmental variables are involved. The objective of this study was to explore the heterogeneity of the affected people and the existence of differentiated subgroups based on different personality traits.

The results suggest that differences in personality can be useful in determining clusters with different levels of comorbidity. These differences imply that the experience of playing online videogames may affect different players, and clinicians and researchers should consider them in order to develop treatment options or assessment instruments. Future research should include more heterogeneous samples in relation to age and sex, and consider treatment outcome as a variable to analyze.

Author Contributions: D.F., V.G.-B. and J.J.S. conceived and planned this article. V.G.-B. and J.J.S. carried out the search and revision of the literature. D.F. and I.O. analyzed the data. V.G.-B. and J.J.S. drafted the study. All authors (V.G.-B., J.J.S., D.F., I.O., E.M., M.B., S.J.-M., A.d.P.-G. and J.R.) revised the article critically for important intellectual content. All authors (V.G.-B., J.J.S., D.F., I.O., E.M., M.B., S.J.-M., A.d.P.-G. and J.R.) commented on and approved the final manuscript and are accountable for all aspects of the work. All authors have read and agreed to the published version of the manuscript.

Funding: This work was funded by an AIS (Atención e Investigación en Socioaddiciones) intramural research program. This research is partially supported by the Marsden grant E2987-3648 (Royal Society of New Zealand), by grant RTI2018-100927-J-I00 Retos Investigación from Ministerio de Ciencia, Innovación y Universidades (Spain), and by grant 2017 SGR 622 (GRBIO) from the Departament d'Economia i Coneixement de la Generalitat de Catalunya (Spain). These partial funders had no role in the study design, data collection, and analysis, decision to publish, or preparation of the manuscript.

Conflicts of Interest: The authors declare no conflict of interest.

References

1. World Health Organization. *International Classification of Diseases for Mortality and Morbidity Statistics 11th Revision*; World Health Organization: Geneva, Switzerland, 2018.
2. Poli, R. Internet addiction update: Diagnostic criteria, assessment and prevalence. *Neuropsychiatry* **2017**, *07*, 4–8. [CrossRef]
3. Fauth-Bühler, M.; Mann, K. Neurobiological correlates of internet gaming disorder: Similarities to pathological gambling. *Addict. Behav.* **2017**, *64*, 349–356. [CrossRef] [PubMed]
4. Kaptsis, D.; King, D.; Delfabbro, P.; Gradisar, M. Withdrawal symptoms in internet gaming disorder: A systematic review. *Clin. Psychol. Rev.* **2016**, *43*, 58–66. [CrossRef] [PubMed]
5. APA. *Diagnostic and Statistical Manual of Mental Disorders*, 5th ed.; (DSM-5); American Psychiatric Association, Ed.; APA: Washington, DC, USA, 2013.
6. Gervasi, A.M.; La Marca, L.; Costanzo, A.; Pace, U.; Guglielmucci, F.; Schimmenti, A. Personality and Internet Gaming Disorder: A Systematic Review of Recent Literature. *Curr. Addict. Rep.* **2017**, *4*, 293–307. [CrossRef]
7. Şalvarlı, Ş.İ.; Griffiths, M.D. Internet Gaming Disorder and Its Associated Personality Traits: A Systematic Review Using PRISMA Guidelines. *Int. J. Ment. Health Addict.* **2019**. [CrossRef]
8. Matthews, G.; Deary, I.J.; Whiteman, M.C. *Personality Traits*, 3rd ed.; Cambridge University Press: Cambridge, UK, 2009; ISBN 9780511812743.
9. Agüera, Z.; Krug, I.; Sánchez, I.; Granero, R.; Penelo, E.; Peñas-Lledó, E.; Jiménez-Murcia, S.; Menchón, J.M.; Fernández-Aranda, F. Personality Changes in Bulimia Nervosa after a Cognitive Behaviour Therapy. *Eur. Eat. Disord. Rev.* **2012**, *20*, 379–385. [CrossRef]
10. Kim, E.J.; Namkoong, K.; Ku, T.; Kim, S.J. The relationship between online game addiction and aggression, self-control and narcissistic personality traits. *Eur. Psychiatry* **2008**, *23*, 212–218. [CrossRef]
11. Kardefelt-Winther, D. The moderating role of psychosocial well-being on the relationship between escapism and excessive online gaming. *Comput. Hum. Behav.* **2014**, *38*, 68–74. [CrossRef]
12. Mehroof, M.; Griffiths, M.D. Online gaming addiction: The role of sensation seeking, self-control, neuroticism, aggression, state anxiety, and trait anxiety. *Cyberpsychol. Behav. Soc. Netw.* **2010**, *13*, 313–316. [CrossRef]

13. Khazaal, Y.; Chatton, A.; Rothen, S.; Achab, S.; Thorens, G.; Zullino, D.; Gmel, G. Psychometric properties of the 7-item game addiction scale among french and German speaking adults. *BMC Psychiatry* **2016**, *16*, 132. [CrossRef]
14. Jiménez-Murcia, S.; Fernández-Aranda, F.; Granero, R.; Chóliz, M.; La Verde, M.; Aguglia, E.; Signorelli, M.S.; Sá, G.M.; Aymamí, N.; Gómez-Peña, M.; et al. Video game addiction in gambling disorder: Clinical, psychopathological, and personality correlates. *BioMed Res. Int.* **2014**, *7*, 105–110. [CrossRef]
15. Müller, K.W.; Beutel, M.E.; Egloff, B.; Wölfling, K. Investigating risk factors for Internet gaming disorder: A comparison of patients with addictive gaming, pathological gamblers and healthy controls regarding the big five personality traits. *Eur. Addict. Res.* **2014**, *20*, 129–136. [CrossRef]
16. Festl, R.; Scharkow, M.; Quandt, T. Problematic computer game use among adolescents, younger and older adults. *Addiction* **2013**, *108*, 592–599. [CrossRef] [PubMed]
17. Wang, C.W.; Ho, R.T.H.; Chan, C.L.W.; Tse, S. Exploring personality characteristics of Chinese adolescents with internet-related addictive behaviors: Trait differences for gaming addiction and social networking addiction. *Addict. Behav.* **2015**, *42*, 32–35. [CrossRef] [PubMed]
18. Montag, C.; Flierl, M.; Markett, S.; Walter, N.; Jurkiewicz, M.; Reuter, M. Internet addiction and personality in first-person-shooter video gamers. *J. Media Psychol.* **2011**, *23*, 163–173. [CrossRef]
19. Goby, V.P. Personality and Online/Offline Choices: MBTI Profiles and Favored Communication Modes in a Singapore Study. *CyberPsychol. Behav.* **2006**, *9*, 5–13. [CrossRef] [PubMed]
20. Jain, A.K.; Jain, A.K.; Murty, M.N.; Flynn, P.J. Data Clustering: A Review. *ACM Comput. Surv.* **1999**, *31*, 264–323. [CrossRef]
21. Noori, H.R.; Helinski, S.; Spanagel, R. Cluster and meta-analyses on factors influencing stress-induced alcohol drinking and relapse in rodents. *Addict. Biol.* **2014**, *19*, 225–232. [CrossRef]
22. Alikhanian, H.; Crawford, J.D.; DeSouza, J.F.X.; Cheyne, D.O.; Blohm, G. Adaptive cluster analysis approach for functional localization using magnetoencephalography. *Front. Neurosci.* **2013**. [CrossRef]
23. Henry, D.; Dymnicki, A.B.; Mohatt, N.; Allen, J.; Kelly, J.G. Clustering Methods with Qualitative Data: A Mixed-Methods Approach for Prevention Research with Small Samples. *Prev. Sci.* **2015**, *16*, 1007–1016. [CrossRef]
24. Gamito, P.S.; Morais, D.G.; Oliveira, J.G.; Brito, R.; Rosa, P.J.; Gaspar De Matos, M. Frequency is not enough: Patterns of use associated with risk of Internet addiction in Portuguese adolescents. *Comput. Hum. Behav.* **2016**, *58*, 471–478. [CrossRef]
25. Álvarez-Moya, E.M.; Jiménez-Murcia, S.; Aymamí, M.N.; Gómez-Peña, M.; Granero, R.; Santamaría, J.; Menchón, J.M.; Fernández-Aranda, F. Subtyping study of a pathological gamblers sample. *Can. J. Psychiatry* **2010**, *55*, 498–506. [CrossRef] [PubMed]
26. de la Fuente-Tomas, L.; Arranz, B.; Safont, G.; Sierra, P.; Sanchez-Autet, M.; Garcia-Blanco, A.; Garcia-Portilla, M.P. Classification of patients with bipolar disorder using k-means clustering. *PLoS ONE* **2019**, *14*. [CrossRef] [PubMed]
27. Vachon, D.D.; Bagby, R.M. Pathological Gambling Subtypes. *Psychol. Assess.* **2009**, *21*, 608–615. [CrossRef]
28. Suomi, A.; Dowling, N.A.; Jackson, A.C. Problem gambling subtypes based on psychological distress, alcohol abuse and impulsivity. *Addict. Behav.* **2014**, *39*, 1741–1745. [CrossRef]
29. Granero, R.; Fernandez-Aranda, F.; Mestre-Bach, G.; Steward, T.; Bano, M.; Agvera, Z.; Mallorqui-Bague, N.; Aymami, N.; Gomez-Pena, M.; Sancho, M.; et al. Cognitive behavioral therapy for compulsive buying behavior: Predictors of treatment outcome. *Eur. Psychiatry* **2017**, *39*. [CrossRef]
30. Billieux, J.; Thorens, G.; Khazaal, Y.; Zullino, D.; Achab, S.; Van der Linden, M. Problematic involvement in online games: A cluster analytic approach. *Comput. Hum. Behav.* **2015**, *43*, 242–250. [CrossRef]
31. Torres-Rodríguez, A.; Griffiths, M.D.; Carbonell, X.; Oberst, U. Internet gaming disorder in adolescence: Psychological characteristics of a clinical sample. *J. Behav. Addict.* **2018**, *7*, 707–718. [CrossRef]
32. Mallorquí-Bagué, N.; Fernández-Aranda, F.; Lozano-Madrid, M.; Granero, R.; Mestre-Bach, G.; Baño, M.; Pino-Gutiérrez, A.D.; Gómez-Peña, M.; Aymamí, N.; Menchón, J.M.; et al. Internet gaming disorder and online gambling disorder: Clinical and personality correlates. *J. Behav. Addict.* **2017**, *6*, 669–677. [CrossRef]
33. González-Bueso, V.; Santamaría, J.; Fernández, D.; Merino, L.; Montero, E.; Ribas, J. Association between Internet Gaming Disorder or Pathological Video-Game Use and Comorbid Psychopathology: A Comprehensive Review. *Int. J. Environ. Res. Public Health.* **2018**, *15*, 668. [CrossRef]

34. Millon, T. *Manual of Millon Adolescent Clinical Inventory*; National Computer Systems: Minneapolis, MN, USA, 1993.
35. Aguirre, G. *Adaptación Español del MACI*; TEA: Madrid, Spain, 2004.
36. Murrie, D.C.; Cornell, D.G. Psychopathy screening of incarcerated juveniles: A comparison of measures. *Psychol. Assess.* **2002**, *14*, 390–396. [CrossRef] [PubMed]
37. Salekin, R.T.; Larrea, M.A.; Ziegler, T. Relationships between the MACI and the BASC in the assessment of child and adolescent offenders. *J. Forensic Psychol. Pract.* **2002**, *2*, 35–50. [CrossRef]
38. Velting, D.M.; Rathus, J.H.; Miller, A.L. MACI personality scale profiles of depressed adolescent suicide attempters: A pilot study. *J. Clin. Psychol.* **2000**, *56*, 1381–1385. [CrossRef]
39. Burton, D.L.; Duty, K.J.; Leibowitz, G.S. Differences between sexually victimized and nonsexually victimized male adolescent sexual abusers: Developmental antecedents and behavioral comparisons. *J. Child Sex. Abus.* **2011**, *20*, 77–93. [CrossRef] [PubMed]
40. Ferrer, L.; Kirchner, T. Suicidal tendency among adolescents with adjustment disorder: Risk and protective personality factors. *Crisis* **2015**, *36*, 202–210. [CrossRef] [PubMed]
41. Grilo, C.M.; Fehon, D.C.; Walker, M.; Martino, S. A comparison of adolescent inpatients with and without wubstance abuse using the Millon Adolescent Clinical Inventory. *J. Youth Adolesc.* **1996**, *25*, 379–388. [CrossRef]
42. Derogatis, L.R. *SCL-90-R. Administration, Scoring and Procedures Manual*; Clinical Psychometric: Baltimore, MD, USA, 1990.
43. Derogatis, L.R. *SCL-90-R. Cuestionario de 90 Síntomas-Manual. [Questionnaire of the 90 Symptoms-Manual]*; TEA Ediciones: Madrid, Spain, 2002.
44. Spielberger, C.; Gorsuch, R.L.; Lushene, R.E. *Manual for the State/Trait Anxiety Inventory*; Consulting Psychologists Press: Palo Alto, CA, USA, 1970.
45. Guillén-Riquelme, A.; Buela-Casal, G. Psychometric revision and differential item functioning in the State Trait Anxiety Inventory (STAI). *Psicothema* **2011**, *23*, 510–515.
46. First, M.; Gibbon, M.; Spitzer, R.; Williams, J. *Users Guide for the Structured Clinical Interview for DSM IV Axis I Disorders—Research Version (SCID-I, Version 2.0)*; New York State Psychiatric Institute: New York, NY, USA, 1996.
47. Scrucca, L.; Raftery, A.E. Clustvarsel: A package implementing variable selection for Gaussian model-based clustering in R. *J. Stat. Softw.* **2018**, *84*. [CrossRef]
48. Kassambara, A.; Mundt, F. Factoextra: Extract and Visualize the Results of Multivariate Data Analyses. *R Package Version* **2017**, *1*, 1–74.
49. Dempster, A.P.; Laird, N.M.; Rubin, D.B. Maximum Likelihood from Incomplete Data via the EM Algorithm. *J. R. Stat. Soc. B* **1977**, *39*, 1–22.
50. McLachlan, G.J.; Krishnan, T. *(Thriyambakam) The EM Algorithm and Extensions*; Wiley-Interscience: New York, NY, USA, 2008; ISBN 9780471201700.
51. Fraley, C.; Raftery, A.E. Model-based methods of classification: Using the mclust software in chemometrics. *J. Stat. Softw.* **2007**, *18*, 1–13. [CrossRef]
52. Mclachlan, G.J.; Lee, S.X.; Rathnayake, S.I. Annual Review of Statistics and Its Application Finite Mixture Models. *Annu. Rev. Stat.* **2019**. [CrossRef]
53. Cohen, J. *Statistical Power Analysis for the Behavioral Sciences*, 2nd ed.; L. Erlbaum Associates: Hillsdale, NJ, USA, 1988; ISBN 9780805802832.
54. Lehenbauer-Baum, M.; Klaps, A.; Kovacovsky, Z.; Witzmann, K.; Zahlbruckner, R.; Stetina, B.U. Addiction and Engagement: An Explorative Study Toward Classification Criteria for Internet Gaming Disorder. *Cyberpsychology. Behav. Soc. Netw.* **2015**, *18*, 343–349. [CrossRef] [PubMed]
55. Gentile, D.A.; Choo, H.; Liau, A.; Sim, T.; Li, D.; Fung, D.; Khoo, A. Pathological video game use among youths: A two-year longitudinal study. *Pediatrics* **2011**, *127*, e319–e329. [CrossRef] [PubMed]
56. Seay, A.F.; Kraut, R.E. Project Massive: Self-Regulation and Problematic Use of Online Gaming. In Proceedings of the Conference on Human Factors in Computing Systems, San Jose, CA, USA, 24–28 April 2007; pp. 829–838.
57. Müller, K.W.; Koch, A.; Dickenhorst, U.; Beutel, M.E.; Duven, E.; Wölfling, K. Addressing the Question of Disorder-Specific Risk Factors of Internet Addiction: A Comparison of Personality Traits in Patients with Addictive Behaviors and Comorbid Internet Addiction. *Biomed Res. Int.* **2013**. [CrossRef]

58. Van Rooij, A.J.; Schoenmakers, T.M.; Vermulst, A.A.; Van Den Eijnden, R.J.J.M.; Van De Mheen, D. Online video game addiction: Identification of addicted adolescent gamers. *Addiction* **2011**, *106*, 205–212. [CrossRef]
59. Baer, S.; Bogusz, E.; Green, D.A. Stuck on screens: Patterns of computer and gaming station use in youth seen in a psychiatric clinic. *J. Can. Acad. Child Adolesc. Psychiatry* **2011**, *20*, 86–94.
60. Cole, S.H.; Hooley, J.M. Clinical and Personality Correlates of MMO Gaming. *Soc. Sci. Comput. Rev.* **2013**, *31*, 424–436. [CrossRef]
61. Starcevic, V.; Berle, D.; Porter, G.; Fenech, P. Problem Video Game Use and Dimensions of Psychopathology. *Int. J. Ment. Health Addict.* **2011**, *9*, 248–256. [CrossRef]
62. Stetina, B.U.; Kothgassner, O.D.; Lehenbauer, M.; Kryspin-Exner, I. Beyond the fascination of online-games: Probing addictive behavior and depression in the world of online-gaming. *Comput. Hum. Behav.* **2011**, *27*, 473–479. [CrossRef]
63. King, D.L.; Delfabbro, P.H.; Zwaans, T.; Kaptsis, D. Clinical features and axis I comorbidity of Australian adolescent pathological Internet and video game users. *Aust. N. Zeal. J. Psychiatry* **2013**, *47*, 1058–1067. [CrossRef] [PubMed]
64. Castellanos-Ryan, N.; Briere, F.N.; O'Leary-Barrett, M.; Banaschewski, T.; Bokde, A.; Bromberg, U.; Biichel, C.; Flor, H.; Frouin, V.; Gallinat, J.; et al. The IMAGEN Consortium The structure of psychopathology in adolescence and its common personality and cognitive correlates. *J. Abnorm. Psychol.* **2016**, *125*, 1039–1052. [CrossRef] [PubMed]
65. Kotov, R.; Gamez, W.; Schmidt, F.; Watson, D. Linking "Big" personality traits to anxiety, depressive, and substance use disorders: A meta-analysis. *Psychol. Bull.* **2010**, *136*, 768–821. [CrossRef] [PubMed]
66. Tackett, J.L.; Kushner, S.C.; De Fruyt, F.; Mervielde, I. Delineating Personality Traits in Childhood and Adolescence: Associations Across Measures, Temperament, and Behavioral Problems. *Assessment* **2013**, *20*, 738–751. [CrossRef] [PubMed]
67. Malouff, J.M.; Thorsteinsson, E.B.; Schutte, N.S. The relationship between the five-factor model of personality and symptoms of clinical disorders: A meta-analysis. *J. Psychopathol. Behav. Assess.* **2005**, *27*, 101–114. [CrossRef]
68. Carragher, N.; Teesson, M.; Sunderland, M.; Newton, N.C.; Krueger, R.F.; Conrod, P.J.; Barrett, E.L.; Champion, K.E.; Nair, N.K.; Slade, T. The structure of adolescent psychopathology: A symptom-level analysis. *Psychol. Med.* **2016**, *46*, 981–994. [CrossRef]
69. McCann, J.T. The MACI: Composition and Clinical Applications. In *The Millon Inventories: Clinical and Personality Assessment*; Millon, T., Ed.; The Guilford Press: New York, NY, USA, 1997; pp. 363–388.
70. Barry, C.T.; Frick, P.J.; Killian, A.L. The relation of narcissism and self-esteem to conduct problems in children: A preliminary investigation. *J. Clin. Child Adolesc. Psychol.* **2003**, *32*, 139–152. [CrossRef]
71. Przybylski, A.K.; Weinstein, N.; Murayama, K. Internet gaming disorder: Investigating the clinical relevance of a new phenomenon. *Am. J. Psychiatry* **2017**, *174*, 230–235. [CrossRef]
72. Rothmund, T.; Klimmt, C.; Gollwitzer, M. Low Temporal Stability of Excessive Video Game Use in German Adolescents. *J. Media Psychol.* **2018**, *30*, 53–65. [CrossRef]
73. Soto, C.J.; John, O.P.; Gosling, S.D.; Potter, J. Age Differences in Personality Traits From 10 to 65: Big Five Domains and Facets in a Large Cross-Sectional Sample. *J. Pers. Soc. Psychol.* **2011**, *100*, 330–348. [CrossRef]
74. Bean, A.M.; Nielsen, R.K.L.; van Rooij, A.J.; Ferguson, C.J. Video game addiction: The push to pathologize video games. *Prof. Psychol. Res. Pract.* **2017**, *48*, 378–389. [CrossRef]
75. Feng, W.; Ramo, D.E.; Steven, R.C.; Bourgeois, J.A. Internet gaming disorder: Trends in prevalence 1998–2016. *Addict. Behav.* **2017**, *75*, 17–24. [CrossRef] [PubMed]
76. Han, D.H.; Kim, S.M.; Lee, Y.S.; Renshaw, P.F. The effect of family therapy on the changes in the severity of on-line game play and brain activity in adolescents with on-line game addiction. *Psychiatry Res. Neuroimaging* **2012**, *202*, 126–131. [CrossRef] [PubMed]
77. Kim, S.M.; Han, D.H.; Lee, Y.S.; Renshaw, P.F. Combined cognitive behavioral therapy and bupropion for the treatment of problematic on-line game play in adolescents with major depressive disorder. *Comput. Hum. Behav.* **2012**, *28*, 1954–1959. [CrossRef]

International Journal of
Environmental Research and Public Health

MDPI

Article

Spanish Validation of the Internet Gaming Disorder Scale–Short Form (IGDS9-SF): Prevalence and Relationship with Online Gambling and Quality of Life

Marta Beranuy [1], Juan M. Machimbarrena [2], M. Asunción Vega-Osés [3], Xavier Carbonell [4], Mark D. Griffiths [5], Halley M. Pontes [6,7] and Joaquín González-Cabrera [1,*]

1. Faculty of Education, Universidad Internacional de la Rioja (UNIR), Avenida de la Paz, 137, 26006 Logroño, Spain; marta.beranuy@unir.net
2. Faculty of Psychology, University of the Basque Country (UPV/EHU), Avenida de Tolosa, 70, 20018 Donostia, Spain; juanmanuel.machimbarrena@ehu.eus
3. Faculty of Health Sciences, Universidad Pública de Navarra (UPNA), Calle Cataluña, s/n, 31006 Pamplona, Navarra, Spain; mvegaose@educacion.navarra.es
4. Faculty of Psychology, Education and Sport Blanquerna, Universitat Ramon Llull. Calle Císter, 34, 08022 Barcelona, Spain; xaviercs@blanquerna.url.edu
5. International Gaming Research Unit, Psychology Department, Nottingham Trent University, Nottingham NG11 8NS, UK; mark.griffiths@ntu.ac.uk
6. University of Tasmania, School of Psychological Sciences, Newnham Campus, Building O, Launceston TAS 7250, Australia; contactme@halleypontes.com
7. The International Cyberpsychology and Addictions Research Laboratory (iCARL), University of Tasmania, Launceston TAS 7250, Australia

* Correspondence: joaquin.gonzalez@unir.net

Received: 8 February 2020; Accepted: 27 February 2020; Published: 28 February 2020

Abstract: Online gaming is a very common form of leisure among adolescents and young people, although its excessive and/or compulsive use is associated with psychological impairments in a minority of gamers. The latest (fifth) edition of the Diagnostic and Statistical Manual of Mental Disorders (DSM-5, Section III) tentatively introduced Internet Gaming Disorder (IGD). Since then, a number of evaluation tools using the DSM-5 criteria have been developed, including the Internet Gaming Disorder Scale–Short Form (IGDS9-SF). The main objective of this study was to translate and adapt the IGDS9-SF into Spanish, as well as to obtain indicators relating to its validity and reliability. The Spanish version of four scales were administered: IGDS9-SF, Mobile Phone-Related Experiences Questionnaire (CERM), Online Gambling Disorder Questionnaire (OGD-Q), and KIDSCREEN-27. The sample comprised 535 Vocational Training students (mean age 18.35 years; SD±2.13; 78.5% males) who reported playing video games in the past 12 months. Confirmatory factor analysis yielded a one-dimensional model with a good fit while the reliability indicators were satisfactory. Findings indicated that 1.9% of gamers were classified with IGD (meeting five or more criteria for more than 12 months). Additionally, another 1.9% were considered gamers 'at-risk' because they endorsed four criteria. Positive and significant relationships were found between the IGDS9-SF, the CERM, and the OGD-Q. Participants classified with IGD had poorer health-related quality of life. In conclusion, the Spanish IGDS9-SF is a valid and reliable instrument to assess IGD according to the DSM-5.

Keywords: Internet Gaming Disorder; gaming disorder; gaming addiction; behavioral addiction; Internet Gaming Disorder Scale-Short Form

1. Introduction

The way in which individuals interact with technology is constantly evolving. New behaviors have emerged, communication and leisure activities have changed, and new psychological problems arose. In the late 1990s, concerns about the addictive use of the internet [1,2] were discussed and, since then, the concept has been extensively studied and debated [3–6]. Although it has been addressed from multiple perspectives and researchers have used different terms, 'internet addiction' has been one of the most commonly used terms, along with 'problematic internet use' [7–9]. Early research focused on internet-related and mobile-related behavior in general terms. However, over the years, studies have focused on more specific uses. This approach has been defined as the move from general problematic internet use (GPIU) to specific problematic internet use (SPIU) (e.g., [5]). Consequently, research has especially focused on internet gaming [10–12], online gambling [13–15], online sex/cybersex [16,17], and social media use [18–20].

Among the aforementioned problems, 'Gaming Disorder' (GD) has recently been introduced in the nosological manuals (American Psychiatric Association [APA], and World Health Organization [WHO],) by being included as a disorder under the heading of addictive behaviors. The DSM-5 [21] places the 'Internet Gaming Disorder' (IGD) in Section III (disorders requiring further investigation) and the International Classification of Diseases 11th (ICD-11) [22] considers GD among non-substance addictions.

IGD is considered an addictive behavior that does not involve the ingestion of a psychoactive substance, and is mainly characterized by recurrent and persistent participation in online video games, leading to clinically significant distress [21]. The nine IGD criteria contain the characteristics indicated in the components model of addiction [23], including salience, mood modification, tolerance, withdrawal, the conflicts it generates (whether interpersonal and/or intrapersonal), and the risk of relapse. The theoretical overlap between the components model of addiction and the nine IGD criteria has been previously ascertained empirically in earlier studies [24]. Despite this definition, the use of different theoretical frameworks has created difficulties in the conceptualization of a problem with worrisome prevalence data and which also produces harmful effects on those who suffer from it, making it a possible public health problem [25].

The video game industry generates millions of Euros in revenue every year (1.530 million Euros in 2018, 12.6% more than the previous year) as video gaming is considered one of the main forms of leisure for many audiences across all stages of life. According to the Spanish Video Game Association [26], the total number of gamers in 2018 amounted to 16.8 million (41% female). Their age ranged from 6 to 64 years old, although the youngest (6–14 years) stand out, and they played for an average of 6.2 h per week. Studies on participation in video games indicate the highest prevalence among younger populations [27], which appears to be an at-risk population due to specific features associated with adolescence (i.e., being at a developmental stage where there is little thought given to the long-term consequences of their actions [28]) and membership of the Z Generation (i.e., born during the early years of the 21st century comprising individuals who have never known a world without the internet and mobile phones [29]).

Adolescent studies indicate a prevalence of IGD ranging from 1.7% to 10% [4,30–33]. According to the meta-analytical study by Fam [27], the prevalence of IGD among male adolescents is 6.8%, and 1.3% among females. In international samples, and without an established age range, the percentage ranges from 0.7% to 15.6% [34] or slightly higher in Chinese studies (from 3.5% to 17% [35]).

In Spain (where the present study was carried out), a study with a school sample of 708 students reported 72.8% online gamers [36] of whom 8.3% met five or more of the nine criteria for IGD (86.44% were male). Another study conducted with Spanish-speaking online gamers reported a prevalence of 2.6% disordered gamers [37]. This same study found that 6.5% were "engaged gamers at high risk", and 11.9% were "engaged gamers at low risk", with the remainder classified as "regular" or "casual" gamers. In a clinical sample of 86 disordered adolescent gamers, it was found that 96.6% were male [38].

In terms of comorbidity, IGD has been associated with a wide spectrum of psychological problems including depression, anxiety, social phobias, poorer school performance, and sleep disorders [36,39–42]. In addition, studies have begun to appear comprising clinical samples demanding psychological treatment, which meet the criteria for the disorder [38,43]. In the study by Martín-Fernández et al. [38], all participants had diagnostic comorbidity, in accordance with other studies [44,45]. The most prevalent comorbidities with IGD were found to be depression, social anxiety, ADHD, and aggressive behaviors [46,47]. Additionally, it is also important to examine the relationship of IGD with other non-substance addictions (particularly gambling) because several studies have established common risk factors such as personality traits [48], levels of impulsivity and compulsivity [49], and similarities in the neurobiological functioning of patients with IGD and patients with pathological gambling [50]. It is relevant to relate IGD with online gambling disorder or other problematic behaviors such as problematic smartphone use, because these three behaviors occur via Information and Communications Technologies (ICTs) and which typically meet addiction criteria similar to IGD (i.e., salience, mood modification, tolerance, withdrawal, conflict, and relapse) when people present with a clinical problem. Currently, there are psychometric tests that evaluate online gambling disorder specifically in adolescent populations [14]. In addition, the most used device to play or connect to the internet is via smartphone. Although a decade ago, gaming consoles and computers were the only technological hardware available to play online, in recent years, the use of the smartphone has significantly increased [51]. Currently smartphones are used by 21% of almost 17 million players in Spain (only surpassed by gaming consoles; 26%). This percentage increases considerably (up to 40%) in the 15–24-year age range. It should also be noted that in the Spanish context, the average time spent weekly with mobile gaming is 5.1 h, compared to 3.9 h on gaming consoles and the 4.9 h on computers [26]).

Many studies have focused on the negative effects of IGD on psychological health and other health-related problems, but fewer studies have linked it to more global general well-being constructs (such as health-related quality of life [HRQoL]). HRQoL is defined as a state of complete physical, mental, and social well-being that is perceived by individuals and by those around them (for more information see the review by Wallander and Koot [52]). The evaluation of HRQoL is complex because it is a polyhedral construct that presents multiple conceptual approaches although one of the best approaches for examining the infant-juvenile stage is with the KIDSCREEN [53], a psychometric test adapted in almost 30 languages. The HRQoL provides general indicators on the impact of a problem in areas relevant to an adolescent's life, such as physical and psychological well-being or the relationship with parents and peers. Several studies indicate that inadequate use of the internet is related to low scores on HRQoL, in addition to a lower self-perceived social support and more friends only known through the internet [54,55]. Similarly, other studies indicate that both HRQoL and social aspects are affected among people presenting problems related to the use of video games in adolescence [31,36].

The APA's [21] operationalization of IGD has arguably reduced the diversity in the assessment instruments as well as the number of items contained in each of these instruments, providing more uniform measures with high internal consistency and adequate criterion validity. A systematic review [56] indicated the nine-item Internet Gaming Disorder Scale-Sort Form (IGDS9-SF) [57] as the most reviewed and translated instrument to assess IGD based on the nine IGD criteria developed by the APA. In fact, since its publication, the IGDS9-SF has been translated into at least nine languages: Chinese [58], German [59], Czech [60], Slovenian [61], Italian [62], Persian [63], Turkish [64,65], Polish [66], European and South American Portuguese [67,68]. Therefore, it is a cross-culturally suitable psychometric tool to assess IGD that allows framing the problem uniformly and inter-culturally with adequate psychometric properties.

This IGDS9-SF is based on the nine criteria suggested by the DSM-5 for IGD that includes: (i) preoccupation with gaming; (ii) withdrawal symptoms; (iii) tolerance; (iv) unsuccessful attempts to reduce or quit gaming; (v) loss of interest in previous activities or entertainments as a result of (and with the exception of) gaming; (vi) continuing to game despite knowing the associated psychosocial problems; (vii) deceiving family members, therapists, or others about the amount of time spent on gaming; (viii) playing video games to evade or relieve negative moods; and (ix) jeopardizing or losing a meaningful relationship, job, or educational or employment opportunity due to gaming.

As aforementioned, the use of different theoretical frameworks to assess IGD generated problems in conceptualization and, given the broad international acceptance of the IGDS9-SF and its robust psychometric properties, the objective of the present study was to extend the cross-cultural psychometric assessment evidence-base related to the assessment of IGD by translating and adapting the IGDS9-SF into Spanish to ascertain its psychometric suitability to this specific cultural context in terms of the validity and reliability of its scores. The validation in this cultural context for the Spanish language is a priority, given that Spanish is the third most spoken language in the world (534 million speakers [69,70]) and there are 21 countries where Spanish is the official language [70]. Consequently, a Spanish version of the most used IGD assessment tool is needed to encourage and improve research investigating IGD in Spanish-speaking countries and to facilitate cross-culturally unified research of this emerging public health issue.

The secondary objectives of this study were to: (a) to obtain indicators of validity and reliability of the Spanish version of IGDS9-SF, including the confirmatory study of its factor structure; (b) to test whether the newly developed psychometric test works equally in both men and women, as well as in adolescences and young adults; (c) establish the prevalence of IGD in a sample of adolescents and young Vocational Training (VT) students aged between 15 and 25 years; and (d) examine the relationship between the IGD and HRQoL. To achieve the aforementioned objective, it was hypothesized that: (i) the IGDS9-SF would show adequate psychometric properties in the sample recruited, similarly to previous IGDS9-SF psychometric validation studies conducted in different countries [58,59,61,67]; (ii) the measurement model would be invariant across both genders [57]; (iii) the prevalence of IGD would be between 2% and 4%, which is referred to in other national and international studies [27,37]; and (iv) those who met the IGD criteria within the sample recruited would present lower scores on the different HRQoL dimensions [31].

2. Materials and Methods

2.1. Design and Participants

The instrument validation study was conducted from February to May 2019. The sample was recruited from 17 VT centers in the Autonomous Community of Navarre by means of non-parametric incidental sampling. The distribution of students by cycles and school stages was as follows: basic VT (152 first-grade students, 14.2%; 70 second-grade students, 6.5%); middle degree VT (433 first-grade students, 40.4%; 56 second-grade students, 5.2%); and higher degree VT (325 first-grade students, 30.3%; 35 second-grade students, 3.4%). The initial number of participants was 1064 (593 males and 471 females), of whom 535 reported playing video games in the last 12 months. Of this final sample, 420 were males (78.5%) and 115 were females (21.5%). The mean and standard deviation of age was 18.35 years (±2.13), with a range of 15–25 years.

2.2. Instruments

The participants provided information about demographic variables including gender, grade, school, and age. They also indicated the name of the video game they spent the most hours on in the past 12 months and whether or not they considered themselves addicted to online video games. In addition, they completed the following assessment tools.

Spanish translation of the IGDS9-SF [57,61] (see the Spanish version in Appendix A). The IGDS9-SF assesses the severity of IGD and its detrimental effects, examining online and/or offline gaming activities that occur over a 12-month period with nine questions based on the DSM-5 IGD criteria that are rated on a five-point Likert scale: 1 (never), 2 (rarely), 3 (sometimes), 4 (often), and 5 (very often). Participants' total scores are obtained by adding the score of each answer (ranging from 9 to 45). Higher scores typically indicate a higher level of IGD symptom-severity and greater incidence of problems related to gaming behaviors. For the Spanish adaptation, three experts evaluated (through a table of test specifications) each of the translations and psychological adaptations of the nine items. High inter-rater reliability was recorded throughout the process (> 0.9) on all the items [71]. The set of items is shown in Table 1, however the whole questionnaire including instructions and response scale can be found in Appendix A. In addition, initial piloting was carried out on a sample of 30 participants, providing adequate indicators of reliability and content and internal validity, and not reporting any comprehension or reading problems. The pilot participants were not included in the final sample.

Table 1. Means, standard deviations, item-total correlation, positive response percentage, and factorial loads of the Internet Gaming Disorder Scale–Short Form (IGDS9-SF) items (*n* = 9).

IGDS9-SF Items	M	SD	IT	%+	CFE
1. ¿Te sientes preocupado por tu comportamiento con el juego? (Algunos ejemplos: ¿Piensas en exceso cuando no estás jugando o anticipas en exceso a la próxima sesión de juego?, ¿Crees que el juego se ha convertido en la actividad dominante en tu vida diaria?)	1.74	0.95	0.53	48.0	0.58
2. ¿Sientes irritabilidad, ansiedad o incluso tristeza cuando intentas reducir o detener tu actividad de juego?	1.45	0.79	0.64	32.5	0.71
3. ¿Sientes la necesidad de pasar cada vez más tiempo jugando para lograr satisfacción o placer?	1.57	0.97	0.67	34.3	0.74
4. ¿Fallas sistemáticamente al intentar controlar o terminar tu actividad de juego?	1.64	0.89	0.63	43.2	0.69
5. ¿Has perdido intereses en aficiones anteriores y otras actividades de entretenimiento como resultado de tu compromiso con el juego?	1.56	0.93	0.61	34.3	0.66
6. ¿Has continuado jugando a pesar de saber que te estaba causando problemas con otras personas? (pareja, amistad o familia)	1.45	0.88	0.65	27.2	0.71
7. ¿Has engañado a alguno de tus familiares, terapeutas o amigos sobre el tiempo que pasas jugando?	1.57	0.99	0.47	33.6	0.51
8. ¿Juegas para escapar temporalmente o aliviar un estado de ánimo negativo (por ejemplo, desesperanza, tristeza, culpa o ansiedad)?	1.94	1.20	0.47	49.0	0.50
9. ¿Has comprometido o perdido una relación importante, un trabajo o una oportunidad educativa debido a tu actividad de juego?	1.23	0.64	0.58	14.8	0.63

Note: For original items see Pontes and Griffiths [57]; M = Arithmetic mean; SD = Standard deviation; IT = corrected item-total correlation; %+ Percentage that has responded positively (at least once); CFE = Standardized factorial loads.

Cuestionario de Experiencias Relacionadas con el teléfono móvil (CERM [Mobile Phone-Related Experiences Questionnaire] [72]). This instrument has 10 items that evaluate two factors: (i) conflicts related to mobile phone abuse and (ii) problems due to the emotional and communicational use of the mobile phone. The items of this instrument are rated on a four-point Likert scale, ranging from 1 (hardly ever) to 4 (almost always). The CERM has been previously shown to have adequate indicators of reliability and validity in Spanish adolescents. In the present sample, the Cronbach's alpha reliability coefficient for the CERM was 0.78 and the Omega coefficient was 0.79.

Online Gambling Disorder Questionnaire (OGD-Q) [14]. This instrument was designed to evaluate online gambling disorder using a total of 11 items that are rated on a five-point Likert scale ranging from 1 (never) to 5 (every day). The total OGD-Q score varies between 11 and 55, with higher scores indicating higher levels of online disordered gambling. The questionnaire has been validated for a Spanish sample of adolescents and presents adequate indicators of reliability and validity. In the present sample, the Cronbach's alpha reliability coefficient for the OGD-Q was 0.91 and Omega coefficient was 0.92.

Spanish version of the KIDSCREEN-27 [53]. This instrument assesses HRQoL in children and adolescents between ages 8 and 18 years. This version assesses five dimensions by using 27 items: physical well-being, psychological well-being, autonomy and relationship with parents, peers and social support, and school environment. The development of the KIDCREEN was based on the probabilistic partial credit model (PCM) which pertains to the family of Rasch models. PCM explains the actual behavior of the responders in the testing situation by the estimated person parameter and the location of the item-answers-category-thresholds. The PCM assumes all items of a scale to be the indicators of a single unidimensional latent trait [53]. For the KIDSCREEN-27, the mean scores varied around 50 (SD = 10) due to T-value standardization. There are standardized data for the Spanish infant-juvenile population. The reliability of the each dimension was as follows: physical well-being (α= 0.86; ω = 0.86); psychological well-being (α= 0.84; ω = 0.84), autonomy and relationship with parents (α= 0.84; ω = 0.84), peers and social support (α= 0.88; ω = 0.88), and school environment (α= 0.80; ω = 0.80). Due to the nature of the KIDSCREEN (designed for children and adolescents), responses from participants over 18 years were not considered in the analyses of this instrument in the present study.

2.3. Procedure

The battery of questionnaires was applied in an online format utilizing Survey Monkey©. Participants completed the questionnaires in the different computer technology classrooms coordinated by the guidance departments of each center, and under the supervision of the classroom tutor. At the outset of the study participants were advised to answer all questions truthfully and to not stop at any particular question for a long time. The overall average time needed to complete the survey ranged between 15 and 25 min, depending on students' age and reading ability.

2.4. Ethical Considerations

The study was conducted with the informed consent of the participants and the directors of the schools. Through the official communication channels with the families, the schools sent a consent form that informed either the legal tutors or the students themselves (if they were at least 18 years old) about the purpose of the study and its characteristics, its promoters, and their right not to participate without penalties. Those parents/tutors who did not wish to allow participation returned the signed consent. This occurred in less than 1% of the sample. The study was approved by the Research Ethics Committee of the research from Universidad Internacional de la Rioja (UNIR) (PI:008/2019). There were no exclusion criteria, except for the refusal to participate by the legal guardians or by the students themselves.

2.5. Data Analysis

Statistical analyses were carried out using the Statistical Package for the Social Sciences (SPSS) [73] program, the R software, the psych package [74], the Lavaan package [75], and the equaltestMI [76]. Firstly, regarding internal validity, an analysis of the psychometric properties of each item was performed, indicating the arithmetic mean, standard deviation, item-total correlation, percentage of positive responses to each item, and the factorial loadings of each item (see Table 1). The multiple criterion for the selection of items without technical deficiencies was that none of them could fail to meet two of the following three statistical indices: (i) a mean between 1.5 and 2.5; (ii) a standard deviation equal to or greater than 1; and (iii) an item-total correlation equal to or greater than 0.35.

To ensure better comparability between the present study and the original IGDS9-SF study, the structure of the IGDS9-SF was initially examined with Exploratory Factor Analysis (EFA) of the items, following the verification of the assumptions (Kaiser-Meyer-Olkin index and Bartlett sphericity test). The factor extraction method used was Principal Axis Factoring with Oblimin rotation. Confirmatory Factor Analysis (CFA) was then performed using Weighted Least Squares Median adjusted method (WLSM). Following the recommendations of Hu and Bentler [77], goodness of fit was assessed using the chi-squared statistic, the comparative fit index (CFI), Tucker-Lewis Index (TLI), the root mean square error of approximation (RMSEA), and the standardized root mean square residual (SRMR). In general, CFI and TLI values of 0.95 or higher reflect a good fit while RMSEA values between 0.06 and 0.08 indicate acceptable fit. Finally, SRMR values lower than 0.08 indicate adequate fit [78]. The hypothesized model was unidimensional, in which all nine items would load on the same latent factor. To determine the internal consistency of the instruments employed, the Cronbach's alpha, McDonald's Omega, greatest Lower Bound (GBL), Gutmann's $\lambda 6$, Average Variance Explained (AVE) and Composite Reliability (CR) coefficients were estimated. To calculate Measurement Invariance (MI) the sample was split by gender (males and females) and age (under 18 years old and 18 years old or older). MI across age and gender was evaluated through the following steps: (a) testing for the invariance of number of factors (configural invariance); (b) testing for the equality of factor loadings (weak or metric invariance); and (c) testing for the equality of indicator intercepts (strong or scalar invariance). Given that chi-square is sensitive to sample size and non-normality conditions, it was assumed that the model is invariant if the ΔCFI is not above 0.01 [79].

Finally, the following analyses were performed in relation to the secondary objectives of the study: (i) analysis of frequencies and central tendency and dispersion measurements of the study variables; (ii) t-test for independent samples (or failing that, Welch's test); (iii) calculation of the effect size with Cohen's d or Hedges' g, as appropriate; (iv) Pearson correlations; (v) analysis of variance with post-hoc Games-Howell comparisons; and (vi) Mann-Whitney U-test for independent samples. A value of less than p=.05 was considered significant.

To obtain the prevalence rate of IGD in the past 12 months, the indications of the APA [21] and Pontes et al. [67] were followed (i.e., endorsing five or more items in classifying individuals with IGD). To establish item endorsement, the items of the IGDS9-SF were dichotomized so that response categories as 4 (often) and 5 (very often) were used to classify the item as a problem (i.e., endorsement of the specific criterion). The remainder of the responses were classified as 'no problem' (i.e., no endorsement of the specific criterion). In addition, participants who endorsed four items were considered 'at-risk' of IGD while participants' preferred video game genre was classified in the following categories: action/shooter (FPS (First-Person Shooter), TPS (Three-Person Shooter), etc.), strategy (4x, RTS (Real-Time Strategy), etc.), role-playing (ARPG (Action Role-Playing Games), JRPG (Japanese Role-Playing Game), RPG (Role-Playing Game), etc.), fighting, MOBA (Multiplayer Online Battle Arena), MMORPG (Massive Multiplayer Online Role-Playing Game), simulators, letters, sports, musical, and casual.

3. Results

The scores of Items 1, 4, 6, 7, and 9 showed significant differences as a function of gender, with higher scores in males than in females ($p < 0.05$). The rest of the items did not present significant gender differences. The effect sizes were small in most cases ($d < 0.3$), except for Item 1 ($d = 0.31$). Additionally, in relation to the differences according to class year, significant differences were only found on Item 6 ($p < 0.05$) between the students of first basic VT and first middle degree, with a small effect size ($d < 0.3$). No significant differences were observed in the remaining items for the class year variable.

3.1. Evidence of Validity of the IGDS9-SF Scores

Table 1 shows the different psychometric indicators for each of the IGDS9-SF items, namely the mean, standard deviation, item-total correlation, percentage of positive response on each item, and factor loadings for each item. At the psychometric level, the scores obtained revealed problems in the mean and standard deviation of all the items, although the item-total correlations in all the items were satisfactory. Items with at least one positive value ranged from 14.8% for Item 9 to 49% for Item 1.

With regards to the EFA results, the data of the Kaiser-Meyer-Olkin index and the Bartlett sphericity test produced values of 0.909 and $\chi^2 = 1629.36$, $p < 0.001$. The correlation matrix between the items was appropriate for the EFA. The results further indicated a single latent factor explaining 47.49% of the total sample variance. Regarding the CFA, the hypothesized unidimensional model yielded adequate fit indices: χ^2 (27, n = 532) = 9.908., $p < 0.001$, RMSEA = 0.019 (95% CI [0.000, 0.026], CFI = 0.995, NNFI = 0.993, and SRMR = 0.035. The standardized factor loadings (see Table 1) were statistically significant and notable in all items, ranging from 0.50 to 0.74. The Cronbach's alpha was 0.85 and Omega coefficients for the IGDS9-SF were both 0.85 (IC: [0.83, 0.87]). The Greatest Lower Bound was 0.88 and Gutmann's λ was 0.85. Finally, the average variance extracted was 0.5 and the Composite Reliability was 0.88.

3.2. Measurement Invariance

To evaluate the generalizability of the model across males and females, participants under 18 years old and 18 years old or older, two multi-group CFAs were performed. For each analysis, an unconstrained model with factor loadings free to vary between subgroups was compared with a more constrained model, in which the factor loadings were held constant across subgroups. Before conducting multi-group analyses, separate CFAs were performed for gender and age subgroups. Regarding gender, the model for females offered a lower fit in general than that of males or the overall model. However, the indicators still presented adequate threshold and the MI analyses were performed. The MI of the single-factor solution was supported at the configural and metric levels. However, the increase in the CFI and RMSEA prevented testing the model any further for gender. Regarding age, both subgroups showed a good fit for the data, in the case of age the MI supported the structure of the single-factor solution across all three levels (configural, metric, and scalar). The results obtained for the different models are displayed in Table 2.

Table 2. Invariance analyses across gender and age.

Model	χ^2	df	Com. Md	ΔSB χ^2	Δdf	p	CFI	ΔCFI	RMSEA	ΔRMSEA	SRMR
1.Overall model	9.91	27	–	–	–	–	0.995	–	0.019	–	0.035
Gender											
2. Men Model	9.08	27	–	–	–	–	0.994	–	0.02	–	0.037
3. Women Model	12.95	27	–	–	–	–	0.984	–	0.035	–	0.072
4. Configural Model	22.03	54	–	–	–	–	0.992	–	0.024	–	0.044
5. Metric Model	39.05	62	4–5	9.57	8	0.296	0.989	−0.003	0.026	0.002	0.058
6. Scalar Model	57.11	70	5–6	26.81	8	>0.001	0.976	−0.013	0.036	0.010	0.062
Age											
7. ≤17 Model	5.38	27	–	–	–	–	0.998	–	0.012	–	0.035
8.≥18 Model	11.93	27	–	–	–	–	0.988	–	0.029	–	0.049
9. Configural Model	17.33	54	–	–	–	–	0.992	–	0.023	–	0.043
10. Metric Model	28.85	62	9–10	11.99	8	0.151	0.988	−0.004	0.026	0.003	0.056
11. Scalar Model	37.40	70	10–11	14.60	8	0.464	0.983	−0.005	0.029	0.003	0.058

Note: n for Men's model = 419; n for women's model = 113; n for ≤ 17's Model = 296; n for ≥ 18's model = 236; χ^2 = Chi-Squared; df = Degrees of freedom; Comp. Md = Compared models; Δ SBχ^2 = differences in Satorra-Bentler Scaled Chi-Squared; Δdf = difference in number of degrees of freedom; p = significance value for the Scaled Chi-Squared Difference Test; CFI: Comparative Fit Index; ΔCFI = differences in Comparative Fit Index; RMSEA = Root Mean Square Error of Approximation; SRMR = Standardized Root Mean Square Residual;.

3.3. Convergent Validity

In relation to convergent validity, the Pearson's bivariate correlation carried out between the total IGDS9-SF scores and the OGD-Q scores had a value of r = 0.440, $p < 0.001$ (n = 101); with the CERM, it was r = 0.553, p = 0.001. Additionally, IGDS9-SF correlated with the five dimensions of the KIDSCREEN-27 as follows: physical well-being (r = −0.164, p = 0.001), psychological well-being (r = −0.315, p = 0.001), autonomy and relationship with parents (r = −0.167, p = 0.001), peers and social support (r = −0.257, p = 0.001), and school environment (r = −0.176, p = 0.001).

It was also found that gamers who preferably played the MOBA-, RPG-, or MMORPG-type game genres reported higher scores on the IGDS9-SF (15.21 ± 6.14) than those who played other genres (FPS, action/platforms, musical, sports simulators, fighting, or casual) (13.81 ± 4.55) (t = 2.679, $p < 0.008$, d = 0.26). Additionally, in response to the question '*I am addicted to video games*', 65 answered 'yes', and 473 replied 'no'. Those who self-reported that they were addicted had a significantly higher mean score on the IGDS9-SF (19.19 ± 8.30) compared to those who did not (13.43 ± 4.73) (t = 8.233, $p < 0.001$, d = 0.85).

3.4. Prevalence and Psychological Involvement of Internet Gaming Disorder

Following the diagnostic approach suggested by Pontes et al., [61], the participants who were classified with IGD (i.e., endorsing at least five of the criteria within the last 12 months) accounted for 1.9% of the sample of gamers (n = 10; see Table 3) and almost 1% of the total study sample. Of these 10 participants, nine were male and one was female. In addition, 1.9% endorsed four diagnostic criteria (n = 10) and were classed as 'at-risk' of developing IGD. It should also be noted that 76.2% (n = 410) did not endorse any diagnostic criteria.

Table 3. Number of participants who meet between 1 and 9 of the Internet Gaming Disorder criteria (adapted from Pontes, et al. [61]).

Number of Criteria Endorsed	Number of Participants	Total % of the Sample (n = 1064)	Total % of Online Gamers (n = 535)
1	71	6.67	13.2
2	27	2.54	5
3	10	0.94	1.9
4	10	0.94	1.9
5	1	0.09	0.2
6	1	0.09	0.2
7	2	0.19	0.4
8	2	0.19	0.4
9	4	0.38	0.8

Table 4 shows the psychological involvement of players with IGD (those with five or more symptoms) compared to those with four or fewer symptoms in relation to HRQoL. The loss of physical (p = 0.011) and psychological well-being (p = 0.018) is especially notable. There was also a loss of autonomy and relationship with parents (p = 0.047) and worse school environment (p = 0.038). There were no differences for the dimension of peers and social support (p = 0.080). The effect sizes for all contrasts were greater than $g > 0.42$.

Table 4. Comparison between gamers with Internet Gaming Disorder (IGD) (endorsing five or more criteria) versus those endorsing four criteria or fewer with respect to the five dimensions of the KIDSCREEN-27.

Instrument	n (< 5 symptoms)	$M \pm SD$ (< 5 symptoms)	n (≥ 5 symptoms)	$M \pm SD$ (≥ 5 symptoms)	Mann-Whitney U (p)	Effect Size (*Hedges' g*)
KD Phy-Wb	341	45.40 ± 11.39	7	33.92 ± 10.51	2.529 (.011)	1.00
KD Psy-Wb	341	46.70 ± 9.47	7	38.00 ± 8.88	2.370 (.018)	0.92
KD A and Pr	341	48.51 ± 10.91	7	39.75 ± 21.70	1.749 (.080)	0.78
KD SS and P	341	50.97 ± 11.31	7	37.81 ± 20.06	1.988 (.047)	1.14
KD SE	341	46.19 ± 9.10	7	42.31 ± 13.26	2.076 (.038)	0.42

Note: n (< 5 symptoms): participants with fewer than four problem criteria/items; n (≥ 5 symptoms): participants with four or more problem criteria/items; KD Phy-Wb = Physical well-being; KD Psy-Wb = Psychological well-being; KD A and Pr = Autonomy and relations with parents; KD SS and P = Social Support and Peers; KD; SE = School Environment; M = arithmetic mean; SD = standard deviation.

4. Discussion

The IGDS9-SF is a sound psychometric test that assesses IGD, and it is one of the most frequently used instruments as it had the greatest number of adaptations to different languages and cultural contexts [56]. In the present study, the development of the Spanish IGDS9-SF was carried out through a rigorous conceptual and methodological procedure that followed conventional international standards [71]. Appropriate indicators of validity and reliability were obtained in a sample of adolescents and young people. Factor analysis confirmed a single-factor solution with adequate goodness of fit, the item-total correlations were also high, and the factor loadings of all the items were satisfactory. Furthermore, the unidimensional factor model was found to be gender and age invariant across the metric level, which is considered a prerequisite for meaningful cross-group comparisons [80]. The present study adopted a similar procedure to studies conducted for different constructs, such as nomophobia [81–83], online gambling disorder [14], and previous validations of the IGDS9-SF in other languages such as Portuguese [67], Slovenian [61], and Italian [62], among others.

The prevalence rate of IGD in the present sample was 1.9%, which is similar to the 2.6% found in another Spanish sample by Fuster et al. [37] but noticeably lower than that of 8.3% reported by Buiza-Aguado et al. [36]. These discrepancies may be due to the use of different psychometric tools in assessing IGD. In other studies that have utilized the IGDS9-SF, the prevalence rates of disordered gaming were reported to be between 3% to 5% [59,61]. The results of the present study fall within the prevalence rate range reported by other international studies [31]. Finally, it also corroborates the fact that males are more frequently classified with IGD than females [31,37,38].

In relation to other validity indicators, the present study sought to evaluate the relationship between the IGDS9-SF and instruments assessing conceptually similar psychological problems such as the CERM [72] and the OGD-Q [14]. The results obtained indicated high correlations between these constructs, suggesting a convergent relationship with other relevant problems related to maladaptive use of technology and the internet such as problematic smartphone use and online gambling disorder. In addition, indicators relating to the relationship between IGDS9-SF and HRQoL allowed the examination of five key dimensions in adolescence (i.e., physical well-being; psychological well-being; autonomy and relationship with parents; peers and social support; and the school environment). Inverse and significant relationships were found between all five aforementioned dimensions, which means that higher level problems with online gaming associate with poorer self-reported quality of life. It is especially interesting to compare participants categorized as gamers with IGD or who are at risk with those who are not, because there was a significant decrease in quality of life scores in the former. Overall, the effects sizes of these correlations were high (i.e., most were greater than 0.80). These results are consistent with those of other studies using the 20-item Internet Gaming Disorder Test (IGD-20 Test [24]) and the KIDSCREEN-27 [31]. This finding also has a theoretical-conceptual relationship with the components model of addiction by Griffiths [23], which highlights the importance relating to the negative effects of the symptoms of addiction. In addition, it also supports the notion of other more general conceptualizations of problematic use of the internet [8], in which problems are related to poorer social and personal functioning, as well as to compulsive use and negative consequences.

Limitations and Future Lines of Research

The study conducted presents with several potential limitations worth discussing. Firstly, the IGDS9-SF is a self-report psychometric tool, so the potential for confounding effects stemming from response biases and social desirability by the adolescents and young people who completed it cannot be completely ruled out. This could be improved in the future by developing complementary measures combining behavioral tracking data (e.g., actual time spent playing and in-game preferences) to enhance self-report data. Secondly, the sample recruited was not randomly selected. However, the sample size of the present study was significantly large, especially in the context of a psychometric study. Nevertheless, caution is suggested when extrapolating the prevalence rates reported in the present study and considering them as a first approximation to the problem. Thirdly, although the diagnostic approach of the APA [21] and the recommendation of the original authors of the IGDS9-SF were followed when classifying disordered gamers, the authors have sought to establish a less conservative diagnostic approach which requires further discussion and analyses. Fourthly, whereas adequate indicators of validity and reliability were obtained, other important measures such as test-retest were not considered due to the imperatives of fieldwork. Fifthly, the KIDSCREEN-27 is a tool designed to evaluate HRQoL in the infantile-juvenile population, and the sample here included some participants over age 18 years. Thus, participants over 18 years were not considered in the analyses related to this construct, which reduced the sample size. It would be of interest for future research to use developmentally specific quality of life assessment tools for those over 18 years. It would also be fruitful to explore in future research other processes that favor the diagnostic accuracy of this scale, such as Receiver Operating Characteristic (ROC) curves (as demonstrated by Severo et al. [68] and Monacis et al. [62]), and other diagnostic elements, such as interviewing or complementary measures, should be used in the future in order to establish a robust clinically-driven gold standard diagnosis. Finally, according to the data obtained, the possible relationship between IGD and online gambling disorder with the advent of gambling-type elements in video games (e.g., loot boxes) should be noted for future lines of research to be explored. Legally, loot boxes are not considered online gambling, but at a psychosocial level, they meet the characteristics to be defined as a type of gambling [84].

5. Conclusions

The present study corroborates the psychometric properties of the scores obtained on the IGDS9-SF. In addition, preliminary data on the prevalence of the disordered gaming were obtained, which are useful for knowledge of an emerging global health challenge. The findings reported here will be particularly useful to pediatric and psychological care units, as well as for those in charge of school orientation at schools. All the above is also of special interest to parents, because education and parental supervision can play a very important role in the prevention of these problems associated with the internet, and in particular, internet gaming.

Author Contributions: Conceptualization, M.B, X.C, and J.G.-C.; Data curation, M.B. and J.M.M.; Formal analysis, J.M.M. and J.G.-C.; Funding acquisition, J.G.-C.; Investigation, M.B., J.M.M., M.A.V.-O., X.C., M.D.G., H.M.P, and J.G.-C.; Methodology, J.M.M. and J.G.-C.; Project administration, J.G.-C.; Resources, M.A.V.-O.; Supervision, X.C., M.D.G., H.M.P., and J.G.-C.; Writing—original draft, M.B., J.M.M., and J.G.-C.; Writing—review and editing, M.B., J.M.M., M.A.V.-O., X.C., M.D.G., H.M.P., and J.G.-C. All authors have read and agreed to the published version of the manuscript.

Funding: This research was funded by Spanish Ministry of Economy, Industry and Competitiveness, RTI2018-094212-B-I00: (CIBER-AACC), and by the International University of la Rioja, Project "Cyberpsychology (Trienio 2017-20)".

Conflicts of Interest: The authors declare no conflict of interest.

Appendix A. The Spanish version of IGDS9-SF

Los siguientes ítems hacen referencia a tu actividad con los videojuegos durante el último año (es decir, los últimos 12 meses). Por actividad en los videojuegos entendemos cualquier acción relacionada con los mismos (jugar desde un ordenador/portátil o desde una videoconsola) o desde cualquier otro tipo de dispositivo (por ejemplo, teléfono móvil, tablet, etc.) tanto conectado a Internet como sin estarlo y a cualquier tipo de juego

Preguntas	Nunca	Raramente	Ocasionalmente	A menudo	Muy a menudo
1.¿Te sientes preocupado por tu comportamiento con el juego? (Algunos ejemplos: ¿Piensas en exceso cuando no estás jugando o anticipas en exceso a la próxima sesión de juego?, ¿Crees que el juego se ha convertido en la actividad dominante en tu vida diaria?)					
2. ¿Sientes irritabilidad, ansiedad o incluso tristeza cuando intentas reducir o detener tu actividad de juego?					
3. ¿Sientes la necesidad de pasar cada vez más tiempo jugando para lograr satisfacción o placer?					
4. ¿Fallas sistemáticamente al intentar controlar o terminar tu actividad de juego?					
5. ¿ Has perdido intereses en aficiones anteriores y otras actividades de entretenimiento como resultado de tu compromiso con el juego?					
6. ¿ Has continuado jugando a pesar de saber que te estaba causando problemas con otras personas? (pareja, amistad o familia)?					
7. ¿ Has engañado a alguno de tus familiares, terapeutas o amigos sobre el tiempo que pasas jugando?					
8. ¿ Juegas para escapar temporalmente o aliviar un estado de ánimo negativo (por ejemplo, desesperanza, tristeza, culpa o ansiedad)?					
9. ¿ Has comprometido o perdido una relación importante, un trabajo o una oportunidad educativa debido a tu actividad de juego?					

References

1. Griffiths, M.D. Technological addictions. *Clin. Psychol. Forum* **1995**, *76*, 14–19.
2. Young, K.S. Internet addiction: The emergence of a new clinical disorder. *Cyberpsychol. Behav.* **1998**, *1*, 237–244. [CrossRef]
3. Carbonell, X.; Guardiola, E.; Beranuy, M.; Bellés, A. A bibliometric analysis of the scientific literature on Internet, video games, and cell phone addiction. *J. Med. Libr. Assoc. JMLA* **2009**, *97*, 102–107. [CrossRef] [PubMed]

4. Kuss, D.; Griffiths, M.; Karila, L.; Billieux, J. Internet addiction: A systematic review of epidemiological research for the last decade. *Curr. Pharm. Des.* **2014**, *20*, 4026–4052. [CrossRef] [PubMed]
5. Lopez-Fernandez, O. Generalised versus specific internet use-related addiction problems: A mixed methods study on internet, gaming, and social networking behaviours. *Int. J. Environ. Res. Public. Health* **2018**, *15*, 2913. [CrossRef] [PubMed]
6. Paulus, F.W.; Ohmann, S.; von Gontard, A.; Popow, C. Internet gaming disorder in children and adolescents: A systematic review. *Dev. Med. Child Neurol.* **2018**, *60*, 645–659. [CrossRef] [PubMed]
7. Davis, R.A. A cognitive-behavioral model of pathological Internet use. *Comput. Hum. Behav.* **2001**, *17*, 187–195. [CrossRef]
8. Caplan, S.E. Theory and measurement of generalized problematic Internet use: A two-step approach. *Comput. Hum. Behav.* **2010**, *26*, 1089–1097. [CrossRef]
9. Caplan, S.E. Problematic Internet use and psychosocial well-being: Development of a theory-based cognitive–behavioral measurement instrument. *Comput. Hum. Behav.* **2002**, *18*, 553–575. [CrossRef]
10. Kuss, D.J.; Griffiths, M.D. Internet gaming addiction: A systematic review of empirical research. *Int. J. Ment. Health Addict.* **2012**, *10*, 278–296. [CrossRef]
11. Pearcy, B.T.D.; McEvoy, P.M.; Roberts, L.D. Internet gaming disorder explains unique variance in psychological distress and disability after controlling for comorbid depression, OCD, ADHD, and anxiety. *Cyberpsychol. Behav. Soc. Netw.* **2017**, *20*, 126–132. [CrossRef] [PubMed]
12. van Rooij, A.J.; Schoenmakers, T.M.; van de Eijnden, R.J.J.M.; van de Mheen, D. Compulsive internet use: The role of online gaming and other internet applications. *J. Adolesc. Health* **2010**, *47*, 51–57. [CrossRef] [PubMed]
13. Gainsbury, S.M.; Russell, A.; Hing, N.; Wood, R.; Lubman, D.; Blaszczynski, A. How the internet is changing gambling: Findings from an Australian prevalence survey. *J. Gambl. Stud.* **2015**, *31*, 1–15. [CrossRef]
14. González-Cabrera, J.; Machimbarrena, J.M.; Beranuy, M.; Pérez-Rodríguez, P.; Fernández-González, L.; Calvete, E. Design and measurement properties of the Online Gambling Disorder Questionnaire (OGD-Q) in Spanish adolescents. *J. Clin. Med.* **2020**, *9*, 120. [CrossRef] [PubMed]
15. Wood, R.T.; Williams, R.J. A comparative profile of the Internet gambler: Demographic characteristics, game-play patterns, and problem gambling status. *New Media Soc.* **2011**, *13*, 1123–1141. [CrossRef]
16. Griffiths, M.D. Internet sex addiction: A review of empirical research. *Addict. Res. Theory* **2012**, *20*, 111–124. [CrossRef]
17. Meerkerk, G.-J.; Eijnden, R.J.J.M.V.D.; Garretsen, H.F.L. Predicting compulsive internet use: It's all about sex! *Cyberpsychol. Behav.* **2006**, *9*, 95–103. [CrossRef]
18. Hormes, J.M.; Kearns, B.; Timko, C.A. Craving Facebook? Behavioral addiction to online social networking and its association with emotion regulation deficits: Online social networking addiction. *Addiction* **2014**, *109*, 2079–2088. [CrossRef]
19. Ryan, T.; Chester, A.; Reece, J.; Xenos, S. The uses and abuses of Facebook: A review of Facebook addiction. *J. Behav. Addict.* **2014**, *3*, 133–148. [CrossRef]
20. Carbonell, X.; Panova, T. A critical consideration of social networking sites' addiction potential. *Addict. Res. Theory* **2017**, *25*, 48–57. [CrossRef]
21. American Psychiatric Association. *Diagnostic and Statistical Manual of Mental Disorders*, 5th ed.; American Psychiatric Publishing: Washington, DC, USA, 2013.
22. World Health Organization. *The ICD-11 Classification of Mental and Behavioural Disorders*; WHO: Geneva, Switzerland, 2018; Available online: https://icd.who.int/browse11/l-m/en (accessed on 18 January 2020).
23. Griffiths, M. A 'components' model of addiction within a biopsychosocial framework. *J. Subst. Use* **2005**, *10*, 191–197. [CrossRef]
24. Pontes, H.M.; Király, O.; Demetrovics, Z.; Griffiths, M.D. The conceptualisation and measurement of DSM-5 internet gaming disorder: The development of the IGD-20 Test. *PLoS ONE* **2014**, *9*, e110137. [CrossRef]
25. Rumpf, H.-J.; Achab, S.; Billieux, J.; Bowden-Jones, H.; Carragher, N.; Demetrovics, Z.; Higuchi, S.; King, D.L.; Mann, K.; Potenza, M.; et al. Including gaming disorder in the ICD-11: The need to do so from a clinical and public health perspective: Commentary on: A weak scientific basis for gaming disorder: Let us err on the side of caution (van Rooij et al., 2018). *J. Behav. Addict.* **2018**, *7*, 556–561. [CrossRef] [PubMed]
26. Asociación Española de Videojuegos La industria del videojuego en españa. *Anuario 2018*; Asociación Española de Videojuegos: Madrid, España, 2018.

27. Fam, J.Y. Prevalence of internet gaming disorder in adolescents: A meta-analysis across three decades. *Scand. J. Psychol.* **2018**, *59*, 524–531. [CrossRef] [PubMed]
28. Salmela-Aro, K. Stages of Adolescence. In *Encyclopedia of Adolescence*; Elsevier: Oxford, UK, 2011; pp. 360–368. ISBN 978-0-12-373951-3.
29. Mascó, A. *Entre Generaciones [between Generations]*; Temas: Buenos Aires, Argentina, 2013.
30. Petry, N.M.; Rehbein, F.; Gentile, D.A.; Lemmens, J.S.; Rumpf, H.-J.; Mößle, T.; Bischof, G.; Tao, R.; Fung, D.S.S.; Borges, G.; et al. An international consensus for assessing internet gaming disorder using the new DSM-5 approach: Internet gaming disorder. *Addiction* **2014**, *109*, 1399–1406. [CrossRef] [PubMed]
31. Wartberg, L.; Kriston, L.; Thomasius, R. The prevalence and psychosocial correlates of internet gaming disorder. *Dtsch. Aerzteblatt Online* **2017**. [CrossRef]
32. Yau, Y.; Potenza, M. Internet gaming disorder. *Psychiatr. Ann.* **2014**, *44*, 379–383. [CrossRef]
33. Yu, H.; Cho, J. Prevalence of internet gaming disorder among Korean adolescents and associations with non-psychotic psychological symptoms, and physical aggression. *Am. J. Health Behav.* **2016**, *40*, 705–716. [CrossRef]
34. Feng, W.; Ramo, D.E.; Chan, S.R.; Bourgeois, J.A. Internet gaming disorder: Trends in prevalence 1998–2016. *Addict. Behav.* **2017**, *75*, 17–24. [CrossRef]
35. Long, J.; Liu, T.; Liu, Y.; Hao, W.; Maurage, P.; Billieux, J. Prevalence and correlates of problematic online gaming: A systematic review of the evidence published in Chinese. *Curr. Addict. Rep.* **2018**, *5*, 359–371. [CrossRef]
36. Buiza-Aguado, C.; Alonso-Canovas, A.; Conde-Mateos, C.; Buiza-Navarrete, J.J.; Gentile, D. Problematic video gaming in a young Spanish population: Association with psychosocial health. *Cyberpsychol. Behav. Soc. Netw.* **2018**, *21*, 388–394. [CrossRef] [PubMed]
37. Fuster, H.; Carbonell, X.; Pontes, H.M.; Griffiths, M.D. Spanish validation of the Internet Gaming Disorder-20 (IGD-20) Test. *Comput. Hum. Behav.* **2016**, *56*, 215–224. [CrossRef]
38. Martín-Fernández, M.; Matalí, J.L.; García-Sánchez, S.; Pardo, M.; Lleras, M.; Castellano-Tejedor, C. Adolescentes con trastorno por juego en internet (IGD): Perfiles y respuesta al tratamiento. *Adicciones* **2016**, *29*, 125. [CrossRef] [PubMed]
39. Gentile, D.A.; Choo, H.; Liau, A.; Sim, T.; Li, D.; Fung, D.; Khoo, A. Pathological video game use among youths: A two-year longitudinal study. *Pediatrics* **2011**, *127*, e319–e329. [CrossRef] [PubMed]
40. Kim, E.J.; Namkoong, K.; Ku, T.; Kim, S.J. The relationship between online game addiction and aggression, self-control and narcissistic personality traits. *Eur. Psychiatry* **2008**, *23*, 212–218. [CrossRef]
41. Lemmens, J.S.; Valkenburg, P.M.; Gentile, D.A. The Internet Gaming Disorder Scale. *Psychol. Assess.* **2015**, *27*, 567–582. [CrossRef] [PubMed]
42. Thomée, S.; Härenstam, A.; Hagberg, M. Mobile phone use and stress, sleep disturbances, and symptoms of depression among young adults-a prospective cohort study. *BMC Public Health* **2011**, *11*, 66. [CrossRef]
43. Torres-Rodríguez, A.; Griffiths, M.D.; Carbonell, X.; Oberst, U. Treatment efficacy of a specialized psychotherapy program for Internet Gaming Disorder. *J. Behav. Addict.* **2018**, *7*, 939–952. [CrossRef]
44. Ha, J.H.; Yoo, H.J.; Cho, I.H.; Chin, B.; Shin, D.; Kim, J.H. Psychiatric comorbidity assessed in Korean children and adolescents who screen positive for internet addiction. *J. Clin. Psychiatry* **2006**, *67*, 821–826. [CrossRef]
45. Ferguson, C.J.; Coulson, M.; Barnett, J. A meta-analysis of pathological gaming prevalence and comorbidity with mental health, academic and social problems. *J. Psychiatr. Res.* **2011**, *45*, 1573–1578. [CrossRef]
46. Fernández-Villa, T.; Alguacil Ojeda, J.; Almaraz Gómez, A.; Cancela Carral, J.M.; Delgado-Rodríguez, M.; García-Martín, M.; Jiménez-Mejías, E.; Llorca, J.; Molina, A.J.; Ortíz Moncada, R.; et al. Uso problemático de internet en estudiantes universitarios: Factores asociados y diferencias de género. *Adicciones* **2015**, *27*, 265. [CrossRef] [PubMed]
47. Ko, C.-H.; Yen, J.-Y.; Yen, C.-F.; Chen, C.-S.; Chen, C.-C. The association between Internet addiction and psychiatric disorder: A review of the literature. *Eur. Psychiatry* **2012**, *27*, 1–8. [CrossRef] [PubMed]
48. Müller, K.W.; Beutel, M.E.; Egloff, B.; Wölfling, K. Investigating risk factors for internet gaming disorder: A comparison of patients with addictive gaming, pathological gamblers and healthy controls regarding the big five personality traits. *Eur. Addict. Res.* **2014**, *20*, 129–136. [CrossRef] [PubMed]
49. Choi, S.-W.; Kim, H.; Kim, G.-Y.; Jeon, Y.; Park, S.; Lee, J.-Y.; Jung, H.; Sohn, B.; Choi, J.-S.; Kim, D.-J. Similarities and differences among Internet gaming disorder, gambling disorder and alcohol use disorder: A focus on impulsivity and compulsivity. *J. Behav. Addict.* **2014**, *3*, 246–253. [CrossRef] [PubMed]

50. Fauth-Bühler, M.; Mann, K. Neurobiological correlates of internet gaming disorder: Similarities to pathological gambling. *Addict. Behav.* **2017**, *64*, 349–356. [CrossRef]
51. Liu, C.-H.; Lin, S.-H.; Pan, Y.-C.; Lin, Y.-H. Smartphone gaming and frequent use pattern associated with smartphone addiction. *Medicine* **2016**, *95*, e4068. [CrossRef]
52. Wallander, J.L.; Koot, H.M. Quality of life in children: A critical examination of concepts, approaches, issues, and future directions. *Clin. Psychol. Rev.* **2016**, *45*, 131–143. [CrossRef]
53. Kidscreen Group Europe. *The KIDSCREEN Questionnaires-Quality of Life Questionnaires for Children and Adolescents, Handbook*; Pabst Science Publishers: Lengerich, Germany, 2006.
54. Takahashi, M.; Adachi, M.; Nishimura, T.; Hirota, T.; Yasuda, S.; Kuribayashi, M.; Nakamura, K. Prevalence of pathological and maladaptive Internet use and the association with depression and health-related quality of life in Japanese elementary and junior high school-aged children. *Soc. Psychiatry Psychiatr. Epidemiol.* **2018**, *53*, 1349–1359. [CrossRef]
55. Barayan, S.; Al Dabal, B.; Abdelwahab, M.; Shafey, M.; Al Omar, R. Health-related quality of life among female university students in Dammam district: Is Internet use related? *J. Fam. Community Med.* **2018**, *25*, 20. [CrossRef]
56. Bernaldo-de-Quirós, M.; Labrador-Méndez, M.; Sánchez-Iglesias, I.; Labrador, F.J. Instrumentos de medida del trastorno de juego en internet en adolescentes y jóvenes según criterios DSM-5: Una revisión sistemática [Measuring instruments for internet gambling disorder in adolescents and young people according to DSM-5 criteria: A systematic review]. *Adicciones* **2019**. [CrossRef]
57. Pontes, H.M.; Griffiths, M.D. Measuring DSM-5 internet gaming disorder: Development and validation of a short psychometric scale. *Comput. Hum. Behav.* **2015**, *45*, 137–143. [CrossRef]
58. Yam, C.-W.; Pakpour, A.H.; Griffiths, M.D.; Yau, W.-Y.; Lo, C.-L.M.; Ng, J.M.T.; Lin, C.-Y.; Leung, H. Psychometric testing of three Chinese online-related addictive behavior instruments among Hong Kong university students. *Psychiatr. Q.* **2019**, *90*, 117–128. [CrossRef] [PubMed]
59. Montag, C.; Schivinski, B.; Sariyska, R.; Kannen, C.; Demetrovics, Z.; Pontes, H.M. Psychopathological symptoms and gaming motives in disordered gaming—A psychometric comparison between the WHO and APA diagnostic frameworks. *J. Clin. Med.* **2019**, *8*, 1691. [CrossRef] [PubMed]
60. Suchá, J.; Dolejš, M.; Pipová, H.; Maierová, E.; Cakirpaloglu, P. *Hraní digitálních her českými adolescenty*, 1st ed.; Univerzita Palackého v Olomouci: Křížkovského: Olomouc, Czech, 2018; ISBN 978-80-244-5424-5.
61. Pontes, H.M.; Macur, M.; Griffiths, M.D. Internet gaming disorder among Slovenian primary schoolchildren: Findings from a nationally representative sample of adolescents. *J. Behav. Addict.* **2016**, *5*, 304–310. [CrossRef]
62. Monacis, L.; Palo, d.V.; Griffiths, M.D.; Sinatra, M. Validation of the Internet Gaming Disorder Scale—Short-Form (IGDS9-SF) in an Italian-speaking sample. *J. Behav. Addict.* **2016**, *5*, 683–690. [CrossRef]
63. Wu, T.-Y.; Lin, C.-Y.; Årestedt, K.; Griffiths, M.D.; Broström, A.; Pakpour, A.H. Psychometric validation of the Persian nine-item Internet Gaming Disorder Scale–Short Form: Does gender and hours spent online gaming affect the interpretations of item descriptions? *J. Behav. Addict.* **2017**, *6*, 256–263. [CrossRef]
64. Aricak, O.T. Psychiatric Symptomatology as a predictor of cyberbullying among university students. *Eurasian J. Educ. Res.* **2009**, *34*, 167–184.
65. Evren, C.; Dalbudak, E.; Topcu, M.; Kutlu, N.; Evren, B.; Pontes, H.M. Psychometric validation of the Turkish nine-item Internet Gaming Disorder Scale–Short Form (IGDS9-SF). *Psychiatry Res.* **2018**, *265*, 349–354. [CrossRef]
66. Schivinski, B.; Brzozowska-Woś, M.; Buchanan, E.M.; Griffiths, M.D.; Pontes, H.M. Psychometric assessment of the internet gaming disorder diagnostic criteria: An item response theory study. *Addict. Behav. Rep.* **2018**, *8*, 176–184. [CrossRef]
67. Pontes, H.M.; Griffiths, M.D. Portuguese validation of the Internet Gaming Disorder Scale–Short-Form. *Cyberpsychol. Behav. Soc. Netw.* **2016**, *19*, 288–293. [CrossRef]
68. Severo, R.B.; Barbosa, A.P.P.N.; Fouchy, D.R.C.; Coelho, F.M.d.C.; Pinheiro, R.T.; de Figueiredo, V.L.M.; de Siqueira Afonso, V.; Pontes, H.M.; Pinheiro, K.A.T. Development and psychometric validation of Internet Gaming Disorder Scale-Short-Form (IGDS9-SF) in a Brazilian sample. *Addict. Behav.* **2020**, *103*, 106191. [CrossRef] [PubMed]
69. Ethnologue What are the top 200 most spoken languages? Available online: https://www.ethnologue.com/guides/ethnologue200 (accessed on 18 January 2020).

70. Instituto Cervantes. El español: Una lengua viva. Informe 2019 [Spanish: A Living Language. 2019 report]. Available online: https://www.cervantes.es/imagenes/File/espanol_lengua_viva_2019.pdf (accessed on 20 February 2020).
71. Muñiz, J.; Fonseca-Pedrero, E. Diez pasos para la construcción de un test. *Psicothema* **2019**, 7–16. [CrossRef]
72. Beranuy, M.; Chamarro, A.; Graner, C.; Carbonell, X. Validación de dos escalas breves para evaluar la adicción a Internet y el abuso de móvil [Validation of two short scales to assess Internet addiction and mobile abuse]. *Psicothema* **2009**, *21*, 480–485.
73. *IBM Corp Statistical Package for the Social Sciences for Windows*; IBM Corp: Armonk, NY, USA, 2015.
74. Revelle, W. *Psych: Procedures for Personality and Psychological Research*, R package version 1.9.12; Northwestern University: Evanston, IL, USA, 2015; Available online: https://CRAN.R-project.org/package=psych (accessed on 27 February 2020).
75. Rosseel, Y. lavaan: An *R* Package for structural equation modeling. *J. Stat. Softw.* **2012**, *48*. [CrossRef]
76. Giang, M.; Mai, Y.; Yuan, K.-H. *equaltestMI: Examine measurement invariance via equivalence testing and projection method*, R package version 0.1.0; 2017. Available online: https://CRAN.R-project.org/package=equaltestMI (accessed on 27 February 2020).
77. Hu, L.; Bentler, P.M. Cutoff criteria for fit indexes in covariance structure analysis: Conventional criteria versus new alternatives. *Struct. Equ. Model. Multidiscip. J.* **1999**, *6*, 1–55. [CrossRef]
78. Byrne, B.M. *Structural equation modeling with EQS: Basic concepts, application, and programming*; Lawrence Elbaum Associates: Mahwah, NJ, USA, 2006.
79. Cheung, G.W.; Rensvold, R.B. Evaluating goodness-of-fit indexes for testing measurement invariance. *Struct. Equ. Model. Multidiscip. J.* **2002**, *9*, 233–255. [CrossRef]
80. Bollen, K.A. *Structural equations with latent variables: Bollen/structural equations with latent variables*; John Wiley & Sons, Inc.: Hoboken, NJ, USA, 1989; ISBN 978-1-118-61917-9. [CrossRef]
81. Yildirim, C.; Correia, A.-P. Exploring the dimensions of nomophobia: Development and validation of a self-reported questionnaire. *Comput. Hum. Behav.* **2015**, *49*, 130–137. [CrossRef]
82. León-Mejía, A.; Calvete, E.; Patino-Alonso, C.; Machimbarrena, J.M.; González-Cabrera, J. Factor structure and sex and age cut-off points for the Spanish version of the Nomophobia Questionnaire. *Adicciones* **2020**. [CrossRef]
83. González-Cabrera, J.; León-Mejía, A.; Pérez-Sancho, C.; Calvete, E. Adaptación al español del cuestionario Nomophobia Questionnaire (NMP-Q) en una muestra de adolescentes [Spanish adaptation of the Nomophobia Questionnaire (NMP-Q) in a sample of adolescents]. *Actas Esp. Psiquiatr.* **2017**, *45*, 137–144.
84. Griffiths, M.D. Is the buying of loot boxes in video games a form of gambling or gaming? *Gaming Law Rev.* **2018**, *22*, 52–54. [CrossRef]

International Journal of
Environmental Research and Public Health

MDPI

Article

The Associations between Family-Related Factors and Excessive Internet Use in Adolescents

Anna Faltýnková [1], Lukas Blinka [1,*], Anna Ševčíková [1] and Daniela Husarova [2]

[1] Faculty of Social Studies, Masaryk University, Brno 601 77, Czech Republic; anafaltynkova@gmail.com (A.F.); asevciko@fss.muni.cz (A.Š.)

[2] Department of Health Psychology and Methodology Research, Faculty of Medicine, P.J. Safarik University in Kosice, Košice 040 01, Slovakia; daniela.husarova@upjs.sk

* Correspondence: lukasblinka@gmail.com

Received: 7 February 2020; Accepted: 6 March 2020; Published: 8 March 2020

Abstract: This study examined the relationship between Excessive Internet Use (EIU) in adolescents and their family environment, namely the family type, the family economic status, the effect of parental care, the level of parental control, the amount of parental monitoring, the quality of communication, and the time spent together. The study was based on data from an international survey, Health Behaviour in School Aged Children (HBSC), conducted in Slovakia. The sample representative for adolescents included 2547 participants (51% boys) aged 13–15. Multiple-step linear regression revealed that higher parental care and parental monitoring predicted lower EIU, while higher parental overprotection and lower socioeconomic status predicted higher EIU. The results suggest that both so-called optimal parenting (i.e., the balance of emotional warmth and protection) and the adolescent's autonomy lower the risk of EIU. Family factors explained about 14% of the variance, which suggests that aside from personal, cognitive and affective factors, a close social environment also plays an important role in adolescence EIU.

Keywords: adolescent internet use; excessive internet use; internet addiction; family factors; parenting styles

---

1. Introduction

The internet and the use of various digital devices are among the most important socialization factors and leisure activities in the lives of adolescents. The extent and intensity of internet usage has spawned research on both the benefits and the risks of online activities, including their addictive potential. This facilitated the inclusion of Internet Gaming Disorder in the International Classification of Diseases 11th Revision (ICD-11, World Health Organization). Generalized internet addiction has not reached such a consensus in the scientific community and various terms have been proposed (e.g. internet addiction, compulsive internet use [1]; pathological internet use [2]; problematic internet use [3]; internet use disorder [4]. In this paper, we work with the term Excessive Internet Use (EIU) and define it as a problem that manifests as preoccupation, mood changes, difficulties with limiting time online, and subsequent conflicts, including difficulties that belong among the symptoms of addictive behaviour. In our study, EIU represents an umbrella term for any excessive online behaviour. Although some older literature suggests the existence of generalised internet addictions [2], more current literature leans towards specific internet use disorders [4]. This has been supported by empirical findings that generalised and specific internet addiction mostly overlap [5], and that some online applications, especially online gaming [6] and social networking [7], contribute to this phenomenon more than others.

Adolescents are often shown to be the most vulnerable group in terms of developing EIU [8]. Their vulnerability stems from specific immaturity that is characteristic for the developmental period and includes increased impulsivity, impaired self-regulation, and higher tendencies toward reward-seeking behaviours [9,10]. Given the fact that the internet is a readily available tool that is thoroughly embedded in their everyday routines [11], EIU may become an issue for some adolescents. Higher levels of internet use have been linked to a range of problems, such as nervousness and irritability due to lack of sleep [12], increased depression [13], health problems caused by sedentary behaviour [14,15], and neglecting academic and other responsibilities [16,17].

Prior research has mostly investigated various individual risk factors that range from low self-esteem and self-efficacy [18,19], self-directedness [20], neuroticism [21], and loneliness [22] to attention difficulties [23]. As such, the role of social environment is neglected in models that describing both generalised and specific disordered internet use-only the role of social cognition (e.g., feelings of isolation, perceived lack of social support) is mentioned in the Interaction of Person-Affect-Cognition-Execution (I-PACE) model [4] and the Cognitive-Behavioral Model of Pathological Internet Use [2]. However, theories of problematic adolescent behaviour, e.g., the Problem Behavior Theory [24] and theories for media effects, e.g., the Differential Susceptibility to Media Effects Model [25] propose stronger consideration of the environmental influence, notably the family, which may both predictand moderate adolescent behaviour, including patterns of internet use.

The family environment is the closest social environment for adolescents and it plays an essential role in development [26]. Previous research suggests that the presence of a dysfunctional and high-conflict family environment may increase the likelihood of developing various forms of adolescent pathological behaviour, like alcohol use and gambling [27,28], as well as developing excessive online behaviour [29]. Specifically, dysfunctional parent–child communication [30], the existence of family conflict [31], and the lower emotional availability of parents [32] have been found to be associated with EIU in adolescents. More parental emotional warmth and care have been shown to be protective factors against EIU [32,33].

Parents face challenges as their adolescent grows. Adolescents develop their own individuality and learn to take responsibility for their own actions. Their needs for privacy and independence gain importance [34], but at the same time compete with needs for closeness and relatedness, which still remains significant [35]. Insufficient social support in this developmental stage can intensify feelings of loneliness or social isolation, which often appear in adolescence [36]. Indeed, the levels of social support have been found to be associated with EIU [37]. Therefore, parents have to find a balance between promoting autonomy and monitoring behaviour, which also applies to internet use. According to existing literature, restriction and control are associated with negative outcomes, while parental monitoring promotes healthy and positive psychological development [38]. With respect to internet use, the adequate monitoring of an adolescent´s whereabouts and the strategies of behavioural control have been linked to lower levels of EIU [39,40], while intrusive parental control and autonomy restriction are likely to manifest as risk factors for EIU [41].

The present overview suggests that family environmental factors might play an important role in moderating internet use in adolescence. However, to our knowledge, none of the available studies have comprehensively examined these factors in relation to EIU. Therefore, the aim of this study is to analyse the associations between EIU and an array of family-related factors—family composition, socioeconomic status, parenting strategies (e.g., care, monitoring, overprotection), quality of communication, and time spent together, while controlling for adolescent gender, age, and digital screen time. We assume that adolescents, who experience sufficient social support, care, and personal autonomy will report lower scores for EIU than those who perceive their parents as unloving, too controlling, and restrictive of their personal freedom.

2. Materials and Methods

2.1. Data Collection

We used data from the Health Behaviour in School-aged Children (HBSC) study, which was a cross-national survey conducted in Slovakia in 2014. A two-step procedure was used for data collection in order to obtain a nationally representative sample of adolescents. In the first step, 151 elementary schools were randomly selected from the register of all eligible schools in Slovakia, as provided by the Slovak Institute of Information and Prognosis for Education. In the second step, every school that agreed to participate had one randomly selected class per grade within the target group of students (i.e., fifth to ninth grades). Standardized self-reporting questionnaires were administered by trained administrators and completed by individual students during a classroom session. The study was approved by the Ethics Committee of the Medical Faculty at P.J. Safarik University in Kosice (No: 9/2012). All participation was fully voluntary and anonymous. Parental consent was obtained before administration.

2.2. Sample

A total of 10,179 Slovak adolescents aged 11–15 participated in the survey and completed the self-report questionnaire. Measures listed below were offered only to these children. Items to measure excessive internet use were included only in version A of the questionnaire, which was designed for students aged 13–15 (only 3135 adolescents were given version A). We also eliminated participants who did not provide 50% of the values for any of the variables of interest. The final sample consisted of 2547 adolescents from 13 to 15 years old (M = 14.32, SD = 0.91), with 50.9% being boys.

2.3. Measures

HBSC followed the strict methodology of the survey, which also applied to the translation process. All of the measures used in the study were translated from English to Slovak using the back-translation procedure. Moreover, back-translations are also independently checked by a translation team within the HBSC network, which was specifically established for this task.

EIU was measured using the *Excessive Internet Use* scale (EU Kids Online network; eukidsonline.net). This scale was validated in 25 European countries [42]. It consists of five items that cover five of the six factors of the components for the model of behavioural addiction [43]: salience ("I have gone without eating and sleeping because of the internet"); withdrawal symptoms ("I have felt bothered when I cannot be on the internet"); tolerance ("I have caught myself surfing when I am not really interested"); relapse ("I have tried unsuccessfully to spend less time on the internet"); and conflict ("I have spent less time than I should with either family, friends, or doing schoolwork because of the time I spend on the internet"). Using a four-point scale (ranging from 1 = "never" to 4 = "very often"), participants rated how often they had experienced the symptoms in the preceding12 months. The final variable was created as a mean of the five items, with higher numbers representing more excessive internet use. The scale was reliable. Cronbach's alpha equaled 0.79.

Digital screen time was assessed as the total time spent on the computer during an average weekday. The scale had nine options that ranged from zero hours a day to seven and more hours a day. Digital screen time was previously associated with the greater duration of internet use per week and with internet addiction [44].

Socioeconomic status was assessed by a question about the perceived financial situation of the family (1 = "our family is not at all well off" to 5 = "our family is very well off").

Family composition was constructed as a dichotomous variable, with 0 representing an incomplete family (e.g., the child lives with only one parent and/or step-parent, or with none of the parents) and 1 representing a complete family (e.g., the child lives with both parents in one household).

Parental care was assessed with a composite score that combines information from two scales that measure identical construct. Eight items were taken from the Parental Bonding Instrument: A Brief

Current Form (PBI-BC [45]) and four items came from the Multidimensional Scale of Perceived Social Support (MSPSS [46]). Items in both subscales measure parental warmth, interest, and supportive behaviour from the adolescents' perspective (e.g., "My mother/my father helps me as much as she/he can"; "My family is willing to help me make decisions."). The scale had a single-factor structure and the composite score for the variable was constructed as a mean score of 12items for the mother and father. For children who had only one parent, the mean score was computed for one parent. This allowed us to avoid reducing the statistical power because of the exclusion of children from single-parent families. The scale was reliable and Cronbach's alpha equaled 0.88.

Parental overprotection was measured with eight items from the Parental Bonding Instrument: A Brief Current Form (PBI-BC [45]). The scale is designed to measure parental control and overprotective behaviour from the perspective of the adolescent (e.g., "Mother/father likes me to make my own decisions"). Participants completed scales for mothers and fathers separately, and each item was rated on a three-point scale according to the perceived frequency of certain behaviour (1 = "never"; 2 = "sometimes"; 3 = "almost always"). The variable Parental Overprotection was constructed as a mean score of items for paternal overprotection and maternal overprotection. For children who had only one parent, the mean score was computed for one parent. The scale was reliable and Cronbach's alpha equaled 0.65.

Parental monitoring was measured with a five-item scale for how much the parents know about the following aspects of their adolescent's life: how they spend their free time; what they do after school; where they go in the evening; how they spend their money; and who their friends are. Response options were 1 = "she/he knows a lot"; 2 = "she/he knows a little"; and 3 = "she/he knows nothing". The mean score was computed from all 10items for the mother and father. For children who had only one parent, the mean score was computed for one parent. The scale was reliable and Cronbach's alpha equaled 0.88.

Time spent together was measured by eight items that assessed how often adolescents engaged in certain common activities with their parents, like playing indoor games, visiting relatives, and watching television. For each item, participants were asked to choose the most appropriate of the following five response categories: 1 = "never"; 2 = less often"; 3 = "about once a week"; 4 = "most days"; and 5 = "every day". The variable was constructed as a mean score of all eight items. The scale was reliable and Cronbach's alpha equaled 0.85.

Quality of communication with parents was measured by asking respondents how easy was for them to talk to their parents about things that bothered them (1 = "very easy" to 4 = "very difficult"). The mean score was computed from both items for the mother and father. For children with only one parent we used the single score they chose for that parent. The scale was reliable and Cronbach's alpha equaled 0.68.

Descriptive statistics and Cronbach's alpha values for all of the scales are presented in Table 1.

Table 1. Descriptive statistics for study variables.

	M	*SD*	Skewness	Kurtosis	Cronbach´s Alpha
Excessive internet use	1.62	0.61	1.37	1.95	0.79
Age	14.32	0.91	0.18	−0.46	
Incomplete family			−1.57	0.47	
Digital screen time	4.43	2.17	0.56	−0.41	
Socioeconomic status	3.84	0.83	−0.50	0.29	
Parental care	3.73	0.66	−1.04	1.18	0.88
Parental overprotection	1.64	0.35	0.24	−0.07	0.65
Parental monitoring	2.47	0.46	−0.81	0.06	0.88
Time spent together	2.82	0.79	0.30	0.05	0.85
Communication	2.81	0.81	−0.34	−0.44	0.68

2.4. Data Analysis

The data were analysed using IBM SPSS Statistics 25 software. Spearman´s correlation coefficients were calculated for all of the variables. Gender differences in the level of excessive internet use and other variables were assessed using the Mann–Whitney U-test. Nonparametric tests were used because the majority of variables were not normally distributed (values of skewness and kurtosis are presented in Table 1). To determine the associations between excessive internet use and family factors, we conducted a hierarchical multiple regression that controlled for the gender and age of the participants and for the frequency of computer use. Variables were entered into the regression in two steps with the Enter method. In the case of sufficiently large sample sizes, regression analysis provided valid results, even for non-normally distributed variables [47]. Control variables were entered in the regression in the first step. All of the family factors were entered in the second step. The assumptions of regression analysis were met: the VIF score for any variable was below 1.94 (tolerance above 0.52), indicating no problem with multicollinearity, and the residuals were uncorrelated (Durbin–Watson coefficient = 1.88) and approximately normally distributed.

3. Results

The means, standard deviations, and Spearman´s rank correlation coefficients of the studied variables are presented in Table 2. On average, participants reported relatively low scores on excessive internet use ($M = 1.62$, $SD = 0.61$). The majority of the sample obtained the minimum possible score in the EIU scale; thus, the distribution of the dependent variable was positively skewed. Excessive internet use was positively related to adolescent age, frequency of computer use, and parental overprotection, and negatively related to all other family variables.

Table 2. Spearman´s rank correlations of all study variables.

	1	2	3	4	5	6	7	8	9	10
1. Excessive internet use	-									
2. Age	0.10 **	-								
3. Digital screen time	0.23 **	0.12 **	-							
4. Socioeconomic status	−0.13 **	−0.08 **	−0.02	-						
5. Family composition	−0.03	−0.01	−0.05 *	0.14 **	-					
6. Parental care	−0.24 **	−0.10 **	−0.05 *	0.27 **	0.03	-				
7. Parental overprotection	0.23 **	−0.03	0.05 **	−0.12 **	−0.07 **	−0.34 **	-			
8. Parental monitoring	−0.28 **	−0.10 **	−0.17 **	0.17 **	0.14 **	0.52 **	−0.28 **	-		
9. Time spent together	−0.13 **	−0.16 **	−0.08 **	0.26 **	0.06 **	0.46 **	−0.11 **	0.37 **	-	
10. Parental communication	−0.17 **	−0.09 **	−0.03	0.23**	−0.01	0.51 **	−0.22 **	0.31 **	0.33 **	-

* $p < 0.05$, ** $p < 0.01$.

A Mann–Whitney U test was conducted to compare the level of excessive internet use in boys ($M = 1.62$, $SD = 0.64$) and girls ($M = 1.62$, $SD = 0.59$). There was no significant difference ($U = 806670.5$, $p > 0.05$, $\eta^2 = 0.01$). There were also no differences between boys and girls in parental care ($U = 807552.5$, $p > 0.05$, $\eta^2 = 0.01$), parental overprotection ($U = 783925.0$, $p > 0.05$, $\eta^2 = 0.06$), parental monitoring ($U = 794794.0$, $p > 0.05$, $\eta^2 = 0.03$), communication ($U = 806933.0$, $p > 0.05$, $\eta^2 = 0.01$), time spent together ($U = 809466.5$, $p > 0.05$, $\eta^2 = 0.002$), socioeconomic status ($U = 800034.5$, $p > 0.05$, $\eta^2 = 0.02$), family composition ($U = 791531.0$, $p > 0.05$, $\eta 2 = 0.04$), and digital screen time ($U = 795816.5$, $p > 0.05$, $\eta^2 = 0.03$).

Regression coefficients for each predictor are presented in Table 3.

Table 3. Regression coefficients.

	Model 1			Model 2		
	B	*SE*	*β*	*B*	*SE*	*β*
Constant	0.65	0.19		1.01	0.23	
Age	0.05	0.01	0.07 ***	0.04	0.01	0.06 ***
Gender	0.01	0.02	0.01	0.02	0.02	0.02
Digital screen time	0.06	0.01	0.22 ***	0.05	0.01	0.18 ***
Socioeconomic status				−0.04	0.02	−0.05 *
Family composition				0.02	0.03	0.01
Parental care				−0.08	0.02	−0.08 ***
Parental overprotection				0.29	0.04	0.17 ***
Parental monitoring				−0.20	0.03	−0.15 ***
Time spent together				0.06	0.02	0.08 ***
Communication				0.01	0.02	0.01
R		0.23			0.37	
R^2		0.06			0.14	
Adjusted R^2		0.05			0.14	

* $p < 0.05$, *** $p < 0.001$.

The first model, including age, gender, and frequency of computer use, significantly predicted excessive internet use ($F(3, 2544) = 49.14$; $p < 0.001$) and accounted for 5.5% of the variance of the dependent variable. Both age ($t(2544) = 3.69$; $p < 0.001$) and digital screen time ($t(2544) = 11.09$; $p < 0.001$) significantly predicted EIU.

After entering family-related variables in the second step, the proportion of explained variance increased to 14% and this model also significantly predicted EIU ($F(10, 2537) = 41.37$; $p < 0.001$). Our analysis revealed that higher socioeconomic status ($t(2537) = −2.44$; $p < 0.05$), parental care ($t(2537) = −3.19$; $p < 0.01$), and parental monitoring ($t(2537) = -6.62$; $p < 0.001$) were significant negative predictors (i.e., protective factors) of EIU. Parental overprotection ($t(2537) = 8.14$; $p < 0.001$) and time spent together ($t(2537) = 3.75$; $p < 0.001$) significantly predicted EIU in the opposite direction, indicating that those variables figure as risk factors. The associations between family composition ($t(2537) = 0.53$; $p > 0.05$), parental communication ($t(2537) = 0.23$; $p > 0.05$), and excessive internet use were not significant. Age and frequency of computer use remained significant predictors, whereas association between gender and excessive internet use remained insignificant.

4. Discussion

This study examined family environmental factors in relation to adolescent Excessive Internet Use. The findings show that the strongest protective factor was the parental monitoring of the adolescent's activities (e.g., knowing about friends, knowing how they spend free time), followed by parental care (e.g., emotional warmth, overall support within the family environment). On the other hand, the strongest risk factor associated with a higher score of EIU was parental overprotection (e.g., parental behaviour that decreases adolescents' independence), followed by increased time spent together and lower socioeconomic status of the family.

These findings point to two important issues. First, our study supports the so-called optimal parenting that seems to be an ideal parenting strategy that is associated with lower scores for EIU [48]. Parents who use this strategy might be successful at balancing their care for their adolescents and their protection, while, at the same time, respecting their personal autonomy [49]. As the need for autonomy in adolescence increases with age, autonomy restrictions may lead to diverse problems, including lower self-esteem or self-efficacy. Sufficient support for autonomy has a protective function, which has also been shown in previous studies [50].

Second, from the perspective of associations among family environmental factors and problematic behaviour in youth, our results show a mix of similarities and dissimilarities between EIU and other traditional problematic adolescent behaviours. In the case of traditional antisocial adolescent behaviours, such as drinking alcohol and delinquency, parental monitoring and care have generally been identified as protective factors, while parenting styles that harm the confidence and independence of children were found to be risk factors [51,52]. On the other hand, antisocial behaviours were often explained decreased family monitoring and involvement [52,53]. In this respect, EIU seems to be distinct from traditional antisocial and problematic behaviours—our results showed that overprotection and increased time at home and in the family environment (in other words, decreased time in peer groups) increase EIU. The distinction between EIU and problematic adolescent behaviour is further supported by De Looze and his colleagues [54], who found that the overall trend in the decrease of substance use in recent adolescents is not related to increased use of digital media; thus, EIU is not simply a replacement for other forms of problematic behaviour. Despite the fact that some studies found an association between EIU and various conduct problems [55,56], and that problematic adolescent behaviours and EIU share some underpinnings with ADHD [57], our study points to the need for specific parenting approaches when addressing EIU and when addressing traditional problematic behaviours. The mutual relatedness of EIU and other forms of problematic behaviours should be the subject of a more thorough investigation.

It must be noted that family factors themselves explain only about 8% (14% if we include digital screen time and age) of the variance of EIU. This is in line with previous research that showed that family factors explained 6% of the variance of EIU through parental mediation strategies [58]. It is likely that the strongest influences for EIU are personality and psychological factors, like ADHD, lower self-esteem, and emotional difficulties in general [42], which is also suggested by theoretical models like the I-PACE [4], and cognitive-behavioural models [2,59]. Since environmental factors play a non-negligible role in adolescent lives and they should be incorporated into theoretical models of EIU and specific internet disorders as has been done in general media-effect models (e.g., Differential Susceptibility to Media-Effects Model [25]).

The strength of associations was rather weak in our study. Except for parental monitoring and overprotection, all other significant predictors were negligible. Whether they remained significant only due to the large sample size or because they indeed play some role in EIU should be the focus of further research. For example, lower social economic status, although correlated with EIU, lost most of its strength in the factor analysis. However, the literature has identified lower socioeconomic status as a risk factor of EIU [60]. In our research, lower socioeconomic status worked in the same direction as parental over protection, the time that the adolescent spent at home, and the time that the adolescent spent with digital devices. This may indicate that, in such families, there is a lack resources for meaningful structured afterschool activities (e.g., sports, music lessons) and that digital devices are used as cheap alternatives for filling free time. Such a hypothesis needs further investigation.

The results of this study should be considered in light of several limitations. First, since the research design is cross-sectional, causal relationships between family factors and EIU could not be inferred based on our results. In this case, longitudinal studies would better examine the direction of the observed effects. Second, the data were obtained solely from adolescent self-report questionnaires. For future studies, it would be useful to complement this method of data collection with reports from parents to provide a comparison for parent–adolescent dyads. Third, the data did not include variables to measure specific internet-related mediation strategies, so we were only able to analyse general attitudes in parenting and the atmosphere in the family. In addition, digital screen time did not measure internet use, per se. Nonetheless, prior research has clearly shown that digital screen time predominantly overlaps with time spent online [61]. And, lastly, the main variable, EIU, although used in many previous studies and validated in most European countries, consists of only five items and does not allow for an in-depth investigation of the excessive use of various applications. This may be problematic from the perspective of the ongoing discussion of generalised vs. specific internet-based disorders. In this respect, our study lacked the differentiation of the applications that contribute to EIU Also, the main variable has not been validated for discriminative purposes, and it is a continuum without differentiation between healthy and pathological internets users. Despite these limitations, the main advantages of this study are the use of the national representative sample of adolescents and a several family variables that allow for their mutual control in the regression analysis.

5. Conclusions

Our study revealed several important factors associated with adolescent Excessive Internet Use. The strongest protective factor was the parental monitoring of their adolescent's activities (i.e., parents being aware of what, where, when, and with whom their adolescent children spend their leisure time). This was supported by higher care (i.e., emotional warmth, atmosphere in the family). The main risk factor of EIU appeared to be the parental overprotection (i.e., parental behaviour that harmed the respondent's independence and confidence). Certain risk factors, although very small, were also brought by the lower socioeconomic status of the family and increased time spent at home, which were also supported by increased digital screen time. These results indicate that optimal parenting is a balance among parental care, protection, and allowing adolescents to build independence and competence. It also points to the potential role of family lifestyles associated with limited resources.

Author Contributions: A.F. conducted the statistical analysis. A.F. and L.B. conducted the literature review and wrote the draft of the article. A.Š. and D.H. helped with the interpretation of the findings and draft editing. All authors have read and agreed to the published version of the manuscript.

Funding: This research was supported by the Slovak Research and Development Support Agency under Contract No. APVV-18-0070.

Conflicts of Interest: The authors declare no conflicts of interest.

References

1. Van Rooij, A.J.; Schoenmakers, T.M.; van den Eijnden, R.J.; van de Mheen, D. Compulsive Internet use: The role of online gaming and other internet applications. *J. Adolesc. Health* **2010**, *47*, 51–57. [CrossRef]
2. Davis, R.A. A cognitive-behavioral model of pathological Internet use. *Comput. Hum. Behav.* **2001**, *17*, 187–195. [CrossRef]
3. Shapira, N.A.; Goldsmith, T.D.; Keck, P.E.; Khosla, U.M.; McElroy, S.L. Psychiatric features of individuals with problematic Internet use. *J. Affect. Disord.* **2000**, *57*, 267–272. [CrossRef]
4. Brand, M.; Young, K.S.; Laier, C.; Wölfling, K.; Potenza, M.N. Integrating psychological and neurobiological considerations regarding the development and maintenance of specific Internet-use disorders: An Interaction of Person-Affect-Cognition-Execution (I-PACE) model. *Neurosci. Biobehav. Rev.* **2016**, *71*, 252–266. [CrossRef] [PubMed]
5. Montag, C.; Bey, K.; Sha, P.; Li, M.; Chen, Y.F.; Liu, W.Y.; Reuter, M. Is it meaningful to distinguish between generalized and specific Internet addiction? Evidence from a cross-cultural study from Germany, Sweden, Taiwan and China. *Asia Pac. Psychiatry* **2015**, *7*, 20–26.
6. Blinka, L.; Škařupová, K.; Ševčíková, A.; Wölfing, K.; Müller, K.W.; Dreier, M. Excessive internet use in European adolescents: What determines differences in severity? *Int. J. Public Health* **2015**, *60*, 249–256. [CrossRef]
7. Lopez-Fernandez, O. Generalized Versus Specific Internet Use-Related Addiction Problems: A Mixed Methods Study on Internet, Gaming, and Social Networking Behaviours. *Int. J. Environ. Res. Public Health* **2018**, *15*, 2913. [CrossRef]
8. Kuss, D.J.; van Rooij, A.J.; Shorter, G.W.; Griffiths, M.D.; van de Mheen, D. Internet addiction in adolescents: Prevalence and risk factors. *Comput. Hum. Behav.* **2013**, *29*, 1987–1996. [CrossRef]
9. Boucher, M.; Sandhu, P. The Neurobiology of Adolescent Addiction. In *Clinical Handbook of Adolescent Addiction*; Rosner, R., Ed.; Wiley: Hoboken, NJ, USA, 2007; pp. 78–87.
10. Spear, L. The Teenage Brain: Adolescents and Alcohol. *Curr. Dir. Psychol. Sci.* **2013**, *22*, 152–157. [CrossRef]
11. Livingstone, S.; Haddon, L.; Görzig, A.; Ólafsson, K. Risk and safety on the internet: The role of online skills and internet self-efficacy. *New Media Soc.* **2011**, *12*, 309–329. [CrossRef]
12. Marino, C.; Vieno, A.; Lenzi, M.; Borraccino, A.; Lazzeri, G.; Lemma, P. Computer Use, Sleep Difficulties, and Psychological Symptoms Among School-Aged Children: The Mediation Role of Sleep Difficulties. *Int. J. Sch. Health* **2016**, *4*, 50–58. [CrossRef]

13. Ha, J.H.; Kim, S.; Bae, S.; Bae, S.; Kim, H.; Sim, M.; Lyoo, I.; Cho, S. Depression and Internet Addiction in Adolescents. *Psychopathology* **2007**, *402*, 424–430. [CrossRef] [PubMed]
14. Vandelanotte, C.; Sugiyama, T.; Gardiner, P.; Owen, N. Associations of Leisure-Time Internet and Computer Use With Overweight and Obesity, Physical Activity and Sedentary Behavior: Cross-Sectional Study. *J. Med. Internet Res.* **2009**, *11*, e28. [CrossRef] [PubMed]
15. Brindova, D.; Dankulincova Veselska, Z.; Klein, D.; Hamrik, Z.; Sigmundova, D.; van Dijk, J.P.; Reijneveld, S.A.; Madarasova Geckova, A. Is the association between screen-based behavior and health complaints among adolescents moderated by physical activity? *Int. J. Public Health* **2015**, *60*, 139–145. [CrossRef] [PubMed]
16. Kim, S.Y.; Kim, M.S.; Park, B.; Kim, J.H.; Choi, H.G. The associations between internet use time and school performance among Korean adolescents differ according to the purpose of internet use. *PLoS ONE* **2017**, *12*, e0174878. [CrossRef]
17. Camerini, A.L.; Chulz, P.J.; Jeannet, A.M. The social inequalities of Internet access, its use, and the impact on children´s academic performance: Evidence from a longitudinal study in Switzerland. *New Media Soc.* **2017**, *20*. [CrossRef]
18. De Berardis, D.; D´Albenzio, A.; Gambi, F.; Sepede, G.; Valchera, A.; Conti, C.M.; Fulcheri, M.; Cavuto, M.; Ortolani, C.; Salerno, R.M.; et al. Alexithymia and its relationships with dissociative experiences and Internet addiction in a nonclinical sample. *Cyberpsychol. Behav.* **2009**, *12*, 67–69. [CrossRef]
19. Montag, C.; Jurkiewicz, M.; Reuter, M. Low self-directedness is a better predictor for problematic internet use than high neuroticism. *Comput. Hum. Behav.* **2010**, *26*, 1531–1535. [CrossRef]
20. Servidio, R.; Gentile, A.; Boca, S. The Mediational Role of Coping Strategies in the Relationship Between Self-Esteem and Risk of Internet Addiction. *Eur. J. Psychol.* **2018**, *14*, 176–187. [CrossRef]
21. Wu, C.Y.; Lee, M.B.; Liao, S.C.; Chang, L.R. Risk Factors of Internet Addiction among Internet Users: An Online Questionnaire Survey. *PLoS ONE* **2015**, *10*, e0137506. [CrossRef]
22. Xinxin, S.; Wang, J.; Zou, H. Family Functioning and Internet Addiction among Chinese Adolescents: The Mediating Roles of Self-esteem and Loneliness. *Comp. Hum. Behav.* **2017**, *76*, 201–210.
23. Tateno, M.; Teo, A.R.; Shirasaka, T.; Tayama, M.; Watabe, M.; Kato, T.A. Internet addiction and self-evaluated attention-deficit hyperactivity disorder traits among Japanese college students. *Psychiatry Clin. Neurosci.* **2016**, *70*, 567–572. [CrossRef] [PubMed]
24. Jessor, R.; Jessor, S.L. *Problem Behavior and Psychological Development: A Longitudinal Study of Youth*; Academic Press: New York, NY, USA, 1977.
25. Valkenburg, P.M.; Peter, J. The Differential Susceptibility to Media Effects Model. *J. Commun.* **2013**, *63*, 221–243. [CrossRef]
26. Bronfenbrenner, U.; Moris, P.A. The Bioecological Model of Human Development. In *Handbook of Child Psychology*, 6th ed.; Damon, W., Lerner, R.M., Eds.; Wiley: Hoboken, NJ, USA, 2007; pp. 793–828.
27. McComb, J.L.; Sabiston, C.M. Family Influences on Adolescent Gambling Behavior: A Review of the Literature. *J. Gambl. Stud.* **2010**, *26*, 503–520. [CrossRef]
28. Swaim, R.C.; Stanley, L.R. Effects of Family Conflict and Anger on Alcohol Use among American Indian Students: Mediating Effects of Outcome Expectancies. *J. Stud. Alcohol Drugs* **2018**, *79*, 102–110. [CrossRef]
29. Schneider, L.A.; King, D.L.; Delfabbro, P.H. Family factors in adolescent problematic Internet gaming: A systematic review. *J. Behav. Addict.* **2017**, *6*, 321–333. [CrossRef]
30. Kim, Y.H.; Chong, Y.S. Parent-Child Communication, Peer-Relationship and Internet Addiction in Children. *J. Korean Home Econ. Assoc.* **2005**, *43*, 103–114. [CrossRef]
31. Wu, C.; Wong, H.T.; Yu, K.F.; Fok, K.W.; Yeung, S.M.; Lam, C.H.; Liu, K.M. Parenting approaches, family functionality, and internet addiction among Hong Kong adolescents. *BMC Pediatrics* **2016**, *16*, 130. [CrossRef]
32. Karaer, Y.; Akdemir, D. Parenting styles, perceived social support and emotion regulation in adolescents with internet addiction. *Compr. Psychiatry* **2019**, *92*, 22–27. [CrossRef]
33. Yao, M.Z.; He, J.; Ko, D.M.; Pang, K. The influence of personality, parental behaviors, and self-esteem on internet addiction: A study of Chinese college students. *Cyberpsychol. Behav. Soc. Netw.* **2014**, *17*, 104–110. [CrossRef]
34. Branje, S. Development of Parent-Adolescent Relationships: Conflict Interactions as a Mechanism of Change. *Child Dev. Perspect.* **2018**, *12*, 171–176. [CrossRef]
35. Reis, O.; Buhl, H.M. Individuation during adolescence and emerging adulthood–Five German studies. *Int. J. Behav. Dev.* **2008**, *32*, 369–371. [CrossRef]

36. Laursen, B.; Hartl, A.C. Understanding loneliness during adolescence: Developmental changes that increase the risk of perceived social isolation. *J. Adolesc.* **2013**, *36*, 1261–1268. [CrossRef] [PubMed]
37. Wang, E.S.T.; Wang, M.C.H. Social support and social interaction ties on internet addiction: Integrating online and offline contexts. *Cyberpsycholg. Behav. Soc. Netw.* **2013**, *16*, 843–849. [CrossRef] [PubMed]
38. Bleakley, A.; Ellithorpe, M.; Romer, D. The Role of Parents in Problematic Internet Use among US Adolescents. *Media Commun.* **2016**, *4*, 24–34. [CrossRef]
39. Lin, C.H.; Lin, S.L.; Wu, C.P. The effects of parental monitoring and leisure boredom on adolescents´ internet addiction. *Adolescence* **2009**, *44*, 993–1004.
40. Shek, D.T.L.; Zhu, X.; Ma, C.M.S. The Influence of Parental Control and Parent-Child Relational Qualities on Adolescent Internet Addiction: A 3-Year Longitudinal Study in Hong Kong. *Front. Psychol.* **2018**, *9*, 1–16. [CrossRef]
41. Li, C.; Dang, J.; Zhang, X.; Zhang, Q.; Guo, J. Internet addiction among Chinese adolescents: The effect of parental behavior and self-control. *Comput. Hum. Behav.* **2014**, *41*, 1–7. [CrossRef]
42. Škařupová, K.; Ólafsson, K.; Blinka, L. Excessive Internet Use and its association with negative experiences: Quasi-validation of a short scale in 25 European countries. *Comput. Hum. Behav.* **2015**, *53*, 118–123. [CrossRef]
43. Griffiths, M. A 'components' model of addiction within a biopsychosocial framework. *J. Subst. Use* **2005**, *10*, 191–197. [CrossRef]
44. Gupta, A.; Khan, A.M.; Rajoura, O.P.; Srivastava, S. Internet addiction and its mental health correlates among undergraduate college students of a university in North India. *J. Family Med. Prim. Care* **2018**, *7*, 721–727. [PubMed]
45. Klimidis, S.; Minas, I.H.; Ata, A.W. The PBI-BC: A Brief Current Form of the Parental Bonding Instrument for Adolescent Research. *Compr. Psychiatry* **1992**, *33*, 374–377. [CrossRef]
46. Zimet, G.D.; Dahlem, N.W.; Zimet, S.G.; Farley, G.K. The Multidimensional Scale of Perceived Social Support. *J. Personal. Assess.* **1988**, *52*, 30–41. [CrossRef]
47. Lumley, T.; Diehr, P.; Emerson, S.; Chen, L. The importance of the normality assumption in large public health data sets. *Annu. Rev. Public Health* **2000**, *23*, 151–169. [CrossRef]
48. Floros, G.; Siomos, K. The relationship between optimal parenting, Internet addiction and motives for social networking in adolescence. *Psychiatry Res.* **2013**, *209*, 529–534. [CrossRef]
49. Parker, G. Parental rearing style: Examining for links with personality vulnerability factors for depression. *Soc. Psychiatry Psychiatr. Epidemiol.* **1990**, *28*, 97–100. [CrossRef]
50. Shek, D.T.L.; Zhu, X.; Dou, D. Influence of Family Processes on Internet Addiction Among Late Adolescents in Hong Kong. *Front. Psychiatry* **2019**, 10. [CrossRef]
51. Baumrind, D. The influence of parenting style on adolescent competence and substance use. *J. Early Adolesc.* **1991**, *11*, 56–95. [CrossRef]
52. Ary, D.V.; Duncan, T.E.; Duncan, S.C.; Hops, H. Adolescent problem behavior: The influence of parents and peers. *Behav. Res. Ther.* **1999**, *37*, 217–230. [CrossRef]
53. Dishion, T.J.; Patterson, G.R.; Stoolmiller, M.; Skinner, M.L. Family, school, and behavioral antecedents to early adolescent involvement with antisocial peers. *Dev. Psychol.* **1991**, *27*, 172–180. [CrossRef]
54. De Looze, M.; van Dorsselaer, S.; Stevens, G.W.J.M.; Boniel-Nissim, M.; Vieno, A.; Van den Eijnden, R.J.J.M. The decline in adolescent substance use across Europe and North America in the early twenty-first century: A result of the digital revolution? *Int. J. Public Health* **2019**, *64*, 229–240. [CrossRef] [PubMed]
55. Lim, J.A.; Gwak, A.R.; Park, S.M.; Kwon, J.G.; Lee, J.Y.; Jung, H.Y.; Sohn, B.K.; Kim, J.W.; Kim, D.J.; Choi, J.S. Are adolescents with internet addiction prone to aggressive behavior? The mediating effect of clinical comorbidities on the predictability of aggression in adolescents with internet addiction. *Cyberpsychol. Behav. Soc. Netw.* **2015**, *18*, 260–267. [CrossRef] [PubMed]
56. Blinka, L.; Šmahel, D. Predictors of adolescents' excessive internet use: A comparison across European countries. In Proceedings of the 15th European Conference on Developmental Psychology, Bergen, Norway, 23–27 August 2011; pp. 337–341.
57. Wang, B.Q.; Young, K.S.; Zhou, X.; Liu, J.; Lv, Z.T. The association between attention deficit/hyperactivity disorder and internet addiction: A systematic review and meta-analysis. *BMC Psychiatry* **2017**, *17*, 260. [CrossRef] [PubMed]
58. Kalmus, V.; Blinka, L.; Ólafsson, K. Does It Matter What Mama Says: Evaluating the Role of Parental Mediation in European Adolescents' Excessive Internet Use. *Child. Soc.* **2015**, *29*, 122–133. [CrossRef]

59. Dong, G.; Potenza, M.N. A cognitive-behavioral model of Internet gaming disorder: Theoretical underpinnings and clinical implications. *J. Psychiatr. Res.* **2014**, *58*, 7–11. [CrossRef]
60. Bitto Urbanova, L.; Holubcikova, J.; Madarasova Geckova, A.; Reijneveld, S.A.; van Dijk, J.P. Does Life Satisfaction Mediate the Association between Socioeconomic Status and Excessive Internet Use? *Int. J. Environ. Res. Public Health* **2019**, *16*, 3914. [CrossRef]
61. Vaala, S.E.; Bleakley, A. Monitoring, Mediating, and Modeling: Parental Influence on Adolescent Computer and Internet Use in the United States. *J. Child. Media* **2015**, *9*, 40–57. [CrossRef]

International Journal of
Environmental Research and Public Health

MDPI

Article

Relationships between Severity of Internet Gaming Disorder, Severity of Problematic Social Media Use, Sleep Quality and Psychological Distress

Hiu Yan Wong [1], Hoi Yi Mo [1], Marc N. Potenza [2,3,4], Mung Ni Monica Chan [1], Wai Man Lau [1], Tsz Kwan Chui [1], Amir H. Pakpour [5,6,*] and Chung-Ying Lin [1,*]

1 Department of Rehabilitation Sciences, The Hong Kong Polytechnic University, Hung Hom, Hong Kong; anniehiuyan@gmail.com (H.Y.W.); elainemohoiyi@gmail.com (H.Y.M.); monicachan0710@gmail.com (M.N.M.C.); sunny313971048@gmail.com (W.M.L.); chuitszkwan1996@gmail.com (T.K.C.)
2 Departments of Psychiatry and Neuroscience and the Child Study Center, School of Medicine, Yale University, New Haven, CT 06511, USA; marc.potenza@yale.edu
3 Connecticut Council on Problem Gambling, Wethersfield, CT 06109, USA
4 Connecticut Mental Health Center, New Haven, CT 06519, USA
5 Social Determinants of Health Research Center, Research Institute for Prevention of Non-Communicable Diseases, Qazvin University of Medical Sciences, Qazvin 3419759811, Iran
6 Department of Nursing, School of Health and Welfare, Jönköping University, 55511 Jönköping, Sweden
* Correspondence: pakpour_amir@yahoo.com (A.H.P.); cylin36933@gmail.com (C.-Y.L.); Tel.: +98-283-323-9259 (A.H.P.); +852-2766-6755 (C.-Y.L.)

Received: 31 December 2019; Accepted: 10 March 2020; Published: 13 March 2020

Abstract: Internet gaming and social media use are prevalent and integral to many people's lives. However, excessive engagement in either could lead to negative health impacts. This study aimed to investigate relationships between severities of internet gaming disorder (IGD) and problematic social media use (operationalized as social media addiction; SMA) with sleep quality and psychological distress among young adults. A cross-sectional study with snowball sampling was conducted among Hong Kong university students in 2019. All participants (n = 300; mean (SD) age = 20.89 (1.48); 122 males (40.67%)) responded to an online survey that included Chinese versions of the Internet Gaming Disorder Scale-Short Form (IGDS9-SF), Bergen Social Media Addiction Scale (BSMAS), Pittsburgh Sleep Quality Index (PSQI), and Depression Anxiety Stress Scales (DASS-21). Multiple linear regressions demonstrated that IGDS-SF9 scores demonstrated associations with psychological distress measures (standardized coefficient (β) = 0.295 for depression, 0.325 for anxiety, 0.339 for stress, all $p < 0.001$). BSMAS scores showed similar albeit numerically less robust associations ($\beta = 0.235$ for depression, $p < 0.001$; 0.219 for anxiety, $p = 0.001$; 0.262 for stress, $p < 0.001$). BSMAS scores demonstrated associations with poorer sleep quality ($\beta = 0.292$; $p < 0.001$) and IGDS9-SF scores ($\beta = 0.157$; $p = 0.024$) showed a significantly less robust association ($p = 0.01$ for comparing the two βs). These findings suggest that both severities of IGD and SMA associate with more psychological distress and poorer sleep quality, although the strengths of associations may differ.

Keywords: gaming; social media; behavior addiction; sleep quality; psychological distress

1. Introduction

Serious health concerns have been linked to specific types and patterns of engagement in internet usage, leading to the concept of "internet addiction" [1,2]. The availability, accessibility and affordability of internet usage worldwide may lead to worsening health problems. According to the International Telecommunication Union (ITU) [3], global internet usage increased from 1990 million in 2010 to 3385 million in 2016. In Hong Kong and Macau, it is estimated that 23.3% of young adults aged 18 to 30 years experience problems with online gaming [4]. Therefore, problematic gaming is a significant concern for young adults in these Asian jurisdictions, and findings suggest similar concerns in other (American, European) countries. In addition to online gaming, internet-based social media platforms (e.g., YouTube, Snapchat, and Instagram) are frequently used, with nearly three fourths (71%) of young adults aged 18–24 years in the US using these media platforms doing so multiple times per day [5]. As such, many young adults may be vulnerable to experiencing Internet gaming and/or social media addictions.

Definitions for addictions have changed over time. Recently, the American Society of Addiction Medicine defined addiction as, "a treatable, chronic medical disease involving complex interactions among brain circuits, genetics, the environment, and an individual's life experiences. People with addiction use substances or engage in behaviors that become compulsive and often continue despite harmful consequences" [6]. Indeed, *behavioral addictions* have been included in the latest edition of the *Diagnostic and Statistical Manual of Mental Disorders (DSM-5)*, with gambling disorder and internet gaming disorder (IGD) defined, with the latter as a provisional diagnostic entity [7]. Recently, *gaming disorder* has been added to the *International Classification of Diseases 11th Revision (ICD-11)* as a disorder due to addictive behaviors [8]. With poor control over repetitive, persistent, and dysfunctional behaviors, behavioral addictions can result in negative health consequences, including social relationship problems, impairments in functional roles and activities, sleep problems, and mental disorders [9]. Therefore, assisting people, especially young adults, in tackling behavioral addictions is important for healthcare providers.

Debates have focused on whether generalized internet addiction or specific internet addictions are diagnosable entities, and if the latter, which disorders constitute specific internet addictions [10,11]. In the current study, we focused on gaming and social media addictions given their potential impacts. Both social media addiction (SMA) and IGD have been associated with sleep disturbances, including sleep quality deterioration and reductions in sleep duration [12,13]. A potential reason for both SMA and IGD leading to sleep disturbances involves portable smartphones and other portable digital devices that may be brought to bed [13].

With portable smartphones and Wi-Fi access, internet (including of social media) may constitute an unstructured activity, one not limited in time and one that can displace time [14]. As such, SMA may lead to lack of sleep. Delayed bedtimes and late waking times may further lead to the rhythm desynchronization and impact school performance [15]. Other potential mechanisms in disrupting sleep may include psychological stimulation (i.e., excited mood due to use of social media) [16], light-emitting screens (i.e., light may suppress sleep-promoting hormones like melatonin that are typically elevated before bedtime) [17], and shortened sleep duration [18]. Similar considerations exist for online gaming. Gamers from different time zones often participate in multiplayers games [12]. Therefore, some gamers may avoid logging off or may arise in the middle of the night to continue gaming [19], leading to irregular or disorganized sleep-wake patterns. Individuals may become sleep-deprived in order to engage in virtual activities. Consistently, playing online games has been associated with longer sleep latencies and shorter total rapid eye movement (REM) sleep [20].

Apart from negative effects on sleep, SMA and IGD may impact psychological well-being, with associations with depression, anxiety, and stress [2,21]. "Fear of missing out" may keep individuals on social media sites and lead to stress and burnout [4,21]. Internet addiction, as a general concept, may lead to poor sleep quality and psychological problems [2,22–24]. Both SMA [25–29] and IGD [30–32] have been associated with poor psychological health. However, differential associations of IGD and SMA with health outcomes have been rarely examined and compared in the same studies. To our best knowledge, only one study has examined relationships between IGD/SMA and symptoms of psychiatric disorders (i.e., attention-deficit/hyperactivity disorder, obsessive- compulsive disorder, anxiety, and depression) [33] and another has examined relationships between IGD/SMA and psychological distress (i.e., anxiety, stress, and depression) [32]. Therefore, more research is needed. Specifically, the activity demands and behavioral patterns of the two addictions are different. As internet gaming and use of social media may have different activity demands and behavioral patterns, with gaming typically requiring more rapid reactions and sustained attention over a continuous time period than using social media use, IGD and SMA may differentially impact sleep quality and psychological distress.

The current preliminary study was designed to explore how severities of IGD and SMA relate to sleep quality and emotional distress in a population at risk of internet addiction (i.e., young adults) [34] in order to complement prior studies of primary and secondary school students [1,35]. We hypothesized that both severities of IGD and SMA would relate to poorer sleep quality and more emotional distress.

2. Materials and Methods

A cross-sectional study using convenience and snowball sampling was conducted among university students aged 18–24 in Hong Kong in 2019.

2.1. Participants and Procedure

Data were collected from April to June 2019 using *Google Form*. The inclusion criteria were: (1) currently studying as an undergraduate or a postgraduate in a university or college in Hong Kong; (2) able to understand written Chinese in traditional characters and spoken Chinese in Cantonese; (3) currently possessing a smartphone with Internet access. The exclusion criterion was having a self-reported diagnosis of psychiatric problems or neurological disease. Finally, 306 participants agreed to participate in this study and data from 300 were analyzed after removing data from those who were not eligible (the six participants were removed because they were not university students or not studying in Hong Kong).

An invitation message with an online link to the survey was sent to the existing social network of the present study's Hong Kong researchers. The recipients were invited to complete an online survey and spread the message to their university peers. The message explained the present study's objectives and the estimated time for completing the survey. Before survey commencement, a project information sheet explaining the study aims, survey structure, and confidentiality assurance was provided to the participants. Participants could enter the survey only after they provided an online consent form that expressed their willingness to participate. The survey included a background information sheet and the Chinese versions of the following questionnaires: the Internet Gaming Disorder Scale-Short Form (IGDS9-SF), Bergen Social Media Addiction Scale (BSMAS), Pittsburgh Sleep Quality Index (PSQI) and Depression Anxiety Stress Scales (DASS-21). The background information sheet included questions relating to age, gender, education level, height, weight, smoking history, monthly income, time spent on smartphone per day in the past month, time spent on internet gaming per day in the past month, and time spent on social media per day in the past month. Five university students completed the survey to help ensure the survey readability before recruiting study participants. The study was approved by the Human Subjects Ethics Subcommittee of The Hong Kong Polytechnic University (Ref No. HSEARS20190130006).

2.2. Instruments

2.2.1. Internet Gaming Disorder Scale- Short Form (IGDS9-SF)

The 9-item IGDS9-SF assesses the severity of Internet Gaming Disorder (IGD) and its detrimental effects over a 12-month period [36–42]. The nine questions correspond to the nine DSM-5 criteria for IGD [7]. The items are rated using a 5-point Likert scale from 1 (*Never*) to 5 (*Very often*). A total score can be obtained by summing responses with a range from 9 to 45 points, with higher scores reflecting greater IGD severity. The cut-off score over 21 indicates problematic internet gaming [43]. A sample question is, "Do you feel the need to spend increasing amount of time engaged gaming in order to achieve satisfaction or pleasure?" The instrument has demonstrated validity and reliability for assessing IGD [36–42]. IGDS9-SF scores have been correlated with weekly time spent gaming (r = 0.325, $p < 0.0001$) and IGD-20 Test scores (r = 0.842, $p < 0.0001$), providing evidence for validity [44]. The IGDS9-SF has demonstrated high internal consistency ($\alpha = 0.87$) and low floor or ceiling effects [44]. The IGDS9-SF has also been validated among Hong Kong participants [45,46]. The internal consistency of the IGDS9-SF in the present study was 0.91.

2.2.2. The Bergen Social Media Addiction Scale (BSMAS)

The BSMAS is a self-report scale for assessing SMA severity [33,47–49]. It contains six items relating to six core addiction components (salience, mood modification, tolerance, withdrawal, conflict, and relapse) [50]. Each item examines the experience of using social media over the last year and is rated on a five-point Likert scale ranging from 1 (*very rarely*) to 5 (*very often*), producing a composite score ranging from 6 to 30. A sample question from the BSMAS is, "How often during the last year have you used social media so much that it has had a negative impact on your job/studies?" Higher BSMAS scores indicate greater SMA severity. A cut-off score over 19 indicates problematic social media use [51]. The BSMAS has satisfactory psychometric properties with high internal consistency ($\alpha = 0.819$) and low floor and ceiling effects [34]. BSMAS scores correlated with IGDS9-SF scores and assessments of anxiety and depression [34]. The BSMAS has been previously validated among Hong Kong participants [45,46]. The internal consistency of the BSMAS in the present study was 0.83.

2.2.3. Depression Anxiety Stress Scales (DASS-21)

The DASS-21 includes three self-reported subscales to measure negative emotional states, including depression, anxiety, and stress [52]. Each scale consists of seven items [52]. By using a four-point severity/ frequency scale, each participant is asked to rate the extent to the experiences relate to each negative emotion. The rating scale for each item is comprised of 0 *(did not apply to me at all)*, 1 (applied *to me to some degree, or some of the time*), 2 *(applied to me to a considerable degree, or a good part of time*) and 3 *(applied to me very much, or most of the time*). A sample question from the DASS-21 depression subscale is, "I felt that I had nothing to look forward to." A sample question from the DASS-21 anxiety subscale is, "I was aware of dryness of my mouth." A sample question from the DASS-21 stress subscale is, "I found it hard to wind down." The scores for depression, anxiety, and stress are summed, respectively, to generate scores in each domain [52]. The high internal consistency (α = 0.94, 0.88, and 0.93 for depression, anxiety, and stress, respectively) of the DASS-21 reflects its reliability [53]. The DASS-21 has been validated among Hong Kong participants [52]. The internal consistency of the DASS-21 in the present study was 0.88 (Depression domain), 0.83 (Anxiety domain), and 0.85 (Stress domain).

2.2.4. The Pittsburgh Sleep Quality Index (PSQI)

The PSQI assesses sleep quality in clinical populations using nineteen questions that can be classified into seven domains that include subjective sleep quality (Q6), sleep latency (Q2 and Q5a), sleep duration (Q4), habitual sleep efficiency (Q1, Q3 and Q4), sleep disturbances (Q5b-Q5j), use of sleeping medication (Q7), and daytime dysfunction (Q8 and Q9) [54]. The items examine sleep quality and disturbances over a 1-month period. A sample question from the PSQI is, "How often have you had trouble sleeping because you have bad dreams?" Scores for each question range from 0 (*no difficulty*) to 3 (*severe difficulty*). The summed component scores generate a global score ranging from 0 to 21, where higher scores represent poorer sleep quality. A cutoff score of 5 for sleep impairment has been proposed for distinguishing good sleep and poor sleep [54]. PSQI scores have correlated with severity of Internet addiction [55]. The PQSI has demonstrated high internal consistency ($\alpha = 0.83$). Both global scores (T1-T2 correlation coefficient = 0.85, $p < 0.001$) and component scores such as sleep latency (T1-T2 correlation coefficient = 0.84, $p < 0.001$) show high test-retest reliability [54]. The PSQI has been suggested as a good screening test for insomnia in college students, with a recommended cut-off ≥ 6 due to its high diagnostic accuracy [56]. The Chinese PSQI has been validated among general and clinical populations in Hong Kong [57,58].

2.3. Statistical Analysis

Statistical analysis was performed using Statistical Package for the Social Sciences (SPSS) version 24.0 (IBM Inc., Armonk, NY, USA). Pearson's correlations were applied to examine the bivariate correlations between IGDS9-SF, BSMAS and PSQI scores and DASS-21 domains. Fisher's *r*-to-*z* transformation on dependent samples (http://quantpsy.org/corrtest/corrtest2.htm) was used to examine for significant differences in magnitudes of bivariate correlations. The associations of severities of IGD and SMA and health outcomes were further analyzed using regression models, which controlled for potential confounders. Specifically, DASS-21 domain scores (i.e., depression, anxiety, and stress domains) and PSQI score were entered as dependent variables; IGDS9-SF and BSMAS were used as independent variables; age, gender, time spent on smartphones, time spent on gaming on the internet, time spent on social media, and smoking history were treated as confounding variables. Further statistical testing on the differences of regression coefficients was conducted using a t-test statistic: dividing the difference between the two regression coefficients by the square root of their sum of the squared standard errors. All statistics were tested using two-tailed with significance level set at p-value < 0.05.

3. Results

Participants (N = 300) had a mean (SD) age of 20.89 (1.48) years. Less than half of the participants were male (40.67%), with small proportion of current smokers (0.36%). On average, participants spent 5.29 h (2.34) on smartphones, 3.09 h (1.80) on social networks and 1.19 h (1.53) gaming per day.

Participants had a mean (SD) score of 16.98 (6.48) on the IGDS9-SF, 15.37 (4.11) on the BSMAS, 5.16 (4.47) for Depression; 4.18 (3.83) for Anxiety; 6.72 (4.25) for Stress; and 6.63 (2.14) on the PSQI (Table 1). Additionally, all scores on the IGDS9-SF, BSMAS, DASS-21, and PSQI distributed normally or nearly normally as their skewness values were between 0.34 and 1.15 and kurtosis values were between −0.32 and 2.24. According to the cutoffs on the IGDS9-SF (score of 21 or above) and BSMAS (score of 19 or above), 77 (25.7%) of participants had IGD and 67 (22.3%) had SMA.

Table 1. Participant characteristics.

Characteristic	M ± SD or *n* (%)
Age (year)	20.89 ± 1.48
Gender (Male)	122 (40.67)
Height (cm)	165.57 ± 8.75
Weight (kg)	56.54 ± 10.24
Body mass index (kg/m^2)	20.52 ± 2.64
Smoke (No) [1]	279 (99.64)
Monthly income (HKD)	
≤ 1000	34 (12.27)
1001–2000	52 (18.77)
2001–3000	59 (21.30)
3001–5000	88 (31.77)
≥5001	44 (15.88)
Time on smartphones (hours/day)	5.29 ± 2.34
Time spent gaming on the internet (hours/day)	1.19 ± 1.53
Time spent on social media (hours/day)	3.09 ± 1.80
IGDS9-SF score (range: 9–45)	16.98 ± 6.45
BSMAS score (range: 6–30)	15.37 ± 4.11
Depression (range: 0–21)	5.16 ± 4.47
Anxiety (range: 0–21)	4.18 ± 3.83
Stress (range: 0–21)	6.72 ± 4.25
PSQI score (range: 0–21)	6.63 ± 2.14

[1] With 10 missing values. IGDS9-SF = Internet Gaming Disorder Scale-Short Form; BSMAS = Bergen Social Media Addiction Scale; DASS-21= Depression Anxiety Stress Scales; PSQI = Pittsburgh Sleep Quality Index; HKD = Hong Kong Dollar, 1 HKD ≈ 7.8 UDS. Depression, Anxiety, and Stress were subdomains of the DASS-21.

Table 2 shows that all the studied variables were significantly correlated ($p < 0.001$). More specifically, IGDS9-SF, BSMAS, depression, anxiety, stress, and PSQI scores were all positively correlated (r = 0.221 to 0.363). The magnitudes of correlations between DASS-21 subscales and IGDS9-SF (r = 0.331 to 0.363) were similar to those between DASS-21 subscales and BSMAS (r = 0.336 to 0.384). Indeed, the Fisher's r-to-z transformation on dependent samples showed non-significant findings (z = 0.227 to 0.807; p = 0.42 to 0.82). Although the correlation between BSMAS and PSQI scores (r = 0.351) showed a numerically larger magnitude than that between IGDS9-SF and PSQI (r = 0.249), the Fisher's r-to-z transformation on dependent samples showed no significant difference in the correlation comparison (z = 1.509; $p = 0.13$).

Table 2. Correlation matrix between IGDS9-SF, BSMAS, DASS-21, and PSQI scores.

	r					
	IGDS9-SF	**BSMAS**	**Depression**	**Anxiety**	**Stress**	**PSQI**
IGDS9-SF	–					
BSMAS	0.221	–				
Depression	0.351	0.336	–			
Anxiety	0.363	0.344	0.792	–		
Stress	0.331	0.384	0.792	0.815	–	
PSQI	0.249	0.351	0.351	0.411	0.386	–

IGDS9-SF = Internet Gaming Disorder Scale-Short Form; BSMAS = Bergen Social Media Addiction Scale; DASS-21 = Depression Anxiety Stress Scales; PSQI = Pittsburgh Sleep Quality Index. Depression, anxiety, and stress were subdomains of the DASS-21. All p-values < 0.001.

Multiple linear regressions in Table 3 (F-values ranged between 6.916 and 10.323; all $p < 0.001$) showed that both IGDS9-SF and BSMAS were associated with all outcome variables (i.e., depression, anxiety, stress, and sleep quality) after controlling for age, gender, time spent on smartphones, time spent on social media, and time spent gaming. As shown by the standardized coefficient (β), IGDS9-SF had a higher association with depression (β = 0.295), anxiety (β = 0.325), and stress (β = 0.339) as compared to BSMAS (β = 0.235 for depression; 0.219 for anxiety; 0.262 for stress) without significance (t = 0.116 to 0.577; p = 0.56 to 0.91). In contrast, BSMAS scores had a significantly higher association with poor sleep quality (β = 0.292) as compared to IGDS9-SF scores (β = 0.157; t = 2.436; $p = 0.01$).

Table 3. Regression models assessing associations between problematic use of Internet and health outcomes.

	Coefficient (SE)/Standardized Coefficient			
	Depression	**Anxiety**	**Stress**	**Sleep Quality**
IGDS9-SF score	0.203 (0.047)/0.295 ***	0.188 (0.039)/0.325 ***	0.220 (0.043)/0.339 ***	0.052 (0.023)/0.157 *
BSMAS score	0.250 (0.067)/0.235 ***	0.196 (0.057)/0.219 **	0.264 (0.063)/0.262 ***	0.150 (0.033)/0.292 ***
Age	−0.356 (0.167)/−0.121 *	−0.144 (0.140)/−0.058	−0.171 (0.154)/−0.061	−0.037 (0.082)/−0.026
Gender	0.251 (0.528)/0.028	0.375 (0.444)/0.050	1.030 (0.489)/0.121 *	0.069 (0.259)/0.016
Time on smartphone	0.215 (0.169)/0.102	0.094 (0.143)/0.053	0.125 (0.157)/0.062	−0.021 (0.083)/−0.020
Time gaming on the internet	−0.249 (0.216)/−0.085	−0.165 (0.182)/−0.067	−0.354 (0.200)/−0.128	0.052 (0.106)/0.037
Time on social media	0.039 (0.187)/0.016	0.117 (0.157)/0.057	0.043 (0.173)/0.018	0.121 (0.092)/0.103
Smoking status	1.731 (4.131)/0.023	2.42 (3.479)/0.039	6.44 (3.827)/0.092	−2.055 (2.027)/−0.058
Model summary				
F-value	8.501 ***	8.459 ***	10.323 ***	6.916 ***
R^2	0.202	0.201	0.235	0.171
Adjusted R^2	0.178	0.177	0.212	0.146

IGDS9-SF = Internet Gaming Disorder Scale- Short Form; BSMAS= Bergen Social Media Addiction Scale; PSQI = Depression, anxiety, and stress were subdomains of the Depression Anxiety Stress Scales; sleep quality was assessed using Pittsburgh Sleep Quality Index. * $p < 0.05$; ** $p < 0.01$; *** $p < 0.001$.

4. Discussion

This study found that participants with greater IGD and/or SMA severities had more depression, anxiety, and stress. Moreover, participants' sleep quality was poorer with greater severities of

IGD or SMA. According to our Pearson's correlation and multiple regression results, sleep quality demonstrated a numerically more robust association with SMA severity as compared to IGD severity. On the other hand, similar to findings from Pontes [32], our findings showed that psychological distress demonstrated a numerically more robust association with IGD severity, as compared SMA severity. Although no significant differences were found in the comparisons of Pearson's correlation coefficients, regression analyses revealed that some relationships were significantly different in the association magnitudes and some were not. Specifically, the sleep quality had a significantly stronger association with SMA severity ($\beta = 0.292$) than with IGD severity ($\beta = 0.157$); psychological distress had similar magnitudes of association with both SMA severity ($\beta = 0.219$ to 0.262) and IGD severity ($\beta = 0.295$ to 0.339). We believe that the results of comparing regression coefficients are more robust than those of comparing Pearson's correlation coefficients, because Pearson's correlation did not control confounders' effects, whereas regression analysis did. Therefore, we tentatively conclude that severities of IGD and SMA share similar associations with psychological distress; SMA severity had stronger association with sleep quality than did IGD severity.

Our sample had comparable or higher IGDS9-SF scores (Mean score = 16.98) than those from the general population in Italy (Mean score = 15.93; $n = 687$) [43]. Our sample also had comparable or higher BSMAS scores (Mean score = 15.37) than those from the general population in Italy (Mean score = 14.20; $n = 734$) [48]. When comparing scores to similar ethnic populations, our sample had similar IGDS9-SF scores relative to a sample comprised of young adults from Hong Kong and Taiwan (Mean score = 16.92; $n = 640$) [46] and comparable or slightly higher BSMAS scores (Mean score = 14.46; $n = 640$) [46].

Associations between IGD severity and psychological distress found in the present study are in line with previous studies [2,4,22]. Adolescents with anxiety, depression, or stress may spend excessive time on gaming on the internet as a coping mechanism, to escape from worries and difficulties in reality [22,59]. Unfortunately, gaming may not help people cope with psychological distress in a positive way. Spending more time gaming may provoke more emotional difficulties, and subsequently lead to IGD and related problems [59]. This cycle could further exacerbate symptoms of IGD and psychological distress [59]. Additionally, people with IGD have exhibited relatively less cognitive reappraisal and more expressive suppression, and these tendencies may prevent emotional difficulties from being resolved and the potential for depression to develop or worsen [59].

The present study also found a positive correlation between IGD severity and poor sleep quality, consistent with prior studies. Achab et al. suggested that people with IGD tend to have sleep deprivation or diurnal sleepiness [12], in line with findings that people who engage frequently in gaming are likely to delay their bedtimes and shorten their total time spent in bed [14]. Internet gaming may also arouse central and autonomic nervous systems, which may subsequently lead to increases in sleep latency [20]. Prolonged gaming may lead to physical discomfort, such as muscular pain and headache, which may impact sleep quality [13]. As a result, engaging more frequently in internet gaming may lead to poorer sleep.

The present study found a significantly positive association between SMA severity and psychological distress, in line with previous studies that people with SMA are prone to depressive and anxiety symptoms [4,21]. A sense of loss of control may underlie this phenomenon [9]. Individuals with more problematic smartphone use may experience more functional impairment, and this may interfere with their school/work and family life and eventually lead to increased stress [9]. Individuals with SMA may also be less physically active and have poorer sleep, which may predispose to psychological distress. Social networking provides a platform for persons to escape from negative emotions. However, excessive escapism may disrupt sleep and eventually induce more stress and depressive symptoms [21]. Negative social comparison may also contribute. Feelings of inferiority may be elicited if individuals engage in self-comparison with idealized representations often communicated on social media [4].

The present study found that SMA severity was positively correlated with sleep disturbance. This observation supports prior findings that suggest that time spent on social media is positively associated to sleep problems, such as insomnia, daytime sleepiness, and evening chronotype [13]. A phase delay in the circadian clock could be a reason for the association between SMA severity and sleep quality [13]. Using social media, especially before bed, may lead to difficulties falling asleep and bedtime delay [13]. The relationship between SMA severity and sleep quality may reflect cognitive and emotional arousal when using social media before sleep, with increased arousal preventing sleep [60]. Additionally, blue light from the screen may activate circadian clocks and generate delays in sleep latency [15].

Although both severities of both IGD and SMA were associated with poor sleep, SMA severity was more strongly related. Different strengths of associations could involve multiple possible mechanisms. First, the use of social media may constitute a more on-going process than internet gaming, which may have more well-defined beginnings and endings for each gaming task or component. The sleep patterns of users may be disrupted when they are preoccupied and waiting for replies from others on social media, which may not be well limited in time [14]. Being offline from interactions on social media may generate greater preoccupation and withdrawal when compared to gaming, potentially leading to greater disturbance in sleep quality [35]. Second, different levels and natures of interactions between social media and internet gaming may also explain differences in strengths of associations. Internet gaming may contain similar stimuli within games with main goals involving winning. In contrast, interactions on social media may be more complex, although this may be debatable given the complexities and varieties of games. Individuals may have a variety of goals and aims when using social media, involving multiple interpersonal relations which may require more mental or emotional effort that could lead to poorer sleep. However, as these factors were not assessed in the present study, further investigation is needed to test these presently speculative possibilities.

IGD severity showed associations with psychological distress that were comparable to those between SMA severity and psychological distress. Aspects of gaming and social media warrant consideration in relationships to psychological distress. Internet gaming often involves activities in multi-player chat rooms. Online gamers may thus form a team with individuals they have never physically contacted before when they engage in online games [61]. The teammates may share a goal in the game, and they may discuss their teamwork strategy to win the game. With the mindset of winning and within a virtual competitive environment, these individuals may become aggressive as they focus on winning [61]. Therefore, online gaming may not be an ideal platform for disclosing inner feelings, while social media may be used as a platform to express emotions [62]. Indeed, people with psychological distress can upload a new post on social media to share their current feelings or have an online discussion regarding their work or academic issues in *WhatsApp*. Therefore, the reasons may be different for the associations between IGD/SMA severity and psychological distress; however, both may share similar magnitudes of association with psychological distress. Additional research is needed to examine possible mechanisms linking IGD and SMA severities with psychological distress.

The present findings suggest that teaching faculty and healthcare providers in universities should attend to students' online gaming behaviors and social media use, especially as IGD and SMA may interact in mutually reinforcing ways [63]. Future research should investigate overlapping and unique factors relating to IGD and SMA. For example, interventions related to SMA may wish to focus on sleeping habits, although given the associations between SMA and IGD with sleep disturbances, the specificity may not be robust.

Study Limitations and Future Research

This study has several limitations. First, convenience and snowball sampling were used to collect data. Therefore, the representativeness of the collected sample is limited. Specifically, as participants were approached on the basis of availability and questionnaires were distributed to familiar people who may share similar characteristics, the potential bias of self-selection exists and it reduces the representativeness of sample. Thus, generalizability is reduced. Second, the sample size was small, especially as we recruited participants from the general population instead of clinical settings (i.e., those who with diagnosed IGD or SMA); therefore, the findings may not generalize to clinical samples. Third, all study variables (i.e., psychological distress, sleep, and behavioral addiction) were assessed using self-report instruments. Therefore, effects of social desirability or recall biases cannot be excluded. Fourth, the present study cannot make causal inferences given its cross-sectional design. Fifth, we did not explore gender-related differences. Given that males and females may have different propensities for IGD, SMA, and other domains assessed in the present study (e.g., psychological distress that is affective in nature), future studies should systematically examine these relationships in females and males. In such studies, examining motivations (e.g., fear of missing out for SMA) will be important [64]. Lastly, diverse findings in significance testing were observed when we compared the associations between sleep quality and severities of IGD and SMA. Pearson correlation did not support that sleep quality had different levels of associations with severities of IGD and SMA. However, the regression coefficient between sleep quality and SMA severity was significantly stronger than that between sleep quality and IGD severity. Although we believed that the findings from regression analyses are more robust than those from Pearson correlations, future studies are needed to investigate further the robustness of potential differences.

5. Conclusions

In conclusion, the present study assessed associations between severities of IGD and SMA and two health outcomes (i.e., psychological health and sleep quality) among Hong Kong university students. The results showed that both severities of IGD and SMA were positively associated with psychological distress and sleep disruption. In addition, although severities of IGD and SMA show similar associations with psychological distress, SMA severity was more strongly associated with poor sleep relative to IGD severity (Table 4). Thus, healthcare providers should consider both domains (psychological distress and sleep) when encountering IGD or SMA, with consideration of the potentially greater impact of SMA on sleep. More research is needed into specific interventions to improve health and sleep in individuals with features of IGD or SMA.

Table 4. Summary of associations between severities of internet gaming disorder (IGD)/social media addiction (SMA) and health measures.

	Depression	Anxiety	Stress	Sleep Quality
Severities of IGD	o	o	o	−
Severities of SMA	o	o	o	+
t (p) [a]	0.574 (0.57)	0.116 (0.91)	0.577 (0.56)	2.436 (0.01)

+ indicates a stronger association; − indicates a weaker association; o indicates a similar association. [a] T-tests that compared unstandardized coefficient of IGD severity on health outcomes and that of SMA severity on health outcomes.

Author Contributions: Conceptualization, C.-Y.L. and A.H.P.; methodology, C.-Y.L.; software, H.Y.W., H.Y.M., M.N.M.C., W.M.L., and T.K.C.; validation, H.Y.W., H.Y.M., M.N.M.C., W.M.L. and T.K.C.; formal analysis, C.-Y.L., H.Y.W., and H.Y.M.; investigation, C.-Y.L.; resources, C.-Y.L.; data curation, H.Y.W., H.Y.M., M.N.M.C., W.M.L., and T.K.C.; writing—original draft preparation, H.Y.W., H.Y.M., M.N.M.C., W.M.L. and T.K.C.; writing—review and editing, C.-Y.L., H.Y.W., M.N.P. and A.H.P.; visualization, M.N.M.C. and W.M.L.; supervision, C.-Y.L. and A.H.P.; project administration, C.-Y.L.; funding acquisition, C.-Y.L. All authors have read and agreed to the published version of the manuscript.

Funding: This research was funded by the internal funding of the Hong Kong Polytechnic University, and The APC was funded by the Hong Kong Polytechnic University.

Acknowledgments: We thank all the participants.

Conflicts of Interest: The authors declare no conflict of interest. Dr. Potenza has received financial support or compensation for the following: Dr. Potenza has consulted for and advised RiverMend Health, Opiant Pharmaceuticals, the Addiction Policy Forum and AXA; has received research support from the Mohegan Sun Casino and the National Center for Responsible Gaming; and has consulted for or advised law offices and gambling entities on issues related to addictive disorders and behaviors.

References

1. Van den Eijnden, R.J.; Meerkerk, G.J.; Vermulst, A.A.; Spijkerman, R.; Engels, R.C. Online communication, compulsive Internet use, and psychosocial well-being among adolescents: A longitudinal study. *Dev. Psychol.* **2008**, *44*, 655–665. [CrossRef] [PubMed]
2. Weinstein, A.; Dorani, D.; Elhadif, R.; Bukovza, Y.; Yarmulnik, A.; Dannon, P. Internet addiction is associated with social anxiety in young adults. *Ann. Clin. Psychiatry* **2015**, *27*, 4–9. [PubMed]
3. Internet Users by Region and Country, 2010–2016. Available online: https://www.itu.int/en/ITUD/Statistics/Pages/stat/Treemap.aspx (accessed on 20 December 2019).
4. Tang, C.S.K.; Wu, A.M.S.; Yan, E.C.W.; Ko, J.H.C.; Kwon, J.H.; Yogo, M.; Gan, Y.Q.; Koh, Y.Y.W. Relative risks of Internet-related addictions and mood disturbances among college students: A 7-country/region comparison. *Public Health* **2018**, *165*, 16–25. [CrossRef] [PubMed]
5. Smith, A.; Anderson, M. *Social Media Use in 2018: A Majority of Americans Use Facebook and YouTube, but Young Adults Are Especially Heavy Users of Snapchat and Instagram*; Pew Reserch Center: Washington, DC, USA, 2018.
6. American Society of Addiction Medicine (ASAM). Available online: https://www.asam.org/Quality-Science/publications/magazine/public-policy-statements/2019/10/21/short-definition-of-addiction (accessed on 10 March 2020).
7. American Psychiatric Association. *Diagnostic and Statistical Manual of Mental Disorders DSM-5®*, 5th ed.; American Psychiatric Association Publishing: Washington, DC, USA, 2013.
8. ICD-11 for Mortality and Morbidity Statistics—Gaming Disorder (6C51). Available online: https://icd.who.int/browse11/l-m/en#/http://id.who.int/icd/entity/1448597234 (accessed on 20 December 2019).
9. Alhassan, A.A.; Alqadhib, E.M.; Taha, N.W.; Alahmari, R.A.; Salam, M.; Almutairi, A.F. The relationship between addiction to smartphone usage and depression among adults: A cross sectional study. *BMC Psychiatry* **2018**, *18*, 148. [CrossRef] [PubMed]
10. Griffiths, M.D.; Pontes, H.M. Internet Addiction Disorder and Internet Gaming Disorder are Not the Same. *J. Addict. Res.* **2014**, *5*, e124. [CrossRef]
11. Griffiths, M.D.; Szabo, A. Is excessive online usage a function of medium or activity? An empirical pilot study. *J. Behav. Addict.* **2013**, *3*, 74–77. [CrossRef]
12. Achab, S.; Nicolier, M.; Mauny, F.; Monnin, J.; Trojak, B.; Vandel, P.; Sechter, D.; Gorwood, P.; Haffen, E. Massively multiplayer online role-playing games: Comparing characteristics of addict vs non-addict online recruited gamers in a French adult population. *BMC Psychiatry* **2011**, *11*, 144. [CrossRef]
13. Fossum, I.N.; Nordnes, L.T.; Storemark, S.S.; Bjorvatn, B.; Pallesen, S. The association between use of electronic media in bed before going to sleep and insomnia symptoms, daytime sleepiness, morningness, and chronotype. *Behav. Sleep Med.* **2014**, *12*, 343–357. [CrossRef]
14. Van den Bulck, J. Television viewing, computer game playing, and Internet use and self-reported time to bed and time out of bed in secondary-school children. *Sleep* **2004**, *27*, 101–104. [CrossRef]
15. Touitou, Y.; Touitou, D.; Reinberg, A. Disruption of adolescents' circadian clock: The vicious circle of media use, exposure to light at night, sleep loss and risk behaviors. *J. Physiol. Paris* **2016**, *110*, 467–479. [CrossRef]
16. Hale, L.; Kirschen, G.W.; LeBourgeois, M.K.; Gradisar, M.; Garrison, M.M.; Montgomery-Downs, H.; Kirschen, H.; McHale, S.M.; Chang, A.M.; Buxton, O.M. Youth Screen Media Habits and Sleep: Sleep-Friendly Screen Behavior Recommendations for Clinicians, Educators, and Parents. *Child Adolesc. Psychiatr. Clin. N. Am.* **2018**, *27*, 229–245. [CrossRef] [PubMed]
17. van der Lely, S.; Frey, S.; Garbazza, C.; Wirz-Justice, A.; Jenni, O.G.; Steiner, R.; Wolf, S.; Cajochen, C.; Bromundt, V.; Schmidt, C. Blue blocker glasses as a countermeasure for alerting effects of evening light-emitting diode screen exposure in male teenagers. *J. Adolesc. Health* **2015**, *56*, 113–119. [CrossRef] [PubMed]
18. Alimoradi, Z.; Lin, C.-Y.; Broström, A.; Bülow, P.H.; Bajalan, Z.; Griffiths, M.D.; Ohayon, M.M.; Pakpour, A.H. Internet addiction and sleep disorders: A systematic review and meta-analysis. *Sleep Med. Rev.* **2019**, *47*, 51–61. [CrossRef] [PubMed]
19. Rehbein, F.; Kliem, S.; Baier, D.; Mößle, T.; Petry, N.M. Prevalence of internet gaming disorder in German adolescents: Diagnostic contribution of the nine DSM-5 criteria in a state-wide representative sample. *Addiction* **2015**, *110*, 842–851. [CrossRef] [PubMed]

20. Higuchi, S.; Motohashi, Y.; Liu, Y.; Maeda, A. Effects of playing a computer game using a bright display on presleep physiological variables, sleep latency, slow wave sleep and REM sleep. *J. Sleep Res.* **2005**, *14*, 267–273. [CrossRef] [PubMed]
21. Elhai, J.D.; Dvorak, R.D.; Levine, J.C.; Hall, B.J. Problematic smartphone use: A conceptual overview and systematic review of relations with anxiety and depression psychopathology. *J. Affect. Disord.* **2017**, *207*, 251–259. [CrossRef]
22. Ho, R.C.; Zhang, M.W.; Tsang, T.Y.; Toh, A.H.; Pan, F.; Lu, Y.; Cheng, C.; Yip, P.S.; Lam, L.T.; Watanabe, H.; et al. The association between internet addiction and psychiatric co-morbidity: A meta-analysis. *BMC Psychiatry* **2014**, *14*, 183. [CrossRef]
23. Chan, W.S.; Fu, K.W.; Wong, P.W.; Yip, P.S. Internet addiction: Prevalence, discriminant validity and correlates among adolescents in Hong Kong. *Br. J. Psychiatry* **2010**, *196*, 486–492.
24. Huang, C. Internet use and psychological well-being: A meta-analysis. *Cyberpsychol. Behav. Soc. Netw.* **2010**, *13*, 241–249. [CrossRef]
25. Griffiths, M.D.; Kuss, D.J.; Demetrovics, Z. Behavioral addictions: Criteria, evidence and treatment. In *Social Networking Addiction: An Overview of Preliminary Findings*; Rosenberg, K., Feder, L., Eds.; Elsevier Inc.: New York, NY, USA, 2014; pp. 119–141.
26. Griffiths, M.D.; Kuss, D. Adolescent social media addiction (revisited). *Educ. Health* **2017**, *35*, 49–52.
27. Kuss, D.J.; Griffiths, M.D. Excessive online social networking: Can adolescents become addicted to Facebook? *Educ. Health* **2011**, *29*, 68–71.
28. Kuss, D.J.; Griffiths, M.D. Online social networking and addiction—A review of the psychological literature. *Int. J. Environ. Res. Public Health* **2011**, *8*, 3528–3552. [CrossRef] [PubMed]
29. Kuss, D.J.; Griffiths, M.D. Social networking sites and addiction: Ten lessons learned. *Int. J. Environ. Res. Public Health* **2017**, *14*, 311. [CrossRef] [PubMed]
30. Griffiths, M.D.; Kuss, D.J.; King, D.L. Video game addiction: Past, present and future. *Curr. Psychiatry Rev.* **2012**, *8*, 308–318. [CrossRef]
31. Kuss, D.J.; Griffiths, M.D. Internet gaming addiction: A systematic review of empirical research. *Int. J. Ment. Health Addict.* **2012**, *10*, 278–296. [CrossRef]
32. Pontes, H.M. Investigating the differential effects of social networking site addiction and Internet gaming disorder on psychological health. *J. Behav. Addict.* **2017**, *6*, 601–610. [CrossRef]
33. Andreassen, C.S.; Billieux, J.; Griffiths, M.D.; Kuss, D.J.; Demetrovics, Z.; Mazzoni, E.; Pallesen, S. The relationship between addictive use of social media and video games and symptoms of psychiatric disorders: A large-scale cross-sectional study. *Psychol. Addict. Behav.* **2016**, *30*, 252–262. [CrossRef]
34. Yam, C.W.; Pakpour, A.H.; Griffiths, M.D.; Yau, W.Y.; Lo, C.M.; Ng, J.M.T.; Lin, C.Y.; Leung, H. Psychometric testing of three Chinese online-related addictive behavior instruments among Hong Kong university students. *Psychiatr. Q.* **2019**, *90*, 117–128. [CrossRef]
35. Leung, L. Predicting Internet risks: A longitudinal panel study of gratifications-sought, Internet addiction symptoms, and social media use among children and adolescents. *Health Psychol. Behav. Med.* **2014**, *2*, 424–439. [CrossRef]
36. Arıcak, O.T.; Dinç, M.; Yay, M.; Griffiths, M.D. Adapting the Short Form of the Internet Gaming Disorder Scale into Turkish: Validity and reliability. *Addicta Turk. J. Addict.* **2019**, *6*, 15–22.
37. de Palo, V.; Monacis, L.; Sinatra, M.; Griffiths, M.D.; Pontes, H.M.; Petro, M.; Miceli, S. Measurement invariance of the nine-item Internet Gaming Disorder Scale across Albania, United States of America, United Kingdom, and Italy. *Int. J. Ment. Health Addict.* **2019**, *17*, 935–946. [CrossRef]
38. Evren, C.; Dalbudak, E.; Topcu, M.; Kutlu, N.; Evren, B.; Pontes, H.M. Psychometric validation of the Turkish nine-item Internet Gaming Disorder Scale-Short Form (IGDS9-SF). *Psychiatry Res.* **2018**, *265*, 349–354. [CrossRef] [PubMed]
39. Pontes, H.M.; Griffiths, M.D. Portuguese validation of the Internet Gaming Disorder Scale-Short-Form. *Cyberpsychol. Behav. Soc. Netw.* **2016**, *19*, 288–293. [CrossRef] [PubMed]
40. Pontes, H.M.; Stavropoulos, V.; Griffiths, M.D. Measurement Invariance of the Internet Gaming Disorder Scale–Short-Form (IGDS9-SF) between the United States of America, India and the United Kingdom. *Psychiatry Res.* **2017**, *257*, 472–478. [CrossRef] [PubMed]

41. chivinski, B.; Brzozowska-Woś, M.; Buchanan, E.M.; Griffiths, M.D.; Pontes, H.M. Psychometric assessment of the Internet Gaming Disorder diagnostic criteria: An Item Response Theory study. *Addict. Behav. Rep.* **2018**, *8*, 176–184. [CrossRef]
42. Wu, T.-Y.; Lin, C.-Y.; Årestedt, K.; Griffiths, M.D.; Broström, A.; Pakpour, A.H. Psychometric validation of the Persian nine-item Internet Gaming Disorder Scale—Short Form: Does gender and hours spent online gaming affect the interpretations of item descriptions? *J. Behav. Addict.* **2017**, *6*, 256–263. [CrossRef]
43. Monacis, L.; Palo, V.; Griffiths, M.D.; Sinatra, M. Validation of the Internet Gaming Disorder Scale-Short-Form (IGDS9-SF) in an Italian-speaking sample. *J. Behav. Addict.* **2016**, *5*, 683–690. [CrossRef]
44. Pontes, H.M.; Griffiths, M.D. Measuring DSM-5 Internet gaming disorder: Development and validation of a short psychometric scale. *Comput. Hum. Behav.* **2015**, *45*, 137–143. [CrossRef]
45. Leung, H.; Pakpour, A.H.; Strong, C.; Lin, Y.C.; Tsai, M.C.; Griffiths, M.D.; Lin, C.Y.; Chen, I.H. Measurement invariance across young adults from Hong Kong and Taiwan among three internet-related addiction scales: Bergen Social Media Addiction Scale (BSMAS), Smartphone Application-Based Addiction Scale (SABAS), and Internet Gaming Disorder Scale-Short Form (IGDS-SF9) (Study Part A). *Addict. Behav.* **2020**, *101*, 105969.
46. Chen, I.-H.; Strong, C.; Lin, Y.-C.; Tsai, M.-C.; Leung, H.; Lin, C.-Y.; Pakpour, A.H.; Griffiths, M.D. Time invariance of three ultra-brief internet-related instruments: Smartphone Application-Based Addiction Scale (SABAS), Bergen Social Media Addiction Scale (BSMAS), and the nine-item Internet Gaming Disorder Scale- Short Form (IGDS-SF9) (Study Part B). *Addict. Behav.* **2020**, *101*, 105960. [CrossRef]
47. Lin, C.-Y.; Broström, A.; Nilsen, P.; Griffiths, M.D.; Pakpour, A.H. Psychometric validation of the Persian Bergen Social Media Addiction Scale using classic test theory and Rasch models. *J. Behav. Addict.* **2017**, *6*, 620–629. [CrossRef] [PubMed]
48. Monacis, L.; De Palo, V.; Griffiths, M.D.; Sinatra, M. Social networking addiction, attachment style, and validation of the Italian version of the Bergen Social Media Addiction Scale. *J. Behav. Addict.* **2017**, *6*, 178–186. [CrossRef] [PubMed]
49. Pontes, H.M.; Andreassen, C.S.; Griffiths, M.D. Portuguese validation of the Bergen Facebook Addiction Scale: An empirical study. *Int. J. Ment. Health Addict.* **2016**, *14*, 1062–1073. [CrossRef]
50. Griffiths, M. A 'components' model of addiction within a biopsychosocial framework. *J. Subst. Use* **2005**, *10*, 191–197. [CrossRef]
51. Bányai, F.; Zsila, Á.; Király, O.; Maraz, A.; Elekes, Z.; Griffiths, M.D.; Andreassen, C.S.; Demetrovics, Z. Problematic social media use: Results from a large-scale nationally representative adolescent sample. *PLoS ONE* **2017**, *12*, e0169839. [CrossRef] [PubMed]
52. Taouk, M.; Lovibond, P.F.; Laube, R. *Report for New South Wales Transcultural Mental Health Centre*; Cumberland Hospital: Sydney, Australia, 2001.
53. Nieuwenhuijsen, K.; De Boer, A.; Verbeek, J.; Blonk, R.; Van Dijk, F. The Depression Anxiety Stress Scales (DASS): Detecting anxiety disorder and depression in employees absent from work because of mental health problems. *Occup. Environ. Med.* **2003**, *60*, I77–I82. [CrossRef]
54. Buysse, D.J.; Reynolds, C.F., III; Monk, T.H.; Berman, S.R.; Kupfer, D.J. The Pittsburgh Sleep Quality Index: A new instrument for psychiatric practice and research. *Psychiatry Res.* **1989**, *28*, 193–213. [CrossRef]
55. Khayat, M.A.; Qari, M.H.; Almutairi, B.S.; Rambo, M.Z.; Alrogi, M.J.; Alkhattabi, S.Z.; Alqarni, D.A. Sleep Quality and Internet Addiction Level among University Students. *Egypt J. Hosp. Med.* **2018**, *73*, 7042–7047.
56. Dietch, J.R.; Taylor, D.J.; Sethi, K.; Kelly, K.; Bramoweth, A.D.; Roane, B.M. Psychometric evaluation of the PSQI in US college students. *J. Clin. Sleep Med.* **2016**, *12*, 1121–1129. [CrossRef]
57. Ho, R.T.; Fong, T.C. Factor structure of the Chinese version of the Pittsburgh Sleep Quality Index in breast cancer patients. *Sleep Med.* **2014**, *15*, 565–569. [CrossRef]
58. Wong, W.S.; Fielding, R. Prevalence of insomnia among Chinese adults in Hong Kong: A population-based study. *J. Sleep Res.* **2011**, *20*, 117–126. [CrossRef] [PubMed]
59. Yen, J.; Lin, H.; Chou, W.; Liu, T.; Ko, C. Associations Among Resilience, Stress, Depression, and Internet Gaming Disorder in Young Adults. *Int. J. Environ. Res. Public Health* **2019**, *16*, 318. [CrossRef]
60. Levenson, J.C.; Shensa, A.; Sidani, J.E.; Colditz, K.B.; Primack, B.A. The association between social media use and sleep disturbance among young adults. *Prev. Med.* **2016**, *85*, 36–41. [CrossRef] [PubMed]

61. Ko, C.H.; Yen, J.Y.; Liu, S.C.; Hunag, C.F.; Yen, C.F. The associations between aggressive behaviours and Internet addiction and online activities in adolescents. *J. Adolesc. Health* **2009**, *44*, 598–605. [CrossRef] [PubMed]
62. Steinkuehler, C.A.; Williams, D. Where Everybody Knows Your (Screen) Name: Online Games as "Third Places". *J. Comput. Mediat. Commun.* **2006**, *11*, 885–909. [CrossRef]
63. Ostovar, S.; Allahyar, N.; Aminpoor, H.; Moafian, F.; Mariani, M.N.; Griffiths, M.D. Internet addiction and its psychosocial risks among Iranian adolescents and young adults: A structural equation model in a cross-sectional study. *Int. J. Ment. Health Addict.* **2016**, *14*, 257–267. [CrossRef]
64. Wegmann, E.; Oberst, U.; Stodt, B.; Brand, M. Online-specific fear of missing out and Internet-use expectancies contribute to symptoms of Internet-communication disorder. *Addict. Behav. Rep.* **2017**, *5*, 33–42. [CrossRef]

International Journal of
Environmental Research and Public Health

MDPI

Article

Pathological Traits Associated to Facebook and Twitter among French Users

Élodie Verseillié, Stéphanie Laconi * and Henri Chabrol

Centre d'Etudes et de Recherches en Psychopathologie et Psychologie de la Santé (CERPPS) EA7411, Université de Toulouse, 5, Allées Antonio Machado, 31058 Toulouse Cedex, France; elodie.verseillie@univ-tlse2.fr (É.V.); henri.chabrol@univ-tlse2.fr (H.C.)

* Correspondence: stephanielaconi.b@gmail.com

Received: 23 February 2020; Accepted: 24 March 2020; Published: 26 March 2020

Abstract: Background: With a growing number of users, social networking sites have been the subject of numerous recent studies, but little investigation has been given to their problematic use. Objectives: Our main objective was to study the relationship between psychopathological variables (i.e., personality traits, depressive and anxiety symptoms, and stress) and problematic Facebook and Twitter use. Participants and method: A sample of 1068 Internet users (Mage = 26.64; SD = 9.5) has been recruited online. Participants completed scales exploring problematic Facebook and Twitter use, and psychopathological variables. Results: Problematic Facebook and Twitter use were predicted by different pathological personality traits, regrouped in clusters in our study. Depressive and anxiety symptoms were also predictive of problematic Facebook and Twitter use but only stress explained problematic Facebook use. Gender differences have been observed. Discussion: This study highlights the relationship between depression, anxiety, stress, pathological personality traits, and problematic Facebook and Twitter use. Significant differences have been retrieved between these two uses and their relationship to psychopathology. Future research should also explore the causal relationship between social networking sites use and psychopathology and consider gender.

Keywords: problematic use; addiction; Twitter; Facebook; psychopathology; personality

---

1. Introduction

Facebook and Twitter are among the most popular Social Networking Sites (SNS) worldwide [1]. All SNS facilitate social interactions, even if they do not have the same finalities. Twitter is more about sharing information and opinions [2] than about developing social interactions like Facebook [3]. Besides, Twitter restores anonymity sought initially through SNS [3] and seems more focused on what users say than on what they are or look like.

Several researches highlight the existence of addictive-like symptoms related to SNS use [4,5]. In recent years, the use of some SNS has been described as pathological, problematic, or even addictive. In this paper, the term of problematic use will be preferred because of the lack of consensual definition and designation [6]. Moreover, this term allows us to avoid the unreasonable pathologization of every SNS use. Problematic Facebook Use (PFU) has been defined as a "problematic behavior that is likely to be characterized by addictive-like symptoms and/or self-regulation difficulties leading to negative consequences in personal and social life" [5] (p. 263). Given the risks associated to SNS use, a growing interest has been put on PFU [7,8] while few studies focused on Problematic Twitter Use (PTU) [9–11]. Some authors [10] found patterns of addiction in PTU through withdrawal symptoms, salience, and desire to control the use.

As for Problematic Internet Use (PIU) or Internet addiction, problematic SNS uses seem to be more frequent among adolescents and young adults. However, a prevalence of these problematic use is complicated to establish given methodological differences which are partially related to the lack of standardized definitions and assessment tools [12]. In occidental Europe, prevalence estimates of PFU are approximately ranged from 4.5% to 10% [13–15]. Regarding Twitter use, one French study revealed that 3.1% of participants (n = 822; Mage = 21.6; SD = 2.8; Sex Ratio = 0.59) have PTU [14].

Few studies focused on the relationship between PFU and psychological disorders. A French researcher [16] showed that among French Internet users with a PFU, 69% had anxiety symptoms and 48% had depressive symptoms. Depression has been frequently linked to PFU [16–19]. As for anxiety, mood disorders would have a causal relationship with PIU, as risk factors and consequences [20]. Besides, some studies highlighted the significant relationships between several pathological personality traits and PIU [21–25], yet no study focused on their relationships with SNS addictions, except for particular pathological traits. For example, borderline personality traits have been shown to be a risk factor for PFU [17], such as narcissistic personality traits, which also appeared to be related to Facebook and Twitter use [26,27].

In order to complete a lack of empirical data on PFU and PTU, particularly on French samples, our main objective was to explore the relationship between psychopathological variables (i.e., personality traits, depressive and anxiety symptoms, and stress) and problematic Facebook and Twitter use. Another aim of this research was to look at gender differences.

2. Materials and Methods

2.1. Participants

The sample consisted of 1068 participants aged from 18 to 64 years old (M = 26.64; SD = 9.5). There were 37% (n = 398) of men for 63% (n = 670) of women. The majority of participants were students (56%; n = 598). More detailed results are presented in Table 1.

Table 1. General characteristics of the study subjects.

Samples	Total		Facebook Users (n = 679)		Twitter Users (n = 389)	
Variables	n	%	n	%	n	%
Gender						
Women	670	62.7	462	68	208	53.5
Men	398	37.3	217	32	181	46.5
Working status						
Students	598	56	383	56.4	215	55.3
Workers	397	37.2	255	37.6	142	36.5
Unemployed	73	6.8	41	6	32	8.2
Marital status						
Single	608	57	381	56	227	58.4
In relationship	460	43	298	44	162	41.6
Academic level						
School certificate	47	4.4	27	4	20	5.1
High school degree	164	15.4	92	13.5	72	18.5
Bachelor degree	496	46.4	327	48.2	169	43.4
Master	309	28.9	198	29.2	111	28.5
Doctoral degree	52	4.9	35	5.2	17	4.4
Number of visits per day						
Less than once a day	-	-	34	5	27	6.9
Between 1 and 5 times a day	-	-	289	42.6	148	38
Between 6 and 10 times a day	-	-	164	24.2	106	27.2
More than 10 times a day	-	-	192	28.3	108	27.8
Time per connection						
Less than 10 min	-	-	329	48.5	218	56
Between 10 and 30 min	-	-	236	34.8	125	32.1
Between 31 and 60 min	-	-	58	8.5	22	5.7
More than an hour	-	-	56	8.2	24	6.2

Table 1. *Cont.*

Samples	Total		Facebook Users (n = 679)		Twitter Users (n = 389)	
Hours per day						
One hour or less	-	-	346	51	216	55.5
Two hours	-	-	180	26.5	93	23.9
Three hours	-	-	68	10	44	11.3
Four hours	-	-	46	6.8	17	4.4
Five hours or more	-	-	39	5.6	19	5

We spread our survey online with Survey Monkey through Facebook groups. These groups were randomly selected and covered a variety of topics (education, online games, sports, etc.) to reach the widest possible range of French users. For Twitter, the link of our survey has been shared with the most popular hashtags at different time periods. Participants were informed of the anonymity and confidentiality of their responses and gave their free and informed consent. Our study follows the ethical principles of the Declaration of Helsinki. Inclusion criteria were being at least 18 years old, understanding and speaking French, and having an account on Facebook and/or Twitter.

2.2. Measures

The Bergen Facebook Addiction Scale (BFAS) [28] has been used to assess PFU. Given there was no French validation, we used a previous translated version [14]. The BFAS had six items rated on a 5-point scale ranged from 1 "Very rarely" to 5 "Very often" (e.g., How often have you used Facebook in order to forget about personal problems?). Scores varied from 5 to 30, and high scores suggest a plausible PFU. The internal consistency was $\alpha = 0.83$ [28], $\alpha = 0.81$ [14], and $\alpha = 0.78$ in our study.

Given the lack of scale to measure PTU, we chose to use an adapted version of the BFAS [28] by replacing the term "Facebook" by "Twitter" (BFAS-T), similar to a previous study [14]. The Cronbach coefficient for the BFAS-T was $\alpha = 0.90$ [14]. In this study, it was $\alpha = 0.87$.

The Depression Anxiety and Stress Scale (DASS-12) [29] has been used to assess depressive and anxiety symptoms, and stress. This scale is composed of 12 items rated on a 4-point scale from 0 "Rarely or never" to 3 "Most of the time or always". Scores are ranged from 0 to 36. Cut off scores suggesting a severe symptomatology were respectively of 4, 4, and 6. Cronbach' alphas were ranged between $\alpha = 0.72$ and $\alpha = 0.76$ [29], and between $\alpha = 0.67$ and $\alpha = 0.78$ in the present study.

The Personality Diagnostic Questionnaire 4+ (PDQ4+) [30,31] measures the ten pathological personality traits of the DSM-IV with 99 items on a dichotomous scale. For this study, we used a Likert scale from 1 "Not typical to me" to 5 "very typical to me" as in previous studies [32,33]. Cronbach' alphas for each subscale were ranged between $\alpha = 0.47$ and $\alpha = 0.83$ [32,33]. In our study, it varied from $\alpha = 0.38$ to $\alpha = 0.94$. Given this, we decided to use clusters of pathological personalities as classified in the DSM for statistical analyses. According to the DSM, Cluster A is composed of eccentric personalities (paranoid, schizoid, schizotypal), Cluster B regroups dramatic personalities (antisocial, borderline, histrionic, narcissistic), and Cluster C anxious personalities (avoidant, dependent, obsessive-compulsive). Cronbach' alphas were $\alpha = 0.85$ for Cluster A and $\alpha = 0.87$ for Clusters B and C [28]. In our study, it was $\alpha = 0.78$ for Cluster A, $\alpha = 0.79$ for Cluster B, and $\alpha = 0.84$ for Cluster C.

2.3. Statistical Analyses

Descriptive analyses were conducted in the entire sample, and among Facebook and Twitter users. Pearson correlation analyses have been performed to assess the relationship between Facebook and Twitter use (including problematic use) and psychopathological variables (i.e., personality traits, depression, anxiety, stress). Multiple linear regressions helped to assess the influence of these last on Facebook and Twitter use. These analyses have also been performed separately for men and women.

3. Results

Among participants, 679 (64%) had a Facebook account only and 389 (36%) also had a Twitter account. No participant had only a Twitter account. As show in Table 1, the majority of participants (41%; n = 437) go to Facebook between one and five time a day and do not stay connected for long: Less than 10 min by connection (51%; n = 547). For Twitter, people connect once a day (50%; n = 193) and stay connected also less than 10 min by connection (67%; n = 262).

Results of correlation analyses between independent variables (depression, anxiety, stress, and cluster A, B, and C of personality) are presented in Table 2 for Facebook users only and in Table 3 for Facebook and Twitter users. There is a poor but significant correlation between dependent variables PFU and PTU (r = 0.11; $p < 0.05$).

Table 2. Bivariate Pearson correlations between variables among users with an account on Facebook (n = 679).

Variables	1	2	3	4	5	6
Depression (1)	-	0.61	0.51	0.40	0.34	0.50
Anxiety (2)		-	0.59	0.36	0.32	0.47
Stress (3)			-	0.29	0.35	0.39
Cluster A (4)				-	0.59	0.60
Cluster B (5)					-	0.53
Cluster C (6)						-

Note. All correlations were significant at $p < 0.01$.

Table 3. Bivariate Pearson correlations between variables among users of an account on Facebook and on Twitter (n = 389).

Variables	1	2	3	4	5	6
Depression (1)	-	0.63 **	0.56 **	0.54 **	0.41 **	0.67 **
Anxiety (2)		-	0.65 **	0.43 **	0.43 **	0.59 **
Stress (3)			-	0.43 **	0.46 **	0.51 **
Cluster A (4)				-	0.60 **	0.69 **
Cluster B (5)					-	0.57 **
Cluster C (6)						-

Note. * = $p < 0.05$; ** = $p < 0.01$.

Regression model assumptions were checked by examining the plot of predicted values of the dependent variable against residuals. It showed that assumptions were sufficiently met. Thus, among Facebook users only (n = 679), PFU has been explained by depressive symptoms ($\beta = 0.14$; $p < 0.01$), anxiety ($\beta = 0.12$; $p < 0.05$) and stress ($\beta = 0.14$; $p < 0.01$). Among Twitter users (n = 389), PTU has been explained by depressive symptoms ($\beta = 0.17$; $p < 0.01$) and anxiety ($\beta = 0.19$; $p < 0.01$).

For personality, PFU has been explained by cluster B personality traits ($\beta = 0.17$; $p < 0.001$) and cluster C personality traits ($\beta = 0.26$; $p < 0.001$) contrary to Twitter, which has been explained by cluster A personality traits ($\beta = 0.16$; $p < 0.05$).

Correlation performed separately for men and women revealed few significant differences (see Table 4). Among women there was no significant correlation between PFU and PTU (r = 0.05; $p = 0.50$) compared to men (r = 0.22; $p < 0.01$).

Regression analysis performed separately for men and women revealed significant differences for psychopathological variables. Among women, PFU was explained by depressive symptoms ($\beta = 0.15$; $p < 0.05$) whereas PTU was explained by anxiety ($\beta = 0.24$; $p < 0.05$). However, among men, no significant result has been retrieved for PFU and PTU, and psychopathological variables. Regarding personality, PFU was explained by cluster B and cluster C personality traits among women ($\beta = 0.11$; $p < 0.05$ and $\beta = 0.15$; $p < 0.05$) and men ($\beta = 0.22$; $p < 0.01$ and $\beta = 0.21$; $p < 0.05$), whereas PTU has been explained by cluster A personality traits (β = 0.23; $p < 0.05$) for women. PTU have not been explained by personality variables among men.

Table 4. Bivariate Pearson correlations between variables among users of an account on Facebook and on Twitter (n = 389) regarding gender.

Variables	1	2	3	4	5	6
Depression (1)	-	0.67 **	0.56 **	0.55 **	0.51 **	0.70 **
Anxiety (2)	0.58 **	-	0.64 **	0.46 **	0.48 **	0.60 **
Stress (3)	0.55 **	0.66*	-	0.39 **	0.47 **	0.48 **
Cluster A (4)	0.52 **	0.39 **	0.47 **	-	0.65 **	0.73 **
Cluster B (5)	0.32 **	0.40 **	0.47 **	0.55 **	-	0.65 **
Cluster C (6)	0.64 **	0.55 **	0.52 **	0.64 **	0.52 **	-

Note. Grey = Men; * = $p < 0.05$; ** = $p < 0.01$.

4. Discussion

4.1. Relationships with Depressive and Anxiety Symptoms

Our results are similar to those retrieved in previous studies in PFU regarding the relationship with depressive and anxiety symptoms [16–19,34,35] and PIU [36,37]. These results are not surprising and it is still necessary to fill the lack of studies on the causal relationship between these variables. The cognitive and behavioral model of problematic Internet use [20] suggests that depressive or socially anxious individuals are more prone to use Internet in order to relieve negative feelings. Young adults also experience a lot of stress in their lives due to their academic responsibilities, uncertainty about professional life or financial difficulty, which can lead to anxiety and depressive symptoms increasing the risk of PFU [34]. Using SNS in particular could induce more depressive and anxiety symptoms [38].

As expected, PTU was also significantly correlated to psychopathology in our study, but with a lower coefficient with stress than with PFU, among men and women. This should be explored and confirmed in other studies.

4.2. Relationships with Pathological Personality Traits

Our results showed that pathological traits of Clusters B and C predicted PFU, with no significant differences among gender, as it has already been shown [34,39]. Previous works highlighted the relationship between PFU and Cluster B symptomatology such as borderline [35,40] and narcissism traits [41–43]. This is similar with results found with PIU for Cluster B [21,23,25,28] and Cluster C [21,25]. Borderline and antisocial traits have been frequently associated to addictive disorders, with impulsivity in common, which could be a supplementary evidence of the high similarity between PFU and other addictive disorders [44]. Besides, we can assume that borderline personalities spend a lot of time on Facebook, especially adding/searching new friends to avoid true or imagined abandonment, while narcissistic and histrionic personalities used their time on posting content hoping to receive the expected attention and compliments. Antisocial personalities might use SNS to fulfill a lack of harmonious relationships in real life with less close individuals, but also to find a satisfaction by watching others' profiles. Cluster C traits, such as the social anxiety and the fear of being criticized or rejected in social situations, may lead to a preference for SNS with a strong control on one's public image. The relationship between PFU and social anxiety has already been shown [45]. PFU has an effect on social anxiety and depression that decrease the level of life satisfaction [46]. Other studies showed links between PFU and obsessive-compulsive traits [38,40]. Users seems to experience craving—the urge desire to check Facebook (assimilated to compulsions), which can be explain by the fear of missing out [38]. We can suppose that dependent personality spent a lot of time on Facebook with an important Facebook involvement to maintain social relationships because of their fear of the separation and their need to be reassured. Avoidant personality traits can be easily related to PFU; avoidant individuals might feel more prone to look for online communication rather than face-to-face interaction [47]. However, similarly to what we have found, other studies found a significant relationship between PIU and Clusters B and C personalities [25,47]. Facebook appears as more similar to Internet than Twitter, which can explain some similarities.

Contrarily to PFU, PTU was predicted by Cluster A traits. This is not surprising given that Twitter restores anonymity, compared to Facebook [48]. People with Cluster A' traits are more detached from social relationships and intimacy, and present more social anxiety and suspiciousness. Anonymity appears as initially looked for in SNS use [3] but the reduction of the social pressure and paranoid fears induced by anonymity can be the reason why Twitter use is preferred by individuals with Cluster A personality traits. Future researches should continue to study these relationships to try to confirm these results.

4.3. Gender Differences

An interesting result is the lack of significant relationship between PTU and PFU among women, which highlights the specificity of these problematic SNS use as those found in gender. It could be interesting to explore if women are indeed more prone to have PFU or PTU than both as men. This result should also be interpreted given the association between other variables, such as personality traits.

Correlation analyses revealed small differences between men and women. Among men, the correlation between Cluster B and PTU was lower than among women. PFU was also more highly related to Cluster B and C in women than men. This could suggest that problematic SNS use, in women, was related to a higher level of pathological personality traits. This is confirmed by the results of regression where no psychopathological variables explained PFU or PTU among men. Their use of SNS seems to be less related to psychopathology than for women.

Our regression analyses suggested that the relationship between PTU and cluster A is only significant among women, and that among them, PTU was explained by anxiety (and PFU by depressive symptoms). Cluster A personalities experienced more stress than other personality and it is known that women are also more at risk for anxiety symptoms [49]. Therefore, this could explain the significant relationship between PTU and cluster A, particularly among women.

4.4. Implications

Even if SNS received increasing attention worldwide, the lack of data on problematic SNS use or their relationship with psychopathological variables makes this research particularly relevant. Given the increasing number of SNS users, it seems necessary to better understand SNS use and the characteristics of their users. Prevention campaign adapted for specific audience (different for Twitter and Facebook users, and probably for users of other SNS) as early detection would have a positive impact on the health of youth and could prevent people to develop a problematic use but also depressive and anxiety symptoms, as a constant pathological personality even if there is no study on the causality of these relationships. This could improve the care of these specific problematic users, with a different orientation and different therapeutic objectives according to the SNS use, its characteristics, and the presence of psychopathological symptoms and traits.

4.5. Limits

Some methodological bias might have an influence on our results, such as sampling method and online recruitment. However, it seems that the use of Internet recruitment compared to the paper-and-pencil methodology has little influence on the consistency of the results [50]. Given the significant number of Facebook users, it was impossible to create a sample with only Twitter users which can make it difficult to read our results and can have an influence on them. The BFAS, which evaluates the PFU, does not have a cut off score. Thus, we have been unable to distinguish those with PFU and/or PTU from those without which did not give us the opportunity to make more precise statistics. Therefore, our assessment tools seem to be a limitation for the interpretation of our results, especially in the absence of diagnostic value. Besides, there are no consensual diagnostic criteria or definition of problematic SNS use, therefore our study lacks of a validated and diagnostic assessment tool to confirm the diagnosis of PFU and PTU such as in any study focused on this topic. Consistent in most studies, the criteria used to assess problematic SNS use are based on those of PIU even if many differences have been observed between specific and generalized Internet use [33]. This lack of specific evaluation increases the need for a unified definition and well-established and validated tools in the field of problematic SNS use. Besides, the lack of similar studies impedes our interpretation.

5. Conclusions

Our objectives were to explore the relationship between PFU, PTU, and psychopathological variables. This study confirmed the relationship between PFU and psychopathological variables but also brings new data about PTU. More importantly, pathological personality traits of cluster A were related to Twitter use and PTU whereas those of cluster B and C were related to Facebook use and PFU. In light of a lack of data in this area, another aim of this research was to look at gender differences in the context of PFU and PTU. Depressive and anxiety symptoms were related to PFU, whereas stress was related to PTU in both gender but with a lower intensity. Women have more anxiety symptoms and PTU than men. As a result, women seem now at a higher risk of developing a problematic SNS use. This research is one of the first to explore problematic Facebook and Twitter use and pathological personality traits. It would deserve to be replicated and deepened.

Author Contributions: Conceptualization, É.V. and S.L.; methodology, É.V. and S.L.; formal analysis, É.V. and S.L.; investigation, É.V.; * writing—original draft preparation, É.V.; writing—review and editing, É.V., S.L. and H.C.; supervision, H.C. All authors have read and agreed to the published version of the manuscript.

Funding: This research received no external funding.

Conflicts of Interest: The authors declare no conflict of interest.

References

1. Blog du Modérateur. Available online: http://www.blogdumoderateur.com/50-chiffres-medias-sociaux-2020/ (accessed on 23 February 2020).
2. Kwak, H.; Lee, C.; Park, H.; Moon, S. What is Twitter, a social network or a news media? In Proceedings of the 19th International Conference on World Wide Web, Raleigh, NC, USA, 26-30 April 2010.
3. Huberman, B.A.; Romero, D.M.; Wu, F. Social networks that matter: Twitter under the micro-scope. *First Monday* **2009**, *14*, 1–9.
4. Kuss, D.J.; Griffiths, M.D. Online social networking and addiction—A review of the psychological literature. *Int. J. Environ. Res. Public Health* **2011**, *8*, 3528–3552. [CrossRef] [PubMed]
5. Marino, C.; Gini, G.; Vieno, A.; Spada, M.M. A comprehensive meta-analysis on Problematic Facebook Use. *Comput. Hum. Behav.* **2018**, *83*, 262–277. [CrossRef]
6. Caci, B.; Cardaci, M.; Scrima, F.; Tabacchi, M.E. The Dimensions of Facebook Addiction as Measured by Facebook Addiction Italian Questionnaire and Their Relationships with Individual Differences. *Cyberpsychol. Behav. Soc. Netw.* **2017**, *20*, 251–258. [CrossRef]
7. Marino, C.; Finos, L.; Vieno, A.; Lenzi, M.; Spada, M.M. Objective Facebook behaviour: Differences between problematic and non-problematic users. *Comput. Hum. Behav.* **2017**, *73*, 541–546. [CrossRef]
8. Moretta, T.; Buodo, G. Modeling Problematic Facebook Use: Highlighting the role of mood regulation and preference for online social interaction. *Addict. Behav.* **2018**, *87*, 214–221. [CrossRef]
9. Kircaburun, K. Self-Esteem, Daily Internet Use and Social Media Addiction as Predictors of Depression among Turkish Adolescents. *J. Educ. Pract.* **2016**, *7*, 64–72.
10. Miles, P.C.; Zhang, L.G. China turns to Tweeting: Exploring the problematic use of Tweeting in China. *Int. J. Bus. Soc. Sci.* **2012**, *3*, 91–94.
11. Saaid, S.A.; Al-Rashid, N.A.A.; Abdullah, Z. The impact of addiction to Twitter among university students. In *Future Information Technology*; Springer: Berlin/Heidelberg, Germany, 2014; pp. 231–236.

12. Griffiths, M.D.; Kuss, D.J.; Demetrovics, Z. Social networking addiction: An overview of preliminary findings. In *Behavioral Addictions: Criteria, Evidence and Treatment*, 1st ed.; Rosenberg, K., Feder, L., Eds.; Elsevier: New York, NY, USA, 2014; pp. 119–141.
13. Brailovskaia, J.; Margraf, J. Facebook addiction disorder (FAD) among German students—A longitudinal approach. *PLoS ONE* **2017**, *12*, e0189719. [CrossRef]
14. Laconi, S.; Verseillié, E.; Chabrol, H. Exploration of the Problematic Twitter, Facebook Uses and Their Relationships with Psychopathological Symptoms among Facebook Users. *Int. J. High Risk Behav.* **2018**, *7*, e61775. [CrossRef]
15. Walburg, V.; Mialhes, A.; Moncla, D. Does school-related burnout influence problematic Facebook use? *Child. Youth Serv. Rev.* **2016**, *61*, 327–331. [CrossRef]
16. Couderc, E. Search for an addiction to social networks and study of the concerned user profile [Recherche d'une addiction aux réseaux sociaux et étude du profil d'utilisateur concerné]. Ph.D. Thesis, Université de Limoge, Limoges, France, 2012.
17. Delfour, M.; Moreau, A.; Laconi, S.; Goutaudier, N.; Chabrol, H. Problematic Facebook use in adolescents and young adultsProblematic Facebook use in adolescents and young adults [Utilisation problématique de Facebook à l'adolescence et au jeune âge adulte]. *Neuropsychiatrie de L'enfance de et L'adolescence* **2014**, *63*, 244–249. [CrossRef]
18. Herrera-Harfuch, M.F.; Pacheco Murguia, M.P.; Lever, J.P.; Andrade, D.Z. La adiccion a Facebook relacionada con la baja autoestima, la depression y la falta de habilitades sociales. *Psicol. Iberoam.* **2010**, *18*, 6–18.
19. Hong, F.Y.; Huang, D.H.; Lin, H.Y.; Chiu, S.L. Analysis of the psychological traits, Facebook usage, and Facebook addiction model of Taiwanese University students. *Telemat. Inform.* **2014**, *31*, 597–606. [CrossRef]
20. Caplan, S.E. Theory and measurement of generalized problematic Internet use: A two-step approach. *Comput. Hum. Behav.* **2010**, *26*, 1089–1097. [CrossRef]
21. Bernardi, S.; Pallanti, S. Internet addiction: A descriptive clinical study focusing on comorbidities and dissociative symptoms. *Compr. Psychiatry* **2009**, *50*, 510–516. [CrossRef]
22. Laconi, S.; Andreoletti, A.; Chauchard, E.; Rodgers, R.; Chabrol, H. Problematic Internet Use, time spent online and personality traits [Utilisation problématique d'Internet, temps passé en ligne et traits de personnalité]. *Encéphale* **2016**, *42*, 214–218. [CrossRef]
23. Leménager, T.; Hoffmann, S.; Dieter, J.; Reinhard, I.; Mann, K.; Kiefer, F. The links between healthy, problematic, and addicted Internet use regarding comorbidities and self-concept-related characteristics. *J. Behav. Addict.* **2018**, *7*, 31–43. [CrossRef]
24. Sepehrian, F.; Lotf, J.J. Study of the relationship between Internet addiction with anxiety and personality types A and B. *Aust. J. Basic Appl. Sci.* **2011**, *5*, 928–934.
25. Zadra, S.; Bischof, G.; Besser, B.; Bischof, A.; Meyer, C.; John, U.; Rumpf, H.J. The association between Internet addiction and personality disorders in a general population-based sample. *J. Behav. Addict.* **2016**, *5*, 691–699. [CrossRef]
26. Davenport, S.W.; Bergman, S.M.; Bergman, J.Z.; Fearington, M.E. Twitter versus Facebook: Exploring the role of narcissism in the motives and usage of different social media platforms. *Comput. Hum. Behav.* **2014**, *32*, 212–220. [CrossRef]
27. Walters, N.T.; Horton, R. A diary of the influence of Facebook use on narcissism among male college students. *Comput. Hum. Behav.* **2015**, *52*, 326–330. [CrossRef]
28. Andreassen, C.S.; Torsheim, T.; Brunborg, G.C.; Pallesen, S. Development of Facebook addiction scale. *Psychol. Rep.* **2012**, *110*, 501–517. [CrossRef] [PubMed]
29. Yussof, M.S.B. Psychometric properties of the Depression Anxiety Stress Scale in a sample of medical degree applicants. *Int. Med. J.* **2013**, *20*, 295–300.
30. Bouvard, M. *Questionnaires et Echelles D'évaluation de la Personnalité*, 3rd ed.; Masson: Paris, France, 2002.
31. Hyler, S.E. *Personality Diagnostic Questionnaire (PDQ-4 Plus)*; New York State Psychiatric Institute: New York, NY, USA, 1994.
32. Chabrol, H.; van Leeuwen, N.; Rodgers, R.; Séjourné, N. Contributions of psychopathic, narcissistic, Machiavellian, and sadistic personality traits to juvenile delinquency. *Personal. Individ. Differ.* **2009**, *47*, 734–739. [CrossRef]

33. Laconi, S.; Tricard, N.; Chabrol, H. Differences between specific and generalized problematic Internet uses according to gender, age, time spent online and psychopathological symptoms. *Comput. Hum. Behav.* **2015**, *48*, 236–244. [CrossRef]
34. Koc, M.; Gulyagci, S. Facebook Addiction among Turkish college students: The role of psychological health, demographic, and usage characteristics. *Cyberpsychol. Behav. Soc. Netw.* **2013**, *16*, 279–284. [CrossRef]
35. Moreau, A.; Laconi, S.; Delfour, M.; Chabrol, H. Psychopathological profiles of adolescent and young adult problematic Facebook users. *Comput. Hum. Behav.* **2015**, *44*, 64–69. [CrossRef]
36. Ostovar, S.; Allahyar, N.; Aminpoor, H.; Moafian, F.; Nor, M.B.M.; Griffiths, M.D. Internet addiction and its psychosocial risks (depression, anxiety, stress and loneliness) among Iranian adolescents and young adults: A structural equation model in a cross-sectional study. *Int. J. Ment. Health Addict.* **2016**, *14*, 257–267. [CrossRef]
37. Stavropoulos, V.; Gomez, R.; Steen, E.; Beard, C.; Liew, L.; Griffiths, M.D. The longitudinal association between anxiety and Internet addiction in adolescence: The moderating effect of classroom extraversion. *J. Behav. Addict.* **2017**, *6*, 237–247. [CrossRef]
38. Andreassen, C.S.; Billieux, J.; Griffiths, M.D.; Kuss, D.J.; Demetrovics, Z.; Mazzoni, E.; Pallesen, S. The relationship between addictive use of social media and video games and symptoms of psychiatric disorders: A large-scale cross-sectional study. *Psychol. Addict. Behav.* **2016**, *30*, 252–262. [CrossRef] [PubMed]
39. Beyens, I.; Frison, E.; Eggermont, S. "I don't want to miss a thing": Adolescents' fear of missing out and its relationship to adolescents' social needs, Facebook use, and Facebook related stress. *Comput. Hum. Behav.* **2016**, *64*, 1–8. [CrossRef]
40. Chabrol, H.; Laconi, S.; Delfour, M.; Moreau, A. Contributions of psychopathological and interpersonal variables to problematic Facebook use in adolescents and young adults. *Int. J. High Risk Behav. Addict.* **2017**, *6*, e32773. [CrossRef]
41. Atroszko, P.A.; Balcerowska, J.M.; Bereznowski, P.; Biernatowska, A.; Pallesen, S.; Andreassen, C.S. Facebook addiction among Polish undergraduate students: Validity of measurement and relationship with personality and well-being. *Comput. Hum. Behav.* **2018**, *85*, 329–338. [CrossRef]
42. Błachnio, A.; Przepiórka, A. Facebook intrusion, fear of missing out, narcissism, and life satisfaction: A cross-sectional study. *Psychiatry Res.* **2018**, *259*, 514–519. [CrossRef] [PubMed]
43. Casale, S.; Fioravanti, G. Why narcissists are at risk for developing Facebook addiction: The need to be admired and the need to belong. *Addict. Behav.* **2018**, *76*, 312–318. [CrossRef]
44. Echeburua, E. Addiction to new technologies and to online social networking in young people: A new challenge. *Adicciones* **2010**, *22*, 91–95.
45. Lee-Won, R.J.; Herzog, L.; Park, S.G. Hooked on Facebook: The role of social anxiety and need for social assurance in problematic use of Facebook. *Cyberpsychol. Behav. Soc. Netw.* **2015**, *18*, 567–574. [CrossRef]
46. Foroughi, B.; Iranmanesh, M.; Nikbin, D.; Hyun, S.S. Are depression and social anxiety the missing link between Facebook addiction and life satisfaction? The interactive effect of needs and self-regulation. *Telemat. Inform.* **2019**, *43*, e101247. [CrossRef]
47. Laconi, S.; Vigouroux, M.; Lafuente, C.; Chabrol, H. Problematic Internet Use, Psychopathology, Personality, Defense and Coping. *Comput. Hum. Behav.* **2017**, *73*, 47–54. [CrossRef]
48. Petrocchi, N.; Asnaani, A.; Martinez, A.P.; Nadkarni, A.; Hofmann, S.G. Differences between people who use only Facebook and those who use Facebook plus Twitter. *Int. J. Hum. Comput. Interact.* **2015**, *31*, 157–165. [CrossRef] [PubMed]
49. Donner, N.C.; Lowry, C.A. Sex differences in anxiety and emotional behavior. *Pflügers Arch. Eur. J. Physiol.* **2013**, *465*, 601–626. [CrossRef] [PubMed]
50. Brock, R.L.; Barry, R.A.; Lawrence, E.; Dey, J.; Rolffs, J. Internet administration of paper-and-pencil questionnaires used in couple research: Assessing psychometric equivalence. *Assessment* **2012**, *19*, 226–242. [CrossRef] [PubMed]

International Journal of
Environmental Research and Public Health

MDPI

Article

Psychological Characteristics and Addiction Propensity According to Content Type of Smartphone Use

Jinhee Lee [1,2], Joung-Sook Ahn [1,2], Seongho Min [1] and Min-Hyuk Kim [1,*]

1 Department of Psychiatry, Yonsei University Wonju College of Medicine, 20 Ilsan-ro, Gangwon-do, Wonju 26426, Korea; jinh.lee95@yonsei.ac.kr (J.L.); jsahn@yonsei.kr (J-S.A.); mchorock@yonsei.ac.kr (S.M.)
2 Division of Child and Adolescent Psychiatry, Yonsei University Wonju College of Medicine, 20 Ilsan-ro, Gangwon-do, Wonju 26426, Korea
* Correspondence: mhkim09@yonsei.kr

Received: 29 February 2020; Accepted: 27 March 2020; Published: 29 March 2020

Abstract: The aim of this study was to evaluate the association between content type of smartphone use and psychological characteristics and addiction propensity, including the average time of smartphone use and problematic smartphone use. Data were obtained from the 2017 Korea Youth Risk Behavior Web-Based Survey, a nationally representative survey of middle- and high-school students (n = 62,276). The content type of smartphone use was divided into four categories: (1) Study, (2) Social-Networking Services (SNS), (3) Game, and (4) Entertainment. The association of depressive mood and suicidal ideation with content type of smartphone use was analyzed, using multiple and binary logistic regression analyses, respectively. The relationship between content type of smartphone use and time spent on smartphone use and problematic smartphone use was analyzed by using multiple logistic regression, adjusted for related covariables. The results of this study revealed that depressive mood and suicidal ideation were significantly associated with the SNS smartphone use group, compared with the other groups. Our results also indicate that the SNS group showed higher addiction propensity, such as overuse and experiencing adverse consequences of smartphone use.

Keywords: suicide; suicide attempts; intervention; case management; smartphone use; addiction

1. Introduction

Smartphones are perceived as indispensable information and communication tools in daily life for many people and are now the most frequently used technology worldwide [1]. For adolescents, in particular, who are sensitive to new technology and media use, smartphones have become an important part of their life. Recent studies estimate that 84% of adolescents in Japan [2] and 97% of adolescents in Switzerland [3] have their own smartphone. Furthermore, similar to substance or other types of behavioral addictions, adolescents are known to be vulnerable to smartphone addiction. A prior study reported that 60% of adolescents in the UK are highly addicted to their smartphones [2], and the rate of smartphone addiction among adolescents was double for adults in South Korea [2]. Prior studies have suggested some neurobiological evidences of the vulnerability toward smartphone addiction among adolescents, such as the dual processing model and an imbalance between the go and stop networks [4,5].

Smartphones are useful for multiple purposes, including study, information searching, social communication, and entertainment [6]. Compared with the traditional forms of computer and internet use, the portability and connectivity of smartphones give users easier access to information and entertainment content—nearly anytime and anywhere. These characteristics can also make people more vulnerable to behavioral addiction [7] in the form of habitual checking or excessive use of smartphones. Previous studies have reported that excessive smartphone use in adolescents is associated with psychopathologies (i.e., depression, anxiety, high-stress levels, and low mood) and behavioral problems [8,9], because adolescents are easily affected by external stimulus, interpersonal issues, and emotional changes. Another study on young adults suggested that excessive smartphone use is related to high stress, and it is also inversely related to academic performance, as well as life satisfaction [10].

Previous studies have used the smartphone addiction scale, smartphone usage time, or the frequency of use to clarify the relationship between addictive smartphone use and adverse effects on physical and mental health. However, despite the possible associations between the purpose of smartphone use and the risk of addictive behaviors, little is known regarding the relative impact of the content type of smartphone use on addiction and adverse consequences of smartphone use, including adolescent mental health. Andreassen and her colleagues have suggested that addictive online behaviors, including both addictive social networking and video gaming, are associated with underlying psychiatric disorders, such as ADHD, OCD, anxiety, and depression [11]; however, the differences in specific associations according to the purpose of its use have not been clarified. Another study has indicated that social-networking addiction and internet-gaming disorder can augment the symptoms of each other and simultaneously contribute to deterioration of overall psychological health in a similar fashion [12]. Some evidence shows that internet addiction comprises strongly directed internet activities, such as excessive online-video-game playing, excessive use of online pornography, or online shopping, and there are indeed different forms of internet addiction [13–15].

Thus, we aim to investigate the association between content type of smartphone use and adolescent mental health, with the hypothesis that psychological characteristics and addiction propensity are related with content type of smartphone use. This study examines the average time of smartphone use and problematic smartphone use, using a school-based, nationally representative dataset of the Korean adolescent population.

2. Materials and Methods

2.1. Methods

Study Population and Source of Data

Data on the study population were obtained from the 13th Korea Youth Risk Behavior Web-Based Survey (KYRBS), which was administered in 2017 by the Korean Ministry of Education, Science and Technology; the Ministry of Health and Welfare; and the Korea Centers for Disease Control and Prevention. KYRBS is a self-reported anonymous online survey of a nationally representative sample of Korean adolescents (aged 12–18 years) [16]. The sample design of this survey used a stratified multistage cluster strategy with 123 questions divided into 15 sections inquiring about health-related behaviors and mental and physical health. In the 13th KYRBS, 64,991 students from 800 middle and high schools were randomly selected, and 62,276 (31,636 boys and 30,640 girls) students (95.8% response rate) from 799 schools responded to the survey [17]. Participants were provided with identification numbers and were guaranteed anonymity, and written informed consent was obtained from each participant after the survey had been fully explained. All data used in this study were fully anonymized before we accessed them. This consent procedure was approved by the Institutional Review Board of the Korea Centers for Disease Control and Prevention (2014-06EXP-02-P-A).

2.2. Measures

2.2.1. Content Type of Smartphone Use

The exposure variable, content type of smartphone use, was assessed by the question "In the last 30 days, please select only one service that you used mainly, when using your smartphone", and the answers were classified into four categories: (1) Study; (2) Social-Networking Services (SNS) (e.g., messaging and chat, communities, and social networks); (3) Game; and (4) Entertainment (e.g., watching movies, reading comics and fiction, listening to music, creating User-Created Content and videos).

2.2.2. Sociodemographic and General Characteristics

The sociodemographic characteristics reported included age, sex, residential area, and family economic status of the participant. Respondents who lived in the country or rural areas were categorized as "Rural"; those who lived in small, middle-sized or large cities were categorized as "Urban". Family economic status was assessed by the question "What is your family economic status?" The five possible response categories, very high/high/middle/low/very low, were grouped into three categories, for the purpose of our analysis: high (very high or high), middle (middle), and low (very low or low) [18]. Sleep hours were divided into two categories: under 6 h; and 6 or more hours. Physical activity was divided into two categories, Yes/No, from the question "In the last 7 days, did you have physical activity with higher heart rate than usual?"

2.2.3. Psychological Characteristics

Subjective stress was measured by the question "How much stress do you usually feel?" The five possible response categories of very high//high/middle/low/very low were grouped into three categories: high (very high or high), middle (middle), and low (very low or low). Current alcohol consumption was assessed by the question "How many days during the past 30 days did you drink more than one cup of alcohol? (None/1–2 days/3–5 days/6–9 days/10–19 days/20–29 days/Every day)" Respondents who responded "None" were classified as not current alcohol drinkers, and those who responded between "1–2 days" and "Every day" were classified as current alcohol drinkers. The current cigarette smoking was assessed by using the following question: "How many days during the past 30 days did you smoke a cigarette? (None/1–2 days/3–5 days/6–9 days/10–19 days/20–29 days/Every day)". Respondents who responded "No" to the question were classified as not current cigarette smokers, and those who responded between "1–2 days" and "Every day" were classified as current cigarette smokers.

Depressed mood among the subjects was assessed by the question "In the past year, have you ever felt so sad or despaired that your feelings disturbed everyday life for two whole weeks?" Subjects responded with the following: (1) "No, I never felt it" or (2) "Yes, I have felt it". We also examined whether the subjects had suicidal ideations with the question "In the past year, did you ever seriously consider attempting suicide?" Subjects responded with the following: (1) "No, I never thought of it" or (2) "Yes, I have thought of it".

2.2.4. Addiction-Propensity-Related Factors

The average time spent using a smartphone was assessed by the question "On an average school day, how many hours do you use a smartphone?" According to the results of the previous study [3], the use of a smartphone over 5 h a day was defined as "smartphone overuse". Participants were also asked the following: "In the last 30 days, have you experienced severe conflicts with family due to your smartphone usage?" and "In the last 30 days, have you experienced severe conflicts with friends due to your smartphone usage?", (Yes/No), which would suggest a tolerance that is one of the important factors of smartphone addiction. They were also asked whether they had experienced poor academic performance due to smartphone use (Yes/No), by the question "In the last 30 days, were there any difficulties in your academic performance due to your smartphone usage?", which would suggest a daily disturbance due to smartphone addiction [1].

2.2.5. Statistical Analyses

The participants' general characteristics according to each content type of smartphone use were summarized by using either a one-way analysis of variance for continuous variables or a chi-squared test with Bonferroni correction for categorical variables. The relationships of the four different groups with psychological factors and problematic smartphone use were analyzed by using Pearson's chi-square. Subsequently, a multiple logistic regression analysis was performed to identify the associations between content type of smartphone use with depressed mood, suicidal ideation, and overuse of smartphones. General characteristics that showed a significant difference in the chi-square test were mutually adjusted for the analysis. Two-tailed analyses were conducted, and p-values lower than 0.05 were considered significant. Adjusted odds ratios (AORs) and 95% confidence intervals (CIs) were calculated. All statistical analyses were performed by using SPSS software (version 23.0, IBM Corp., Armonk, NY, USA).

3. Results

The descriptive characteristics according to the content type of smartphone use are presented in Table 1. Results of chi-square analysis revealed that there were significant differences depending on age, sex, residential area, family economic status, sleep hours, and physical activity by content type of smartphone use. The "Study" group was more likely to be older, live in large cities, and have a higher family economic status. The "SNS" group had a higher prevalence of female respondents and a lower prevalence of physical activity. The "Game" group was more likely to be younger, boys, living in rural areas, sleeping less than 6 h, and less physically active. The "Entertainment" group had a higher prevalence of low family economic status compared to other groups (all $p < 0.001$).

Table 1. General characteristics of participants, according to content type of smartphone use.

General Characteristics		Study (n = 4202, 10.7%)	SNS (n = 10,192, 25.9%)	Game (n = 7282, 18.5%)	Entertainment (n = 17,642, 44.9%)	p
Age (years)	Mean (SD)	15.58 ± 1.78	14.92 ± 1.66	14.31 ± 1.71	15.07 ± 1.76	< 0.001
Sex (%)						< 0.001
	Boys	62.9	28.9	87.9	50.8	
	Girls	37.1	71.1	12.1	49.2	
Region (%)						< 0.001
	Rural Area	7.3	7.2	8.5	7.8	
	Small City	45.6	47.3	50.3	47.3	
	Large City	47.1	45.5	41.2	44.9	
Family Economic Status (%)						< 0.001
	High	47.2	39.0	40.4	37.0	
	Middle	40.3	46.8	46.2	47.1	
	Low	12.5	14.2	13.4	15.9	
Sleep Hours (%)						< 0.001
	<6hrs	33.2	38.6	62.4	42.9	
	≥6hrs	66.8	61.4	37.6	57.1	
Physical Activity (%)						< 0.001
	No	34.2	38.9	32.0	36.5	
	Yes	65.8	61.1	68.0	64.0	

SNS: Social-Networking Services.

The psychological characteristics of participants according to content type of smartphone use are presented in Table 2. The "SNS" group had a higher prevalence of high subjective stress level, current cigarette smoking, and current alcohol drinking. The "SNS" group also had significantly higher prevalence of depressive mood and suicidal ideation compared to other groups. The "Game" group had the lowest proportion of depressive mood and suicidal ideation among groups (all $p < 0.001$).

The average amount of time spent using a smartphone was greater in the "SNS" group (322.17 ± 228.90 min/day) than in other groups and lower in the "Study" group (176.97 ± 173.06 min/day). The proportion of adverse consequences of smartphone use, including conflicts with family, conflicts with friends, and poor academic performance due to smartphone use, were higher in the "SNS" group (59.2%, 27.6%, and 58.4% respectively), whereas the "Study" group had a lower prevalence of adverse consequences (41.8%, 20.1%, and 43.0% respectively) (Table 3).

Table 2. Psychological characteristics of participants, according to content type of smartphone use.

Psychological Characteristics		Study (n = 4202, 10.7%)	SNS (n = 10,192, 25.9%)	Game (n = 7282, 18.5%)	Entertainment (n = 17,642, 44.9%)	p
Subjective Stress						< 0.001
	High	37.1	41.1	30.3	38.0	
	Middle	40.9	42.0	43.2	42.1	
	Low	22.0	16.6	26.5	19.8	
Current Cigarette Smoking						0.694
	No	93.4	93.2	93.6	93.3	
	Yes	6.6	6.8	6.4	6.7	
Current Alcohol Drinking						< 0.001
	No	77.7	74.7	77.4	75.5	
	Yes	22.3	25.3	22.6	24.5	
Depressive Mood						< 0.001
	No	76.2	71.1	82.5	76.2	
	Yes	23.8	28.9	17.5	23.8	
Suicidal Ideation						< 0.001
	No	89.5	85.3	90.6	87.7	
	Yes	10.5	14.7	9.4	12.3	

SNS: Social-Networking Services.

Table 3. Adverse consequences of smartphone use, according to content type of smartphone use.

Adverse Consequences		Study (n = 4202, 10.7%)	SNS (n = 10,192, 25.9%)	Game (n = 7282, 18.5%)	Entertainment (n = 17,642, 44.9%)	p
Average Spent Time Using a Smartphone (min/day)						< 0.001
	Mean (SD)	176.97 ± 173.06	322.17 ± 228.90	243.87 ± 191.73	255.10 ± 187.00	
Conflict with Family Members due to Smartphone Use (%)						< 0.001
	No	58.2	40.8	41.5	45.8	
	Yes	41.8	59.2	58.5	54.2	
Conflict with Friends due to Smartphone Use (%)						< 0.001
	No	79.9	72.4	74.1	78.4	
	Yes	20.1	27.6	25.9	21.6	
Poor Academic Performance due to Smartphone Use (%)						< 0.001
	No	57.0	41.6	51.9	47.4	
	Yes	43.0	58.4	48.1	52.6	

SNS: Social-Networking Services.

Compared to the "Study" group, the "SNS" group was significantly more likely to report a depressive mood (AOR 1.36; 95% CI 1.24–1.49) and suicidal ideation (AOR 1.49; 95% CI 1.32–1.69). The "Entertainment" group also showed a positive association with suicidal ideation (AOR 1.20; 95% CI 1.06–1.35), and the "Game" group showed a negative association with depressive mood (AOR 0.77; 95% CI 0.69–0.85) (Table 4).

Table 4. Multivariable logistic regression analysis of content type of smartphone use, for depressive mood and suicidal ideation.

Content Type	Depressive Mood		Suicidal Ideation	
	Crude OR	Adjusted OR *	Crude OR	Adjusted OR *
	OR (95% CI)	OR (95% CI)	OR (95% CI)	OR (95% CI)
Study	Ref.	Ref.	Ref.	Ref.
SNS	1.30 (1.20–1.41)	1.36 (1.24–1.49)	1.46 (1.31–1.64)	1.49 (1.32–1.69)
Game	0.68 (0.62–0.74)	0.77 (0.69–0.85)	0.87 (0.77–0.99)	0.95 (0.82–1.09)
Entertainment	1.00 (0.92–1.08)	1.02 (0.94–1.12)	1.19 (1.07–1.32)	1.20 (1.06–1.35)

* Adjustment for age, sex, region of residence, family economic status, sleep hours, and physical activity; SNS: Social-Network Services, OR: odds ratio.

The AORs for smartphone overuse (> 5 h per day) were 4.57 (95% CI, 4.20–4.98), 2.24 (95% CI, 2.05–2.45) and 2.60 (95% CI, 2.40–2.81) in the "SNS" group, "Game" group, and "Entertainment" group, respectively (Table 5).

Table 5. Multivariable logistic regression analysis of content type of smartphone use, for smartphone overuse (more than 5 h per day).

Content Type	Crude OR	Adjusted OR
	OR (95% CI)	OR (95% CI)
Study	Ref.	Ref.
SNS	4.67 (4.32–5.05)	4.57 (4.20–4.98)
Game	2.22 (2.04–2.40)	2.24 (2.05–2.45)
Entertainment	2.68 (2.49–2.88)	2.60 (2.40–2.81)

* Adjustment for age, sex, region of residence, family economic status, sleep hours, and physical activity; SNS: Social-Networking Services, OR: odds ratio

4. Discussion

This study examined the association of psychological characteristics and addiction propensity with content type of smartphone use, within a relatively large convenience sample of adolescents in Korea. The results of this study revealed that depressive mood and suicidal ideation are significantly associated with higher SNS use, compared with smartphone use for games, study, and entertainment. Our results also suggested that the "SNS" group showed higher addiction propensity, including overuse and adverse consequences of smartphone use.

The results of this study expanded upon and shared similarities with previous findings on the relationship between mental health and SNS use. A prior systematic review by Frost et al. reported associations between SNS use (i.e., Facebook) and mental-health outcomes, such as alcohol use, addiction, anxiety, and depression [19]. Several studies have indicated that the prolonged use of SNS may be related to signs and symptoms of depression, and some authors have indicated that certain SNS activities might be associated with low self-esteem, especially in children and adolescents [20–23]. On the other hand, our results were contrary to a previous study that reported that the use of non-social smartphone features (i.e., news consumption, entertainment, and relaxation) were most related to depression and problematic smartphone use [24]. Furthermore, a prior study by our research group reported a potential protective effect from moderate use (1–2 h) of smartphones for social purposes (i.e., SNS and messaging) in regard to suicide attempts [1]. In these contexts, we should also consider the positive psychological effects of SNS use. In this study, we did not simply divide content type of smartphone use as social and non-social, but instead we compared detailed non-social uses: study, game, entertainment, and SNS. According to our results, content type of smartphone use should not be classified simply as social and non-social use but should also take into account the detailed characteristics of SNS use and various other tasks, including differences in the effects of mental health on adolescents.

There has been wide discussion on the potential causes for depressive mood resulting from increased time on SNS. The most commonly used mediator to explain the association between SNS use and depression is self-esteem. It is an important factor in developing and maintaining mental health and overall quality of life, and low self-esteem is associated with numerous mental illnesses, including depression and addiction [25,26]. Some authors have presented that individuals higher in narcissism and lower in self-esteem also showed more online activity, including self-promotional content such as SNS [27]. On the other hand, there is the that hypothesis feelings of depression can be predicted indirectly by SNS addiction [21]. Authors have indicated that SNS allows the user to get virtual community gratification and gain gratification from creating a self-image online. Based on the uses and gratifications theory, SNS use can lead to SNS addiction, as the functions available to the users allow them to gain instant gratification from using the service, which in turn could lead to excessive use.

Contrary to previous research that indicated a negative association between online-game use and adolescent mental health [28,29], the current study did not find that smartphone-game use was associated with depressive mood and suicidal ideation. The results might reflect the characteristics of categorization and reference group of study. The "Game" group of this study included those who enjoyed "gaming" more than other contents of the smartphone, but it does not mean that they had a "gaming addiction". Specifically, if a person performs gaming in a regular pattern, the person may relieve his/her stress. However, if a person overly performs gaming, he or she may have increased psychological problems, as shown in the literature. On the other hand, because of the statistically low number of "non-smartphone users", we used "Study" as a reference group. Studying does not mean the person cannot be addicted to it, and using "Study" as the reference can create some biases. For example, a person who is over-studying may have increased distress. Therefore, we cannot capture whether gaming is related to increased distress if studying is associated with high distress.

Furthermore, smartphone-game use predicted problematic smartphone use compared to the "Study" group, but showed a weak association compared to the "SNS" or "Entertainment" groups. Smartphone games are somewhat different from computer-based online games, allowing users to access them anywhere, anytime, but there is a limit to the use of tools for the games. There have been a number of studies on problematic game use, and recently, a WHO ICD-11 proposal for a new category named "Gaming Disorder" [30]. However, most of the studies so far have been limited to computer-based online games [31–33]. Furthermore, considering the recent trend that the use of entertainment, such as the use of YouTube, is particularly popular among adolescents and has become dominant in the media market worldwide [34,35], the results of the current study indicated the necessity for further studies about game and entertainment on the smartphone. Smartphones, which are relatively simple tools compared to conventional computers, may be better suited for simple functions, such as watching videos, than for more complex tasks, such as playing games, which may result in adolescents indulging in media instead.

The present study has a number of limitations that should be considered when interpreting the findings. First, due to the cross-sectional nature of national surveys, the present findings have limitations in explaining the causal inferences between content type of smartphone use and psychological characteristics. Further studies with sufficient time for investigation are needed to develop a clear understanding about the association of psychological characteristics and addiction propensity with content type of smartphone use. Second, the psychological characteristics and internet use were measured through the ad hoc questions rather than mental-health experts' assessment or validated scales, because the data were collected through the participants' self-reports, and therefore, some reporting bias could have occurred. Moreover, because the group was divided only for one main purpose of smartphone use, we could not distinguish those who performed two or more content types. Third, in our study, the addictive propensity was estimated only by time spent using a smartphone, not by the scales for smartphone addiction. In addition, most variables in the study, including conflicts with family/friends, and poor academic performance due to smartphone use, were surveyed on the basis of a self-reported questionnaire, which has inherent limitations regarding the validity of the data and the recall bias. However, in the previous study, excessive smartphone use was validated as the most powerful independent predictor of smartphone addiction [8], and we can use this to estimate the propensity to addiction. Fourth, our data lacked information regarding the familiarity or personological profile of the participants that might affect individuals with mood disorder and/or addiction. Despite the limitations of this cross-sectional survey, the present study has some strengths. We used a multilevel multinomial logistic modeling approach based on a nationally representative sample of Korean adolescents, who have the highest smartphone ownership rate in the world. Moreover, the response rate to the survey was very high. To the best of our knowledge, this study is the first to report on the association of psychological characteristics and addiction propensity with the content type of smartphone use in adolescents.

5. Conclusions

In the present nationally representative sample of Korean adolescents aged 12–18 years, the significant and specific association between content type of smartphone use and the prevalence of depressed mood, suicidal ideation, and addiction propensity has been confirmed. Our findings indicate that SNS use was not only a stronger predictor of smartphone addiction than other content types, but also had stronger associations with negative psychological characteristics, including depressive mood and suicidal ideation. The results of this study suggest that careful consideration should be given to improving screening for the risk of smartphone addiction and mental health problems in adolescents, with a focus on SNS use. Further research is needed to understand the longitudinal impact of specific content type of smartphone use on the addictive behavior of adolescents, allowing for the development of effective prevention strategies and to help strengthen the current smartphone-use guidelines.

Author Contributions: Conceptualization, J.L. and M.-H.K.; methodology, J.L.; validation, J.-S.A. and S.M..; formal analysis, J.L.; investigation, J.L.; writing—original draft preparation, J.L.; writing—review and editing, M.-H.K.; supervision, J.-S.A. and S.M. All authors have read and agreed to the published version of the manuscript.

Funding: This research received no external funding.

Conflicts of Interest: The authors declare no conflict of interest.

References

1. Kim, M.H.; Min, S.; Ahn, J.S.; An, C.; Lee, J. Association between high adolescent smartphone use and academic impairment, conflicts with family members or friends, and suicide attempts. *PLoS ONE* **2019**, *14*, e0219831. [CrossRef] [PubMed]
2. Davey, S.; Davey, A. Assessment of smartphone addiction in Indian adolescents: A mixed method study by systematic-review and meta-analysis approach. *Int. J. Prev. Med.* **2014**, *5*, 1500–1511. [PubMed]
3. Haug, S.; Castro, R.P.; Kwon, M.; Filler, A.; Kowatsch, T.; Schaub, M.P. Smartphone use and smartphone addiction among young people in Switzerland. *J. Behav. Addict.* **2015**, *4*, 299–307. [CrossRef] [PubMed]
4. Adisetiyo, V.; Gray, K.M. Neuroimaging the neural correlates of increased risk for substance use disorders in attention-deficit/hyperactivity disorder—A systematic review. *Am. J. Addict.* **2017**, *26*, 99–111. [CrossRef] [PubMed]
5. Sussman, C.J.; Harper, J.M.; Stahl, J.L.; Weigle, P. Internet and video game addictions: Diagnosis, epidemiology, and neurobiology. *Child Adolesc. Psychiatr. Clin.* **2018**, *27*, 307–326. [CrossRef]
6. Van Deursen, A.J.; Bolle, C.L.; Hegner, S.M.; Kommers, P.A. Modeling habitual and addictive smartphone behavior: The role of smartphone usage types, emotional intelligence, social stress, self-regulation, age, and gender. *Comput. Hum. Behav.* **2015**, *45*, 411–420. [CrossRef]
7. Demirci, K.; Orhan, H.; Demirdas, A.; Akpinar, A.; Sert, H. Validity and reliability of the Turkish Version of the Smartphone Addiction Scale in a younger population. *Klin. Psikofarmakol. Bülteni-Bull. Clin. Psychopharmacol.* **2014**, *24*, 226–234. [CrossRef]
8. Matar Boumosleh, J.; Jaalouk, D. Depression, anxiety, and smartphone addiction in university students-A cross sectional study. *PLoS ONE* **2017**, *12*, e0182239. [CrossRef]
9. Cerniglia, L.; Griffiths, M.D.; Cimino, S.; De Palo, V.; Monacis, L.; Sinatra, M.; Tambelli, R. A latent profile approach for the study of internet gaming disorder, social media addiction, and psychopathology in a normative sample of adolescents. *Psychol. Res. Behav. Manag.* **2019**, *12*, 651–659. [CrossRef]
10. Samaha, M.; Hawi, N.S. Relationships among smartphone addiction, stress, academic performance, and satisfaction with life. *Comput. Hum. Behav.* **2016**, *57*, 321–325. [CrossRef]
11. Andreassen, C.S.; Billieux, J.; Griffiths, M.D.; Kuss, D.J.; Demetrovics, Z.; Mazzoni, E.; Pallesen, S. The relationship between addictive use of social media and video games and symptoms of psychiatric disorders: A large-scale cross-sectional study. *Psychol. Addict. Behav.* **2016**, *30*, 252–262. [CrossRef] [PubMed]
12. Pontes, H.M. Investigating the differential effects of social networking site addiction and Internet gaming disorder on psychological health. *J. Behav. Addict.* **2017**, *6*, 601–610. [CrossRef] [PubMed]

13. Montag, C.; Bey, K.; Sha, P.; Li, M.; Chen, Y.F.; Liu, W.Y.; Zhu, Y.K.; Li, C.B.; Markett, S.; Keiper, J. Is it meaningful to distinguish between generalized and specific Internet addiction? Evidence from a cross-cultural study from Germany, Sweden, Taiwan and China. *Asia-Pac. Psychiatry* **2015**, *7*, 20–26. [CrossRef] [PubMed]
14. Davis, R.A. A cognitive-behavioral model of pathological Internet use. *Comput. Hum. Behav.* **2001**, *17*, 187–195. [CrossRef]
15. Wong, H.Y.; Mo, H.Y.; Potenza, M.N.; Chan, M.N.M.; Lau, W.M.; Chui, T.K.; Pakpour, A.H.; Lin, C.-Y. Relationships between Severity of Internet Gaming Disorder, Severity of Problematic Social Media Use, Sleep Quality and Psychological Distress. *Int. J. Environ. Res. Public Health* **2020**, *17*, 1879. [CrossRef]
16. Kim, Y.; Choi, S.; Chun, C.; Park, S.; Khang, Y.H.; Oh, K. Data Resource Profile: The Korea Youth Risk Behavior Web-based Survey (KYRBS). *Int. J. Epidemiol.* **2016**, *45*, 1076-1076e. [CrossRef]
17. Korea centers for disease control & prevention call center (KCDC). The 13th Korea Youth Risk Behavior Web-Based Survey. Available online: http://www.cdc.go.kr/yhs/home.jsp (accessed on 23 January 2020).
18. Lee, J.; Jang, H.; Kim, J.; Min, S. Development of a suicide index model in general adolescents using the South Korea 2012–2016 national representative survey data. *Sci. Rep.* **2019**, *9*, 1846. [CrossRef]
19. Frost, R.L.; Rickwood, D.J. A systematic review of the mental health outcomes associated with Facebook use. *Comput. Hum. Behav.* **2017**, *76*, 576–600. [CrossRef]
20. Pantic, I. Online social networking and mental health. *Cyberpsychol. Behav Soc. Netw.* **2014**, *17*, 652–657. [CrossRef]
21. Donnelly, E.; Kuss, D. Depression among users of social networking sites (SNSs): The role of SNS addiction and increased usage. *J. Addict. Prev. Med.* **2016**, *1*, 107. [CrossRef]
22. Jelenchick, L.A.; Eickhoff, J.C.; Moreno, M.A. "Facebook depression?" Social networking site use and depression in older adolescents. *J. Adolesc. Health* **2013**, *52*, 128–130. [CrossRef] [PubMed]
23. Cerniglia, L.; Guicciardi, M.; Sinatra, M.; Monacis, L.; Simonelli, A.; Cimino, S. The Use of Digital Technologies, Impulsivity and Psychopathological Symptoms in Adolescence. *Behav. Sci.* **2019**, *9*, 82. [CrossRef] [PubMed]
24. Elhai, J.D.; Levine, J.C.; Dvorak, R.D.; Hall, B.J. Non-social features of smartphone use are most related to depression, anxiety and problematic smartphone use. *Comput. Hum. Behav.* **2017**, *69*, 75–82. [CrossRef]
25. Mann, M.; Hosman, C.M.; Schaalma, H.P.; de Vries, N.K. Self-esteem in a broad-spectrum approach for mental health promotion. *Health Educ. Res.* **2004**, *19*, 357–372. [CrossRef]
26. J Kuss, D.; D Griffiths, M.; Karila, L.; Billieux, J. Internet addiction: A systematic review of epidemiological research for the last decade. *Curr. Pharm. Des.* **2014**, *20*, 4026–4052. [CrossRef]
27. Mehdizadeh, S. Self-presentation 2.0: Narcissism and self-esteem on Facebook. *Cyberpsychol. Behav. Soc. Netw.* **2010**, *13*, 357–364. [CrossRef]
28. Grüsser, S.M.; Thalemann, R.; Griffiths, M.D. Excessive computer game playing: Evidence for addiction and aggression? *Cyberpsychol. Behav.* **2006**, *10*, 290–292. [CrossRef]
29. Mentzoni, R.A.; Brunborg, G.S.; Molde, H.; Myrseth, H.; Skouverøe, K.J.M.; Hetland, J.; Pallesen, S. Problematic video game use: Estimated prevalence and associations with mental and physical health. *Cyberpsychol. Behav. Soc. Netw.* **2011**, *14*, 591–596. [CrossRef]
30. Aarseth, E.; Bean, A.M.; Boonen, H.; Colder Carras, M.; Coulson, M.; Das, D.; Deleuze, J.; Dunkels, E.; Edman, J.; Ferguson, C.J. Scholars' open debate paper on the World Health Organization ICD-11 Gaming Disorder proposal. *J. Behav. Addict.* **2017**, *6*, 267–270. [CrossRef]
31. Petry, N.M.; O'Brien, C.P. Internet gaming disorder and the DSM-5. *Addiction* **2013**, *108*, 1186–1187. [CrossRef]
32. Petry, N.M.; Rehbein, F.; Gentile, D.A.; Lemmens, J.S.; Rumpf, H.J.; Mößle, T.; Bischof, G.; Tao, R.; Fung, D.S.; Borges, G. An international consensus for assessing internet gaming disorder using the new DSM-5 approach. *Addiction* **2014**, *109*, 1399–1406. [CrossRef] [PubMed]
33. Ferguson, C.J. Do angry birds make for angry children? A meta-analysis of video game influences on children's and adolescents' aggression, mental health, prosocial behavior, and academic performance. *Perspect. Psychol. Sci.* **2015**, *10*, 646–666. [CrossRef] [PubMed]

34. Cranwell, J.; Murray, R.; Lewis, S.; Leonardi-Bee, J.; Dockrell, M.; Britton, J. Adolescents' exposure to tobacco and alcohol content in YouTube music videos. *Addiction* **2015**, *110*, 703–711. [CrossRef] [PubMed]
35. Chau, C. YouTube as a participatory culture. *New Dir. Youth Dev.* **2010**, *2010*, 65–74. [CrossRef]

International Journal of
Environmental Research and Public Health

MDPI

Article

A Marketing Approach to a Psychological Problem: Problematic Smartphone Use on Adolescents

Adnan Veysel Ertemel [1,*] and Ela Ari [2]

[1] Istanbul Commerce University, Faculty of Business Administration, International Trade Department, 34445 Istanbul, Turkey

[2] Istanbul Medipol University, Faculty of Humanities and Social Sciences, Psychology Department, 34810 Istanbul, Turkey; eari@medipol.edu.tr

* Correspondence: avertemel@ticaret.edu.tr

Received: 28 February 2020; Accepted: 1 April 2020; Published: 4 April 2020

Abstract: *Background:* Smartphones have become an indispensable part of the daily lives of adolescents in the 21st century, which is characterized by a highly digitized modern world. Besides their many advantages, smartphones might pave the way to compulsive usage and addictive experiences. To remedy this problem, this study proposes an authentic approach which integrates consumer behavior theories and techniques such as unhook and gamification. An education program has been designed based on these approaches to decrease the problematic smartphone use. *Method:* The participants of the education program consisted of 305 students (48.2% girls and 51.8% boys) with a mean age of 14.57 (SD = 0.74). The Demographic Form and Smartphone Addiction Scale for Adolescents (SASA) were conducted before the education program and three weeks after the education. *Results:* The results of the paired sample *t*-test analysis before and after the education program revealed that the SASA total scores decreased significantly ($p < 0.01$). There are significant differences in terms of gender, mothers' education and class levels. *Conclusion:* This research emphasizes the role of an interdisciplinary approach to the addiction problem. The content used in the education program includes strategies that originally aimed at increasing consumption. The effectiveness of the program can be enhanced further in the future along with self-regulatory additions.

Keywords: problematic smartphone use; adolescence; marketing; unhook; gamification

---

1. Introduction

With more than 3.1 [1] billion worldwide users as of 2019, which is expected to reach 3.8 billion by 2021 [2], smartphones have permanently changed our daily routines. They have become the de facto means for communication, socialization and access to information; and hence become inseparable in our lives [3]. When separated, individuals may encounter the following incidents: physiological withdrawal-like symptoms [4], increased anxiety [5] and even phantom vibration syndrome which can be described as the sensation that one's smartphone is vibrating even though there is no notification coming [6], a syndrome experienced by the majority of users [7].

Not surprisingly, problematic smartphone use, which can be categorized as a form of technological addiction [8], has become a major global concern across societies [3]. Technology addiction may be characterized as a non-chemical behavioral addiction involving human-computer interaction [9]. On the other hand, problematic smartphone use, which is also interchangeably called as smartphone addiction or smartphone use disorder [10], can be defined as the overuse of smartphones in a manner that is difficult to control and has harmful effects on the other areas of life [11].

As smartphones are devices that can be carried anywhere and are available anytime, problematic smartphone use causes various serious potential problems. Among those problems pointed in the literature are their negative effect in interpersonal relations, [12], personal health [13] well-being [14,15], including decreased academic performance and decreased physical activity [16] as well as depression symptoms [17]. Validating this phenomenon, American Psychiatric Association addressed this behavior in Diagnostic and Statistical Manual of Mental Disorders-5 (DSM-5) by introducing the non-substance addiction as a psychiatric diagnosis, although there are still controversies on the topic [18].

Contemporary research on behavioral psychology points out that addictions arise out of problematic habit-driven behavior, which is a natural outcome of ineffective self-regulation and self-control [19]. As such, this repetitive habitual behavior increases the risk of personal and social problems [20]. Frequent interruptions were found to result in loss of productivity [21] and higher tendency of attention deficit hyperactivity disorder [22]. Even the mere existence of a smartphone on a table when studying was found to result in reduced cognitive capacity and up to a 10% decrease in IQ [23,24]. Studies show that it takes 23 min on average for a fully concentrated individual to come back to the previous state after being interrupted with a notification message [25]. Worsening the case, the probability that the individual will self-interrupt increases by 8%.

As devices are being carried around all the time, Oulasvirta et al. [26] distinguish smartphone usage behavior as a series of 'short duration, isolated, reward-based' (SIRB) sessions as opposed to other technological devices like tablets and laptops. In this sense, informational rewards reinforce the checking behavior ultimately, performing it out of habit. As smartphones give access to a broad variety of sources for seeking information, socializing and entertainment, the overall reward value of "checking" habits increases as well.

Until recent times, severe criticisms of technology platforms due to their addictive-by-design experiences have gone unnoticed. However, the last couple of years witness a change in this case. As a result of the increasing awareness of these addictive-by-design techniques and the lobbying activities, technology platforms have started to adopt a more responsible code of conduct. As an example, Instagram and some other social media platforms changed their infinite scroll feature so that the users can scroll back up to two days. Again, as of 2019, Instagram started to hide the number of 'likes' to a post to avoid a meaningless competition of having more likes.

Problematic smartphone use can be attributed to many factors. In the last decade, however, it is argued that digital platforms' architected, finely tuned experiences that are designed to be addictive to deliberately make consumers spend prolonged periods of time on screens result in more problematic smartphone use [27–31]. Therefore, as opposed to looking from a mere psychological perspective, the need to adopt a multidisciplinary approach to prevent problematic smartphone use is brought into consideration.

Looking from a marketing perspective, we propose that using the same techniques that increase consumption—but in reverse manner—can effectively work in preventing problematic smartphone use. In the remainder of the paper, the defining characteristics of the modern digital marketing environment are discussed first. Afterwards, the mechanics of deliberately designed habit-forming experiences are explored. Finally, a consciousness-based preventive educational program on adolescents is proposed. The results of the study conducted among 305 students in 14 schools located in Istanbul, Turkey are also analyzed and discussed.

Adolescents were chosen as the target group since their acquaintance with smartphones is relatively new as opposed to their older counterparts, which implies that their smartphone behavior is not ossified yet. A study conducted late in 2019 in Turkey revealed that the most problematic smartphone use is seen in young adolescents and this problematic behavior rate decreases as the age of the users increases [32]. Another study conducted in Turkey published that 90.7% of the youngest age group (16–24). Internet use is listed as the heaviest compared to all other age groups. This group partially includes adolescents (10–19 years) which implies that this age group is the most susceptible to problematic smartphone use [33].

A recent global mobile consumer survey conducted in 28 countries across six continents revealed that Turkish consumers rank the highest in hourly checking of social networks and also highest in playing mobile games. Turkish consumers also ranked highest (26%) indicating physical health problems related with heavy smartphone usage [32]. These figures justify that adolescents are more prone to the risks associated with problematic smartphone use when compared to adults. Nevertheless, adolescents are more open to guidance for change. Undertaking such a study in Turkey is especially important since Turkey is classified as one of the countries with the highest risk potential of problematic smartphone use. Recent figures highlight that the percentage of Turkish users using WhatsApp heavily is 94% and Instagram with 80% is way higher than that of the European Union averages which is reported as 55% and 35%, respectively [32]. These rates place Turkey either to the first or the second place all over the world and above the averages of European countries.

In recent years, digital health related smartphone apps specifically designed to help individuals achieve positive behavior change started to emerge. These apps frequently use persuasive technology itself as a tool to reduce unhealthy, problematic smartphone use. The techniques used incorporate gamification, self-monitoring and goal setting [34].

However, to our knowledge, there is no structured education program that uses gamification, hook and the like techniques, in reverse manner, to decrease problematic smartphone use among adolescents.

1.1. New Marketing Strategies Appealing to the Unconscious

The highly competitive modern marketing environment of the 21st century imposes unique challenges. Whether online or offline, the abundance of marketing messages coming from a variety of sources has caused consumers to have difficulty in devoting their time and attention to those messages. With the advent of the Internet of things (IoT) and big data, consumers currently witness an even higher level of digitalization, which takes the aforementioned phenomenon to a different level.

In such a world, digital ecosystems are characterized by attention economy [35] in which marketers offer their services in exchange for the consumers' attention. Monetization occurs indirectly through advertisers. Facebook, Instagram, YouTube, Twitter are typical examples which do not charge consumers but monetize their attention in this attention economy.

Since appealing to the conscious mind in the attention economy becomes increasingly difficult marketers have started to resort to alternative strategies that appeal rather to the unconscious. As such, digital platforms, particularly in the last decade, are in an arms race to grab and keep the consumers' attention using sophisticated techniques in today's always on and connected world.

The flow theory, which has the aim of achieving the flow state, an emotional state embracing perceptual distortion and enjoyment [36] is one such strategy. Built on flow theory, new marketing strategies like gamification and user experience design which are collectively named as illusional marketing strategies [37,38] provide shortcuts to consumers by enabling them to take action without effortful thinking. The following sections provide further details on the marketing strategies, namely gamification and hook strategy, which make up the main focus of this paper, and are developed to design addictive experiences by appealing to the unconscious.

1.1.1. Gamification

Gamification can be described as the use of game design features in non-game contexts [39]. Through a gamified process, game elements and mechanics like achievements, progress bars, badges, leader boards, levels and points are employed to engage users [40]. In the context of marketing, gamification refers to a new marketing strategy that aims at changing the consumers' behavior in predetermined ways to foster engagement and ultimately to buy products [41,42]. In order to attain this goal, designation of an engaging experience plays a key role. Previous studies show that gamified and engaging experiences may result in brand recall and buying behavior [43].

Marketing scholars classify gamification as a different kind of persuasive marketing [44], which raises more ethical concerns than other persuasive marketing strategies like product packaging, personalized digital marketing, etc. Some other scholars define gamification as a form of stealth marketing which refers to configuring a product in such a way that people are not aware that the marketer is trying to persuade them to buy it [45]. These concerns arise out of the degree and type of engagement between the brand and consumers due to the highly engaging gamified experience. Deep sense of participatory engagement in gamification can also be compared to co-creation marketing [46,47] in which consumers' co-produce products by actively engaging in the design process [48]. However, gamification can have subversive results.

1.1.2. Hook Strategy

Hook is a recent marketing strategy similar to gamification, based on behaviorist thinking related to marketing software products. It has its roots in Persuasive Technology Lab in Stanford University, founded by BJ Fogg, back in 1998. This relatively new field has been named 'captology', which is an acronym for 'computers as persuasive technologies'. The ultimate goal in captology is to provide meaningful input into how software products can be designed to change consumer beliefs and behaviors [49].

Habitual behavior is central to such strategies. Habit-driven, automatic behaviors occupy the majority of the daily life and are performed with little, or no conscious thinking at all. Previous studies show that following a trigger-action-reward loop, a desirable habit can be imposed on an individual [50–52].

Impressed with Fogg's work, Eyal [53] developed the Hook model, which is grounded on the reinforcement theory in forming habitual behavior, which goes back to reinforcement experiments [54]. The Hook model advises a four-step loop to make the desired behavior a habitual one. The first step is the trigger, which refers to an external or internal stimulus that tells the user what to do next and how to behave next. The second step, action, refers to the actions of the user based on the information provided by the stimulus. The third stage, namely reward, refers to the acquisition of a variable reward as a consequence of the actions mentioned above. The last step, investment, means having the users make a deliberate effort to invest in the platform as in the form of time, effort, social capital investment to increase the likelihood of the repetitive behavior.

The Hook model can be illustrated through a typical Instagram usage habit. The process starts with a notification message, which is an external trigger. The model aims at starting with an external trigger and gradually making it an internal one. Internal triggers are states of mind and emotions. For the behavior to be habitual, the emotion should be somewhat vague and negative like loneliness, boredom, uncertainty, etc. Vague emotions allow the desired behavior to become a habit by avoiding reaching the consciousness level while negative emotions cause the individual to take action more easily [28].

Action refers to the behavior of entering the Instagram application. The key in the action step is to make the desired behavior as effortless as possible. This is because the level of effort required in the behavior directly affects the habitual nature of the behavior [53]. In this context, instead of requiring the user to enter a password, smartphones now enable a mindless sort of authentication by fingerprint or retina scans. This allows consumers to take the action step literally in an effortless manner. Auto-activation of the smartphone camera by a single swipe and the similar features serve the same goal.

The next step is the variable reward. It is what the notification is all about. For example, the user might have uploaded a photo album that consists of many photos. The variable reward is the discovery of which photo got the most likes, how many likes are received in total, who have commented and what they said about the posts and so on. The fact that the reward is variable allows the unconscious to be stimulated by surprise, causing dopamine release in the brain and regarding the perception of the reward as reinforcing [55]. Immediate variable rewards work perfectly well when the brain is in the unconscious mode [55] by gradually associating the reward with the trigger.

Additionally, it has been documented that this state of mind causes an individual's long run self to have no control over his or her decisions [56], which in turn causes a compulsive and addictive behavioral pattern.

The last step, investment to the platform, refers to invest in a small effort on the platform in such a way that calls for the repetition of the behavior over time, and finally makes it impossible to break off from the cycle. In Instagram application, this investment is the number of followers and the photo archive accumulated so far. The more investment is made in the platform, the more addictive it becomes. The system aims to replace the external trigger with the internal trigger in the long term. In Instagram application, the internal trigger is the desire not to miss the opportunity to capture a photo of the moment. An individual's Spotify list, for example, is perceived as valuable since he/she has spent a considerable amount of effort creating that list.

The addictive nature of various digital platforms was confirmed in various studies [31,57]. Some authors who investigated the gamification strategy also warn that the application of variable rewards may create gambling-like addictive experiences, which may lead to consumers being unable to exercise their free will [58]. The flow state is a state of mind characterized with enjoyment and perceptual distortion in time and space contributing to the addictive behavior [36]. The Hook model could be classified as a more sophisticated form of gamification since it targets the unconscious, applying gamification elements such as variable rewards and is designed to be addictive in nature. In this study, both hook and gamification strategies will be included in the education program considering their potential to create new habitual behavior [30].

Finally, this study first sought to explore the demographic factors influence on the adolescents' problematic smartphone use, namely gender, age, grades and parental education [59–62]. Afterwards, the main focus of the study is to find out how does a marketing based education program reduces the problematic smartphone use of adolescents. Further, and last of all, the study investigates how the smartphone problematic use changed after the education according to the above stated demographic factors as gender, age, and grades except parents' education. For all these research aims, the following hypotheses are formulated under three main categories, which are before the education program, about the education program and after the education program. Thus, the following hypotheses guided the present study.

Before the education program;

Hypothesis 1 (H1). *The problematic smartphone use differs according to demographic factors.*

Hypothesis 1 (H1a). *The problematic smartphone use differs on the basis of gender.*

Hypothesis 1 (H1b). *Age and problematic smartphone use are related.*

Hypothesis 1 (H1c). *The problematic smartphone use differs on the basis of the adolescents' grades.*

Hypothesis 1 (H1d). *The problematic smartphone use of adolescents differs on the basis of their parents' education level.*

About the education program;

Hypothesis 2 (H2). *The adolescents' problematic smartphone use will decrease after receiving a marketing based education program including gamification and unhook strategies.*

After the education program,

Hypothesis 3 (H3). *The problematic smartphone use differs according to demographic factors after the education program.*

Hypothesis 3 (H3a). *The problematic smartphone use decrease differs on the basis of gender after the education.*

Hypothesis 3 (H3b). *Age and the problematic smartphone use after the education are related.*

Hypothesis 3 (H3c). *The problematic smartphone use differs on the basis of the adolescents' grades after the education.*

2. Method

2.1. Participants

The sample consisted of 337 students from 14 different schools. Out of 337 participants, 32 were excluded since they were outliers or had missing data. In total, 144 girls (47.2%) and 161 boys (52.8%) participated in the study. Their mean age was 14.57 (SD = 0.74).

2.2. Procedure

2.2.1. Before the Education Program

The researchers have promoted the free education program to decrease problematic smartphone use on the social media platforms of the associated university. After that, the university corporate communications and public relations department e-mailed about the education program to school counselors of various schools residing in Istanbul. Following this, the education program was scheduled with the school counselors who requested it. Afterwards, school counselors have chosen the class sections randomly. Then, the school counselors sent to their families two weeks before the parental permission form including an information sheet about the research and a certificate of consent about the adolescent's participation to the education program and research.

On the day of the education after the education, those students who volunteer to participate signed an assent form. If the adolescents did not sign the assent form they were not included in the study. Thus, only 337 out of 1055 students signed the assent form. Then, the researcher explained the procedures to the participants to complete the surveys. The participants were allowed to ask questions if they encountered problems while completing the questionnaires, and the research assistants helped them to resolve their problems.

2.2.2. Description of the Education Program

The intervention included a PowerPoint presentation accompanied by several striking videos that highlight the non-neutral characteristic of technology. The adolescents were also given flyers that highlight the key points as reminders of the education and a check-list of to-do items to be performed throughout the three-week observation period.

The whole program was delivered in one day using two breaks. The education program consisted of eight modules. Each module of the program roughly lasted about 15 min. The first module consisted of creating awareness about the problematic use of smartphones. The educator presented famous technology figures' examples and introduced statistics about the problematic smartphone use.

The module included evidence based striking facts and insights on how Silicon Valley-based technology platforms create addictive experiences. For example, Twitter, Instagram and Facebook platforms create a "craving" effect on users by making them wait (on variable time) to get their liking feedback.

Once the participants were oriented and motivated to be a conscious smartphone user, then the next module was implemented, which showed how to measure and keep track of the participants' screen time. The educator explained in detail the hook strategy that marketers employ to engage users.

The next module was about presenting the participants how to unhook from their smartphones. Snapchat's Snap Streaks feature could be regarded as a deceptive example to hook strategy. This feature shows the number of consecutive days someone has messaged to his or her respective friend as a badge. Adolescents see it as an indication of how strong their relationships are with their acquaintances and hence they do not want to break the streak. During the education program, some participants reflected by stating that when they were to be offline for a certain period of time (during a holiday with their family, etc.) they gave their user credentials to one of their close friends and asked them to send dummy messages to their stated friends so as not to break their snap streaks.

In order to practice unhook strategies, the participants were first asked to identify the applications that made them feel unhappy. These applications were removed with the guidance of the educator. Additionally, in this module, the participants were persuaded to hide the social media applications that hooked them inside a folder. The home screen was cleared up and the time consuming applications were saved in another folder.

During the third module, participants learned how to turn off all the notifications. Also the educator instructed to mute the group messages and to remove the last seen or read information from the account menu under the privacy settings. Once all the sources of distractions were removed, the fourth module was initiated.

It was in the fourth module that a digital diet program was introduced. The fifth module integrated mindfulness applications as well as attention focusing techniques to control unconscious use. The sixth module included strategies to increase the quality of the participants' social relationships. The adverse effects of being available all the time and its damaging impact on personal productivity were explained along with supportive research findings. Afterwards, steps that could be taken to prevent the problem were introduced. On WhatsApp, for example, the adolescents were advised to avoid immediate reply to not-so-urgent incoming messages and to put a deliberate delay so that the other party's anticipation for an immediate reply was lowered. Then, the educator taught the participants to build the communication rules and facilitate the quality of social communication without phone. Recent studies showed that 72% of adolescents (as opposed to 48% of adults) feel an urge to respond to an incoming call or a message immediately [63]. This phenomenon can be attributed to the wrong interpersonal communication etiquette as well as low self-control and peer pressure. Therefore, the education module also incorporated the proper etiquette for communication that relieves the individual from feeling the urge to respond to everything immediately.

All through the seventh module a gamification strategy was used to engage the adolescents with the use of #iamincontrol hashtag use. This module of the education was about adolescents' controlling their lives. The participants learned how to protect an area they identified as a phone-free zone. The educator taught when, where and how to leave their phone in their daily routines.

The participants were also offered many activities they could plan together with their loved ones. Along the modules, the participants were asked to do homework and were supported with manuals, digital diet planners, activities list and handy calendars. Once all the content was delivered, the researcher scheduled a follow-up meeting three weeks after with the same school.

2.2.3. After the Education Program

Three weeks later, the scales and checklist were collected from the participants with the help of the school counselors. The participants who conducted the demographic questionnaires and the smartphone addiction scale before the education were distributed an id number so that it matched the collected surveys after the program. The researchers prepared an evaluation form to test the effectiveness of each task in the education program. Then, the participants were asked to rate about the effectiveness of each step in the program. Finally, they rated these steps according to their own usage benefits.

2.3. Measures

Demographic Form: The form includes gender, age, parents' education, parents' occupation, social media use of the family and the adolescent, purpose of use for the smartphone and the time they started using it and how much time they spent on it on a daily basis.

The Smartphone Addiction Scale for Adolescents

The Smartphone Addiction Scale for Adolescents (SASA) measures smartphone addiction. SASA was developed by Kwon et al. [60]. The Turkish validation study conducted by Demirci et al. (2014) consisted of 6 point Likert scales ranging from 1 (Definitely No) to 6 (Definitely Yes), 36-items and 7 factors as disturbing daily life and tolerance, withdrawal symptoms, positive anticipation, overuse, social network dependency, physical symptoms and cyberspace-oriented relationships [61].

The correlations between the subscales ranged from 0.22 to 0.51 The highest total scales values indicated a higher risk for problematic smartphone addiction. The Cronbach's α's of the seven factors were found respectively as 0.85, 0.81, 0.74, 0.62, 0.59, 0.64, 0.34 and 0.89 for the total scale in this study, which are very close to the validated scale. The Pearson correlations coefficients varied between the SASA factors and total varied between 0.18 and 0.75, indicating statistically significant positive relations (all p's < 0.01). Since it is a reliable and valid tool of measurement, SASA was conducted to assess the change in the smartphone usage before and after the education program.

2.4. Statistical Analysis

The data analysis was conducted using SPSS 21 for Windows. Descriptive statistics were calculated to investigate the demographic data. In order to test first and third set of hypotheses, the demographics of the participants were compared by independent sample *t*-tests analysis of variance tests. Pearson moment correlation analysis was run to evaluate the relationships between continuous demographics (e.g., age) and the factors of SASA. In order to test Hypothesis 2 which aims to evaluate of the effectiveness of education, paired sample *t*-test was used to compare before and after SASA scores.

2.4.1. Demographic Results

Three hundred and five participants were included in this study. 144 (47.2%) were female, and 161 (52.8%) were male. Only 44 (14.4%) of the participants indicated that they had cyber friendship, which they had met online and then later got to know in person. Table 1 listed gender, grades, electronics they possess, purpose of use, applications they have and parents' education.

2.4.2. Difference Tests Results before the Education

Smartphone use variable was the reported use of smartphone time as minutes per day ($X_{use} =$ 139.84, SD = 83.68). The daily use of smartphones was declared to be less than 180 minutes for 75%, between 180 to 300 min for 21.5% and more than 300 min for 3.5% of the sample.

The first set of hypotheses explored the problematic smartphone use difference in terms of gender (H1a), age (H1b), adolescents' grades (H1c) and parents' education (H1d). In order to test gender differences, an independent sample *t*-test was conducted. As a result, there were no significant difference between boys ($X_{boys} = 81.06$) and girls ($X_{girls} = 84.99$) in terms of total SASA before the education ($p > 0.05$).

Then, to test the age and SASA relation (H1b), a Pearson moment correlation analysis was conducted. The findings revealed no significant relationship between age and SASA scores (r = 0.01, $p > 0.05$).

In order to test the problematic smartphone use difference on the basis of the adolescents' grades (H1c) an analysis of variance test showed that SASA scores before the education do differ significantly according to the students' grade levels. A Tukey post hoc test indicated that the SASA scores increased significantly at 10th grade ($X_{Pep} = 16.19$ $X_{9th} = 17.26$ $X_{10th} = 19.33$ X_{11th} 17.06 $F_{(305)} = 3.59$, $p < 0.05$ d = 0.3).

Table 1. Demographics of the Participants.

N = 305	Frequency	Percentage
Gender		
Female	144	47.2
Male	161	51.8
Grades		
Prep (14 yrs)	44	14.4
9th Grade (15 yrs)	210	68.9
10th Grade (16 yrs)	31	10.2
11th Grade (17 yrs)	20	6.6
The distribution of electronics they do possess		
Smartphone	305	100
Computer	187	61.3
Pad	82	26.9
Console	57	18.7
Smartphone purpose of use		
Education/Research	275	90.2
Entertainment/Game	239	78.4
Free Time	207	67.9
Social Media	238	78.0
Communication	217	71.1
Shopping	124	40.7
Applications participants have		
WhatsApp	302	99.0
Instagram	247	81.0
Facebook	75	24.6
TicToc	10	3.3
Snapchat	102	33.4
Twitter	58	19.0
Pinterest	67	22.0
YouTube	222	72.8
Twitch	43	14.1
Mother's Education		
Middle School	125	41.0
High School	78	25.6
College	102	33.4
Father's Education		
Middle School	89	29.2
High School	91	29.8
College	125	41.0

In order to measure the problematic smartphone difference on the basis of the parents' education (H1d) an analysis of variance test indicated that the adolescents' SASA scores vary according to their mothers' education level. A Tukey post hoc test showed that there was a significant difference between high school graduates and college graduates' mothers in terms of their children's SASA scores ($X_{MiddleSchool}$ = 17.39 $X_{HighSchool}$ = 18.13 $X_{College}$ = 16.52; $F_{(305)} = 3.62$, $p < 0.05$). Another analysis of variance test revealed that fathers' education also differed in terms of their children's SASA scores. A Tukey post hoc test showed that there was a significant difference between high school graduates and college graduates fathers' in terms of their children's SASA scores ($X_{MiddleSchool}$ = 16.95 $X_{HighSchool}$ = 18.22 $X_{College}$ = 16.89; $F_{(305)} = 3.21$, $p < 0.05$ d = 0.3).

2.4.3. Difference Tests Results after the Education

In order to test the second hypothesis, which evaluated the influence of the education program, a paired sample *t*-test analysis before and after the education program was conducted. As shown in Table 2, the results revealed a significant decrease in disturbing daily life and tolerance ($X_{pretest}$ = 2.93 $X_{posttest}$ = 2.68), withdrawal symptoms ($X_{pretest}$ = 2.09 $X_{posttest}$ = 1.97), positive anticipation ($X_{pretest}$ =

3.18 $X_{posttest} = 2.84$), overuse ($X_{pretest} = 2.61$ $X_{posttest} = 2.44$), physical symptoms ($X_{pretest} = 2.67$ $X_{posttest} = 2.51$) dimensions and in the total SASA scale ($X_{pretest} = 17.30$ $X_{posttest} = 16.19$). Cyberspace oriented relationships and social network dependence factors did not change significantly after the education program (p 's > 0.05). Thus, the effectiveness of the education program has been proven for total and five factors except cyberspace oriented relationships and social network dependence factors as can be seen in Table 2.

Table 2. Paired sample t test results before and after the education program.

Factors	Mean Difference	St. Dev	t	p
1. Disturbing daily life and tolerance	0.25	0.81	5.29	<0.01 **
2. Withdrawal symptoms	0.12	0.75	2.56	<0.01 **
3. Positive anticipation	0.34	0.88	6.73	<0.01 **
4. Cyberspace oriented relationships	0.03	0.74	0.75	0.45
5. Overuse	0.17	0.89	3.32	<0.01 **
6. Social network dependence	0.06	1.05	1.06	0.29
7. Physical symptoms	0.15	0.89	2.97	<0.01 **
Total SASA	1.11	3.45	5.61	<0.01 **

Notes: ** deemed significant at the 0.01 level.

In order to test the third set of hypotheses that explore the difference of the problematic smartphone use in terms of gender (H3a), age (H3b) and adolescents' grades (H3c). Firstly, an independent sample t-test was conducted to compare boys' and girls' SASA scores after the education program (H3a). The results revealed that male adolescents' total SASA dropped significantly more than those of the girls with a medium effect size ($X_{female} = 17.17$ $X_{male} = 15.31$; $t_{(305)} = 3.63$, $p < 0.01$ d = 0.42). Additionally, the SASA scores of the participants after the education have been subtracted from the SASA scores before the education. Boys' change is significantly higher than that of girls' change ($X_{girls} = -0.58$ $X_{boys} = -1.58$; $t_{(305)} = 2.55$, $p < 0.01$ d = 0.30).

Following this, to test the age and SASA relation after the education program (H3b), a Pearson moment correlation analysis was conducted. The findings revealed no significant relationship between age and SASA scores after the education program (r = −0.01, $p > 0.05$).

After the completion of the education, an analysis of variance test indicated that SASA scores varied significantly according to the adolescent's grades. A Tukey post hoc test showed that only the SASA scores of preparation grade and 10th graders were significantly different ($X_{Pep} = 14.56$ $X_{9th} = 16.36$ $X_{10th} = 17.33$ $X_{11th} = 16.23$; $F_{(305)} = 2.66$, $p < 0.05$ d = 0.02).

2.5. Feedback Results about the Program

The percentage of the participants who practice no-tolerance rule to phubbing friends, those who prefer to look to their smartphone screens instead of themselves in a physical conversation were 62% and 56% of them have found this method to be somewhat or absolutely useful. The corresponding percentages for turning-off all notifications (except from real people) were 55% and 58%, putting all WhatsApp group messages to silent mode were 68% and 65% and turning off the last seen time and read receipt of WhatsApp messages were 58% and 56% respectively.

Initially, many adolescents had reported that they had hard time trying them. After experiencing these methods, however, many of them reflected that they feel more in control of technology and can now focus more easily on their tasks.

3. Discussion

This study first aims to investigate the role of demographic factors, namely gender, age, grades and parents' education level of adolescents in assessing their tendency for smartphone addiction before the education. The first hypothesis related to demographics' differentiation in terms of problematic smartphone use is examined separately for gender, age, grades and parental education. The problematic smartphone use did not differ on the basis of gender nor has been associated with age. Only the problematic smartphone use differs for grades and parental education. The second hypothesis, which constitutes the main objective of this study, is about inquiring the effectiveness of a novel education program for reducing problematic smartphone use in adolescents. In order to achieve this, a marketing based education program has been developed and then later measured the effect before and three weeks after the program. The change in problematic smartphone use showed a significant decrease. Moreover, the last hypothesis is related to the demographic factors, specifically gender, age and grades. These demographics are critical in order to understand which ones differed after the education for future references. Gender and grades differed in terms of decreases in problematic smartphone addiction.

Concerning the first hypothesis, the current study revealed that there is no statistically significant difference between girls and boys in terms of their SASA scores before the education. Thus, the first hypothesis of this study is rejected confirming the previous studies [59,60]. There are many studies that indicated female addiction is higher than males [64–66]. These studies attribute females' higher tendency for mobile addiction to females' being more likely to engage in interpersonal communication and social relationships via their mobile phone [67,68]. Moreover, Demirci et al.'s (2014) study conducted in Turkey indicated higher SASA scores for female participants. The fact that the difference between boys and girls is not significant in this present study can be explained first by age factor. The participants of this study consisted of adolescents while Demirci et al.'s (2014) study consisted of college students. Second, not only the age group might have brought out the discrepancy between these studies but also the six years' difference (2014–2020) might have changed the rates. Indeed, gender gap is predicted to decrease over the years with the advances in technology spreads, which also explains the current gender similarity in terms of addiction [69]. The total SASA scores of the adolescents in this study (girls were 84.99 and boys were 81.06) are higher but the gender difference has closed compared to previous findings [60] as (females were 78.7 and males were 72.2). As a matter of fact, this finding confirm that the gender gap is closing and problematic smartphone use has gradually become more prevalent among adolescents in recent years [32].

This study could not find a statistically significant relationship between age and problematic smartphone use. As Demirci et al. (2014) study [60] also found non-significant negative relationship between age and smartphone addiction, the results are in line with the previous study. Some studies, particularly in the Internet addiction domain, find such negative but significant relation between adolescents' age and addiction level stating that this risky behavior is temporary in the adolescence period. This phenomenon is said to wear off with increased familiarity with the technology [70,71].

The problematic smartphone use differed significantly on the basis of the adolescents' grades. Beginning around age 10 adolescent risk taking, reward sensitivity behaviors start increasing, then peak between 13 and 16, and then decline [72–74]. The problematic smartphone use according to grades is in line with the risk preference trend. The earlier grades' SASA scores are lower than the later ones, then there is a peak at 10th graders and then it drops again.

The higher the maternal and paternal education level the lower were the SASA scores except for college graduates. The college graduate parents are mostly white collar workers who might spend less time and control over the adolescents which might have caused more problematic smartphone use. Thus, this finding of the present study partially confirms the previous studies in the literature [66,67]. In the light of these findings, we can conclude that, both parents' education and attention, which ultimately contributes to a more beneficial attitude towards adolescents [66].

The main aim this study is to reduce the problematic smartphone use after a marketing based education program. The education program started with explaining the negative consequences of problematic smartphone use in terms of its harm on cognitive capacity, academic performance and social relationships. Afterwards, the internal mechanics of addictive design elements of digital products are explained in detail. The program followed by going through the steps in the hook model aimed at exercising the opposite behavior in each step to prevent unconscious, compulsive smartphone usage and ensure conscious and healthy usage behavior. The education program was coupled with a follow-up meeting three weeks after the education. The education module aimed at building awareness on the specifics of addictive mechanisms that ultimately hook users.

Concerning the second hypothesis, the results are very promising since the comparison of SASA scores before and after the education program has proven significant decrease in problematic smartphone use. The disturbing daily life and tolerance, withdrawal symptoms, positive anticipation, overuse, physical symptoms dimensions and the total SASA scores decreased significantly after the education program. The cyberspace-oriented relationships and social network dependence factors did not change significantly after the education program. When the cyberspace-oriented relationship factors' questions items are checked closely, there is a distinction between cyber friends and friends. Only a small percent of the participants did make cyber friendship (14.1%). This fact might be due to the more conservative nature of families in Turkey [75] in which adolescents are still protected closely. Finally, the social network dependence factor did not change significantly after the education. The two items that measured social dependence did include social applications as Facebook and Twitter. The participants' reports indicated that the rate of use for Facebook (24.6%) and Twitter (18.5%) is considerably low. Since most of the participants do not use these applications, these questions might have misdirected them.

Concerning the third hypothesis, the boys' SASA scores decreased more compared to those of the girls. The addiction literature provides evidence of that women had 2.7 more risk at smartphone addiction than men [76]. When women try to be abstinent, they show greater anxiety and stress compared to men [77,78]. First, it might be less stressful to reduce the problematic smartphone use for boys than girls. Furthermore, girls tend to strengthen their friendships, which keeps them on the smartphones and receive more emotional support, while boys aim to use smartphones to extend their weak social relations [79]. The education program's emphasis on how to build social relationship without the use of smartphones might have reduced boys' problematic smartphone use more than the girls. Girls define smartphones as central component of their personal existence [80] which could bring more resistance to reduce their smartphone use.

It was found that there is no significant relationship with adolescents' age and smartphone addiction levels after the education program. There are in the literature longitudinal and cross sectional studies where the expected behavior and age were not correlated after the intervention either [81,82]. A meta-analysis about the intervention programs applied to subjects under 18 years of age revealed that age did not have an influence on the success of the programs [83]. However, the later study mentioned about the possibility of cognitive developmental level, which might be related to grades.

The problematic smartphone use differed significantly on the basis of the adolescents' grades after the education. As a matter of fact, this finding pattern is very similar to the case before education. This illustrates that adolescents at earlier grades are still more open to guidance for healthier and more proper communication suggestions. This can be also explained by the fact that the education could be more influential when the habitual behavior of using a smartphone has recently started.

The results also reveal that, with a 97.2% usage rate, smartphone use is much more prevalent than the use of any other technology products. The results also show that among the responding adolescents, 57.6% check their social media from their smartphone more than 10 times a day. WhatsApp was found to be the most frequently installed mobile app; almost all the responding adolescents (98.2%) use it. Instagram (80.5%) was found to be the second mostly used app among the adolescents.

Overall, the results of the study suggest that educational guidance for healthy usage of technology is missing or not adequate in the visited schools. Unfortunately, the current educational system fails to provide even the basic communication etiquette to adolescents. That is, adolescents do not know and are eager to learn the proper way to communicate with their peers (importance of face to face communication etc.). This is especially valid for younger adults as opposed to their older counterparts.

Limitations and Implications for Future Research

Due to the peculiar characteristics of this special age group, namely adolescents, self-reported inventories like the one in this study have self-assessment bias risk. The results may also differ in terms of participants' self-involvement level, sincerity, beliefs etc. In addition, this study has been conducted in a relatively more conservative culture, which might have driven participants to give desirable answers.

All in all, the program is distinguished in that an improvement is achieved using a structured program. However, in the future, longer and more thorough educational programs could be designed to achieve more lasting effect. In addition, such an educational program could have much more impact if coupled with a digital health app usage that tracks individual behavior.

4. Conclusions

Technology platforms have been creating increasingly more addictive experiences that result in spending prolonged periods of time on screens, which ultimately leads to more profitability in the attention economy. Some of these addictive features include the lack of stop sign facilitated by infinite scroll feature, the hooking technique that relies mainly on the unconscious and employs mechanisms like variable rewards.

Problematic smartphone use increasingly becomes more of a concern globally especially among adolescents. In this study, a marketing approach was adopted in an education program. Consumer-engaging methodologies are used to increase awareness on specifics of hooking strategies and to disengage adolescents from their phones. The results are significant in that an apparent improvement in problematic smartphone use is accomplished with a short and structured education program. Although there are some mindfulness-based intervention studies that have proven to reduce smartphone use [84], to our knowledge, this is the first study in which a marketing-focused education program has been developed to approach smartphone addiction. The program also offered practical, ready-to-use tips to restrain oneself from unconscious, compulsive smartphone use.

This study makes significant contributions to the existing literature in several aspects. Firstly, looking from a marketing perspective, it uses the strategies employed to increase compulsive, unconscious smartphone usage, but in reverse direction to prevent excessive and hence problematic smartphone use. In addition, an educational program was introduced with a special emphasis placed on the conscious usage of smartphones.

Living without digital technologies has become almost impossible. This is particularly true as global pandemics (i.e., Covid-19 virus a.k.a. Corona virus) and the like developments are poised to change our daily habits radically. However, the key question in our relationship with technology is about which party will be in control. Unconscious, compulsive smartphone usage lets technology control users. It will be only by educating the society with the proper way of smartphone use that we can turn the situation the other way around.

The conducted study was scheduled as a one-time education. However, if these eight modules could be spread across the eight-week period with more supportive homework and with more detailed monitoring, the benefits could be improved. Future studies could integrate cognitive behavioral approaches and more gamified activities in order to hook adolescents back to their own future.

Author Contributions: Conceptualization, project administration, organization of the schools, as well as education content and delivery were prepared by A.V.E. Methodology, scale and demographics selections, data preparation and analysis were done by E.A. Writing, reviewing and editing were distributed equally between the two authors. All authors have read and agreed to the published version of the manuscript.

Funding: This project was funded by Istanbul Commerce University.

Acknowledgments: We would like to acknowledge Istanbul Commerce University for funding to our project and Oya Özkardeş for her guidance. Special thanks to project fellows Dilara Güler and Zehra Özalp for their technical support in data analysis, Ahu Dereli Dursun and Ipek Sarıgöz for her rigorous work in editing process.

Conflicts of Interest: The authors declare no conflict of interest.

References

1. We Are Social. Global Report. 2019. Available online: https://wearesocial.com/global-digital-report-2019 (accessed on 10 September 2019).
2. Ericsson, A.B. *Mobility Report 2015*; Ericsson: Stockholm, Sweden, 2017.
3. Cheever, N.A.; Rosen, L.D.; Carrier, L.M.; Chavez, A. Out of sight is not out of mind: The impact of restricting wireless mobile device use on anxiety levels among low, moderate and high users. *Comput. Hum. Behav.* **2014**, *37*, 290–297. [CrossRef]
4. Clayton, R.B.; Leshner, G.; Almond, A. The extended iSelf: The impact of iPhone separation on cognition, emotion, and physiology. *J. Comput. Mediat. Commun.* **2015**, *20*, 119–135. [CrossRef]
5. Wise, R.A.; Koob, G.F. The development and maintenance of drug addiction. *Neuropsychopharmacology* **2014**, *39*, 254. [CrossRef] [PubMed]
6. Kruger, D.J.; Djerf, J.M. High ringxiety: Attachment anxiety predicts experiences of phantom cell phone ringing. *Cyberpsychol. Behav. Soc. Netw.* **2016**, *19*, 56–59. [CrossRef]
7. Drouin, M.; Kaiser, D.H.; Miller, D.A. Phantom vibrations among undergraduates: Prevalence and associated psychological characteristics. *Comput. Hum. Behav.* **2012**, *28*, 1490–1496. [CrossRef]
8. Lin, Y.H.; Chang, L.R.; Lee, Y.H.; Tseng, H.W.; Kuo, T.B.; Chen, S.H. Development and validation of the Smartphone Addiction Inventory (SPAI). *PLoS ONE* **2014**, *9*, e98312. [CrossRef]
9. Griffiths, M.D. Behavioural addiction and substance addiction should be defined by their similarities not their dissimilarities. *Addiction* **2017**, *112*, 1718–1720. [CrossRef]
10. Griffiths, M. Gambling on the Internet: A brief note. *J. Gambl. Stud.* **1996**, *12*, 4–471. [CrossRef]
11. Park, N.; Lee, H. Social implications of smartphone use: Korean college students' smartphone use and psychological well being. *Cyberpsychol. Psychol. Behav. Soc. Netw.* **2012**, *15*, 491–497. [CrossRef]
12. Augner, C.; Hacker, G.W. Associations between problematic mobile phone use and psychological parameters in young adults. *Int. J. Public Health* **2012**, *57*, 437–441. [CrossRef]
13. Chóliz, M. Mobile-phone addiction in adolescence: The test of mobile phone dependence (TMD). *Prog. Health Sci.* **2012**, *2*, 33–44.
14. Leung, L. Linking psychological attributes to addiction and improper use of the mobile phone among adolescents in Hong Kong. *J. Child. Media* **2008**, *2*, 93–113. [CrossRef]
15. Samaha, M.; Hawi, N.S. Relationships among smartphone addiction, stress, academic performance, and satisfaction with life. *Comput. Hum. Behav.* **2016**, *57*, 321–325. [CrossRef]
16. Thomee, S.; Harenstam, A.; Hagberg, M. Mobile phone use and stress, sleep disturbances, and symptoms of depression among young adults a prospective cohort study. *BMC Public Health* **2011**, *11*, 66. [CrossRef] [PubMed]
17. David, P.; Kim, J.H.; Brickman, J.S.; Ran, W.; Curtis, C.M. Mobile phone distraction while studying. *New Media Soc.* **2015**, *17*, 1661–1679. [CrossRef]
18. American Psychiatric Association. *Diagnostic and Statistical Manual of Mental Disorders*, 5th ed.; American Psychiatric Association: Washington, DC, USA, 1994.
19. LaRose, R.; Lin, C.A.; Eastin, M.S. Unregulated Internet usage: Addiction, habit, or deficient self-regulation. *Media Psychol.* **2003**, *5*, 225–253. [CrossRef]
20. Marlatt, G.A.; Baer, J.S.; Donovan, D.M.; Kivlahan, D.R. Addictive behaviors: Etiology and treatment. *Annu. Rev. Psychol.* **1988**, *39*, 223–252. [CrossRef]
21. Duke, É.; Montag, C. Smartphone addiction, daily interruptions and self-reported productivity. *Addict. Behav. Rep.* **2017**, *6*, 90–95. [CrossRef]
22. Lachmann, B.; Sindermann, C.; Sariyska, R.Y.; Luo, R.; Melchers, M.C.; Becker, B.; Montag, C. The role of empathy and life satisfaction in internet and smartphone use disorder. *Front. Psychol.* **2018**, *9*, 398. [CrossRef]
23. Ward, A.F.; Duke, K.; Gneezy, A.; Bos, M.W. Brain drain: The mere presence of one's own smartphone reduces available cognitive capacity. *J. Assoc. Consum. Res.* **2017**, *2*, 140–154. [CrossRef]
24. Krznaric, R. *Carpe Diem: Reclaiming Seize the Day in a Distracted World*, 1st ed.; Penguin: London, UK, 2017.
25. Mark, G.; Gudith, D.; Klocke, U. The Cost of Interrupted Work: More Speed and Stress. In Proceedings of the SIGCHI Conference on Human Factors in Computing Systems 2008, Florence, Italy, 5–10 April 2008; pp. 107–110.

26. Oulasvirta, A.; Rattenbury, T.; Ma, L.; Raita, E. Habits make smartphone use more pervasive. *Pers. Ubiquitous Comput.* **2012**, *16*, 105–114. [CrossRef]
27. Ertemel, A.V.; Aydın, G. Technology addiction in the digital economy and suggested solutions. *Addicta Turk. J. Addict.* **2018**, *5*, 665–690.
28. Alter, A. *Irresistible: The Rise of Addictive Technology and the Business of Keeping Us Hooked*, 1st ed.; Penguin: London, UK, 2017.
29. Harris, T. How a Handful of Tech Companies Control Billions of Minds Every Day. 2017. Available online: https://www.ted.com/talks/tristan_harris_the_manipulative_tricks_tech_companies_use_to_capture_your_attention (accessed on 10 December 2019).
30. Chou, T.J.; Ting, C.C. The role of flow experience in cyber-game addiction. *Cyberpsychol. Behav.* **2003**, *6*, 663–675. [CrossRef] [PubMed]
31. Alrobai, A.; Phalp, K.; Ali, R. Digital Addiction: A Requirements Engineering Perspective. In *International Working Conference on Requirements Engineering: Foundation for Soft-Ware Quality*; Springer: Berlin/Heidelberg, Germany, 2014; pp. 112–118.
32. Deloitte Global Mobile User Survey. Available online: https://www2.deloitte.com/tr/tr/pages/technology-media-and-telecommunications/articles/TR-GMCS-2019.html (accessed on 12 December 2019).
33. TurkStat. *Annual Report on Information and Communication Technology (ICT) Usage in Households and by Individuals*; Turkish Statistical Institute: Ankara, Turkey, 2018.
34. Bychkov, D.; Young, S.D. Facing Up to Nomophobia: A Systematic Review of Mobile Phone Apps that Reduce Smartphone Usage. In *Big Data in Engineering Applications*; Springer: Singapore, 2018; pp. 161–171.
35. Davenport, T.H.; Beck, J.C. *The Attention Economy: Understanding the New Currency of Business*; Harvard Business Press: Harvard, MA, USA, 2001.
36. Csikszentmihalyi, M.; Csikszentmihalyi, I.S. *Optimal Experience: Psychological Studies of Flow in Consciousness*; Cambridge University Press: Cambridge, UK, 1992.
37. Ertemel, A.V. *Dijital çağda Illüzyonel Pazarlama*, 1st ed.; Abaküs Kitap: İstanbul, Turkey, 2016.
38. Ertemel, A.V. The Importance of Unconscious Mind for Consumer Engagement in the Digital Age: A Model Proposal. In Proceedings of the 23th National Marketing Conference, Trabzon, Turkey, 9 September 2017.
39. Deterding, S.; Dixon, D.; Khaled, R.; Nacke, L. From Game Design Elements to Gamefulness: Defining Gamification. In Proceedings of the 15th International Academic Mindtrek Conference: Envisioning Future Media Environments, Tampere, Finland, 28–20 September 2011.
40. Xu, F.; Weber, J.; Buhalis, D. The Gamification of Tourism. In *Information and Communication Technologies in Tourism*; Xiang, Z., Tussyadiah, I., Eds.; Springer: Wien, Austria, 2014.
41. Rafttopoulos, M. Toward gamification transparency: A conceptual framework for the development of responsible gamified enterprise systems. *J. Gaming Virtual Worlds* **2014**, *6*, 159–178. [CrossRef]
42. Shahri, A.; Hosseini, M.; Phalp, K.; Taylor, J.; Ali, R. Toward a Code of Ethics for Gamification at Enterprise. In *The Practice of Enterprise Modeling*; Frank, V., Loucoploulos, P., Pastor, O., Petrounias, I., Eds.; Springer: Berlin, Germany, 2014; pp. 235–245.
43. Vashisht, D.; Sreejesh, S. Impact of Game Speed and Persuasion Knowledge on Brand Recall and Brand Attitude. In Proceedings of the Twelfth AIMS International Conference on Management, Kerala, India, 2–5 January 2015; pp. 231–234.
44. Thorpe, A.S.; Roper, S. The ethics of gamification in a marketing context. *J. Bus. Ethics* **2019**, *155*, 597–609. [CrossRef]
45. Kaikati, A.M.; Kaikati, J.G. Stealth marketing: How to reach consumers surreptitiously. *Calif. Manag. Rev.* **2004**, *46*, 6–23. [CrossRef]
46. Gamble, J.; Gilmore, A. A new era of consumer marketing An application of co-creational marketing in the music industry. *Eur. J. Mark.* **2013**, *47*, 1859–1888. [CrossRef]
47. Etgar, M. A descriptive model of the consumer co-production process. *J. Acad. Mark. Sci.* **2008**, *36*, 97–108. [CrossRef]
48. Black, I.; Veloutsou, C. Working consumers: Co-creation of brand identity, consumer identity and brand community identity. *J. Bus. Res.* **2017**, *70*, 416–429. [CrossRef]
49. Fogg, B.J. *Persuasive Technology: Using Computers to Change What We Think and Do*; Morgan Kaufmann: Boston, MA, USA, 2003.

50. Duhigg, C. *The Power of Habit: Why We Do What We Do in Life and Business*, 1st ed.; Random House Trade Paperbacks: New York, NY, USA, 2012.
51. Judah, G.; Gardner, B.; Aunger, R. Forming a flossing habit: An exploratory study of the psychological determinants of habit formation. *Br. J. Health Psychol.* **2013**, *18*, 338–353. [CrossRef]
52. Lally, P.; Van Jaarsveld, C.H.; Potts, H.W.; Wardle, J. How are habits formed: Modelling habit formation in the real world. *Eur. J. Soc. Psychol.* **2010**, *40*, 998–1009. [CrossRef]
53. Eyal, N. *Hooked: How to Build Habit-Forming Products*, 1st ed.; Portfolio Books: London, UK, 2014.
54. Ferster, C.B.; Skinner, B.F. *Schedules of Reinforcement*, 1st ed.; Appleton-CenturyCrofts: New York, NY, USA, 1957.
55. Metcalfe, J.; Mischel, W. A hot/cool-system analysis of delay of gratification: Dynamics of willpower. *Psychol. Rev.* **1999**, *106*, 3–19. [CrossRef] [PubMed]
56. Bernheim, B.D.; Rangel, A. Behavioral public economics: Welfare and policy analysis with non-standard decision-makers. *Natl. Bur. Econ. Res.* **2005**, 11518. [CrossRef]
57. Cover, R. Digital addiction: The cultural production of online and video game junkies. *Media Int. Aust. Inc. Cult. Policy* **2004**, *113*, 110–123. [CrossRef]
58. Marczewski, A. The ethics of gamification. XRDS: Crossroads. *ACM Mag. Stud.* **2017**, *24*, 56–59.
59. Kwon, M.; Lee, J.Y.; Won, W.Y.; Park, J.W.; Min, J.A.; Hahn, C.; Kim, D.J. Development and validation of a smartphone addiction scale (SAS). *PLoS ONE* **2013**, *8*, e56936. [CrossRef]
60. Demirci, K.; Orhan, H.; Demirdas, A.; Akpinar, A.; Sert, H. Validity and reliability of the Turkish Version of the Smartphone Addiction Scale in a younger population. *Bull. Clin. Psychopharmacol.* **2014**, *24*, 226–234. [CrossRef]
61. Fırat, S.; Gül, H.; Sertçelik, M.; Gül, A.; Gürel, Y.; Kılıç, B. The relationship between problematic smartphone use and psychiatric symptoms among adolescents who applied to psychiatry clinics. *Psychiatry Res.* **2018**, *270*, 97–103. [CrossRef]
62. Rosen, L.; Carrier, L.M.; Miller, A.; Rokkum, J.; Ruiz, A. Sleeping with technology: Cognitive, affective, and technology usage predictors of sleep problems among college students. *Sleep Health J. Natl. Sleep Found* **2016**, *2*, 49–56. [CrossRef]
63. Commonsense. Available online: https://www.commonsensemedia.org/technology-addiction-concern-controversy-and-finding-balanceinfographic (accessed on 10 October 2019).
64. Frangos, C.C.; Fragkos, K.C.; Kiohos, A. Internet addiction among Greek university students: Demographic associations with the phenomenon, using the Greek version of Young's Internet Addiction Test. *Int. J. Econ. Sci. Appl. Res.* **2010**, *3*, 49–74.
65. Weiser, E.B. Gender Differences in Internet Use Patterns and Internet Application Preferences: A Two-Sample Comparison. *Cyberpsychol. Behav.* **2000**, *3*, 167–178. [CrossRef]
66. Billieux, J.; Van Der Linden, M.; Rochat, L. The role of impulsivity in actual and problematic use of the mobile phone. *Appl. Cogn. Psychol.* **2008**, *22*, 1195. [CrossRef]
67. Hong, F.Y.; Chiu, S.I.; Lin, H.Y. The development and current states of private college student mobile phone addiction scale. *Chung Cheng Educ. Stud.* **2012**, *11*, 87–116.
68. Chiu, S.I.; Hong, F.Y.; Chiu, S.L. An Analysis on the Correlation and Gender Difference between College Students' Internet Addiction as well as Mobile Phone Addiction in Taiwan. *ISRN Addict.* **2013**. [CrossRef]
69. Sherman, R.C.; End, C.; Kraan, E.; Cole, A.; Campbell, J.; Birchmeier, Z.; Klausner, J. The Internet gender gap among college students: Forgotten but not gone. *Cyberpsychol. Behav.* **2000**, *3*, 885–894. [CrossRef]
70. Shek, D.T.; Yu, L. Adolescent internet addiction in Hong Kong: Prevalence, change, and correlates. *J. Pediatric Adolesc. Gynecol.* **2016**, *29*, S22–S30. [CrossRef]
71. Widyanto, L.; McMurran, M. The psychometric properties of the internet addiction test. *Cyberpsychol. Behav.* **2004**, *7*, 443. [CrossRef]
72. Steinberg, L. A social neuroscience perspective on adolescent risk-taking. *Dev. Rev.* **2008**, *28*, 78–106. [CrossRef]
73. Steinberg, L. A dual systems model of adolescent risk-taking. *Dev. Psychobiol.* **2010**, *52*, 216. [CrossRef]
74. Steinberg, L.; Albert, D.; Cauffman, E. Age differences in sensation seeking and impulsivity as indexed by behavior and self-report: Evidence for a dual systems model. *Dev. Psychobiol.* **2008**, *44*, 1764. [CrossRef]
75. Engin, C.; Pals, H. Patriarchal attitudes in Turkey 1990–2011: The influence of religion and political conservatism. *Soc. Politics Int. Stud. Gend. State Soc.* **2018**, *25*, 383–409. [CrossRef]

76. Lee, H.; Kim, J.W.; Choi, T.Y. Risk factors for smartphone addiction in Korean adolescents: Smartphone use patterns. *J. Korean Med. Sci.* **2017**, *32*, 1674–1679. [CrossRef] [PubMed]
77. Hogle, J.M.; Curtin, J.J. Sex differences in negative affective response during nicotine withdrawal. *Psychophysiology* **2006**, *43*, 344–356. [CrossRef] [PubMed]
78. Becker, J.B.; McClellan, M.L.; Reed, B.G. Sex differences, gender and addiction. *J. Neurosci. Res.* **2017**, *95*, 136–147. [CrossRef] [PubMed]
79. Park, N. Gender difference in social networking on smartphones: A case study of Korean college student smartphone users. *Int. Telecommun. Policy Rev.* **2014**, *21*, 1–18.
80. Geser, H. Are Girls (Even) More Addicted? Some Gender Patterns of Cell Phone Usage. In *Sociology in Switzerland: Sociology of the Mobile Phone*; Online Publications: Zuerich, Switzerland, 2006.
81. Reinehr, T.; Temmesfeld, M.; Kersting, M.; De Sousa, G.; Toschke, A.M. Four-year follow-up of children and adolescents participating in an obesity intervention program. *Int. J. Obes.* **2007**, *31*, 1074–1077. [CrossRef]
82. Shochet, I.M.; Dadds, M.R.; Holland, D.; Whitefield, K.; Harnett, P.H.; Osgarby, S.M. The efficacy of a universal school-based program to prevent adolescent depression. *J. Clin. Child Psychol.* **2001**, *30*, 303–315. [CrossRef]
83. Durlak, J.A.; Wells, A.M. Primary prevention mental health programs for children and adolescents: A meta-analytic review. *Am. J. Commun. Psychol.* **2001**, *25*, 115–152. [CrossRef]
84. Li, L.; Niu, Z.M.; Mei, S.L. The mindfulness cognitive behavioral group therapy of medical student'smartphone addiction in group counseling course. *China High. Med. Educ.* **2017**, *5*, 37–38.

International Journal of
Environmental Research and Public Health

MDPI

Article

Problematic Social-Networks-Use in German Children and Adolescents—The Interaction of Need to Belong, Online Self-Regulative Competences, and Age

Sina Ostendorf [1], Elisa Wegmann [1] and Matthias Brand [1,2,*]

[1] Department of General Psychology: Cognition and Center for Behavioral Addiction Research (CeBAR), University of Duisburg-Essen, 47057 Duisburg, Germany; sina.ostendorf@uni-due.de (S.O.); elisa.wegmann@uni-due.de (E.W.)
[2] Erwin L. Hahn Institute for Magnetic Resonance Imaging, 45141 Essen, Germany
* Correspondence: matthias.brand@uni-due.de

Received: 28 February 2020; Accepted: 3 April 2020; Published: 7 April 2020

Abstract: Adolescents nowadays spend much time communicating via social networks. Recent investigations also report a noticeable proportion showing a problematic usage behavior, underlining the importance of better understanding its development and maintenance in young individuals. Theoretical views on Internet-use disorders assume that specific predispositions and needs can contribute to addictive behaviors in interaction with further aspects including Internet-related cognitive biases. This study focuses on vulnerable individuals due to their age and investigates interactions between possible risk (need to belong, NTB) and protective factors (online self-regulative competences, OSRC). Participants (N = 466) between 10 and 17 years answered questionnaires assessing social-networks-use disorder symptoms, NTB, and OSRC. Moderated regression analysis revealed significant effects of age, NTB, and OSRC. Three-way interaction was also significant (potentially mainly caused by females), with highest social-networks-use disorder symptoms found for individuals with high NTB and low OSRC, especially when older. With high OSRC, symptoms were significantly lower for both younger and older individuals having high NTB. However, even if NTB was low, older individuals showed high social-networks-use disorder symptoms if their OSRC were low. The results highlight the importance of improving specific competences to prevent problematic usage behaviors, which should be considered in youth-tailored prevention and intervention programs.

Keywords: social-networks-use disorder; internet addiction; social media use; social networking sites; protective competences; self-regulation; social needs; behavioral addiction

1. Introduction

Given the easy and almost unrestricted access to social media applications including social networking sites such as Facebook, video/photo sharing platforms such as Instagram or TikTok, and instant messengers such as WhatsApp [1,2], social interactions are largely taking place online. By sharing personally relevant information via text, voicemails, photos, and videos, individuals can communicate with others, express and present themselves, build and maintain relationships, or perceive social support and connectedness [3–6]. Especially, young people seem to value the possibilities to socially interact via online applications as this can strengthen their sense of belonging [3,7]. Following research on media use of more than 1200 German children and adolescents between 12 and 19 years, 95% stated that they owned a smartphone [8]. In the new reports of EU Kids Online [9], the authors also point out that the majority of young people state that they use the smartphone almost all the time, with adolescents aged between 15 and 16 years spending about twice as much time online as children aged between nine and 11 years. Furthermore, during their leisure time, the usage of the Internet in general, the usage of the smartphone and listening to music were the most important media activities stated by German adolescents in 2018 [8]. Regarding the Internet-usage time, communication accounted for the largest share, followed by entertainment and games, whereby the most frequently used communication applications were WhatsApp, Instagram, and Snapchat [8]. Communicating and staying in contact with others is generally an important motive for using social networks, as highlighted by several authors [10–15]. Following Pertegal, Oliva, and Rodríguez-Meirinhos [16], research also identified two important motives for very young individuals: on the one hand obtaining social recognition, and on the other hand belonging to an online community.

Besides several advantages of such applications, research has also critically pointed out that the use of social networks can have undesired and disadvantageous consequences as well. These include, for example, becoming a cyberbullying victim, being threatened by other people, or experiencing problems in everyday life due to an uncontrolled usage [17–19]. Based on recent developments in the International Classification of Diseases (11th revision; ICD-11) including the classification of gaming disorder as a disorder due to addictive behaviors, researchers also discuss the problematic use of social networks as potential addictive behavior [20,21]. We therefore use the term social-networks-use disorder (although this term has not been included in classification systems, so far) for the definition of an uncontrolled, problematic use of online-communication applications. This describes the experiences of negative consequences due to, and the diminished control over the use of, social networks, whereby the communicative and social aspect is considered as the key element the users are addicted to rather than a certain device or platform [21,22].

With respect to age, being younger was associated with a more problematic social-networks-use [23] and an addictive/problematic smartphone behavior [24,25]. A recent study with a representative sample of German adolescents reported an estimated prevalence of 2.6% for a problematic social-networks-use [26], which is in line with the European-wide prevalence rates ranging from 0% to 2.1% [9]. The results emphasize the relevance of investigating this phenomenon in children and adolescents. They also highlight the importance of investigating potential mechanisms that contribute to the development and maintenance of such behaviors especially in younger ages in order to derive possible implications for prevention measures. Brand and colleagues [27,28] proposed a theoretical model (Interaction of Person-Affect-Cognition-Execution; I-PACE) to specify mechanisms involved in addictive behaviors, such as the problematic use of social networks. The I-PACE model is based on the assumption that specific motives, needs and further personal predispositions including social deficiencies interact with affective and cognitive processes (e.g., Internet-related cognitive biases such as specific expectancies) as well as executive components (e.g., reduced inhibitory control) and experiences of gratification and compensation in the development and maintenance of addictive behaviors. Considering the development of a (potential) social-networks-use disorder, the aforementioned processes may lead individuals to use specific social network services, resulting in experienced gratification and/or compensation (supposedly depending on individual's addiction stage) and possible reinforcement processes, potentially intensifying a problematic/addictive usage [28]. As one further extension of the development and maintenance of a social-networks-use disorder, Wegmann and Brand [22] specify the addiction process and assume that especially individuals with specific needs (e.g., high need to belong) or social deficits may be vulnerable to developing a problematic use of social networks, since the experience of gratified social needs may reinforce specific expectancies and coping styles, and may lead to the recurring decision to use those applications to strengthen a sense of belonging.

Need to belong in general is considered as a fundamental human motivation [29] and was previously assumed to be a driving motivator for the use of social networks, such as Facebook [5]. The authors propose a dual-factor model of Facebook use including two primary social needs: the need to belong, representing an intrinsic drive to get in touch and connect with others, and the need for self-presentation, corresponding to impression management processes [5]. A study by Beyens, Frison, and Eggermont [30] confirmed the proposed role of need to belong and found that it is associated with an increased Facebook usage. Besides, Martin [31] reported a positive correlation between individual's need to belong and the frequency of checking social media applications as well as intensity of social-networks-use. Wang and colleagues [32] even found that the need to belong depicts a positive predictor for what the authors called "adolescent smartphone addiction" (note that this term has been criticized recently [33]), and it further moderated the relation between self-esteem and addiction. The results show that the effect of need to belong as a risk factor interacts with further variables, which is consistent with the I-PACE model by Brand and colleagues [28]. Brand and colleagues [28] also highlight that the relevance of predisposing variables, such as usage motives, on the tendency of an additive behavior is related to further affective and cognitive responses such as Internet-related cognitive biases (e.g., specific Internet use expectancies, Internet-related competences). Several studies have already shown that the interaction of those Internet-related cognitive biases and predisposing variables is related to addictive behavior online [2,34].

Focusing on the relevance of Internet-related competences, Wegmann, Stodt, and Brand [35], for example, found that self-regulative competences as a subdomain of Internet literacy had a negative effect on the symptom severity of a social-networks-use disorder and, moreover, partially mediated the relationship between psychopathological symptoms and symptom severity. The authors conclude that the ability to regulate one's own online behavior and to be able to manage time online might be a preventive factor for developing a social-networks-use disorder [35]. This conclusion is consistent with further studies, since Stodt, Wegmann, and Brand [36] also found a preventive effect of self-regulation when considering an unspecific Internet-use disorder. Additionally, Błachnio and Przepiorka [37] concluded that a dysfunction of self-regulation and self-control might be a risk factor for a problematic social-networks-use and van Deursen, Bolle, Hegner, and Kommers [24] found a negative effect of general self-regulation on addictive usage behaviors, but a positive effect of age on self-regulation. This underlines, on the one hand, that self-regulation can be considered as a protective factor preventing a problematic usage, and, on the other hand, implies that especially young individuals seem to have a low level of general self-regulation. Going one step further, online-specific self-regulative competences could play an even more relevant role when investigating problematic social-networks-use in children and adolescents.

Based on the aforementioned studies and theoretical assumptions, we argue that there is strong need for investigating not only possible factors in isolation, but especially their interactions in order to better understand the development and maintenance of a problematic social-networks-use. This appears particularly important in order to support prevention and intervention measures especially for young individuals. We hypothesize that the effect of need to belong on symptoms of a social-networks-use disorder is moderated by the ability to self-regulate one's own online behavior. However, when keeping in mind that the online behavior of children and adolescents is notably changing during childhood and adolescence [9], the interaction effects have to be controlled for a possible age effect. This means that we include age as first predictor to investigate the effect age may have on symptoms of a social-networks-use disorder. We then include need to belong and online self-regulative competences to examine their incremental validity beyond age. Overall, we aim at gaining deeper insights into the protective power of online self-regulative competences by focusing on individuals at risk due to their age and specific needs. Our hypothesized model is depicted in Figure 1.

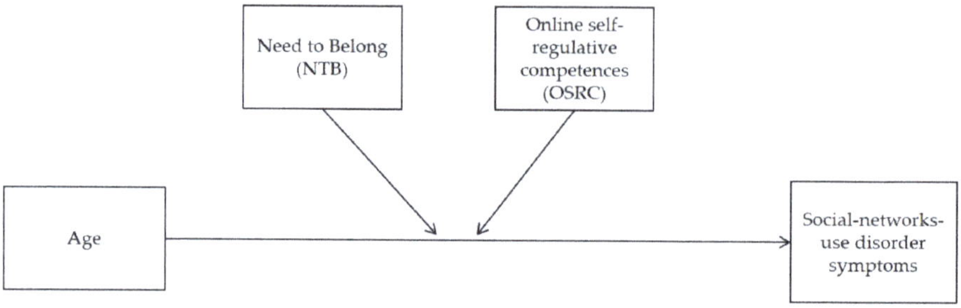

Figure 1. Hypothesized moderated regression model with age, need to belong (NTB), and online self-regulative competences (OSRC) as (moderating) predictors for a social-networks-use disorder.

2. Materials and Methods

2.1. Participants and Recruitment

In total, 466 children and adolescents between 10 and 17 years ($M = 13.05$, $SD = 1.99$) took part in the current study. Of these, 239 were female and 227 were male. The frequency of individual ages was as follows: 10.5% (49 participants) were 10 years old, 18.2% (85) 11 years, 12.2% (57) 12 years, 18.9% (88) 13 years, 13.1% (61) 14 years, 11.4% (53) 15 years, 13.5% (63) 16 years, and 2.2% (10) 17 years. The modal value and the median are both 13 years. Of all participants, 85 stated that they used the Internet actively up to one hour a day, 186 between two and three hours, 105 participants four to five hours, 79 more than five hours a day, and 11 did not report their daily Internet usage time. All participants used at least one social media application including social networking sites, instant messaging services, and photo/video sharing platforms. The most frequently mentioned app was WhatsApp (used by 99.4% of all participants), followed by Instagram (55.4%) and Snapchat (55.2%). Facebook was used by 18.7% of all participants, the Facebook Messenger by 12.9% and Twitter by 15.7%. Besides, 37.1% of the participants mentioned that they used further services, for example Musical.ly (now known as TikTok), YouTube, or Threema. Of all 466 participants, 463 stated to have either their own or a shared smartphone and/or computer. Another three participants reported to not have their own or a shared device but stated that they used at least one social media application. Since the usage of specific applications is the more relevant condition, we decided to include these three participants as well.

The current sample was recruited in 2017 at a local secondary school in a large city in Western Germany within a period of two weeks. Parents and legal guardians gave their informed consent before children and adolescents voluntarily took part. All persons involved (including parents/legal guardians, teaching staff, school administration, and pupils) were informed about the purpose of the study and participants were surveyed in the school premises. The processing of all questionnaires (including those that are not relevant for the current manuscript) took about 40 to 60 min. The study was conducted in accordance with the Declaration of Helsinki and was approved by the local ethics committee of the University of Duisburg-Essen (date: 21st March, 2017).

2.2. Instruments

2.2.1. Short Internet Addiction Test Modified for Social-Networks-Use Disorder

In order to measure tendencies of a social-networks-use disorder, a modified version [35] of the short Internet Addiction Test [38] was answered by the participants. This modified version measures subjectively perceived problems in an individual's everyday life due to the use of online-communication applications, including active (e.g., creating posts/content) and passive use (e.g., browsing/reading content) of social networking sites, (micro-)blogs, and instant messengers. In the course of this study, 4 out of originally 12 items were applied in order to keep the study as short as possible. These four items are considered as key items to represent the scale's two factors: loss of control/time management and social problems/craving [39,40]. In detail, these were "How often do you find that you spend more time with online-communication applications than you intended?" and "How often do you neglect household chores to spend more time with online-communication applications?" for the subscale loss of control/time management and "How often do you feel preoccupied with online-communication applications when offline, or fantasize about online-communication applications?" and "How often do you choose to spend more time with online-communication applications over going out with others?" for social problems/craving. All items were rated from 1 = "never" to 5 = "very often". The sum score for the four items ranges from 4 to 20. The overall internal consistency was $\alpha = 0.844$.

2.2.2. Single-Item Need to Belong Scale

The extent of participants' need to belong was assessed by applying the Single-Item Need to Belong Scale by Nichols and Webster [41], to which a good reliability and validity is attributed. Here, participants are asked to rate from 1 = "disagree" to 4 = "agree" to what extent the statement "I have a strong need to belong" applies to them. Thus, higher scores indicate a higher need to belong.

2.2.3. Internet Literacy Questionnaire

To measure online self-regulative competences we used the Internet Literacy Questionnaire developed by Stodt and colleagues [42]. The scale assesses four dimensions of individuals' competences in using the Internet in an adequate way, namely technical expertise, reflection and critical analysis, production and interaction, and self-regulation. The subscale self-regulation consists of five items, which are answered on a 6-point Likert scale ranging from 0 = "strongly disagree" to 5 = "strongly agree". For the purpose of this manuscript, we only used the subscale self-regulation with an internal consistency of $\alpha = 0.752$. A mean score was used to test the hypothesis.

2.3. Statistical Analyses

We used SPSS 24.0 for Windows [43] for the statistical analyses. To test for bivariate correlations, we calculated Pearson's correlations, with $|r| \geq 0.10$ indicating a small, $|r| \geq 0.30$ a medium, and $|r| \geq 0.50$ a large effect [44]. To test our hypothesis, we calculated a hierarchical moderated regression analysis with age as first predictor, need to belong in the second step, and online self-regulative competences in the third step, followed by the respective interactions and the three-way interaction in the last step. All independent variables were Fisher's z-transformed beforehand. Here, standardized beta-coefficients serve as measure of the effect sizes (with similar indicators for a small, medium, and large effect as for correlations). Simple Slopes analyses are calculated for significant three-way interactions.

3. Results

3.1. Descriptive Values and Multivariate Statistics

Mean values with corresponding standard deviations of all variables are depicted in Table 1. Table 1 also shows bivariate correlations between symptoms of a social-networks-use disorder, age, need to belong, and online self-regulative competences. Social-networks-use disorder symptoms were significantly positively correlated with need to belong and age, and significantly negatively with online self-regulative competences. Need to belong was not significantly correlated with both age and online self-regulative competences, whereas age and online self-regulative competences were significantly negatively correlated. Moreover, age correlated significantly positively with participants' Internet usage time ($r = 0.369$, $p < 0.001$; $n = 455$ due to missing data) and social-networks-use disorder symptoms also correlated significantly positively with Internet usage time ($r = 0.465$, $p < 0.001$, $n = 455$).

Table 1. Mean values, standard deviations, range, and bivariate correlations between the study's variables.

Variables	*M*	*SD*	Range	1	2	3
1. Social-networks-use disorder symptoms	7.56	3.55	4–20			
2. Age	13.05	1.98	10–17	0.310 ***		
3. NTB	2.58	0.89	1–4	0.202 ***	−0.073	
4. OSRC	3.02	1.20	0–5	−0.374 ***	−0.136 **	−0.077

Note. NTB = Need to belong; OSRC = Online self-regulative competences. $^{**}p < 0.01$; $^{***}p < 0.001$.

3.2. Hypothesis Testing

In accordance with our hypothesized model, we calculated a hierarchical moderated regression analysis including age, need to belong (NTB), and online self-regulative competences (OSRC), corresponding interactions and the three-way interaction, in the order as stated. The results revealed significant main effects for age, NTB, and OSRC in the prediction of social-networks-use disorder symptoms (see Figure 2). Furthermore, the interaction effect of age and OSRC was significant ($\Delta R^2 = 0.011$, $\Delta F = 6.88$, $p = 0.009$), as well as the three-way interaction ($\Delta R^2 = 0.009$, $\Delta F = 5.36$, $p = 0.021$). The overall model also proved to be significant, explaining 27.1% of the criterion's variance ($R^2 = 0.271$, F (7, 458) $= 24.35$, $p < 0.001$) and supporting our assumptions. Investigating the interaction effects in more detail, simple slopes analyses were calculated. The results showed that, especially when older, individuals with high NTB and low OSRC showed the highest tendency towards a social-networks-use disorder (grey line, Figure 3). With higher OSRC, social-network-use disorder tendency was significantly lower in both younger and older individuals with high NTB (yellow line, Figure 3), whereby younger individuals showed even less a social-networks-use disorder tendency than older ones. However, older individuals even had a high social-networks-use disorder tendency when their NTB was low, if they also had low OSRC (blue line, Figure 3). Finally, especially older individuals showed the lowest social-networks-use disorder tendency when they had a low NTB level and also high OSRC. Corresponding statistical values including beta-coefficients are depicted in Table 2.

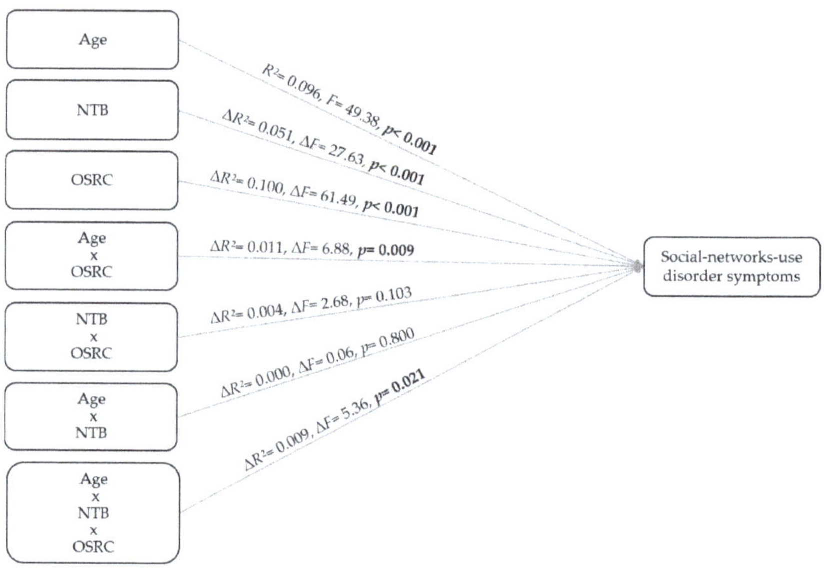

Figure 2. Results of the hierarchical moderated regression analysis with symptoms of a social-networks-use disorder as dependent variable and age, need to belong (NTB), and online self-regulative competences (OSRC) as well as their respective interactions as predictors.

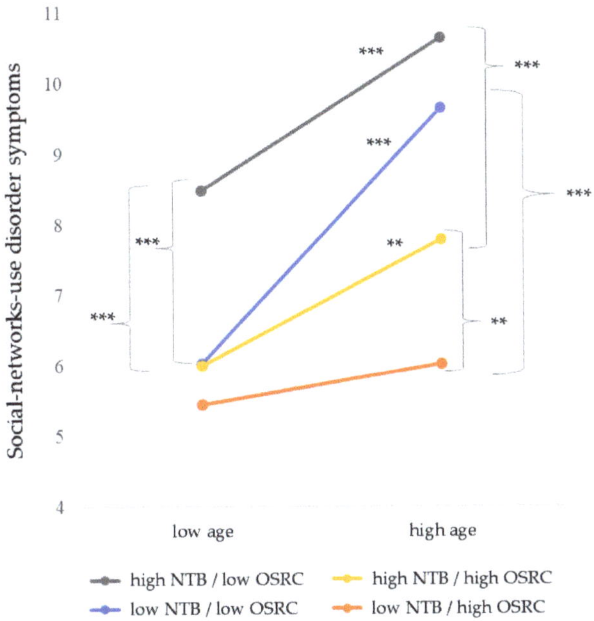

Figure 3. Simple Slopes visualizing the significant three-way interaction between age, need to belong (NTB), and online self-regulative competences (OSRC) in the prediction of social-networks-use disorder symptoms with ** $p < 0.01$, *** $p < 0.001$ and for all significant t: $2.65 \leq t \leq 6.78$.

Table 2. Regression coefficients of the hierarchical moderated regression analysis with age, need to belong (NTB), and online self-regulative competences (OSRC) predicting symptoms of a social-networks-use disorder.

Predictors	*B*	*SE (B)*	ß	*t*	*p*
Age	0.51	0.07	0.287	7.00	<0.001
NTB	0.81	0.16	0.203	4.98	<0.001
OSRC	−1.00	0.12	−0.337	−8.11	<0.001
Age × OSRC	−0.18	0.06	−0.115	−2.81	0.005
NTB × OSRC	−0.14	0.13	−0.044	−1.05	0.295
Age × NTB	−0.02	0.09	−0.009	−0.21	0.834
Age × NTB × OSRC	0.17	0.07	0.097	2.32	0.021

Note. Significant values depicted in bold.

3.3. Additional Analyses

As an additional analysis, we controlled our results for a potential gender effect. We firstly calculated t-tests for independent samples. We found no significant differences between male and female participants for NTB, OSRC, and age (all $p \geq 0.768$). For tendencies towards a social-networks-use disorder, we found a significant difference (males: $M = 6.62$, $SD = 3.14$; females: $M = 8.45$, $SD = 3.68$; $t\,(458.76) = -5.77$, $p < 0.001$, $d = -0.54$). We therefore calculated our hierarchical moderated regression analysis in a second step again separately for male and female participants. For female participants, the effects of age ($\beta = 0.403$, $p < 0.001$), NTB ($\beta = 0.214$, $p < 0.001$), and OSRC ($\beta = -0.418$, $p < 0.001$) on symptoms of a social-networks-use disorder remained significant, as well as the interaction effect of age and OSRC ($\beta = -0.180$, $p < 0.001$) and the three-way interaction ($\beta = 0.151$, $p = 0.003$). The overall model also proved to be significant and explained 46.0% of the criterion's variance ($R^2 = 0.460$, $F(7, 231)$

= 28.11, $p < 0.001$). For male participants we also found significant effects of age (β = 0.197, $p = 0.003$), NTB (β = 0.188, $p = 0.004$), and OSRC (β = −0.244, $p < 0.001$), but no significant interaction effect of age and OSRC (β = −0.045, $p = 0.488$) and no significant three-way interaction (β = 0.052, $p = 0.440$). The overall model remained significant, explaining 14.4% of the criterion's variance ($R^2 = 0.144$, F (7, 219) = 5.25, $p < 0.001$). The subsequently calculated Chow Test comparing both hierarchical moderated regression analyses yielded a significant difference (F (8, 450) = 9.37, $p < 0.001$).

4. Discussion

In this study, we aimed at contributing to a better understanding of a social-networks-use disorder by considering on the one hand specific needs that can increase the risk of developing a problematic use and on the other hand Internet-related cognitive biases (i.e., self-regulative competences) that may play a protective role. Further, we focused on a specific group of individuals that can be particularly vulnerable to developing a social-networks-use disorder due to their young age (here individuals between 10 and 17 years) and investigated interaction effects between the respective variables. The calculated hierarchical moderated regression analysis revealed that 27.1% of the variance of social-networks-use disorder symptoms were explained by the included variables: age, need to belong, and online self-regulative competences. All variables showed a significant main effect and the three-way interaction was significant as well. Controlling for possible gender effects, the results illustrate that female participants showed higher symptoms of social-networks-use disorder compared to male participants, which is in line with previous studies [45,46]. When calculating the hierarchical moderated regression analysis separately for male and female participants, we found significant main effects of age, need to belong, and online self-regulative competences for both males and females, but the interaction effects only remained significant for females.

The results illustrate that high need to belong may be a vulnerability factor for the development and maintenance of a problematic social-networks-use. The findings support previous reports regarding the important role of need to belong [32] and further strengthen that this fundamental social need is also an important factor for both female and male children and adolescents as potentially contributing to a problematic use of social networks. The findings therefore extend current research, illustrating the relevance of specific needs and motives for using social networks for children and adolescents. It can be assumed that online social networks are perceived as a possible means to feel socially integrated and to gratify individual's need for belonging [22]. Especially, young individuals may perceive a strong desire to fulfill their need to belong, since they find themselves in an important developmental part of life, in which their respective peer group occupies an important place [30,47,48] and depicts an essential source of experiencing social support [49]. Thus, particularly younger individuals seem to be a vulnerable group that might use social networks in an intense and potentially critical way to gratify their need to be socially connected and accepted. This is in line with the I-PACE model [27,28], assuming that the aim to fulfill personal needs and motives can lead individuals to use specific applications, resulting in gratification and reinforcement processes, which potentially ends up in an uncontrolled or even problematic usage.

The current results also indicate that high online self-regulative competences are associated with significantly lower symptoms of a problematic use of social-networks in both males and females, underlining the protective power of specific competences and abilities [24,35,36]. These results strengthen previously found empirical results and again broaden current research since the ability to regulate one's own online behavior appears to be an important competence for very young individuals too.

When investigating the interaction between both factors (need to belong and online self-regulative competences), we found no significant effect on the symptom severity of a social-networks-use disorder, neither for females, males, nor when taking both together. This result appears unexpected at first sight, but when considering that age had a significantly positive direct effect (for both males and females), and the interaction between online self-regulative competences and age was (except for males) also significant, it becomes apparent that the investigation of interactions between potential risk and protective factors for individuals, and especially for females, in such a young phase of life additionally needs to consider age. As the three-way interaction illustrates, children and adolescents with a high need to belong and low online self-regulative competences had a high problematic usage behavior, especially when older within the age range of 10 to 17 years. Higher online self-regulative competences, in contrast, were related to significantly lower social-networks-use disorder symptoms. Moreover, especially younger adolescents seem to benefit from online self-regulative competences when having a high need to belong level, as they showed even fewer symptoms of a social-networks-use disorder compared to older ones. However, even if the need to belong is low, particularly older individuals seem to need good self-regulative competences in order to reduce the risk of developing social-networks-use disorder symptoms. Since these interaction effects have been observed in the overall sample and in the female sample, but not in the male sample, they may be mainly caused by female participants. However, in accordance with the I-PACE model, interactions between different variables are assumed to contribute to the development and maintenance of problematic behavior generally for both men and women [28]. Thus, theoretically assumed interactions may be valid for both genders, although gender-specific differences can occur. Regarding the for females significant interaction between online self-regulative competences and age, which was not significant for males, one possible explanation could be that female participants reported higher symptoms of problematic behavior compared to male participants and thus the interaction effect may mainly occur along with a certain symptom severity. Moreover, the development of specific competences in such a young phase of life could be assumed to proceed differentially in females and males and thus potential mechanisms may also be different at this age [50]. The development of these competences with regard to age and gender and their relevance as protective mechanisms for a problematic behavior must specifically be addressed in further, preferably longitudinal studies.

Overall, the current results emphasize the importance of improving specific competences, particularly the ability to regulate oneself when using social networks, in order to prevent problematic usage behaviors. They further underline that in a time span of only seven years, young individuals undergo many changes, and that specific needs and also competences can vary in their importance for the development and maintenance of a social-networks-use disorder depending on the respective age. Since we, moreover, found that age and symptoms of a social-networks-use disorder were positively correlated with Internet usage time within our sample, it seems that using the Internet is rapidly increasing in a relatively short age span and is additionally related to the risk of a problematic social-networks-use. This is also in line with the investigation of Smahel and colleagues [9], illustrating that time online is significantly higher in young adults aged between 15 and 16 years, compared to children aged between nine and 11 years.

Based on the current results and statistics, it appears important to investigate the developmental trends of specific needs as well as competences and strategies especially in youths, in order to better understand respective problematic and uncontrolled behaviors. Self-regulation is likely to be learnable and research has previously pointed out that, when growing older, not only personal interests or goals change, but individuals also seem to be more settled [24,51]. However, loneliness and the need to be socially integrated can also increase over lifetime, probably leading people to use media even more [24,52]. Thus, possible changes and developments over time need to be considered more strongly when investigating the interplay of different factors in the context of problematic online behaviors. In addition, possible differences between male and female children and adolescents should also be kept in mind. Along with changes in people's communicative behavior, future research also needs to continue addressing the question of when the usage of social networks in order to fulfill personal needs becomes a significant problem. Using specific applications to keep social contact might in general be beneficial, but if individuals, and particularly youths, for example, lose control over their usage and perceive those applications as the only way to feel socially integrated, this might increase negative consequences in their everyday lives.

To sum up, we strongly recommend that prevention programs should focus specifically on the improvement in self-regulative competences from early on, including children and adolescents, to sensitize for a more functional use of social networks. Individuals should be supported in developing competence to better evaluate an appropriate extent of their social-networks-use, which is meant as an individually suitable amount not causing problems in everyday life. Teaching specific competences in school could help individuals not to satisfy their personal needs exclusively online, but to focus on a more purposeful behavior. Totally restricting youths' access to the Internet and its various platforms to communicate is certainly not an effective strategy [53], particularly since young individuals use social networks not only to avoid parental control, but also to handle their friendships and to explore new identities in an important stage of life [54,55]. In contrast, finding ways to empower users, especially children and adolescents, seems to be a promising approach, which is also noted by other authors [56]. Besides the communication context, specific competences and needs might also play an important role in other specific Internet-use disorders since there is, at least for German youths, a slow but noticeable increase in using social networks for entertainment and gaming purposes [8]. Thus, the interplay between potential risk and protective factors should also be further investigated in other specific Internet-use disorders.

For the current study, some limitations have to be mentioned. The current sample only includes children and adolescents aged between 10 and 17 years from one school in Germany and is not representative for the whole German population in this age range. The information given is based on self-reports, which did not include questions on participants' primary time of day to use social networks and other applications. However, this information can be very important for prevention and intervention measures and should be included in future studies [57]. Furthermore, we were not able to assess epidemiological characteristics, such as social class or rural/urban area, but since these aspects can give valuable insights, they should be considered in future studies. Overall, it should be noted that conclusions about possible changes in self-regulation or other specific competences over lifespan from childhood to adolescence are only based on the cross-sectional investigation and should be examined with a longitudinal study. However, the results are a first hint about the relevance of specific needs, specific competences such as online self-regulation, and age on a problematic use of social networks. In addition, younger individuals should also be taken further into account, since certain needs and the use of the smartphone have already been observed in even younger ages. In order to be able to generalize the results, future studies are still needed.

5. Conclusions

Childhood and adolescence depict an important phase of life in which communicating and socializing is increasingly realized via social networks. The current study revealed that in particular female children and adolescents aged between 10 and 17 years with high need to belong and low online self-regulative competences showed a high tendency towards a social-networks-use disorder, especially when they were older. High online self-regulative competences were associated with significantly lower symptom severity. Moreover, even if the need to belong was low, older individuals were found to need high online self-regulative competences in order to prevent a problematic usage. Accordingly, prevention and intervention programs should try to improve individuals' competences, in particular self-regulation, to support individuals in using social networks in a purposeful and appropriate way.

Author Contributions: Conceptualization, S.O., E.W. and M.B.; Formal analysis, S.O.; Methodology, S.O. and E.W.; Project administration, M.B.; Supervision, M.B.; Visualization, S.O.; Writing – original draft, S.O.; Writing – review & editing, S.O., E.W. and M.B. All authors have read and agreed to the published version of the manuscript.

Funding: We acknowledge support by the Open Access Publication Fund of the University of Duisburg-Essen. No further external funding was received for this research.

Acknowledgments: We thank Marco Fileccia for actively supporting us in conducting this study. Additionally, we thank Benjamin Stodt for the assistance regarding the administrative support and study design as well as Lydia Harbarth and Dario Stolze for the assistance regarding data preparation.

Conflicts of Interest: The authors declare no conflict of interest.

References

1. Aichner, T.; Jacob, F. Measuring the degree of corporate social media use. *Int. J. Mark. Res.* **2015**, *57*, 257–275. [CrossRef]
2. Wegmann, E.; Ostendorf, S.; Brand, M. Is it beneficial to use Internet-communication for escaping from boredom? Boredom proneness interacts with cue-induced craving and avoidance expectancies in explaining symptoms of Internet-communication disorder. *PLoS ONE* **2018**, *13*, e0195742. [CrossRef]
3. Allen, K.A.; Ryan, T.; Gray, D.L.; McInerney, D.M.; Waters, L. Social media use and social connectedness in adolescents: The positives and the potential pitfalls. *Aust. Educ. Dev. Psychol.* **2014**, *31*, 18–31. [CrossRef]
4. Cheung, C.; Lee, Z.W.Y.; Chan, T.K.H. Self-disclosure in social networking sites: The role of perceived cost, perceived benefits and social influence. *Internet Res.* **2015**, *25*, 279–299. [CrossRef]
5. Nadkarni, A.; Hofmann, S.G. Why do people use Facebook. *Pers. Individ. Dif.* **2012**, *52*, 243–249. [CrossRef] [PubMed]
6. Taddicken, M.; Jers, C. The uses of privacy online: Trading a loss of privacy for social web gratifications? In *Privacy Online*; Trepte, S., Reinecke, L., Eds.; Springer: Berlin, Germany, 2011; pp. 143–156.
7. Davis, K. Friendship 2.0: Adolescents' experiences of belonging and self-disclosure online. *J. Adolesc.* **2012**, *35*, 1527–1536. [CrossRef]
8. Feierabend, S.; Rathgeb, T.; Reutter, T. *JIM-Studie 2018—Jugend, Information, Medien—Basisuntersuchung zum Medienumgang 12- 19-Jähriger*; Medienpädagogischer Forschungsverbund Südwest (LFK, LMK): Stuttgart, Germany, 2018; pp. 1–80.
9. Smahel, D.; Machackova, H.; Mascheroni, G.; Dedkova, L.; Staksrud, E.; Ólafsson, K.; Livingstone, S.; Hasebrink, U. *EU Kids Online 2020: Survey Results from 19 Countries*; EU Kids Online: London, UK, 2020; pp. 1–156.
10. Guedes, E.; Sancassiani, F.; Carta, M.G.; Campos, C.; Machado, S.; King, A.L.S.; Nardi, A.E. Internet addiction and excessive social networks use: What about Facebook? *Clin. Pract. Epidemiol. Ment. Heal.* **2016**, *12*, 43–48. [CrossRef]
11. Ryan, T.; Chester, A.; Reece, J.; Xenos, S. The uses and abuses of Facebook: A review of Facebook addiction. *J. Behav. Addict.* **2014**, *3*, 133–148. [CrossRef]
12. Giannakos, M.N.; Chorianopoulos, K.; Giotopoulos, K.; Vlamos, P. Using Facebook out of habit. *Behav. Inf. Technol.* **2013**, *32*, 594–602. [CrossRef]
13. Lee, E.; Lee, J.A.; Moon, J.H.; Sung, Y. Pictures speak louder than words: Motivations for using Instagram. *Cyberpsychology Behav. Soc. Netw.* **2015**, *18*, 553–556. [CrossRef]
14. Alzougool, B. The impact of motives for Facebook use on Facebook addiction among ordinary users in Jordan. *Int. J. Soc. Psychiatry* **2018**, *64*, 528–535. [CrossRef] [PubMed]
15. Pertegal-Vega, M.A.; Oliva-Delgado, A.; Rodríguez-Meirinhos, A. Systematic review of the current state of research on online social networks: Taxonomy on experience of use. *Communicar* **2019**, *27*, 81–91. [CrossRef]
16. Pertegal, M.A.; Oliva, A.; Rodríguez-Meirinhos, A. Development and validation of the Scale of Motives for Using Social Networking Sites (SMU-SNS) for adolescents and youths. *PLoS ONE* **2019**, *14*, 1–21. [CrossRef] [PubMed]
17. Aharony, N. Relationships among attachment theory, social capital perspective, personality characteristics, and Facebook self-disclosure. *Aslib J. Inf. Manag.* **2016**, *68*, 362–386. [CrossRef]
18. Debatin, B.; Lovejoy, J.P.; Horn, A.K.; Hughes, B.N. Facebook and online privacy: Attitudes, behaviors, and unintended consequences. *J. Comput. Commun.* **2009**, *15*, 83–108. [CrossRef]
19. Andreassen, C.S. Online social network site addiction: A comprehensive review. *Curr. Addict. Reports* **2015**, *2*, 175–184. [CrossRef]
20. Kuss, D.J.; Griffiths, M.D. Social networking sites and addiction: Ten lessons learned. *Int. J. Environ. Res. Public Health* **2017**, *14*, 311. [CrossRef]

21. Wegmann, E.; Müller, S.M.; Ostendorf, S.; Brand, M. Highlighting Internet-communication disorder as further Internet-use disorder when considering neuroimaging studies. *Curr. Behav. Neurosci. Reports* **2018**, *5*, 295–301. [CrossRef]
22. Wegmann, E.; Brand, M. A narrative overview about psychosocial characteristics as risk factors of a problematic social networks use. *Curr. Addict. Reports* **2019**, *6*, 402–409. [CrossRef]
23. Andreassen, C.S.; Billieux, J.; Griffiths, M.D.; Kuss, D.J.; Demetrovics, Z.; Mazzoni, E.; Pallesen, S. The relationship between addictive use of social media and video games and symptoms of psychiatric disorders: A large-scale cross-sectional study. *Psychol. Addict. Behav.* **2016**, *30*, 252–262. [CrossRef]
24. Van Deursen, A.J.A.M.; Bolle, C.L.; Hegner, S.M.; Kommers, P.A.M. Modeling habitual and addictive smartphone behavior: The role of smartphone usage types, emotional intelligence, social stress, self-regulation, age, and gender. *Comput. Human Behav.* **2015**, *45*, 411–420. [CrossRef]
25. Lopez-Fernandez, O.; Honrubia-Serrano, L.; Freixa-Blanxart, M.; Gibson, W. Prevalence of problematic mobile phone use in British adolescents. *Cyberpsychology Behav. Soc. Netw.* **2014**, *17*, 91–98. [CrossRef] [PubMed]
26. Wartberg, L.; Kriston, L.; Thomasius, R. Internet gaming disorder and problematic social media use in a representative sample of German adolescents: Prevalence estimates, comorbid depressive symptoms and related psychosocial aspects. *Comput. Human Behav.* **2020**, *103*, 31–36. [CrossRef]
27. Brand, M.; Young, K.S.; Laier, C.; Wölfling, K.; Potenza, M.N. Integrating psychological and neurobiological considerations regarding the development and maintenance of specific Internet-use disorders: An Interaction of Person-Affect-Cognition-Execution (I-PACE) model. *Neurosci. Biobehav. Rev.* **2016**, *71*, 252–266. [CrossRef] [PubMed]
28. Brand, M.; Wegmann, E.; Stark, R.; Müller, A.; Wölfling, K.; Robbins, T.W.; Potenza, M.N. The Interaction of Person-Affect-Cognition-Execution (I-PACE) model for addictive behaviors: Update, generalization to addictive behaviors beyond internet-use disorders, and specification of the process character of addictive behaviors. *Neurosci. Biobehav. Rev.* **2019**, *104*, 1–10. [CrossRef]
29. Baumeister, R.F.; Leary, M.R. The need to belong: Desire for interpersonal attachments as a fundamental human motivation. *Psychol. Bull.* **1995**, *117*, 497–529. [CrossRef]
30. Beyens, I.; Frison, E.; Eggermont, S. 'I don't want to miss a thing': Adolescents' fear of missing out and its relationship to adolescents' social needs, Facebook use, and Facebook related stress. *Comput. Human Behav.* **2016**, *64*, 1–8. [CrossRef]
31. Martin, P. *Social Media and the Need to Belong: Self-Esteem, Narcissism and Life Satisfaction on Facebook and Instagram*; Higher Diploma Final Year Project, Dublin Business School: Dublin, Ireland, 2018.
32. Wang, P.; Zhao, M.; Wang, X.; Xie, X.; Wang, Y.; Lei, L. Peer relationship and adolescent smartphone addiction: The mediating role of self-esteem and the moderating role of the need to belong. *J. Behav. Addict.* **2017**, *6*, 708–717. [CrossRef]
33. Montag, C.; Wegmann, E.; Sariyska, R.; Demetrovics, Z.; Brand, M. How to overcome taxonomical problems in the study of Internet use disorders and what to do with 'smartphone addiction'? *J. Behav. Addict.* **2019**, 1–7. [CrossRef]
34. Laier, C.; Wegmann, E.; Brand, M. Personality and cognition in gamers: Avoidance expectancies mediate the relationship between maladaptive personality traits and symptoms of Internet-gaming disorder. *Front. Psychiatry* **2018**, *9*, 304. [CrossRef]
35. Wegmann, E.; Stodt, B.; Brand, M. Addictive use of social networking sites can be explained by the interaction of Internet use expectancies, Internet literacy, and psychopathological symptoms. *J. Behav. Addict.* **2015**, *4*, 155–162. [CrossRef] [PubMed]
36. Stodt, B.; Wegmann, E.; Brand, M. Predicting dysfunctional Internet use: The role of age, conscientiousness, and Internet literacy in Internet addiction and cyberbullying. *Int. J. Cyber Behav. Psychol. Learn.* **2016**, *6*, 28–43. [CrossRef]
37. Błachnio, A.; Przepiorka, A. Dysfunction of self-regulation and self-control in Facebook addiction. *Psychiatr. Q.* **2016**, *87*, 493–500. [CrossRef]
38. Pawlikowski, M.; Altstötter-Gleich, C.; Brand, M. Validation and psychometric properties of a short version of Young' s Internet Addiction Test. *Comput. Human Behav.* **2013**, *29*, 1212–1223. [CrossRef]
39. Müller, M.; Brand, M.; Mies, J.; Lachmann, B.; Sariyska, R.Y.; Montag, C. The 2D:4D marker and different forms of Internet use disorder. *Front. Psychiatry* **2017**, *8*, 213. [CrossRef]

40. Wegmann, E.; Oberst, U.; Stodt, B.; Brand, M. Online-specific fear of missing out and Internet-use expectancies contribute to symptoms of Internet-communication disorder. *Addict. Behav. Reports* **2017**, *5*, 33–42. [CrossRef] [PubMed]
41. Nichols, L.; Webster, G.D. The single-item need to belong scale. *Pers. Individ. Dif.* **2013**, *55*, 189–192. [CrossRef]
42. Stodt, B.; Brand, M.; Sindermann, C.; Wegmann, E.; Li, M.; Zhou, M.; Sha, P.; Montag, C. Investigating the effect of personality, Internet literacy, and use expectancies in Internet-use disorder: A comparative study between China and Germany. *Int. J. Environ. Res. Public Health* **2018**, *15*, 579. [CrossRef]
43. IBM. *SPSS Statistics*; IBM: Armonk, NY, USA, 2017.
44. Cohen, J. *Statistical Power Analysis for the Behavioral Sciences*; Lawrence Erlbaum Associates: Hillsdale, MI, USA, 1988.
45. Cudo, M.; Torój, M.; Demczuk, M.; Francuz, P. Dysfunction of self-control in Facebook addiction: Impulsivity is the key. *Psychiatr. Q.* **2020**, *91*, 91–101. [CrossRef]
46. Xie, W.; Karan, K. Predicting Facebook addiction and state anxiety without Facebook by gender, trait anxiety, Facebook intensity, and different Facebook activities. *J. Behav. Addict.* **2019**, *8*, 79–87. [CrossRef]
47. Brown, B.B.; Larson, J. Peer relationships in adolescence. In *Handbook of Adolescent Psychology*; Lerner, R.M., Steinberg, L., Eds.; Wiley: New York, NY, USA, 2009; pp. 74–103.
48. Pombeni, M.L.; Kirchler, E.; Palmonari, A. Identification with peers as a strategy to muddle through the troubles of the adolescent years. *J. Adolesc.* **1990**, *13*, 351–369. [CrossRef]
49. Bokhorst, C.L.; Sumter, S.R.; Westenberg, P.M. Social support from parents, friends, classmates, and teachers in children and adolescents aged 9 to 18 years: Who is perceived as most supportive? *Soc. Dev.* **2010**, *19*, 417–426. [CrossRef]
50. Montroy, J.J.; Bowles, R.B.; Skibbe, L.E.; McClelland, M.M.; Morrison, F.J. The development of self-regulation across early childhood. *Dev. Psychol.* **2016**, *52*, 1744–1762. [CrossRef]
51. Diehl, M.; Semegon, A.B.; Schwarzer, R. Assessing attention control in goal pursuit: A component of dispositional self-regulation. *J. Pers. Assess.* **2006**, *86*, 306–317. [CrossRef] [PubMed]
52. Pearson, J.C.; Carmon, A.; Tobola, C.; Fowler, M. Motives for communication: Why the millennial generation uses electronic devices. *J. Commun. Speech Theatr. Assoc. North Dakota* **2010**, *22*, 45–56.
53. Staksrud, E.; Ólafsson, K.; Livingstone, S. Does the use of social networking sites increase children's risk of harm? *Comput. Human Behav.* **2013**, *29*, 40–50. [CrossRef]
54. Boyd, D. *It's Complicated: The Social Lives of Networked Teens*; Yale University Press: New Haven, CT, USA, 2014; pp. 1–281.
55. Jöckel, S.; Wilhelm, C. Everything under control? The role of habit strength, deficient self-regulation and media literacy for the use of social network sites among children and adolescents. In *Youth and Media: Current Perspectives on Media Use and Effects*, 1st ed.; Kühne, R., Baumgartner, S.E., Koch, T., Hofer, M., Eds.; Nomos: Baden-Baden, Germany, 2018; pp. 55–74.
56. Livingstone, S.; Ólafsson, K.; Staksrud, E. Risky social networking practices among 'underage' users: Lessons for evidence-based policy. *J. Comput. Commun.* **2013**, *18*, 303–320. [CrossRef]
57. Adan, A. A chronobiological approach to addiction. *J. Subst. Use* **2013**, *18*, 171–183. [CrossRef]

International Journal of
Environmental Research and Public Health

MDPI

Article

The Effect of Solution-Focused Group Counseling Intervention on College Students' Internet Addiction: A Pilot Study

Xinhe Zhang [1], Xiaoxuan Shi [1], Shuowei Xu [1], Jingwen Qiu [1], Ofir Turel [2] and Qinghua He [1,3,4,5,*]

1. Faculty of Psychology, Ministry of Education Key Laboratory of Cognition and Personality, Southwest University, Chongqing 400715, China; guhong15@126.com (X.Z.); shixx@email.swu.edu.cn (X.S.); adson2688@gmail.com (S.X.); qjw990130@email.swu.edu.cn (J.Q.)
2. Information Systems and Decision Sciences, California State University, Fullerton, CA 92831, USA; oturel@fullerton.edu
3. Chongqing Collaborative Innovation Center for Brain Science, Chongqing 400715, China
4. Southwest University Branch, Collaborative Innovation Center of Assessment toward Basic Education Quality at Beijing Normal University, Chongqing 400715, China
5. Institute of Psychology, Key Laboratory of Mental Health, Chinese Academy of Sciences, Beijing 100101, China

* Correspondence: heqinghua@swu.edu.cn

Received: 22 February 2020; Accepted: 2 April 2020; Published: 7 April 2020

Abstract: This pilot study aimed to explore the effect of solution-focused group counseling intervention on Internet addiction among college students. Eighteen college students participated in this study, out of which nine subjects were assigned into the experimental group and the rest (n = 9) to a control group. The experimental group received group counseling for five weeks, while the control group did not receive any intervention. The revised version of the Chinese Internet Addiction Scale (CIAS-R) was used to capture pre-test and post-test excessive use in the two groups. The experimental group was also subjected to a follow-up test and self-reported Internet addiction scores six months after the end of group counseling. Results showed that after the five-week solution-focused group counseling, the scores of four dimensions of the CIAS-R in the experimental group had CIAS-R decreased, and the reduction trend of the total score of CIAS-R was similar across all subjects in this group. The treatment effect was larger than the placebo reduction in the control group in two dimensions: compulsive and withdrawal (Sym-C & Sym-W) and tolerance (Sym-T) symptoms. Qualitative research confirmed the conclusions from the quantitative data, showing that the experimental group reduced its Internet addiction symptoms. Overall, the findings suggested that solution-focused group counseling had positive intervention effects on Internet addiction.

Keywords: Internet addiction; solution-focused group counseling; intervention; college students

1. Introduction

The Internet can serve as a double-edged sword. On the one hand, it provides many benefits and increases convenience. On the other, its use can be highly alluring, which may result in excess consumption, which manifests in behavioral addiction-like symptoms [1]. Since Goldberg [2] proposed the concept of "Internet Addiction Disorder", many researchers have used similar concepts, such as Pathological Internet Use, Problematic Internet Use, Excessive Internet Use, and Internet Addiction [3].

Although these concepts are slightly different, they all have two common characteristics: (1) addiction-like symptoms: compulsive use of Internet, withdrawal reactions, and tolerance, combined with inability of users to effectively control themselves; (2) as a result, their daily functions such as academic and social activities, as well as sleep and food intake, are impaired [4]. We believe that the concept of Internet Addiction (IA) can best reflect the above characteristics and is widely used. While the term is controversial [5], for convenience reasons, we will use in this study the term Internet Addiction to describe this phenomenon, while acknowledging the need of future research to fine tune its definition and boundaries. IA has been recognized as a public health concern [6], which can correlate with poor health outcomes such as higher likelihood of being overweight [7], impaired sleep [8], increased suicidal ideation and attempt [9], and increased depressive symptoms [10].

College students can be highly susceptible to IA [11]. Once freed from the strict control of high school and family, they tend to have free time at their disposal, which when coupled with easy access to the Internet can lead to IA [12]. In addition, these students are experiencing changes in the environment with which they must learn to cope. For example, they face pressures on handling new interpersonal relationships and balancing learning and social activities [13]. Consequently, students may seek to relieve these pressures via Internet use, which can result in IA [14]. Among Chinese college students, IA is prevalent. Between 6.56% and 13.5% of Chinese college students may be classified as addicted to the Internet [15]. Therefore, paying close attention to issues related to college students' use of the Internet, and intervening if necessary, is critical for preventing IA and its adverse impacts [16].

Various schools of psychotherapy have been applied to alleviate IA. Among them, solution-focused brief therapy (SFBT), as a relatively new paradigm of post-modern constructivism, has also been applied to IA cases. Specifically, Yang [17] provided psychotherapy with an emphasis on SFBT and comprehensive family intervention to teenagers with IA, and reported positive results showing that SFBT can be an effective therapy for treating IA. SFBT evolved out of the clinical practice of Steve de Shazer, Insoo Kim Berg, and colleagues at the Brief Family Therapy Center in Milwaukee, Wisconsin, in the early 1980s [18]. SFBT change processes were originally grounded in the constructivist approaches to communication and social interactional theories [19]. Over time, SFBT also became associated with social constructionism and the philosophical, post-structural views of language such as in Wittgenstein's language games [20,21]. The main components of SFBT include looking for previous solutions, acknowledging problems but identifying that exceptions to the problem are key to the solution, focusing on present and future, as opposed to past-orientated questions, using validation, and utilizing techniques such as miracle questions, scaling questions, and coping questions [22].

Due to the high cost of psychotherapy, people strive for a faster and more effective treatment, which has led to the widespread application of SFBT, such as for treating students' emotional behavior problems [23], substance abuse [24], adolescent crisis events, suicide and self-harm [25]. SFBT is not only applied to psychotherapy and counseling, but also applied in school class management, enterprise management and other fields [26]. SFBT emphasizes using a positive perspective to comprehend people, expressing their affirmation, and fully trusting that people will understand the methods that suit them. It is complementary to traditional Confucian culture [27] and compatible with it [28]. As such, it is highly suitable for application in China, as demonstrated by large effects in Chinese populations [29].

Comparing SFBT with Cognitive Behavioral Therapy (CBT), Jordan, Froerar, & Brovelas [30] argue that professionals using CBT assume the expert role, aiming to identify problems in people's thinking and behavior. But professionals using SFBT regard their patients as already having all the resources they need, defining their own role as supportive in eliciting the individuals' strengths, and empowering people to articulate these into achievable goals for the future. Therefore, SFBT enables patients to look for possible behaviors to change, and puts patients in control as regulators of their behavioral changes. It is not only beneficial for patients to maintain behavioral change, but also helps them to improve self-efficacy, thus making it more suitable for college students' self-exploration and continuous transformation. In addition, some studies have shown that SFBT used more of the client's exact words, and used more positive language, than other therapies such as CBT and Motivational Interviewing (MI) [30,31]. Therefore, SFBT may be more beneficial to relieve patients' anxiety and their concerns towards the issue, thereby making it easier for college students who are sensitive and care about others' evaluation and social pressures to accept this method and reduce their resistance.

In addition, group counseling is an effective intervention for IA [32]. Compared to solution-focused individual counseling, applying SFBT to group counseling makes the process lively, encourages group members to share their experiences and promotes psychological growth with mutual support and encouragement from the group. Nevertheless, the efficacy of SFBT in group counseling to reduce IA in Chinese populations is not fully established. We aim at ameliorating this gap in this pilot study.

Specifically, we focused on applying SFBT in group counseling and explored the intervention effects of solution-focused group counseling on IA in college students. By examining immediate and long-term post-intervention effects, we aim at clarifying if solution-focused group counseling improves IA in college students. Based on the above mentioned findings of prior intervention studies applying SFBT in different contexts, we hypothesized that participants in the experimental group (i.e., those who will accept five-week solution-focused group counseling aimed at alleviating Internet Addiction core symptoms, such as compulsive use, withdrawal and tolerance, and Internet Addiction related problems such as interpersonal, health-related, and time management problems) will show significant changes in these aspects, beyond the placebo effects observed in a control group.

2. Materials and Methods

2.1. Study Design and Participants

Participants were screened from a sample of 27 college students who presented high IA symptomology, recruited through flyers on campus. Based on an initial interview and the Revised Chinese Internet Addition Scale (CIAS-R), we selected 18 subjects with scores ≥53. The other 9 had score <53 and were excluded. The final 18 participants were randomly assigned to experimental or control groups with inclusion criteria CIAS-R score ≥53, which represents moderate to high levels of IA. There were 9 in the experimental group, (male = 3, female = 6; age = 20.11 ± 1.45) and 9 in the control group (male = 3, female = 6; age = 20 ± 1.56). Non-parametric Mann-Whitney U tests showed that there was no significant difference in CIAS-R score between the two groups [$Z = -0.71$, $p > 0.05$]. We also used a self-reported Symptom Checklist-90 (SCL-90) questionnaire as an auxiliary selection tool, and the scores of each factor of the students in the two groups were less than 5, showing no symptoms of serious mental disorders. The study was given ethical approval by the Institutional Review Board of the Southwest University (No. 20180030).

2.2. Procedures

Closed semi-structured group counseling was conducted on the experimental group once a week for five weeks. Every counseling session lasted about two and a half hours. There were one leader and two assistants in the group. The leader is a certified psychotherapist, and the assistants are psychology graduate students trained for group counseling. Their roles followed basic procedures that were designed in advance. Participants were given free interaction space. The leader regulated and navigated the group process according to the needs of the members. The control group did not receive any intervention. Both groups completed surveys at t1 and t2 (t1 + 5 weeks). The intervention group also completed a t1 + 6 months survey. Instruments included in the survey are outlined in the next section.

2.3. Instruments

2.3.1. Revised Chinese Internet Addiction Scale (CIAS-R)

The scale [33,34] consists of 19 items, on a four-point Likert scale. Two subscales are Internet Addiction Core Symptoms (IA-Sym) and Internet Addiction Related Problems (IA-RP). Core symptoms include four dimensions: Compulsive and Withdrawal Symptoms of Internet addiction (Sym–C&Sym-W), Tolerance Symptoms of Internet Addiction (Sym-T), Interpersonal and Health-Related Problems of Internet Addiction (RP-IH), and Time Management Problems (RP-TM).

The experimental group and the control group completed the scale before and after group counseling, and the experimental group also carried out follow-up measurement six months after group counseling. The higher the score, the greater the risk of IA is. CIAS-R is widely used in the measurement of IA. Several studies show that CIAS-R has good reliability, with Cronbach's alpha ranging from 0.85 to 0.90 [34–36]. In this study, Cronbach's alpha was 0.86, 0.89 and 0.83 in the three rounds of administration.

2.3.2. Symptom Checklist-90 (SCL-90)

The Symptom Checklist-90 (SCL-90) has a total of 90 items rated on a five-point Likert scale. It includes 9 factors, such as depression and anxiety. Studies have found that IA may be comorbid with depression and anxiety disorders [37,38]. If comorbidity exists, it is very important to evaluate and treat related mental disorders. However, our group counseling mainly focused on Internet addiction and had no strong pertinence to other mental disorders. Therefore, SCL-90 is used to screen whether the subjects had major comorbid mental disorders.

2.3.3. Scaling Questions Form

Scaling questions are representative questions of SFBT. Members of the experimental group are required to evaluate the current state of Internet use with a 10-point score, which increases successively from 0 ("the worst state of Internet use") to 10 ("the most ideal state of Internet use"). The experimental group completed the evaluation at the pre- and post-group counseling.

2.3.4. Satisfaction Survey

The experimental group was asked to evaluate the overall satisfaction with group counseling at the end of each session on a 10-point Likert scale (from 0 = "very unsatisfied" to 10 = "very satisfied").

2.3.5. Change Questionnaire after Group Counseling

Six months after group counseling, the experimental group members were tracked by a change questionnaire, to examine the sustainability of changes after receiving the group counseling. The questionnaire employed open-ended questions that asked whether there was any change in the behavior of members after they participated in group counseling compared with before, such as Internet use time, attitude towards Internet use, behaviors related to Internet use, and changes in other life aspects beyond Internet use.

2.4. Data Analysis

Quantitative data were analyzed with SPSS 23.0 (IBM Corp. Released, Armonk, NY, USA). We used the non-parametric Mann-Whitney U test and Friedman test for analyses. Qualitative data were sorted based on the Kawakita Jiro (KJ) method which is used to organize data into useful categories. It adopts the bottom-up sorting process and is very useful for classifying data. Based on the statement of significance of "Change after group counseling", researchers summarized, coded and recorded responses with concise, independent and clear phrases, extracted representative topics, and formed a comprehensive database of self-reported parsed statements from group members.

2.5. Solution-Focused Group Counseling Programs

The Solution Focused Brief Therapy Association listed three elements of SFBT [39]. First, there are the overall topics. SFBT conversations are centered on client concerns: who and what are important to the clients; a vision of a preferred future; clients' exceptions, strengths, and resources related to that vision; scaling of clients' motivational level and confidence in finding solutions; and ongoing scaling of clients' progress toward reaching the preferred future. Second, SFBT conversations involve a therapeutic process of co-constructing altered or new meanings in clients. Third, therapists use a number of specific responding and questioning techniques that invite clients to construct a vision of a preferred future and draw on their past successes, strengths, and resources to make that vision a reality. Based on the above factors we designed five focused-solution group counseling sessions (Table 1), with the purpose of assisting college students to establish the goal of alleviating IA (e.g., Ask group members: what changes about IA do you want to make after our group counseling sessions?), reconstruct the problem (e.g., Let group members reflect on the pros and cons of the internet through the group counseling sessions, discussing and sharing with each other), expand positive resources and exceptional experience (e.g., Ask group members: in your past life, to some extent, did you ever think that your internet usage has reached the target state?), and actively seek and enrich solutions (e.g., Let the group members list obstacles they may encounter when using the internet in the future, proving better solutions, after brainstorms.). In group counseling, the leader encouraged and assisted group members to share their experiences (e.g., All members were invited to participate in the group counseling sessions, freely expressing their views and feelings.), and actively adopt a variety of questions (e.g., When group members share twice regarding their changes, the leader will ask more details about what caused these changes.), compliments (e.g., At the end of each session, praise is given to the group members on their efforts to solve and conquer the problems of IA.), other techniques in SFBT (e.g., miracle question: When you go back to sleep in the dormitory tonight, a miracle will happen and all the problems that you brought here have all been solved. (Pauses) But because you are sleeping, you don't know that a miracle has happened. When you wake up the next day, what will you pay attention to, in order to know that a miracle has happened?).

Table 1. Solution-Focused Group Counseling Programs.

Theme	Objective	Activity Process
You and I together	Preliminary understanding between group members; Clarify and establish a group contract to be observed by all members of the group; Arouse the interest of individuals in participating groups; Clarify members' expectations of the group and define members' personal goals.	1. Warm up: The wind blows 2. Serial self-introduction 3. Form a group contract 4. Clarify members' personal goals 5. Complete the scaling question form 6. Feedback and homework: try to improve your score on scaling questions by 1 point; Think about and document the positive aspects of the Internet 7. Complete the satisfaction survey
When miracles happen	Reconstruct the problem of IA, discuss the change between group counseling interventions and a vision of a preferred future, and dig out the materials to solve the problem.	1. Members share their change 2. Warm up: In the same boat 3. Share homework, reconstruct the problem of IA 4. Progressive muscle relaxation and miracle question 5. Feedback and homework: Think about and document the positive aspects of the Internet use 6. Complete the satisfaction survey
Look for a successful self	Explore exceptional experiences and successful experiences that have been achieved, expand positive resources, and gain confidence.	7. Members share changes based on homework 8. Warm up: Trip to trust 9. Explore the exceptional experiences associated with internet use by group members 10. Feedback and homework: Replicate the exceptional experience 11. Complete the satisfaction survey

Table 1. *Cont.*

Theme	Objective	Activity Process
Make friends with the future	Discuss potential obstacles in the future and find and enrich solutions.	1. Members share changes based on homework 2. Warm up: Break through the dilemma 3. Brainstorming: Discuss solutions to obstacles in the future 4. Leader gives a keynote speech: The content aims at the common trouble of the members of the previous three sessions—procrastination 5. Feedback and homework: Replicate the exceptional experience 6. Complete the satisfaction survey
Common agreement	Deal with the parting mood, discuss and summarize the changes brought by the group counseling to the members, praise the members and end the group.	1. King and angel: assign roles 2. Warm up: Copy the human body 3. Share the overall changes from five sessions of group counseling 4. The angel wanted to talk to the king: Members compliment each other 5. End group counseling: Invite members to finish outcome questionnaire after six months 6. Complete the satisfaction survey and scaling question form

3. Results

3.1. Effects of Group Counseling Programs

The change in scores between t1 and t2 was used to compare the treatment and the control group changes (△= post-test score − pre-test score). We compared the scores with the non-parametric Mann-Whitney U test. The changes in both groups are shown in Table 2. As can be seen, both groups showed reduction in IA symptoms, but the reductions in the experimental group were larger. Specifically, the intervention produced significantly larger reductions, compared to the control group, in two IA symptoms: compulsive use and withdrawal (Sym-C & Sym-W) and tolerance (Sym-T), as well as in the total IA score. The small reduction in the control group may manifest from learning and social desirability.

Table 2. Comparison between the Experimental and Control Groups.

Measures	Experimental Group (n = 9)	Control Group (n = 9)	Z
Sym-C & Sym-W	−2.67 ± 2.17	−0.88 ± 1.05	−1.99 *
Sym-T	−2.78 ± 1.72	−0.78 ± 1.20	−2.48 *
RP-IH	−2.00 ± 2.83	−0.11 ± 2.15	−1.48
RP-TM	−2.00 ± 2.45	−0.89 ± 1.54	−0.82
Total IA	−9.44 ± 6.13	−2.67 ± 4.24	−2.26 *

Note: Sym-C: compulsive use; Sym-W: withdrawal; Sym-T: tolerance; RP-IH: interpersonal and health-related problems; RP-TM: time management problems; * $p < 0.05$.

3.2. Long-term Effects of the Intervention

To examine long-term effects of the intervention, beyond the immediate post-intervention effects, we followed up with participants six months after t1. The results of the non-parametric Friedman test showed that significant effects existed in all dimensions of IA (Table 3). Statistical analysis of the data using the Wilcoxon signed-rank test revealed that the IA scores in the pre-test were significantly higher than those in the post-test and follow-up tests for three dimensions of IA: compulsive and withdrawal

(Sym-C & Sym-W), tolerance (Sym-T) and time management problems (RP-TM), as well as for the total IA score. In the dimension of interpersonal and health-related problems (RP-IH), the scores of the pre- and post-test were significantly higher than those of the follow-up test. As shown in Figure 1, the change trend of the total score of the CIAS-R was similar for all subjects in the experimental group.

Table 3. Comparison of the Differences among Pre-test, Post-test and Tracking Scores in the Experimental Group (x ± s).

Measures	Pre-Test (T1)	Post-Test (T2)	Follow-up Test (T3)	χ2	Multiple Comparisons
Sym-C & Sym-W	17.44 ± 1.67	14.78 ± 1.86	13.89 ± 2.67	11.46 **	T1 > T2, T1 > T3
Sym-T	12.78 ± 1.72	10.00 ± 1.94	9.89 ± 2.20	13.15 **	T1 > T2, T1 > T3
RP-IH	15.89 ± 2.09	13.89 ± 2.57	11.67 ± 2.78	13.24 **	T1 > T3, T2 > T3
RP-TM	12.67 ± 1.73	10.67 ± 2.29	10.33 ± 1.32	11.53 **	T1 > T2, T1 > T3
Total IA	58.78 ± 6.12	49.33 ± 6.63	45.78 ± 7.12	14.80 **	T1 > T2, T1 > T3

Note: Sym-C: compulsive use; Sym-W: withdrawal; Sym-T: tolerance; RP-IH: interpersonal and health-related problems; RP-TM: time management problems; ** $p < 0.01$.

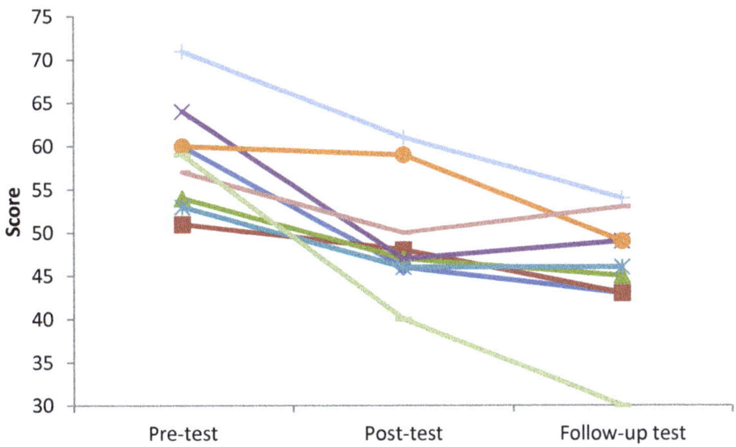

Figure 1. The Total IA Score in the Experimental Group ($n = 9$).

3.3. *Survey Score in Group Counseling of Experimental Group*

As shown in Table 4, the satisfaction degree of the experimental group for each group counseling was above 8.5. Paired-samples t-test revealed that the score of the experimental group on the scaling question at the last group counseling was significantly improved, compared with the first group counseling session ($t = 8.00$, $p < 0.01$).

Table 4. Satisfaction Survey Score of Experimental Group.

Measures	Week 1	Week 2	Week 3	Week 4	Week 5
Satisfaction Score	8.78 ± 0.63	8.75 ± 0.83	9.67 ± 0.47	9.33 ± 0.47	9.78 ± 0.42
scaling question Score	3.11 ± 0.57	/	/	/	5.78 ± 0.92

3.4. *The Change Questionnaire after Group Counseling of Experimental Group*

Based on the change questionnaire after group counseling, the changes of college students can be summarized as "changes related to Internet use" (i.e., changes are directly related to Internet use, including Internet use time, Internet use attitude, Internet use form, etc.) and "changes related to daily life" (i.e., changes in other aspects of daily life that are not directly related to Internet use, such as schedule, interpersonal communication, studying, etc.) (Table 5).

Table 5. Benign Changes after Group Counseling of Experimental Group Members.

Third-Level Coding	Second-Level Coding	First-Level Coding (Number of People Mentioned & Examples of Group Members' Changes)
Changes related to the Internet use	Changes in Internet use time	Time spent on Internet reduced (4; e.g., Time reduced by about an hour or two)
		Reasonable planning and control of Internet use time (2; e.g., Able to put the phone down in time)
		Reasonably allocate time for online entertainment and learning (3; e.g., Study for about 40 minutes, and take a 10-minute break to play on the phone)
		Time spent on games reduced (1)
	Changes in Internet use attitude	Objective view on Internet use (4; e.g., Understand my own Internet use behavior)
		When surfing the Internet, my mentality is more peaceful (2; e.g., Not experiencing dilemmas like before on using internet)
		The appeal of my mobile phone declined (1)
		More focused on other things (1)
		It doesn't matter if I don't surf the Internet (2; e.g., It's no big deal not to be online)
	Changes in Internet use form	The Internet becomes a relaxation tool (3 e.g., Occasionally watch videos to relax)
		The Internet becomes a learning tool (1)
		The Internet becomes a tool of daily life (2; e.g., Now usually use Internet to chat)
		More meaningful content browsed on the Internet (1)
		Dialectical view of Internet point (1)
Changes related to daily life	Changes in schedule	Schedule more regular (2; e.g., Meals on times)
		Internet use will not overly affect sleep (1)
		More days to get up early (1)
		The number and timing of getting up lately adjusted (1)
	Changes in work and study	Spent time on reflection (1)
		More time spent on study (1)
		The number of visits to the library increased (1)
	Changes in interpersonal communication	More activities with friends (2; e.g., Spend more time out with friends for studying and playing)
		Internet use will not affect interpersonal communication (1)
	Changes in life planning	Re-adjust life planning (1)
		Life plan is clearer (1)
	Changes in whole life	Use time to adjust life (1)
		Life is more organized (1)
	Changes in other aspects	The amount of exercise increased (1)
		Internet use will not affect walking (1)
		Began to pay attention to clothing collocation (1)
		Pay more attention to observing emotions (1)
		Continuous counseling (1)

4. Discussion

This pilot study explored the intervention effect of solution-focused group counseling for college students' IA. The results of the comparison between the two groups found that the change in total IA scores as well as in the dimensions of compulsive use, withdrawal (Sym-C & Sym-W) and tolerance

(Sym-T) were significantly larger in the intervention compared to the control group. Within the experimental group, the scores in the four IA dimensions were significantly decreased and the change trend of the total IA score was similar for all subjects in the experimental group. In the dimension of time management problems (RP-TM), the follow-up test was significantly lower than the scores in the pre- and post-tests. In the other dimensions, the post-test was significantly lower than the pre-test, and the follow-up test was significantly lower than the pre-test. The score on the scaling question in the post-test in the experimental group was significantly higher than that in the pre-test, indicating that the members of the group subjectively felt improvements in IA after the group counseling. In addition, the open-ended responses revealed that Internet use time, attitudes toward Internet use and Internet use form, as well as many aspects of daily life of group members, have changed. Moreover, the satisfaction score of the experimental group for every-week group counseling was above 8.5 points, which showed that members valued the provided group counseling. Overall, the results described above indicated that solution-focused group counseling intervention has a positive effect on college students' IA and can lead to healthier Internet use and lifestyles.

After group counseling, the experimental group experienced more significant changes than the control group in IA-Sym, including but not limited to compulsion and withdrawal symptoms (Sym-C & Sym-W), and tolerance symptoms (Sym-T). However, there were no significant differences between the experimental group and the control group with regards to interpersonal and health-related problems (RP-IH), time management problems (RP-TM), and other related IA issues. These results indicated that the characteristics and advantages of SFBT make this therapeutic approach highly targeted, symptom specific, and efficient. Specifically, SFBT helps group members to clarify their goals and find clues to realize their goals by helping them visualize themselves after they use the Internet in a healthier and responsible fashion and avoid IA symptoms. SFBT is effective for several reasons. It guides group members to actively think about the function of the Internet, and to restructure their negative attitudes about the Internet towards more positive views of Internet use. SFBT encourages members to actively explore positive experiences and coping strategies for rational uses of the internet, and to search for past successful experiences to help discover and replicate effective methods of overcoming IA and uncontrolled use. All of this is done in combination with brainstorming, interpersonal communication and sharing, and other related processes to enrich the treatment milieu. SFBT stimulates group members to put various methods into practice through homework, thus gradually boosting changes in members' core IA symptoms. It should be noted here that the change in the experimental group is more significant and may be related to the effect of social desirability. Specifically, in the process of participating in group counseling, the experimental group establishes relationships with leaders and other members, and become eager for recognition and praise. This, coupled with clear goals of group counseling, may stimulate members to actively perform in line with group goals. Of course, this is also one of the motivations of the group. From the perspective of SFBT, the socially desirable responding of the parties is an important psychological resource, which can be used for achieving better results. Through asking questions, the parties can express their abilities and efforts in the process of answering, so that the parties can have a clearer understanding of how to make progress.

Research findings within the groups showed that the experimental group experienced positive significant changes in IA. The classification we employed is based on a study of a sample of 388 college students who completed the CIAS-R revised. They found that a cutoff of 46 separates normal from at-risk groups, and a cutoff of 53 separates at-risk from the Internet addiction group [32]. According to this standard, the average Internet addiction score of the experimental group in this study belongs to the range of Internet addiction before intervention, the range of Internet addiction risk after intervention, and the normal range after six months. Although it is only a rough comparison of the average value, it reflects the clinical significance of the intervention effect to a certain extent. However, the experimental group showed different changes in different aspects of Internet addiction. In terms of core IA symptoms (IA-Sym), such as compulsion (Sym-C), withdrawal (Sym-W) and tolerance symptoms (Sym-T), and time management problems (RP-TM), the immediate effect of group counseling was quite positive and lasted to the follow-up test, six months later. In the Interpersonal and health-related problems (RP-IH) domain, there was no significant change after group counseling, but there was a significant improvement six months later. In addition, the open-ended reports after the follow-up period supported the quantitative data, indicating that experimental group members had experienced positive changes in Internet use and daily life. An earlier study [29] found that the application of SFBT in China can have positive immediate effect. Extending this view, the current study demonstrated that in the form of group counseling SFBT can have both short-term and long-term effects. Additionally, the solution-focused group counseling of this study first stimulated changes in core problems, and those changes were mostly maintained after group counseling. While changes in other dimensions were not as rapid and obvious as changes in core problems, there was still the possibility of change after the group counseling. De Jong and Berg [40] suggested that the professional value of SFBT is multidimensional. It encourages clients' involvement, improves clients' self-determination, maximizes their sense of empowerment, and promotes transferability. Specifically, SFBT helps clients to help themselves, maximizes client potential and resources in solving problems, and makes clients responsible for their own lives by applying what they got from the counseling to other scenarios through the process of creating solutions with counselors. Therefore, solution-focused group counseling not only achieved significant outcomes in treating core IA symptoms, but also had positive longer-lasting impacts on interpersonal relationships, learning, work and other life domains.

Many researchers have also conducted intervention studies on college students' Internet addiction, but follow-up tests are often conducted within six or eight weeks after the intervention [34,41]. We extend such studies in this paper, by focusing on long-term follow-up changes. Moreover, Cognitive-Behavioral Therapy has been widely applied to treatment of IA [42], but the existing cognitive behavioral therapies mostly draw on results of well-developed studies in substance addiction and intervention treatments of other mental disorders. These treatments may not be as efficacious with college students, which is a specific clinical group that needs to use the Internet, but needs to learn to do so responsibly, even under conditions of total freedom. These reasons may account for the limited success of these therapies with IA cases [43]. Besides shedding factors that have controlled their life, many college students lack self-control and behavior-selection strategies. Even if students make positive changes, these can be difficult to maintain. Moreover, each relapse to an addictive behavior will further undermine their self-efficacy [44], thus leading to increased engagement in addictive behaviors and further loss of control. Based on the concept of SFBT, the current pilot study respected the subjectivity of college students and helped them dig deeper into their own resources and strengths to solve the problem. In addition, the form of group counseling promoted positive feedback among members. This is not only conducive to the improvement in the core IA symptoms, but also helps to maintain the change and to generalize the change to other life domains.

Several limitations of the current study are noteworthy. First, the sample size is small. The reliability and validity of major measurement tools such as CIAS-R cannot be fully verified, and the generalizability of the results may be limited. Future studies can consider increasing the sample size to verify the reliability and validity of the measurement tools and improve generalization of the research results. Second, the variables in this study are all self-reported, subjective and easy to be influenced by the effect of social desirability. Future research can consider obtaining more objective data from different approaches such as multi-subject evaluation by consultants, classmates, etc., and/or control for social desirability bias. Third, the sample included college students. The design of the focus solution group counseling program is based on college students' cognitive development level and other psychological characteristics, and the Internet addiction level of the subjects is controlled. In the future, when the group counseling program is applied to other age, identity or Internet addiction level groups, it needs to be revised according to the specific situation. Fourth, changes brought by group counseling may be subtle and even unconscious. More sensitive indicators, such as using neurophysiology related technologies to explore brain structure and functions following SFBT [43], can be used in future research to lay a foundation for improving IA interventions.

5. Conclusions

In sum, this pilot study used solution-focused group counseling to treat college students with moderate to high levels of IA symptoms. The results suggested that solution-focused group counseling had positive effects on IA and some of its core dimensions, and was generally positively valued by participants. SFBT has the characteristics of high-efficiency, simplicity and durability in intervening with college students' Internet use. We call for further studies to apply this approach and examine its merits.

Author Contributions: Conceptualization, X.Z., O.T., and Q.H.; methodology, X.Z. and S.X.; formal analysis, X.Z. and X.S.; investigation, X.Z., X.S., S.X., and J.Q.; resources, X.Z., S.X., and J.Q.; data curation, J.Q.; writing—original draft preparation, X.Z., X.S., S.X., J.Q.; writing—review and editing, O.T. and Q.H.; supervision, X.Z. and Q.H.; project administration, X.Z.; funding acquisition, Q.H. All authors have read and agreed to the published version of the manuscript.

Funding: X.Z. was supported by research grants from the Fundamental Research Funds for the Central Universities (SWU1709618), the Planning Subject for the 13th Five Year Plan of National Education Sciences (EBA180462), and the Chongqing Social Science Planning Project (2016BS104). Q.H. was supported by research grants from the National Natural Science Foundation of China (31972906), Open Research Fund of the Key Laboratory of Mental Health, Institute of Psychology, Chinese Academy of Sciences (KLMH2019K05), and High-end Foreign Expert Introduction Program (G20190022029).

Acknowledgments: We are grateful to all the volunteers for participating in the present study.

Conflicts of Interest: The authors declare no conflict of interest.

References

1. Young, K. *Internet Addiction Test (IAT)*; Stoelting: Wood Dale, IL, USA, 2017.
2. Goldberg, I. Internet Addiction Disorder: Diagnostic Criteria, 1996. Available online: http://users.rider.edu/~{}suler/psycyber/supportgp.html (accessed on 3 April 2020).
3. Turel, O.; Serenko, A.; Giles, P. Integrating technology addiction and use: An empirical investigation of online auction users. *Mis. Q.* **2011**, *35*, 1043–1061. [CrossRef]
4. Turel, O.; Romashkin, A.; Morrison, K.M. Health Outcomes of Information System Use Lifestyles among Adolescents: Videogame Addiction, Sleep Curtailment and Cardio-Metabolic Deficiencies. *PLoS ONE* **2016**, *11*, e0154764. [CrossRef] [PubMed]
5. He, Q.; Turel, O.; Bechara, A. Brain anatomy alterations associated with Social Networking Site (SNS) addiction. *Sci. Rep.* **2017**, *7*, 1–8. [CrossRef] [PubMed]
6. Kumar, N.; Kumar, A.; Mahto, S.K.; Kandpal, M.; Deshpande, S.N.; Tanwar, P. Prevalence of excessive internet use and its correlation with associated psychopathology in 11th and 12th grade students. *Gen. Psychiatry* **2019**, *32*, 84–91. [CrossRef] [PubMed]
7. Ding, Y.J.; Lau, C.H.; Sou, K.L.; Abraham, A.A.; Griffiths, S.M.; Kim, J.H. Association between internet addiction and high-risk sexual attitudes in Chinese university students from Hong Kong and Macau. *Public Health* **2016**, *132*, 60–63. [CrossRef]
8. Canan, F.; Yildirm, O.; Sinani, G.; Ozturk, O.; Ustunel, T.Y.; Ataoglu, A. Internet addiction and sleep disturbance symptoms among adolescents. *Eur. Psychiatry* **2013**, *28*, 1. [CrossRef]
9. Lin, I.H.; Ko, C.H.; Chang, Y.P.; Liu, T.L.; Wang, P.W.; Lin, H.C.; Huang, M.F.; Yeh, Y.C.; Chou, W.J.; Yen, C.F. The association between suicidality and Internet addiction and activities in Taiwanese adolescents. *Compr. Psychiatry* **2014**, *55*, 504–510. [CrossRef]
10. Ozgul, O.; Ozlem, O.; Alaettin, U.; Sinan Ozalp, S. Evaluation of Internet Addiction and Depression among University Students. *Procedia-Soc. Behav. Sci.* **2013**, *82*, 445–454.
11. Turel, O.; Serenko, A. Cognitive biases and excessive use of social media: The facebook implicit associations test (FIAT). *Addict. Behav.* **2020**, *105*, 106328. [CrossRef]
12. Rui, G.L.; Wang, W.; Liu, X.M. An epidemiological survey of Internet addiction among College Students. *Chin. J. Behav. Med. Sci.* **2004**, *13*, 333.
13. Li, Q. The Research about the Relationship of college student Psychological Capital, Stress and creativity. Master's Thesis, Zhejiang University of science and technology, Hangzhou, China, 2014.

14. Ye, B.J.; Zheng, Q. The Effects of Stress on College Students' Internet Addiction. *J. Psychol. Sci.* **2016**, *39*, 621–627.
15. Deng, L.Y.; Fang, X.Y.; Wan, J.J.; Zhang, J.T.; Xia, C.C. The Relationship of Psychological Needs and Need Gratification with Internet Addiction Among College Students. *J. Psychol. Sci.* **2012**, *35*, 123–128.
16. Turel, O.; Mouttapa, M.; Donato, E. Preventing problematic Internet use through video-based interventions: A theoretical model and empirical test. *Behav. Inform. Technol.* **2015**, *34*, 349–362. [CrossRef]
17. Yang, F.R.; Hao, W. The Effect of Integrated Psychosocial Intervention on 52 Adolescents with Internet Addiction Disorder. *Chin. J. Clin. Psychol.* **2005**, *13*, 343–352.
18. Gingerich, W.J.; Eisengart, S. Solution-focused brief therapy: A review of the outcome research. *Fam. Process.* **2000**, *39*, 477–498. [CrossRef] [PubMed]
19. de Shazer, S. *Putting Difference to Work*; Norton: New York, NY, USA, 1991.
20. de Shazer, S. *Words were Originally Magic*; WW Norton &, Co.: New York, NY, USA, 1994.
21. Bavelas, J.B.; De Jong, P.; Jordan, S.; Korman, H. The theoretical and research basis of co-constructing meaning in dialogue. *Solut.-Focus. Brief Ther.* **2014**, *2*, 1–24.
22. Shazer, S.D. *Clues: Investigating Solutions in Brief Therapy*; W.W. Norton Polishers: New York, NY, USA, 1988.
23. Wilmshurst, L.A. Treatment programs for youth with emotional and behavioral disorders: An outcome study of two alternate approaches. *Ment. Health Serv. Res.* **2002**, *4*, 85–96. [CrossRef]
24. Froeschle, J.G.; Smith, R.L.; Ricard, R. The efficacy of a systematic substance abuse program for adolescent females. *Prof. Sch. Couns.* **2007**, *10*, 498–505. [CrossRef]
25. Lamprecht, H.; Laydon, C.; Mcquillan, C.; Wiseman, S.; Reilly, J. Single-session solution-focused brief therapy and self-harm: A pilot study. *J. Psychol. Ment. Health Nurs.* **2010**, *14*, 601–602. [CrossRef]
26. Franklin, C.; Trepper, T.S.; Gingerich, W.J.; McCollum, E.E. *Solution-Focused Brief Therapy: A Handbook of Evidence-Based Practice*; Oxford University Press: Oxford, UK, 2012.
27. Li, Y.C.; Lin, S.L.; Hong, L.Z.; Zhang, D.C.; Zhang, J.W.; Ye, Z.P.; Fan, X.C. *Multiple Applications of Focus Solution Consultation*; Teach. Zhang Cult: Taipei, China, 2006.
28. Xu, W.S. The representative of paradigm shift — the philosophy of Solution-focused Brief Therapy. *Psycho: Tech. Appl.* **2014**, *13*, 41–44.
29. Gong, H.L.; Xu, W.S. A Meta-analysis on the Effectiveness of Solution-focused Brief Therapy. *Psychol. Behav. Res.* **2015**, *13*, 799–803.
30. Jordan, S.S.; Froerer, A.S.; Bavelas, J.B. Microanalysis of positive and negative content in solution focused brief therapy and cognitive behavioral therapy expert sessions. *J. Syst. Ther.* **2013**, *32*, 46–59. [CrossRef]
31. Korman, H.; Bavelas, J.B.; De Jong, P. Microanalysis of formulations in solution-focused brief therapy, cognitive behavioral therapy, and motivational interviewing. *J. Syst. Ther.* **2013**, *32*, 31–45. [CrossRef]
32. Bai, Y.; Fan, F.M. The Effects of Group Counseling on Internet Dependent College Students. *Chin. J. Ment. Health* **2007**, *21*, 247–250.
33. Chen, S.H. Development of Chinese Internet Addiction Scale and Research on Psychometric Characteristics. *Chin. Psychol.* **2003**, *45*, 279–294.
34. Bai, Y.; Fan, F.M. A Study on the Internet Dependence of College Students: The Revising and Applying of a Measurement. *Psychol. Dev. Educ.* **2005**, *4*, 99–104.
35. Gao, F.Q.; Geng, J.Y.; Ren, Y.Q.; Yang, H.Y.; Han, L. Mediating effect of perceived social support and relative deprivation among life events and internet addiction in college students. *Chin. Ment. Health J.* **2017**, *31*, 734–738.
36. Zhang, J.; Zhang, Y.J.; Shen, L.J. Attentional bias characteristics of different levels of over-users in mobile network under different time-course conditions. *Chin. Ment. Health J.* **2019**, *33*, 545–549.
37. Ha, J.H.; Yoo, H.J.; Cho, I.H.; Chin, B.; Shin, D.; Kim, J.H. Psychiatric comorbidity assessed in korean children and adolescents who screen positive for internet addiction. *J. Clin. Psychiatry* **2006**, *67*, 821–826. [CrossRef]
38. Luo, K.L.; Meng, H.Q.; Fu, Y.X. Depression and Anxiety as well as the comorbid of them in adolescents with internet addiction disorder. *Chin J. Clin. Rehabil.* **2005**, *8*, 5–6.

39. Trepper, T.S.; McCollum, E.E.; De Jong, P.; Korman, H.; Gingerich, W.; Franklin, C. *Solution-Focused Therapy Treatment Manual for Working with Individuals*; Research Committee of the Solution Focused Brief Therapy Association, 2010; Available online: https://www.andrews.edu/sed/gpc/faculty-research/coffen-research/trepper_2010_solution.pdf (accessed on 3 April 2020).
40. De Jong, P.D.; Berg, I.K. Interviewing for solutions. *J. Marital. Fam. Ther.* **1998**, *26*, 117.
41. Fang, H.Z.; Dai, Y.Z. Psychological intervention of positive psychology on internet addiction of college students. *Chin. J. Health Psychol.* **2019**, *27*, 1906–1909.
42. Young, K.S. CBT-IA: The first treatment model for internet addiction. *J. Cogn. Psychother.* **2011**, *25*, 304. [CrossRef]
43. Fang, X.Y.; Liu, L.; Deng, L.Y.; Liu, Q.X.; Su, W.L.; Lan, J. The Prevention and Intervention on Adolescent Internet Addiction. *Psychol. Dev. Educ.* **2015**, *31*, 100–107.
44. He, J.F. Effects of group counseling on internet using and study management in college students. *Chin. J. Ment. Health* **2013**, *27*, 496–501.

International Journal of
Environmental Research and Public Health

MDPI

Review

Internet Addiction in the Web of Science Database: A Review of the Literature with Scientific Mapping

Antonio-José Moreno-Guerrero, Gerardo Gómez-García, Jesús López-Belmonte and Carmen Rodríguez-Jiménez *

Department of Didactics and School Organization, University of Granada, 18071 Granada, Spain; ajmoreno@ugr.es (A.-J.M.-G.); gomezgarcia@ugr.es (G.G.-G.); jesuslopez@ugr.es (J.L.-B.)
* Correspondence: carmenrj@ugr.es

Received: 27 February 2020; Accepted: 15 April 2020; Published: 16 April 2020

Abstract: Information and communication technologies (ICT) is a major element of today's society with great potential that can offer both advantages and disadvantages. Addiction to the Internet and social networks is a growing problem in all age groups. Education is the context in which to work and train in the correct use of these media. The objective of the study focuses on knowing the scientific production and the performance of the concepts "addiction" and "internet" (ADIN). A bibliometric methodology complemented with the scientific mapping technique was followed. Different processes related to the quantification, analysis, evaluation, and estimation of scientific documents were carried out. The literature was analyzed by specific programs such as SciMAT, Analyze Results, and Creation Citation Report. The unit of analysis was specified in 5644 scientific publications extracted from Web of Science (WoS), belonging to the period of years between 1996 and 2019. The results showed that the evolution in the study of the addiction to the Internet is constant and continuous, with articles in English being the most used means to present the information on the part of the investigators. In addition, the subject of study was based on time, given that the coincidence of key words between the periods analyzed was high. In conclusion, the importance of promoting healthy living habits that include responsible use of the Internet are discussed.

Keywords: scientific production; bibliometric analysis; scientific mapping; addiction; internet; Web of Science

1. Introduction

Today's society is increasingly focused on technology and therefore people are progressively more dependent on it, particularly with regards to the Internet [1,2]. Mobile phones and different technological devices have made the world of video games and online shopping more accessible to everyone, making markets even more desirable and broad than before [3].

The Internet and the different social networks have become popular instruments that are currently used to communicate and exchange information, among the most outstanding actions [4]. The way in which people interact with each other, entertain themselves, and consume through the media has been modified through the Internet [5]. However, as more people connect to multimedia devices to share information and make use of the possibilities they are offered, cases of media addiction are growing at a dangerous pace [6,7]. These addictions have grown significantly over the last decade [8], surpassing other types of addictions.

There is a great deal of scientific evidence showing how there is a serious risk associated with excessive use of the Internet and everything that derives from it [9], where, in addition, personal characteristics, gender, socio-cultural differences [10], and biological factors could play an important role in the appearance of an addiction to these media. The literature reveals an association between the Internet and social networks and an increase in so-called techno-stress [11–13].

The constant use of the Internet and social networks leads to an addiction and obsession, which results in negative behavior [14]. The consequences of these habits regarding the excessive use of different smart devices and the Internet can affect both the physical and mental health of users [15]. Issues such as anxiety, depression, impulsiveness, and aggressive behavior can also result or be increased from uncontrolled use [16]. Those with internet addiction have higher levels of depression, anxiety, and impulsivity than those without such an addiction [9].

Numerous studies address internet and smartphone addiction, regardless of the age of those suffering from it [17] or even the field in which they are trained or work [18]. In addition, the literature contains relevant research that addresses this type of addiction and its relationship to the educational setting [8,19]. Technology, the Internet, and the impact of the digital world have a great influence on the student body and the development of their personality [20] and health [21].

At present, students, regardless of their educational stage, are one of the groups most closely related to everything related to ICT and its use for different purposes, such as socialization or proper academic development [22]. Experts on the field state that those students who use the Internet for purposes that go beyond simple socialization through the different social networks and video games have a better level with regard to problems related to the use of this technology [23]. The exclusive use of these media for playful or socializing purposes without an educational perspective or previous training in the correct use and handling can lead to problems such as sexting [24], cyberbullying [25], or video game addiction [26]. The consequences derived from these phenomena can affect the academic, social, and emotional spheres.

In this respect, from the educational field, mechanisms of prevention and action should be established in the face of these new existing trends. Therefore, it is necessary to educate in terms of the correct handling of the devices at a technical level, but also at a regulatory level, where the self-regulation of certain behaviors towards these devices is promoted, always extracting the maximum potential for all the areas of the life of the students, including their academic life [27]. The objective from the educational point of view should be to form digitally competent citizens who have positively developed their ethical, literacy, participatory, and critical aspects towards technology and the use of the Internet [28,29].

In recent times, there have been constant publications that analyze ICT, as well as the Internet and its relationship with education [30,31], as the tendency of these topics is to increase and acquire more relevance in terms of research. On the other hand, studies approached from bibliometrics, such as the one presented in this paper, are in recent times a constant that addresses a multitude of topics of incidence, such as medical research [32], virtual reality [33,34], or the area of sports [35].

2. Justification of Study

In the research presented, the concepts "addiction" and "internet" (ADIN) are analyzed in the impact scientific literature collected from the Web of Science (WoS) database. Web of Science is an online platform containing databases of bibliographic information and information analysis resources that allow the evaluation and analysis of research performance, especially in the field of social sciences. In addition, it presents a series of analysis tools that allow specific and concrete searches to be carried out [36]. This study acquires a bibliometric nature with the purpose of analyzing the publications by means of a scientific mapping developed as a result of various bibliometric indicators and a dynamic and structural evolution of the aforementioned terms.

To carry out and follow a study model validated by experts in this field of research, other impact investigations taken from the Journal Citation Reports were taken as a reference (JCR) [36,37].

This study focused on the evaluation of the trajectory and evolution of the mentioned concepts of the main WoS collection. Initially, a search was carried out in such a database to know the status of the terms issue and, in addition, to check if there were any studies of similar characteristics to the one presented in this work. After the search phase, no publication was found that analyzed both terms with a scientific mapping, and thus this study acquired an exploratory nuance about the state of the art in impact literature. Therefore, this study helps to increase the findings and reduce the gap found in the literature concerning the relationship between "addiction" and "internet". Therefore, the results obtained will acquire a relevant value as they will contribute to the advancement of science, in addition to serving as a basis for successive research.

3. Materials and Methods

3.1. Research Objectives

The general objective of this study was to analyze the scientific literature on ADIN indexed in the WoS database. This general objective is articulated in the following specific objectives:

- To determine the performance of publications on ADIN in WoS.
- To decree the scientific evolution of ADIN in WoS.
- To find out the most incident issues about ADIN in WoS.
- To identify the most influential authors in ADIN in WoS.

The research questions posed are

- RQ1: What is the performance of ADIN publications in WoS?
- RQ2: What is the scientific evolution of ADIN in WoS?
- RQ3: What are the most frequent problems of ADIN in WoS?
- Q4: Who are the most influential authors of ADIN in WoS?

3.2. Research Design

The research methodology adopted in this study to achieve the proposed objectives was bibliometry. This choice is justified in the scientometric potentialities linked to the actions of searching, recording, analyzing, and predicting scientific literature [38]. Similarly, the guidance and considerations of experts in this type of study were followed [39].

Specifically, this work was based on the analysis of co-words [40] and various bibliometric indicators such as the h-index, taken as a reference, in addition to other indices (g, hg, q2) [41]. This gave rise to the creation of node maps whose interpretation determined the performance and position of conceptual subdomains alluding to ADIN. In addition, the analytical treatment delimited the thematic development of these concepts in WoS [42]. From this database, we used the complete record in order to obtain as much information as possible, in order to have the most relevant data to perform the co-word analysis [43].

3.3. Procedure and Data Analysis

This investigation was carried out following various actions that specified the following procedures: (a) selection of the database (WoS), (b) determination of the keywords ("addiction" and "internet") after consultation in specialized thesauri (ERIC and UNESCO) and in the information provided in the Special Issue of the *International Journal of Environmental Research and Public Health*, and (c) construction of a precise search equation to obtain significant results ("addiction*" [TOPIC] AND "internet" [TOPIC]) and report documents containing both terms in different metadata (title, abstract, and keywords).

The first search action compiled a scientific volume of 5776 publications. In addition, the entire report was analyzed to eliminate repeated documents, those belonging to the year 2020 and the poorly indexed ones. After this, an analysis unit of 5644 publications remained, as a result of the configuration of various production indicators with their respective inclusion criteria (Table 1).

Table 1. Production indicators and inclusion criteria.

Indicators	Criteria
Year of publication	All production except 2020
Language	x ≥ 50
Publication area	x ≥ 300
Type of documents	x ≥ 300
Organizations	x ≥ 70
Authors	x ≥ 60
Sources of origin	x ≥ 100
Countries	x ≥ 400
Citation	The four most cited documents

To improve the understanding and visualization of the different actions carried out, the following flow chart was established following the protocols of the PRISMA-P matrix (Figure 1).

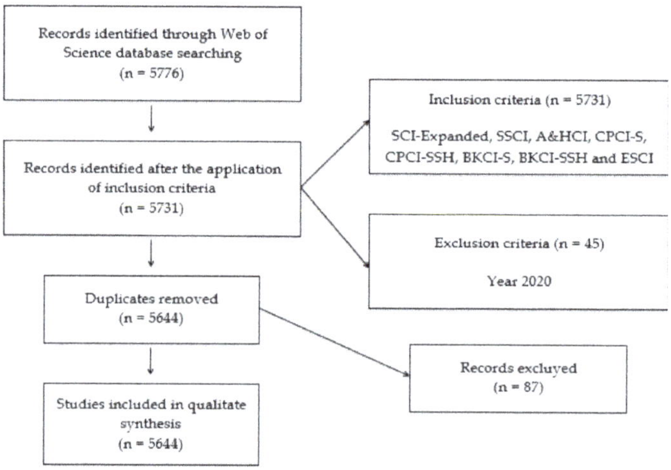

Figure 1. Flowchart according to the PRISMA Declaration.

The documents obtained were analyzed with various programs (Analyze Results, Creation Citation Report, and SciMAT). The first two tools are provided by WoS and were used to obtain data concerning the year, authorship, country, type of document, institution, language, means of publication, and most cited documents. For the structural and dynamic development of a longitudinal nature of the scientific volume, SciMAT was used, under the considerations of the experts [44].

With SciMAT, a co-word analysis of the topics was carried out through the following processes:

- Recognition: In this process, various actions were carried out. The keywords of the literature obtained were analyzed (n = 10,555). A map of co-occurrence nodes was generated. A standardized network of co-words was developed. The most significant keywords were detected (n = 9842), and the themes and concepts with greater predominance were represented with a clustering algorithm.
- Reproduction: In this process, a strategic diagram and a thematic network based on the principles of centrality and density were prepared. In the generated graphs, four regions can be seen: (1) top right (motor themes and highlights), (2) upper left (rooted and isolated themes); (3) lower left (themes to disappear or project), and (4) lower right (underdeveloped and transversal themes).
- Determination: In this process, the evolution of the nodes in certain periods or intervals of time was analyzed. In particular, four periods were defined (P_1 = 1996-2010; P_2 = 2011-2014; P_3 = 2015-2017; P_4 = 2018-2019), except in the authorship that one that covered the entire documentary volume

was established (P_X = 1996-2019). The number of matching keywords determined the strength of association between the configured periods.

- Performance: In this process, a set of production indicators associated with inclusion criteria that determine the scientific works that become part of the study were delimited (Table 2).

Table 2. Production indicators and inclusion criteria.

Configuration	Values
Analysis unit	Keywords authors, keywords WoS
Frequency threshold	Keywords: P_1 = (5), P_2 = (7), P_3 = (7), P_4 = (7) Authors: P_X = (10)
Network type	Co-occurrence
Co-occurrence union value threshold	Keywords: P_1 = (2), P_2 = (4), P_3 = (5), P_4 = (5) Authors: P_X = (6)
Normalization measure	Equivalence index
Clustering algorithm	Maximum size: 9; minimum size: 3
Evolutionary measure	Jaccard index
Overlapping measure	Inclusion rate

4. Results

4.1. Performance and Scientific Production

The evolution of the 5644 documents in the scientific production on ADIN has been constant and continuous over time, showing an exponential growth from its beginnings until the year 2019 (Figure 2). In this sense, and according to the Price law about growth of scientific information, visualization of the distribution of publications on internet addiction doubled in its development 10 years after its first publication. At present, productivity on the subject is in a phase of linear growth.

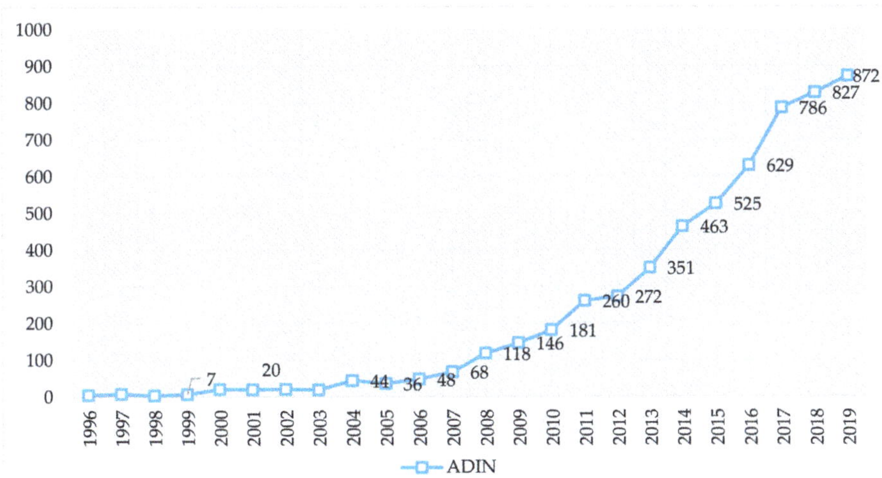

Figure 2. Evolution of scientific production of the keywords "addiction" and "internet" (ADIN) in the Web of Science (WoS).

The language chosen by the various authors for the presentation of the academic results was English (Table 3).

Table 3. Scientific language used in ADIN.

Language	n
English	5265
Spanish	119
German	95
French	56
Russian	53

The main area of knowledge in ADIN studies is psychiatry, distanced considerably from other areas of knowledge (Table 4).

Table 4. Research areas.

Research Area	n
Psychiatry	1649
Psychology (multidisciplinary)	801
Substance abuse	678
Psychology (clinical)	522
Psychology (experimental)	374
Education educational research	313

The type of document mainly used by the various scientists are articles (Table 5).

Table 5. Document types.

Document Types	n
Article	4467
Meeting abstract	383
Proceedings paper	347
Review	320

Nottingham Trent University is the leading organization in ADIN studies (Table 6).

Table 6. Institutions.

Institution	n
Nottingham Trent University	242
Seoul National University (SNU)	107
Catholic University of Korea	105
Yale University	98
Kaohsiung Medical University	87
Chung Ho Memorial Hospital	76
Johannes Gutenberg Univesrity of Mainz	72

The most prolific author is Griffinths, M.D., with considerable scientific output on ADIN (Table 7).

Table 7. Authors.

Authors	n
Griffinths, M.D.	183
Kim, D.J.	72
Ko, C.H.	67
Yen, C.F.	66

Table 7. *Cont.*

Authors	n
Potenza, M.N.	65
Billieux, J.	63
Brand, M.	60
Wolfling, K.	60

The main source of presentation of studies on ADIN are, with similar productions, *Computers in Human Behavior* and the *Journal of Behavioral Addictions* (Table 8).

Table 8. Source titles.

Source Titles	n
Computers in Human Behavior	356
Journal of Behavioral Addictions	331
Cyberpsychology Behavior and Social Networking	168
Addictive Behaviors	105
International Journal of Mental Health Addiction	105

The country with the greatest interest in production on ADIN is the United States (Table 9).

Table 9. Countries.

Country	n
United States	1263
China	773
England	511
South Korea	469
Turkey	406
Germany	405

The reference author for the scientific community in relation to the subject of ADIN, in terms of number of citations, is Davis (2001), followed at a short distance by Morahan-Martin and Schumacher (2000) and Block (2008) (Table 10).

Table 10. Most cited articles and main conclusions.

Reference	Citations	Main Conclusion
[45]	881	The model implies a more important role of cognitions in pathological internet use (PIU), and describes the means by which PIU is both developed and maintained. Furthermore, it provides a framework for the development of cognitive-behavioral interventions for PIU.
[46]	593	Pathological users, for pathological internet use (PIU), scored significantly higher on the UCLA Loneliness Scale, and were socially disinhibited online.
[47]	579	This document refers to how the DSM-IV considers internet addiction.
[48]	535	Among the findings are the urgent need to develop research on best practices for treating pain in adolescents, as well as the development of prevention strategies to reduce diversion and abuse.

4.2. Structural and Thematic Development

The evolution of keywords shows information about the number of keywords in each of the established time intervals, the number of matching keywords between the periods, and the number of keywords leaving and entering a certain period with respect to another. The studies on ADIN, taking into account the evolution of keywords in scientific production, mark a settled and consolidated line of research, given that the level of coincidence between periods was between 40% and 50% (Figure 3).

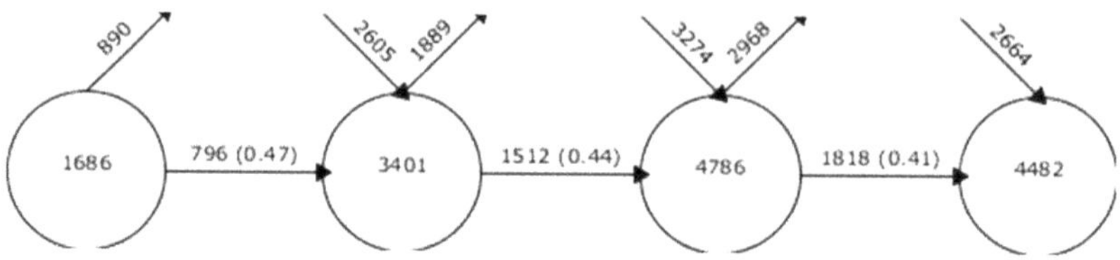

Figure 3. Continuity of keywords between contiguous intervals.

The academic performance in the established periods offers the subjects with the greatest bibliometric indicators, using the h-index as the main reference, and completing this information with the g-index, hg-index and q2-index, in addition to the number of citations. On the subject of ADIN, the subjects with the highest bibliometric indicators varied between periods. In the first time period (1996–2010), the topic with the highest bibliometric indicator was "addiction"; in the second time period (2011–2014) it was "prevalence"; in the third time period (2015–2017) it was "depression" and "addiction"; and in the last time period (2018–2019) it was "adolescents". This shows changes in the interests of researchers in this field of study. We also observed the changes between periods of the topics of greatest interest to researchers, showing the main lines of research in each of the established time intervals (Table 11).

Table 11. Thematic performance.

Interval 1996–2010						
Denomination	**Works**	**Index h**	**Index g**	**Index hg**	**Index q^2**	**Citations**
Health-care	12	8	10	8.94	13.86	210
Addiction	238	76	130	99.4	108.19	18,177
Impulse-control-disorder	13	11	12	11.46	25.48	1203
Prevalence	62	37	61	47.51	65.8	5423
Symptoms	20	15	20	17.32	48.68	2216
Computer-games	13	11	11	11	29.1	1328
Sensation-seeking	15	14	14	14	31.3	1547
Technology	22	12	20	15.49	35.16	2060
Online	42	26	38	31.43	43.86	3061
Students	17	13	17	14.87	24.45	848
Intervention	11	8	10	8.94	22.09	523
Scale	21	17	19	17.97	31.4	1488
Alcohol-use	5	5	5	5	10.72	491
Video-games	6	6	6	6	21.77	489
Support	6	6	6	6	21.21	591
Model	5	4	5	4.47	25.14	611
Risk	4	4	4	4	21.82	466
Interval 2011–2014						
Denomination	**Works**	**Index h**	**Index g**	**Index hg**	**Index q^2**	**Citations**
Prevalence	519	69	105	85.12	83.89	19,238
College-students	95	39	71	52.62	54.44	5256
Children	44	23	37	29.17	33.57	1384
Behavior	115	38	61	48.15	54.79	4168
Loneliness	81	30	56	40.99	46.8	3339
Alcohol	30	17	30	22.58	28.57	1026
Motivations	28	20	27	23.24	36.06	2048
Gender	27	13	23	17.29	19.42	596

Table 11. *Cont.*

Interval 2011–2014						
Denomination	**Works**	**Index h**	**Index g**	**Index hg**	**Index q²**	**Citations**
FMRI	15	13	14	13.49	28.62	870
Treatment	26	17	25	20.62	29.15	1038
Validation	16	13	16	14.42	22.8	681
Health	15	10	15	12.25	15.81	228
Pornography	12	10	12	10.95	17.03	406
TDAH	8	7	8	7.48	12.69	481
DSM-5	18	11	17	13.67	26.12	759
Play	10	8	9	8.49	14.7	296
Computer-use	20	15	19	16.88	24.49	724
Anterior-cingulate-cortex	5	5	5	5	16.28	414
Aggression	7	7	7	7	20.32	579
Interval 2015–2017						
Denomination	**Works**	**Index h**	**Index g**	**Index hg**	**Index q²**	**Citations**
Funcional-connectivity	54	21	29	24.68	25.51	998
Depression	571	42	63	51.44	51.44	8642
Addiction	680	42	63	51.44	51.03	9259
Self-esteem	108	22	33	26.94	27.75	1492
Decision-making	35	17	29	22.2	22.58	866
DSM-5	73	21	38	28.25	29.7	1627
Smartphone-addiction	44	17	26	24.74	29.44	1314
Validation	85	16	25	20	21.54	937
Substance-use	52	15	26	19.75	21.21	821
Children	53	15	24	18.97	18.57	763
Facebook	44	11	15	12.85	12.85	331
Students	45	15	22	18.17	18.97	596
Model	47	13	23	17.29	18.73	605
Internet-pornography	43	13	26	18.38	17.66	736
Play	19	11	16	13.27	16.58	361
Smoking	15	6	13	8.83	12.73	179
Technology	10	8	10	8.94	13.56	244
Computer-games	12	6	9	7.35	11.22	126
Substance-use-disorders	8	6	7	6.48	10.39	207
Gaming-disorder	9	5	7	5.92	7.07	68
Interval 2018–2019						
Denomination	**Works**	**Index h**	**Index g**	**Index hg**	**Index q²**	**Citations**
Adolescents	860	17	22	19.34	19.34	2433
Internet-pornography	48	8	13	10.2	11.66	231
Mobile-phone	66	8	14	10.58	12.33	262
Self-esteem	138	11	17	13.97	15.56	544
Facebook	73	7	8	7.48	7.94	200
Impulsivity	68	9	11	9.95	9.95	229
Video-games	82	6	10	7.75	8.49	180
Validation	74	7	9	7.94	7.94	177
Children	112	8	10	8.94	9.8	263
Problematic-internet-use	119	10	13	11.4	11.83	399
University-students	45	8	13	10.2	10.95	222
Prefrontal-cortex	20	5	7	5.92	6.71	60
Internet-use	25	5	7	5.92	7.42	83
Life	16	4	7	5.29	6.93	55
Cue-reactivity	7	3	4	3.46	3.46	20
International-consensus	12	5	7	5.92	6.71	59
Stress	12	3	5	3.87	3.46	33
Risk	16	3	4	3.46	4.58	29

The diagrams of the intervals developed show data on the importance of each of the themes in the different periods. For this purpose, a grouping process was developed, according to Callon's indicators, which assesses the degree of interaction of a network with respect to other networks, from two axes: centrality, which analyses the strength of the relationship of external links with other topics, where it shows the importance of the development of a topic in a field of research; and density, which assesses the internal strength of the network, analyzing the internal links between the key words that are grouped around a specific topic, giving information on the degree of development of a field of study. In the first period (1996–2010), the driving themes were "addiction", "prevalence", "symptoms", "health-care", and "sensation-seeking". In the second period (2011–2014), the themes were "prevalence", "children", "computer-use", and "loneliness". In the third period (2015–2017), themes were "depression", "addiction", "DSM-5", "functional-connectivity", "self-esteem", "children", and "validation". In the last period (2018–2019), themes were "adolescents", "self-esteem", "impulsivity", "children", and "video-games". From this last period, we should also highlight the themes "international-consensus", "stress", "risk", "life", and "internet-use", given that their location in the diagram makes them unknown themes, as they may be the next references in the next few years in this field of study, or they may disappear. Generally speaking, the interests of ADIN researchers are focused on aspects related to adolescents, children, frequency, and addiction, as these are the issues that most appear in the motor themes of the diagram (Figure 4).

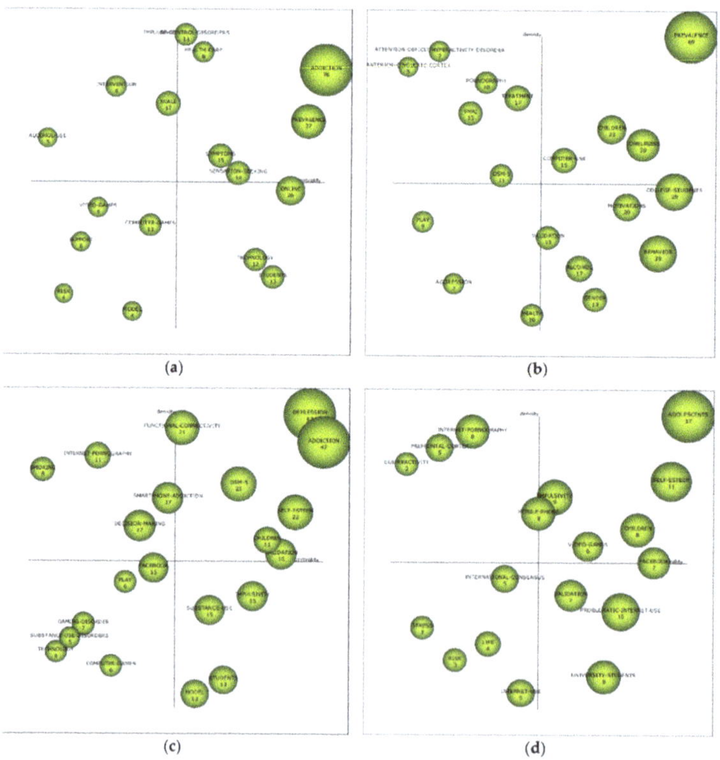

Figure 4. ADIN's strategic diagram by h-index. Note: (**a**) interval 1996–2010; (**b**) interval 2011–2014; (**c**) interval 2015–2017; (**d**) interval 2018–2019.

4.3. Thematic Evolution of the Terms

The thematic evolution offers the strength of the relationship maintained between the themes of the various intervals generated according to the Jaccard index. The development is generated if a theme of a particular interval shares keywords or themes with the subsequent or previous periods. The greater the number of keywords between the two themes of consecutive intervals, the more solid its evolution will be. The possible connections are continuous line, where its connection is thematic, and discontinuous line, whose connection is by keywords. The thickness of the lines shows the strength of the relationship between the themes. The data showed a conceptual gap in the thematic evolution in the ADIN study, as there was no keyword that was repeated in all the established periods. Even so, there were themes that were repeated in at least two or more periods. The timeline with the greatest connecting force was the one established between "prevalence-prevalence-depression-adolescent", being the basis of the studies on the research topic. There were other strong connections, such as those established between "play-play", from the second and third period, and "impulsivity-impulsivity" or "self-esteem" from the third and fourth period. In addition, Figure 5 shows that there were an even number of thematic connections and keyword connections. The studies on ADIN had an established basis, focusing mainly on people under 18 years of age and the different pathologies caused by internet use addiction.

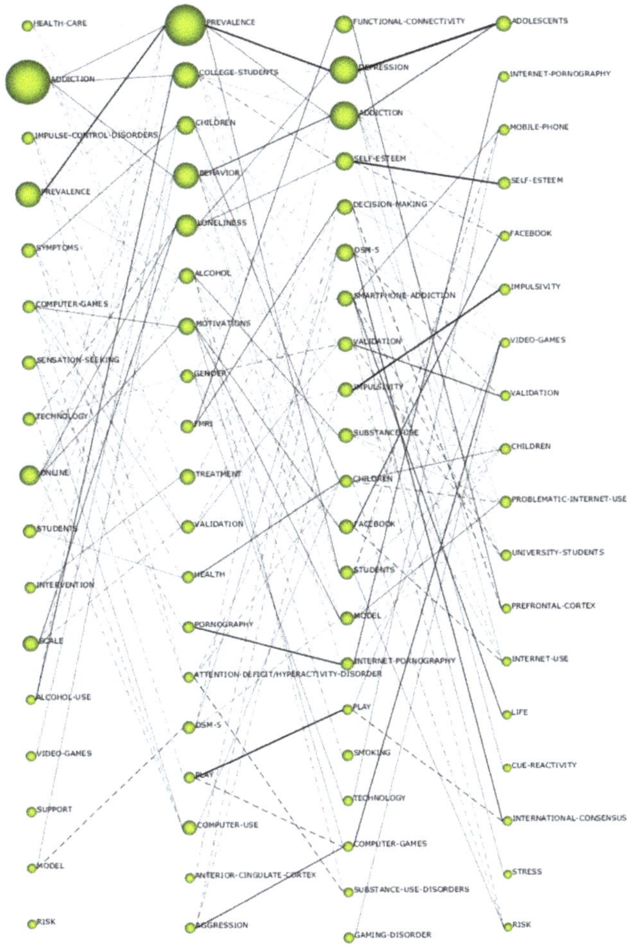

Figure 5. Thematic evolution by h-index.

4.4. *Authors with a Higher Relevance Index*

The driving authors on the field of study of ADIN are Mihara, S.; Marino, C.; and Levine, S. In addition, the author that have a great relevance for the scientific community, bearing in mind their h-index, are Yen, J.Y., and Griffiths, M.D., although their location in the diagram placed them as isolated or cross-sectional authors. For the next few years, we should bear in mind the authors Brand, M.; Dong, G.H.; and Montag, C., because their location in the diagram placed them as unknown authors, and they may disappear or be the reference authors in this field of study (Figure 6).

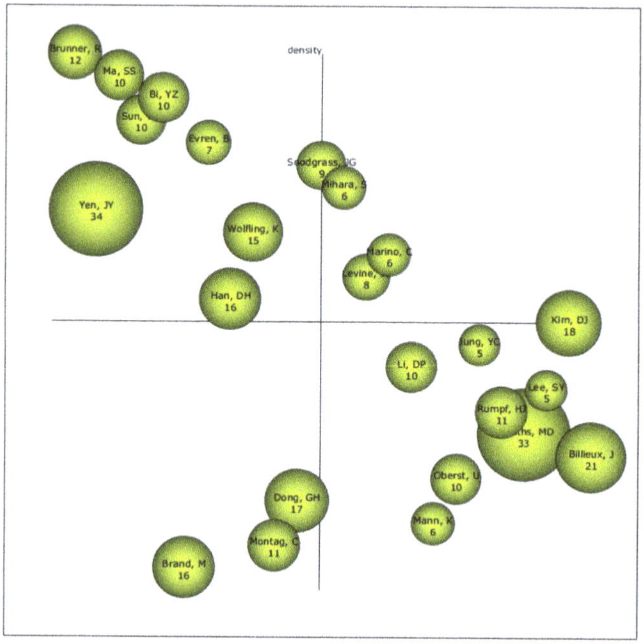

Figure 6. Strategic authoring diagram.

5. Discussion

ADIN has become one of the most frequented topics in the scientific panorama over the last years. The arrival of the Internet in people's lives has brought many benefits, but also risks that citizens must be aware of in order to take preventive measures [9].

The present work aimed to establish a state of the art study on scientific productivity on ADIN in one of the most prestigious databases at present (WoS). The analysis of the data allowed the elucidation of the exponential growth in which the discipline of knowledge finds itself at present, doubling in 10 years from its origin the number of publications and being nowadays in a phase of linear growth [44]. The large percentage of manuscripts found belonged mainly to the psychiatric and psychological area, linked to the treatment of addictions, drugs, and harmful substances, which denotes that this is a global concern that particularly affects the compendium of the health discipline [32]. However, there is also an emerging amount of research concerning the educational branch, which allows us to see that ADIN is beginning to be a concern for experts in the field of education, in order to promote a practice of a prophylactic nature [33].

Likewise, the analysis on the most prolific institutions and authors in the area allowed us to know that the research on ADIN is an object of study by multiple researchers from different parts of the world, with numerous countries being among the biggest producers of scientific works on the subject, and therefore configuring one of the most important risks that reside in the society [1].

In this sense, the co-word analysis provided an understanding of which of the descriptors had the highest academic performance over the years about ADIN, in addition to being able to visualize its evolution over the years. By way of this, it was possible to distinguish how there was an evolution in this issue, from the origin of ADIN to the present, starting with descriptors such as "sensation-seeking" or "symptoms", to "loneliness", "depression", "self-esteem", "stress", "video-games", or "impulsivity". Undoubtedly, a negative progression is visualized that began with descriptors that alluded to minor symptoms and disorders, and ended with descriptors referring to risks related to serious disorders or diseases such as depression, social isolation, or serious self-esteem problems. This indicates that there has been an increase in ADIN in society and, therefore, there are publications that refer to this phenomenon associated with this type of disease [9,16].

In sum, the results indicated that, to a greater extent, risks associated with the Internet, such as the consumption of pornography and young people's addiction to video games and the behavioral disorders associated with them, obtained a considerable density of publication, and therefore these are risks that affect society today [23,24]. In a similar vein, the analysis of the relationship between issues through the Jaccard index allowed us to discern relevant connections between issues, as was the case with the one maintained between "prevalence-prevalence-depression-adolescent", "addiction-behavior-addiction-adolescent", or "online-loneliness-self-esteem". These links allowed us to corroborate the interest of the research community in the risks associated with ADIN over the years, an idea similar to those expressed in previous research [7,17,33,34]. Specifically, the main population under study, according to the works analyzed, are young people, especially adolescents.

Finally, the scientific mapping of authors with a higher index of relevance allowed us to know a considerable compendium of experts in the present on ADIN, which indicates the importance that this topic is gaining within the current scientific panorama, as well as the need for its study and deepening by the research community.

6. Conclusions

ADIN is among the most threatening risks to society today and in the future, especially among the younger population, who live more with technological devices continuously, which can lead to severe consequences associated with serious behavioral disorders and psychological diseases.

Through this work, we intended to establish a general state on the productivity and scientific performance of the subject, trying to contribute interesting ideas about the evolution of the subject and about which are the most frequent aspects to be developed by the researchers, who focus their attention on the study of ADIN. All of this information can be used in order to be able to elucidate the working path in the near future.

Among the limitations that comprise this work is the debugging of the data presented in WoS, where in some cases some repeated or unrelated documents are presented. This was also the case in the establishment of the intervals, this being a question of equity, given that the researchers tried to maintain a similar density of manuscripts in each of them. This study focused solely on the WoS database to show the scientific community the state of the art research on the existing literature on the subject. This may also be a limitation of the study. Therefore, as a future line of research, it is proposed that the analysis be replicated in other impact databases such as Scopus. On the other hand, as future lines of research, it is recommended that future authors investigate the search for new risks associated with the excessive use of the Internet, as well as the establishment of effective educational research to help combat and prevent this dangerous phenomenon for present and future generations.

In conclusion, the idea of continuing on the path of promoting healthy living habits with respect to digital health, especially among the youngest people, through a responsible practice to promote a healthy society that avoids coming into contact with the risks associated with the digital world, is stressed.

Author Contributions: Conceptualization, C.R.-J. and A.-J.M.-G.; methodology, A.-J.M.-G. and J.L.-B.; software, A.-J.M.-G.; validation, J.L.-B. and G.G.-G.; formal analysis, J.L.-B.; investigation, A.-J.M.-G.; resources, G.G.-G.;

data curation, A.-J.M.-G.; writing—original draft preparation, C.R.-J.; writing—review and editing, C.R.-J. and G.G.-G.; visualization, C.R.-J. and G.G.-G.; supervision, A.-J.M.-G. and J.L.-B. All authors have read and agreed to the published version of the manuscript.

Funding: Ministry of Education and Vocational Training of the Government of Spain (project reference: FPU18/01595).

Acknowledgments: To the researchers of the research group Análisis de la Realidad Educativa (AREA) (HUM-672). Research group by belonging to the Ministry of Education and Science of the Junta de Andalucía and based in the Department of Didactics and School Organization of the Faculty of Education Sciences of the University of Granada.

Conflicts of Interest: The authors declare no conflict of interest.

References

1. Brooks, S.; Wang, X.; Schneider, C. Technology Addictions and Technostress: An Examination of the US and China. *J. Organ. End User Comput. (JOEUC)* **2020**, *32*, 1–19. [CrossRef]
2. Sahu, M.; Gandhi, S.; Sharma, M.K. Mobile Phone Addiction Among Children and Adolescents: A Systematic Review. *J. Addictions Nurs.* **2019**, *30*, 261–268. [CrossRef] [PubMed]
3. San-Martín, S.; Jimenez, N.; Camarero, C.; San-José, R. The Path between Personality, Self-Efficacy, and Shopping Regarding Games Apps. *J. Theor. Appl. Electron. Commer. Res.* **2020**, *15*, 59–75. [CrossRef]
4. Aparicio-Martínez, P.; Ruiz-Rubio, M.; Perea-Moreno, A.J.; Martínez-Jiménez, M.P.; Pagliari, C.; Redel-Macías, M.D.; Vaquero-Abellán, M. Gender differences in the addiction to social networks in the Southern Spanish university students. *Telemat. Inform.* **2020**, *46*, 1–12. [CrossRef]
5. Firth, J.; Torous, J.; Stubbs, B.; Firth, J.A.; Steiner, G.Z.; Smith, L.; Álvarez-Jiménez, M.; Gleeson, J.; Vancampfort, D.; Armitage, C.J.; et al. The "online brain": How the Internet may be changing our cognition. *World Psychiatry* **2019**, *18*, 119–129. [CrossRef]
6. Molavi, P.; Mikaeili, N.; Ghaseminejad, M.A.; Kazemi, Z.; Pourdonya, M. Social Anxiety and Benign and Toxic Online Self-Disclosures: An Investigation into the Role of Rejection Sensitivity, Self-Regulation, and Internet Addiction in College Students. *J. Nerv. Ment. Dis.* **2018**, *206*, 598–605. [CrossRef]
7. Rodríguez-García, A.M.; Moreno-Guerrero, A.J.; López Belmonte, J. Nomophobia: An Individual's Growing Fear of Being without a Smartphone—A Systematic Literature Review. *Int. J. Environ. Res. Public Health* **2020**, *17*, 580. [CrossRef]
8. Carbonell, X.; Chamarro, A.; Oberst, U.; Rodrigo, B.; Prades, M. Problematic use of the internet and smartphones in university students: 2006–2017. *Int. J. Environ. Res. Public Health* **2018**, *15*, 475. [CrossRef] [PubMed]
9. Bisen, S.S.; Deshpande, Y.M. Prevalence, predictors, psychological correlates of internet addiction among college students in India: A comprehensive study. *Anatol. J. Psychiatry.* **2020**, *21*, 117–123. [CrossRef]
10. Urbanova, L.B.; Holubcikova, J.; Madarasova Geckova, A.; Reijneveld, S.A.; van Dijk, J.P. Does Life Satisfaction Mediate the Association between Socioeconomic Status and Excessive Internet Use? *Int. J. Environ. Res. Public Health* **2019**, *16*, 3914. [CrossRef] [PubMed]
11. Cuervo-Carabel, T.; Orviz-Martínez, N.; Arce-García, S.; Fernández-Suárez, I. Tecnoestrés en la Sociedad de la Tecnología y la Comunicación: Revisión bibliográfica a partir de la Web of Science. *Arch. Prevención Riesgos Labor.* **2018**, *21*, 18–25. [CrossRef]
12. Kim, Y.J.; Jang, H.M.; Lee, Y.; Lee, D.; Kim, D.J. Effects of internet and smartphone addictions on depression and anxiety based on propensity score matching analysis. *Int. J. Environ. Res. Public Health* **2018**, *15*, 859. [CrossRef] [PubMed]
13. Xie, J.Q.; Zimmerman, M.A.; Rost, D.H.; Yin, X.Q.; Wang, J.L. Stressful life events and problematic smartphone usage among Chinese boarding-school adolescents: A moderated mediation model of peer support and depressive symptoms. *Addict. Res. Theory* **2019**, 1–8. [CrossRef]
14. Masi, L.; Herba, C.; Garel, P. Projet pilote: Exploration de l'utilisation d'Internet et des médias sociaux chez un groupe d'adolescents ayant participé à Espace Transition. *Ann. Médico-Psychologiques Rev. Psychiatr.* **2019**, *177*, 319–326. [CrossRef]
15. Savolainen, I.; Oksanen, A.; Kaakinen, M.; Sirola, A.; Paek, H.J. The Role of Perceived Loneliness in Youth Addictive Behaviors: Cross-National Survey Study. *JMIR Men. Health* **2020**, *7*, e14035. [CrossRef]

16. Obeid, S.; Saade, S.; Haddad, C.; Sacre, H.; Khansa, W.; Al Hajj, R.; Kheir, N.; Hallit, S. Internet addiction among Lebanese adolescents: The role of self-esteem, anger, depression, anxiety, social anxiety and fear, impulsivity, and aggression—A cross-sectional study. *J. Nerv. Ment. Dis.* **2019**, *207*, 838–846. [CrossRef]
17. Lyvers, M.; Cutinho, D.; Thorberg, F.A. Alexithymia, impulsivity, disordered social media use, mood and alcohol use in relation to facebook self-disclosure. *Comput. Hum. Behav.* **2020**, *103*, 174–180. [CrossRef]
18. Rafiee, F.; Ghamari Zare, Z.; Adib-Hajbaghery, M.; Bakhtiari Zade, S. Internet addiction and psychological strain in nursing and midwifery students. *Nurs. Midwifery Stud.* **2020**, *9*, 36–42. [CrossRef]
19. Bhatt, S.; Gaur, A. Psychological risk factors associated with internet and smartphone addiction among students of an Indian dental institute. *Indian J. Public Health* **2019**, *63*, 313–317. [CrossRef]
20. Al Majali, S.A. The Digital World for Children and its Relationship with Personality Disorders: Exploring Emerging Technologies. *Int. J. Emerg. Technol. Learn. (iJET)* **2020**, *15*, 213–221. [CrossRef]
21. Lai, H.Y.; Cheah, W.L. Stress and its association with sociodemographic characteristics, coping and internet use among late adolescent university students in East Malaysia. *J. Indian Assoc. Child. Adolesc. Ment. Health* **2020**, *16*, 13–32.
22. Ergun-Basak, B.; Aydin, M. Problematic Internet Use in Terms of the Purposes of Internet Use, Irrational Beliefs, Feelings of Inferiority, and Gender. *Addicta: The Turkish J. Addict.* **2019**, *6*, 469–494. [CrossRef]
23. Truzoli, R.; Viganò, C.; Galmozzi, P.G.; Reed, P. Problematic internet use and study motivation in higher education. *J. Comput. Assist. Learn.* **2019**, 1–7. [CrossRef]
24. Cooper, K.; Quayle, E.; Jonsson, L.; Svedin, C.G. Adolescents and self-taken sexual images: A review of the literature. *Comput. Hum. Behav.* **2016**, *55*, 706–716. [CrossRef]
25. Cardozo, G.; Dubini, P.; Lorenzino, L. Bullying y ciberbullying: Un estudio comparativo con adolescentes escolarizados. *Rev. Mex. Psicol.* **2017**, *34*, 101–109.
26. Griffiths, M.D.; Nuyens, F. An Overview of Structural Characteristics in Problematic Video Game Playing. *Curr. Addict. Rep.* **2017**, *4*, 272–283. [CrossRef]
27. Caro-Amada, C.; Plaza, J. Intervención educativa familiar y terapia sistémica en la adicción adolescente a internet: Fundamentación teórica. *Rev. Española Orientac. Psicopedag. (REOP)* **2016**, *27*, 99–113. [CrossRef]
28. Choi, M. A concept analysis of digital citizenship for democratic citizenship education in the internet age. *Theory Res. Soc. Educ.* **2016**, *44*, 565–607. [CrossRef]
29. Hartley, J. *The Uses of Digital Literacy*; Routledge: New York, NY, USA, 2017.
30. Van Laar, E.; Van Deursen, A.J.; Van Dijk, J.A.; De Haan, J. The relation between 21st-century skills and digital skills: A systematic literature review. *Comput. Hum. Behav.* **2017**, *72*, 577–588. [CrossRef]
31. Rodríguez-Jiménez, C.; Gómez-García, G.; Romero-Rodríguez, J.M. Los retos tecnológicos del docente del siglo XXI/Technological challenges of the 21st century teacher. *Educ. Soc.* **2019**, *17*, 44–56.
32. Liao, H.; Tang, M.; Luo, L.; Li, C.; Chiclana, F.; Zeng, X.J. A bibliometric analysis and visualization of medical big data research. *Sustainability* **2018**, *10*, 166. [CrossRef]
33. Aznar, I.; Trujillo, J.M.; Romero-Rodríguez, J.M. Estudio bibliométrico sobre la realidad virtual aplicada a la neurorrehabilitación y su influencia en la literatura científica. *Rev. Cuba. Inf. Cienc. Salud* **2018**, *29*, 1–10.
34. Salmeron-Manzano, E.; Manzano-Agugliaro, F. The Higher Education Sustainability through Virtual Laboratories: The Spanish University as Case of Study. *Sustainability* **2018**, *10*, 4040. [CrossRef]
35. Sánchez-Santos, J.M.; Castellanos-García, P.A. Bibliometric analysis of sports economics research. *Int. J. Sport Financ.* **2011**, *6*, 222–244.
36. López, J.; Moreno-Guerrero, A.J.; López, J.A.; Pozo, S. Analysis of the Productive, Structural, and Dynamic Development of Augmented Reality in Higher Education Research on the Web of Science. *Appl. Sci.* **2019**, *9*, 1–21.
37. Rodríguez-García, A.M.; López, J.; Agreda, M.; Moreno-Guerrero, A.J. Productive, Structural and Dynamic Study of the Concept of Sustainability in the Educational Field. *Sustainability* **2019**, *11*, 5613. [CrossRef]
38. Martínez, M.A.; Cobo, M.J.; Herrera, M.; Herrera, E. Analyzing the scientific evolution of social work using science mapping. *Res. Soc. Work Pract.* **2015**, *25*, 257–277. [CrossRef]
39. Moral-Muñoz, J.A.; Herrera-Viedma, E.; Santisteban-Espejo, A.; Cobo, M.J. Software tools for conducting bibliometric analysis in science: An up-to-date review. *Prof. Inf.* **2020**, *29*, 1–20. [CrossRef]
40. Hirsch, J.E. An index to quantify an individual´s scientific research output. *Proc. Natl. Acad. Sci. USA* **2005**, *102*, 16569–16572. [CrossRef]

41. Cobo, M.J.; López, A.G.; Herrera, E.; Herrera, F. Science mapping software tools: Review, analysis, and cooperative study among tools. *J. Am. Soc. Inf. Sci. Technol.* **2011**, *62*, 1382–1402. [CrossRef]
42. López-Robles, J.R.; Otegi-Olaso, J.R.; Porto, I.; Cobo, M.J. 30 years of intelligence models in management and business: A bibliometric review. *Int. J. Inf. Manag.* **2019**, *48*, 22–38. [CrossRef]
43. Belli, S.; Baltá, J. Stocktaking scientific publication on bi-regional collaboration between Europe 28 and Latin America and the Caribbean. *Scientometrics* **2019**, *121*, 1447–1480. [CrossRef]
44. Montero-Díaz, J.; Cobo, M.J.; Gutiérrez-Salcedo, M.; Segado-Boj, F.; Herrera-Viedma, E. Mapeo científico de la Categoría «Comunicación» en WoS (1980–2013). *Comunicar* **2018**, *26*, 81–91. [CrossRef]
45. Davis, R.A. A cognitive-behavioral model of pathological Internet use. *Comput. Hum. Behav.* **2001**, *17*, 187–195. [CrossRef]
46. Morahan-Martin, J.; Schumacher, P. Incidence and correlates of pathological Internet use among college students. *Comput. Hum. Behav.* **2000**, *16*, 13–29. [CrossRef]
47. Block, J.J. Issues for DSM-V: Internet addiction. *Am. J. Psychiatry* **2008**, *165*, 306–307. [CrossRef]
48. Compton, W.M.; Volkow, N.D. Major increases in opioid analgesic abuse in the United States: Concerns and strategies. *Drug Alcohol Depend.* **2006**, *81*, 103–107. [CrossRef]

International Journal of
Environmental Research and Public Health

MDPI

Article

Adolescents' Addictive Phone Use: Associations with Eating Behaviors and Adiposity

Sarah E. Domoff [1], Emma Q. Sutherland [2], Sonja Yokum [3] and Ashley N. Gearhardt [4,*]

[1] Department of Psychology, Central Michigan University, Mount Pleasant, MI 48859, USA; domof1se@cmich.edu
[2] School of Public Health, University of Michigan, Ann Arbor, MI 48109, USA; emmasuth@umich.edu
[3] Oregon Research Institute, Eugene, OR 97403, USA; sonjas@ori.org
[4] Department of Psychology, University of Michigan, Ann Arbor, MI 48109, USA
* Correspondence: agearhar@umich.edu;

Received: 27 February 2020; Accepted: 16 April 2020; Published: 21 April 2020

Abstract: Concerns have been raised about excessive or "addictive" phone use among adolescents, and the impact that addictive phone use (APU) can have on adolescent development and health. Most research on the physical health correlates of smartphone use has been limited to sleep health, whereas other outcomes, such as eating behaviors and obesity risk have not received as much attention. To address this gap in the literature, we examined the association between APU and emotion regulation difficulties, impulsivity, maladaptive eating behaviors, and adiposity in a sample of 111 adolescents. We found that APU is associated with greater emotion regulation difficulties, dysregulated eating, restrained eating, food addiction, and higher percent body fat. Further, we found that emotion regulation difficulties mediated the association between APU and dysregulated eating, restrained eating, and food addiction. Findings suggest that addictive phone use may confer increased risk for obesogenic eating behaviors and food addiction via challenges in regulating emotions.

Keywords: phone; adolescents; addiction; BMI percentile; food addiction; emotional eating; impulsivity; emotion regulation

1. Introduction

Smartphone access and ownership continues to increase among adolescents in the United States. Indeed, a majority of youth (53%) own a smartphone by age 11—this increases to over 80% of youth by age 14 [1]. In addition to gaining access to smartphones at younger ages, shifts have also occurred in frequency of smartphone use. Social media applications (apps) are more frequently used by teens, with 38% of teens stating they use social media multiple times per hour, and 16% indicating they use social media nearly constantly [2].

Given the massive shift in adolescents' consumption of social media and other apps on mobile devices, concern has been expressed regarding whether adolescents can become "addicted" to screens, particularly smartphones. Nearly half of parents in the United States report that their children are addicted to mobile devices [3]. Although the term "addiction" may be used colloquially to describe too much use, the construct of addictive phone use is conceptualized in the scientific literature as a behavioral addiction, wherein excessive use leads to dysfunction or impairment [4]. Similar to the measurement of specific types of screen media addiction (e.g., problematic media use in children: [5]; social media dependence: [6]), symptoms of addictive phone use are based on the DSM-5 criteria for Internet Gaming Disorder (IGD [7]), but applied to smartphone use [4]. The nine criteria include indicators of both excessive use and dysfunction related to use, such as tolerance (needing to use phone for longer amounts of time), withdrawal (i.e., feeling irritable when unable to use or access phone), use to manage/modulate mood, and psychosocial impairment due to use.

Most research on addictive phone use has been conducted in college student and other adult populations [8]. Although a growing body of research has examined addictive phone use in adolescents in Europe and Asia [9], very limited research has examined addictive phone use among adolescents in the United States. One recent exception described the development of the Addictive Patterns of Use measure (APU measure) and its association with academic functioning in US adolescents. Domoff, Foley, and Ferkel (2019) found that greater addictive phone use (using the aforementioned DSM-5 criteria) is associated with poorer academic functioning, over and above the amount of time spent using social media on school days [4]. This study suggests that how adolescents use smartphones, and not necessarily the time they spend on smartphones, confers risks to adolescent development.

Traditional screen time (i.e., time spent watching TV) has long been associated with physical health outcomes, such as obesity and poorer sleep, in children and adolescents [10]. A recent review on the literature on physical health correlates of mobile devices noted major gaps in our understanding of the impact of excessive smart phone use on the physical health of youth [11]. Although research has most consistently linked smartphone use with disrupted sleep and shorter sleep duration [12], we are unaware of research examining addictive phone use with other health concerns, such as problematic eating behaviors, food addiction, and adiposity. The limited research that has considered smartphone use and health risks examines overall time spent on smartphones [13–15], but not addictive phone use. Given the high rates of phone use and concerns about phone addiction, it is critical to clarify whether addictive phone use may contribute to problematic eating behaviors and adiposity in US adolescents.

Furthermore, consideration of the underlying mechanisms accounting for the relationship between addictive phone use and physical health should be explored to clarify potential targets for clinical intervention. In particular, theoretical and empirical support exists for the role of impulsivity and emotion regulation difficulties on problematic phone use (see the Billieux et al. (2015) pathway model for the emergence of problematic mobile phone use [16]). Specifically, Billieux et al. (2015) outline three pathways that lead to problematic phone use. Two of these pathways, the excessive reassurance pathway and the impulsive pathway, describe both emotion regulation challenges (i.e., emotional instability) and poor impulse control (i.e., lack of premeditation, low self-control, and ADHD symptoms), as risk factors for addictive patterns of phone use. Billieux et al., (2015)'s theoretical pathways have been supported in studies of adolescents in Turkey and South Korea [17]. Similar correlates have also been defined as likely contributors to maladaptive eating behaviors, food addiction, and obesity in adolescents [18,19]. As such, impulsivity and emotion regulation difficulties will be explored as potential mediators.

In summary, smartphone addiction and its impact on adolescent health is largely understudied [11]. This is a significant area of research as a vast majority of US teens have their own smartphones, and nearly half of US parents are concerned about their adolescents' addiction to mobile devices [3]. Although countless studies have examined additive phone use and adolescent mental health [20], no studies have examined addictive phone use and obesity risk in US adolescents. The United States has a high rate of childhood obesity, with one in five adolescents meeting criteria for obesity [21]. To our knowledge, no research has considered whether smartphone addiction confers obesogenic risk among US adolescents. Given that obesity prevention is critical to the health of adolescents, examining whether and how addictive phone use may increase obesity risk fits within the priorities of the National Institute of Child Health and Human Development's Strategic Plan [22]. Preliminary research is necessary to investigate obesity-related health correlates of addictive phone use.

To address the limitations of previous research and set the foundation for future clinical research, our study aims to: (1) examine differences by gender, race/ethnicity, and parental education in addictive phone use; (2) investigate whether adolescents who report greater addictive phone use have greater emotion regulation difficulties, impulsivity, dysregulated eating, restrained eating, food addiction, and adiposity (objectively measured by body mass index (BMI) percentile and percent body fat); and (3) explore whether emotion regulation difficulties and impulsivity mediate the associations between addictive phone use and eating behaviors and adiposity.

2. Materials and Methods

2.1. Participants

This study is drawn from a larger parent study (N = 193) examining adolescents' neural responsiveness to fast food advertisements [23]. Administration of the Addictive Patterns of Use measure (APU measure) occurred part-way through the study, for a final N of 111 adolescents (see Table 1 for demographic characteristics). Data collection occurred between July 2015 and August 2017.

2.2. Procedure

Study procedures were approved by the Institutional Review Board at the University of Michigan (UM IRB board: HUM00095596). After obtaining parent/legal guardian informed consent and adolescent written assent, participants completed two visits to the lab. These visits consisted of survey completion, anthropometric measurement, and a functional magnetic resonance imaging paradigm (see Gearhardt et al., 2020, for additional information) [23]. Participants received compensation for their participation.

2.3. Measures

2.3.1. Demographic Characteristics

Information about the participants' demographics (e.g., age, gender, ethnicity, race, and highest education level of parent in the household) was provided. We categorized race/ethnicity as White or non-White, and parental education was categorized as college degree or higher or no college degree attained.

2.3.2. Addictive Patterns of Use (APU) Scale

Participants were administered the APU scale [4], which consists of nine items assessing excessive use of Smartphones (α = 0.85). Sample items are as follows: "During the last year, how often have you felt restless or tense when you were unable to use your phone?" and "During the last year, how often have there been times when all you could think about was using your phone?" Participants responded to items on a 5-point Likert scale, ranging from 1 (Never) to 5 (Very often). Mean scores were calculated to provide a total score of addictive phone use, with higher scores indicating greater addictive phone use.

2.3.3. Difficulties in Emotion Regulation Scale (DERS)

The DERS is a 36-item scale (α = 0.90) which assesses adolescents' perceptions of difficulties in regulating emotions [24]. Sample items include, "I have difficulty making sense out of my feelings" and "When I'm upset, I have difficulty controlling my behaviors." Adolescents respond on a Likert scale ranging from Almost Never (1) to Almost Always (5). Responses were summed to calculate a total score, with higher scores indicating greater difficulties in emotion regulation.

2.3.4. Barratt Impulsiveness Scale-Brief (BIS-Brief)

The BIS-Brief is a shortened version of the BIS-11, a widely used measure of trait impulsivity [25]. The BIS-Brief consists of eight items (α = 0.76), including "I do things without thinking" and "I am self-controlled." Participants respond to questions on a scale from rarely/never (1) to almost always/always (4). Responses are summed to provide a total score, with higher scores indicating greater trait impulsivity.

2.3.5. Eating Behaviors

The Dutch Eating Behaviour Questionnaire (DEBQ) is a widely used self-report measure of three types of eating behaviors: restrained eating, emotional eating, and external eating [26]. Participants responded to items on a 5-point Likert scale from (1) never to (5) very often (33 items). Restrained eating is measured by 10 items ($\alpha = 0.93$) such as "Do you try to eat less at mealtimes than you would like to eat?" The mean of these items was calculated to provide an overall restrained eating score. Emotional eating is measured by 13 items ($\alpha = 0.91$), such as "Do you get the desire to eat when you are anxious, worried or tense?" and external eating is measured by responses to 10 items ($\alpha = 0.86$), such as "If you see or smell something delicious, do you have a desire to eat it?" Given the high inter-correlation of emotional and external eating scales ($r = 0.58$, $p < 0.01$), the mean of the emotional and external eating items was calculated for a score of dysregulated eating.

2.3.6. The Dimensional Yale Food Addiction Scale for Children 2.0 (dYFAS-C 2.0)

The dYFAS-C 2.0 [27] consists of 16 items ($\alpha = 0.91$), adapted from the adult version of the Yale Food Addiction Scale 2.0 [28]. Items include: "When I started to eat certain foods, I found it hard to stop" and "When I cut down or stopped eating certain foods, I craved them a lot more." Participants responded to questions on a 5-point Likert-scale, ranging from Never (0) to Always (4). The sum of items was calculated to provide a total score. Higher scores indicate higher levels of food addiction.

2.3.7. Body Mass Index (BMI) Percentile

Objectively measured height and weight via standardized anthropometric measurements occurred, with participants instructed to wear light clothing. Height was measured using an O'Leary Acrylic Stadiometer (Ellard Instrumentation LLD, Monroe, Washington, USA) height (in centimeters, to the nearest tenth); weight was measured using a Detecto Portable Scale (Cardet, Webb City, Missouri, USA) (in kilograms, to the nearest tenth). Sex, age, height, and weight were used to calculate BMI percentiles [29].

2.3.8. Percent Body Fat

After participants completed the standard anthropometric measurements, we also objectively measured their percent body fat (PBF) via InBody 570 scale (InBody USA, Cerritos, California, USA), with an 8-point Bioelectrical Impedance Analyzer (Frequencies 5, 50, 500 kHz). Consistent with the measurement of height and weight, participants were instructed to wear light clothing and to remove coats, shoes, and socks for PBF measurement.

2.4. Statistical Analyses

All variables used in this study were investigated for normality. Given that some of the variables were skewed, non-parametric analyses that do not require normal distributions were used. For Aim 1, Mann-Whitney tests were used to compare mean APU scores across demographic categories. For Aim 2, Spearman rho correlations were used to investigate the association of APU with the outcome variables. For Aim 3, Hayes' PROCESS macro (Model 4) was used to explore whether emotion dysregulation and impulsivity mediated the associations examined in Aim 2 [30]. Due to identified sex differences in APU, sex was entered as a covariate in the mediation analyses.

3. Results

3.1. Participant Characteristics:

The mean age of the participants was 14.57 ($SD = 1.08$) years. The average BMI percentile was 78.59 ($SD = 21.84$) and the average percent body fat (PBF) was 26.84 ($SD = 11.60$). The sample was

44.1% male (n = 49); 71.2% identified as White (n = 79), followed by 17.1% identifying as Black/African American (n = 19; see Table 1 for additional characteristics of the sample).

Table 1. Sample Characteristics.

Characteristics	M (*SD*)/*n* (%)
Age	14.57 (1.08)
Gender	
Female	62 (55.9%)
Male	49 (44.1%)
Ethnicity	
Hispanic	9 (8.1%)
Race	
White	79 (71.2%)
Black	19 (17.1%)
Biracial	8 (7.2%)
Other	5 (4.5%)
Parent education level	
Less than high school or less	12 (10.8%)
High school diploma	7 (6.3%)
Some college courses	18 (16.2%)
Associate's degree	10 (9.0%)
Bachelor's degree	30 (27.0%)
Advanced degree	34 (30.6%)
BMI Percentile	78.59 (21.84)
Percent body fat	26.84 (11.60)
Dysregulated eating	1.97 (0.52)
Restrained eating	2.06 (0.87)
DERS	74.52 (17.37)
BIS-Brief	15.96 (3.66)
dYFAS-C 2.0	26.89 (8.89)
APU	1.83 (0.60)

Note: BMI—Body Mass Index; Dysregulated eating was measured with the external and emotional eating subscales of the Dutch Eating Behaviour Questionnaire (range 1–5); DERS—Difficulties in Emotion Regulation Scale (range: 36–180); BIS-Brief—Barratt Impulsiveness Scale, Brief version (range: 8–32); dYFAS-C 2.0—Dimensional Yale Food Addiction Scale for Children, 2.0 (range: 0–64); APU—Addictive Patterns of Use Scale (range: 1–5).

3.2. Aim 1

Demographic differences in APU were examined by gender, race/ethnicity, and parent education level. Girls (M = 2.05, SD = 0.49) reported greater APU than boys (M = 1.55, SD = 0.60; t (109) = −4.69, $p < 0.01$). No significant differences emerged by race/ethnicity and parent education level (p's > 0.05).

3.3. Aim 2

To investigate whether adolescents who report greater APU have greater emotion regulation difficulties, impulsivity, dysregulated eating, restrained eating, food addiction, and adiposity (as measured by BMI percentile and percent body fat), we conducted Spearman's Rho correlations. We found significant associations between APU and emotion dysregulation, dysregulated eating, restrained eating, food addiction and percent body fat (see Table 2). APU was not significantly associated with impulsivity, and was thus not included in the exploratory mediation analyses. APU did not significantly associate with BMI percentile.

Table 2. Associations of Addictive Phone Use (APU) with Emotion Regulation Difficulties (DERS), Impulsivity (BIS-Brief), Eating Behaviors, Food Addiction (dYFAS-C 2.0), Body mass index (BMI) Percentile, and Percent Body Fat (PBF).

Variables	1	2	3	4	5	6	7	8
1. APU	-							
2. DERS	0.29 **	-						
3. BIS-Brief	0.18 +	0.20 **	-					
4. Dysregulated eating	0.53 **	0.51 **	0.16 *	-				
5. Restrained eating	0.39 **	0.33 **	0.07	0.68 **	-			
6. dYFAS-C 2.0	0.49 **	0.35 **	0.07	0.56 **	0.39 **			
7. BMI percentile	0.08	0.10	0.00	0.25 **	0.43 **	0.32 **	-	
8. PBF	0.22 *	0.04	−0.04	0.27 **	0.37 **	0.40 **	0.76 **	-

Note: APU—Addictive Patterns of Use Scale; DERS—Difficulties in Emotion Regulation Scale; BIS-Brief—Barratt Impulsiveness Scale, Brief version; dYFAS-C 2.0—Dimensional Yale Food Addiction Scale for Children, 2.0; BMI—Body Mass Index; PBF—Percent Body Fat; **. $p < 0.01$; *. $p < 0.05$.; +. $p < 0.10$.

3.4. Aim 3

To explore whether emotion regulation difficulties mediated the associations between addictive phone use and maladaptive eating behaviors and adiposity, we used Hayes' (2017) PROCESS macro (Model 4). We found that emotion regulation difficulties mediated the association between APU and dysregulated eating (see Table 3), restrained eating (see Table 4), and food addiction (see Table 5).

Table 3. Emotion Regulation Difficulties (DERS) as a Mediator of the Relation Between Addictive Phone Use (APU) and Dysregulated Eating.

Model	*B*	*SE*	*t*	*p*, Cis	*F*	R^2
Model summary				0.00	*F* (3, 107) = 24.47	0.41
Direct effects						
APU	0.33	0.07	4.43	0.00		
DERS	0.01	0.00	4.480	0.00		
Sex	0.11	0.08	1.32	0.19		
Indirect effect						
DERS	0.10	0.03 [a]		(0.04, 0.17) [b]		
Completely Standardized Indirect Effect						
DERS	0.11	0.04 [a]		(0.05, 0.19) [b]		

Note: APU—Addictive Patterns of Use Scale; DERS—Difficulties in Emotion Regulation Scale. Estimates are unstandardized. [a] Bootstrap-derived estimate of standard error of indirect effect, 10,000 bootstrapped samples; [b] 95% lower and upper level confidence intervals.

Table 4. Emotion Regulation Difficulties (DERS) as a Mediator of the Relation Between Addictive Phone Use (APU) and Restrained Eating.

Model	*B*	*SE*	*t*	*p*, CIs	*F*	R^2
Model summary				0.00	*F* (3, 107) = 11.69	0.25
Direct effects						
APU	0.45	0.14	3.21	0.00		
DERS	0.01	0.00	3.05	0.00		
Sex	0.12	0.16	0.71	0.48		
Indirect effect						
DERS	0.12	0.05 [a]		(0.03, 0.24) [b]		
Completely Standardized Indirect Effect						
DERS	0.09	0.04 [a]		(0.02, 0.16) [b]		

Note: APU—Addictive Patterns of Use Scale; DERS—Difficulties in Emotion Regulation Scale. Estimates are unstandardized. [a] Bootstrap-derived estimate of standard error of indirect effect, 10,000 bootstrapped samples; [b] 95% lower and upper level confidence intervals.

Table 5. Emotion Regulation Difficulties (DERS) as a Mediator of the Relation Between Addictive Phone Use (APU) and Food Addiction.

Model	*B*	*SE*	*t*	*p*, CIs	*F*	R^2
Model summary				0.00	*F* (3, 86) = 11.25	0.28
Direct effects						
APU	3.97	1.55	2.57	0.01		
DERS	0.13	0.05	2.58	0.01		
Sex	3.29	1.80	1.83	0.07		
Indirect effect						
DERS	1.33	0.62 [a]		(0.18, 2.64) [b]		
Completely Standardized Indirect Effect						
DERS	0.09	0.04 [a]		(0.01, 0.18) [b]		

Note: APU—Addictive Patterns of Use Scale; DERS—Difficulties in Emotion Regulation Scale. Estimates are unstandardized. [a] Bootstrap-derived estimate of standard error of indirect effect, 10,000 bootstrapped samples; [b] 95% lower and upper level confidence intervals.

4. Discussion

A major aim of this study was to examine the physical health correlates of APU that are under-studied in the literature and that are major areas of public health concern in the United States (see Domoff, Borgen, Foley, & Maffett, 2019 [11] for a review). Specifically, we sought to investigate whether addictive phone use associated with objectively measured indicators of obesity (i.e., percent body fat and body mass index, BMI), as well as adolescents' eating behaviors which have consistently been linked to obesity risk [31]. Consistent with prior research on non-mobile media (e.g., TV, computer, gaming [14]), we similarly found that addictive phone use associates with various obesogenic risk factors, including dysregulated eating, food addiction and percent body fat. We, however, did not find an association between addictive phone use and BMI percentile. It is possible that percent body fat might be a more accurate measure of adiposity. BMI percentile scores may be biased for individuals with elevated lean muscle mass.

We also sought to explore whether emotion regulation difficulties mediated the association of addictive phone use with dysregulated eating, restrained eating, food addiction, and adiposity. We found that emotion regulation difficulties mediated the association between addictive phone use and dysregulated eating, restrained eating, and food addiction, but not adiposity. It is possible that emotion regulation difficulties may be an underlying or common factor for both maladaptive eating behaviors (e.g., dysregulated eating, food addiction) and excessive phone use. As has recently been suggested [32], overlap in the experience of multiple addictions (including behavioral addiction) may indicate an underlying phenotype for addiction proneness. Future research may seek to test out other common factors across addictions, as has been conducted specific to screen-based behavioral addictions (e.g., [24]). Though preliminary, these findings suggest that clinicians assisting youth with emotion regulation challenges should assess for the presence of co-occurring addictions, such as phone addiction and food addiction.

Consistent with prior research in countries outside of the US [9,33], we found that girls had higher rates of addictive phone use and no differences emerged by race/ethnicity or parent education level. Given the social features of several popular smartphone apps (e.g., social media, texting or communication apps), the higher utilization of social media by girls [1], and the salience of peer relations in girls during adolescence, it is understandable that they would be at heightened risk for excessive phone use and dysfunction related to phone use. Although prior research on access to mobile devices and amount of screen time differs across various demographic factors [1], we examined addictive phone use and not hours of smartphone use. When prior research has examined problematic smartphone use, similar findings have emerged in adolescents in other Western countries [9].

Limitations

There are limitations of the study that should be addressed in future research. First, our sample size was small; future research should seek to replicate our findings in a larger, more representative sample of adolescents. It is possible that, with a larger, more representative sample, impulsivity may significantly associate with addictive phone use, as has been found in research on young adults in the US [34]. We should also note that a larger sample size would be beneficial to conduct more sensitive analyses to understand how the associations may be moderated by demographic factors (e.g., race, ethnicity, gender). As has been found in prior research on problematic phone use (e.g., [9,33]), girls may be at greater risk for compulsive use. Additional research should consider gender and stage of development (early adolescent versus older adolescent/emerging adult) in future research. Additionally, the current cross-sectional study does not permit unambiguous conclusions regarding temporal precedence and the directions of effects. It is important to emphasize, however, that given this is the first study to examine the links between addictive phone use and obesity, and underlying mechanisms (i.e., emotion regulation) connecting the constructs, it would have been premature to carry out a longitudinal design as preliminary research was lacking. The current study has thus set the foundation for future research on physical health correlates of addictive phone use, which should utilize a longitudinal design, as some of the physical health outcomes (e.g., obesity) may not emerge until later in adolescence or early adulthood, and to demonstrate temporal precedence and direction of effects.

5. Conclusions

This study provides preliminary evidence demonstrating the physical health risks associated with addictive phone use in US adolescents. Further, we found that emotion regulation difficulties appear to be a mechanism through which youths' addictive phone use may associate with dysregulated eating, restrained eating, and food addiction. On the other hand, it may also be possible that youth with greater emotion regulation concerns are at heightened risk for multiple addictions (e.g., food addiction, addictive phone use). Future research should seek to replicate our findings in a larger, more diverse sample of adolescents in order to inform clinical recommendations and intervention targets.

Author Contributions: Conceptualization, S.E.D. and A.N.G.; methodology, S.E.D., S.Y., and A.N.G.; formal analysis, S.E.D., E.Q.S., and A.N.G.; resources, S.Y. and A.N.G.; data curation, S.E.D., E.Q.S., and A.N.G.; writing—original draft preparation, S.E.D., E.Q.S., and A.N.G.; writing—review and editing, S.E.D., E.Q.S., S.Y. and A.N.G.; supervision, S.E.D. and A.N.G.; project administration, A.N.G.; funding acquisition, S.Y. and A.N.G. All authors have read and agreed to the published version of the manuscript.

Funding: This work was supported by a grant from the National Institute on Diabetes and Digestive and Kidney Diseases: R01 DK102532 (PI: AG).

Conflicts of Interest: The authors declare no conflict of interest.

References

1. Rideout, V.; Robb, M. The Common Sense Census: Media Use By Tweens and Teens. *Common Sense Media* **2019**, 1–104. [CrossRef]
2. Rideout, V.; Robb, M.B. *Social Media, Social Life: Teens Reveal Their Experiences*; Common Sense Media: San Francisco, CA, USA, 2018. [CrossRef]
3. Common Sense Media. Common Sense and SurveyMonkey Poll Parents on YouTube and Technology Addiction. Available online: https://www.surveymonkey.com/newsroom/common-sense-and-surveymonkey-poll-parents-on-youtube-and-technology-addiction/ (accessed on 10 January 2020).
4. Domoff, S.E.; Foley, R.P.; Ferkel, R. Addictive Phone Use and Academic Performance in Adolescents. *Hum. Behav. Emerg. Technol.* **2019**. [CrossRef]
5. Domoff, S.E.; Harrison, K.; Gearhardt, A.N.; Gentile, D.A.; Lumeng, J.C.; Miller, A.L. Development and Validation of the Problematic Media Use Measure: A Parent Report Measure of Screen Media "Addiction" in Children. *Psychol. Pop. Media Cult.* **2019**, *8*, 2–11. [CrossRef]

6. van den Eijnden, R.J.J.M.; Lemmens, J.S.; Valkenburg, P.M. The Social Media Disorder Scale. *Comput. Hum. Behav.* **2016**, *64*, 478–487. [CrossRef]
7. American Psychiatric Association. *Diagnostic and Statistical Manual of Mental Disorders*, 5th ed.; American Psychiatric Association: Washington, DC, USA, 2013.
8. Yu, S.; Sussman, S. Does Smartphone Addiction Fall on a Continuum of Addictive Behaviors? *Int. J. Environ. Res. Public Health* **2020**, *19*, 422. [CrossRef]
9. López Fernández, O.; Losada López, J.; Honrubia Serrano, M. Predictors of Problematic Internet and Mobile Phone Usage in Adolescents. *Aloma Rev. Psicol. ciències l'educació i l'esport* **2015**, *33*, 49–58.
10. Robinson, T.N.; Banda, J.A.; Hale, L.; Lu, A.S.; Fleming-Milici, F.; Calvert, S.L.; Wartella, E. Screen Media Exposure and Obesity in Children and Adolescents. *Pediatrics* **2017**, *140* (Suppl. 2), S97–S101. [CrossRef]
11. Domoff, S.E.; Borgen, A.L.; Foley, R.P.; Maffett, A. Excessive Use of Mobile Devices and Children's Physical Health. *Hum. Behav. Emerg. Technol.* **2019**, *1*, 169–175. [CrossRef]
12. Carter, B.; Rees, P.; Hale, L.; Bhattacharjee, D.; Paradkar, M.S. Association Between Portable Screen-Based Media Device Access or Use and Sleep Outcomes: A Systematic Review and Meta-Analysis. *JAMA Pediatr.* **2016**, *170*, 1202–1208. [CrossRef]
13. Bradbury, K.M.; Turel, O.; Morrison, K.M. Electronic Device Use and Beverage Related Sugar and Caffeine Intake in US Adolescents. *PLoS ONE* **2019**. [CrossRef]
14. Kenney, E.L.; Gortmaker, S.L. United States Adolescents' Television, Computer, Videogame, Smartphone, and Tablet Use: Associations with Sugary Drinks, Sleep, Physical Activity, and Obesity. *J. Pediatr.* **2017**, *182*, 144–149. [CrossRef] [PubMed]
15. Wada, K.; Yamakawa, M.; Konishi, K.; Goto, Y.; Mizuta, F.; Koda, S.; Uji, T.; Tamura, T.; Nakamura, K.; Tsuji, M.; et al. Associations of Cell Phone Use and Screen Viewing with Overweight in Children. *Child. Obes.* **2019**, *15*, 417–425. [CrossRef] [PubMed]
16. Billieux, J.; Maurage, P.; Lopez-Fernandez, O.; Kuss, D.J.; Griffiths, M.D. Can Disordered Mobile Phone Use Be Considered a Behavioral Addiction? An Update on Current Evidence and a Comprehensive Model for Future Research. *Curr. Addict. Reports* **2015**, *2*, 156–162. [CrossRef]
17. Yıldız, M. Emotion Regulation Strategies as Predictors of Internet Addiction and Smartphone Addiction in Adolescents. *J. Educ. Sci. Psychol.* **2017**, *7*, 66–78.
18. Miller, A.L.; Gearhardt, A.N.; Fredericks, E.M.; Katz, B.; Shapiro, L.F.; Holden, K.; Kaciroti, N.; Gonzalez, R.; Hunter, C.; Lumeng, J.C. Targeting Self-Regulation to Promote Health Behaviors in Children. *Behav. Res. Ther.* **2018**, *101*, 71–81. [CrossRef]
19. Eirini, T.; Kamal, A.; Turvill, A.; Holler, R. Emotion Dysregulation and Loneliness as Predictors of Food Addiction. *J. Heal. Soc. Sci.* **2019**, *4*, 43–48. [CrossRef]
20. Odgers, C.L.; Jensen, M.R. Annual Research Review: Adolescent Mental Health in the Digital Age: Facts, Fears, and Future Directions. *J. Child Psychol. Psychiatry Allied Discip.* **2020**, *61*, 336–348. [CrossRef]
21. Hales, C.M.; Carroll, M.D.; Fryar, C.D.; Ogden, C.L. *Prevalence of Obesity Among Adults and Youth: United States, 2015–2016*; NCHS Data Brief; No 288; National center for Health Statistics: Hyattsville, MD, USA, 2017; pp. 2–7.
22. Eunice Kennedy Shriver National Institute of Child Health and Human Development, NIH, D. *NICHD Strategic Plan 2020*; Government Printing Office: Washington, DC, USA, 2019; pp. 1–44.
23. Gearhardt, A.N.; Yokum, S.; Harris, J.L.; Epstein, L.H.; Lumeng, J.C. Neural Response to Fast Food Commercials in Adolescents Predicts Intake. *Am. J. Clin. Nutr.* **2020**, *3*, 493–502. [CrossRef]
24. Gratz, K.L.; Roemer, L. Multidimensional Assessment of Emotion Regulation and Dysregulation: Development, Factor Structure, and Initial Validation of the Difficulties in Emotion Regulation Scale. *J. Psychopathol. Behav. Assess.* **2004**, *26*, 41–54. [CrossRef]
25. Mathias, C.W.; Stanford, M.S.; Liang, Y.; Goros, M.; Charles, N.E.; Sheftall, A.H.; Mullen, J.; Hill-Kapturczak, N.; Acheson, A.; Olvera, R.L.; et al. A Test of the Psychometric Characteristics of the BIS-Brief among Three Groups of Youth. *Psychol. Assess.* **2018**, *30*, 847–856. [CrossRef]
26. van Strien, T.; Frijters, J.E.R.; Bergers, G.P.A.; Defares, P.B. The Dutch Eating Behavior Questionnaire (DEBQ) for Assessment of Restrained, Emotional, and External Eating Behavior. *Int. J. Eat. Disord.* **1986**, *5*, 295–315. [CrossRef]
27. Schiestl, E.T.; Gearhardt, A.N. Preliminary Validation of the Yale Food Addiction Scale for Children 2.0: A Dimensional Approach to Scoring. *Eur. Eat. Disord. Rev.* **2018**, *26*, 605–614. [CrossRef] [PubMed]

28. Gearhardt, A.N.; Corbin, W.R.; Brownell, K.D. Development of the Yale Food Addiction Scale Version 2.0. *Psychol. Addict. Behav.* **2016**, *30*, 113–121. [CrossRef] [PubMed]
29. Centers for Disease Control and Prevention. BMI Percentile Calculator for Child and Teen. Available online: https://www.cdc.gov/healthyweight/bmi/calculator.html. (accessed on 8 August 2017).
30. Hayes, A.F. *Introduction to Mediation, Moderation, and Conditional Process Analysis, Second Edition: A Regression-Based Approach*, 2nd ed.; Little, T.D., Ed.; The Guilford Preess: New York, NY, USA, 2018.
31. Boutelle, K.N.; Braden, A.; Knatz-Peck, S.; Anderson, L.K.; Rhee, K.E. An Open Trial Targeting Emotional Eating among Adolescents with Overweight or Obesity. *Eat. Disord.* **2018**, *26*, 79–91. [CrossRef]
32. Mak, K.K.; Lai, C.M.; Watanabe, H.; Kim, D.I.; Bahar, N.; Ramos, M.; Young, K.S.; Ho, R.C.; Aum, N.R.; Cheng, C. Epidemiology of Internet Behaviors and Addiction among Adolescents in Six Asian Countries. *Cyberpsychol. Behav. Soc. Netw.* **2014**, *17*, 720–728. [CrossRef]
33. Lee, S.Y.; Lee, D.; Nam, C.R.; Kim, D.Y.; Park, S.; Kwon, J.G.; Kweon, Y.S.; Lee, Y.; Kim, D.J.; Choi, J.S. Distinct Patterns of Internet and Smartphone-Related Problems among Adolescents by Gender: Latent Class Analysis. *J. Behav. Addict.* **2018**, *7*, 454–465. [CrossRef]
34. Grant, J.E.; Lust, K.; Chamberlain, S.R. Problematic Smartphone Use Associated with Greater Alcohol Consumption, Mental Health Issues, Poorer Academic Performance, and Impulsivity. *J. Behav. Addict.* **2019**, *8*, 335–342. [CrossRef]

International Journal of
Environmental Research and Public Health

MDPI

Article

The Association of Problematic Internet Shopping with Dissociation among South Korean Internet Users

Young-Mi Ko [1], Sungwon Roh [2] and Tae Kyung Lee [3,*]

1 Mental Health Research Institute, National Center for Mental Health, Seoul 04933, Korea; daquin07@korea.kr
2 Department of Psychiatry, Hanyang University, College of Medicine, Seoul 04763, Korea; swroh@hanyang.ac.kr
3 Department of Mental Health, ChunCheon National Hospital, ChunCheon City 24409, Korea
* Correspondence: TKLeeMD@gmail.com; Tel.: +82-33-260-3200

Received: 19 March 2020; Accepted: 1 May 2020; Published: 6 May 2020

Abstract: *Background*: This study examined patterns of problematic shopping behavior by South Korean internet users to investigate the association between problematic internet shopping (PIS) and dissociative experiences.; *Methods*: Five hundred and ninety eight participants from 20–69 years old were recruited through an online panel survey. We gathered information about sociodemographic characteristics, alcohol use, caffeine intake, and online shopping behaviors. Psychopathological assessments included Korean version of dissociative experience scale (DES-K), Canadian Problem Gambling Index (CPGI-K), the modified Stress Response Inventory (SRI-MF), the Barratt Impulsive Scale-11-Revised (BIS-K). We used multiple logistic regression analysis with the Richmond compulsive buying scale (RCBS-K) as the dependent variable.; *Results*: The prevalence of shoppers with internet-based problem shopping was 12.5%. The amount of time spent on online shopping was correlated with PIS severity (OR = 1.008, $p < 0.01$). The risk of PIS was related to an increased tendency toward dissociation (OR = 1.044, $p < 0.001$) and impulsivity (OR = 1.046, $p < 0.05$). *Conclusions*: PIS participants with dissociation showed higher levels of perceived stress, gambling problems, and impulsivity than did PIS participants without dissociation. This study suggests that dissociation was associated with a higher burden of PIS as it was connected to poor mental health problems.

Keywords: compulsive buying; addictive shopping; impulsivity; online shopping; dissociation

1. Introduction

With the spread of the internet, online shopping is also showing steady growth. The convenience offered by the internet contributes to its popularity. The internet provides a huge diversity of shopping information and simultaneous access to many online stores, thereby living up to expectations for immediate rewards and emotional enhancement.

South Korea is one of the best-connected countries in the world. In 2018, the internet penetration rate was at around 96 percent, while the internet usage rate was at 91.6 percent [1]. The South Korean government reported that online or internet-based shopping transaction value was 9.926 billion USD in July 2019, rising by 15.4% from July 2018 [2]. Because of an increase in high-speed internet access connections, lower connection costs, and increasing consumer competence, e-commerce activity has been rising.

For most people, buying is a normal and routine part of everyday life. Today, shopping is considered both a utilitarian and social or leisure activity with hedonistic features [3]. However, excessive shopping leads to distress or impairment. Compulsive and excessive spending is controversial, and such consumption and buying behaviors have been researched in both the business and medical literature.

The phenomenon of compulsive buying or an uncontrolled shopping spree has been described for 100 years in the literature [4,5]. Still, the phenomenon has not been included as a formal mental health disorder in the most recent, 5th edition of the Diagnostic and Statistical Manual of Mental Disorders (DSM-5) and the recently released 11th revision of the International Classification of Diseases (ICD-11) [6–8]. Uncontrolled buying behavior has been referred to in the literature as uncontrolled buying [9], compulsive buying [10], compulsive shopping [11], addictive buying [12], excessive buying [13], and pathological buying [14].

Given the rapid growth of online shopping, researchers have concerns about issues surrounding repetitive problematic behaviors associated with internet-based shopping. According to Rose and Dhandayudham [15], addictive online shopping may negatively influence not only an individual's daily and social life but also their economic status. Zhao et al. [16] described a tendency to engage in excessive, compulsive, and problematic shopping behavior via the internet that results in consequences associated with economic, social, and emotional problems such as online shopping addiction. Müller et al. use the term "online buying-shopping disorder" [17]. In the current paper, we use "Problematic Internet Shopping (PIS)", a more neutral expression in accordance with Lam and Lam [18].

Investigating the clinical characteristics of PISis of great importance as it will provide insights for developing diagnostic criteria. Improvement of understanding for PIS will enhance our capacity to recognize and define their presence. However, PIS is a complex and highly debated concept. Previous studies suggest that compulsive or pathological forms of buying behavior fit well into the behavioral addiction spectrum [19–21]. Disordered gambling, the only behavioral addiction being included in the main section of DSM-5 [6], is associated with problematic shopping [22,23]. Problematic shopping is associated with comorbid psychiatric conditions, such as mood disorders, and other behavioral addictions [14,24].

Behavioral addictions have been hypothesized as having similarities to substance addictions. Substance use disorders (SUD) co-occur with behavioral addictions. Higher caffeine intake has been associated with impulsivity and gambling [25]. In the case of Compulsive Buying, the co-morbidity rate with SUD was estimated as 21–46% [26]. Maraz et al. reported that Compulsive Buyers showed high impulsivity, high levels of distress, and substance use (alcohol, smoking, and illicit drug) in their shopping mall visitor sample [27]. These findings suggest that behavioral addictions share a common pathophysiology with SUD.

Some research has noted a correlation between behavioral addiction and dissociation [20,21]. Berancy and colleagues (2013) suggested that addiction is an absorbent relation established with an object that determines a person's cognitive, emotional, and behavioral states, causing significant damage in different areas of life [28].

Ludwig [29] said, "Dissociation represents a process whereby certain mental functions which are ordinarily integrated with other functions presumably operate in a more compartmentalized or automatic way, usually outside the sphere of conscious awareness or memory recall."

Maldonado and Spiegel [30] and DSM-5 [6] defines dissociation as the "disruption of and/or discontinuity in the normal integration of consciousness, memory, identity, emotion, perception, body representation, motor control, and behavior. Dissociative symptoms in mental disorders are of high clinical relevance".

Researchers have reported that symptoms of dissociation have been linked to maladaptive functioning, symptom severity, and poor treatment response in substance-related and addictive disorders [30,31].

The importance of dissociation in the psychopathology of addiction has been confirmed [32–34]. Somer [35] found that levels of dissociation negatively contributed to the prediction of abstinence. Craparo et al. [32] reported that addictive behaviors have a dissociative nature that allows individuals to manage negative and unregulated emotions. They suggested that dissociation is a predictor of addiction.

Thus, the present exploratory study examined clinical characteristics and psychological aspects of PIS among South Korean Internet users. Based on the extant literature, we expected that PIS would be associated with elevated impulsivity, emotional distress, depression, more severe gambling problems, and substance abuse (alcohol and caffeine). Secondly, we investigate the relationship between PIS and dissociative symptoms. We hypothesized that dissociation would be positively associated with higher PIS severity and mental health problems.

2. Methods

2.1. Participants

The total number of participants was 598, and we applied random sampling through proportionate allocation in accordance with the sex and age distribution of South Korea. Participants from 20–69 years were recruited through an online research service, ZINNOS R&C (ZINNOS R&C Co. Ltd., Seoul, Korea). The cross-sectional study was performed from August to September 2015. As this study aimed to collect research materials from dedicated, internet-based shoppers, this study was done through an online panel survey conducted by ZINNOS R&C, which, at the time of the survey, had 75,000 or more members.

Email invitations to complete an online survey were sent out to a random sample of potential respondents in the ZINNOS R&C panel. Of the 8977 Korean internet users who received an invitation, 598 (6.66%) visited the survey webpage. Participants completed a self-administered questionnaire that gathered information about sociodemographic characteristics, alcohol and caffeine use, online shopping behaviors, problematic buying, dissociation experiences, gambling problems, depression, stress perception, and impulsivity. Problematic shopping was measured with the Korean version of the Richmond Compulsive Buying Scale (RCBS). This study was approved by the National Center for Mental Health Institutional Review Board (NCMH 2015-06) and adhered to ethical policies.

2.2. Measures

2.2.1. Sociodemographic Characteristics

We examined the subjects' sex, age, and marital status. Participants aged 20–69 years were recruited, and for analysis were divided into age groups by decade: 20–29, 30–39, 40–49, 50–59, and 60–69. Marital status was divided into three groups: married, separated/widowed/divorced, and single.

2.2.2. Internet Shopping Behaviors

The total time that participants had spent internet shopping, the amount of money they had spent on internet shopping in the past month, the time spent on activities related to shopping during the day, and the average number of days per week they engaged in activities related to shopping in the past month were surveyed and measured as continuous variables. Additionally, participants were asked whether they had the experience of making an online purchase in excess of their income.

2.2.3. Alcohol and Caffeine Use

Alcohol use was assessed with the question: "How many glasses do you drink?" For the question, "How many times have you drunk four or more glasses of an alcoholic beverage in the past year?" answers of "not at all," "once a month," "twice a month," "once a week," and "two or three times a week" were given. The question, "How many caffeine drinks (including coffee) do you drink a day on average?" was asked, and the answers were measured as a continuous variable.

2.2.4. The Korean Version of the Richmond Compulsive Buying Scale (RCBS-K)

The scale was developed by Ridgway et al. [36] and was used to measure the severity of PIS in this study. The scale consists of six questions that use a seven-point Likert scale. A score of 25 points or higher indicates that a respondent's online shopping is a source of significant problems [36]. The RCBS has been reported to have high internal consistency, with a Cronbach's alpha coefficient of 0.84, and evidence of validity [36]. For this study, the Cronbach's alpha coefficient of RCBS-K was 0.906.

2.2.5. The Korean Version of the Dissociative Experiences Scale (DES-K)

The Dissociative Experiences Scale (DES) is a 28-item self-report instrument. The DES was developed by Bernstein and Putman [37] and has adequate test-retest reliability, good split-half reliability, and good clinical validity. It can be completed in 10 min and scored in less than 5 min. It is easy to understand, and the questions are framed in a normative way that does not stigmatize the respondent for positive responses. The respondent clicks along a line anchored at 0% on the left and 100% on the right to show how often they have this experience. The overall DES score is obtained by adding up the answers of the 28 items and then dividing by 28; this yields an overall score ranging from 0 to 100. Scores higher than 20 on the Korean version of the DES (DES-K) may indicate the presence of a dissociative disorder. The DES-K has good validity and reliability, and overall good psychometric properties, exhibiting a Cronbach's alpha coefficient of 0.94 [38]. The Cronbach's alpha of DES-K for this study was 0.985.

2.2.6. The Korean Version of the Canadian Problem Gambling Index (CPGI-K)

This scale was developed by Ferris and Wynne [39] to measure participant degree of gambling addiction. Scores for a total of nine questions are measured on a 4-point Likert-type scale (0: never, 1: sometimes, 2: frequently, 3: always). The range of the total score is from 0–27. According to the total score, the degree of gambling addiction is divided into nonproblematic gambling (0 points), low-danger gambling (1–2 points), mid-danger gambling (3–7 points), and problematic gambling (8 points or higher). The Korean version of the scale was standardized by Kim et al. [40], and the Cronbach's alpha coefficient was 0.94. The Cronbach's alpha of CPGI-K for this study was 0.947.

2.2.7. The Korean Version of the Zung Self-Rating Depression Scale (ZDS-K)

The ZDS is a 20-item self-report measure of the symptoms associated with depression. Subjects rate each item with regard to how they have felt during the preceding week using a four-point Likert scale, with 4 representing the most unfavorable response. The sum of the 20 items, after transposing the 10 items that are reverse-scored, produces a raw score from 20–80. Previous studies have pointed out that scores are not meant to offer strict diagnostic criteria but rather denote levels of depressive symptoms that might be clinically significant [41,42]. The Korean version of the ZDS (ZDS-K) was used in this study and has high internal consistency (i.e., Cronbach's alpha for the SDS = 0.79) [43]. The Cronbach's alpha coefficient for ZDS-K in this study was 0.876

2.2.8. The Modified Form of the Stress Response Inventory (SRI-MF)

SRI-MF is the short form of the Stress Response Inventory (SRI) was developed by Choi and colleagues [44] to score mental and physical symptoms occurring during the previous two weeks that might influence the current status of mental stress levels [45]. SRI scores can be categorized into seven stress factors: tension, aggression, somatization, anger, depression, fatigue, and frustration. The SRI assesses stress severity based on the stress symptoms or the effects of stressors. The SRI consists of 39 items that focus on emotional, somatic, cognitive, and behavioral stress responses. The SRI-MF consists of 22 items employing three factors: somatization, depression, and anger. The SRI-MF has good validity and reliability, exhibiting a Cronbach's alpha coefficient of 0.93 [44]. The Cronbach's alpha coefficient for SRI-MF in this study was 0.958.

2.2.9. The Barratt Impulsive Scale-11-Revised (BIS-K).

The Barratt Impulsive Scale was developed to evaluate the degree of impulsiveness [46]. Sora Lee et al. [47] conducted a study on the reliability and validity of the Korean version of the scale. The scale has a total of 30 questions that are scored based on a four-point Likert scale. Sora Lee et al.'s (2012) study analyzed three sub-factors (cognitive impulsiveness, motor impulsiveness, and unplanned impulsiveness). Cronbach's alpha for all questions was 0.78 (for cognitive impulsiveness: 0.623, for motor impulsiveness: 0.626, for unplanned impulsiveness: 0.580) [47]. For this study, the Cronbach's alpha coefficient was 0.855.

2.3. Data Analysis

Investigators classified the participants into two groups (i.e., PIS and NPIS) based on their RCBS score. Byeon et al. [48] suggested that a score of 25 points or higher on the Korean version of the RCBS indicates a problematic shopper. We classified those whose RCBS score was 25 or more as the PIS group. Those whose RCBS-K score was below 25 were assigned to the non-PIS group (NPIS).

Chi-square tests of independence were employed to analyze the sociodemographic differences between the PIS and NPIS groups. For continuous variables, we used two-tailed t-tests to compare group mean differences. Mann–Whitney U test was adapted for non-normal distributed data. Pearson correlation analysis was used for data analysis. A multiple logistic regression analysis was used for four models. We estimated odds ratios (OR), adjusting for sex, age, and marital status (Model 1); adjusting for online shopping duration, online shopping amount, online shopping time, online shopping days, and experience of buying in excess of income (Model 2); adjusting for drinking and caffeine (Model 3); and adjusting for DES-K, CPGI-K, ZDS-K, SRI-MF, BIS-K (Model 4).

Model 4 is a full model. We considered statistical tests to be significant at an alpha level of 0.05 using a two-tailed test. We performed our data analyses using IBM SPSS Statistics version 21.0 (SPSS Inc., Chicago, IL, USA).

3. Results

Descriptive Statistics

The sample size of the online panel survey was 598 adults. Their sociodemographic characteristics included 50.7% (n = 303) men and 49.3% (n = 295) women. Seventyfive (12.5%) participants were classified as PIS. Thirty-four (45.3%) men and 41 (54.7%) women scored 25 or above, with no difference between them.

By age group, there was no statistically significant difference between PIS and non-PIS (NPIS) participants. As for marital status, the PIS participants were more often married than NPIS participants (Table 1). The PIS participants spent more money on shopping than did NPIS participants in the previous month. The PIS participants reported that they bought things more often in excess of their income than did NPIS participants.

With respect to shopping behaviors, PIS participants spent more money on shopping, more of their time on shopping-related activities, and had more buying experiences in excess of their income than did NPIS participants (Table 2).

Compared to the NPIS participants, the PIS participants reported higher alcohol use. The PIS group also had higher scores on measures of dissociation, gambling severity, depression, perceived stress, and impulsivity (Table 3). Pearson's correlation analysis revealed that the RCBS-K scores were positively related to DES-K, CPGI-K, ZDS-K, SRI-MF, and BIS-K ($p < 0.01$) (Table 4).

Table 1. Comparison of online shopper with and without problematic shopping behavior.

Variable		Demographics			Group Difference	
		PIS n = 75 (%), Mean (± SD)	NPIS n = 523 (%), Mean (± SD)	Total n = 598 (%), Mean (± SD)	χ2 /t	p Value
Gender	Male	34 (45.3)	269 (51.4)	303 (50.7)	0.977	0.323
	Female	41 (54.7)	254 (48.6)	295 (49.3)		
Age (years)	20–29	18 (24.0)	87 (16.6)	105 (17.6)	3.090	0.543
	30–39	16 (21.3)	114 (21.8)	130 (21.7)		
	40–49	17 (22.7)	131 (25.0)	148 (24.7)		
	50–59	17 (22.7)	120 (22.9)	137 (22.9)		
	60–69	7 (9.3)	71 (13.6)	78 (13.0)		
Marital status	Married	41 (54.7)	329 (62.9)	370 (61.9)	6.476	0.039
	Separated/ widowed/divorced	2 (2.7)	39 (7.5)	41 (6.9)		
	single	32 (42.7)	155 (29.6)	187 (31.3)		

PIS, Problematic Internet Shopping; NPIS, Non-Problematic Internet Shopping.

Table 2. Comparisons of online shopping pattern between PIS and NPIS participants.

		Shopping Pattern			Group Differences	
		PIS Mean or n (SD or %)	NPIS Mean or n (SD or %)	Total Mean or n (SD or %)	χ2 /t	p Value
Online shopping duration	year	11.06 (5.19)	10.37 (4.56)	10.46 (4.64)	1.202	0.230
Online shopping amount of budget	10k Won	46.69 (56.30)	27.84 (44.96)	30.21 (46.90)	2.776	0.007
Time spent shopping per day	minutes	109.84 (91.10)	47.26 (55.49)	55.12 (64.43)	5.796	0.000
The number of days shopping per week	days	4.37 (1.92)	3.33 (1.87)	3.46 (1.91)	4.477	0.000
Buying in the excess over the income	No	37 (49.3)	427 (81.6)	464 (77.6)	39.386	0.000
	Yes	38 (50.7)	96 (18.4)	134 (22.4)		

PIS, Problematic Internet Shopping; NPIS, Non-Problematic Internet Shopping.

Table 3. Mental health problems in participants with and without PIS.

Variables		Mental Health			Group Difference	
		PIS Mean or n (SD)	NPIS Mean or n (SD)	Total	χ2 /t	p Value
Alcohol use	Not at all	8 (10.7)	96 (18.4)	104 (17.4)	11.968	0.018
	Once a month	10 (13.3)	134 (25.6)	144 (24.1)		
	Twice a month	16 (21.3)	99 (18.9)	115 (19.2)		
	Once a week	17 (22.7)	88 (16.8)	105 (17.6)		
	Two or three times a week	24 (32.0)	106 (20.3)	130 (21.7)		
Caffeine use	cups	2.71 (1.84)	2.66 (2.01)	2.66 (1.98)	0.199	0.842
DES-K		41.78 (24.43)	14.30 (14.50)	17.75 (18.46)	9.502	0.000
CPGI-K		6.16 (6.79)	1.14 (2.96)	1.77 (4.02)	6.319	0.000
ZDS-K		50.47 (8.09)	42.33 (9.13)	43.35 (9.39)	8.017	0.000
SRI-MF		41.24 (17.09)	21.84 (15.57)	24.27 (17.02)	9.963	0.000
BIS-K		69.69 (10.30)	58.46 (9.14)	59.87 (10.00)	9.791	0.000

DES-K, The Korean version of the Dissociative Experiences Scale; CPGI-K, The Korean version of the Canadian Problem Gambling Index; ZDS-K, The Korean version of ZungSelf-Rating Depression Scale; SRI-MF, The modified form of the Stress Response Inventory; BIS-K, The Korean version of The Barratt Impulsive Scale-11-Revised.

Table 4. Reliabilities and Correlation between RCBS-K and other psychological scales.

Psychological Scales	Cronbach's Alpha	95% CI	r(p)
RCBS-K	0.906	0.894–0.917	1
DES-K	0.985	0.984–0.987	0.559 **
CPGI-K	0.947	0.941–0.954	0.422 **
ZDS-K	0.876	0.861–0.890	0.356 **
SRI-MF	0.958	0.953–0.963	0.461 **
BIS-K	0.855	0.837–0.871	0.484 **

** $p < 0.01$. RCBS-K, The Korean version of the Richmond Compulsive Buying Scale; DES-K, The Korean version of the Dissociative Experiences Scale; CPGI-K, The Korean version of the Canadian Problem Gambling Index; ZDS-K, The Korean version of Zung Self-Rating Depression Scale; SRI-MF, The modified form of the Stress Response Inventory; BIS-K, The Korean version of the Barratt Impulsive Scale-11-Revised.

Table 5 presents the result of the multivariate logistic regression analysis. The dependent variable was the RCBS-K, and the reference group was NPIS. The OR was calculated from Model 1 with explanatory variables of sex, age, and marital status that are sociodemographic aspects, but no variable was statistically significant. The variable of online shopping behaviors was added to Model 2, while the variables of Model 1 were controlled. Among them, the OR value for the time spent on shopping-related activities was 1.010, showing a tendency toward an increased risk of PIS following an increase in the time spent on shopping-related activities ($p < 0.001$). Those reporting the experience of buying in excess of their income showed a tendency toward a greater risk for PIS than those who did not report such an experience (OR = 2.961, $p < 0.001$). Model 3 evaluated the impact of alcohol and caffeine consumption. The risk of PIS rose by frequency of alcohol consumption. Notably, those who consumed alcohol two or three times a week showed a higher OR of 2.88 than those who did not consume alcohol at all, and the value was statistically significant ($p < 0.05$). However, there is not significant relation for the amount of caffeine consumption with PIS.

Model 3 showed statistical significance between the time spent on shopping-related activities during a day (OR = 1.010, $p< 0.001$) and the experience of buying in excess of income (OR = 2.860, $p < 0.001$), both of which were still variables that were associated with PIS. Model 4 was a full model that assessed the impact of DES-K, CPGI-K, ZDS-K, SRI-MF, and BIS-K while all the variables of Model 3 were controlled. The result showed that the risk of PIS increased when the tendency toward pathological dissociation was higher (OR = 1.044, $p < 0.001$) and that the risk of PIS was higher when the level of impulsiveness was higher (OR = 1.046, $p < 0.05$). In addition, time spent on shopping-related activities was correlated with PIS severity (OR = 1.008, $p < 0.01$), as well as with the duration of online shopping, indicating that the higher the online shopping duration was, the higher the risk of PIS (OR = 1.093, $p < 0.05$).

When we compared mental health problems between PIS participants with dissociation and without dissociation, PIS participants with dissociation showed higher levels of perceived stress, gambling problems, and impulsivity than did PIS participants without dissociation (Table 6).

Table 5. Contributing factors to RCBS-K identified in the multiple logistic regression.

Variable	Model 1				Model 2				Model 3				Model 4			
	B	S.E.	OR	p Value	B	S.E.	OR	p Value	B	S.E.	OR	p-Value	B	S.E.	OR	p Value
Gender (ref:male)																
female	0.253	0.250	1.288	0.310	−0.168	0.285	0.845	0.555	0.111	0.304	1.118	0.714	0.090	0.387	1.094	0.817
Age (ref:60~69)																
20~29	0.138	0.580	1.148	0.813	0.092	0.655	1.096	0.889	−0.002	0.658	0.998	0.997	−0.539	0.844	0.584	0.523
30~39	0.027	0.512	1.028	0.958	−0.436	0.574	0.647	0.448	−0.694	0.582	0.499	0.233	−1.116	0.753	0.328	0.138
40~49	0.121	0.482	1.128	0.802	0.266	0.531	1.304	0.617	0.048	0.536	1.050	0.928	−0.253	0.676	0.777	0.709
50~59	0.333	0.476	1.395	0.484	0.385	0.527	1.470	0.465	0.295	0.525	1.343	0.574	0.093	0.668	1.098	0.889
Marital status(ref:married)																
seperated/widowed/divorced	−0.915	0.747	0.400	0.221	−1.098	0.795	0.333	0.167	−1.097	0.790	0.334	0.165	−1.245	0.892	0.288	0.163
single	0.552	0.361	1.737	0.126	0.478	0.385	1.613	0.214	0.628	0.396	1.874	0.113	0.980	0.499	2.664	0.050
Online shopping duration					0.039	0.034	1.039	0.253	0.054	0.035	1.055	0.119	0.089	0.040	1.093	0.026
Online shopping amount					0.003	0.002	1.003	0.177	0.002	0.002	1.002	0.336	0.000	0.003	1.000	0.964
Time spent shopping per day					0.010	0.002	1.010	0.000	0.010	0.002	1.010	0.000	0.008	0.002	1.008	0.001
The number of days shopping per week					0.099	0.079	1.104	0.210	0.101	0.080	1.106	0.211	0.180	0.097	1.197	0.063
Buying exceeding the income (ref:No)																
Yes					1.086	0.287	2.961	0.000	1.051	0.294	2.860	0.000	0.608	0.372	1.837	0.102
Alcohol use (ref:Not at all)																
Once a month									−0.293	0.539	0.746	0.586	−0.275	0.601	0.759	0.647
Twice a month									0.552	0.505	1.737	0.274	0.111	0.617	1.117	0.858
Once a week									0.972	0.505	2.643	0.054	0.614	0.636	1.847	0.335
Two or three times a week									1.058	0.488	2.880	0.030	1.017	0.579	2.765	0.079
Caffeine consumption									−0.042	0.073	0.959	0.569	−0.018	0.100	0.982	0.854
DES-K													0.043	0.010	1.044	0.000
CPGI-K													0.075	0.042	1.078	0.073
ZDS-K													0.006	0.029	1.006	0.829
SRI-MF													0.021	0.014	1.021	0.131
BIS-K													0.045	0.022	1.046	0.035
-2LL			442.969			371.731				359.010				247.218		

RCBS-K, The Korean version of the Richmond Compulsive Buying Scale; DES-K, The Korean version of the Dissociative Experiences Scale; CPGI-K, The Korean version of the Canadian Problem Gambling Index; ZDS-K, The Korean version of ZungSelf-Rating Depression Scale; SRI-MF, The modified form of the Stress Response Inventory; BIS-K, The Korean version of the Barratt Impulsive Scale-11-Revise.

Table 6. Comparison of mental health problems in PIS participants with and without dissociation.

Variables	Mental Health		Group Difference	
	Normal (n = 19) Mean Rank	Dissociation (n = 56) Mean Rank	M-W U	p Value
CPGI-K	25.24	42.33	289.50	0.003
SRI-MF	25.42	42.27	293.00	0.004
BIS-K	26.18	42.01	307.50	0.006

PIS, Problematic Internet Shopping, CPGI-K, The Korean version of the Canadian Problem Gambling Index; SRI-MF, The modified form of the Stress Response Inventory; BIS-K, The Korean version of the Barratt Impulsive Scale-11-Revised; M-W U Mann–Whitney U.

4. Discussion

In line with the rapid increase in e-commerce activities, there have been growing concerns about PIS. However, little is known about the clinical and psychopathological aspects of PIS. In this cross-sectional study, we found out a significant prevalence rate (12.5%) of PIS among South Korean internet users. The number was lower than in those reported in previous studies (17.7% [49] and 33.6% [17]). This rate difference might be due to the use of a different survey method: The participants in the Kukar-Kinney et al. [49] study were women. Muller et al. [17] analyzed the pooled data of treatment-seeking patients with shopping disorders.

Contrary to previous findings which indicating that young people and women were more prone to manifest PIS [17,50], the results indicated no link between PIS and gender, age, and monthly income. Of the demographic variables, only marital status distinguished between the group, with PIS participants were more often single than NPIS members. The result is not consistent with past findings regarding link between online buying disorder and partnership status [17]. Some studies have argued that loneliness is an important reason why people are developing addictive behaviors. Andreassen et al. (2017) suggested that individuals who were no in a personal relationship were more prone to developing addictive social media use than people who had partners [51]. Elton-Marshall et al. (2018) presented that gambling to escape feeling of loneliness was linked with problem gambling severity. They suggested that being married was a protective factor against problem gambling severity [52].

The results showed that PIS participants connected to online shopping sites longer and more frequently, and they spent more money when online shopping. This result was consistent with previous studies [53,54]. Increasing the time spent using the internet has been considered an index of problematic use and possible addiction. Lemmens and Hendriks [55] indicated that the time spent playing online games was strongly related to internet addiction.

Although the amount of coffee intake did not distinguish PIS from NPIS, the results showed that individuals with PIS had more mental health predicaments as they presented with an increase in depression, perceived stress, impulsiveness, gambling, alcohol use, and dissociative experiences than the NPIS group. This result provides support for PIS as a behavioral addiction requiring clinical recognition and treatment.

In the current study, PIS participants' gambling severity was higher than NPIS participants. Although disordered gambling is the only behavioral addiction classified in DSM-5, problematic gambling was categorized into the Impulse-Control Disorder section in DSM-IV. In this regard, the association of PIS with gambling indicates that PIS is a behavioral addiction.

The relationship of PIS with impulsivity was consistent with previous studies. Impulsivity has been found to play an important role in the occurrence of addiction-related disorders [56,57]. Billieux et al. [58] reported that compulsive buying was positively correlated with impulsivity, and impulsivity was the most significant predictor of compulsive buying. Black [4] argued that pathological or compulsive buying should be classified as an impulse control disorder.

The key finding here was that the best-fit logistic regression model identified dissociation and impulsivity as being associated with PIS. The importance of dissociation in the psychopathology of addiction has been confirmed [32–34]. The literature indicates that symptoms of dissociation are present in a variety of mental disorders and connect to higher-burden illnesses and poorer treatment response [31,33,59–61].

As hypothesized, participants with PIS and dissociation exhibited poorer mental health, including higher stress levels, gambling, and impulsivity. According to Jacobs [34], dissociative symptoms resemble detachment states accompanying the acting-out phase of impulse control disorder. Maldonado and Spiegel (2019) indicated that dissociation represents more a disturbance in the organization or structure of mental contents than a disturbance in the mental contents themselves [30]. Lyssenko et al. (2018) suggested that the experience of dissociation can induce stress itself because it not only disrupts neurocognitive functioning but can also be perceived as losing control [31]. Kianpoor and Bakhshani (2012) reported that dissociation related with high-risk behaviors such as violence, heavy drinking, the use of illicit drugs, and dangerous driving [62]. From a clinical perspective, this finding underlines the importance of careful evaluation of dissociation symptoms. It will help health professionals to have recognition of people prone to high-risk behaviors as well as implement more effective strategies to prevent high-risk behaviors among at-risk populations.

The current research has important implications for the prevention of problematic internet shopping. As the results were reviewed, our data pointed to the importance of interventions aimed at helping internet-using shoppers increase their level of awareness to prevent dissociation. For example, introducing a time or monetary limit pop-up reminder into the internet-based shopping mall sites might help control impulsiveness. Stewart and Wohl [63] suggested that a monetary limit pop-up reminder was effective for internet gamblers to facilitate adherence to monetary limits.

Limitations

Despite this being one of the first studies to explore the clinical and psychopathological characteristics of PIS, this study has several limitations that might influence the generalizability of the findings. First, the use of cross-sectional, self-reported data in this study might have influenced our results through common method bias (e.g., causality problems; a prevalence-incidence bias) [64]. Second, because we used self-report measures instead of structured interviews, we cannot make a clinical diagnosis of dissociative disorders or clinically assess the relationship of PIS with different types of dissociative disorders (e.g., dissociative identity disorder, dissociative amnesia, depersonalization/derealization disorder). Third, we did not screen participants for the presence of traumatic psychological experiences, which can play an important role as the cause of dissociative disorders. Hence, we could not evaluate whether the dissociative disorder is a consequence of psychological trauma-related dissociative disorders. Finally, it is worth noting that we recruited participants through an online research service, ZINNOS R&C, which operates its own independent consumer panel in which participants are preregistered. The sample might not be representative of South Korean internet-based online shoppers or the general public. Besides, the response rate of this online panel study was 6.6%. The low rate may reflect the fact that only the panel members who have interested in the topic may have answered the invitation. Callegaro et al. [65] indicated that completion rates of online panel studies had a large variability going from 3%–91% with an average of 18%.

5. Conclusions

This study suggests that increasing awareness to prevent dissociation is important for addressing PIS. Individuals with PIS and dissociation showed a more severe mental health status than the PIS participants without dissociation. PIS participants with dissociation showed higher levels of perceived stress, gambling problems, and impulsivity than did PIS participants without dissociation. Introducing measures to prevent PIS and triggering of the state of detachment might increase a person's ability to tolerate negative affect.

Author Contributions: Writing—original draft, Y.-M.K. and T.K.L.; Writing—review & editing, Y.-M.K., S.R. and T.K.L. All authors have read and agreed to the published version of the manuscript.

Funding: This research was supported by the 2015 intramural research fund of National Center for Mental Health (NCMH 2015-05) in South Korea.

Conflicts of Interest: The authors declare no conflict of interest.

References

1. Statistical Research Department. Internet Usage in South Korea-Statistics & Facts. Available online: https://www.statista.com/topics/2230/internet-usage-in-south-korea/ (accessed on 16 February 2020).
2. Statistics Korea. Online Shopping in July 2019. Available online: http://kostat.go.kr/portal/eng/pressReleases/1/index.board?bmode=read&aSeq=377546 (accessed on 4 September 2019).
3. Hirschman, E.C.; Holbrook, M.B. Hedonic consumption: Emerging concepts, methods and propositions. *J. Mark.* **1982**, *46*, 92–101. [CrossRef]
4. Black, D.W. A review of compulsive buying disorder. *World Psychiatry* **2007**, *6*, 14–18. [PubMed]
5. Müller, A.; Brand, M.; Claes, L.; Demetrovics, Z.; de Zwaan, M.; Fernández-Aranda, F.; Frost, R.O.; Jimenez-Murcia, S.; Lejoyeux, M.; Steins-Loeber, S. Buying-shopping disorder—is there enough evidence to support its inclusion in ICD-11? *CNS Spectr.* **2019**, 1–6. [CrossRef] [PubMed]
6. American Psychiatric Association. *Diagnostic and statistical manual of mental disorders (DSM-5®)*; American Psychiatric Pub: Arlington, VA, USA, 2013.
7. WHO. ICD-11 for Mortality and Morbidity Statistics. Available online: https://icd.who.int/browse11/l-m/en#/http%3a%2f%2fid.who.int%2ficd%2fentity%2f826065555 (accessed on 21 April 2020).
8. Gaebel, W.; Zielasek, J.; Reed, G.M. Mental and behavioural disorders in the ICD-11: Concepts, methodologies, and current status. *Psychiatr. Pol.* **2017**, *51*, 169–195. [CrossRef] [PubMed]
9. McElroy, S.L.; Keck, P.E., Jr.; Pope, H.G., Jr.; Smith, J.M.; Strakowski, S.M. Compulsive buying: A report of 20 cases. *J. Clin. Psychiatry* **1994**, *55*, 242–248.
10. O'Guinn, T.C.; Faber, R.J. Compulsive buying: A phenomenological exploration. *J. Consum. Res.* **1989**, *16*, 147–157. [CrossRef]
11. Black, D.W. Compulsive shopping. In *Clinical Manual of Impulse-control Disorders*; Hollander, E., Stein, D.J., Eds.; American Psychiatric Pub: Washington, DC, USA, 2006; pp. 203–228.
12. Scherhorn, G.; Reisch, L.A.; Raab, G. Addictive buying in West Germany: An empirical study. *J. Consum. Policy* **1990**, *13*, 355–387. [CrossRef]
13. Dittmar, H.; Beattie, J. Impulsive and excessive buying behaviour. In *Choice and public policy: The limits to welfare markets*; Taylor-Gooby, P., Ed.; Macmillan: London, UK, 1998; pp. 123–144.
14. Muller, A.; Mitchell, J.E.; de Zwaan, M. Compulsive buying. *Am. J. Addict.* **2015**, *24*, 132–137. [CrossRef]
15. Rose, S.; Dhandayudham, A. Towards an understanding of Internet-based problem shopping behaviour: The concept of online shopping addiction and its proposed predictors. *J. Behav. Addict.* **2014**, *3*, 83–89. [CrossRef]
16. Zhao, H.; Tian, W.; Xin, T. The Development and Validation of the Online Shopping Addiction Scale. *Front. Psychol* **2017**, *8*, 735. [CrossRef]
17. Muller, A.; Steins-Loeber, S.; Trotzke, P.; Vogel, B.; Georgiadou, E.; de Zwaan, M. Online shopping in treatment-seeking patients with buying-shopping disorder. *Compr. Psychiatry* **2019**, *94*, 152120. [CrossRef] [PubMed]
18. Lam, L.T.; Lam, M.K. The association between financial literacy and Problematic Internet Shopping in a multinational sample. *Addict. Behav. Rep.* **2017**, *6*, 123–127. [CrossRef] [PubMed]
19. Brewer, J.A.; Potenza, M.N. The neurobiology and genetics of impulse control disorders: Relationships to drug addictions. *Biochem. Pharmacol.* **2008**, *75*, 63–75. [CrossRef] [PubMed]
20. Granero, R.; Fernandez-Aranda, F.; Mestre-Bach, G.; Steward, T.; Bano, M.; Del Pino-Gutierrez, A.; Moragas, L.; Mallorqui-Bague, N.; Aymami, N.; Gomez-Pena, M.; et al. Compulsive Buying Behavior: Clinical Comparison with Other Behavioral Addictions. *Front. Psychol* **2016**, *7*, 914. [CrossRef] [PubMed]
21. Brand, M.; Wegmann, E.; Stark, R.; Muller, A.; Wolfling, K.; Robbins, T.W.; Potenza, M.N. The Interaction of Person-Affect-Cognition-Execution (I-PACE) model for addictive behaviors: Update, generalization to addictive behaviors beyond internet-use disorders, and specification of the process character of addictive behaviors. *Neurosci. Biobehav. Rev.* **2019**, *104*, 1–10. [CrossRef]
22. Ford, M.; Hakansson, A. Problem gambling, associations with comorbid health conditions, substance use, and behavioural addictions: Opportunities for pathways to treatment. *PLoS ONE* **2020**, *15*, e0227644. [CrossRef]

23. Moore, T.J.; Glenmullen, J.; Mattison, D.R. Reports of pathological gambling, hypersexuality, and compulsive shopping associated with dopamine receptor agonist drugs. *JAMA Intern. Med.* **2014**, *174*, 1930–1933. [CrossRef]
24. Aboujaoude, E. Compulsive buying disorder: A review and update. *Curr. Pharm. Des.* **2014**, *20*, 4021–4025. [CrossRef]
25. Grant, J.E.; Chamberlain, S.R. Caffeine's influence on gambling behavior and other types of impulsivity. *Addict. Behav.* **2018**, *76*, 156–160. [CrossRef]
26. Grant, J.E.; Potenza, M.N.; Weinstein, A.; Gorelick, D.A. Introduction to behavioral addictions. *Am. J. Drug Alcohol Abuse* **2010**, *36*, 233–241. [CrossRef]
27. Maraz, A.; van den Brink, W.; Demetrovics, Z. Prevalence and construct validity of compulsive buying disorder in shopping mall visitors. *Psychiatry Res.* **2015**, *228*, 918–924. [CrossRef] [PubMed]
28. Beranuy, M.; Carbonell, X.; Griffiths, M.D. A qualitative analysis of online gaming addicts in treatment. *Int. J. Ment. Health Addict.* **2013**, *11*, 149–161. [CrossRef]
29. Ludwig, A.M. The psychobiological functions of dissociation. *Am. J. Clin. Hypn.* **1983**, *26*, 93–99. [CrossRef] [PubMed]
30. Maldonado, J.R.; Spiegel, D. Dissociative disorders. In *The American Psychiatic Association Publishing Textbook of Psychiatry*, 7th ed.; Roberts, L.W., Ed.; American Psychiatric Association Publishing: Washington DC, USA, 2019; pp. 437–474.
31. Lyssenko, L.; Schmahl, C.; Bockhacker, L.; Vonderlin, R.; Bohus, M.; Kleindienst, N. Dissociation in Psychiatric Disorders: A Meta-Analysis of Studies Using the Dissociative Experiences Scale. *Am. J. Psychiatry* **2018**, *175*, 37–46. [CrossRef] [PubMed]
32. Craparo, G.; Ardino, V.; Gori, A.; Caretti, V. The relationships between early trauma, dissociation, and alexithymia in alcohol addiction. *Psychiatry Investig.* **2014**, *11*, 330–335. [CrossRef]
33. Karadag, F.; Sar, V.; Tamar-Gurol, D.; Evren, C.; Karagoz, M.; Erkiran, M. Dissociative disorders among inpatients with drug or alcohol dependency. *J. Clin. Psychiatry* **2005**, *66*, 1247–1253. [CrossRef]
34. Jacobs, D.F. Evidence for a common dissociative-like reaction among addicts. *J. Gambl. Behav.* **1988**, *4*, 27–37. [CrossRef]
35. Somer, E. Prediction of abstinence from heroin addiction by childhood trauma, dissociation, and extent of psychosocial treatment. *Addict. Res. Theory* **2003**, *11*, 339–348. [CrossRef]
36. Ridgway, N.M.; Kukar-Kinney, M.; Monroe, K.B. An expanded conceptualization and a new measure of compulsive buying. *J. Consum. Res.* **2008**, *35*, 622–639. [CrossRef]
37. Bernstein, E.M.; Putnam, F.W. Development, reliability, and validity of a dissociation scale. *J. Nerv. Ment. Dis.* **1986**, *174*, 727–735. [CrossRef]
38. Park, J.; Choe, B.; Kim, M.; Han, H.; Yoo, S.; Kim, S. Standardization of dissociative experiences scale-Korean version. *Korean J. Psychopathol* **1995**, *4*, 105–125.
39. Ferris, J.A.; Wynne, H.J. *The Canadian problem gambling index*; Final report; Canadian Centre on Substance Abuse: Ottawa, Canada, 2001.
40. Kim, A.Y.; Cha, J.E.; Kwon, S.J.; Lee, S.M. Construction and Validation of Korean Version of CPGI. *Korean J. Psychol. Gen.* **2011**, *30*, 1011–1038.
41. Zung, W.W. A self-rating depression scale. *Arch. Gen. Psychiatry* **1965**, *12*, 63–70. [CrossRef] [PubMed]
42. Passik, S.D.; Kirsh, K.L.; Donaghy, K.B.; Theobald, D.E.; Lundberg, J.C.; Holtsclaw, E.; Dugan, W.M., Jr. An attempt to employ the Zung Self-Rating Depression Scale as a "lab test" to trigger follow-up in ambulatory oncology clinics: Criterion validity and detection. *J. Pain Symptom Manag.* **2001**, *21*, 273–281. [CrossRef]
43. Lee, Y.H.; Song, J.Y. A study of the reliability and the validity of the BDI, SDS, and MMPI-D scales. *Korean J. Clin. Psychol.* **1991**, *10*, 98–113.
44. Choi, S.-M.; Kang, T.-Y.; Woo, J.-M. Development and validation of a modified form of the stress response inventory for workers. *J. Korean Neuropsychiatr. Assoc.* **2006**, *45*, 541–553.
45. Koh, K.B.; Park, J.K.; Kim, C.H.; Cho, S. Development of the stress response inventory and its application in clinical practice. *Psychosom. Med.* **2001**, *63*, 668–678. [CrossRef]
46. Patton, J.H.; Stanford, M.S.; Barratt, E.S. Factor structure of the Barratt impulsiveness scale. *J. Clin. Psychol.* **1995**, *51*, 768–774. [CrossRef]

47. Lee, S.-R.; Lee, W.-H.; Park, J.-S.; Kim, S.-M.; Kim, J.-W.; Shim, J.-H. The Study on Reliability and Validity of Korean Version of the Barratt Impulsiveness Scale-11-Revised in Nonclinical Adult Subjects. *J. Korean Neuropsychiat. Assoc.* **2012**, *51*, 378–386. [CrossRef]
48. Byeon, G.H.; Kim, R.; Han, J.H.; Ko, Y.M.; Roh, S.; Lee, T.K. Reliability and validity of the Korean version of Richmond Compulsive Buying Scale. *J. Korean Neuropsychiatr. Assoc.* **2017**, *56*, 35–44. [CrossRef]
49. Kukar-Kinney, M.; Ridgway, N.M.; Monroe, K.B. The relationship between consumers' tendencies to buy compulsively and their motivations to shop and buy on the Internet. *J. Retail.* **2009**, *85*, 298–307. [CrossRef]
50. Sorce, P.; Perotti, V.; Widrick, S. Attitude and age differences in online buying. *Int. J. Retail. Distrib. Manag.* **2005**, *33*, 122–132. [CrossRef]
51. Andreassen, C.S.; Pallesen, S.; Griffiths, M.D. The relationship between addictive use of social media, narcissism, and self-esteem: Findings from a large national survey. *Addict. Behav.* **2017**, *64*, 287–293. [CrossRef] [PubMed]
52. Elton-Marshall, T.; Wijesingha, R.; Sendzik, T.; Mock, S.E.; van der Maas, M.; McCready, J.; Mann, R.E.; Turner, N.E. Marital Status and Problem Gambling among Older Adults: An Examination of Social Context and Social Motivations. *Can. J. Aging* **2018**, *37*, 318–332. [CrossRef]
53. Lejoyeux, M.; Mathieu, K.; Embouazza, H.; Huet, F.; Lequen, V. Prevalence of compulsive buying among customers of a Parisian general store. *Compr. Psychiatry* **2007**, *48*, 42–46. [CrossRef]
54. Duroy, D.; Gorse, P.; Lejoyeux, M. Characteristics of online compulsive buying in Parisian students. *Addict.Behav* **2014**, *39*, 1827–1830. [CrossRef]
55. Leeman, R.F.; Potenza, M.N. A targeted review of the neurobiology and genetics of behavioural addictions: An emerging area of research. *Can. J. Psychiatry* **2013**, *58*, 260–273. [CrossRef]
56. Cudo, A.; Toroj, M.; Demczuk, M.; Francuz, P. Dysfunction of Self-Control in Facebook Addiction: Impulsivity Is the Key. *Psychiatr. Q.* **2019**, *91*, 91–101. [CrossRef]
57. Rømer Thomsen, K.; Callesen, M.B.; Hesse, M.; Kvamme, T.L.; Pedersen, M.M.; Pedersen, M.U.; Voon, V. Impulsivity traits and addiction-related behaviors in youth. *J. Behav. Addict.* **2018**, *7*, 317–330. [CrossRef]
58. Billieux, J.; Rochat, L.; Rebetez, M.M.L.; Van der Linden, M. Are all facets of impulsivity related to self-reported compulsive buying behavior? *Personal. Individ. Differ.* **2008**, *44*, 1432–1442. [CrossRef]
59. Schimmenti, A.; Bifulco, A. Toward a Better Understanding of the Relationship between Childhood Trauma and Psychiatric Disorders: Measurement and Impact on Addictive Behaviors. *Psychiatry Investig.* **2015**, *12*, 415–416. [CrossRef] [PubMed]
60. Schimmenti, A.; Guglielmucci, F.; Barbasio, C.P.; Granieri, A. Attachment disorganization and dissociation in virtual worlds: A study on problematic Internet use among players of online role playing games. *Clin. Neuropsychiatry* **2012**, *9*, 195–202.
61. Moskowitz, A. Dissociation and violence: A review of the literature. *Trauma Violence Abus.* **2004**, *5*, 21–46. [CrossRef] [PubMed]
62. Kianpoor, M.; Bakhshani, N.M. Trauma, dissociation, and high-risk behaviors. *Int. J. High. Risk Behav. Addict.* **2012**, *1*, 9–13. [CrossRef]
63. Stewart, M.J.; Wohl, M.J. Pop-up messages, dissociation, and craving: How monetary limit reminders facilitate adherence in a session of slot machine gambling. *Psychol. Addict. Behav.* **2013**, *27*, 268–273. [CrossRef]
64. Carlson, M.D.; Morrison, R.S. Study design, precision, and validity in observational studies. *J. Palliat. Med.* **2009**, *12*, 77–82. [CrossRef]
65. Callegaro, M.; Baker, R.P.; Bethlehem, J.; Göritz, A.S.; Krosnick, J.A.; Lavrakas, P.J. *Online panel research: A data quality perspective*; John Wiley & Sons: Hoboken, NJ, USA, 2014.

International Journal of
Environmental Research and Public Health

MDPI

Article

Aligning the "Manifesto for a European Research Network into Problematic Usage of the Internet" with the Diverse Needs of the Professional and Consumer Communities Affected by Problematic Usage of Pornography

Darryl Mead [1,2,*] and Mary Sharpe [1,3]

1 The Reward Foundation, The Melting Pot, Edinburgh EH2 2PR, UK; mary@rewardfoundation.org
2 Digital Humanities, Information Studies, University College London, London WC1E 6BT, UK
3 Lucy Cavendish College, University of Cambridge, Cambridge CB3 0BU, UK
* Correspondence: info@rewardfoundation.org

Received: 29 February 2020; Accepted: 7 May 2020; Published: 15 May 2020

Abstract: The Manifesto for a European research network into Problematic Usage of the Internet was published in May 2018. It was written from the perspective of the COST Action Network, a programme of the European Cooperation in Science and Technology CA16207 and is expected to have significant influence on research funding priorities over the next decade. The Manifesto identified nine key research priorities to advance understanding in the field. Our analysis shows that while at the most general level it identified problematic usage of pornography (PUP) as a key research priority, it then barely mentioned it again within the body of the report. This paper uses the Manifesto's framework to suggest research areas into the problematic usage of pornography which are of particular relevance to clinicians and other professionals working in the field who want to develop approaches to assist individuals and target groups affected by PUP. It also looks at potential research opportunities inspired by the lived-experience of users withdrawing from PUP. A large number of opportunities are identified for new work on PUP across all nine key research areas of the Manifesto.

Keywords: problematic usage of pornography; manifesto; problematic usage of the internet; COST action network; behavioural addiction research.

1. Introduction

The publication of the Manifesto for a European research network into Problematic Usage of the Internet (PUI) [1] in May 2018 was a milestone in planning the roadmap for the behavioural addictions' community. It provided an international focus for developing policy across different elements of the research landscape for behavioural addiction over the next decade. The Manifesto was written from the perspective of the framework of the COST Action Network, a programme of the European Cooperation in Science and Technology CA16207.

The Manifesto identified nine key research priorities to advance the understanding of PUI (Table 1).

The Manifesto identifies pornography usage as a potential PUI. We have called this activity problematic usage of pornography (PUP).

Table 1. Summary of key research priorities to advance the understanding of PUI [1]

1.	*"Reliable consensus-driven conceptualisation of PUI (defining main phenotypes and specifiers, related comorbidities and brain-based mechanisms)*
2.	*Age- and culture-appropriate assessment instruments to screen, diagnose and measure the severity of different forms of PUI*
3.	*Characterise the impacts of different forms of PUI on health and quality of life*
4.	*Define the clinical courses of different forms of PUI*
5.	*Reduce obstacles to timely recognition and interventions*
6.	*Clarify the possible role of genetics and personality features in different forms of PUI*
7.	*Consider the impact of social factors in the development of PUI*
8.	*Generate and validate effective interventions, both to prevent PUI, and to treat its various forms once established*
9.	*Identify biomarkers, including digital markers, to improve early detection and intervention".*

Problematic Use of Pornography

Since 2008 pornography consumption has transitioned from the model where a market supplied physical media such as magazines and DVDs to consumers through retail networks, to a transnational, Internet-based system mainly operating on a freemium trading model [2–4]. In the process it has scaled from predominantly national businesses reaching an audience of millions through shops and mail-order, to a world-wide online one with perhaps a billion customers [5–8]. Today, pornography is generally accessed on smartphones and other devices. Its consumption has the potential to develop into a large-scale form of PUI [9–11]. This technology-driven progression has massively increased the number of people potentially exposed to PUP, while simultaneously removing barriers moderating individual levels of consumption.

Some consumers uncomfortable with the apparent impacts of their pornography use have now gathered together in large-scale, online recovery communities [12,13]. Many of these are self-help groups who try to support individuals to manage their consumption, or to end it all together. A growing community of professional therapists [14–17], coaches [18] and mental health experts [19] has developed to support these self-help groups, as well as people accessing mental and physical health services through more mainstream medical, psychological and psychiatric routes.

Academic research and public policy to support consumers, recovery communities and professionals is still at an early stage of development. Much of the motivation for writing the current paper comes from the opportunity the Manifesto offers when it says:

> *"Additionally, through an interactive Dissemination Plan (including website, social media, blogs) we are reaching out to relevant stakeholders at international and national levels with an emphasis on encouraging people with the lived experience of PUI to become involved in the Action"* [1] p. 1235.

This paper is a result of the authors heeding this call to action. First, it discusses the extent to which the topic of pornography usage is carried through in the nine key research themes. Next it considers how the Manifesto tries to address the diverse needs of the professional and consumer communities affected by PUP. It then goes on to suggest additional areas of research to bridge any identified gaps.

Our research question is "what topics should be included within future research proposals under the Manifesto to meet the diverse needs of consumers, recovery communities and professionals impacted by the problematic use of pornography"?

2. Materials and Methods

The current study has used the content of the Manifesto [1] as its target document. Its authors created the Manifesto both in the hope of influencing future research directions around PUI and to improve the availability of funding to support those aspirations. As far as the authors are aware, the Manifesto is the only research planning and policy document of its type currently available, which is relevant to the field of PUI.

Rather than developing a mechanism to choose to include some parts of the Manifesto and to exclude other portions, the target document is sufficiently short for it to be analysed as a single entity. Each and every paragraph referencing pornography within the Manifesto has been identified and is reproduced in this paper to provide context to allow consideration of both what has been said, and just as importantly, what has been omitted.

The Manifesto began by setting a level playing field for different problematic Internet usage behaviours. It also introduced the central focus on health impacts.

> *"We use the umbrella term PUI to encompass all potentially problematic Internet related behaviours, including those relating to gaming, gambling, buying, pornography viewing, social networking, "cyber-bullying," "cyberchondria" among others. PUI may have mental and physical health consequences"* [1] p. 1234.

The nature of the health impacts is then developed further.

> *"Disordered online behaviours, such as excessive video gaming, pornography viewing, buying, gambling, or streaming and social networks use (Ioannidis et al., 2018) have been associated with marked functional impairment including loss of productivity (or reduced scholastic achievement), and mental health sequelae including mood and anxiety disorders (Derbyshire et al., 2013; Ho et al., 2014)"* [1] p. 1234.

The next step was to contextualise the behaviours according to their form, their place within the international medical diagnostic frameworks and as issues affecting identifiable user populations.

> *"As noted, PUI envelops a wide range of activities including video gaming, pornography viewing (and other compulsive sexual behaviours), buying, gambling, web-streaming, social media use and other behaviours. Some of these behaviours may fall into an existing mental disorder in psychiatric nomenclature (e.g., gambling disorder), whereas others are likely to be formally recognized in future DSM/ICD revisions, notably Internet Gaming Disorder (Kim et al., 2016b). Different types of PUI often start in childhood or adolescence (Volpe et al., 2015), but broad age ranges can be affected (Ioannidis et al., 2018). Age and gender relate importantly to PUI behaviours, with younger people typically having problems with gaming and media streaming, males with gaming, gambling and pornography viewing and females with social media and buying (Andreassen et al., 2016)"* [1] p. 1234.

The manifesto then moves into a process of exploring the individual research priorities according to the list above.

> ***"1.1. Reliable consensus-driven conceptualisations of different forms of PUI (phenomenologies, comorbidities and brain-based mechanisms)***
>
> *The clinical aspects of some Internet-related behaviours appear phenomenologically much like addiction (e.g., gambling or viewing pornography), and demonstrate impaired control (unsuccessful attempts to reduce or cease the behaviour), preoccupation (craving), associated functional impairment (neglect of other areas of life), and persistence despite damaging effects (Billieux et al., 2015; Ioannidis et al., 2016; Kardefelt-Winther, 2017). However, it remains less clear whether, apart from gambling disorder, these other forms of PUI meet the physiological criteria relating to addiction (tolerance, withdrawal). (Fineberg et al., 2018)"* [1] p. 1235.

This section starts to indicate at least one area of potential future research. Is viewing pornography a PUI that meets the physiological criteria for addiction? It then adds:

> *"Interestingly, obsessive-compulsive personality traits are common in excessive Internet users and are associated with problematic Internet use (Chamberlain et al., 2017b), hinting that compulsive*

behaviours contribute to some forms of PUI. Some forms of online shopping or cybersex, on the other hand, may closely resemble ICD-10 or DSM-IV impulse control or sexual disorders (Volpe et al., 2015)" [1] p 1236.

However, after this fairly strong opening, placing pornography viewing into the PUI research context, we do not hear about pornography again until Research Priority 7.

"7. Consider the impact of social factors in the development of PUI

For problematic online sexual behaviour (e.g., cybersex), three structural elements have been highlighted as being important contributors per the so-termed Triple A Model involving: accessibility, affordability, and anonymity (Cooper, 1998; Cooper et al., 1999), though more research is needed on this topic (Brand et al., 2016a; Wery and Billieux, 2017). Another similar proposed framework is the ACE Model (anonymity, convenience, and escape) (Young, 2008). For excessive streaming (watching videos excessively), important structural features may include the ability of given programmes to grab attention by activating a biological 'orientating response', mediated through techniques including the use of attention-grabbing noises, zooming/panning, and presentation of rewarding stimuli (e.g., of a sexual or thrilling nature) (Flayelle et al., 2017, 2018). Collectively, public research into the structural elements that may promote PUI in different contexts is lacking" [1] p. 1241.

However, this is the only mention of pornography after Section 1. So what is missing and what areas of research would be useful to the consumer, recovery and professional communities affected by problematic usage of pornography?

3. Results

This paper sets out to determine "what topics should be included within future research proposals under the Manifesto to meet the diverse needs of consumers, recovery communities and professionals impacted by the problematic use of pornography"?

The Manifesto tells us that PUP is a part of PUI and provides some discussion on phenomenologies, comorbidities and brain-based mechanisms. It introduces pornography in the context of how to classify it as a disorder. Some early models of pornography consumption are discussed in Research Priority 7.

Beyond these initial pointers, direct reference to the Manifesto provides little specific guidance for policy-makers considering the research environment for PUP. Unlike other behavioural disorders such as gambling and gaming, which were referred to repeatedly in the Manifesto, with pornography the Manifesto was mostly silent. It did not try linking PUP to assessment instruments, characterisation, defining clinical forms and removing obstacles to recognition and treatment. Equally it did not recommend directly relevant research on issues around genetics, personality features, interventions for prevention and/or treatments, or the field of biomarkers.

Pornography and the Manifesto

Is the absence of more mentions of pornography viewing as a PUI in the Manifesto accidental or driven by other factors? We will not attempt to answer this. We recognise that the Manifesto's authors covered a great deal of ground in 7000 words and some omissions were always likely. Several of the authors have published widely in the field of PUP. While still limited in comparison with research in areas such as gambling and gaming, the scale of literature investigating pornography viewing as a PUI is growing rapidly. During 2018 we read over 200 papers on pornography usage, about 90% of them published in 2017–2018.

New work which has appeared since the Manifesto was published revealed previously unquantified consumer patterns. The diagnostic category of compulsive sexual behaviour disorder (CSBD) was published in ICD-11, two months after the Manifesto appeared [20]. Over 80% of people now seeking treatment for CSBD have a pornography-use related issue, a statistic not widely known when the Manifesto was published [21]. Subsequent pornography-consumer behaviour has also been significantly influenced by entirely new factors such as the COVID-19 pandemic [22].

4. Discussion

So, where does an analysis of the Manifesto leave policy-makers, funders, researchers and other interested parties? Our discussion now approaches this from two separate, though linked, directions. First, we consider how PUP can be researched to meet the needs of the different stakeholder communities. We then consider the potential for future research, as seen through the lens of the nine priorities set out in the Manifesto.

4.1. Community-Focused Issues

Pornography accessed for free via the Internet is now used on a frequent basis by hundreds of millions of people. For most users any problems only emerge slowly over time. However, the scale of the consumer population suggests that problematic pornography usage has the potential to impact the health of millions of people across the nine key research priorities set out in the Manifesto. To make wise choices for research resource allocation, policy-makers and funders need to have sight of a research landscape that addresses each priority. What might it look like?

The following is a pragmatic assessment based on a consideration of the existing literature, mediated by the constraints set within the Manifesto. This assessment is supported by the author's wide-ranging dialogues with professional, recovery and consumer communities. Since 2016 we have had face-to-face discussions on pornography consumption with over 9000 people in the United Kingdom, the Republic of Ireland, France, Germany, Croatia, Hungary, the Ukraine, Turkey, Japan, Australia and the United States [23,24].

4.1.1. Professional Communities Affected by Problematic Usage of Pornography

What else do therapists, medical practitioners, counsellors and sex educators want to know about the mental and physical health implications of problematic pornography usage? What are the issues that their clients, the problematic pornography users, want to have investigated? It is unhelpful that the diagnostic manuals do not mention the word "pornography" when over 80% of people seeking treatment for compulsive sexual behaviour have a pornography-related issue [21]. Further, the nature of the condition for many meets the criteria for addictive disorders and should be clearly categorised as such to enable healthcare professionals to respond with appropriate treatments. Both groups have an interest in distinguishing between classifying pornography usage as a potentially addictive disorder versus it being considered as an impulse control disorder. At any rate, there is clear desire in both lay people and professionals to have the word "pornography" appear in any given classification of a future edition of the International Classification of Diseases [20] and the Diagnostic and Statistical Manual of Mental Health Disorders [25].

It is apparent from our experience in meeting researchers from all parts of the world [23,24] that the therapists' own knowledge, attitudes and prejudices are very important issues in terms of delivering access to appropriate treatment services. While the Internet brings PUP to all parts of the world, different cultures have different taboos and even blind spots about what is PUP, what is unacceptable behaviour and what sorts of treatment might be considered appropriate. The legal systems of different countries are also relevant, and there are no universal standards for most sexual behaviours.

For the last four years the authors have been teaching healthcare professionals about Internet pornography and behavioural addiction. These audiences have consistently expressed a desire for the health and social implications arising from pornography usage to be taught as a component of general medical education. Clinicians have also expressed interest in integrating the management of compulsive sexual behaviour disorder into the regular functioning of national healthcare systems.

4.1.2. Issues within the Pornography Recovery Communities

It would be desirable for the Manifesto to address the concerns of the online pornography recovery communities and members of 12-step programmes such as Sex Addicts Anonymous [18,19,26]. So far, we know of no quantitative research that has investigated elimination of digital porn use to reverse sexual dysfunction and a variety of mental health disorders reported by those who quit [27].

The focus within recovery communities begins with recognition/identification/diagnosis of problematic pornography usage. The first question is "do I have a problem?" If they do have a problem, this quickly becomes "how do I stop?" with a focus on tools, techniques and support. It then becomes "how do I maintain my desired level of sexual sobriety?" Here we need more longitudinal research on the mechanisms and techniques for overall programmes of quitting pornography usage, on abstaining from pornography viewing, and the issue of managing masturbation. Unlike most PUIs such as gambling, shopping or gaming, pornography viewing is linked to very deep biological sexual and reproductive drives, with the option of the interactive reinforcement by masturbation. The pornography, masturbation and orgasm (PMO) cycle reinforces the viewing behaviour, making simple abstinence strategies difficult to maintain for many former users. When you stop gaming you can get rid of your game console or you can have yourself banned from gambling venues, but you cannot stop being a sexual being.

There are some specific characteristics of problematic pornography viewing which merit more research, particularly issues around triggering and flatlining. A deeper understanding of PUP triggering factors would be useful to help individuals and clinicians build better strategies and models for avoiding problematic usage. Flatlining is the term for the severe and sustained, short to medium-term loss of libido some users report upon quitting a cycle of problematic pornography viewing. While widely reported in community forums such as NoFap [12] and RebootNation [13], it is not well-covered by the academic literature [28].

The medical establishment has an ongoing interest in the development of new approaches to support quitting pornography viewing by identifying effective drug treatments. Naltrexone is a medication that is used often in drug and alcohol addiction. It blocks the effects of drugs known as opiates. It competes with these drugs for opioid receptors in the brain. Naltrexone can reduce a patient's desire for drinking thereby helping them remain abstinent. Three papers have looked at the effect of naltrexone on problematic pornography use [29–31]. Given the similarity of brain changes in substance abuse disorders and behavioural addictions, this could prove a fruitful area for further research. Very recently transcranial magnetic stimulation has been suggested as a novel treatment [32]. Investment in research to try to demonstrate the value of such approaches to clinical populations at scale would be very helpful.

Therapists operating within the context of organisations such as the Association for the Treatment of Sexual Addiction and Compulsivity (ATSAC) in the United Kingdom [16] and the Society for the Advancement of Sexual Health (SASH) in the United States [15] would benefit from more research into behavioural modification techniques to help their clients stop Internet-based behaviours. This might include the creation of apps and other Internet tools. Very recently metacognitive therapy has claimed to be three-times as effective as traditional cognitive behavioural therapy. It deals successfully with 'desire thinking' and rumination, a behaviour that often predicts relapse. Extending new work on this to clinical populations of recovering pornography users would be helpful [33,34].

4.1.3. Consumer Communities Affected by Problematic Usage of Pornography

Pornography viewing is now a widespread behaviour in most countries where there is good quality access to the Internet [8]. To have a protective effect on health and wellbeing, consumer-facing academic research on PUP needs to have impact in the wider community. This is a science communication issue. Little published research has looked at the ecology of spreading this type of message. There is scope for developing a very diverse array of effective communication strategies and toolkits. More research is needed on what would help adults to avoid situations and behaviours that might lead to PUP.

There is more active research focused on schools and adolescents. The Manifesto could support the sort of educational work The Reward Foundation does in schools to help equip young people with the knowledge about the reward system of the brain and the skills they need to avoid not just

PUP, but PUI in all its forms [35]. Other educational programmes are available from an array of non-governmental organisations such as Culture Reframed, eChildhood, and the Naked Truth Project. Common to all programmes is a lack of longitudinal studies evaluating their effectiveness over time, across cultures and among varying age groups.

We also need to recognise that PUP is not just a first-world issue. The Reward Foundation website operates in English, but uses the machine translations system GTranslate to make all pages available in over 100 languages. As a result, a significant proportion of our traffic is now in African languages such as Hausa and Somali. These people have a need for information about PUP, but it is not readily available from within their own language communities.

World-wide data from Pornhub shows that heaviest usage of pornography is by people in the 18 to 34 years age brackets. It is strongly tilted towards male consumers, but supply-side drivers are slowly pushing consumption towards gender parity [5–8].

Research over the past decade has shown that pornography viewing correlates with higher levels of sexual assaults and dating violence, pathways into domestic violence and reduced levels of bystander intervention [36–39]. The next generation of research should help move us from noting these as correlations to investigating questions of causation (or not) in these fields.

4.2. Future Research Seen Through the Lens of the Nine Priorities

The categories in the following discussion are directly based on the Manifesto's nine research priorities, simply substituting 'problematic usage of pornography' for the wider PUI category. They map out some of the ways the Manifesto could respond to the needs of the professional and consumer communities.

- Research area 1: Reliable consensus-driven categorization of Problematic Usage of Pornography (defining main phenotypes and specifiers, related comorbidity and brain-based mechanisms) [1].

Separate and interlocking strands of research would be useful in the two main fields analysing sexual behaviour; compulsive sexual behaviour towards people and the usage of pornography. In particular, the overlap between them needs more work. There would also be considerable value in research on the overlap of pornography viewing with other PUI's involving sexual behaviour-related activity, especially dating apps and social media.

- Research area 2: Age- and culture-appropriate assessment instruments to screen, diagnose and measure the severity of different forms of Problematic Usage of Pornography [1].

The number of screening, diagnostic and measuring tools for pornography usage keeps expanding. There would be considerable value in testing and calibrating the best of the existing crop of tools, such as the Brief Pornography Screener and Problematic Pornography Use Scale, across genders, age groups and societies [14,40]. Extending field trials would materially improve the ease with which these tools can then be rolled out for widespread adoption by clinicians and other people who are in the first line of contact by individuals concerned about PUP. Even among the experts the Brief Pornography Screener is not universally known [41]. Separately, major international reviews of erectile dysfunction have not referred in any way to the behavioural addiction field for data [42].

Tolerance and escalation are classic behavioural symptoms of the journey towards an addictive disorder. Each contributes to some consumers developing problematic usage. More research would be useful in relating the roles of tolerance and escalation in the PUP of viewers who escalate to viewing child abuse material. This seems to have become a wide and growing aspect of cybercrime around the world [43,44]. More knowledge of effective crime-fighting strategies could lead to better policy-making on prevention and to better risk-based advice for the courts and judges around sentencing and disposal. Standardised and tested risk assessment tools would be particularly useful.

Problematic usage of pornography seems to be creating a new separate sub-population of consumers who choose to view child abuse material without having any particular sexual interest in children in real life. They appear to be a separate clinical population to individuals with a paedophilic sexual orientation.

Another area where screening tools are woefully lacking is in the categorisation of the use of pornography by people on the autistic spectrum. Whereas about 1% of the population have autism spectrum disorders, crime statistics suggest that a disproportionately large number of the people (circa 30%) who escalate to viewing child abuse material come from specific vulnerable communities such those with as autism spectrum disorders, ADHD and those with learning difficulties [35]. These vulnerable populations are often considered 'asexual', and not requiring any sex education. This is wrong as many use the Internet as a substitute for relationships that they find so difficult to negotiate in real life. People with ASD are rarely represented in other crime categories. Their emotional immaturity and inability to interpret social rules can result in an attraction to younger people online who match their emotional age better. More research needs to be done on this group to improve prevention of access to child abuse material and to consider their special needs in terms of disposal following a conviction. For example, they do not function well in groups, which are the typical types of setting for sex offender rehabilitation [45].

- Research Area 3: Characterise the impacts of problematic Usage of Pornography on health and quality of life [1].

4.2.1. Is Problematic Pornography Usage a Primary Disorder?

This question can be explored by research into the pornography, masturbation and orgasm cycle. In turn, this line of work could lead to consideration of the extent to which PMO may be driving depression, social isolation and other mental health issues in the wider population. Is it a bi-directional process – to what extent are these mental health issues powering the PMO cycle? Work in this space to test causation was recommended as far back as 2016 [27], but has yet to gain funding.

Bi-directional work would also be helpful for studies exploring health links or problematic pornography usage and the Big-5 personality factors. First, factors such as narcissism are regularly shown as correlates of heavy pornography usage [46], but few studies look at what happens to personality factors after people end problematic pornography use. Do they remain constant or is there an individual and/or population tendency for them to revert to what they were before the individual began engaging in PUP [27]. Other recent studies have highlighted the need for longitudinal studies in this area [47].

Similarly, conference reports [23,24] and our informal discussions with both individual problematic pornography users in the community [26] and among sex therapists [15,16], suggest that there are co-morbidities with other addictions, both to substances and behaviours. Developing new and/or improved strategies for treating complex webs of problematic usage are always of value.

More literature would be helpful on the impact of problematic pornography use at different points in the human life-cycle. Most research published to date has focused on teens or university students, though around half of all pornography viewers are older [8]. This makes case studies or more specific work on older people very rare [48,49]. It would also allow us to look at the potentially very different trajectories of younger populations who began with access to unlimited, free streaming videos in or around 2008, and older users who may have transitioned through several successive modes of access to pornography such as magazines and DVDs, to Internet use.

4.2.2. Individual, Relationship and Community Impacts

The many roles of pornography in forming, sustaining and ending couple relationships would benefit from further research. The majority of research to date has been on users, and to a lesser extent their sexual partners. There would be value in unpacking the factors relating to pornography consumption as they impact on various aspects of an individual's health and quality of life in the context of a dyad [18,19,28]. This focus on couple relationships naturally tends to exclude work on the lack of relationships among young heavy users, around 50% of whom are virgins and not in relationships [12]. The other area needing more work is on the way pornography creates expectations of "how partnered sex will be" [35,46]. The literature of wider family and community impacts of pornography use across the human lifespan is also limited [48].

Additional research into the impact of accessibility would be valuable, as problematic pornography usage would seem to be a case where the PUI strongly links to private viewing. This is especially true for mobile platforms which can be used alone or with sexual partners. Smartphones are also now used widely for technology-enabled flirting including "sexting" with diverse relationship, social and potentially legal implications.

How does pornography consumption relate to the health and quality of life in a couple or a family? The authors frequently train social workers and criminal justice professionals who report that consumption of pornography, particularly violent pornography, is a consistent negative factor in the out-of-control domestic situations they encounter in their work. Unlike mainstream television where intimate partner violence in programmes tends to be simulated, most violence depicted in pornography is real [37,50]. It is not simulated and may offer viewers ethically challenging issues while they are viewing. It may also lead to post traumatic stress disorder and related symptoms after they stop. Once seen, real violence and intimate partner violence shown in pornography cannot be unseen. Where the real rape of minors is being streamed as entertainment by major commercial pornography sites, the legal and ethical challenges can be even more complex [51].

To what extent does acting out violent, coercive or controlling pornography scripts with real partners drive domestic violence and a growing appetite for dangerous practices such as sexual strangulation [52]? These questions are key in many different cross-disciplinary approaches to understanding PUI and PUP.

- Research Area 4: Define the Clinical Courses of Different Forms of Problematic Usage of Pornography [1].

The Manifesto noted that:

> *"Remarkably little prospective research has been conducted on the courses of different forms of PUI and we remain relatively ignorant of key factors affecting long-term out-comes. Such data are of crucial importance in understanding aetiology, planning treatment and improving prognostication. For example, for some individuals PUI may represent a temporary phenomenon and spontaneously resolve (e.g., in some young people as brain systems mature), whereas for others PUI may become chronic."* [1] p. 1238.

These observations apply very strongly to PUP. There are many areas where we have little data.

In simple terms, PUP can lead to the individual failing to stop their use when faced with consequences, following the model set out in the diagnosis for compulsive sexual behaviour disorder [20]. However, ICD-11 is careful not to set out the details of the specific sexual behaviours that are causing the distress. Equally it is vague about the progression of the disorder from the perspectives of screening, assessment and diagnosis. This gives considerable scope to clearly define the clinical courses of each and every form of PUP.

We suggest that there are two fundamental classes of PUP, and each needs more research. The first is where the consumption of pornography leads to problems directly attributable to the consumption process. The user cannot stop their use and it is leading to negative consequences. This is applicable to CSBD as an impulse control disorder [20] and would also apply if a future version of the International Classification of Diseases gave it an entry in the addictive disorders section.

We propose that the second category is where the development of a PUP comes from the ideas it introduces into a viewer's mind and life. These can range across a whole spectrum from requiring to replay sexual scripts from pornography in your head to get aroused during partnered sex, to learning to ignore the need for consent in sexual activities, to promoting rape myths, the diminution of bystander intervention and the promotion of practicing dangerous activities such as autoerotic asphyxiation. Other problematic uses can encompass escalation to criminal activities around viewing or creating images of child sexual abuse. It can also lead individuals to engage in sextortion, 'capping', and other illegal activities towards minors [35]. Other new populations are the technology-enabled voyeurs whose illegal activities are also directed at adults, encompassing upskirting, hidden cameras, revenge porn and sharing sexual material without permission. Pornography has also appeared that was secretly recorded on baby monitors lacking password protection.

Pornography-induced erectile dysfunction (PIED) [28,53] has been identified as a key factor that causes male sufferers of PUP to seek treatment [54], but this area still lacks clear before and after brain-imaging studies [27]. PIED has also been linked to escalation to child abuse material. As desensitisation, tolerance and hypofrontality develop, a problematic pornography user needs more shocking material to feel any arousal.

Therapists need to know how long it can take for a PUP to develop and what external or internal factors can act as triggers. Reports from the recovery communities talk about the triggering influence of everything from life events and rites of passage, to social networks and the manipulation of viewer's taste by the algorithms used by the commercial pornography suppliers.

The withdrawal process from PUP, both physical and mental, seems generally similar to that of other PUIs. The only unique characteristic of PUP identified so far is 'flatlining'. The literature does not adequately characterise the features of flatlining as a symptom across the population of users withdrawing from PUP. This means that when it happens, problematic users can relapse to try to feel aroused again and overcome their feeling that their sexuality might be permanently 'broken'.

More studies of co-morbidity, both between PUP and other behavioural or substance addictions, and with other compulsive sexual behaviours, would also be welcome.

There are other elements research could reveal about the general characteristics of PUP. What are the characteristics of the incubation periods of PUP at both the individual and population levels? Do they have characteristics driven by gender or other factors? This in turn leads to the need for wider studies looking at gender, lifestyle and sexuality-based approaches to tease apart key elements in the PUP process. What commonalties and differences exist across the LGBTQI++ spectrum? In parallel, is PUP in the chemsex community driving addiction, or is its influence bi-directional [55]?

Existing brain-imaging studies [56,57] hint that both the quantity of pornography viewed and the period the viewing is spread across can have impacts on the development of PUP. However, the current crop of brain studies on pornography are restricted to people in their 20s and 30s and they also draw on very narrow demographic and cultural samples. In a world where an average age for boys' first exposure is 12 years or younger, brain development issues need investigation [2].

More women-focused studies would be beneficial. How is PUP the same for women as for men, and how is it different? Are the genders the same in the etiology of compulsive use? To what extent is sex toy used [58] a component of women's PUP? Some research has suggested that pornography usage by women has influenced their behaviour in couple relationships [59,60]. Are there differences according to ages and points in the lifecycle? We are unaware of any PUP studies in women controlling for stages across the menopause as a variable.

Age and stage of life should also be addressed from other perspectives. Virgins make up perhaps 50% of the recovery community [12]. What are the longer-term implications of early PUP across their lives as sexual beings? Separately, how do the challenges faced by older addicts, who used a lot of pornography before the Internet, compare to those of purely Internet-based consumers?

What are the impacts of PUP on relationship formation, sustainability and sexual health? A study of Korean men in stable relationships suggested they had developed a preference for viewing violent pornography as an alternative to real sex with their dyadic partner [61]. This also raises the issue of the nature of the content of the pornography being viewed. In line with the self-reports from the recovery community, a Swedish and Italian study of university students demonstrated that unusual sexual content creates more excitement. It also concluded that lots of the acts considered "unusual" by the researchers in 2015, already had become mainstream to many viewers and did not seem "unusual" at all [62].

- Research Area 5: Reduce Obstacles to Timely Recognition and Interventions [1].

Obstacles around the timely recognition of PUP and in delivering interventions exist both within the user population and among professionals. The interests of the groups overlap where there is a desire to develop, test and deploy more public health intervention strategies to draw the attention of pornography users across society to indicators of problematic usage.

At the widest level there is value in developing strategies to have the issue of PUP accepted by national healthcare systems. It can now be diagnosed as compulsive sexual behaviour disorder, coded 6C72 under the parent category of impulse control disorders [20]. Some clinicians are using this categorisation, but there will be considerable lags before it is adopted by nations world-wide. In some countries such as the United States, the need for clinicians to have a code for CSBD to allow patients to claim on medical insurance is paramount.

A separate issue revolves around making effective training for professionals more widely available. There is a need for the right balance of availability of practitioners (numbers, skills, geographical spread) to the scale of needs of people with issues.

Training can be built into continuing education programmes for professionals, as it is in the United States through the Society for the Advancement of Sexual Health and the International Institute for Trauma and Addiction [15,17]. In the UK this is done through the Association for the Treatment of Sexual Addiction and Compulsivity and the College of Sexual and Relationship Therapists [11,16]. However, CSBD and PUP are very new ideas and they have not become part of general medical training in most tertiary institutions.

- Research Area 6: Clarify the Possible Role of Genetics and Personality Features in Different Forms of Problematic Usage of Pornography [1].

This area of PUP research is still very immature. There would be value in developing a broad literature looking at the relationship between genetic factors and the natural history of problematic pornography usage. Are particular individuals or populations especially vulnerable? Key dimensions might include gender, sexual orientation and exploring health links to Big-5 personality factors.

Proof of concept was demonstrated by a 2017 paper [63] examining links between gene methylation and hypersexual disorder. The study found that the epigenetic state in the CRH gene may contribute to explaining the biological mechanisms of hypersexual disorder. Thus, some genes appear to play a role in hypersexual disorder.

Within the recovery community there is a strong desire to find answers to the 'why me' question. What is the relationship between predisposition to PUP versus acquiring it through consumer behaviour in a changed technological environment? Epigenetic factors need to be further researched. Informal evidence from the online recovery communities indicates that certain character traits or personality types thought to be fixed actually change when regular use of a supernormal stimulus such as hardcore pornography ceases. For example, in some users out of control hypersexual or narcissistic or sensation seeking or aggressive behavioural traits remit or disappear when the stressor is removed.

- Research Area 7: Consider the Impact of Social Factors in the Development of Problematic Usage of Pornography [1].

Several separate, but overlapping strands of research around social factors would be useful. Five are highlighted here as requiring more investigation: the technologies of the attention economy; women's changing role as consumers; environmental impacts; social acceptability; and the potential for early exposure to pornography to be considered as an Adverse Childhood Experience.

First, the Manifesto in Research area 7, in Section 2 above, identified

"For excessive streaming (watching videos excessively), important structural features may include the ability of given programmes to grab attention by activating a biological 'orientating response', mediated through techniques including the use of attention-grabbing noises, zooming/panning, and presentation of rewarding stimuli (e.g., of a sexual or thrilling nature)" [1] p. 1241.

This neatly summarises the core behaviour of people engaged in PUP, where watching is generally accompanied by the viewer masturbating. This is absolutely an area where public research is required into the structural elements of commercial pornography sites as a part of the attention economy. Commercial sites seem to have adopted the structural concepts pioneered by B.J. Fogg at Stanford University and the behavioural design work of Nir Eyal [64,65]. This allows them to build websites and apps that change users' thoughts and behaviour without their knowing. In particular, the pornographers target the unconscious mind via dopamine pathways in the reward centre to stimulate cravings that keep users coming back for more, resulting in PUP for increasing numbers of consumers. Artificial intelligence is being deployed by the commercial pornography suppliers to learn about user's sexuality to deliver the most engaging experience possible. This has implications for privacy. As some research suggests that pornography viewing can change a person's sexual tastes, this is a very slippery area. Do we really want a big-porn machine algorithm to reshape our sexual templates [66]?

Another area ripe for research is the issue of intentional manipulation of consumption patterns by commercial interests in the pornography industry. In particular, since 2016 there has been a sustained increase of usage by women as a world-wide phenomenon [5–8,67]. To what extent has this resulted from deliberate female-focused promotional strategies of the commercial pornography suppliers and what are its implications for dyadic relationships, family structures and society in general?

Pornography sites get heavy, repeat traffic from users. The largest supplier, Pornhub, streamed 115 billion videos in 2019 to a world-wide audience of perhaps 1 billion individual viewers [8]. It would be extremely helpful to have reliable calculations for the world-wide use of pornography. The industry is now so large that it has been calculated to consume 27% of the energy used by the Internet and is currently responsible for 0.2% of all greenhouse gas emissions world-wide, roughly equivalent to the domestic energy use of every household in France [68]. The environmental impact is large enough for researchers to calculate a quantifiable contribution to rising sea levels.

Across cultures and within cultures, the phenomenon of "normalisation" of pornography consumption as a socially acceptable activity has grown rapidly. Since the arrival of the iPhone in June 2007, supported by high-speed network services, the smartphone supplying "free" commercial pornography has removed barriers and increased temptation through its Triple-A model of effectively unlimited access, affordability and anonymity [69]. The speed of this adoption has outstripped the capacity of researchers to place it in context and to include it in theoretical models. It is clear that the arrival of smartphones, Wi-Fi and 4G networks leads to most societies adopting pornography as a more acceptable form of recreation, especially among the young. The trajectory of this change can be unique to each country [2].

Many stories circulate within the recovery community suggesting that childhood exposure to pornography in general, and Internet pornography in particular, should be considered an adverse childhood experience (ACE). However, at present this cannot happen. The list of definitions of ACEs pre-dates the arrival of problematic usage of pornography from the Internet. It would mean revisiting the research field of defining, testing and verifying the validity of PUP as an ACE. This would be a very worthwhile project.

- Research Area 8: Generate and Validate Effective Interventions, both to prevent Problematic Usage of Pornography, and to Treat its Various Forms once Established [1].

4.2.3. Supply-Side Preventative Interventions

These tend to revolve around whole-population approaches to changing the ease of access to Internet pornography. After three false starts, will age verification legislation in the UK prove to be a successful intervention? How and where might it be tried in other countries? How will it be validated and will it achieve the intended outcomes?

There is an established and diverse commercial ecosystem of blocking technologies now available for individuals and groups to prevent access to Internet pornography. Their characteristics tend to be specific to their local geographies. They can operate at the individual level, usually by subscription. Many Internet Service Providers (ISPs) offer filtering on either an opt-in or opt-out basis. Academic analysis of the effectiveness of these measures would be helpful.

PUP research could also look to public health models used for other areas of addiction, particularly online gambling, where the business models and delivery mechanisms are often quite similar. Policy research into new models of control, monitoring and product delivery limitation involving the government, commercial and voluntary sectors could be developed. This remains minimally charted territory.

A recent development in pornography delivery (and sometimes production) is the increasing role of social media as a point of access. While games delivery platforms are now a significant source of pornography for some consumers, we have also encountered adolescents getting around parental controls via the Internet of Things. A poorly protected smart refrigerator could get a grounded teen online.

4.2.4. Supply-Side Treatment Interventions

Individual or community approaches to treatment from counsellors, schools, churches or other local units can be successful, especially when linked to demand-side reduction measures. Accountability partners are popular in some communities [26]. Again, more external validation of success, or not, of different approaches would be valuable.

4.2.5. Demand-Side Preventative Interventions

Are all pornography consumers at risk of developing PUP? The simple answer is probably not, but we do not know for certain. If the general characteristics of PUP are broadly similar to other better studied behaviours with a propensity to lead to behavioural addiction, the answer is 'no'. A proportion of users will be likely to move first into problematic use and then addictive use. Some studies suggest that all pornography consumption generates brain changes, even at quite low durations [56,57]. So far, the World Health Organization only considers the research base is strong enough to support the diagnosis of compulsive sexual behaviour disorder under the parent category of impulse control [20]. At the same time, few pornography consumers have been exposing themselves to large volumes of online erotica for longer than two decades, so we have only explored it as a factor for, at most, half of the duration of a human beings' sexual life-cycle. We do know that the more pornography some people consume, the more likely they are to develop compulsive behaviours as a result. It is the sort of issue that could respond well to the development of a risk-managed framework. Such a framework could balance benefits of pornography consumption at the individual and society levels with the inevitable side-effects from some individuals being unhappy with the consequences arising from their use.

Where consumers do develop concerns, many practitioners believe that the solution to PUP is rooted in consumer education. As a society we can choose to try to reduce the strength of demand for access to pornography. Some individuals who see pornography as a valuable part of a sex positive culture would be uncomfortable about this, but it is up to each society to consider what restrictions may or may not be appropriate if some individuals may be harmed by excessive consumption.

More systematic assessments of the success of educational strategies would show how effective they can be as risk-reduction strategies. What role will learning about the addictive potential of PUP play in reducing demand? Does learning about porn-induced erectile dysfunction reduce consumption? Thousands of anonymous self-reports on the recovery websites show that the availability of knowledge of the effects of Internet pornography on the brain was sufficient to encourage users to experiment with quitting porn [12,13]. When they did, a range of mental and physical health problems remitted or cleared up. Only when heavy users quit porn did they realise that those conditions were related to their porn use. Formal research into this phenomenon would be valuable.

A separate investigation into the issues of prevention versus cure would be interesting. We are suggesting research to compare the benefits of never beginning pornography usage versus treating the problem when it has been 'cured'. At a population level, are people who quit ultimately as mentally and physically healthy as those who never engaged in PUP in the first place? What methodologies might help unpack this question?

4.2.6. Demand-Side Treatment Interventions

The development of possible drug treatments is still at an early stage. In theory adjusting brain chemistry with substances could reduce libido and cravings for pornography. The most common approach is to look at substances that are already approved for clinical use and to begin the process of taking them through stages from case reports and on to small and then large-scale clinical trials. What drugs might discourage PUP and what pathways should be targeted? A major review was published by Sniewski and colleagues in 2018 [70]. Early work focused on possible similarities to obsessive compulsive disorder and suggested trying drugs known as selective serotonin re-uptake inhibitors (SSRIs). Recent case reports have been published for Paroxetine [71], Nalmefene [49] and Aripiprazole [72], with mixed results.

- Research Area 9: Identify Biomarkers, Including Digital Markers, to Improve Early Detection and Intervention [1].

This is a new area, but it may have real potential to help improve health at the individual and population levels. As far as we are aware, biomarkers are not currently in use for detecting or diagnosing PUP. Would there be value in their introduction?

Research might be able to separate out the underlying tendency of some individuals or groups to be susceptible to problematic usage. This could be contrasted to the potential for Internet pornography to lead to problematic usage as a natural characteristic in and of itself. The issue here is "are you the problem?" versus "is pornography the problem". Co-morbidity studies involving other PUIs would also be helpful—many users find themselves trying to unhook from gaming or recreational drugs and pornography at the same time.

Biomarkers are a high-tech intervention. In face-to-face situations, such as a doctor's surgery or a counsellor's office, are they a better solution than simply asking a person about their level of pornography consumption? The brief pornography screener with five questions on a single A4 sheet of paper can be administered in three minutes and seems to be both sensitive and reliable [14].

A different line of research would be to look at the potential for widescale interventions using Internet data. It might even extend to artificial intelligence algorithms at commercial pornography suppliers or the ISP level. It would mean monitoring people for signs of problematic usage and then invoking a warning or even a cut-off protocol. This has privacy and data management implications, but the reality seems to be that monitoring rather like this is already at the core of the business model of major commercial pornography suppliers. Could governments consider introducing national level intervention programmes? How would they operate and what would be the success criteria?

At the same time, the pornography supply companies are very reluctant to share their data. In this they follow the lead of other industries supplying addictive substances like tobacco and alcohol. The commercial pornography suppliers are quite vulnerable to a 'smoking gun' which showed that were trying to induce addictive behaviour as a part of their business model.

5. Conclusions

When you consider the conclusions in the Manifesto [1] in the light of the current paper, there is nothing recommended for research into PUI that does not equally apply to the sub-discipline of PUP.

To come back to our research question, "what topics should be included within future research proposals under the Manifesto to meet the diverse needs of consumers, recovery communities and professionals impacted by the problematic use of pornography"?

The comments in this section have been grouped to answer these questions individually, but there is, and should be, overlap between them, so we begin by looking at general questions about the field of research. The management and treatment of PUP requires the right approaches to meet the specific needs of everybody in this mental and physical health ecosystem.

5.1. General Research

Given the massive scale of use of pornography and its potentially negative outcomes for some users, it is important for the European Union's Problematic Usage of the Internet (EU-PUI) Network to have a visible and trusted champion for PUP, just as it should for other behaviours. Linked to this, Europe should hold a planning conference under the umbrella of COST Action to design its response. The opportunity for PUP data to be made accessible to researchers through shared multinational databases is extremely attractive.

To assist the World Health Organization in improving the classification of compulsive sexual behaviour disorder in future to include the word pornography, research is required to separate out the natural history of PUP from compulsive sexual behaviours focused on people.

Experimental work to investigate causation through longitudinal studies would provide a better strategic foundation for PUP. This would be stronger if it included large-scale brain imaging studies to gather evidence before and after people have quit or have been given treatment. It is essential to develop a wider evidence-base covering all ages and stages of the human life-cycle for people engaging in PUP. There should be much more research focus on women as they are now the main sector of growth for PUP. Equally our data should embrace sexual diversity and investigate the usage among LGBTQI++ communities where levels of mental health disorders and chemical dependencies are higher than in the mainstream heterosexual population. Could PUP be a contributing factor?

5.2. Research for the Professional and Therapist Community

The 2020s is the decade when training of health and social care professionals should begin to incorporate a deep and broad understanding of all PUIs as a matter of course. This will be a communication and public relations programme. It should be recognised that within the field of PUP we are starting from a low-base with modest levels of training or resources available for people working as professionals and therapists. New research should be leading towards giving professionals a wider toolkit to support treatment of all the different forms of PUP.

Ideally the next ten years will see a rising quality of research into PUP, moving us towards a landscape with more double-blind placebo trials for drugs and longitudinal studies covering time periods that allow us to take a broader view on PUP. It is a disorder that is slow to develop and potentially also slow to heal.

We should encourage better and deeper validation of a few assessment tools across diverse populations. Standardisation is helpful and surely two proven ones are sufficient. Link this to a strong communication programme to make them known widely in communities from general practitioners (MDs) and urologists to sex therapists, relationship counsellors, school counsellors and parents.

The next decade should see the debate around pornography use and erectile health fully resolved. Dialogues about the impact of pornography use in the context of violence against women and children also require more experimental approaches if the discussion is going to be able to move from correlation to causation.

The next few years will clearly be a time of great advances in linking genetic data and behaviour and we will monitor the contributions that this form of high-technology can make to epigenetic changes in humans.

5.3. Research for Consumers

At the low-tech end, the most important thing is to initiate a research and development programme to determine if early exposure to Internet pornography can be recognised as an adverse childhood experience (ACE). Success in that project could have a profound impact on raising awareness of the need to reduce early exposure and thereby reduce the risk of subsequent PUP.

Work monitoring the success, or otherwise, of age verification legislation around the world will be important for policy development. For individuals and communities, it is important that we learn which interventions are sustainable and cost effective.

Could commercial pornography suppliers be encouraged to take a pro-active role in heading off the development of PUP, rather than following their current laissez faire view that a customer can never have too much pornography? It will be important to have research to boost the accountability of suppliers. Accountability can be achieved through government intervention or industry self-regulation. So far the latter does not seem to be working in the consumer's favour. In general the access to pornography is free, but the consumption risks are all transferred to the user.

The powerful position of commercial pornography suppliers within the context of the attention economy leaves the average consumer poorly informed about how their tastes and behaviour are being influenced by 'free' pornography. Commercial pornography suppliers use the data that gather to develop a deep understanding of consumer behaviour. Commercial players also benefit from the access to artificial intelligence and other data-mining tools to build market share and develop new markets.

5.4. Research for the Recovery Community

Good academic studies of the recovery communities, be they NoFap, Sex Addicts Anonymous or others, could be revolutionary. Recovery communities are a cheap solution to a potentially very expensive medical problem. However, often the communities do not really want to be studied. Members just want to be healed or to help others to heal. There is some acceptance that good research could help. There is also a recognition that for some PUP members, recovery communities are only part of the answer and that they may also need help from therapists and other professionals.

Quantitative research looking at the elimination of pornography is the highest priority. Does it really reverse sexual dysfunctions and have a positive impact on a variety of mental health disorders? The field currently has extensive anecdotal evidence to support these suggestions, but that is not a substitute for large-scale, well-grounded studies.

5.5. Moving Forward

It is now nearly two years since the Manifesto was published. In that time there has been a continuing growth world-wide in the volume of research appearing on pornography use in general and PUP in particular. However, most of the work being done tends to be at the easier, cheaper end of the scale. As far as we can tell as frequent observers of this field, there remains a general lack of vision and large-scale co-ordination between the many research groups around the world. The future needs to be more about designing investigations to test causation, not just to demonstrate yet more correlations.

The time is right for the COST Action team and the EU-PUI Network to create, fund and promote a visible and trusted champion for PUP research. Europe has sufficient resources to support such a role. Perhaps the research ideas sketched in this paper could be a positive contribution towards COST Action accepting it should deliver a leadership role in the field of PUP.

5.6. Limitations

Focusing on the Manifesto is a limiting strategy, in that many researchers in the pornography field around the world are unlikely to be familiar with it. Our approach has been to work with what is already available to build a more coherent research environment. At the same time, focusing on the Manifesto is essential. It is the only major published document which deals with the whole policy context of problematic use of the Internet. It is Europe-wide in its reach and is designed to endure for a decade.

By writing a response homing in on the issues around PUP, we hope to encourage researchers to have conversations to advance the field in ways that will benefit the continent's taxpayers and publicly funded research programmes.

Our methodology of simply analysing all of the Manifesto as a whole through the particular lens of PUP is a limitation. This approach was taken to balance achieving the clearest level of understanding of what readers could infer from the document with an open agenda to build the strongest possible set of recommendations to shape the future of PUP research in Europe over the next decade.

Author Contributions: Conceptualisation, D.M. and M.S.; Writing—original draft preparation, D.M.; Writing—review and editing, D.M. and M.S. All authors have read and agreed to the published version of the manuscript.

Funding: This research received no external funding.

Acknowledgments: The authors would like to thank Julius Burkauskas, the Science Communications Manager of the COST Action Dissemination Plan for encouraging the preparation of this paper.

Conflicts of Interest: The authors declare no conflict of interest.

References

1. Fineberg, N.; Demetrovics, Z.; Stein, D.; Ioannidis, K.; Potenza, M.; Grünblatt, E.; Brand, M.; Billieux, J.; Carmi, L.; King, D.; et al. Manifesto for a European Research Network into Problematic Usage of the Internet. *Eur. Neuropsychopharmacol.* **2018**, *28*, 1232–1246. [CrossRef] [PubMed]
2. Mead, D. The Risks Young People Face as Porn Consumers. *Addicta- Turk. J. Addict.* **2016**, *3*, 387–400. [CrossRef]
3. Döring, N.M. The Internet's impact on sexuality: A critical review of 15 years of research. *Comput. Hum. Behav.* **2009**, *25*, 1089–1101. [CrossRef]
4. Döring, N.; Daneback, K.; Shaughnessy, K.; Grov, C.; Byers, E.S. Online sexual activity experiences among college students: A four-country comparison. *Arch. Sex. Behav.* **2017**, *46*, 1641–1652. [CrossRef] [PubMed]
5. Pornhub Insights. 2016 Year in Review. Available online: https://www.pornhub.com/insights/2016-year-in-review (accessed on 27 February 2020).
6. Pornhub Insights. 2017 year in Review. Available online: https://www.pornhub.com/insights/2017-year-in-review (accessed on 27 February 2020).
7. Pornhub Insights. 2018 Year in Review. Available online: https://www.pornhub.com/insights/2018-year-in-review (accessed on 27 February 2020).
8. Pornhub Insights. The 2019 Year in Review. Available online: https://www.pornhub.com/insights/2019-year-in-review (accessed on 27 February 2020).
9. Brand, M.; Blycker, G.R.; Potenza, M. When pornography becomes a problem: Clinical insights. *Psychiatric Times* **2019**, *36*, 12.
10. Stark, R.; Klucken, T.; Potenza, M.N.; Brand, M.; Strahler, J. A current understanding of the behavioral neuroscience of compulsive sexual behavior disorder and problematic pornography use. *Curr. Behav. Neurosci. Rep.* **2018**, *5*, 218–231. [CrossRef]
11. College of Sexual and Relationship Therapists. Available online: https://www.cosrt.org.uk/ (accessed on 30 April 2020).
12. NoFap.com. Available online: https://www.nofap.com (accessed on 27 February 2020).
13. RebootNation.org. Available online: http://www.rebootnation.org/ (accessed on 27 February 2020).
14. Kraus, S.W.; Hoff, R.A.; Gola, M.; Kowalewska, E.; Potenza, M.N. Clinical Characteristics of Compulsive Pornography Users: A Military Sample. *J. Behav. Addict.* **2018**, *7*, 98.
15. Society for the Advancement of Sexual Health. Available online: https://www.sash.net/ (accessed on 30 April 2020).
16. Association for the Treatment of Sexual Addiction and Compulsivity. Available online: https://atsac.org.uk/ (accessed on 30 April 2020).

17. International Institute for Trauma and Addiction Professionals. Available online: https://iitap.com/default.aspx (accessed on 30 April 2020).
18. Church, N. *Wack: Addicted to Internet Porn*; Bvrning Qvestions LLC: Portland, OR, USA, 2014.
19. Hall, P. *Understanding and Treating Sex and Pornography Addiction: A Comprehensive Guide for People Who Struggle with Sex Addiction and Those Who Want to Help Them*; Routledge: Abingdon-on-Thames, UK, 2018.
20. World Health Organization. International Classification of Diseases, 11th Revision (ICD-11). Available online: https://www.who.int/classifications/icd/en/ (accessed on 15 February 2020).
21. Bőthe, B.; Tóth-Király, I.; Potenza, M.N.; Griffiths, M.D.; Orosz, G.; Demetrovics, Z. Revisiting the Role of Impulsivity and Compulsivity in Problematic Sexual Behaviors. *J. Sex Res.* **2019**, *56*, 166–179. [CrossRef]
22. Mestre-Bach, G.; Blycker, G.R.; Potenza, M.N. Pornography use in the setting of the COVID-19 pandemic. *J. Behav. Addict.* **2020**. Available online: https://akjournals.com/view/journals/2006/aop/article-10.1556-2006.2020.00015/article-10.1556-2006.2020.00015.xml (accessed on 27 April 2020). [CrossRef]
23. Mead, D.; Sharpe, M. Pornography and Sexuality Research Papers at the 4th International Conference on Behavioral Addictions. *Sex. Addict. Compulsivity* **2017**, *24*, 217–223. [CrossRef]
24. Mead, D.; Sharpe, M. Pornography and Sexuality Research Papers at the 5th International Conference on Behavioral Addictions. *Sex. Addict. Compulsivity* **2018**, *25*, 248–268. [CrossRef]
25. American Psychiatric Association. *Diagnostic and Statistical Manual of Mental Disorders (DSM-5®)*; American Psychiatric Pub: Washington, DC, USA, 2013.
26. Anonymous. *Sex Addicts Anonymous*; International Service Organization of SAA: Houston, TX, USA, 2017.
27. Wilson, G. Eliminate Chronic Internet Pornography Use to Reveal Its Effects. *Addicta Turk. J. Addict.* **2016**, *3*, 1–13. [CrossRef]
28. Park, B.Y.; Wilson, G.; Berger, J.; Christman, M.; Reina, B.; Bishop, F.; Klam, W.P.; Doan, A.P. Is Internet Pornography Causing Sexual Dysfunctions? A Review with Clinical Reports. *Behav. Sci.* **2016**, *6*, 17. [CrossRef] [PubMed]
29. Bostwick, J.M.; Bucci, J.A. *Internet Sex Addiction Treated with Naltrexone. Mayo Clinic Proceedings*; Elsevier: Amsterdam, The Netherlands, 2008; Volume 83, pp. 226–230.
30. Capurso, N.A. Naltrexone for the Treatment of Comorbid Tobacco and Pornography Addiction. *Am. J. Addict.* **2017**, *26*, 115–117. [CrossRef]
31. Kraus, S.W.; Meshberg-Cohen, S.; Martino, S.; Quinones, L.J.; Potenza, M.N. Treatment of Compulsive Pornography Use with Naltrexone: A Case Report. *Am. J. Psychiatry* **2015**, *172*, 1260–1261. [CrossRef]
32. Blum, A.W.; Grant, J.E. Positive Response of Compulsive Sexual Behavior to Transcranial Magnetic Stimulation. *Prim. Care Companion CNS Disord.* **2020**, *22*, 19l02469. [CrossRef]
33. Caselli, G.; Spada, M.M. Desire Thinking: A New Target for Treatment of Addictive Behaviors? *Int. J. Cogn. Ther.* **2016**, *9*, 344–355. [CrossRef]
34. Wells, A. Advances in Metacognitive Therapy. *Int. J. Cogn. Ther.* **2013**, *6*, 186–201. [CrossRef]
35. The Reward Foundation - Love, Sex and the Internet. Available online: https://rewardfoundation.org (accessed on 27 February 2020).
36. Rodenhizer, K.A.E.; Edwards, K.M. The Impacts of Sexual Media Exposure on Adolescent and Emerging Adults' Dating and Sexual Violence Attitudes and Behaviors: A Critical Review of the Literature. *Trauma Violence Abuse* **2019**, *20*, 439–452. [CrossRef]
37. Bridges, A.J. Pornography and Sexual Assault. In *Handbook of Sexual Assault and Sexual Assault Prevention*; O'Donohue, W.T., Schewe, P.A., Eds.; Springer International Publishing: Cham, Switzerland, 2019; pp. 129–149. [CrossRef]
38. Foubert, J.D.; Blanchard, W.; Houston, M.; Williams, R.R. Pornography and Sexual Violence. In *Handbook of Sexual Assault and Sexual Assault Prevention*; O'Donohue, W.T., Schewe, P.A., Eds.; Springer International Publishing: Cham, Switzerland, 2019; pp. 109–127. [CrossRef]
39. Brem, M.J.; Garner, A.R.; Grigorian, H.; Florimbio, A.R.; Wolford-Clevenger, C.; Shorey, R.C.; Stuart, G.L. Problematic Pornography Use and Physical and Sexual Intimate Partner Violence Perpetration Among Men in Batterer Intervention Programs. *J. Interpers. Violence* **2018**, 088626051881280. [CrossRef]
40. Bőthe, B.; Tóth-Király, I.; Zsila, Á.; Griffiths, M.D.; Demetrovics, Z.; Orosz, G. The Development of the Problematic Pornography Consumption Scale (PPCS). *J. Sex Res.* **2018**, *55*, 395–406. [CrossRef] [PubMed]
41. de Alarcón, R.; de la Iglesia, J.I.; Casado, N.M.; Montejo, A.L. Online Porn Addiction: What We Know and What We Don't—A Systematic Review. *J. Clin. Med.* **2019**, *8*, 91. [CrossRef] [PubMed]

42. Goldstein, I.; Goren, A.; Li, V.W.; Tang, W.Y.; Hassan, T.A. Epidemiology Update of Erectile Dysfunction in Eight Countries with High Burden. *Sex. Med. Rev.* **2020**, *8*, 48–58. [CrossRef] [PubMed]
43. Nowak, M. Promotion and Protection of All Human Rights, Civil, Political, Economic, Social and Cultural Rights, Including the Right to Development. Report of the Special Rapporteur on Torture, and Other Cruel, Inhuman or Degrading Treatment or Punishment. Available online: https://www.refworld.org/docid/47c2c5452.html (accessed on 15 February 2020).
44. Knack, N.; Holmes, D.; Fedoroff, J.P. Motivational Pathways Underlying the Onset and Maintenance of Viewing Child Pornography on the Internet. *Behav. Sci. Law* **2020**, *37*, 100–116. [CrossRef]
45. Hackett, S.; Phillips, J.; Masson, H.; Balfe, M. Individual, Family and Abuse Characteristics of 700 British Child and Adolescent Sexual Abusers. *Child Abuse Rev.* **2013**, *22*, 232–245. [CrossRef]
46. Efrati, Y.; Amichai-Hamburger, Y. Are Adolescents Who Consume Pornography Different from Those Who Engaged in Online Sexual Activities? *Child. Youth Serv. Rev.* **2020**, *111*, 104843. [CrossRef]
47. Borgogna, N.C.; Aita, S.L. Problematic Pornography Viewing from a Big-5 Personality Perspective. *Sex. Addict. Compulsivity* **2019**, *26*, 293–314. [CrossRef]
48. Herbenick, D.; Fu, T.-C.; Wright, P.; Paul, B.; Gradus, R.; Bauer, J.; Jones, R. Diverse Sexual Behaviors and Pornography Use: Findings From a Nationally Representative Probability Survey of Americans Aged 14 to 60 Years. *J. Sex. Med.* **2020**, S1743609520300473. [CrossRef]
49. Yazdi, K.; Fuchs-Leitner, I.; Gerstgrasser, N.W. Nalmefene in the Treatment of Internet Pornography Addiction—A Case Report and Review of Literature. *J. Addict. Med.* **2020**, 1. [CrossRef]
50. Bridges, A.J.; Wosnitzer, R.; Scharrer, E.; Sun, C.; Liberman, R. Aggression and Sexual Behavior in Best-Selling Pornography Videos: A Content Analysis Update. *Violence Women* **2010**, *16*, 1065–1085. [CrossRef]
51. McDonald, S. Florida man arrested after 58 porn videos, photos link him to missing underage teen girl. Available online: https://www.newsweek.com/florida-man-arrested-after-58-porn-videos-photos-link-him-missing-underage-teen-girl-1467413 (accessed on 22 February 2020).
52. Chapple, L. Characteristics of Individuals Who Participate In Autoerotic Asphyxiation Practices: An Exploratory Study. PhD thesis, University of North Dakota, Grand Forks, North Dakota, August.
53. Retzler, K. Erectile Dysfunction: A Review of Comprehensive Treatment Options for Optimal Outcome. *J. Restor. Med.* **2019**, *8*. [CrossRef]
54. Wilson, G. *Your Brain on Porn: Internet Pornography and the Emerging Science of Addiction*, 2nd ed.; Commonwealth Publishing: Margate, UK, 2018.
55. Corkery, J.M.; Loi, B.; Claridge, H.; Goodair, C.; Corazza, O.; Elliott, S.; Schifano, F. Gamma Hydroxybutyrate (GHB), Gamma Butyrolactone (GBL) and 1,4-Butanediol (1,4-BD.; BDO): A Literature Review with a Focus on UK Fatalities Related to Non-Medical Use. *Neurosci. Biobehav. Rev.* **2015**, *53*, 52–78. [CrossRef] [PubMed]
56. Kühn, S.; Gallinat, J. Brain Structure and Functional Connectivity Associated with Pornography Consumption: The Brain on Porn. *JAMA Psychiatry* **2014**, *71*, 827–834. [CrossRef] [PubMed]
57. Voon, V.; Mole, T.B.; Banca, P.; Porter, L.; Morris, L.; Mitchell, S.; Lapa, T.R.; Karr, J.; Harrison, N.A.; Potenza, M.N. Neural Correlates of Sexual Cue Reactivity in Individuals with and without Compulsive Sexual Behaviours. *PloS ONE* **2014**, *9*, e102419. [CrossRef] [PubMed]
58. Davis, C.M.; Blank, J.; Lin, H.; Bonillas, C. Characteristics of Vibrator Use among Women. *J. Sex Res.* **1996**, *33*, 313–320. [CrossRef]
59. Wright, P.J.; Sun, C.; Miezan, E. Individual Differences in Women's Pornography Use, Perceptions of Pornography, and Unprotected Sex: Preliminary Results from South Korea. *Personal. Individ. Differ.* **2019**, *141*, 107–110. [CrossRef]
60. Bélanger Lejars, V.O.; Bélanger, C.H.; Razmak, J. Exploring New Measures of Online Sexual Activities, Device Use, and Gender Differences. *Comput. Hum. Behav.* **2020**, *108*, 106300. [CrossRef]
61. Sun, C.; Miezan, E.; Lee, N.-Y.; Shim, J.W. Korean Men's Pornography Use, Their Interest in Extreme Pornography, and Dyadic Sexual Relationships. *Int. J. Sex. Health* **2015**, *27*, 16–35. [CrossRef]
62. Tripodi, F.; Eleuteri, S.; Giuliani, M.; Rossi, R.; Livi, S.; Petruccelli, I.; Petruccelli, F.; Daneback, K.; Simonelli, C. Unusual online sexual interests in heterosexual Swedish and Italian university students. *Sexologies* **2015**, *24*, e84–e94. [CrossRef]
63. Jokinen, J.; Boström, A.E.; Chatzittofis, A.; Ciuculete, D.M.; Öberg, K.G.; Flanagan, J.N.; Arver, S.; Schiöth, H.B. Methylation of HPA Axis Related Genes in Men with Hypersexual Disorder. *Psychoneuroendocrinology* **2017**, *80*, 67–73. [CrossRef]

64. Fogg, B.J. *Tiny Habits: The Small Changes That Change Everything*; Houghton Mifflin Harcourt: Boston, MS, USA, 2019.
65. Eyal, N. *Hooked: How to Build Habit-Forming Products*; Penguin: New York, NY, USA, 2014.
66. Wéry, A.; Billieux, J. Online Sexual Activities: An Exploratory Study of Problematic and Non-Problematic Usage Patterns in a Sample of Men. *Comput. Hum. Behav.* **2016**, *56*, 257–266. [CrossRef]
67. The Official xHamster 2019 Trend Report. Available online: https://xhamster.com/blog/posts/911001 (accessed on 1 May 2020).
68. Efoui-Hess, M. *Climate Crisis: The Unsustainable Use of Online Video*; The Shift Project: Paris, France, 2019.
69. Cooper, A. Sexuality and the Internet: Surfing into the New Millennium. *Cyberpsychol. Behav.* **1998**, *1*, 187–193. [CrossRef]
70. Sniewski, L.; Farvid, P.; Carter, P. The Assessment and Treatment of Adult Heterosexual Men with Self-Perceived Problematic Pornography Use: A Review. *Addict. Behav.* **2018**, *77*, 217–224. [CrossRef] [PubMed]
71. Gola, M.; Potenza, M.N. Paroxetine Treatment of Problematic Pornography Use: A Case Series. *J. Behav. Addict.* **2016**, *5*, 529–532. [CrossRef] [PubMed]
72. Mousailidis, G.; Mehboob, R.; Papanna, B.; Bhan-Kotwal, S.; Shoka, A. Hypersexuality and New Sexual Orientation Following Aripiprazole Use. *Prog. Neurol. Psychiatry* **2020**, *24*, 14–16. [CrossRef]

International Journal of
Environmental Research and Public Health

MDPI

Review

Preventing Harmful Internet Use-Related Addiction Problems in Europe: A Literature Review and Policy Options

Olatz Lopez-Fernandez [1,*] and Daria J. Kuss [2]

1 Monash Addiction Research Centre, Turning Point, Easter Health Clinical School, Monash University, Clayton, VIC 3800, Australia
2 International Gaming Research Unit, Cyberpsychology Research Group, Psychology Department, Nottingham Trent University, Nottingham NG1 4FQ, UK; Daria.Kuss@ntu.ac.uk
* Correspondence: Olatz.Lopez-Fernandez@monash.edu or Lopez.Olatz@gmail.com; Tel.: +61-03-8413-8509

Received: 25 March 2020; Accepted: 23 May 2020; Published: 27 May 2020

Abstract: Internet use-related addiction problems are increasingly being recognized on a European scale due to international health organizations considering gaming addiction. In April 2013, the American Psychiatric Association recognized Internet Gaming Disorder in the fifth Diagnostic and Statistical Manual of Mental Disorders, and in April 2018, the World Health Organization included Gaming Disorder in the eleventh International Classification of Diseases. However, findings on these problems within this period are lacking in Europe, and a preventive approach is missing globally. A detailed critical literature review was conducted using PsycINFO and Web of Science in this five-year period. A total of 19 studies were reviewed and problems identified were: generalized Internet addiction and online gaming and gambling addictions across seven European countries (i.e., Spain, Germany, France, Italy, Greece, The Netherlands, and Denmark). The individuals with problematic use were found to be educated adolescents, usually young males with comorbid disorders, and gaming and gambling disorders were implicated in the most severe cases. Cognitive behavioral therapy was the main treatment, sometimes combined with a systemic approach for adolescents. Prevalence, high-risk populations, and factors contributing to these addiction problems are discussed, and a set of policy options are developed for this region. The implications for early detection, diagnosis, treatment, and prevention in Europe are considered.

Keywords: Internet addiction; problematic Internet use; generalized Internet addiction; online gaming addiction; online gambling addiction; Europe; policy option; prevention; public health

1. Introduction

Contemporary use of the Internet has led to a number of benefits in the health field (e.g., digital health), but also negative impacts at an individual and psychological level (e.g., gaming addiction).

Excessive Internet use has been classed in the mid-nineties as Internet Addiction (IA) [1], Problematic Internet Use (PIU) [2], or as technological (behavioral) addiction [3]. This broad term, however, has evolved and at present encompasses many types of addiction problems related to generalized Internet addiction (GIA) and a set of specific addictive uses of the Internet [4]. These include online gambling, online gaming, social networking, and cybersex, which are the most prevalent ones that have evolved alongside gaming addiction [5]. These behavioral problems can be engaged in using any device as the Internet is ubiquitous. Accordingly, during the last decade, the Internet has facilitated the development of addiction problems through online technology in many ways, and is associated with health problems (e.g., distress, functional impairment, and comorbidity [6]).

During the last decade, international health bodies, which publish diagnostic manuals for mental health diseases, recognized two associated conditions as behavioral addictions, i.e., gambling and gaming disorders. First, the American Psychiatric Association (APA) proposed Internet Gaming Disorder (IGD) in its fifth Diagnostic and Statistical Manual of Mental Disorders (DSM-5) within its third appendix in April 2013 [7]. Subsequently, the World Health Organization (WHO) included Gaming Disorder (GD) in its first version of the eleventh International Classification of Diseases (ICD-11) in April 2018 [8]. This inclusion has produced the following consequences. Firstly, gambling and gaming disorders have been recognized in the mental health sciences and by health practitioners as behavioral or process addictions, leading to many debates [9,10]. Secondly, the mass media have alerted the general public regarding these emerging online addiction problems which usually affect young populations [11,12]. Thirdly, these addictions are now understood as international public health concern, and the preventive actions undertaken have had limited success and focused on English speaking countries (i.e., in American, European, and Australasian regions [13]), and Asian countries [14].

However, to the authors' knowledge, no literature review has been conducted focusing on the period when gaming addiction was officially recognized by global health organizations, within an intercultural continental region (i.e., Europe), to detect the main concerns, and to propose a set of policy options which are culturally and geographically based. For these reasons, the European Parliament's Scientific Foresight Unit (STOA) endeavored to perform a recent literature review to study the individual and psychological aspects of the harms associated with Internet use, including IA and related harms (e.g., gaming addiction) in the European Union (EU) [15].

To the authors' knowledge, only two reviews exist with similar characteristics, but both with an international scope rather than a regional focus performed in 2016 [6,16].

Kuss and Lopez-Fernandez [6] focused on clinical research on IA and reported characteristics of treatment seekers and online addiction treatments. First, treatment seeker characteristics from various continents included European clinical studies performed in Germany, The Netherlands, and Greece, and focused on both, GIA and gaming addiction problems (among other comorbidities). Second, psychopharmacotherapy was covered, which appeared to have positive effects in decreasing IA symptomatology and Internet gaming addiction problems through antidepressants and anxiolytics, and obsessive–compulsive disorders (OCD) and attention deficit hyperactivity disorder (ADHD) medications for comorbid problems. Third, psychological therapies usually with an individual approach (e.g., cognitive behavioral therapy (CBT)) were applied to outpatients, apart from a few group therapy approaches (e.g., multi-family group therapy (MFGT)). Fourth, combined treatments were researched, which included psychological treatment in combination with pharmacotherapy or electroacupuncture therapy.

Vondráčková and Gabrhelík's review [16] focused on IA prevention. First, they stated some target groups may benefit from prevention (e.g., children and adolescents) when it is indicated (e.g., focusing on psychopathological factors). Second, the need to improve specific skills with the help of professionals and other significant individuals (i.e., counsellors, parents) was emphasized. Third, program characteristics were deemed relevant (e.g., information-provision versus interactive interventions). Fourth, environmental interventions were indicated as being needed in some regions (e.g., in countries in which IA is a public health concern where regulation should be promoted, similar to the approach taken by the Chinese government [17]).

Apart from these reviews, a world-wide meta-analysis on IA performed by Chen and Li in 2014 [18] indicated that the global estimated prevalence rate was approximately 6%, with the lowest numbers found in Northern and Western Europe (2.6%). IA prevalence was inversely associated with self-perception of quality of life regarding subjective (e.g., life satisfaction) and objective indicators (e.g., environmental conditions). Furthermore, many cross-cultural studies on IA have emerged since 2012, especially in intercultural regions, such as Europe [19,20]. These studies were school based with adolescent samples and found between 1%–4% estimated prevalence of GIA (which was higher in males). There has been a continuing increase in the number of these studies in the field [21], including mostly cross-national intercontinental studies (covering Asia, America, and Europe), which have researched GIA and estimated its prevalence with psychometric scales, obtaining higher rates in Asian countries and in young male users.

Considering the above, the objective of the present paper was to present a timely critical review of the literature on Internet use-related addiction and associated problems published in Europe between April 2013 (i.e., when IGD was included in the DSM-5's III appendix) and April 2018 (i.e., when GD was first officially recognized in the ICD-11 beta test version). The aims were to critically analyze online harms by addressing: (i) the cross-cultural approach adopted within the EU, (ii) the users' characteristics based on community and clinical populations, (iii) Internet use-related addiction problems and the interventions to target the resultant harms in Europe, and (iv) its implications at a public health level with an eye towards prevention. Furthermore, we aim to provide the first set of policy options for harm minimization at the level of the individual in Europe.

2. Materials and Methods

A literature review was conducted using the databases PsycINFO and Web of Science between January and April 2018 at Nottingham Trent University (United Kingdom). The rationale to select these two scientific databases was to contain research in Psychology and related disciplines [15]. Initially, PubMed was also selected but the results almost duplicated all outcomes collected via the first two databases, and consequently this search was discarded. PsycINFO and Web of Science are also among the most relevant in the field of Internet addiction covering the majority of current scientific sources targeted in the present paper's aims. Moreover, they offered sufficient information to perform a timely, expeditious, and recent review, and are among the databases which are usually used in literature reviews published in this field from a disciplinary perspective (i.e., PsycINFO) and also using interdisciplinary approaches (i.e., Web of Science), which allowed us to study the individual and psychological aspects of the harms associated with IA.

The review comprised scientific papers published between April 2013 and April 2018 as this is the period between the official recognition of IGD and GD. The following search was undertaken using the following terms, clusters, and Boolean operators: ("Internet" OR "online" OR "game*" OR "gaming" OR "video gam*" OR "videogame*" OR "video-game*" OR "social network*" OR "social media") AND ("Addict*" OR "compuls*" OR "problem*" OR "disorder" OR "pathology*" OR "excess*") AND ("clinic*" OR "treat*" OR "therap*" OR "harm*" OR "risk factor" OR "prevent*"). The search was performed by paper titles as this was the only option available across both search engines.

The inclusion criteria were for studies to: (i) contain empirical data (i.e., data collected using quantitative, qualitative, and mixed methods approaches), (ii) assess online addictions in the EU, (iii) be published between 2013–2018, (iv) include community and clinical samples, (v) provide a full-text article, and (vi) be published in the languages the authors manage (i.e., English, Spanish, French, German, Polish, Italian, and Portuguese).

The literature review was performed as indicated in Figure 1 [22]. Over 390 sources resulted from the initial search. Of these, a great number were filtered out based on the following criteria: (i) duplicates, (ii) meeting and conference abstracts and non-empirical studies (e.g., case studies, anecdotal studies, reviews, editorials, letters, and commentaries), (iii) studies that did not assess IA and related harms in the EU, (iv) studies that were not published between April 2013 and April 2018 (both months included), (v) did not include the population groups targeted (i.e., community and clinical samples), (vi) did not provide a full text article, (vii) were not published in a language the authors manage. Thus, after removing duplicates ($n = 34$), articles in other languages ($n = 32$), conference abstracts ($n = 188$), non-empirical studies ($n = 81$) and non-EU papers ($n = 36$), 19 relevant sources were included in the final analysis.

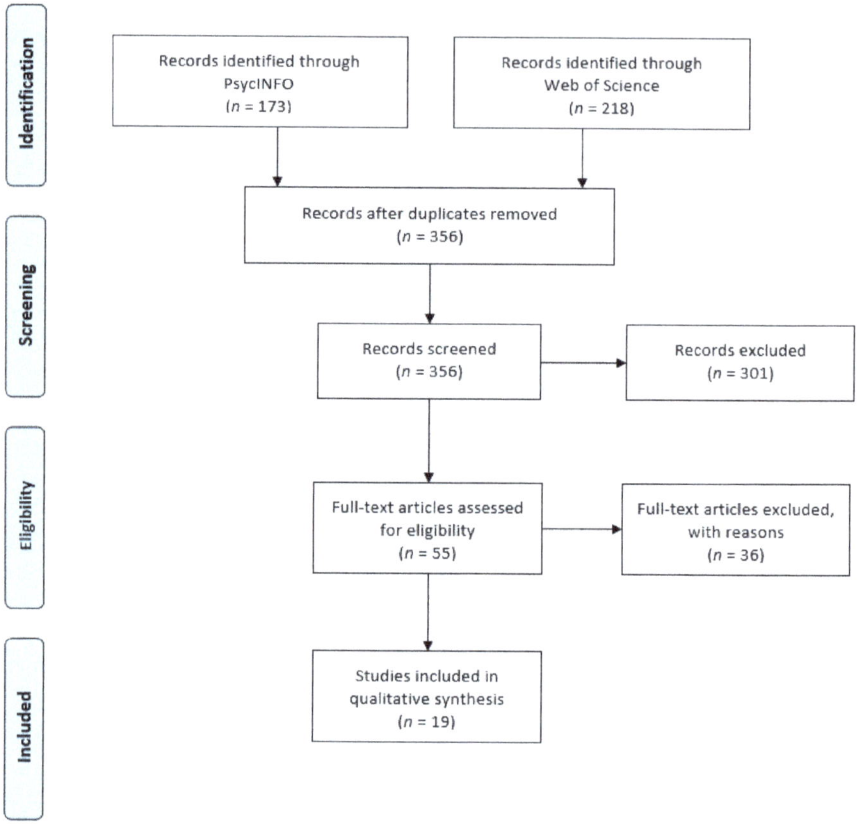

Figure 1. PRISMA Flow diagram of study selection processes.

Two rounds of searches were used: a first round (in January and February 2018) using PsycINFO and subsequently Web of Science, followed by a second round (in April 2018) to ensure all papers were consistently collected and no new paper was published within the specified period and in accordance with the inclusion and exclusion criteria. From the initial pool of 390 papers, after deleting duplicate papers, the remaining 324 results were manually scanned (i.e., title, abstract, key words, and, the paper) to identify the relevant outcomes. Thus, the literature search provided non-exclusive categories of Internet use-related addiction problems as follows: eight Internet addiction papers (i.e., seven by Internet addiction itself, and one including Internet addiction and gaming addiction), 11 online gaming addiction papers (i.e., eight with gaming addiction by itself, and two about gambling and gaming addictions together, and one including Internet addiction and gaming addiction), and three online gambling addiction papers (i.e., own with gambling addiction by itself, and two including gambling and gaming addictions together).

3. Results

Data were initially organized into four main categories which emerged in the qualitative analysis of the 19 European empirical papers undertaken by the two co-authors by categories (see Table 1).

Table 1. Papers selected for the review ($N = 19$).

Authors (Year) [Reference]	Country	Problem
Andrisano, Santoro, De Caro, Palmieri, Capunzo, Venuleo, & Boccia (2016) [23]	Italy	Internet addiction
Beranuy, Carbonell, & Griffiths (2013) [24]	Spain	Gaming addiction
Brand, Laier, & Young (2014) [25]	Germany	Internet addiction
Danet, & Miljkovitch (2016) [26]	France	Internet addiction
Floros, Siomos, Stogniannidou, Giouzepas, & Garyfallos (2014) [27]	Greece	Internet addiction
Frölich, Lehmkuhl, Orawa, BrombaWolf, & Görtz-Dorten (2016) [28]	Germany	Gaming addiction
González, & Orgaz (2014) [29]	Spain	Internet addiction
Holstein, Pedersen, Bendtsen, Madsen, Meilstrup, Nielsen, & Rasmussen (2014) [30]	Denmark	Gaming addiction
Jiménez-Murcia, Fernández-Aranda, Granero, Chóliz, La Verde, Aguglia, & del Pino-Gutiérrez (2014) [31]	Spain	Gambling & Gaming addictions
Lai, Altavilla, Mazza, Scappaticci, Tambelli, Aceto, & Tonioni (2017) [32]	Italy	Internet addiction
Luquiens, Tanguy, Lagadec, Benyamina, Aubin, & Reynaud (2016) [33]	France	Gambling addiction
Mallorquí-Bagué, Fernández-Aranda, Lozano-Madrid, Granero, Mestre-Bach, Baño, & Jiménez-Murcia (2017) [34]	Spain	Gambling & Gaming addictions
Marco, & Chóliz (2017) [35]	Spain	Gaming addiction
Martín-Fernández, Matalí, García-Sánchez, Pardo, Lleras, Castellano-Tejedor (2017) [36]	Spain	Gaming addiction
Müller, Beutel, & Wölfling (2014) [37]	Germany	Internet addiction
Taquet, & Hautekeete (2013) [38]	France	Gaming addiction
Torres-Rodríguez, Griffiths, Carbonell, Farriols-Hernando, & Torres-Jimenez (2017) [39]	Spain	Gaming addiction
Van Rooij, Schoenmaker, & van de Mheen (2017) [40]	The Netherlands	Gaming addiction
Wölfling, Beutel, Dreier, & Müller (2014) [41]	Germany	Internet addiction & Gaming addiction

Therefore, both authors independently first qualitatively analyzed the papers divided by categories (i.e., O.L.-F. performed the examination of gaming and gambling addiction articles [24,28,30,31,33–36,38–40]; and D.J.K. evaluated Internet addiction articles [23,25–27,29,32,37,41]). Therefore, both authors reviewed the contents of the identified articles according to the following categories of analysis: title and journal, authors and country, sample, design, aim(s), measures, results, implications for policy options and prevention, and conclusions for harm minimization [15]. From this categorical analysis the authors proceeded to discuss the main preliminary results, and extracted the information based on the aims with the final purpose of creating a set of preliminary policy options and preventive actions for IA and related harms in Europe. This process included several rounds until theoretical saturation of the contents from all 19 papers was achieved, according to the aims.

The present qualitative and narrative analysis resulted in the division of identified research papers into four categories: (i) the characteristics of problem users (including community and clinical samples); (ii) GIA; (iii) specific IA problems (i.e., gaming, and gambling addictions); and (iv) policy options for preventing Internet use-related harm in Europe. The first category about users' characteristics is subdivided by Internet use-relate addiction problems (i.e., GIA and specific problems), as the literature shows there are differences in Internet users based on typology of disordered behaviour. Therefore, the categories related to GIA and specific problems were analyzed in detail covering both non-clinical and clinical studies, which were researched from a policy implications perspective. Lastly, the fourth category was divided into respective policy options to reduce Internet harm from an individual person perspective.

Regarding geographical location, half of the studies included in this review ($n = 10$; 53%) were from the Southern European region (countries are ordered from higher to lower frequency): Spain ($n = 7$), Italy ($n = 2$), and Greece ($n = 1$); and 42% ($n = 8$) were from the Western European region: Germany ($n = 4$), France ($n = 3$), and The Netherlands ($n = 1$). Finally, only one study (5%) was conducted in the Northern region (i.e., Denmark).

3.1. The Characteristics of Targeted Problem Internet Users

Almost all participants included in the studies were adolescents and young adults from high schools or universities, and the assessed studies dealt with GIA and gaming addiction. Only a few studies assessed online gambling addiction, with participants usually being middle-aged male adults.

3.1.1. Generalized Internet Addiction Users

The main characteristics extracted were:

- Sample sizes: variability depending on the method applied (all research methods were used);
- Age groups: majority of adolescents, and some adults;
- Gender: balanced in adolescent community samples, and more males in clinical samples;
- Regions: In Western and Southern Europe (i.e., Germany, France, Greece, Italy, and Spain).

Sample sizes included studies which varied in number depending on the research methods applied. For instance, samples ranged from 16 Italian Internet-addicted patients investigated through an experiment with a control group to assess the biological causes of IA [32] to a survey with 1,019 German adults to test a new model for GIA [25]. Regarding life stage, participants were usually adolescents and students in high schools [23,27,29], although a few studies included adults [25]. Regarding participant gender, studies on GIA tended to cover both genders in a balanced way. However, when participants were university students, there tended to be more females than males in the sample [25,26,29], and there were significantly more males in clinical samples [27,37]). No study analyzed potential differences between female and male problem Internet users.

3.1.2. Specific Internet Addiction Problem Users: Gamers and Gamblers

The main characteristics extracted from Internet-addicted gamers and gamblers were:

- Sample sizes ranged from one case study to group surveys including mixed methods studies;
- Age groups: majority of adolescents, and a few young adults who were gamers and only adult gamblers;
- Gender: more males, especially in clinical samples and gambling studies;
- Regions: all regions studied (i.e., Spain, France, Germany, The Netherlands, and Denmark).

Sample sizes included clinical case studies of one [38] and nine adolescent patients [24], and online surveys with gaming or gambling participants [30,31] used mixed methods studies combining interviews and surveys on Internet gambling [33,34]. Almost all studies about gaming addiction used adolescent samples [28,38–40], and the clinical studies were conducted with males who usually played Massively Multiplayer Online Role-Playing Games (MMORPGs [24,38,39]) and sometimes Multiplayer Online Battle Arena games (MOBA [39]) or First-Person Shooters (FPS [39]). Interestingly, these studies usually included family members (e.g., the mother, both parents, or a sibling [24,28,38,39]), type of family [41], or type of parenting style [28] to treat existing conflicts (e.g., loneliness and discussions with parents) and measured the impact of environmental factors and interventions [28,38–40].

However, studies on Internet gambling were conducted with patients within a pathological gambling unit [31,34] and explored factors related to IGD, and some participants were invited from an online gambling site (i.e., Winnimax [33]). These studies came from Spain [31,33–36,39], France [33,38], Germany [28,41], The Netherlands [40], and Denmark [30].

3.2. Generalised Internet Addiction Problems

Eight studies (42.1%) assessed GIA, and referred to non-specific Internet use (i.e., not reliant on the engagement with a particular online activity), and few considered prevention of IA [23]. The studies reviewed were from Germany [25,37], France [26], Greece [27], Italy [23,32], and Spain [29].

Findings suggested a wide range of problems could arise from overusing (e.g., difficulty cutting down, lack of sleep, fatigue, irritability, apathy, racing thoughts, declining grades or poor job performance, and neglecting other duties). Thus, the presence of addiction symptoms (e.g., tolerance), impairment in daily functioning, high comorbidity (i.e., anxiety, depression, and OCD), and risk factors (e.g., preoccupied and fearful attachment styles) were identified. Specifically, problem users tended to present psychological characteristics and co-occurring disorders (i.e., when two or more health problems occur at the same time, e.g., an addiction problem and a mental health disorder are present simultaneously). Usually, these other mental health disorders related to personality disorders, mood disorders, or anxiety disorders. For example, those who were affected by GIA also presented with poor coping strategies and low self-esteem [25], and attachment difficulties (e.g., preoccupied and fearful types [26]). Regarding co-occurring disorders, it seems at least half of the samples presented at least two problems [25,27]). The most prevalent associated problems were depression and anxiety disorder (e.g., the latter with the social subtype [25]). However, CBT emerged as effective in leading to significant changes in symptom experience.

The main characteristics of individuals with GIA in Europe were:

- Almost all studies used Young's psychometric tests, and their derivatives, whilst finding higher prevalence in clinical studies;
- Models in IA explain risk factors, which are diverse (cognitive, attachment styles, and comorbidity);
- Peer education programs when used as school interventions have good outcomes;
- Clinical interventions rely on cognitive and emotional components, including CBT approaches.

Prevalence rates were higher in clinical studies than in community studies. For instance, Müller et al. [37] estimated a prevalence of 71% of German treatment seekers with the clinical diagnosis of IA; while Andrisano and colleagues [23] found a prevalence rate of 4% of severe Internet-addicted Italian adolescent users among the community sample they studied, which was similar to the other community samples with young Spanish adults, where the prevalence was 10% according to Gonzalez and Orgaz [29]. The scale that most frequently used to measure IA [23,25,27] was the Internet Addiction Test (IAT [42]) and its short version (s-IAT [43]). However, other valid measures have also been used [27,29] (e.g., Online Cognitions Scale (OCS [44]); Index of Problematic Online Experiences (I-POE [45]), and the Assessment of Internet and Computer game Addiction—Scale (AICA-S [41])).

Brandt et al.'s [25] model on GIA explained 64% of GIA variance based on addiction symptoms, and included associated disorders and IA symptoms experience, suggesting users' cognitions (e.g., poor coping and cognitive expectations) increase the risk of IA. However, comorbidity can also mediate the relationship between symptomatology and factors which seem to act as a cause. Similarly, Danet and Miljkovitch [26] stated fearful and preoccupied attachments can be associated with IA, and Lai et al. [32] suggested a generalized impairment in emotional and cognitive processing abilities in those who suffer from IA, which can be linked to dissociative symptoms. Comorbidities in IA seem, therefore, to be diverse and present in half of Internet-addicted patients [25,27,37,41]. The identified comorbidities include depression, social anxiety, and associated symptoms experienced, such as low self-esteem, low self-efficacy, and high stress vulnerability. According to Müller et al. [27], the majority of treatment seekers present criteria sufficient to be diagnosed with IA, and half of them have comorbidities (i.e., depression, OCD, and dissociative symptoms) and stress. In general, comorbidities include Axis I diagnoses, such as:

- Anxiety Disorders (e.g., panic, social anxiety, and post-traumatic stress disorders);
- Mood Disorders (e.g., major depression, bipolar disorder);
- Eating Disorders (e.g., anorexia nervosa, bulimia nervosa);
- Psychotic Disorders;
- Dissociative Disorders;
- Substance Use Disorders (i.e., drug addictions).

Furthermore, it seems anxiety disorders are associated with the onset of GIA, and mood disorders can be precursors of or follow IA [27].

School interventions which have shown excellent outcomes are the peer education program evaluated by Andrisano and colleagues [23] in Italy, which included brainstorming and video co-creation. In Spain, potential Internet-addicted students [29] also presented with the problem, and this was associated with environmental factors (e.g., family, friends, online interactions, etc.). Both school-based studies came from Southern Europe, suggesting there is a need of educational policies to prevent GIA and related harms in this European region.

Clinical interventions usually aimed to validate tools and cut-off points to estimate the prevalence of GIA [37,41] (e.g., AICA-S [41]).

3.3. *Specific Internet Addiction Problems: Gaming and Gambling*

Twelve papers (63.2%) reported results on online gaming and gambling addictions, nine of which focused only on gaming (44.4%). Thus, these two problems together were more prevalent in comparison with GIA in the assessed samples of European studies.

3.3.1. Internet Gaming Addiction

The main characteristics of Internet gaming addiction in Europe were:

- All studies used different scales and methods (from qualitative to experiments) and addressed prevalence;
- A few models and risk factors emerged (i.e., cognitive, emotional, environmental, and comorbidity);
- Peer education programs used as school interventions have shown contradictory outcomes;
- Clinical studies relied on cognitive, emotional, and personality components and used CBT.

Studies that screened for gaming addiction in community samples were a minority in this section, and usually measured self-perceived problematic video gaming through different devices (e.g., computers and consoles) to assess both offline and online gaming through cross-sectional surveys in Denmark and Spain [30,35]. Regarding their commonalities, males and older adolescents were at a higher risk of gaming addiction problems, non-clinical measures were useful as preventive actions, and programs to reduce gaming were usually effective, and even more so if personality traits (such as impulsivity) were addressed in the interventions.

However, clinical studies were the most common in the samples that were included in the present review, coming from Spain [24,36,39], France [38], The Netherlands [40], and Germany [28]. Patients were brought to health centers by their families, usually by their mother or a sibling [24,38], and in general parent supervision was required [28]. The main factors associated with problematic MMORPG, MOBA and FPS behaviors were dissociation (i.e., a psychological mechanism of stepping out of oneself to be protected from external harm; e.g., bullying or the loss of a loved one), entertainment (e.g., enjoyment and escapism), and virtual friendship (e.g., social relationships in game without any need to personally know one's fellow gamers; the 'clan' or the 'guild') [24]. Therefore, there was a need to assess present motivations [38]: to change (e.g., if you continue gaming like this during a decade, what will happen to you?), and to work therapeutically (e.g., playing time was double the usual adult working time per week). Simultaneously, functional analyses were performed (e.g., to support the patient to treat themselves regarding the co-occurring disorders associated with gaming), treating the gaming behaviour (i.e., psychological gaming experience), while addressing alternative pastime opportunities, and improving other relationships.

In CBT interventions, the emotional component was as relevant as the cognitive component; e.g., using techniques related to empathy, self-esteem, self-control, assertiveness, communication skills, or insight [39]. One of the main aspects in the therapeutic intervention was relapse prevention [24,38,39]. Furthermore, one study [28,38–40] showed that nonspecific psychiatric disorders pose an increased risk for gaming addiction. This supports the argument that Internet gaming addiction might be a discrete psychiatric entity usually combined with emotional and social problems [24,38]. It can be related to ADHD, Asperger's, Autism, and other disorders, such as anxiety and depression, social phobia, pervasive developmental disorders, among other comorbid conditions and problems (e.g., parent–child relationship problems, school relationship problems, obesity, cannabis use, and anhedonia). The prognosis is generally positive at three or six months of treatment [24,36,38] for those patients with an externalized profile (i.e., disruptive behaviour disorder, ADHD, and adaptive disorder) or an internalized profile (i.e., anxiety, mood and personality disorders, social relationship problems, previous mental disorders family histories, and individuals who use gaming to escape discomfort experienced in their daily lives [39]). Furthermore, clinicians have stated that increasing numbers of patients sought help through families in the recent years in European public hospitals and health centers [34,36].

Thus, gaming addiction in Europe during the last decade has required both, the development of new short non-clinical measures to screen for it in young adolescents (e.g., using the computer gaming index, console gaming index, or the Internet use index [30]), while clinical studies usually were case studies and used mixed methods dealing with interventions through tailored CBT, including for instance the 'Individualized Psychotherapeutic Program for the Addiction to the Information and Communication Technologies' (PIPATIC [39]), and new psychometric tools (e.g., Clinical Video game Addiction test second version(C-VAT 2.0 [40]), or Assessment of Internet and Computer Game Addiction (AICA-S [41])). Related to prevalence, according to Martin-Fernandez et al. [36], 69% of Spanish adolescent patients met the DSM-5 criteria for IGD, and 91% of young Dutch patients met the IGD criteria through the C-VAT 0.2 [40]. However, only 37% of gamblers also experienced video game addiction as co-occurring disorder.

Furthermore, the effectiveness of impulsivity techniques to prevent gaming addiction has been demonstrated [35]. The most studied gaming problem is related to using MMORPGs [24,38,39], which has been researched through qualitative and mixed methods approaches to create a theoretical model [24] or to test CBT interventions [38,39]. The MMORPG online gaming addiction phenomenon has been described by Beranuy et al. [24], including use motivations (e.g., entertainment, escapism or disassociation, and virtual friendship) and factors associated with it, its symptomatology, and consequences (e.g., game context, conflict, and loss of control, respectively). Taquet and Hautekeete [38] and Torres-Rodriguez et al. [39] also highlighted good knowledge of the world of video games by the therapist and a balance between emotional and cognitive components in the intervention are positive factors to ensure therapeutic alliance and successful treatment outcome.

The scales used to measure gaming addiction in this European review were diverse and validated in different languages. These instruments include the AICA-S [41], Assessment of Pathological Computer Gaming (CSV-S [46]), Problem Video Game Playing Scale (PVP [47]), and the Video game dependency test (TDV [35]). Consequently, only a few of the assessed studies measured IGD, as stated by the APA [36,39,40] (e.g., the C-VAT 2.0 [40], the Internet Gaming Disorder test with 20 items (IGD-20 [48])).

School interventions have also been studied [30,35], usually to develop non-clinical measures for problematic gaming and Internet use, screen time, and other problems. The study from Denmark [30] did not find any problems regarding GIA or specific Internet uses in their population of study. On the other hand, a similar Spanish study [35] used an intervention for adolescents to prevent video gaming addiction and to treat two intervention groups with a program to prevent addiction to technologies (i.e., "PrevTec 3.1"). In one group, impulsivity management techniques were added to intensify the positive outcomes of the preventive program, in addition to a waitlist control group. They found the preventive program significantly reduced perceived dependence on video games, and the group who received instructions on impulsivity techniques maintained the successful results in the follow-up better than those who did not receive these techniques in the program. Accordingly, personality traits, such as impulsivity, appear to play a role in prevention on a long-term basis.

Ten lessons have been extracted regarding the problem gamer profile in Europe:

1. It appears they are high-school students,
2. Usually males,
3. Who usually play MMORPGs,
4. Spend considerable time at home alone and game for many hours daily,
5. Treatment is usually sought by parents,
6. These patients present distinct addiction symptomatology,
7. With specific comorbidities (internalized versus externalized profiles),
8. Together with problems with social relationships (e.g., social phobia),
9. CBT has positive results after three and six months, which are maintained after six months,
10. Prognosis improves if family support the treatment.

These studies also highlighted that preventative programs are effective over time in reducing gaming. However, in clinical settings, time spent gaming, age and gender, type of games (e.g., MMORPGs), and type of comorbidities were associated with gaming addiction (e.g., individuals with externalizing profiles have the best prognosis after three months and both profiles have a good prognosis at six months of treatment [36]). Moreover, lack of external parental control should be considered as important risk factor. However, one study did not find any specific psychiatric disorder as a risk factor for this addiction problem [28]. Thus, gaming addiction appears to be a unique clinical entity that can be treated by CBT (e.g., with a treatment length of three to six months). Follow-up studies are required to verify its benefits across groups and cultures.

3.3.2. Internet Gambling Addiction

The main characteristics of Internet gambling addiction in Europe were:

- All studies used different measures for gambling addiction;
- Risk factors emerged, especially when related to gaming addiction;
- Severe comorbidity exists when both Internet gambling and gaming problems were present;
- Clinical studies rely on self-seeking treatment and tailored interventions.

Three studies (15.8%) addressed mainly Internet Gambling Disorder in clinical samples, suggesting it is a different clinical entity in comparison to IGD, although both share some sociodemographic characteristics (e.g., both usually affect males) and psychological features (e.g., type of emotional distress, higher harm avoidance, and reward dependence traits).

The measures used to assess gambling addiction can be considered traditional (i.e., Stinchfield's Diagnostic Questionnaire for Pathological Gambling [49], and the Problem Gambling Severity Index (PGSI [50])).

In addition, when gambling was the main addiction and the patient played videogames, the comorbidity was more severe than for Internet gaming addiction itself [31,34,36], specifically if gaming addiction was identified together with gambling addiction (in which case paranoid ideation, distress, OCD, and interpersonal sensitivity were also present [31]). However, inversely, comorbidity did not appear in gamblers who were not gaming addicts, although the reviewed research indicated that gambling addiction appeared to be the more severe behavioral addiction. In other words, both disorders appear to be independent of each other, which is supported by evidence regarding their different clinical profiles [24]. Internet gambling had a higher mean age of disordered onset, disorder severity, somatization and depression symptoms, among other personality traits (i.e., novelty seeking and persistence) and associations with substance use (e.g., tobacco use). Furthermore, patients with both problems are younger, present more dysfunctional personality traits (e.g., lower self-directedness and higher persistence), and general psychopathology (e.g., depression, anxiety, and social phobia), higher body mass index (BMI) and food addiction (FA). In summary, although both addictive online behaviors share some emotional distress and personality traits, gambling disorder appeared to be more severe in the included studies.

Moreover, online interventions seem only to be effective when the gambler seeks treatment, and a commitment with a health professional is made, even if it is short-lived; inversely, if there is no help-seeking the efficacy of any intervention is counter-productive or may have an aversive effect [33]; therefore, 'more is not always better' in terms of prevention. Lastly, CBT should also be personalized to the type of gambling activity (e.g., online poker), which again requires knowledge from the therapist, as highlighted in the case of gaming [38].

3.4. European Policy Options to Prevent Internet Addiction Problems

From the present European literature review encompassing the period between 2013 and 2018, the following four policy options have been developed based on the included 19 studies (see Figure 2).

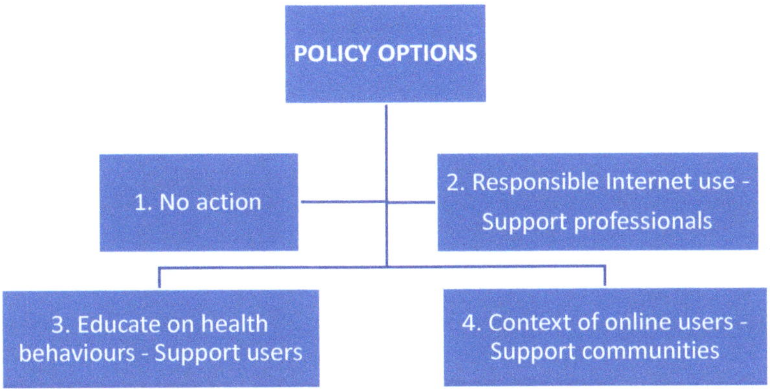

Figure 2. Policy options derived from the present literature review.

3.4.1. No Action

The first policy option concerns all stakeholders involved and should be considered with caution. In a few studies, contradictory findings have emerged which have highlighted it is not worth taking action if treatment is not sought [33] or if the problem does not emerge [30]. Thus, the first policy option is related to the following preventive actions:

- Internet addiction, Internet gaming, and gambling addictions can be hazardous (e.g., they can appear temporarily and last less than a year and then disappear). These respective problems are associated with a variety of internal and individual factors and are not always associated with the same comorbidity or associated symptom experiences. Environmental factors may also contribute to the emergence of different online addictions. Thus, Internet use-related addiction problems can appear temporarily and disappear spontaneously before they can be considered a disorder.
- This option should probably be considered when the period of the problem affecting users is less than a year, with no comorbidity or other associated severe symptoms and problems. However, if there is a suspicion of chronification (e.g., with a length of at least six months), preventive measures should be taken to avoid the development of the disorder and future disorder diagnosis.
- At present, this option seems to be unsustainable because the public health concern is growing and health practitioners are reporting increasing numbers of cases (e.g., more clinical papers than non-clinical) and interventions.

This option refers to natural recovery, which seems to exist but has not been studied much. However, due to the precautionary principle (i.e., which encourages policies that protect human health and the environment in the face of uncertain risks) it may jeopardize Europe's ability to prevent Internet use-related problems from emerging in the first place, and to take advantage of the instruments, therapeutic, and of preventive opportunities that the scientific literature on generalized and specific Internet addiction problems has recently provided.

3.4.2. Promote and Disseminate Applied Research on Responsible Internet Use and Prevention

The second policy option mainly addresses support organizations, professionals, and practitioners. Applied research is present in almost all reviewed studies, as research seeks to find solutions to the problems in European community and clinical settings. Only a couple of studies described models of GIA [25] and Internet gaming addiction [24] to provide an understanding about which factors contribute to the problems, their phenomenology, and therapy components to clinically address them. Moreover, providing information and interventions seem to be the key public health strategies to prevent and address the problem in all settings. Early preventive actions can:

- Promote evidence-based information and tools and are supported by applied research outcomes for professionals who can help to prevent and intervene in Internet use-related addiction problems. For instance, information can be available through an EU webpage with public resources (e.g., containing validated scales which exist for non-clinical measures), contact details for support organizations across the respective European countries, a list of experts, and information provided in the different EU languages.
- Include other scientific European initiatives disseminating the state of the art of research into these problems across Europe, especially if these technological problems have associated risks, such as comorbidity (e.g., anxiety and depression) and other associated psychosocial problems (e.g., cyberbullying).
- When these problems are present, self-screen tools and other test actions can be offered as a package to different professional groups and settings (e.g., clinicians in hospitals and teachers in schools), especially to those who are close to and/or work with children and adolescent populations.

There is a need to improve this research field to diversify the methodologies used and translate them to the professional sector, and to promote joint clinical and educational research. It is also crucial for standardized measures to assess problems and compare them in order to support diagnosis and treatment success (e.g., cross-culturally and trans-diagnostically). These preventive actions can provide support for decision-makers to better understand the problems from a European public health perspective, and to promote responsible Internet use and media literacy.

3.4.3. Promote Education on Offline and Online Health Behaviors in Young Populations

The third policy option addresses all Internet users, especially those who have appeared as more at risk for developing Internet use-related addiction problems. All reviewed studies have highlighted common aspects related to promoting healthy Internet use, especially in adolescents and young adults, as some gaming genres are very demanding regarding competition and social involvement (e.g., MMORPGs). The following options are preventive actions which can be introduced:

- To encourage alternative motivations, engagement in alternative entertainment behaviors (also those including Internet use), new coping skills, cognitive and emotional skills, healthy attachment styles to reduce the risk of Internet use-related problems and to reduce Internet usage, if needed (e.g., through alerts and notifications) and to provide alternative options of relaxation (e.g., reading, meeting people, and engaging in physical activities).
- To detect the risk of experiencing other comorbidities or problems and address them with professional support and the support of significant others (e.g., caregivers in the case of adolescents), and to embrace systemic approaches.
- If a problem with Internet usage is present, all Internet use-related addictions should be simultaneously assessed, as many of the reviewed studies were clinical studies on gaming addiction, which together with gambling in young adults seemed to indicate the worst-case scenario regarding these problems.

In the EU, there is a need for programs and campaigns addressing children and youth to promote awareness of the risks of online behaviors at an individual person's level. Studies have shown school interventions are usually effective, even more so if they include a psychological component (e.g., impulsivity management techniques when gaming). Young individuals should be engaged in conversations and activities concerning offline and online health, potential positive and negative implications of excessive online behaviors, and provided with information on alternative pastime activities and alternative coping strategies not involving Internet use.

The usual problem user is an adolescent who increasingly spends time gaming alone at home, usually plays MMORPGs, and experiences co-occurring problems and negative impacts in their daily life. Thus, problem cases should be treated on a case-by-case basis when detected with tailored psychological and educational interventions, including CBT, whilst ensuring treatment encompasses interventions for comorbidities. The young user should be able to determine which functions the maladaptive Internet use fulfils in his or her life, and which other options are available to him or her with professional support. These actions can be supported by other initiatives in educational and health settings, which require resources and action plans (e.g., providing funding and resources).

3.4.4. Support Communities and Significant Others of Problematic Internet Users

In addition to supporting professionals (i.e., the second policy option) and users (i.e., the third policy option), communities (i.e., the fourth policy option) also need attention. The preventive actions for this group are the following:

- Enhanced family, partner, and peer communication and caretaking (e.g., through parents, siblings, partners or friends) can prevent the problems from emerging when risk indicators appear and develop progressively (e.g., excessive time spent playing online games, lack of sleep due to constant online connection, irritability and mood changes when disconnecting, neglecting school or relationships). Thus, 'keeping an eye' on time spent and having conversations about online uses can be the first measure of prevention in families and friendship groups.
- Information should also be available for users' environments (e.g., for families, schools and communities) about the risks of habitually engaging in online role-playing games or online gambling applications which can be out of control and cause negative health consequences (i.e., functional impairment and distress) or financial problems (i.e., online gambling).

These problems usually affect families, education or workplace organizations, and communities. Thus, basic information, education, social, and clinical support can help individuals in the immediate context of problem users with community support. The EU should consider facilitating information provision to healthcare providers to support general practitioners when taking care of communities in health settings. Moreover, the implementation of actions, programs, and services for information, early detection, and facilitation of support and treatment routes for future problem users and their significant others are options to develop. For instance, at a school and community level, actions to promote prevention can be provided together with those for other related problems (e.g., substance use disorders and cyberbullying).

4. Discussion

This timely European literature review provides an overview of the currently available research on Internet use-related addiction problems in this region in the period between gaming addiction recognition by the APA and the WHO (i.e., April 2013–April 2018). It has used a public health approach and a preventative perspective to offer a set of policy options and preventive actions. The aims, therefore, were to use a cross-cultural approach across the EU to identify the problematic users' profiles for risk management in community and clinical settings; to ascertain how Internet addiction problems have been researched in Europe within the period when gaming disorder was officially recognized by health organizations; to understand the scope of their harm implications; and, at a public health level, which preventive actions can be extracted and policy options proposed.

An update of these problems at an individual level in the EU has been provided as Internet use-related addiction problems seem to have increased world-wide in the past two decades [6,18,19], with an estimated global prevalence of 6% [18]. Low rates have generally been reported in European regions in school community sample studies (e.g., via meta-analyses and cross-cultural studies) published between 2012 and 2015 (with an average of a 2.5% prevalence [18–21]). However, this review has highlighted the prevalence is growing, as GIA in similar adolescent and young community samples is now approximately 4%–10% [23,29]. Indeed, a recent cross-cultural study has indicated that the prevalence of PIU in Europe is relatively higher than previously indicated, although this observation is based on an adult community sample (i.e., where prevalence rates ranged from 14% to 55% [51]). However, caution is needed to be considered a cause for concern in the present general population due to the highlighted conceptual and methodological issues in the respective studies of these addiction problems.

On the other hand, another indicator that requires attention and has appeared in this review is the higher rates of GIA and online gaming addiction in European clinical samples [36,37,40] compared with community samples [23,29], which range between 69%–91% for both addiction problems. This is in line with Carbonell et al.'s [52] and Lopez-Fernandez's [53] bibliometric studies of IA and other specific online addiction problems published in the last two decades. It seems increasing world-wide Internet use is accompanied by an increasing number of publications on Internet addiction problems. However, it also seems clear gaming addiction has surpassed IA, probably due to IGD recognition by the APA [5] within the period studied, which was the starting point for this review and attracted the attention of clinicians and researchers who deal with these health problems and have published their findings since. The fact that the number of publications on gaming disorders in Europe, and internationally, is increasing may be due to the official recognition of gaming addiction as a disorder, which is indicated by the number of recent reviews [54–59]. Furthermore, neurological functions have commonalities and differences across these two behavioral addictions [54].

The number of studies included in this literature review, however, is scarce compared to other previous international reviews on GIA and IGD [5,6,16,21,54–58]. One explanation is that these health concerns are less prevalent in Europe relative to Asian regions, which is supported by the scientific literature. However, this does not mean that the precautionary principle cannot be applied [10], and almost no reviews have analyzed papers cross-culturally using different languages [6]. In the reviewed European samples, no study from Eastern Europe has been identified, and the regions that have seen more publications are both Southern and Western Europe for GIA, and Northern regions for gaming addiction. This is consistent with a previous cross-cultural study on dependent mobile phone use [60]. In that case, the Northern and Southern regions were the ones with heaviest online mobile use (the Northern countries especially for gaming), and the Eastern regions had lower rates. France appeared as one of the countries with the highest problematic mobile phone use, although Spain has seen a larger number of publications on Internet use-related addiction problems in the period studied.

The European scientific evidence reviewed here published between April 2013–April 2018 identified three potential problems: GIA, online gaming, and gambling disorders at community and clinical levels, which usually affected adolescents and young male adults, except for online gambling (middle-aged adults). This distribution corresponds to previous literature [5,6,16,54–57], specifically as gambling requires financial resources. However, recent empirical European and international studies on IA and gaming addiction show that females are increasingly affected, although they have not been the main study group yet [51,61]. To the best of the authors' knowledge, no review on these problems at an individual person level in the EU has been published yet, and the main findings correspond to the results presented in two previous international reviews on clinical issues related to IA [6,55].

Furthermore, comorbidity seems to be the norm [6,59], and usually includes depression, social anxiety disorder, social phobia, OCD, ADHD, hostility, substance use disorders (e.g., gambling, alcohol, marihuana, nicotine, and cocaine use), eating disorders (e.g., binge eating disorder, bulimia, and obesity), and certain personality traits and personality disorders (e.g., impulsivity, borderline, avoidant personality, or antisocial disorders) [6,55]. However, the present review showed different comorbidities depending on the type of Internet use-related addiction problems, a novel finding which reinforces the independent identities of GIA and gaming addiction [5,24,38,53–59]. In GIA, half of the investigated samples present with comorbid Axis I disorders, which is consistent with previous research [25,27,62]. This suggests a complete psychiatric evaluation is needed for these types of problems. Nevertheless, this review also highlights the need of a psychological evaluation as other emotional, cognitive, and behavioral features have emerged, such as the role of self-esteem [25,32,39], attachment or defense styles [26,60], cognitive coping and disassociation [25,32], and other personality traits and mental disorders [6,25,27,37,41,59] in specific developmental stages (i.e., adolescence [58]). In gaming addiction, however, the spectrum of comorbid disorders is more diverse and severe, especially if gambling is one of the co-occurring disorders, including internalizing and externalizing profiles which need to be considered regarding recovery length [24,35,36,38,39,59]. A European gamer profile has also emerged where environmental factors appear for the etiology, development, and recovery of these problems (e.g., CBT with a systemic approach for adolescent gamers).

At present, research has moved the field forward considerably, resulting in clinicians and researchers recognizing Internet use-related addiction problems across different devices [30,53,60,62] as more scientific research is emerging [5,6,52], and so is the demand for diagnosis and treatment [24,34,36,38,39,41]. The present literature is slightly contradictory, as it has been suggested that the device used to engage in gaming can be associated with the occurrence, course, and prognosis of IGD [62], and it has also been stated that the device does not influence gaming addiction problems [30]. The most alarming studies on gaming addiction and the role of gaming devices come from Asia, and those studies which contradict the findings related to the role of gaming devices in gaming addiction were mainly conducted in Europe.

On the other hand, other contradictions regarding gaming addiction, such as comorbidity or associated symptom experience identified, are the reasons why, together with the low number of prevention research studies in the field [13,16,57,59,63], it seems essential to start addressing possible preventive actions regarding IA and related harms at all levels (i.e., by regions and globally). Simultaneously, qualitative work is needed to address the uniqueness of the phenomenological expression of these types of behavioral addiction problems [64] facilitated through the Internet.

Regarding the policy options, the first one (i.e., no action) has also been put forward in similar preventive studies on substance use disorders. This action is based on the available knowledge on natural recovery, which is considered in the context of IGD as well [65]. However, as almost no follow-up studies have produced evidence on long-term relapse and the natural recovery rates, caution must be applied for this first policy option. The second policy option will aid informing and training professionals and practitioners and is aligned with the few international reviews on IA and prevention. For instance, according to Vondráčková and Gabrhelík [16], the improvement of skills in specific professions (i.e., for researchers, counsellors, and teachers) can support preventive action and better intervention plans or, as Kiraly et al. [57] stated, measures are taken to make health services available to gamers who experience problems. The third policy option, related to users, also highlights the need to pay attention to those who are among the highest risk groups, individuals of particular age (e.g., children and youth, especially MMORPG gamers), gender (males), engaged in a general or specific Internet activity, and who experience comorbidity [6,15,16,55,57–59]. The fourth policy option is aimed at supporting communities, including signposting families, schools, institutions, and governments [13,16,57,63,66,67].

Nevertheless, as King et al. [13] highlighted, in Western cultures including Europe, at the moment community-based support derives from non-profit organizations and the private sector; although a few countries are starting to provide support through their national health systems, such as Germany [10]. However, not all measures that have been put into place to prevent Internet and gaming addiction have obtained effective results [13,57], and there is a bias regarding what is known through the current literature, which is dominated by English language publications from English and Asian regions [13,57]. Thus, the current scientific literature base may not sufficiently reflect what other non-English speaking countries are already doing regarding prevention at all levels (e.g., Switzerland [10]).

This review also has its limitations, including the strategy applied to identify the included studies. For instance, the period of five years selected, and the number of databases used can be considered short and small, but both decisions have a rationale (i.e., recognition of gaming addiction, and disciplinary and interdisciplinary scientific search engines associated with the aims, respectively). Preliminary findings are relevant as they have shown, for example, the emergence of clinical research in online addiction problems in Europe with its specificities (e.g., gamer profile and specific comorbidity depending on internet use-related problems). However, in this emergent field, other literature reviews have also been undertaken with even shorter periods of analysis for relevant reasons and with a larger or smaller number of databases (e.g., internet use-related to self-harm and suicidal behaviour using Medline, Cochrane, and PsychINFO, and covering four years [68], or the utility of magnetic resonance imaging to study IA using Scopus, which covered a three year period [69]). The keywords applied did not take into consideration other possible Internet addiction problems which are currently being researched in Europe (e.g., cybersex). However, the identified limited number of studies produced preliminary findings to achieve the present aims to obtain an overview of the status quo in Europe regarding these problems from a cross-cultural and preventive perspectives, with a qualitative analysis using the lens of harm minimization to develop a set of policy options with preventive actions. Future research should first extend a similar procedure to non-European countries and also collect grey literature with non-English language publications to produce a holistic perspective of the policy options and prevention actions and consequences of what kinds of initiatives are already taken in several countries, which can be useful at local and global levels. Thus, this review offers a brief and timely snapshot of scientific studies in the recent period where gaming addiction has been officially recognized. Indeed, it is the first review on these problems at a European level using a preventive approach. However, methodological improvements can aid more robust future research, which should apply methods and procedures to compute other quantitative inter-rater reliability measures to complement the qualitative inter-rated reliability obtained through sharing and comparing coding agreements in iterative rounds until arriving at a consensus and theoretical saturation of findings (e.g., the Cohen's Kappa coefficient [70]), and complementary quality checks of the procedure (e.g., the Critical Appraisal Skills Programme (CASP) [71]). The included studies' findings have been synthesized and analyzed in detail in the present literature review to provide an overview regarding these emerging addiction problems in Europe, which can be used with caution as the present literature review constitutes a qualitative narrative synthesis, for international comparisons, and to translate some of the identified policy options into preventive actions.

5. Conclusions

In summary, the most prevalent Internet addiction problems appeared to be generalized Internet addiction and online gaming addiction in the EU between April 2013 and April 2018, both of which tend to present with specific comorbid disorders. More clinical studies compared to non-clinical studies were identified and analyzed which shows the emergence of and need for action, public health, and prevention. Gaming and gambling addictions were usually more severe problems compared to generalized Internet addiction. In addition, gambling appears to be more severe than gaming. However, the current scientific literature base does not report much prevention work in Europe (and internationally). A set of preventive recommendations and policy options have been formulated, which can support future harm minimization actions.

Author Contributions: Conceptualization, O.L.-F.; methodology, O.L.-F. and D.J.K.; formal analysis, O.L.-F. and D.J.K.; investigation, O.L.-F. and D.J.K.; resources, O.L.-F.; data curation, O.L.-F. and D.J.K.; writing—original draft preparation, O.L.-F.; writing—review and editing, O.L.-F. and D.J.K.; visualization, O.L.-F.; supervision, O.L.-F. and D.J.K.; project administration, O.L.-F.; funding acquisition, O.L.-F. and D.J.K. All authors have read and agreed to the published version of the manuscript.

Funding: This research was funded by the European Parliament through the Scientific Foresight Unit (STOA), Directorate for impact assessment and European added value, grant number SQM 02Y066 (EPRS/STOA/SER/17/226N; ED number 03210-01-00/5218/107321).

Acknowledgments: We would like to acknowledge the contribution of the European Parliament through the Science and Technology Options Assessment (STOA) Panel and thank them for their invitation to develop this research project: Gianluca Quaglio, Damir Plese, and Emilia Bandeira Morais. To Nottingham Trent University for facilitating the development of the project, which included a research assistant in the data collection: Bailey Foster.

Conflicts of Interest: The authors declare no conflict of interest.

References

1. Young, K.S. Psychology of computer use: XL. Addictive use of the Internet: A case that breaks the stereotype. *Psychol. Rep.* **1996**, *79*, 899–902. [CrossRef]
2. Shapira, N.A.; Goldsmith, T.D.; Keck, P.E., Jr.; Khosla, U.M.; McElroy, S.L. Psychiatric features of individuals with problematic internet use. *J. Affect. Disord.* **2000**, *57*, 267–272. [CrossRef]
3. Griffiths, M. Technological addictions. *Clin. Psych. For.* **1995**, *76*, 14–19.
4. Davis, R.A. A cognitive–behavioral model of pathological Internet use. *Comput. Hum. Behav.* **2001**, *17*, 187–195. [CrossRef]
5. Lopez-Fernandez, O. How has internet addiction research evolved since the advent of internet gaming disorder? An overview of cyberaddictions from a psychological perspective. *Curr. Addict. Rep.* **2015**, *2*, 263–271. [CrossRef]
6. Kuss, D.J.; Lopez-Fernandez, O. Internet addiction and problematic Internet use: A systematic review of clinical research. *World J. Psychiatry* **2016**, *6*, 143–176. [CrossRef]
7. American Psychiatric Association (APA). *Diagnostic and Statistical Manual of Mental Disorders 5th Edition (DSM-5)*; American Psychiatric Association Publishing: Washington, DC, USA, 2013.
8. World Health Organization (WHO). ICD-11 Beta Draft—Mortality and Morbidity. StatisticsMental, Behavioural or Neurodevelopmental Disorders. 2018. Available online: https://icd.who.int/dev11/l-m/en#/http://id.who.int/icd/entity/1448597234 (accessed on 29 February 2020).
9. Van Rooij, A.J.; Ferguson, C.J.; Carras, M.C.; Kardefelt-Winther, D.; Shi, J.; Brus, A.; Coulson, M.; Deleuze, J.; Dullur, P.; Dunkels, E.; et al. A weak scientific basis for gaming disorder: Let us err on the side of caution. *J. Behav. Addict.* **2018**, *7*, 1–9. [CrossRef]
10. Rumpf, H.J.; Achab, S.; Billieux, J.; Bowden-Jones, H.; Carragher, N.; Demetrovics, Z.; Higuchi, S.; King, D.L.; Mann, K.; Potenza, M.; et al. Including gaming disorder in the ICD-11: The need to do so from a clinical and public health perspective. *J. Behav. Addict.* **2018**, *7*, 556–561. [CrossRef]
11. NY Times. Video Game Addiction Tries to Move from Basement to Doctor's Office. Available online: https://www.nytimes.com/2018/06/17/business/video-game-addiction.html (accessed on 29 February 2020).
12. Time. 'Gaming Disorder' is Now an Official Medical Condition, According to the WHO. Available online: https://time.com/5597258/gaming-disorder-icd-11-who/ (accessed on 29 February 2020).
13. King, D.L.; Delfabbro, P.H.; Doh, Y.Y.; Wu, A.M.; Kuss, D.J.; Pallesen, S.; Mentzoni, R.; Carragher, N.; Sakuma, H. Policy and prevention approaches for disordered and hazardous gaming and Internet use: An international perspective. *Prev. Sci.* **2018**, *19*, 233–249. [CrossRef]
14. Choi, J.; Cho, H.; Lee, S.; Kim, J.; Park, E.C. Effect of the online game shutdown policy on Internet use, Internet addiction, and sleeping hours in Korean adolescents. *J. Adolesc. Health* **2018**, *62*, 548–555. [CrossRef]
15. Lopez-Fernandez, O.; Kuss, D. *Internet addiction—PartI: Internet addiction and problematic internet use*—Report Scientific Foresight Unit—Panel for the Future of Science and Technology, Science and Technology Options Assessment (STOA). In Proceedings of the Directorate for Impact Assessment and European Added Value, Directorate-General for Parliamentary Research Services, European Parliament, Brussels, Belgium, 31 January 2019; pp. 1–80. Available online: https://www.europarl.europa.eu/stoa/en/document/EPRS_STU(2019)624249 (accessed on 29 February 2020).

16. Vondráčková, P.; Gabrhelík, R. Prevention of Internet addiction: A systematic review. *J. Behav. Addict.* **2016**, *5*, 568–579. [CrossRef]
17. Kuss, D.J. Policy, prevention and regulation for Internet Gaming Disorder: A commentary on Király et al. (2017). *J. Behav. Addict.* **2018**, *7*, 553–555. [CrossRef]
18. Cheng, C.; Li, A.Y. Internet addiction prevalence and quality of (real) life: A meta-analysis of 31 nations across seven world regions. *Cyberpsychol. Behav. Soc. Netw.* **2014**, *17*, 755–760. [CrossRef]
19. Durkee, T.; Kaess, M.; Carli, V.; Parzer, P.; Wasserman, C.; Floderus, B.; Apter, A.; Balazs, J.; Barzilay, S.; Bobes, J.; et al. Prevalence of pathological internet use among adolescents in Europe: Demographic and social factors. *Addiction* **2012**, *107*, 2210–2222. [CrossRef]
20. Tsitsika, A.; Janikian, M.; Schoenmakers, T.M.; Tzavela, E.C.; Ólafsson, K.; Wojcik, S.; Macarie, G.F.; Tzavara, C.; Richardson, C. Internet addictive behavior in adolescence: A cross-sectional study in seven European countries. *Cyberpsychol. Behav. Soc. Netw.* **2014**, *17*, 528–535. [CrossRef]
21. Lopez-Fernandez, O. Cross-cultural research on Internet addiction: A systematic review. *I Arch. Addict. Res. Med.* **2015**, *1*, 1–5.
22. Moher, D.; Liberati, A.; Tetzlaff, J.; Altman, D.G.; The PRISMA Group. Preferred Reporting Items for Systematic Reviews and Meta-Analyses: The PRISMA Statement. *PLoS Med.* **2009**, *6*, e1000097. [CrossRef]
23. Andrisano, R.; Santoro, E.; De Caro, F.; Palmieri, L.; Capunzo, M.; Venuleo, C.; Boccia, G. Internet Addiction: A prevention action-research intervention. *Epidem. Biostat Public Health* **2016**, *13*, e11817-1. [CrossRef]
24. Beranuy, M.; Carbonell, X.; Griffiths, M.D. A qualitative analysis of online gaming addicts in treatment. *Int. J. Ment Health Addict.* **2013**, *11*, 149–161. [CrossRef]
25. Brand, M.; Laier, C.; Young, K.S. Internet addiction: Coping styles, expectancies, and treatment implications. *Front. Psychol.* **2014**, *5*, 1256. [CrossRef]
26. Danet, M.; Miljkovitch, R. Etre soi-même sur le net: Un facteur de risque à l'usage problématique d'Internet chez les personnes insécures. *L'Encéphale* **2016**, *42*, 506–510. [CrossRef] [PubMed]
27. Floros, G.; Siomos, K.; Stogiannidou, A.; Giouzepas, I.; Garyfallos, G. Comorbidity of psychiatric disorders with Internet addiction in a clinical sample: The effect of personality, defense style and psychopathology. *Addict. Behav.* **2014**, *39*, 1839–1845. [CrossRef] [PubMed]
28. Frölich, J.; Lehmkuhl, G.; Orawa, H.; Bromba, M.; Wolf, K.; Görtz-Dorten, A. Computer game misuse and addiction of adolescents in a clinically referred study sample. *Comput. Hum. Behav.* **2016**, *55*, 9–15. [CrossRef]
29. González, E.; Orgaz, B. Problematic online experiences among Spanish college students: Associations with Internet use characteristics and clinical symptoms. *Comput. Hum. Behav.* **2014**, *31*, 151–158. [CrossRef]
30. Holstein, B.E.; Pedersen, T.P.; Bendtsen, P.; Madsen, K.R.; Meilstrup, C.R.; Nielsen, L.; Rasmussen, M. Perceived problems with computer gaming and internet use among adolescents: Measurement tool for non-clinical survey studies. *BMC Pubic Health* **2014**, *14*, 361. [CrossRef]
31. Jiménez-Murcia, S.; Fernández-Aranda, F.; Granero, R.; Chóliz, M.; Verde, M.L.; Aguglia, E.; Signorelli, M.S.; Sá, G.M.; Aymamí, N.; Gómez-Peña, M.; et al. Video game addiction in gambling disorder: Clinical, psychopathological, and personality correlates. *BioMed Res. Int.* **2014**, *315062*, 1–11. [CrossRef]
32. Lai, C.; Altavilla, D.; Mazza, M.; Scappaticci, S.; Tambelli, R.; Aceto, P.; Luciani, M.; Corvino, S.; Martinelli, D.; Alimonti, F.; et al. Neural correlate of Internet use in patients undergoing psychological treatment for Internet addiction. *J. Ment. Health* **2017**, *26*, 276–282. [CrossRef]
33. Luquiens, A.; Tanguy, M.; Lagadec, M.; Benyamina, A.; Aubin, H.J.; Reynaud, M. The efficacy of three modalities of Internet-based psychotherapy for non-treatment-seeking online problem gamblers: A randomized controlled trial. *J. Med. Internet Res.* **2016**, *18*, e36. [CrossRef]
34. Mallorquí-Bagué, N.; Fernández-Aranda, F.; Lozano-Madrid, M.; Granero, R.; Mestre-Bach, G.; Baño, M.; Jiménez-Murcia, S. Internet Gaming Disorder and Online Gambling Disorder: Clinical and personality correlates. *J. Behav. Addict.* **2017**, *6*, 669–677. [CrossRef]
35. Marco, C.; Chóliz, M. Eficacia de las técnicas de control de la impulsividad en la prevención de la adicción a videojuegos. *Ter. Psicológica* **2017**, *35*, 57–69. [CrossRef]
36. Martín-Fernández, M.; Matalí, J.L.; García-Sánchez, S.; Pardo, M.; Lleras, M.; Castellano-Tejedor, C. Adolescents with Internet Gaming Disorder (IGD): Profiles and treatment response. Adolescentes con Trastorno por juego en Internet (IGD): Perfiles y respuesta al tratamiento. *Adicciones* **2017**, *29*, 125–133. [CrossRef]

37. Müller, K.; Beutel, M.E.; Wölfling, K. A contribution to the clinical characterization of Internet addiction in a sample of treatment seekers: Validity of assessment, severity of psychopathology and type of co-morbidity. *Compr. Psychiatry* **2014**, *55*, 770–777. [CrossRef] [PubMed]
38. Taquet, P.; Hautekeete, M. Prise en charge TCC d'une addiction aux jeux vidéo: L'expérience de jeu contribue à la thérapie. *J. Thérapie Comport. Cogn.* **2013**, *23*, 102–112. [CrossRef]
39. Torres-Rodríguez, A.; Griffiths, M.D.; Carbonell, X. The treatment of internet gaming disorder: A brief overview of the pipatic program. *Int. J. Ment. Health Addict.* **2018**, *16*, 1000–1015. [CrossRef] [PubMed]
40. van Rooij, A.J.; Schoenmakers, T.M.; van de Mheen, D. Clinical validation of the C-VAT 2.0 assessment tool for Gaming Disorder: A sensitivity analysis of the proposed DSM-5 criteria and the clinical characteristics of young patients with 'video game addiction'. *Addict. Behav.* **2017**, *64*, 269–274. [CrossRef] [PubMed]
41. Wölfling, K.; Beutel, M.E.; Dreier, M.; Müller, K.W. Treatment outcomes in patients with Internet addiction: A clinical pilot study on the effects of a cognitive-behavioral therapy program. *BioMed Res. Int.* **2014**, *2014*, 425924. [CrossRef]
42. Young, K.S. *Caught in the Net: How to Recognize the Signs of Internet Addiction–and a Winning Strategy for Recovery*; John Wiley & Sons: New York, NY, USA, 1998.
43. Pawlikowski, M.; Altstötter-Gleich, C.; Brand, M. Validation and psychometric properties of a short version of Young's Internet Addiction Test. *Comput. Hum. Behav.* **2013**, *29*, 1212–1223. [CrossRef]
44. Davis, R.A.; Flett, G.L.; Besser, A. Validation of a new scale for measuring problematic Internet use: Implications for pre-employment screening. *CyberPsychology Behav.* **2002**, *5*, 331–345. [CrossRef]
45. Mitchell, K.J.; Sabina, C.; Finkelhor, D.; Wells, M. Index of Problematic Online Experiences: Item characteristics and correlation with negative symptomatology. *CyberPsychology Behav.* **2009**, *12*, 707–711. [CrossRef]
46. Wölfling, K.; Müller, K.W.; Beutel, M. Reliability and validity of the Scale for the Assessment of Pathological Computer-Gaming (CSV-S). *Psychother. Psychosom. Med. Psychol.* **2011**, *61*, 216–224. [CrossRef]
47. Salguero, R.A.T.; Morán, R.M. Measuring problem video game playing in adolescents. *Addiction* **2002**, *97*, 1601–1606. [CrossRef] [PubMed]
48. Pontes, H.M.; Király, O.; Demetrovics, Z.; Griffiths, M.D. The conceptualisation and measurement of DSM-5 Internet Gaming Disorder: The development of the IGD-20 Test. *PLoS ONE* **2014**, *9*, e110137. [CrossRef] [PubMed]
49. Stinchfield, R. Reliability, validity and classification accuracy of a measure of DSM-IV diagnostic criteria for pathological gambling. *Am. J. Psychiatry* **2003**, *160*, 180–182. [CrossRef] [PubMed]
50. Holtgraves, T. Evaluating the problem gambling severity index. *J. Gambl. Stud.* **2009**, *25*, 105. [CrossRef] [PubMed]
51. Laconi, S.; Kaliszewska-Czeremska, K.; Gnisci, A.; Sergi, I.; Barke, A.; Jeromin, F.; Király, O. Cross-cultural study of Problematic Internet Use in nine European countries. *Comput. Hum. Behav.* **2018**, *4*, 430–440. [CrossRef]
52. Carbonell, X.; Guardiola, E.; Beranuy, M.; Bellés, A. A bibliometric analysis of the scientific literature on Internet, video games, and cell phone addiction. *J. Med. Libr. Assoc.* **2009**, *97*, 102–107. [CrossRef]
53. Lopez-Fernandez, O. Generalised Versus specific Internet use-related addiction problems: A mixed methods study on Internet, gaming, and social networking behaviours. *Int. J. Environ. Res. Pubic Health* **2018**, *15*, 2913. [CrossRef]
54. Kuss, D.J.; Griffiths, M.D. Internet and gaming addiction: A literature review of neuroimaging studies. *Brain Sci.* **2012**, *2*, 347–374. [CrossRef]
55. Ko, C.H.; Yen, J.Y.; Yen, C.F.; Chen, C.S.; Chen, C.C. The association between Internet addiction and psychiatric disorder: A review of the literature. *Eur. Psychiatry* **2012**, *27*, 1–8. [CrossRef]
56. Kuss, D.J.; Griffiths, D.M.; Karila, L.; Billieux, J. Internet addiction: A review of epidemiological research for the last decade. *Curr. Pharm. Des.* **2014**, *20*, 4026–4052. [CrossRef]
57. Kiraly, O.; Griffiths, M.D.; King, D.L.; Lee, H.K.; Lee, S.Y.; Banyai, F.; Zsila, Á.; Takacs, Z.K.; Demetrovics, Z. Policy responses to problematic video game use: A review of current measures and future possibilities. *J. Behav. Addict.* **2017**, *1*, 1–15. [CrossRef] [PubMed]
58. Paulus, F.W.; Ohmann, S.; Von Gontard, A.; Popow, C. Internet gaming disorder in children and adolescents: A systematic review. *Dev. Med. Child. Neurol.* **2018**, *60*, 645–659. [CrossRef] [PubMed]

59. González-Bueso, V.; Santamaría, J.J.; Fernández, D.; Merino, L.; Montero, E.; Ribas, J. Association between Internet Gaming Disorder or Pathological Video-Game Use and Comorbid Psychopathology: A Comprehensive Review. *Int. J. Environ. Res. Public Health* **2018**, *15*, 668. [CrossRef] [PubMed]
60. Lopez-Fernandez, O.; Kuss, D.J.; Pontes, H.M.; Griffiths, M.D.; Dawes, C.; Justice, L.V.; Männikkö, N.; Kääriäinen, M.; Rumpf, H.-J.; Bischof, A.; et al. Measurement invariance of the short version of the Problematic Mobile Phone Use Questionnaire (PMPUQ-SV) across eight languages. *Int. J. Environ. Res. Public Health* **2018**, *15*, 1213. [CrossRef] [PubMed]
61. Lopez-Fernandez, O.; Williams, A.J.; Kuss, D.J. Measuring female gaming: Gamer profile, predictors, prevalence, and characteristics from psychological and gender perspectives. *Front. Psychol.* **2019**, *10*, 898. [CrossRef]
62. Paik, S.-H.; Cho, H.; Chun, J.-W.; Jeong, J.-E.; Kim, D.-J. Gaming device usage patterns predict Internet Gaming Disorder: Comparison across different gaming device usage patterns. *Int. J. Environ. Res. Pubic Health* **2017**, *14*, 1512. [CrossRef]
63. Throuvala, M.A.; Griffiths, M.D.; Rennoldson, M.; Kuss, D.J. School-based prevention for adolescent Internet addiction: Prevention is the key. A literature review. *Curr. Neuropharmacol.* **2019**, *17*, 507–525. [CrossRef]
64. Billieux, J.; Rooij, A.J.; Heeren, A.; Schimmenti, A.; Maurage, P.; Edman, J.; Blaszczynski, A.; Khazaal, Y.; Kardefelt-Winther, D. Behavioural addiction open definition 2.0–using the Open Science Framework for collaborative and transparent theoretical development. *Addiction* **2017**, *112*, 1723–1724. [CrossRef]
65. Petry, N.M.; Rehbein, F.; Ko, C.H.; O'Brien, C.P. Internet gaming disorder in the DSM-5. *Curr. Psychiatry Rep.* **2015**, *17*, 72. [CrossRef]
66. Neverkovich, S.D.; Bubnova, I.S.; Kosarenko, N.N.; Sakhieva, R.G.; Sizova, Z.M.; Zakharova, V.L.; Sergeeva, M.G. Students' Internet addiction: Study and prevention. *Eurasia J. Math. Sci. Technol. Educ.* **2018**, *14*, 1483–1495.
67. Bağatarhan, T.; Siyez, D.M. Programs for preventing Internet addiction during adolescence: A review. *Addicta Turk. J. Addict.* **2017**, *4*, 243–265.
68. Marchant, A.; Hawton, K.; Stewart, A.; Montgomery, P.; Singaravelu, V.; Lloyd, K.; Purdy, N.; Daine, K.; John, A. A systematic review of the relationship between internet use, self-harm and suicidal behaviour in young people: The good, the bad and the unknown. *PLoS ONE* **2017**, *12*, e0181722. [CrossRef] [PubMed]
69. Sharifat, H.; Rashid, A.A.; Suppiah, S. Systematic review of the utility of functional MRI to investigate internet addiction disorder: Recent updates on resting state and task-based fMRI. *MJMHS* **2018**, *14*, 21–33.
70. Sun, S. Meta-analysis of Cohen's kappa. *Health Serv. Outcomes Res. Meth.* **2011**, *11*, 145–163. [CrossRef]
71. Critical Appraisal Skills Programme UK. Critical Appraisal Skills Programme Making Sense of Evidence. 2012. Available online: http://www.casp-uk.net (accessed on 14 May 2020).

International Journal of
Environmental Research and Public Health

MDPI

Article

Confirmatory Factor Analysis of the Malay Version of the Smartphone Addiction Scale among Medical Students in Malaysia

Siti Rubiaehtul Hassim [1], Wan Nor Arifin [1,*], Yee Cheng Kueh [1] and Nor Azwany Yaacob [2]

1 Biostatistics and Research Methodology Unit, School of Medical Sciences, Universiti Sains Malaysia, Kubang Kerian 16150, Malaysia; bisyarah_istikharah@yahoo.com (S.R.H.); yckueh@usm.my (Y.C.K.)
2 Department of Community Medicine, School of Medical Sciences, Universiti Sains Malaysia, Kubang Kerian 16150, Malaysia; azwany@usm.my
* Correspondence: wnarifin@usm.my; Tel.: +60-9767-6825; Fax: +60-9765-3370

Received: 20 April 2020; Accepted: 7 May 2020; Published: 28 May 2020

Abstract: Background: At present, the validity and reliability evidence of the Malay version of the Smartphone Addiction Scale (SAS-M) is only available by exploratory factor analysis (EFA). The aim of this study is to validate and determine the psychometric properties of the SAS-M by confirmatory factor analysis (CFA). Methods: A cross-sectional study was conducted among 323 medical students in Universiti Sains Malaysia. The students were given questionnaire forms consisting of socio-demographic information, the SAS-M and the Malay version of the Internet Addiction Test (MVIAT). The CFA was conducted using robust maximum likelihood estimator. The internal consistency reliability was determined by Raykov's rho coefficient. The concurrent validity was assessed by the Pearson's correlations between the factor scores of the SAS-M and the MVIAT. Results: The analysis showed the five-factor model of the SAS-M has an acceptable model fit after the inclusion of 12 correlated errors (SRMR = 0.067, RMSEA 0.059 (90% CI: 0.054, 0.065), CFI = 0.895, TLI = 0.882). The factor loadings ranged from 0.320 to 0.875. The internal consistency reliability was good (Raykov's rho = 0.713 to 0.858) and it showed good concurrent validity with the MVIAT. Conclusions: The CFA showed that the SAS-M is a valid and reliable self-administered questionnaire to measure the level of smartphone addiction among medical students.

Keywords: confirmatory factor analysis; Malay version; medical student; smartphone addiction; validation study

1. Introduction

Nowadays, a smartphone is an essential and popular communication tool that makes our daily life more convenient. As compared to a phone with basic features, a smartphone packs more computing capability and connectivity [1]. A smartphone performs many of the functions of a computer and comes with an operating system that allows it to run mobile applications [2,3]. The hardware capability, connectivity and inclusion of operating systems allow a smartphone to have a wide range of software, Internet and multimedia (music, video, camera and gaming) functionality, alongside main phone functions such as voice calls and text messaging [3]. All these functions come in the form of a pocket-sized device and its popularity is inevitable [4]. Widespread and pervasive smartphone use has become the social norm because of the high accessibility and mobility of the device [5].

In Malaysia, the smartphone penetration rate stands at a staggering 97.9% based on the latest report by the Department of Statistics, Malaysia in 2019 [6]. The smartphone has become the most common device to access the Internet based on the survey by the Malaysian Communications and Multimedia Commission (MCMC) in 2018, which reported 93.1% of Internet users used smartphones to go online [7]. Based on another survey by MCMC in 2017, other common uses of smartphones are text and voice messaging (98.5%), voice calls (93.8%), social media (88.1%), entertainment (83.7%), map/navigation (73.6%), email (60.0%) and video calls (53.2%) [8]. Judging from these nationwide surveys, the findings clearly show that the smartphone permeates the life of all Malaysians.

There is an increasing concern about the negative effects that come from its overuse, and these may interfere with our daily life [4]. De-Sola Gutiérrez et al. [9] define smartphone addiction as an excessive attention and uncontrolled dedication to one's smartphone. It is a behavior that manifests tolerance, withdrawal symptoms and dependence, along with social problems [10]. Studies on the effects of smartphone use on students have revealed many negative sides of smartphone addiction [11]. Smartphone addiction is associated with low academic performance [12,13], poor attention during learning sessions [12,14], poor psychological health [15,16], poor social interaction [17–19] and low life satisfaction [13].

Studies have also attributed problematic smartphone use with the social media use of smartphones [12,20].

A number of local studies have also supported the negative effects of smartphone use among students in Malaysia. In a study among undergraduate students, Ithnain, Ghazali and Jaafar [21] concluded that smartphone addiction is associated with depression and anxiety. In another study among medical students, Hadi et al. [22] found the association of smartphone addiction with poor psychological health. A recent study reported that 60.7% of undergraduate students had problematic smartphone use, and this was found to be associated with anxiety, depression, stress and a deterioration in academic performance [23]. An earlier study also supported the relationship between problematic phone use with lower academic performance [24].

Given the importance of assessing smartphone-related problems, so far nine major scales have been used in studies, which are the Smartphone Addiction Scale (SAS), Smartphone Addiction Scale Short Version (SAS-SV), Smartphone Addiction Proneness Scale (SAPS), Mobile Phone Addiction Index (MPAI), Mobile Phone Addiction Scale (MPAS), Mobile Phone Problem Use Scale (MPPUS), Smartphone Addiction Scale (SPAS), Smartphone Addiction Inventory (SPAI), and SPAI Short Form (SPAI-SF) [5]. Of these scales, most studies used either the SAS-SV or SAS [5]. Kwon et al. [10] developed the SAS which consists of 33 items and six factors. It was the first questionnaire for the measurement and diagnosis of smartphone addiction, and they showed that the SAS has good validity and reliability [10]. Alternatively, the authors of the SAS also developed the short version of the questionnaire (SAS-SV) which clusters 10 items under one factor [25]. The SAS-SV is intended for easy administration among adolescents, while the SAS is mainly for use among adults [10,25]. The SAS-SV showed comparably good validity and reliability to the longer version of the questionnaire [25]. The SAS-SV has been translated and validated in several languages; Arabic [26], Chinese [27], French [28], German [4], Italian [29], Portuguese [30], Spanish [28] and Turkish [31]. The SAS has been translated and validated in Arabic [26], Japanese [32], Romanian [33] and Thai [34]. In order to facilitate further research on smartphone addiction in Malaysia, Ching et al. [35] translated the SAS into the Malay language (SAS-M) and they showed that the SAS-M is valid and reliable for use among students. The SAS-M has been used in a number of research projects [21,23] in Malaysia to study the effects of smartphone use on students.

However, in both of the SAS-M validation studies [10,35], the researchers used exploratory factor analysis (EFA) to analyze the internal factor structure of the SAS. As compared to EFA, confirmatory factor analysis (CFA) provides stronger evidence to support the validity of the factor structure of a measure [36]. Although CFA provides a better option of analysis, both studies [10,365 were exploratory in nature, which justified the use of EFA. CFA requires specialized software, for example Mplus [37] and *lavaan* R package in R software environment [38,39], while EFA can be easily conducted in many kinds of statistical software. This could have hindered the use of CFA in previous SAS validation studies (with exception of [32,33]). Thus, this study aimed to validate and determine the psychometric properties of the SAS-M by CFA to provide stronger validity and reliability evidence to the earlier EFA study [35].

2. Materials and Methods

2.1. Study Setting and Participants

A cross-sectional study was conducted among Year 1 to Year 5 undergraduate medical students from School of Medical Sciences, Universiti Sains Malaysia (USM), Kubang Kerian, Kelantan, Malaysia. The data collection was done between October 2016 and December 2016. The sample size was determined for the CFA, which required at least 300 respondents for a scale with seven or less factors with item communality of <0.45 [40]. After taking into consideration a 20% drop-out rate, the target sample size was 375 respondents.

The students were selected by stratified random sampling based on the list obtained from the academic office, with the group size per year of study as the stratification variable. The selected students were then invited to participate in the study. The nature of the study and the confidentiality of their responses was explained. Those who consented to participate in the study signed informed consent forms and were given questionnaire forms to be completed. Each questionnaire form consists of socio-demographic information (age, gender, race and family income), purpose of using smartphones, the SAS-M and the MVIAT. The students returned the completed questionnaire forms to the researchers on the same day.

2.2. Instruments

2.2.1. The Malay Version of the Smartphone Addiction Scale (SAS-M)

The English version of the SAS questionnaire [10,35] is a self-administered, 6-point Likert type scale consisting of 33 items that belong to six factors, which are: 1. Cyberspace oriented relationship (7 items): S20–S26; 2. Daily Life Disturbance (5 items): S1–S5; 3. Tolerance (3 items): S31–S33; 4. Overuse (4 items): S27–S30; 5. Positive Anticipation (8 items): S6–S13; and 6. Withdrawal (6 items): S14–S19. Each item has a response scale from 1 (strongly disagree) to 6 (strongly agree). Higher scores indicate a higher degree of smartphone addition [10,35]. The SAS showed good internal consistency reliability (Cronbach's alpha = 0.967 (total), 0.825 to 0.913 (subscales)) and concurrent validity ($r = 0.32$ to 0.61) [10].

The Malay version of the SAS (SAS-M) was validated among medical students in Malaysia [35]. It showed good internal consistency reliability (Cronbach's alpha = 0.940 (total), 0.837 to 0.877 (subscales)), test-retest reliability (intraclass correlation (ICC) = 0.85) and concurrent validity ($r = 0.645$) with the Malay version of the Internet Addiction Test (MVIAT). Based on EFA, the number of extracted factors from SAS-M matched the original SAS [35]. As detailed in the study [35], a new factor "Primacy" replaced the "Tolerance" factor from the original SAS, and there were a number of changes to the placement of items under each factor.

In contrast to the original SAS, the SAS-M questionnaire [35] consists of 33 items that belong to six factors, which are: 1. Cyberspace oriented relationship (7 items): S19–S24, S26; 2. Daily Life Disturbance (6 items): S1–S5, S33; 3. Primacy (5 items): S10-S14; 4. Overuse (7 items): S25, S27–S32; 5. Positive Anticipation (4 items): S6–S9; and 6. Withdrawal (4 items): S15–S18.

2.2.2. The Malay Version of the Internet Addiction Test (MVIAT)

The Internet Addiction Test (IAT) is the most commonly used questionnaire to diagnose internet addiction [35]. The questionnaire is a self-administered, 5-point Likert type scale consisting of 20 items, and each item has a response scale from 1 (never) to 5 (always), reflecting the frequency of the symptoms [41]. Higher scores indicate higher degree of pathological use of the internet [41]. The Malay version of the IAT (MVIAT) was validated among a group of medical students in Malaysia; it showed good internal consistency reliability (Cronbach's alpha = 0.91) and parallel reliability with the IAT (intraclass correlation coefficient (ICC) = 0.88, $p < 0.001$) [41]. The MVIAT [41] showed five extracted factors, which are: 1. Lack of control (8 items): T1, T2, T10, T11, T12, T14, T16, T17; 2. Neglect of Duty (7 items): T3, T6, T8, T9, T18, T19, T20; 3. Problematic Use (2 items): T4, T15; 4. Social relationship disruption (2 items): T5, T13; and 5. Email Primacy (1 item): T7.

2.3. Statistical Analysis

The data analysis was performed in R software environment [39]. The CFA was performed using *lavaan* and *semTools* R packages [38,42]. Robust maximum likelihood estimator was used because the data were not normally distributed [36]. The model fit assessment was based on the following fit indices (robust) with their respective cut-off values [36]: χ^2 ($p > 0.05$), comparative fit index (CFI) and Tucker-Lewis fit index (TLI) ≥ 0.95 (good) or ≥ 0.90 (acceptable), root mean square error of approximation (RMSEA) ≤ 0.08, and standardized root mean square residual (SRMR) ≤ 0.08. Model-to-model comparison was based on Akaike information criterion (AIC) and Bayesian information criterion [36]. The model with the lowest values of AIC and BIC was chosen as the best fitting model for the CFA. Items with factor loadings of >0.3 were considered acceptable [40]. Multicollinearity between the factors was identified when factor-to-factor correlation was $r > 0.85$ [36]. The internal consistency reliability was determined by Raykov's rho coefficient [43]. Raykov's rho values of ≥ 0.7 were considered to reflect good reliability [40].

The concurrent validity was assessed by Pearson's correlation coefficient (r) for the correlations between factor scores of the SAS-M and MVIAT. The strength of correlation was interpreted according to Colton [44]: no or little correlation (0.00–0.25), fair correlation (0.26–0.50), moderate correlation (0.51–0.75), and perfect correlation (0.76–1.00).

2.4. Ethical Consideration

The approval for this research was obtained from the Human Research Ethics Committee of Universiti Sains Malaysia (JEPeM Code: USM/JEPeM/16030107). The confidentiality of the data was maintained and only the researchers had access to the data. The permission to conduct the study among the medical students in USM was obtained from the academic office. The permission to use the SAS-M and the MVIAT was obtained from the original authors.

3. Results

A total of 323 medical students agreed to participate in the study. The socio-demographic characteristics of the respondents are displayed in Table 1. The mean (SD) age of the students was 21.39 (1.71) years. The majority of them were female (64.4%), of Malay race (53.6%) and those coming from families with a monthly income of more than RM5000 (38.7%).

The CFA of the six-factor model of SAS-M (Model 1) showed that the model did not fit the data. After the inclusion of 12 correlated errors to the model, the CFA showed an acceptable model fit (Model 2). The factors were not multicollinear because all r values were less than 0.85, which indicated good discrimination between the factors. The model fit details, factor-to-factor correlation and correlated errors are presented in Table 2.

The factor loadings for all items exceeded the cut-off value of 0.3 (0.320 to 0.875). The reliability for each factor was good (Raykov's rho = 0.713 to 0.858). The factor loadings and Raykov's rho values by factor are displayed in Table 3.

The strength and direction of the correlation between the factor scores of the SAS-M and MVIAT are shown in Table 4. The result shows significant correlations between the related SAS-M and MVIAT factors in the range of 0.258 to 0.511, which indicated concurrent validity of the SAS-M.

Table 1. Socio-demographic characteristics of the medical students (N = 323).

Variable		N (%)
Age (Mean [SD])		21.39 (1.71)
Year of Study	Year 1	62 (19.2)
	Year 2	59 (18.3)
	Year 3	63 (19.5)
	Year 4	61 (18.9)
	Year 5	78 (24.1)
Gender	Male	115 (35.6)
	Female	208 (64.4)
Race	Malay	173 (53.6)
	Chinese	77 (23.8)
	India	64 (19.8)
	Others	9 (2.8)
Monthly family income (Ringgit Malaysia; RM)	<RM1000	33 (10.2)
	RM1000–1999	49 (15.2)
	RM2000–2999	39 (12.1)
	RM3000–3999	48 (14.9)
	RM4000–4999	28 (8.7)
	>RM5000	125 (38.7)

Table 2. Model fit indices and correlated errors of the CFA.

Models	χ^2 (df), p	SRMR	RMSEA (90% CI)	CFI	TLI	AIC	BIC
Model 1	1213.1 (480), < 0.001	0.075	0.074 (0.069, 0.079)	0.834	0.817	31,194.3	31,623.5
Model 2 *	931.3 (468), < 0.001	0.067	0.059 (0.054, 0.065)	0.895	0.882	30,895.2	31,369.7

* Correlated errors in Model 2: S3↔S4 (r = 0.410, $p \leq 0.001$), S14↔S15 (r = 0.481, $p \leq 0.001$), S31↔S32 (r = 0.378, $p \leq 0.001$), S33↔S31 (r = 0.336, $p \leq 0.001$), S21↔S23 (r = 0.371, $p \leq 0.001$), S10↔S13 (r = 0.292, $p \leq 0.001$), S23↔S26 (r = 0.353, $p \leq 0.001$), S21↔S26 (r = 0.259, p = 0.001), S20↔S21 (r = 0.218, p = 0.002), S1↔S2 (r = 0.274, p = 0.002), S14↔S17 (r = 0.220, p = 0.001), S22↔S30 (r = 0.282, p = 0.001). Correlation between factors in Model 2: S1↔S2 (r = 0.538, $p \leq 0.001$), S1↔S3(r = 0.762, $p \leq 0.001$), S1↔S4(r = 0.635, $p \leq 0.001$), S1↔S5 (r = 0.422, $p \leq 0.001$), S1↔S6 (r = 0.844, $p \leq 0.001$), S2↔S3 (r = 0.483, $p \leq 0.001$), S2↔S4 (r = 0.694, $p \leq 0.001$), S2↔S5 (r = 0.292, p = 0.004), S2↔S6 (r = 0.623, $p \leq 0.001$), S3↔S4 (r = 0.587, $p \leq 0.001$), S3↔S5 (r = 0.676, $p \leq 0.001$), S3↔S6 (r = 0.808, $p \leq 0.001$), S4↔S5 (r = 0.388, $p \leq 0.001$), S4↔S6 (r = 0.697, $p \leq 0.001$), S5↔S6 (r = 0.513, $p \leq 0.001$). Abbreviations: F = factor; S = SAS item; CFA = confirmatory factor analysis; SRMR = standardized root mean square residual; RMSEA = root mean square error of approximation; CFI = comparative fit index; TLI = Tucker-Lewis fit index; AIC = Akaike information criterion; BIC: Bayesian information criterion.

Table 3. The factor loadings and Raykov's rho values by factor.

Factor	Question No.	Factor Loading	Raykov's Rho
F1. Cyberspace oriented relationship	S19	0.592	0.756
	S20	0.621	
	S21	0.546	
	S22	0.791	
	S23	0.651	
	S24	0.531	
	S26	0.562	
F2. Daily life disturbance	S1	0.592	0.713
	S2	0.637	
	S3	0.543	
	S4	0.485	
	S5	0.654	
	S33	0.582	

Table 3. *Cont.*

Factor	Question No.	Factor Loading	Raykov's Rho
F3. Primacy	S10 S11 S12 S13 S14	0.667 0.783 0.792 0.809 0.735	0.858
F4. Overuse	S25 S27 S28 S29 S30 S31 S32	0.483 0.320 0.585 0.699 0.836 0.786 0.586	0.798
F5. Positive anticipation	S6 S7 S8 S9	0.752 0.875 0.677 0.668	0.831
F6. Withdrawal	S15 S16 S17 S18	0.767 0.743 0.699 0.604	0.800

Table 4. Pearson's correlation coefficients (p-values) between the Malay Version of the Smartphone Addiction Scale (SAS-M) and Malay version of the Internet Addiction Test (MVIAT) factor scores.

	SAS-M					
MVIAT	***Cyberspace oriented Relationship***	***Daily Life Disturbance***	***Primacy***	***Overuse***	***Positive Anticipation***	***Withdrawal***
Lack Of control	0.331 (<0.001)	0.437 (<0.001)	0.338 (<0.001)	0.511 (<0.001)	0.211 (<0.001)	0.466 (<0.001)
Neglect of duty	0.455 (<0.001)	0.454 (<0.001)	0.320 (<0.001)	0.422 (<0.001)	0.116 (0.220)	0.463 (<0.001)
Problematic use	0.400 (<0.001)	0.367 (<0.001)	0.291 (<0.001)	0.310 (<0.001)	0.101 (0.270)	0.393 (<0.001)
Socialrelationship	0.402 (<0.001)	0.383 (<0.001)	0.258 (<0.001)	0.214 (<0.001)	0.051 (0.360)	0.426 (<0.001)
Email primacy	0.243 (<0.001)	0.130 (0.130)	0.097 (0.270)	0.107 (0.270)	0.099 (0.270)	0.158 (0.030)

4. Discussion

EFA was used to analyze the internal structure of the original SAS [10] and the Malay version of the SAS [35]. As CFA can provide a stronger evidence in support of the validity of the factor structure [36], this study provided the psychometric properties and model fit assessment of the SAS-M by CFA.

The results of this study showed that the six-factor model of the SAS-M has an acceptable model fit after the inclusion of 12 correlated errors. These correlated errors can be justified on the basis of method effect that reflects additional item covariation, for example between similarly worded items [36]. In the context of the SAS and the SAS-M, the correlation is possible because all items contained the "smartphone" word, and most items contain the "feeling" word. One notable example in this study is between S20 and S21; both item statements start with "Feeling" (SAS) or "Berasa" (SAS-M).

To the authors' knowledge, this is the first study that reports the CFA of the SAS-M. For the CFA of the SAS, there are a very limited number of studies that used the CFA method [32,33]. The five-factor model is supported mainly by SRMR and RMSEA values that were clearly below the cut-off value of 0.08, while the support from CFI and TLI values were just close enough to the cut-off value of more than 0.90. In contrast, the CFA of the Romanian version of the SAS supported bi-factor model [13], and similar to this study the RMSEA and SRMR values were below 0.08 (0.064 and 0.054 respectively), while the CFI and TLI values were below 0.90 (0.870 and 0.848 respectively). For the Japanese version, they could not confirm the factor structure by CFA and resorted to using EFA which extracted four factors instead of six from the original SAS [32]. In this study, the approach to the CFA was to maintain the six-factor structure found in the earlier EFA [35], to keep all 33 items and to avoid overfitting the model. For these reasons, although it was possible to add more correlated errors to obtain higher CFI and TLI values, the modification to the model was stopped when these values were very close to the cut-off value of 0.90 to qualify as an acceptable model fit. In addition, similar to Vintila et al. [33], the decision to accept the model fit was also based on two other fit indices, which were RMSEA and SRMR. In both studies, the RMSEA and SRMR values were below 0.08 which indicated good model fit.

The factor loadings in each of the factors of the SAS-M showed all items were acceptable in view of the recommended cut-off value of 0.3 [40]. The factor loadings in this study ranged from 0.320 to 0.875. The result is comparable to the factor loadings obtained from the EFA of the SAS-M [35], which ranged from 0.415 to 0.897. The internal consistency reliability for each factor was also good, which ranged from 0.713 to 0.858 (by Raykov's rho). However, these values were lower than the EFA, which ranged from 0.837 to 0.877 (by Cronbach's alpha) [35]. This can be attributed to the Raykov's rho formula, which also takes into account the correlated error covariances in the calculation of the reliability [43].

In this study, most of the factor scores of the SAS-M were correlated to the MVIAT factor scores at fair-to-moderate strength. Similar to the earlier SAS-M validation study [35], there was negligible correlation between the "Positive anticipation" factor score with all MVIAT factor scores. In contrast to the earlier study which examined the correlation between SAS-M factor scores with the total score of the MVIAT [35], this study examined the correlation with each MVIAT factor score. The correlation was low between all factor scores of the SAS-M and "Email primacy" factor score of the MVIAT, and between the "Overuse" factor score of the SAS-M and "Social relationship disruption" factor score of the MVIAT. The poor correlation between the SAS-M and "Email primacy" of the MVIAT could be because email was not the main cause of smartphone addiction. Previous studies indicated that the addiction is better attributed to the social media use of smartphone [12,20]. The poor relationship between "Overuse" of the SAS-M and "Social relationship disruption" of the MVIAT in this study was unexpected because this finding contradicts previous studies which associate problematic smartphone use with poor social interaction [17–19]. This could be a population-specific finding (medical students), which warrants a further investigation in future studies.

There are two important limitations of this study that must be highlighted. First, the sample that was used for this study came from a single medical school in Malaysia, which might not represent all medical students in Malaysia. This limitation also affects the earlier study by Ching et al. [35], which also consists of a sample from a single medical school. Thus, it is recommended to cross-validate the SAS-M among medical students from other universities in Malaysia. Second, similar to the study by Ching et al. [35], all participants were medical students. The homogeneity of the sample makes it difficult to generalize the findings to students from other field of studies. In contrast, the validation of the original SAS was done among adult participants from companies and universities [10]. Thus, the validity of the SAS-M could be limited to a medical student population. To address this limitation, it is recommended to cross-validate the questionnaire in other heterogeneous samples in future studies. Despite these limitations, it was decided that in this study medical students would be selected as the sample to reflect closely the population in the EFA study by Ching et al. [35]. By restricting the sample to medical students, this will allow a direct comparison between the factor structure extracted by EFA and the one confirmed by CFA in the selected population.

5. Conclusions

This study provided the validity evidence and psychometric properties of the SAS-M by CFA. The results showed that the SAS-M is a valid and reliable self-administered questionnaire to measure the level of smartphone addiction among medical students. The generalizability of this validity evidence can be further supported by conducting confirmatory studies among medical students in other universities in Malaysia. In addition, the use of SAS-M in other populations requires further research in the form of cross-validation studies to explore and confirm its psychometric properties in these populations.

Author Contributions: Conceptualization, S.R.H. and N.A.Y.; methodology, S.R.H.; W.N.A.; Y.C.K. and N.A.Y.; software, S.R.H. and W.N.A.; validation, W.N.A.; Y.C.K. and N.A.Y.; formal analysis, S.R.H.; W.N.A.; investigation, S.R.H. and N.A.Y.; resources, W.N.A. and Y.C.K.; data curation, S.R.H. and W.N.A.; writing—original draft preparation, S.R.H; writing—review and editing, W.N.A.; supervision, W.N.A.; Y.C.K. and N.A.Y.; project administration, S.R.H. All authors have read and agreed to the published version of the manuscript.

Funding: This research received no external funding.

Acknowledgments: We are grateful to the medical students in School of Medical Sciences, Universiti Sains Malaysia for their participation in this study. We also thank the academic office of the medical school for allowing use to conduct the study.

Conflicts of Interest: The authors declare no conflict of interest.

References

1. Rashvand, H.F.; Hsiao, K.F. Smartphone intelligent applications: A brief review. *Multimed. Syst.* **2015**, *21*, 103–119. [CrossRef]
2. "Smartphone", Oxford Advanced Learner's Dictionary. Available online: https://www.oxfordlearnersdictionaries.com (accessed on 28 February 2020).
3. Wikipedia Contributors. "Smartphone", Wikipedia, the Free Encyclopedia. Available online: https://en.wikipedia.org/w/index.php?title=Smartphone (accessed on 28 February 2020).
4. Duke, E.; Montag, C. Smartphone addiction, daily interruptions and self-reported productivity. *Addict. Behav. Rep.* **2017**, *6*, 90–95. [CrossRef] [PubMed]
5. Yu, S.; Sussman, S. Does Smartphone Addiction Fall on a Continuum of Addictive Behaviors? *Int. J. Environ. Res. Public Health* **2020**, *17*, 422. [CrossRef] [PubMed]
6. Department of Statistics Malaysia. *Press Release: ICT Use and Access by Individuals and Households Survey Report, Malaysia, 2018*; Department of Statistics Malaysia: Putrajaya, Malaysia, 2019.
7. Malaysian Communications and Multimedia Commission. *Internet Users Survey 2018*; Malaysian Communications and Multimedia Commission: Cyberjaya, Selangor, Malaysia, 2018.
8. Malaysian Communications and Multimedia Commission. *Hand Phone Users Survey 2017*; Malaysian Communications and Multimedia Commission: Cyberjaya, Selangor, Malaysia, 2017.
9. De-Sola Gutiérrez, J.; Rodríguez de Fonseca, F.; Rubio, G. Cell-phone addiction: A review. *Front. Psychiatry* **2016**, *7*, 175. [CrossRef] [PubMed]
10. Kwon, M.; Lee, J.Y.; Won, W.Y.; Park, J.W.; Min, J.A.; Hahn, C.; Gu, X.; Choi, J.H.; Kim, D.J. Development and validation of a smartphone addiction scale (SAS). *PLoS ONE* **2013**, *8*, e56936. [CrossRef]
11. Singh, M.K.K.; Samah, N.A. Impact of Smartphone: A Review on Positive and Negative Effects on Students. *Asian Soc. Sci.* **2018**, *14*, 83–89. [CrossRef]
12. Rozgonjuk, D.; Saal, K.; Täht, K. Problematic smartphone use, deep and surface approaches to learning, and social media use in lectures. *Int. J. Environ. Res. Public Health* **2018**, *15*, 92. [CrossRef]
13. Samaha, M.; Hawi, N.S. Relationships among smartphone addiction, stress, academic performance, and satisfaction with life. *Comput. Hum. Behav.* **2016**, *57*, 321–325. [CrossRef]
14. Froese, A.D.; Carpenter, C.N.; Inman, D.A.; Schooley, J.R.; Barnes, R.B.; Brecht, P.W.; Chacon, J.D. Effects of classroom cell phone use on expected and actual learning. *Coll. Stud. J.* **2012**, *46*, 323–332.
15. Hwang, K.H.; Yoo, Y.S.; Cho, O.H. Smartphone overuse and upper extremity pain, anxiety, depression, and interpersonal relationships among college students. *J. Korea Contents Assoc.* **2012**, *12*, 365–375. [CrossRef]

16. Lee, Y.K.; Chang, C.T.; Lin, Y.; Cheng, Z.H. The dark side of smartphone usage: Psychological traits, compulsive behavior and technostress. *Comput. Human Behav.* **2014**, *31*, 373–383. [CrossRef]
17. Misra, S.; Cheng, L.; Genevie, J.; Yuan, M. The iPhone effect: The quality of in-person social interactions in the presence of mobile devices. *Environ. Behav.* **2014**, *48*, 275–298. [CrossRef]
18. Przybylski, A.K.; Weinstein, N. Can you connect with me now? How the presence of mobile communication technology influences face-to-face conversation quality. *J. Soc. Pers. Relatsh.* **2013**, *30*, 237–246. [CrossRef]
19. Kuss, D.J.; Griffiths, M.D. Online social networking and addiction—A review of the psychological literature. *Int. J. Environ. Res. Public Health* **2011**, *8*, 3528–3552. [CrossRef]
20. Carbonell, X.; Chamarro, A.; Oberst, U.; Rodrigo, B.; Prades, M. Problematic use of the internet and smartphones in university students: 2006–2017. *Int. J. Environ. Res. Public Health* **2018**, *15*, 475. [CrossRef]
21. Ithnain, N.; Ghazali, S.E.; Jaafar, N. Relationship between smartphone addiction with anxiety and depression among undergraduate students in Malaysia. *Int. J. Health Sci. Res.* **2018**, *8*, 163–171.
22. Hadi, A.A.; Nawawi, H.M.M.; Shamsuri, N.; Rahim, N.N.; Pasi, H. Smartphone Addiction and Its Relationship with Psychological Health among Clinical Years Medical Students of International Islamic University Malaysia (IIUM), Kuantan. *IIUM Med. J. Malays.* **2019**, *18*, 10.
23. Nasser, N.S.; Loh, J.L.; Rashid, A.A.; Sharifat, H.; Ahmad, U.; Ibrahim, B.; Mustafa, S.; Hoo, F.K.; Ching, S.M.; Suppiah, S. *A Cross-Sectional Survey on Smartphone Usage Pattern, the Level of Mobile Phone Dependence and Psychosocial Effects among Undergraduate Students in a Malaysian University*; MedRxiv Preprint; Cold Spring Harbor Laboratory Press: New York, NY, USA, 2020. [CrossRef]
24. Ng, S.F.; Hassan, N.S.I.C.; Nor, N.H.M.; Malek, N.A.A. The Relationship between Smartphone Use and Academic Performance: A Case of Students in a Malaysian Tertiary Institution. *Malays. Online J. Educ. Technol.* **2017**, *5*, 58–70.
25. Kwon, M.; Kim, D.J.; Cho, H.; Yang, S. The smartphone addiction scale: Development and validation of a short version for adolescents. *PLoS ONE* **2013**, *8*, e83558. [CrossRef]
26. Sfendla, A.; Laita, M.; Nejjar, B.; Souirti, Z.; Touhami, A.A.O.; Senhaji, M. Reliability of the Arabic smartphone addiction scale and smartphone addiction scale-short version in two different moroccan samples. *CyberPsychol. Behav. Soc. Netw.* **2018**, *21*, 325–332. [CrossRef]
27. Luk, T.T.; Wang, M.P.; Shen, C.; Wan, A.; Chau, P.H.; Oliffe, J.; Viswanath, K.; Chan, S.S.C.; Lam, T.H. Short version of the Smartphone Addiction Scale in Chinese adults: Psychometric properties, sociodemographic, and health behavioral correlates. *J. Behav. Addict.* **2018**, *7*, 1157–1165. [CrossRef] [PubMed]
28. Lopez-Fernandez, O. Short version of the Smartphone Addiction Scale adapted to Spanish and French: Towards a cross-cultural research in problematic mobile phone use. *Addict. Behav.* **2017**, *64*, 275–280. [CrossRef]
29. De Pasquale, C.; Sciacca, F.; Hichy, Z. Italian validation of smartphone addiction scale short version for adolescents and young adults (SAS-SV). *Psychology* **2017**, *8*, 1513–1518. [CrossRef]
30. Andrade, A.L.M.; Kim, D.J.; Caricati, V.V.; Martins, G.D.G.; Kirihara, I.K.; Barbugli, B.C.; Enumo, S.R.F.; De Micheli, D. Validity and reliability of the Brazilian version of the Smartphone Addiction Scale-Short Version for university students and adult population. *Estudos de Psicologia (Campinas)* **2020**, *37*, e190117. [CrossRef]
31. Akın, A.; Altundağ, Y.; Turan, M.E.; Akın, U. The validity and reliability of the turkish version of the smart phone addiction scale-short form for adolescent. *Procedia Soc. Behav. Sci.* **2014**, *152*, 74–77. [CrossRef]
32. Kennedy, O.; Healy, S. Japanese Validation of the Smartphone Addiction Scale. *AI Mach. Learn. Lang. Education* **2019**, *2019*, 75–88.
33. Vintilă, M.; Tudorel, O.I.; Goian, C.; Bărbat, C. Determining the structure of Smartphone Addiction Scale: A bifactor model analysis. *Curr. Psychol.* **2018**, 1–8. [CrossRef]
34. Areeudomwong, P.; Ruanjai, T.; Srimalee, P. Reliability of a Thai Version of Smartphone Addiction Scale in Thai Adolescents. *J. Med Technol. Phys. Ther.* **2019**, *31*, 391–400.
35. Ching, S.M.; Yee, A.; Ramachandran, V.; Lim, S.M.S.; Sulaiman, W.A.W.; Foo, Y.L.; Kee Hoo, F. Validation of a Malay version of the smartphone addiction scale among medical students in Malaysia. *PLoS ONE* **2015**, *10*, e0139337. [CrossRef]
36. Brown, T.A. *Confirmatory Factor Analysis for Applied Research*; The Guilford Press: New York, NY, USA, 2015.
37. Muthén, L.K.; Muthén, B.O. *Mplus User's Guide*, 8th ed.; Muthén & Muthén: Los Angeles, CA, USA, 2017.

38. Rosseel, Y. lavaan: An R Package for Structural Equation Modeling. *J. Stat. Softw.* **2012**, *48*, 1–36. [CrossRef]
39. R Core Team. *R: A Language and Environment for Statistical Computing*; R Foundation for Statistical Computing: Vienna, Austria, 2020.
40. Hair, J.F., Jr.; Black, W.C.; Babin, B.J.; Anderson, R.E. *Multivariate Data Analysis*, 7th ed.; Prentice Hall: Upper Saddle River, NJ, USA, 2010.
41. Chong Guan, N.; Isa, S.M.; Hashim, A.H.; Pillai, S.K.; Harbajan Singh, M.K. Validity of the Malay version of the Internet Addiction Test: A study on a group of medical students in Malaysia. *Asia Pac. J. Public Health* **2015**, *27*, NP2210–NP2219. [CrossRef] [PubMed]
42. Jorgensen, T.D.; Pornprasertmanit, S.; Schoemann, A.M.; Rosseel, Y. *semTools: Useful Tools for Structural Equation Modeling*, R Package Version 0.5-1; 2018. Available online: https://CRAN.R-project.org/package=semTools (accessed on 5 September 2019).
43. Raykov, T. Estimation of congeneric scale reliability using covariance structure analysis with nonlinear constraints. *Br. J. Math. Stat. Psychol.* **2001**, *54*, 315–323. [CrossRef] [PubMed]
44. Colton, T. *Statistics in Medicine*, 1st ed.; Little, Brown, and Company: Boston, MA, USA, 1974.

International Journal of
Environmental Research and Public Health

MDPI

Article

Self-Perception of Dependence as an Indicator of Smartphone Addiction—Establishment of a Cutoff Point in the SPAI–Spain Inventory

María Luisa Ballestar-Tarín [1], Conchín Simó-Sanz [1,2], Elena Chover-Sierra [1,2,*], Carlos Saus-Ortega [1,3], Carmen Casal-Angulo [1,4] and Antonio Martínez-Sabater [1,5]

1 Nursing Department, Facultat d'Infermeria I Podologia, Universitat de València, 46010 Valencia, Spain; M.luisa.ballestar@uv.es (M.L.B.-T.); conchin.simo@uv.es (C.S.-S.); sausor@uv.es (C.S.-O.); m.carmen.casal@uv.es (C.C.-A.); Antonio.Martinez-Sabater@uv.es (A.M.-S.)
2 Hospital General, Universitario de Valencia, 46014 Valencia, Spain
3 Nursing School La Fe, Universitario de Valencia, 46026 Valencia, Spain
4 Emergencies Service, Extra-hospital, 46010 Valencia, Spain
5 Hospital Clínico, Universitario de Valencia, 46010 Valencia, Spain
* Correspondence: elena.chover@uv.es; Tel.:+34-96-3864182

Received: 30 April 2020; Accepted: 25 May 2020; Published: 28 May 2020

Abstract: Background: In recent years, the abusive use of the smartphone has reached a situation that could be considered pathological. In this sense, different instruments to assess this problematic use or addiction to the smartphone are used. One of these instruments is the Smartphone Addition Inventory (SPAI), which has been validated in the Spanish language (SPAI-Spain). The main difficulty of these scales is to establish a cut-off point that determines such mobile addiction. On the other hand, self-perception was used in different addictions as a predictor of the problem. Aim: The objective of this study was to establish the cut-off point in the scores of the SPAI-Spain, using as a reference the self-perception of addiction values. Methods: A receiver operating characteristics (ROC) analysis was carried out, establishing as the cut-off point the one that presented a higher value of Youden J, indicative of its sensitivity and specificity. Results: 2958 participants from the university community completed the SPAI–Spain questionnaire. Differences in SPAI–Spain scores were found among age groups and gender, even though not all of them were statistically significant. When using the self-perception of smartphone addiction as the benchmark value, a score of 44 was established as the cutting point of the SPAI-Spain questionnaire, with a Youden J corresponding to 0.416. Conclusions: The implementation of a cut-off point of the SPAI-Spain questionnaire makes it an instrument that allows early identification of those individuals at risk of addiction, as well as the establishment of preventive and/or intervention measures.

Keywords: smartphones; addiction; self-perceived addiction; ROC analysis; cutoff point; SPAI–Spain

1. Introduction

In the Diagnostic and Statistical Manual of Mental Disorders (DSM–V), the term behavioral addiction is introduced, with the gaming disorder as the only category [1]. It is admitted, therefore, that the "pathological gambling" is an addictive disorder and not a disorder of impulse control, as it was previously classified [2,3]. The World Health Organization (WHO) included gambling and gaming disorder in the International Classification of Diseases (ICD) 11, suggesting that behavioral addictions share some common ground with substance use disorders [4]. Besides, the American Psychiatric Association (APA) recognizes that reward systems are activated in behavioral addictions, and similar behavioral symptoms occur in clinical conditions caused by substance use [1]. This is an open door to other behavioral addictions such as sex, shopping, food, work, physical exercise, mobile, or technologies [5]. However, the DSM–V only points out a small reference to the existence of these excessive behaviors, calling them behavioral syndromes. They are supposed not to have a solid base to be considered as mental disorders [1]. Despite this, many researchers suggest the addictive nature of these behaviors, although they do not share some features with substances' addiction [4,6,7].

This study focuses on the use of the smartphone, to which authors such as Flores et al. [7] catalog it as the consequence of the new "society of autism." The increase in connectivity in different areas of life has led to a behavioral change in people. Moreover, the excessive use of network technology can lead to physical, mental, and social problems [8]. This symptomatology is assimilated to substances addiction, giving rise to the lack of control of impulses, dependence, craving, anxiety, interferences in daily life, in the dream, and/or in personal relationships, among other symptoms [9]. Hence, currently, the problematic use of technologies is considered a social issue, with adolescents and young adults constituting the largest risk group [10–13].

It must be said that adolescence is a period when it is more likely to create dependence on the mobile phone due to the device's almost irresistible characteristics, for instance: autonomy, freedom, identity, social prestige, and entertainment [14]. Although it is true, authors such as Lu et al. [15] suggest that the psychological dependence and abuse created with some of its functionalities are not something exclusive to adolescents. Nevertheless, some other studies indicate that this dependence could affect the adult population [16–19].

Some authors such as De-Sola et al. [20] and Lin [21] have already echoed the need for more studies in other population groups and higher age ranges because there is a lack of awareness about how this device works in them.

On top of that, the difference in use based on gender is another aspect that was analyzed regarding the use of the smartphone. Most studies identify that women use the smartphone much more, and they even have higher levels of addiction [19,22]. Besides, some authors speak of different patterns of use according to gender. Women use the smartphone as an instrument to communicate with others; in contrast, men use it as an instrument for accessing the Internet or gaming [22,23].

Additionally, authors such as Billieux consider the "problematic mobile phones' use" (PMPU) as an inability to regulate the use of mobile phones, which eventually leads to negative consequences in daily life [24]. Studies carried out in Spain find percentages of people with PMPU that vary between 7.99% and 12.5% using different scales [25]. To this end, three different ways were established to define the problematic use of the mobile with its three respective behavior patterns: addictive pattern, antisocial pattern, and risk use pattern. It is necessary to know through validated instruments and semi-structured interviews three aspects of the individual to be catalogued as a user with a real problem: the user profile, the actual use of the device, and the type of problematic use [26]. To achieve the objective of establishing diagnostic criteria for behavioral addictions, it is also necessary to develop reliable and valid measures of these behaviors and regularly assess their psychometric properties, especially because the technological and social trends related to these behaviors change rapidly [8]. Diverse authors designed several instruments to evaluate this problematic use of the mobile phone [25], and even different diagnostic criteria were established for smartphone addiction [27].

Aside from these criteria, self-perception plays an important role in this work. Various authors defined this perception as the process by which people infer their attitudes from their behavior and which refers to the set of evaluations that a person has regarding their abilities [28]. In other words, it is the mental image of oneself. Each person creates it from his or her experiences and needs, and this self-perception is mediated by the circumstances that surround the individual. Escamilla, Córdoba, and Campos [29] consider it a dynamic reality that changes with experience, integrating new data and information, and develops according to social experiences.

In several studies, validated scales asking participants to express their perception regarding the risk indicated or perceived by them are used [30,31]. It is important to emphasize that the degree of problem recognition that anyone has can be related to the valence of the content of their scheme related to consumption [32]. Therefore, self-perception is widely used in different studies as an indicator of disease. In this sense, some factors for it were evaluated, such as the association of self-perception with the dependence on text messages [8], the use/addiction to pornography and personality factors [33], the reasons and barriers in the search for help in gambling problems [34], risk of committing violence [31], alcohol dependence [32,35], and predictors of change in alcohol consumption habits [36]. Even the self-perception of physical health is related as a predictive factor in the use of health services [37] or the degree of participation in endurance sports and self-reported data on self-image, physical and psychological health, and style of life in general [30]. Going further, in other studies, the variable self-perception of health status proves to be a criterion that is strongly related to the presence of chronic diseases [38].

Consequently, this study was performed to establish the cut-off point in the scores of the SPAI–Spain scale [39] related to the self-perceived addiction score indicated by the participants. Besides, the intention is to verify that this criterion variable is better than other measures of dependence, such as the mobile's dedication time.

It is worth highlighting that SPAI is a smartphone addiction screening tool for the adult population, which is already proven to be valid and reliable by research carried out in different countries (China, Italy, Turkey, and Brazil) [40–43]. Its main advantage is its ability to assess aspects as important in other addictions such as compulsive behavior, functional impairment, withdrawal, and tolerance. The authors selected this instrument for its validation in Spanish [39] after carrying out a review of different instruments used to assess mobile addiction [25].

An exploratory study was thereby carried out to establish a cutoff point that allows discrimination between addicted and nonaddicted individuals. Our starting hypothesis is that dependency self-perception is a good indicator of the level of mobile addiction. Therefore, it will be an appropriate criterion to establish this cut-off point.

2. Materials and Methods

2.1. Design

It is a cross-sectional observational study. The data collection instrument (in online format) was distributed among university students by using mailing lists from the University of Valencia's central services as a mean of dissemination after obtaining the permission of the University Rector's office. Data were collected from 5 to 28 April 2017.

The University of Valencia postgraduate studies ethics committee reviewed and approved this study. All the participants gave their written consent to their voluntary and anonymous involvement before completing the inventory.

The inclusion criteria were as follows: age over 17 years old, having a smartphone, and being a student of any discipline at the University of Valencia. The participants were selected using nonprobabilistic convenience sampling. Although the selection of the participants was made by convenience sampling, it must be said that it reflected the reality of the university population.

2.2. Data Collection Instrument

The instrument designed allowed us to collect information about the sociodemographic characteristics of the population, patterns of mobile phone use, and the subjective perception of smartphone dependence, by using a numerical scale (1–10); the SPAI–Spain inventory used for the analysis of smartphone addiction was also included.

The SPAI–Spain instrument [39] shows adequate indices of internal consistency ($\alpha = 0.94$) and consists of 22 items, four less than its original version [41]. Each item is answered by a Likert scale ranging from 1 (strongly agree) to 4 (strongly disagree), obtaining a global score between 22 and 88. The higher the score obtained, the higher the degree of addiction. The original SPAI scale [41] obtained a global Cronbach's alpha of 0.949. Each of its corresponding factors, compulsive behavior, functional impairment, abstinence, and tolerance, obtained the following Cronbach's alpha indexes = 0.856, 0.888, 0.855, and 0.712, respectively. The overall Cronbach's alpha indexes of the SPAI–Spain version [39] were 0.95 for the global scale and 0.87, 0.88, 0.81, and 0.72 for each of its factors.

2.3. Data Analysis

Receiver operating characteristics (ROC) analysis was conducted to examine the diagnostic efficacy of the SPAI–Spain for smartphone addiction by using the area under the ROC curve (AUC). Through this analysis, the probability of correctly classifying a subject as an "achiever" of a specific characteristic (sensitivity of the test) was compared with the probability of classifying as "achiever" somebody who was not (1-specificity). This analysis leads to obtaining an optimal cutoff point, from which it was possible to correctly classify the most significant number of subjects [44].

Thus, to obtain a cut-off score of the SPAI–Spain, two variables that measure dependence on the smartphone, self-perception of the dependency and hours of dedication, were used as criterion variables. These two variables were chosen because both obtained high correlations with the final score of the SPAI–Spain, Spearman's r 0.595, and 0.447, respectively.

Self-perception of dependency was evaluated using a Visual Analogical Scale (VAS). The subject was asked to rate their level of dependence from 0 to 10, with 0 being "I do not consider myself dependent on the smartphone at all" and 10 "I consider myself totally dependent on the smartphone." We decided to use a VAS scale to evaluate dependence, in a similar way as this type of scales are used for assess pain levels, thereby being able to evaluate dependence objectively.

The phone usage time was evaluated by asking the subject the number of hours he or she spent using the smartphone daily (less than two hours, between two and four hours, or more than four hours).

Both criterion variables were dichotomized to conduct ROC analysis, considering as "dependent" those subjects who indicated scores between 8 and 10 in the smartphone's dependence self-perception or those who used their phone 4 h or more a day. "Not dependent" subjects were those with scores lower than 8 in the variable smartphone's dependence self-perception or those who used it less than 4 h a day.

The sensitivity, specificity, and Youden's J were calculated for each SPAI–Spain score. The cutoff point for the SPAI–Spain was optimal for smartphone dependence diagnosis when the score was accompanied by the higher Youden's J.

Youden's J is the maximum vertical distance from the ROC curve to the line between (0, 0) and (1, 1). When sensitivity and specificity are equally weighted, the optimal cutoff point is the point with the highest value of J, calculated according to the formula (sensitivity + specificity − 1) [45].

Subsequently, the percentage of false negatives and true positives was assessed using a cross-tabulated table to determine the best criterion variable. Diagnostic accuracy achieved at each score was also calculated.

Analyses were carried out using SPSS Version 24.0. Armonk, NY: IBM Corp. and Microsoft Excel 2010 spreadsheets for Windows.

3. Results

3.1. Characteristics of the Studied Population

The SPAI–Spain questionnaire was completed by 2,958 participants whose sociodemographic characteristics, as well as the profile of use of the smartphone, are shown in Table 1.

Table 1. Characteristics of the studied population.

	Mean	SD	*n*	%
Age	27.96	12.13		
18–25			1944	65.7
26–35			408	13.8
36–45			253	8.6
Over 46			353	11.9
Gender				
Men			1025	34.7
Women			1933	65.3
Age When Started Using A Smartphone	17.11	9.27		
Smartphone's Dedication Hours				
<4 (Nondependent)			2062	69.7
≥4 (Dependent)			896	30.3
Smartphone's Dependence Perception				
<8 (Nondependent)			2005	67.8
≥8 (Dependent)			953	32.2

SD: Standard Deviation; *n*: number of participants with this characteristic; %: percentage of participants with this characteristic.

3.2. ROC Analysis Results

The area under the curves is 0.78 [0.763–0.798] by using as a criterion variable the perception of dependence, and 0.699 [0.679–0.719] with the variable dedication time (in hours), as shown in Figures 1 and 2, respectively.

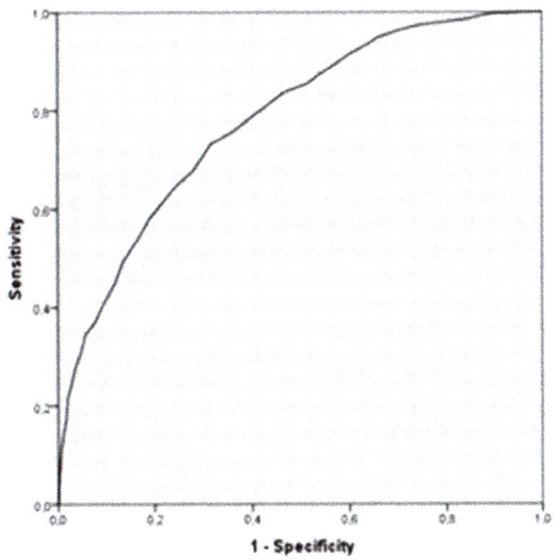

Figure 1. Receiver operating characteristics (ROC) curve. Criterion variable: self-dependence perception.

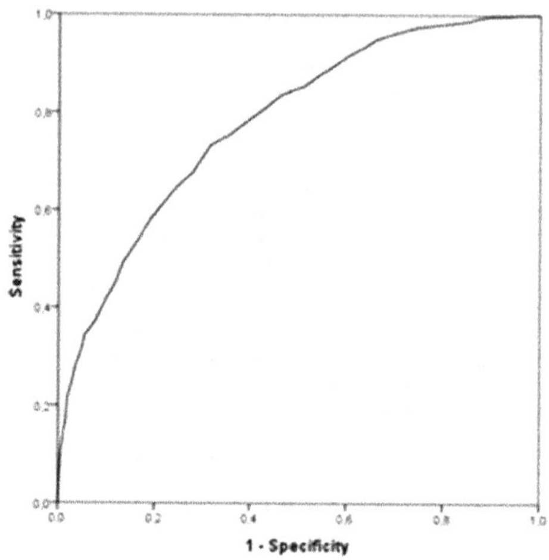

Figure 2. ROC curve. Criterion variable: hours of dedication.

3.3. *SPAI–Spain Cutoff Score Determination*

In order to identify the most suitable cutoff point, the sensitivity and specificity corresponding to each of the scores of the questionnaire for the variable dependence perception and dedication hours, as well as the Youden index, were calculated. It was established that the cutoff point would be the SPAI–Spain score corresponding to the highest value of the Youden index. In the case of dependence self-perception, it was established at score 44 (corresponding to a Youden's J of 0.416). In contrast, in the case of the variable hours of dedication, the cut-off point was established at the value 42 (which corresponds to a Youden's J value of 0.278).

In the supplementary files, we included detailed tables with sensitivity, specificity, positive and negative predictive values, diagnostic accuracy, and Youden's J values for each SPAI–Spain score, according to both external criteria (dependence self-perception and dedication hours).

From these cut-off points identified according to the two established criteria, the percentage of subjects who would be adequately classified with each of said cutoff points was calculated. These results are shown in Table 2.

Table 2. Participant's classification according to their addiction level for the cutoff point for each criterion variable.

		SPAI–Spain: Cut-off point = 44			
		Nondependent		Dependent	
		n	%	*n*	%
Dependence Self-Perception	Nondependent	1371	68.4	632	31.6
	Dependent	255	26.8	698	73.2
		SPAI–Spain: Cut-off point = 42			
		Nondependent		Dependent	
		n	%	*n*	%
Hours of Dedication	Nondependent	1154	56.0	906	44.0
	Dependent	254	28.2	643	71.8

SPAI-Spain: SmartPhone Addiction Inventory Spanish Version; *n*: number of participants with this characteristic.

Although the cutoff point is stricter using the hours of dedication as a criterion variable, the percentage of true positives (dependent–addicts) is higher in the case of the variable self-perception of dependence (73.2% versus 71.8%). Likewise, the percentage of false negatives is higher by using as the variable self-perception of dependence rather than hours of dedication (68.4% versus 56.0%). Subsequently, the score 44 in SPAI–Spain is considered as its cutoff point. It is the one obtained when considering the subjective self-perception of dependence as a criterion since it allows us to classify better both the nondependent subjects and the dependent subjects.

Once the score of 44 (as the cutoff point of the questionnaire) was determined, the percentages of subjects that would be considered as dependent were analyzed, based on characteristics such as the participants' gender and age. These results are shown in Tables 3 and 4.

Table 3. Participant's smartphone dependence according to the variable gender.

		Nondependent		Dependent	
		N	%	N	%
Gender	Female	1041	53.9	891	46.1
	Male	585	57.1	439	42.9

Table 4. Participant's smartphone dependence according to the variable age group.

		Nondependent		Dependent	
		N	%	N	%
Age Group	18–25	956	49.2	988	50.8
	26–35	240	59.0	167	41.0
	36–45	157	62.1	96	37.9
	Over 46	273	77.6	79	22.4

This analysis of the differences by using the chi-square test allowed us to conclude that there were no statistically significant differences between men and women (chi-square = 2.85 degrees of freedom (df) = 1; $p = 0.09$). However, some significant differences were perceived between age groups (chi-square = 106.68; df = 2; $p < 0.001$).

4. Discussion

In this research, we set out to establish the cutoff point of the SPAI scale validated in Spanish (SPAI–Spain), establishing self-perceived addiction as an external criterion.

The perceived addiction to the smartphone refers to the propensity of a person to report feelings of deregulation and a compulsive device use. In this situation, the focus is on the subjective assessment instead of the objectively measured behaviors, and the focus is on perception, which can be considered as a relevant clinical construct and predicts levels of psychological distress [33].

For this reason, it is understood that the self-perception of dependence on the smartphone, evaluated with the SPAI–Spain scale, can also be a predictor of this addiction. Thus, we divided subjects into "dependent or not dependent" according to their subjective perception, which we have considered an optimal parameter to establish a cutoff point in the SPAI–Spain instrument that could allow researchers to identify those subjects with smartphone addiction. This self-perception of mobile dependence was also measured in other research, such as is this European study performed in 2017. In the study, it emerged that young people in southern Europe (including the Spanish) showed the highest phone usage time, being a predictor of their dependency levels, measured in this case with the Problematic Mobile Phone Use Questionnaire (PMPUQ) [46].

In the validation study of the SPAI–Spain version from which this work for the determination of a cut-off point arises [46], we set out to analyze a series of variables that would serve as comparison criteria with the results of the questionnaire. Among these collected data, we find a subjective criterion, such as the self-perception of mobile addiction, and an objective criterion, such as the number of hours an individual uses the mobile phone. For this reason, we tried to use both measures to establish the cutoff point, finding later that the subjective perception of dependence is a criterion that allows us to predict addiction better than the time the smartphone is used.

The methodology used to establish this cutoff point was based on the determination of the AUC in the ROC curve. Subsequently, the sensitivity (capability to detect addicted individuals), the specificity (capability to detect nonaddicted individuals), and the Youden index for each of the cut points were analyzed, choosing as cutoff that with a higher Youden index [44,47]. This methodology was used in other works in which the objective was to establish this cutoff point from which to establish a classification of the population based on the presence or absence of a specific characteristic, evaluated through any instrument [47–49].

When the self-perception of dependence on the smartphone was used as an external criterion, better diagnostic accuracy values were obtained than when using the criterion of daily hours of smartphone use. Therefore, it also points out that self-perception of dependence was a better indicator than usage hours. Thus, considering that the cutoff point based on the self-perception criterion was the one which allowed us to classify the participants in the study more adequately, it was decided to establish the value 44, which was the one with the highest Youden index (and therefore greater sensitivity).

Using this cut-off point, 23.67% of the participants in the study were classified as dependents to the smartphone. Furthermore, when considering their perception as a descriptor of this dependency, 32.24% of these participants considered themselves as a dependent. These results show how self-perception would be a good predictor of the level of dependence measured by the SPAI-Spain.

The establishment of this cut-off point allowed the identification of those subjects addicted or not addicted to the smartphone. This analysis was also established based on descriptive characteristics of the population, such as gender and age.

As for the analysis of the differences according to gender, this work found a higher percentage of women (53.88%) presented this addiction (although the differences were not statistically significant). In other studies that analyzed these differences, this trend of higher levels of smartphone addiction (or PMPU) in females continues.

In a study carried out in Madrid (Spain) among 1,328 young people, it was found that the estimated prevalence of cell phone dependency was 20% (26.1% in females, 13% in males) [50]. In addition to this, the study by Machado Kouri et al., developed in Brazil, identified female gender as a predictor of mobile addiction [42].

In a study carried out in Belgium and Finland by Lopez-Fernandez et al. [51], women also obtained higher scores in terms of smartphone dependence. The women who participated in the work carried out by Roberts et al. [52] reported using the mobile for longer hours and obtained higher scores on the instrument used to assess cell phone addiction. Also, Choliz et al. showed in two different studies [14,16] that females usually had a higher degree of dependency on mobile phones than did males.

In the case of the differences according to the age of the participants, this study showed statistically significant differences, by finding a higher percentage of mainly younger people who can be classified as addicts.

Most of the research about the excessive use of mobile phones focuses on young people. However, in studies that extend the age of the study population, this greater use or even dependence on mobile phones is also observed in younger people [18]. Thus, for example, the work of Machado Khouri et al. [53], in which subjects between 18 and 35 years of age participated, shows an inverse relationship between age and score in SPAI-BR, and that age between 18 and 25 is a predictive factor of addiction to the mobile. Smetaniuk, in a study with subjects between 18 and 75 years old [54], found significant differences in the score of the instrument used to assess mobile addiction between the different age groups, and the highest scores, indicators of higher addiction levels, were obtained by the youngest subjects, as observed in this work.

Several studies that allude to addiction to the Internet and mobile devices focused on the younger population (even at school age) and tried to identify factors that intervene in the development of this addiction. Some research showed that the younger a subject begins to use these devices, the more likely he or she is to develop addictive behaviors. This fact would explain the lower scores on the SPAI–Spain scale that were obtained in this work by older people who started using their devices at an older age.

It was not possible to compare the cut-off point established for the SPAI-Spain with the versions of the instrument in other languages since only two studies were found that proposed to establish this cutoff point. One of them is the study to develop the short version of the SPAI, with only ten items, instead of the 26 that compose the original instrument. In this study, some diagnostic criteria for smartphone addiction were defined. The cut-off point of the SPAI-SF as well as its capability to classify subjects as dependent and non-dependent, were established through ROC analysis based on the prior defined criteria [49]. There is another study that establishes a cutoff point that uses the Brazilian version of the instrument. In that study, which pinpoints the prevalence of smartphone addiction as an external benchmark for establishing the cut-off point, the researchers consider the 26 elements as dichotomous (yes/no) and establish the cutoff point in seven positive responses [42].

In the development of the original instrument the establishment of a cut-off point from which to define a subject as a dependent was not considered. Moreover, the two studies in which it has been established, the versions of SPAI (or the criteria to establish the cut-off point) are not similar, and also they have been adapted to different populations. All in all makes it difficult to compare the cut-off point established in this work with the other two cut-off points established for different versions of SPAI.

In other versions of the SPAI, such as the Turkish or the Italian one, the establishment of a cut-off point was not considered [40,43]. The study to validate the Iranian version of SPAI alludes to the need to establish a cut point in the instrument, in order to make it a diagnostic instrument, even this cut-off point was not established [54].

The establishment of a cutoff point in the SPAI–Spain instrument, to determine from what score a person could be considered as addicted, allows it to become a diagnostic instrument. An instrument useful to identify subjects with levels of dependence on smartphones and the establishment of measures to prevent/manage addresses this problem. It implies that in the daily practice of health professionals they can use an easy, simple, and manageable tool in the detection of mobile addiction. At the same time, it allows the screening of the general population in the existence of a disorder such as addiction to mobile phones, and thus it can establish therapeutic interventions. On the other hand, it is essential to assess the feeling of self-perception of people regarding their addiction, which will allow their involvement in these therapeutic interventions.

As other researchers claim, problematic mobile use is an evolving public health concern that requires more study to determine the boundary between helpful and harmful technology use [12]. The determination of a risk score can allow future establishment of mobile dependence prevalence studies, as well as the establishment of early detection programs and adequate treatment of this current public health problem.

The information provided by the SPAI–Spain instrument can be highly relevant as a support for researchers and clinicians in the diagnosis and treatment of problematic/addictive use of the smartphone in the adult population. It could be included among the instruments used for evaluating addictive disorders, as well as other tools such as the Münchner Alkoholismus Test (MALT) for the detection of problematic alcohol consumption or the Fagerström tobacco dependence test. As it is a self-administered tool, brief and simple, it can be used in any field of research as a screening method.

Finally, we must not forget that the present work has a series of limitations that we would like to consider.

Regarding the methodology used to establish the cutoff point, although it is the most used for this purpose, we must not forget that the result will depend on the external criterion chosen as the definition of "dependency" (it has been self-perception in this case). Besides, the selection of another external criterion could lead to a different cut point. It must be considered the diversity of criteria, methodological approaches, and lack of conceptual delimitation (abuse, misuse, dependency, and addiction) that guided the various studies related to the PMPU (problematic mobile phone use).

The delimitation of the sample to a university setting could have influenced the high level of studies of the sample and the high percentage of women. It would be interesting to have randomized samples from other populations. The same number of male and female participants in each of the age groups was not recruited. Even so, the participants in our work show a faithful reflection of the population of students at the University of Valencia, with a higher percentage of women, which decreases as the participants' age increases. Besides, we found a percentage of subjects over 45 years old and even over 60 years old, because in this university there is a study program specifically aimed at people over 65 years old.

The means used for the diffusion of the questionnaire and data collection (Internet and email) imply a self-selection bias of the individuals that must also be taken into account. This bias entails that people who use technology regularly and those who use their smartphone more frequently are more likely to have responded.

5. Conclusions

The statistical analysis performed allows the establishment of a cut-off value for the SPAI–SP scale at 44, based on the self-perception of addiction.

The cut-off point established for the SPAI-Spain presents adequate sensitivity and specificity values, as well as adequate diagnostic accuracy.

The establishment of its cut-off point allows the use of the scale as a diagnostic tool to improve the detection and early treatment of addicted persons.

Supplementary Materials: The following are available online at http://www.mdpi.com/1660-4601/17/11/3838/s1, Table S1: Sensitivity, specificity, positive and negative predictive values, diagnostic accuracy and Youden Index of cut-off points in SPAI-SP considering self-perception of dependence, Table S2: Sensitivity, specificity, positive and negative predictive values, diagnostic accuracy, and Youden Index of cut-off points in SPAI-SP considering daily hours of smartphone use.

Author Contributions: Conceptualization, M.L.B.-T., A.M.-S., and C.S.-S.; methodology, M.L.B.-T., C.S.-O., and E.C.-S.; formal analysis, M.L.B.-T., and C.S.-O.; data curation, C.C.-A., A.M.-S., M.L.B.-T., C.S.-O., and C.S.-S.; writing—original draft preparation, M.L.B.-T., A.M.-S., and E.C.-S.; writing—review and editing, E.C.-S. and C.C.-A.; supervision, A.M.-S and M.L.B-T. All authors have read and agreed to the published version of the manuscript.

Funding: This research received no external funding.

Acknowledgments: We are grateful for permission from the SPAI's authors for its translation and use as well as for the voluntary participation of the University of Valencia members as respondents in this study.

Conflicts of Interest: The authors declare no conflict of interest.

References

1. American Psychiatric Association. *Diagnostic and Statistical Manual of Mental Disorders*, 5th ed.; American Psychiatric Association: Arlington, VA, USA, 2013.
2. Chóliz, M. Historia De Una Adicción: Del Vicio Al Trastorno Del Juego. Cuadernos de medicina psicosomática y psiquiatría de enlace. *Rev. Iberoam. Psicosom.* **2014**, *111*, 84–88.
3. Muñoz-Rivas, M.; Shih, P.C. Abuso de dispositivos móviles. In *Abuso De Internet: Antesala Para La Adicción Al Juego De Azar Online*; Echeburúa, E., Ed.; Pirámide: Madrid, Spain, 2016.
4. Harris, B.; Regan, T.; Schueler, J.; Fields, S.A. Problematic Mobile Phone and Smartphone use Scales: A Systematic Review. *Front. Psychol.* **2020**, *11*, 672. [CrossRef] [PubMed]
5. Griffiths, M. Gambling on the Internet: A Brief Note. *J. Gambl. Stud.* **1996**, *12*, 471–473. [CrossRef] [PubMed]
6. Echeburúa, E. *Adicciones Sin Drogas? Las Nuevas Adicciones: Juego, Sexo, Comida, Compras, Trabajo, Internet*, 2nd ed.; Editorial Desclée de Brouwer: Bilbao, Spain, 2009.

7. Flores Robaina, N.; Jenaro Río, C.; González Gil, F.; Martín, E.; Poy, R. Adicción Al Móvil En Alumnos De Secundaria: Efectos En La Convivencia. *Eur. J. Investig. Health Psychol. Educ.* **2013**, *3*, 215–225. [CrossRef]
8. Liese, B.S.; Benau, E.M.; Atchley, P.; Reed, D.; Becirevic, A.; Kaplan, B. The Self-Perception of Text-Message Dependency Scale (STDS): Psychometric Update Based on a United States Sample. *Am. J. Drug Alcohol Abus.* **2018**, *45*, 42–50. [CrossRef] [PubMed]
9. Lopez-Fernandez, O.; Kuss, D.J.; Pontes, H.M.; Griffiths, M.D.; Dawes, C.; Justice, L.V.; Männikkö, N.; Kääriäinen, M.; Rumpf, H.; Bischof, A.; et al. Measurement Invariance of the Short Version of the Problematic Mobile Phone use Questionnaire (PMPUQ-SV) Across Eight Languages. *Int. J. Environ. Res. Public Health* **2018**, *15*, 1213. [CrossRef]
10. Sasmaz, T.; Oner, S.; Kurt, A.Ö.; Yapici, G.; Yazici, A.E.; Bugdayci, R.; Sis, M. Prevalence and Risk Factors of Internet Addiction in High School Students. *Eur. J. Public Health* **2014**, *24*, 15–20. [CrossRef]
11. Sohn, S.; Rees, P.; Wildridge, B.; Kalk, N.J.; Carter, B. Prevalence of Problematic Smartphone Usage and Associated Mental Health Outcomes Amongst Children and Young People: A Systematic Review, Meta-Analysis and GRADE of the Evidence. *Bmc Psychiatry* **2019**, *19*, 356.
12. Cimino, S.; Cerniglia, L. A Longitudinal Study for the Empirical Validation of an Etiopathogenetic Model of Internet Addiction in Adolescence Based on Early Emotion Regulation. *Biomed Res. Int.* **2018**, *2018*, 4038541–4038548. [CrossRef]
13. Nakayama, H.; Ueno, F.; Mihara, S.; Kitayuguchi, T.; Higuchi, S. Relationship between Problematic Internet use and Age at Initial Weekly Internet Use. *J. Behave. Addict.* **2020**, *9*, 129–139. [CrossRef]
14. Chóliz Montañés, M.; Villanueva Silvestre, V. *Evaluación De La Adicción Al Móvil En La Adolescencia. Rev. Esp. Drogodep*; Departamento de Psicología Básica. Universidad de Valencia: Valencia, Spain, 2011; pp. 165–183.
15. Lu, X.; Watanabe, J.; Liu, Q.; Uji, M.; Shono, M.; Kitamura, T. Internet and Mobile Phone Text-Messaging Dependency: Factor Structure and Correlation with Dysphoric Mood among Japanese Adults. *Comput. Hum. Behav.* **2011**, *27*, 1702–1709. [CrossRef]
16. Chóliz, M.; Pinto, L.; Phansalkar, S.; Corr, E.; Mujjahid, A.; Flores, C.; Barrientos, P. Development of a Brief Multicultural Version of the Test of Mobile Phone Dependence (TMDbrief) Questionnaire. *Front. Psychol.* **2016**, *7*, 650.
17. Körmendi, A.; Brutóczki, Z.; Petra Végh, B.; Székely, R. Smartphone use can be Addictive? A Case Report. *J. Behav. Addict.* **2016**, *5*, 548–552. [CrossRef] [PubMed]
18. Kwon, M.; Lee, J.; Won, W.; Park, J.; Min, J.; Hahn, C.; Gu, X.; Choi, J.; Kim, D. Development and Validation of a Smartphone Addiction Scale (SAS). *PLoS ONE* **2013**, *8*, e56936. [CrossRef] [PubMed]
19. Long, J.; Liu, T.; Liao, Y.; Qi, C.; He, H.; Chen, S.; Billieux, J. Prevalence and Correlates of Problematic Smartphone use in a Large Random Sample of Chinese Undergraduates. *BMC Psychiatry* **2016**, *16*, 408. [CrossRef] [PubMed]
20. De-Sola Gutiérrez, J.; Rodríguez de Fonseca, F.; Rubio, G. Cell-Phone Addiction: A Review. *Front. Psychiatry* **2016**, *7*, 175. [CrossRef]
21. Lin, J.C. Popularity, Funding for Health-Effect Research and Cell-Phone Addiction. *Ap-M* **2010**, *52*, 164–166. [CrossRef]
22. Ivanova, A.; Gorbaniuk, O.; Błachnio, A.; Przepiórka, A.; Mraka, N.; Polishchuk, V.; Gorbaniuk, J. Mobile Phone Addiction, Phubbing, and Depression among Men and Women: A Moderated Mediation Analysis. *Psychiatr. Q.* **2020**. [CrossRef]
23. Fischer-Grote, L.; Kothgassner, O.D.; Felnhofer, A. Risk Factors for Problematic Smartphone use in Children and Adolescents: A review of Existing Literature. *Neuropsychiatr. Klin. Diagn. Ther. Rehabil. Organ Ges. Osterr. Nervenarzte Psychiater* **2019**, *33*, 179–190. [CrossRef]
24. Billieux, J. Problematic use of the Mobile Phone: A Literature Review and a Pathways Model. *Curr. Psychiatry Rev.* **2012**, *15*, 142. [CrossRef]
25. Simó Sanz, C.; Martinez Sabater, A.; Ballestar Tarín, M.; Domínguez Romero, A. Instrumentos De Evaluación Del Uso Problemático Del Teléfono Móvil/Smartphone. *Health Addict. Salud Y Drog.* **2017**, *17*, 5. [CrossRef]
26. Billieux, J.; Maurage, P.; Lopez-Fernandez, O.; Kuss, D.; Griffiths, M. Can Disordered Mobile Phone use be Considered a Behavioral Addiction? an Update on Current Evidence and a Comprehensive Model for Future Research. *Curr. Addict. Rep.* **2015**, *2*, 156–162. [CrossRef]
27. Lin, Y.; Chiang, C.; Lin, P.; Chang, L.; Ko, C.; Lee, Y.; Lin, S. Proposed Diagnostic Criteria for Smartphone Addiction. *PLoS ONE* **2016**, *11*, e0163010. [CrossRef]

28. Becerra Álvarez, D.R.; Delgado Martínez, A.C.; Villamizar Acevedo, G.A. Autopercepción De Estudiantes De Psicología Sobre Sus Competencias En Los Campos Laboral, Educativo Y Salud. *Rexe Rev. Estud. Exp. Educ.* **2014**, *13*, 151–167.
29. Escamilla Cruz, S.; Campos Castolo, M.; Córdoba Ávila, M.Á. Autopercepción De Competencias Profesionales De Alumnos De La Licenciatura En Enfermería. *Rev. Conamed* **2012**, *17*, 67–75.
30. Kirkcaldy, B.D.; Shephard, R.J.; Siefen, R.G. The Relationship between Physical Activity and Self-Image and Problem Behaviour among Adolescents. *Soc. Psychiatry Psychiatr. Epidemiol.* **2002**, *37*, 544–550. [CrossRef]
31. Lockertsen, Ø.; Procter, N.; Vatnar, S.K.B.; Færden, A.; Eriksen, B.M.S.; Roaldset, J.O.; Varvin, S. Screening for Risk of Violence using Service Users' Self-perceptions: A Prospective Study from an Acute Mental Health Unit. *Int. J. Ment. Health Nurs.* **2018**, *27*, 1055–1065. [CrossRef]
32. Domenico, L.H.; Strobbe, S.; Stein, K.F.; Giordani, B.J.; Hagerty, B.M.; Pressler, S.J. Identifying the Structure and Effect of Drinking-Related Self-Schemas. *West. J. Nurs. Res.* **2017**, *39*, 942–981. [CrossRef]
33. Grubbs, J.B.; Wilt, J.A.; Exline, J.J.; Pargament, K.I.; Kraus, S.W. Moral Disapproval and Perceived Addiction to Internet Pornography: A Longitudinal Examination. *Addiction* **2018**, *113*, 496–506. [CrossRef]
34. Suurvali, H.; Hodgins, D.; Toneatto, T.; Cunningham, J. Motivators for Seeking Gambling-Related Treatment among Ontario Problem Gamblers. *J. Gambl. Stud.* **2012**, *28*, 273–296. [CrossRef]
35. Agrawal, A.; Dick, D.M.; Bucholz, K.K.; Madden, P.A.F.; Cooper, M.L.; Sher, K.J.; Heath, A.C. Drinking Expectancies and Motives: A Genetic Study of Young Adult Women. *Addiction* **2008**, *103*, 194–204. [CrossRef] [PubMed]
36. Zins, M.; Carle, F.; Bugel, I.; Leclerc, A.; Orio, F.D.; Goldberg, M. Predictors of Change in Alcohol Consumption among Frenchmen of the GAZEL Study Cohort. *Addiction* **1999**, *94*, 385–395. [CrossRef] [PubMed]
37. Fleury, M.; Ngui, A.N.; Bamvita, J.; Grenier, G.; Caron, J. Predictors of Healthcare Service Utilization for Mental Health Reasons. *Int. J. Environ. Res. Public Health* **2014**, *11*, 10559–10586. [CrossRef] [PubMed]
38. Acosta, L.D.; Carrizo, E.D.; Pelaez, E. Factores Asociados a La Autopercepción De Salud En Adultos Mayores. *Rev. Cuba. Salud Pública* **2015**, *41*, 638–648.
39. Simó-Sanz, C.; Ballestar-Tarín, M.L.; Martínez-Sabater, A. Smartphone Addiction Inventory (SPAI): Translation, Adaptation and Validation of the Tool in Spanish Adult Population. *PLoS ONE* **2018**, *13*, e0205389. [CrossRef]
40. Arpaci, I.; Esgi, N. Psychometric Properties of the Turkish Version of the Smartphone Addiction Inventory (SPAI). *Curr. Psychol.* **2018**, 1–6. [CrossRef]
41. Lin, Y.; Chang, L.; Lee, Y.; Tseng, H.; Kuo, T.B.J.; Chen, S. Development and Validation of the Smartphone Addiction Inventory (SPAI). *PLoS ONE* **2014**, *9*, e98312. [CrossRef]
42. Machado Khoury, J.; Corrêa de Freitas, A.A.; Valente Roque, M.A.; Rodrigues Albuquerque, M.; Maila de Castro, L.N.; Duarte Garcia, F. Assessment of the Accuracy of a New Tool for the Screening of Smartphone Addiction. *PloS ONE* **2017**, *12*, e0176924. [CrossRef]
43. Pavia, L.; Cavani, P.; Di Blasi, M.; Giordano, C. Smartphone Addiction Inventory (SPAI): Psychometric Properties and Confirmatory Factor Analysis. *Comput. Hum. Behav.* **2016**, *63*, 170–178. [CrossRef]
44. Streiner, D.L.; Cairney, J. What's Under the ROC? an Introduction to Receiver Operating Characteristics Curves. *Can. J. Psychiatry* **2007**, *52*, 121–128. [CrossRef] [PubMed]
45. Perkins, N.J.; Schisterman, E.F. The Inconsistency of "Optimal" Cutpoints obtained using Two Criteria Based on the Receiver Operating Characteristic Curve. *Am. J. Epidemiol.* **2006**, *163*, 670–675. [CrossRef] [PubMed]
46. Lopez-Fernandez, O.; Kuss, D.J.; Romo, L.; Morvan, Y.; Kern, L.; Graziani, P.; Rousseau, A.; Rumpf, H.; Bischof, A.; GÄSSLER, A.; et al. Self-Reported Dependence on Mobile Phones in Young Adults: A European Cross-Cultural Empirical Survey. *J. Behav. Addict.* **2017**, *6*, 168–177. [CrossRef] [PubMed]
47. Cerda, J.; Cifuentes, L. Using ROC Curves in Clinical Investigation: Theoretical and Practical Issues. *Rev. Chil. Infectol. Organo Of. Soc. Chil. Infectol.* **2012**, *29*, 138–141. [CrossRef] [PubMed]
48. Nanishi, K.; Green, J.; Taguri, M.; Jimba, M. Determining a Cut-Off Point for Scores of the Breastfeeding Self-Efficacy Scale–Short Form: Secondary Data Analysis of an Intervention Study in Japan. *PLoS ONE* **2015**, *10*, e0129698. [CrossRef] [PubMed]
49. Lin, Y.; Pan, Y.; Lin, S.; Chen, S. Development of Short-Form and Screening Cutoff Point of the Smartphone Addiction Inventory (SPAI-SF). *Int. J. Methods Psychiatr. Res.* **2017**, *26*, e1525. [CrossRef] [PubMed]
50. Sánchez-Martínez, M.; Otero, A. Factors Associated with Cell Phone use in Adolescents in the Community of Madrid (Spain). *Cyberpsychol. Behav.* **2009**, *12*, 131–137. [CrossRef]

51. Lopez-Fernandez, O.; Männikkö, N.; Kääriäinen, M.; Griffiths, M.D.; Kuss, D.J. Mobile Gaming and Problematic Smartphone use: A Comparative Study between Belgium and Finland. *J. Behave. Addict.* **2018**, *7*, 88–99. [CrossRef]
52. Roberts, J.A.; Yaya, L.H.P.; Manolis, C. The Invisible Addiction: Cell-Phone Activities and Addiction among Male and Female College Students. *J. Behav. Addict.* **2014**, *3*, 254–265. [CrossRef]
53. Machado Khoury, J.; Maila de Castro, L.N.; Valente Roque, M.A.; Correa de Freitas, A.A.; Ralil da Costa, M.; Duarte Garcia, F. Smartphone and Facebook Addictions Share Common Risk and Prognostic Factors in a Sample of Undergraduate Students. *Trends Psychiatry Psychother.* **2019**, *41*, 358–368. [CrossRef]
54. Smetaniuk, P. A Preliminary Investigation into the Prevalence and Prediction of Problematic Cell Phone Use. *J. Behav. Addict.* **2014**, *3*, 41–53. [CrossRef]

International Journal of
Environmental Research and Public Health

MDPI

Article

Stakeholders' Consensus on Strategies for Self- and Other-Regulation of Video Game Play: A Mixed Methods Study

Michelle Colder Carras [1,*], Matthew Carras [2] and Alain B. Labrique [3]

1 Department of International Health, Johns Hopkins Bloomberg School of Public Health, 615 N. Wolfe St. W5501, Baltimore, MD 21205, USA

2 Carras Consulting, 1407 Scanlan Drive, Glen Burnie, MD 21205, USA; matthew.carras@gmail.com

3 Departments of International Health, Epidemiology, and Health Policy and Management, Johns Hopkins Bloomberg School of Public Health, 615 N. Wolfe St. W5501, Baltimore, MD 21205, USA; alabriq1@jhu.edu

* Correspondence: mcarras@jhu.edu

Received: 10 April 2020; Accepted: 25 May 2020; Published: 28 May 2020

Abstract: *Background:* Little is known about strategies or mechanics to improve self-regulation of video game play that could be developed into novel interventions. This study used a participatory approach with the gaming community to uncover insider knowledge about techniques to promote healthy play and prevent gaming disorder. *Methods:* We used a pragmatic approach to conduct a convergent-design mixed-methods study with participants attending a science fiction and education convention. Six participants answered questions about gaming engagement and self- or game-based regulation of gaming which were then categorized into pre-determined (a priori) themes by the presenters during the presentation. The categorized themes and examples from participant responses were presented back to participants for review and discussion. Seven participants ranked their top choices of themes for each question. The rankings were analyzed using a nonparametric approach to show consensus around specific themes. *Results:* Participants suggested several novel potential targets for preventive interventions including specific types of social (e.g., play with others in a group) or self-regulation processes (e.g., set timers or alarms). Suggestions for game mechanics that could help included clear break points and short missions, but loot boxes were not mentioned. *Conclusions:* Our consensus development approach produced many specific suggestions that could be implemented by game developers or tested as public health interventions, such as encouraging breaks through game mechanics, alarms or other limit setting; encouraging group gaming; and discussing and supporting setting appropriate time or activity goals around gaming (e.g., three quests, one hour). As some suggestions here have not been addressed previously as potential interventions, this suggests the importance of including gamers as stakeholders in research on the prevention of gaming disorder and the promotion of healthy gaming. A large-scale, online approach using these methods with multiple stakeholder groups could make effective use of players' in-depth knowledge and help speed discovery and translation of possible preventive interventions into practice and policy.

Keywords: gaming disorder; prevention; video games; mixed methods research; stakeholder engagement; consensus development

1. Introduction

Research on problems with video game play has led to a proposed diagnosis of Internet gaming disorder (IGD) in the Diagnostic and Statistical Manual 5 [1] and inclusion of gaming disorder in ICD-11 [2]. Representative surveys usually show that only a small percentage of the general population has clinical problems related to excessive gaming [3]. Systematic reviews link gaming disorder to correlates such as impulsivity, male gender, impaired mental health and low psychosocial well-being [3–5]. However, the Goldilocks hypothesis (based on the fairytale character Goldilocks, who struggles to find sizes of chair, porridge and bed that are "just right) suggests that engaging with technology in a moderate way is not harmful [6]. The WHO notes that most people are able to play video games without problems [7]. As this suggests that moderate video game play is a reasonable public health goal, investigating how people can play in a healthy way is vital to support the public health objective of preventing gaming disorder.

While many studies examine motivations to play, less is known about how gamers self-regulate their gaming. One qualitative study in children uncovered several reasons for engaging (e.g., motives, habit, and contextual factors) or disengaging (e.g., self-regulation, parental intervention, or disruption/frustration of gaming outcomes or psychological needs; [8]). Contextual factors of games or the environment also contribute to the length of gaming sessions: gaming sessions may be prolonged through losing track of time [9], achievement, winning or game mechanics, including gambling-like mechanisms such as loot boxes [8,10]. Disengaging from play may also be easier with support from real-life friends [11] or when using strategies like setting an alarm, taking regular breaks, or setting goals in the game [9]. The above findings show a range of individual, game and contextual factors that can promote healthy gaming and prevent disorder.

However, policy responses have focused largely on limiting time gaming [12] or regulating loot boxes as gambling [13]. Commentaries suggest that existing tools for self-regulation of gambling, such as self-exclusion, mandatory breaks, or pop-up messaging, might be useful for gaming [14]. A few technology platforms have already implemented feedback mechanisms to alert players about time or allow them to limit their time [15,16]. Games also provide such mechanisms in the form of parental controls. Some also remind players to take breaks, but forcing shutdown or exclusion periods as a preventive mechanism has so far been limited [12]. The current focus on a few limited policy aspects also neglects the need to take into account cultural factors, individual player factors, and how games change over time [12].

With the recent inclusion of gaming disorder in ICD-11, calls for collaboration with industry stakeholders have increased. Many researchers have suggested that the game industry has a responsibility to protect the health of gamers [14,17,18]. Fewer researchers support including gamers as partners in research and policy efforts to understand or prevent gaming disorder [19,20].

Participatory and multi-stakeholder approaches in health research allow for better translation of research into practice [21,22]. Stakeholders are members of diverse groups, including experts and the general public, who can provide unique perspectives on an issue. In the case of gaming disorder, this could mean not just clinicians, researchers, and policy makers, but also gamers, their families, and people who create games. Stakeholder-engaged research that involves group deliberation and decision-making promotes the type of co-learning that allows researchers to better examine all aspects of a problem or question, encourages decision-makers to consider multiple perspectives, and makes science more transparent [23,24]. Using systematic methods that integrate perspectives of those in the know—in this case video game players and developers—with research and theory will be vital to allow research on problematic gaming to address insider perspectives and cultural differences and keep up with changing technologies.

This approach has been used with gamers to explore their unique insights into what it means to be "addicted" [19]. As an illustration of the usefulness of this approach, we discovered that a small sample of gamers at a gaming fan convention were not only interested in discussing game "addiction", their ideas overlapped only partially with criteria for IGD. In fact, their suggestions leaned toward less strict criteria for social impairment and they suggested appointment mechanics (a game feature promoting engagement) as a primary contributor to loss of control. These findings suggest that including gamers in discussions of research into gaming disorder may improve research.

However, a systematic, perspective-integrating approach that examines factors associated with gaming self-regulation would better contribute to intervention development.

To address this gap and provide clinicians, decision-makers, the game industry and the general public with additional insights about gaming self-regulation, we conducted participatory, mixed-methods, convergent-design research with gamers to discuss the challenges of regulating video game play. This research builds on our prior study regarding gamers' insights on game "addiction" [19]. The qualitative objectives of our study were to elicit a wide variety of knowledge from the video gaming stakeholder community about why gamers play, why they stop playing, and how they self-regulate gaming time and to map these to known themes in the literature. The quantitative goal was to numerically rank the consensus on importance of themes. The overall mixed methods goal was to demonstrate a rapid and transparent consensus-development approach that can be used with multiple stakeholders. Such an approach would foster knowledge development and research co-creation in order to develop and prioritize research and prevention interventions for problematic video game play.

2. Materials and Methods

2.1. Overview

Mixed methods research combines qualitative and quantitative data to answer research questions that require a deep understanding of multiple perspectives, contexts, and/or communities/cultures [25]. We used a convergent mixed methods design, i.e., we combined qualitative and quantitative data simultaneously, to develop a rapid consensus on the relative importance of themes related to video gaming engagement and self-regulation. We used a pragmatic approach toward study design and analysis, as these allow researchers to combine diverse approaches to best answer the research question and place equal value on both objective and subjective knowledge [26]. In our previous study [19], for example, we used an emic, free-listing approach, working with our gamer audience at a panel to generate categories and themes for free-listed responses. As this proved difficult within the one-hour time frame of a panel at a convention, we opted to combine deductive (pre-determined themes from the literature) and inductive (free-listing) approaches in the qualitative part of our study design here.

As qualitative research benefits from researchers' self-reflection, we describe our relevant background here. All authors are researchers with experience in conducting mixed methods research in technology and health, particularly in video games. We have conducted a similar study in another setting to address gamers' insights into game "addiction" [19]. The first two authors, who conducted the panel, also identify as members of the gaming community, regularly attend gaming fan conventions, and have personal experience with regulating their gaming time. They use this personal experience to connect with stakeholders and facilitate meaningful discussion of healthy and unhealthy video game play. Author MCC is a researcher specializing in video game play, technology use and mental health; author MC is an information technology specialist with game industry experience; and author AL is a professor specializing in digital health.

To address our mixed method goal of demonstrating a rapid and transparent consensus-development technique, we first used a deductive approach among authors to identify prominent themes around self-regulation and gaming from the literature (Section 2.2). We then used an emic, inductive, free-listing approach to gather data about self- and other-regulation of video game play from audience members at a presentation (Section 2.4). We then facilitated a group decision-making procedure (Delphi process) with the audience to brainstorm, discuss/debate, and finally quantitatively rank concepts related to gaming self-regulation and steps the gaming industry could take to facilitate healthy play. We present the qualitative, quantitative and mixed methods results to better understand self-regulation as gamers see it. Our presentation here was guided by American Psychological Association standards for mixed methods and qualitative research reporting [27,28].

2.2. Selection of Themes for Framework

Prior to the presentation, the authors first briefly reviewed relevant literature on game engagement [8,9,29] and design [29–32]. We extracted themes and frameworks from these into tables and bulleted notes and discussed and combined results to condense themes surrounding game engagement and self-regulation. This deductive approach was not meant to be exhaustive but to present the most salient themes for discussion; other themes could be incorporated during data collection if necessary. This condensed list of themes (Table 1) was used to facilitate rapid thematic analysis during presentation and discussion with audience stakeholders. Themes were transcribed to index cards placed on the table in front of the presenters and used in a deductive, top-down theoretical thematic analysis approach [33] to rapidly classify participant responses during the presentations.

Table 1. Prepared themes and examples used in thematic analysis.

Theme	Example Theme/Concept from Literature [8,9,29–32]
Achievement	Non-competitive. Also includes general success in completing objectives Achievement, challenge, Advancement, Mechanics, Being stronger, Rare items, Meta-game rewards
Behavioral game mechanics	Game-driven regulation, Alerts, Behavioral tracking, Exit points, Timed rewards, Fast loading times
Environmental	Forces outside the scope of the game's design Content-driven, family/parent driven, outside needs & responsibilities
Habit	Habit
Immersion	The feeling of being one with the game Narrative, Look and feel, Interface, Autonomy, Customization, Role-playing, Discovery, Fantasy
Mood regulation	Arousal/enjoyment, seeking diversion or distraction, escapism
Novelty	New or interesting aspects of the game
Self-regulation	Self-regulation
Social	Social interaction, relationships, teamwork

2.3. Participants

A panel was presented at a science fiction and education convention, Escape Velocity, held in Washington, D.C. in 2017. The convention featured topics and events relevant to science and technology, STEAM education, video games, and pop culture. The conference is sponsored by the Museum of Science Fiction and science organizations like NASA, nonprofit organizations and industry groups, and in its first year was attended by over 2000 people [34].

Participants were recruited passively through panel descriptions in the convention schedule and through posters on the door to the panel room. Our panel on Day 1 was entitled "Insights Needed: Video Gaming and the Goldilocks Principle" and our panel on Day 2 was entitled "Insights Needed Wrap-Up: Give Us Your Feedback on Video Games and Engagement Preliminary Analysis." The authors had no prior relationship with the research participants. Each session began with the reading of a consent document where participants were offered the opportunity to leave if they did not want to participate; no audience members left. Participants did not receive compensation for participation. As the pragmatic study design was dependent on the setting of a panel at a convention, we did not identify a needed number of participants ahead of time and our sample was one of convenience. One participant arrived late and two participants did not complete all portions of the questionnaire.

Six respondents completed at least some demographic information and percentages are reported out of those who answered the question. Age ranged from 27 to 49 (Mean = 35, SD 7.9) and one participant was female. Most respondents (66.7%) were white, one identified as mixed race, and one declined to answer. Half had a graduate degree and another 33.3% had graduated from college. All respondents identified as members of the gamer stakeholder group, but also identified with other groups such as family/friend of a gamer (n = 3), researcher (n = 3), educator (n = 1), and industry (n = 1). In response to the question "Do you see yourself as a gamer", all responded either Yes or "Kind of (intermediate/casual)" but one participant also answered "No" to the question "Do you play video games, including mobile games (e.g., Candy Crush, Pokémon GO)?". (This participant identified primarily as a member of the game industry.) All worked full time and the majority (80%) earned over $40,000/year. The group played on average 3.13 h a day (SD 1.31) and 5.33 days per week (SD 1.15). Half played about 1–2 h per session while the other half played 2–3 h per session.

2.4. Data Collection and Study Procedure

Questionnaire responses, sticky notes used in free listing, and field notes made up the qualitative and quantitative data as described here. This approach described is a combination of nominal group technique and Delphi approach that the authors used in a previous study [19]. Prior to data collection the authors printed out questionnaires with questions about demographics, stakeholder group membership, gaming experience and ranking of themes and added a number to each questionnaire. We then created a set of sticky notes with matching participant numbers. Each set of sticky notes contained five sticky notes for each of our four questions. Each note was labeled with a participant number and Q1, Q2, Q3, or Q4.

Data were collected over the course of an hour at a panel discussion. At the beginning of the panel, the presenters (the first two authors) introduced themselves as researchers and gamers and described the presentation as a study using group decision-making techniques to help answer questions about how to achieve just the right amount of engagement with video games. We then asked participants to list ideas for promoting healthy gaming by asking four open-ended questions (Table 2) about gaming engagement and regulation strategies. Participants recorded answers on sticky notes, then completed paper questionnaires asking about demographics and gaming experience. We allowed 10 min for data collection. Participants were asked not to complete the questions about ranking the importance of themes until told to do so.

Table 2. Themes and examples presented for ranking.

Theme	Example 1	Example 2	Example 3
Q1 What features of games make you want to play them or keep playing?			
Competition	*Winning*		
Game type	*Arcade games*		
Immersion	*Known intellectual property that you like*	*Customizing*	
Mood regulation	*Repeatable enjoyment*		
Achievement	*Well-balanced gameplay*	*Rewards*	
Q2 What features of games make you want to stop playing?			
Social	Playing online with terrible people	*Losing*	
Environmental	*Out of time*		
Immersion	*Broken games*	*Games that crash*	*Unclear control of avatar*
Achievement	*Games that are not balanced*	*Hard learning curve*	
Mood regulation	*Not fun anymore*		
Novelty	*Repetition/grinding*		
Q3 What are some strategies people can use to regulate the amount of time they game?			
Self-regulation	*Set an alarm*	*Set goals for gameplay*	
Social	*Play with others to limit time*		
Analyze moves	*Slower-move analytically*		
Q4 What are some things the game industry could do to make it easier to regulate the amount that people play games?			
Game mechanics	*Game-breaking pop-ups*	*Provide break points*	*On screen clock*
Environmental	*Integrate health tips*		
Social	*Local multiplayer only*		
Novelty	*Local multiplayer—AR fun with new tech*		

As data on sticky notes was collected, responses were categorized by the presenters by placing each sticky note under a pre-determined theme's index card. Disagreements about how to categorize responses were resolved through discussion between presenters. Once all responses were categorized (about 10 min after all responses were received) they were transcribed into the slideshow and displayed to all participants. We asked for clarifications of some responses and, as a participant check, discussed whether we had categorized responses for each question appropriately based on our own understanding as gamers and researchers. Field notes were incorporated into the response examples at times for further clarification. Participants elaborated on some responses, disagreed with one another or provided additional explanation, and a robust discussion of specific responses and themes ensued. The authors' goal was to ensure clarity of themes and their examples; this was accomplished with about three minutes of discussion for each question.

Participants were then asked to rank the top three themes they felt best answered each question by listing their choices for most important themes in order on the last page of the questionnaire, e.g., Question 1: #1 = immersion, #2 = game type, #3 = achievement. The numeric ranking participants gave to themes for a particular question constituted the quantitative data. Participants were invited to return in two days for the results. After the panel, one author transcribed questions, responses, and themes into an Excel spreadsheet to record qualitative data and the results of thematic analysis. Another author entered rankings and questionnaire responses directly into a Stata do-file to create and analyze a quantitative analysis database. To validate data quality, data entry was double-checked and analyses re-run prior to manuscript revision.

At a second panel two days later, we discussed the study's organizing frameworks and procedures and presented the results of the ranking. We again asked for clarifications of some responses and further discussed difficult-to-categorize responses. Finally, we presented the limitations and strengths we felt the process provided and invited participants to provide additional feedback. Data are available as Supplementary Materials.

2.5. *Qualitative and Quantitative Analysis*

2.5.1. Categorization of Responses

We categorized responses as the unit of analysis during the presentation using a deductive approach based on the a priori themes. We chose theory-based thematic analysis as a practical way to facilitate rapid categorization in keeping with our pragmatic approach. Initial thematic analysis was performed at the panel and final categorizations and themes for each question were arrived at after discussion with participants; we used participant checks to facilitate categorization of some responses. Themes and examples (Table 2) were then projected onscreen to allow participants to rank them.

2.5.2. Calculation of Ranks

We used an observational design to gather participants' perspectives on the importance of themes for each question. Ranked themes were assigned a numerical value from one to eleven to represent the eleven total themes (the nine a priori themes from Table 1 and two themes emerging during group discussion). We validated data entry against collected data by listing database results for each question in a table and checking against each participant's questionnaire.

For each question, we transformed a participant's ranked themes into coded rankings by replacing a placeholder variable for each theme with the numeral 1, 2, or 3 if a participant ranked it among their top three choices. This resulted in a unique subset of themes as the domain for each question. If a respondent did not rank a particular theme for that subset, the rank for that theme for that participant was coded as 0. We conducted a quantitative consensus analysis of participants' ranking of the most important themes for each question by calculating the weighted sum of centered ranks (WSCR) in Stata 13 IC [35] using the Skillings–Mack statistic (skilmack.ado, [36]). This flexible approach aggregates rankings across participants by combining the ranking for a specific question provided by each participant with the consensus on ranking for that question across all participants. With this approach, high consensus themes (those that are ranked more often or ranked very highly) will have a high and positive WSCR while low consensus themes, i.e., those that are ranked fewer times, ranked as less important, or show greater disagreement among respondents will have a low or negative WSCR. As we used rankings rather than raw numerical values in the analysis and replaced non-ranked themes with 0, standard errors are the same for all resulting WSCR for each question.

2.6. Presentation of Results

Qualitative results from the rapid thematic analysis are presented first, then qualitative and quantitative results are presented together to reflect the convergent design. Table 3 combines ranked themes with examples and important topics of discussion to illustrate how group processes affected participants' choices. We used these results to suggest ways that multiple stakeholder groups could help promote healthy gaming (Figure 1). This report follows standards for reporting qualitative, quantitative, and mixed methods research established by the American Psychological Association [27,28].

For game developers and industry

- Provide clear chapters or break points, even in open world games
- Force or reward play breaks
- Offer in-game reminders (game-breaking pop-ups are very effective)
- Allow autosave or quick save at any time
- Reward short gaming sessions
- Punish long gaming sessions (e.g., XP decay)
- Provide an on-screen clock
- Remove features that may promote excessive play (e.g., online multiplayer where players feel obligated to continue to support their team)

For players, family members and friends

- Play social games that depend on team members being present
- Use alarm clocks to keep track of and limit time
- Set limits for money spent on games
- Set clear activity goals for a gaming session (e.g., 3 quests, 1 hour)
- Schedule game play for after work or other life responsibilities

For clinicians and policymakers

- Discuss self-regulation with gamers and familiesSupport gamers and family/friends in setting appropriate goals around gaming
- Require games to offer features that promote self-regulation

Figure 1. Suggestions for regulating video game play.

Table 3. Ranked themes with examples of responses (n = 7).

Theme as Categorized by Authors	Times Ranked [a]	WSCR [b]	WSCR/SE	Response Examples
Q1. What features of games make you want to play them or keep playing?				
Immersion	5	7.78	1.47	*Visual/audio aesthetics, exploration of dynamic universe, known theme or IP, player/playstyle customization, great stories, color, sound*
Achievement	5	3.54	0.67	*Rewards, either literal ones in the form of items, unlocks, in-game money, OR psychological rewards such as a rewarding story pay off or resolution to a puzzle; well-balanced gameplay*
Game type	4	0	0	*Arcade-based games like racers or mech games.*
Mood	3	−0.71	−0.13	*Repeatable enjoyment*
Competition [c]	1	−10.61	−2.00	*Winning*
Q2. What features of games make you want to stop playing?				
Social	4	5.24	0.89	*Online interaction with terrible people (most)*
Novelty	5	5.24	0.89	*Repetition/grinding with little variety*
Immersion	4	1.96	0.33	*Unclear control of avatar, Games that are crashy or slow*
Environment	2	−3.93	−0.66	*Out of time*
Achievement	2	−3.93	−0.66	*Unfair or unreliable game mechanics that result in "cheap" deaths/failures, games that are not balanced, games that are broken/not working*
Mood	3	−4.58	−0.77	*Not fun anymore*
Q3. What are some strategies people can use to regulate the amount of time they game?				
Self-regulation	7	5.2	1.39	2x[d,c]-*Limit time … (60 min/session)*,2x[d]-*Use an alarm clock*, 2x-*Set firm goals for gameplay ex. "Just 3 quests", set a schedule for gaming after other responsibilities*
Social	6	0	0	*Play with others in a game that requires everyone to be present … since the other players can effectively force you to stop*
Analyzing moves [c]	5	−5.2	−1.39	*Analyzing moves, moving slower*
Q4. What are some things the game industry could do to make it easier to regulate the amount that people play games?				
Mechanics	6	10.07	2.2	*Clear delineations between missions/chapters which offer clear "break points" to the player, as opposed to open-ended/open world games which don't give an "excuse" to stop, autosave/quicksave, rewards that emphasize short sessions or have negative reinforcement mechanisms like XP delay, time allowed*
Environment	5	−0.77	−0.17	*Integrated health tips into game that use health points that players can use themselves*
Social	5	−2.32	−0.51	*Local multiplayer only- take away features of MO*
Novelty	3	−6.97	−1.52	*Augmented Reality Gaming—local multiplayer fun with new tech*

[a] Number of times an item was ranked as one of a participant's top three choices. [b] Weighted sum of centered ranks statistic from the Skillings–Mack statistic, where high/positive ranks correspond to agreed-upon items and low/negative ranks correspond to items that were chosen less often or ranked as less important. SE were the same for each question: $SE_{Q1} = 5.29$, $SE_{Q2} = 5.92$, $SE_{Q3} = 3.74$, $SE_{Q4} = 4.58$. [c] New theme emerging from discussion. [d] More than one version/wording submitted by different respondents.

2.7. Methodological Integrity

To ensure fidelity and utility of our qualitative findings and address our research objectives, we address the adequacy, grounding, and context of data and how we used it to develop meaningful and coherent findings. First, we collected data from cultural insiders—those who had an interest in video gaming and self-regulation—grounding our data in their perspective. We used an emic approach to data collection, free listing, to elicit their cultural understanding and insider knowledge. We further grounded findings in evidence by using free-listed responses to illustrate themes. We included all audience members in our analysis, since all members identified as gamers. However, audience members also identified as members of other stakeholder groups. Although our sample was small, we aimed to gather insights/knowledge from gamers and other stakeholders; we feel the collected data are therefore adequate to capture a broad range of perspectives. Researchers' perspectives were used to determine an initial set of themes, facilitate rapid data analysis and guide group discussion.

We demonstrate consistency in our analysis by describing consensus development and discussion and when and how they were used in the analytic process. We used data displays (tables), author consensus on a priori themes, group discussion with participants, formal quantitative consensus, and participant feedback as checks to promote rigor in the qualitative analysis. We discuss the context of data collection and present findings coherently, reflecting on discrepancies and reporting according to mixed methods standards. Our results further the understanding of how gamers and other stakeholders can make insightful and meaningful contributions to a clinical- and policy-relevant research.

2.8. Ethics

The study protocol was approved by the institutional review board of the Johns Hopkins Bloomberg School of Public Health (IRB No. 00006931) and carried out in accordance with the Declaration of Helsinki. The need for written informed consent was waived by the institutional review board. Participants were fully informed about the study at the beginning of each study session and offered the opportunity not to participate.

3. Results

Six people participated in the free listing round and submitted between one and three answers per question. Seven people participated in the ranking round and six people completed demographic and gameplay questionnaires. Four questionnaires were complete, including rankings, while two were partially complete.

3.1. Qualitative

We initially identified nine themes from our brief review of several articles related to gaming engagement and design [8,9,29–32]. Our selection of themes was not meant to be exhaustive of all themes in the gaming literature, but rather a starting point for our discussion-based study design. Themes were chosen based on the authors' knowledge of game engagement and game design and how these factors contribute to problematic or self-regulated gaming. Categorization of responses to our four questions resulted in three to six themes for each question (Table 2). Some responses transcended themes, such as pop-up windows being a behavioral game mechanic that was "game-breaking", thus primarily having to do with the theme of immersion (Q4). Through discussion, some submissions categorized by the authors as having to do with one theme were reassigned to a different theme and one novel theme, analyzing moves, was suggested but did not end up being ranked highly by participants. Another theme, competition, was suggested as being different from achievement. The pre-determined theme of habit was not seen to match any of the responses. The full database of responses and assigned themes are available as Supplementary Materials.

3.2. Mixed Methods Synthesis

Table 3 presents results of nonparametric quantitative analysis that describe consensus on themes as they were ranked by participants. Immersion (e.g., *exploration of dynamic universe*) and achievement (e.g., *rewards*) emerged as important features of games that motivated playing (Q1), but there was less agreement on game type, mood and competition. Social factors (e.g., *online interactions with terrible people*) and novelty (e.g., *repetition/grinding with little variety*) were considered equally important reasons to stop playing (Q2). Immersion (e.g., *games that are crashy or slow*) was also agreed to be somewhat important, but achievement (*games that are not balanced*), environment (*out of time*), and mood (*not fun anymore*) were considered less important.

Social factors showed up again in Q3 as a strategy people could use to self-regulate (e.g., *play with others in a game that requires everyone to be present*). One response categorized as social described how online group play could limit gaming time. Participants debated this initially, but after deliberation agreed that choosing to play games online in groups might be an effective way to reduce time as other group members' leave, disrupting the group. Self-regulation strategies were also considered important. Responses focused on setting firm goals (*just three quests*) or limiting time (*setting an alarm, set a schedule for gaming time that is after your other work/responsibilities have been finished*) or money spent on gaming.

For Q4, there was strong agreement that the game industry could make it easier for players to self-regulate by changing or adding behavioral game mechanics (e.g., *clear break points, rewards for short sessions, negative reinforcement like XP decay*). Participants also offered social (*taking away features of online multiplayer and have local play only*) or environment-related (e.g., *integrated health tips into game that use health points that players can use themselves*) strategies, but these were less often ranked. The group also suggested benefits to new technology such as augmented reality games (e.g., Pokémon GO) that would offer novelty (*local multiplayer fun with new tech*), but there was low consensus on this theme. The group discussed this with regard to the benefits of staying grounded in the real world, especially when playing with others, which was thought to be useful for reducing the feeling of immersion that could lead to time loss.

4. Discussion

This study used a systematic, consensus-developing participatory approach with members of the gaming community to explore the importance gamers assigned to themes about self-regulating gaming. We found that even a small group (n = 7) was able to offer many specific suggestions that could form the basis of preventive interventions (Figure 1). These results support the emphasis on game mechanics and self-regulation identified in previous research [8,9,19], but add an additional dimension, social factors, that presents a new potential focus of prevention research. The findings also contribute to the importance of incorporating cultural insider knowledge about gaming and self-regulation into research and intervention development across disciplines.

The fact that gamers endorse the theme of self-regulation so strongly suggests that integrating specific techniques for promoting self-regulation (Figure 1) into preventive interventions may be a worthwhile approach. These strategies may be more palatable when endorsed by the gaming community itself. One possibility would be to have celebrity gamers (e.g., popular Twitch streamers) suggest or even discuss such strategies with their gaming audience. Hearing about risk from trusted individuals is important, and recent research shows that the general public may place greater trust in social media personalities than experts who may offer conflicting or confusing messages [37]. Streamers do address mental health topics in Twitch [38], so this may become an important mode for delivering preventive interventions relating to video game play.

Our small sample had many specific suggestions about features to include or leave out of games, such as exit points—places where they could leave the game without feeling that more had to be done. Exit points have been specifically mentioned as a factor of ethical game design that can be promoted in the video game industry [39]. Exit points may be unique to games and were not included in a recent expert-developed list of recommended tools for the prevention of problematic gambling that could be revised to address gaming disorder [14].

Another specific suggestion was to incentivize taking breaks. Our sample mentioned how games sometimes punish players for limiting their play, for example temporarily blocking players from team play if the leave a team abruptly. While games may do this to foster engagement, this also makes it more difficult for gamers to leave when competing opportunities or demands arise, such as a friend's request for a visit or the dawning recognition that time has passed and the dishes still need to be washed. Games can also reward breaks by manipulating experience gains, either by allowing players to progress faster when they return from a break or limiting progress after long sessions. While interventions related to experience gains have been implemented in several countries (experience decay), this approach may be less appealing to game developers [12]. In contrast, allowing faster progress when returning from a break (e.g., rested experience) may be a more palatable way to reward healthy amounts of gaming.

Social factors are an area brought up by our participants that seems to have little mention in suggestions for prevention interventions. Our sample found them important as a driver of disengaging (negative online interactions) and self-regulating (playing with others in a game that requires everyone to be present). In fact, this theme is a good example of the strength of a group reflection process as it reflected a single participant's suggestion. At the beginning of the discussion, most participants didn't understand or agree with the suggestion that choosing to "play with others in a game that requires everyone to be present ... since the other players can effectively force you to stop" would be a good way to self-regulate, but after a few minutes of discussion, six out of seven people ranked that as one of the most effective ways to limit time. This reveals the unique benefit to conducting a Delphi method approach with gamers as stakeholders—the perspective of an individual can be clarified within a group and ultimately recognized as a ground truth: Yes, if we choose to play games with that require a group, our time may be limited whether we choose to limit it (or are able to effectively limit it) or not.

Our findings suggest that focusing research and prevention efforts without involving insights from community members may unnecessarily limit the range of potential interventions decision-makers consider and impede progress toward implementation. The current epistemological approach toward research on gaming disorder places greater emphasis on clinical expertise rather than multiple perspectives [40]. The findings here challenge this traditional approach by considering gamers as vital co-creators of knowledge. For example, loot boxes have been the focus of much public health activity related to gaming disorder to date [13,41], yet loot boxes or other gambling-like mechanics were not mentioned by our sample. Other targets may be more palatable to multiple stakeholders, potentially easier to achieve, or more effective at preventing problematic gaming. Gamers are also useful in pointing out nuances of how we think about interventions. For example, while research may concentrate on immersion as a motivating factor for problematic gaming, our applied research here allowed gamers themselves to make a clear link between an example intervention (pop-ups) and the experience of losing immersion (pop-ups as game-breaking).

While this study was limited in sample size, it does reflect the gender (male) and age of gamers in the United States [42,43]. The small size does limit generalizability, and other sample makeups may suggest different responses. For example, our sample was self-selected for attending a science fiction and education convention and participating in a panel focused on gaming research. Other samples may prioritize different themes when ranking questions related to self-regulation. Our sample also felt the theme of Immersion to be more important than achievement—had we used a sample of gamers at a gaming tournament, for example, this would likely be different. In addition, our sample was English-speaking and based in the US, so findings may not transfer to other countries or cultures. However, one goal of this study was to demonstrate that the combination of perspective seeking, deliberation and systematic quantitative consensus analysis provides additional information beyond that of a survey, interview or focus group. Further research would benefit from systematic recruitment of a large sample of various stakeholders and a formal, multi-round consensus development technique in an online format to encourage transparency.

5. Conclusions

Formal consensus development techniques with gamers produced many specific examples of ways gamers can self-regulate and also emphasized the importance of game mechanics in fostering healthy gaming and preventing gaming disorder. These examples are supported by previous research and could serve as the basis for interventions to promote healthy gaming (Figure 1). Going forward, decision makers could increase the value of research on the prevention of problematic or disordered gaming by involving gamers, developers, clinicians and other stakeholders in co-learning about self-regulation strategies, game mechanics, and other features of game engagement.

Supplementary Materials: The following are available online at http://www.mdpi.com/1660-4601/17/11/3846/s1, data and themes; analysis code.

Author Contributions: Conceptualization, M.C.C. and A.B.L.; data curation, M.C.; formal analysis, M.C.C., M.C. and A.B.L.; funding acquisition, M.C.C.; investigation, M.C.C.; methodology, M.C.C., M.C. and A.B.L.; project administration, M.C.C., M.C. and A.B.L.; resources, M.C. and A.B.L.; supervision, M.C.C. and A.B.L.; validation, M.C.C.; visualization, M.C.; writing—original draft, M.C.C.; writing—review and editing, M.C.C., M.C. and A.B.L. All authors have read and agreed to the published version of the manuscript.

Funding: This work was supported by the National Institute of Mental Health Training Grant 5T32MH014592-39.

Acknowledgments: The authors wish to thank Nico Pandi of the Museum of Science Fiction in Washington, D.C. for his extensive support.

Conflicts of Interest: M.C.C. has spoken to the World Health Organization about this topic at the request of the video game industry but has received no funding, honorarium, fees, meals, accommodations, donations or reimbursement. She is currently seeking funding from industry sources to conduct game research. She has acted as a consultant with a nonprofit organization that provides mental health and suicide prevention support through online video game play. She is the CEO and founder of the Gaming and Wellness Association, Inc., a nonprofit video game research organization. Author M.C. has worked for a video game developer as a game quality tester and heads an IT and research consulting company. Author A.B.L. declares no conflicts of interest.

References

1. American Psychiatric Association. *Diagnostic and Statistical Manual of Mental Disorders*, 5th ed.; American Psychiatric Association: Arlington, VA, USA, 2013.
2. World Health Organization WHO. Gaming Disorder. Available online: http://www.who.int/features/qa/gaming-disorder/en/ (accessed on 28 May 2020).
3. Mihara, S.; Higuchi, S. Cross-sectional and longitudinal epidemiological studies of internet gaming disorder: A systematic review of the literature. *Psychiatry Clin. Neurosci.* **2017**, *71*, 425–444. [CrossRef] [PubMed]
4. González-Bueso, V.; Santamaría, J.J.; Fernández, D.; Merino, L.; Montero, E.; Ribas, J. Association between Internet Gaming Disorder or Pathological Video-Game Use and Comorbid Psychopathology: A Comprehensive Review. *Int. J. Environ. Res. Public Health* **2018**, *15*, 668. [CrossRef]
5. Männikkö, N.; Ruotsalainen, H.; Miettunen, J.; Pontes, H.M.; Kääriäinen, M. Problematic gaming behaviour and health-related outcomes: A systematic review and meta-analysis. *J. Health Psychol.* **2017**, *71*, 425–444. [CrossRef] [PubMed]
6. Przybylski, A.K.; Weinstein, N. A Large-Scale Test of the Goldilocks Hypothesis: Quantifying the Relations Between Digital-Screen Use and the Mental Well-Being of Adolescents. *Psychol. Sci.* **2017**, *28*, 204–215. [CrossRef] [PubMed]
7. World Health Organization WHO. Gaming behavior. Available online: https://www.who.int/substance_abuse/activities/gaming_disorders/en/ (accessed on 28 May 2020).
8. Van Rooij, A.J.; Daneels, R.; Liu, S.; Anrijs, S.; Looy, J.V. Children's Motives to Start, Continue, and Stop Playing Video Games: Confronting Popular Theories with Real-World Observations. *Curr. Addict. Rep.* **2017**, *4*, 323–332. [CrossRef]
9. Wood, R.T.A.; Griffiths, M.D.; Parke, A. Experiences of time loss among videogame players: An empirical study. *Cyber Psychol. Behav.* **2007**, *10*, 38–44. [CrossRef]

10. King, D.L.; Delfabbro, P.H. Predatory monetization schemes in video games (e.g. 'loot boxes') and internet gaming disorder. *Addiction* **2018**, *113*, 1967–1969. [CrossRef]
11. Snodgrass, J.G.; Lacy, M.G.; Francois Dengah, H.J., II; Fagan, J. Enhancing one life rather than living two: Playing MMOs with offline friends. *Comput. Hum. Behav.* **2011**, *27*, 1211–1222. [CrossRef]
12. Kiraly, O.; Griffiths, M.D.; King, D.L.; Lee, H.-K.; Lee, S.-Y.; Banyai, F.; Zsila, A.; Takacs, Z.K.; Demetrovics, Z. Policy responses to problematic video game use: A systematic review of current measures and future possibilities. *J. Behav. Addict.* **2018**, *7*, 503–517. [CrossRef]
13. Zendle, D.; Bowden-Jones, H. Loot boxes and the convergence of video games and gambling. *Lancet Psychiatry* **2019**, *6*, 724–725. [CrossRef]
14. Griffiths, M.D.; Pontes, H.M. The Future of Gaming Disorder Research and Player Protection: What Role Should the Video Gaming Industry and Researchers Play? *Int. J. Ment. Health Addict.* **2019**, 1–7. [CrossRef]
15. Booth, C. Facebook and Instagram officially announce new tools to fight social media addiction. Available online: https://thenextweb.com/facebook/2018/08/01/social-media-addiction-facebook-instagram/ (accessed on 17 February 2020).
16. Apple Use Screen Time on your iPhone, iPad, or iPod touch. Available online: https://support.apple.com/sl-si/HT208982 (accessed on 17 February 2020).
17. Derevensky, J.L.; Richard, J. Response to Commentary: The Future of Gaming Disorder Research and Player Protection: What Role Should the Video Gaming Industry and Researchers Play. *Int. J. Ment. Health Addict.* **2019**, 1–9. [CrossRef]
18. Kuss, D.J. Policy, prevention, and regulation for Internet Gaming Disorder. *J. Behav. Addict.* **2018**, *7*, 553–555. [CrossRef]
19. Colder Carras, M.; Porter, A.M.; Van Rooij, A.J.; King, D.; Lange, A.; Carras, M.; Labrique, A. Gamers' insights into the phenomenology of normal gaming and game "addiction": A mixed methods study. *Comput. Hum. Behav.* **2018**, *79*, 238–246. [CrossRef] [PubMed]
20. Shi, J.; Potenza, M.N.; Turner, N.E. Commentary on: "The Future of Gaming Disorder Research and Player Protection: What Role Should the Video Gaming Industry and Researchers Play?". *Int. J. Ment. Health Addict.* **2020**, 1–9. [CrossRef]
21. Claassen, C.A. The agenda development process of the United States' National Action Alliance for Suicide Prevention Research Prioritization Task Force. *Crisis* **2013**, *34*, 147–155. [CrossRef]
22. Khodyakov, D.; Savitsky, T.D.; Dalal, S. Collaborative learning framework for online stakeholder engagement. *Health Expect.* **2015**, *19*, 868–882. [CrossRef] [PubMed]
23. Dalal, S.; Khodyakov, D.; Srinivasan, R.; Straus, S.; Adams, J. ExpertLens: A system for eliciting opinions from a large pool of non-collocated experts with diverse knowledge. *Technol. Forecast. Soc. Chang.* **2011**, *78*, 1426–1444. [CrossRef]
24. Jones, L.; Wells, K. Strategies for academic and clinician engagement in community-participatory partnered research. *JAMA* **2007**, *297*, 407–410. [CrossRef]
25. Creswell, J.; Klassen, A.; Plano Clark, V.; Smith, K. *Best Practices for Mixed Methods Research in the Health Sciences*; National Institutes of Health: Rockville, MD, USA, 2011.
26. Morgan, D.L. Paradigms Lost and Pragmatism Regained: Methodological Implications of Combining Qualitative and Quantitative Methods. *J. Mix. Methods Res.* **2007**, *1*, 48–76. [CrossRef]
27. Levitt, H.M.; Bamberg, M.; Creswell, J.W.; Frost, D.M.; Josselson, R.; Suárez-Orozco, C. Journal article reporting standards for qualitative primary, qualitative meta-analytic, and mixed methods research in psychology: The APA Publications and Communications Board task force report. *Am. Psychol.* **2018**, *73*, 26–46. [CrossRef] [PubMed]
28. APA Style Journal Article Reporting Standards (APA Style JARS). Available online: https://apastyle.apa.org/jars/index (accessed on 4 May 2020).
29. King, D.; Delfabbro, P.; Griffiths, M. Video game structural characteristics: A new psychological taxonomy. *Int. J. Ment. Heal. Addict.* **2009**, *8*, 90–106. [CrossRef]
30. MDA: A Formal Approach to Game Design and Game Research. Available online: https://users.cs.northwestern.edu/~{}hunicke/MDA.pdf (accessed on 28 May 2020).
31. Ido, S. Using the Agential Structure Model to Classify Fun. Available online: https://www.gamasutra.com/blogs/SatoshiIdo/20170828/304503/Using_the_Agential_Structure_Model_to_classify_fun.php (accessed on 28 May 2020).

32. VandenBerghe, J. The 5 Domains of Play: Applying Psychology's Big 5 Motivation Domains to Games. In Proceedings of the 2012 Game Developers Conference (GDC), San Francisco, CA, USA, 25–29 March 2012.
33. Braun, V.; Clarke, V. Using thematic analysis in psychology. *Qual. Res. Psychol.* **2006**, *3*, 77–101. [CrossRef]
34. Museum of Science Fiction 2017 Museum of Science Fiction: Escape Velocity Prospectus and Program Guide. 2017. Available online: https://static1.squarespace.com/static/52615e54e4b051d1915a8890/t/5a15da39652dea2e1ac54cfb/1511381569778/MOSF+Prospectus+and+EV2017+Program+Guide+Web+Res.pdf (accessed on 1 May 2020).
35. StataCorp. *Stata Statistical Software: Release 13*; StataCorp LP.: College Station, TX, USA, 2013.
36. Chatfield, M.; Mander, A. The Skillings–Mack test (Friedman test when there are missing data). *Stata J.* **2009**, *9*, 299–305. [CrossRef] [PubMed]
37. World Health Organization. *Communicating Risk in Public Health Emergencies: A WHO Guideline for Emergency Risk Communication (ERC) Policy and Practice*; World Health Organization: Geneva, Switzerland, 2018; ISBN 92-4-155020-1.
38. Halberstadt, B. YouTubers and Streamers Speak up about Mental Health | GEMH Lab. Available online: https://gemhlab.com/blog/depression/lets-talk-about-mental-health-youtubers-and-twitch-streamers-speak/ (accessed on 15 February 2020).
39. Extra Credits Exit Points-Putting Down the Game-Extra Credits. 2015. Available online: https://www.youtube.com/watch?v=GqjkWec61gQ (accessed on 17 September 2019).
40. Rumpf, H.-J.; Achab, S.; Billieux, J.; Bowden-Jones, H.; Carragher, N.; Demetrovics, Z.; Higuchi, S.; King, D.L.; Mann, K.; Potenza, M.; et al. Including gaming disorder in the ICD-11: The need to do so from a clinical and public health perspective. *J. Behav. Addict.* **2018**, *7*, 556–561. [CrossRef]
41. King, D.L.; Delfabbro, P.H. Video Game Monetization (e.g., 'Loot Boxes'): A Blueprint for Practical Social Responsibility Measures. *Int. J. Ment. Health Addict.* **2018**, *17*, 166–179. [CrossRef]
42. Duggan, M. Gaming and Gamers. Pew Research Center: Internet, Science & Tech. 2015. Available online: http://www.pewinternet.org/2015/12/15/gaming-and-gamers/ (accessed on 30 December 2015).
43. Entertainment Software Association Essential Facts about the Computer and Video Game Industry: 2018 Sales, Demographic and Usage Data. 2018. Available online: https://www.theesa.com/wp-content/uploads/2019/03/ESA_EssentialFacts_2018.pdf (accessed on 30 April 2019).

International Journal of
Environmental Research and Public Health

MDPI

Article

Psychological Risk Factors that Predict Social Networking and Internet Addiction in Adolescents

Montserrat Peris [1], Usue de la Barrera [2], Konstanze Schoeps [3] and Inmaculada Montoya-Castilla [2,*]

[1] Department of Personality, Evaluation and Psychological Treatments, University of the Basque Country, 20018 San Sebastian, Spain; montserrat.peris@ehu.eus
[2] Department of Personality, Assessment and Psychological Treatment, Faculty of Psychology, University of Valencia, 46010 Valencia, Spain; usue.barrera@uv.es
[3] Department of Psychology, Faculty of Health Sciences, European University of Valencia, 46010 Valencia, Spain; konstanze.schoeps@universidadeuropea.es
* Correspondence: inmaculada.montoya@uv.es; Tel.: +34-963983433

Received: 22 May 2020; Accepted: 24 June 2020; Published: 26 June 2020

Abstract: Adolescents' addictive use of social media and the internet is an increasing concern among parents, teachers, researchers and society. The purpose was to examine the contribution of body self-esteem, personality traits, and demographic factors in the prediction of adolescents' addictive use of social media and the internet. The participants were 447 Spanish adolescents aged 13–16 years (M = 14.90, SD = 0.81, 56.2% women). We measured gender, age, body self-esteem (body satisfaction and physical attractiveness), personality traits (extraversion, neuroticism, disinhibition and narcissism) and social networking and internet addiction (internet addiction symptoms, social media use, geek behaviour, and nomophobia). The effects of gender, age, body self-esteem and personality on the different dimensions of internet addiction were estimated, conducting hierarchical linear multiple regression analysis and a fuzzy-set qualitative comparative analysis (fsQCA). The results evidenced different pathways explaining four types of adolescents' internet addiction: gender and disinhibition were the most relevant predictors of addiction symptoms; gender combined with physical attractiveness best explained social media use; narcissism and neuroticism appear to be the most relevant predictors of geek behaviour; and narcissism was the variable that best explained nomophobia. Furthermore, the advantages and differences between both methodologies (regressions vs. QCA) were discussed.

Keywords: adolescents; internet addiction; social networking; body self-esteem; personality traits; fsQCA models

1. Introduction

1.1. Risk of Internet Addiction and Social Media Use

The use of social networking sites and the internet has grown in popularity over the last few decades and new technological tools such as smartphones may have become indispensable today [1,2]. Prevalence rates vary considerably in internet addiction research. Across Europe, recent studies reported prevalence ranging between 4.4% to 13.5% for pathological internet use and between 14.3% and 54.9% for problematic internet use [3]. In Spain, the prevalence of problematic internet users has been estimated to be between 18.5% and 4.9% of pathological internet users [4]. Adolescence is an especially vulnerable period of change and teenagers face the risk of suffering symptoms of addiction as a result of their daily social network use [2]. Traditionally, internet addiction has been analysed in a generalised and global way, however, recent studies suggest the relevance of investigating each subtype of addiction in each particular population [5]. For instance, Peris et al. [6] consider four types of risk behaviour regarding social networking and internet addiction in adolescents on a subclinical level: internet addiction symptoms, social media use, geek behaviour, and nomophobia.

Internet addiction symptoms express the users' urge to continue being connected despite the desire to stop, experiencing unpleasant emotions when they do not succeed. Adolescents, who perceive the constant need to be online, usually present an increased use of social networking sites and the internet. In extreme cases, such behaviour may produce psychological problems that traditionally correspond to substance-related addictions [2]. Addiction symptoms would include sleep disturbance, angry or agitated reaction when forced to disconnect, and loosing track of time while online [6]. Personal characteristics such as gender and age appear to play an essential role in internet addiction [7], however, the results are not conclusive. On the one hand, recent studies report that girls experience higher amount of internet addiction symptoms than boys [4,8,9]. On the other hand, some other studies reveal that boys are more susceptible to addictive online behaviour than girls [10]. With regard to age difference, the results are also inconclusive. On the one hand, some studies report a progressive increase in addiction to the internet with age, enhancing also the negative consequences [11], while other studies report a progressive decrease [12]. On the other hand, there are studies that do not observe any age differences at all [9]. Hence, research investigating the influence of gender and age on internet addiction during adolescence seems warranted.

The aspect of social media use refers to a virtual way of relating with peers, reducing face-to-face interaction. Social networking sites have become a new environment of group socialization for adolescents. Such virtual environments enable the users to create public profiles in an interactive space where they communicate with friends but also with strangers, people they never have contact with outside the network [13]. Excessive use of social networking sites, however, can have negative implications such as negative mood states, concentration problems, and less interest in spending time with friends and family [14]. In relation to gender, it appears that women tend to use smartphones primarily for communication and social media such as Instagram and Facebook [15,16]. On the contrary, other studies have reported that boys are more likely to present problematic social media use than girls [17]. With regard to age difference, studies reveal that the risk and frequency of addictive use of social networking sites decrease as age increases [18].

Geek behaviour refers to a certain habit, which is commonly ascribed to an intense level of interest in the technological field. The term *geek* is widely used among young people in the digital era that is why there is still little literature about the phenomenon. In this study, the term geek behaviour refers to the quality of young people who are passionate about information and communication technologies (ICT), including online social networking. In addition, they usually prefer interacting with their peers through the internet, given the easy access to websites where you can make friends online or look for an online date [19]. Three characteristics have been identified: relating through online games, searching for similar interest groups and/or online eroticism [20]. With regard to gender, research consistently shows that boys engage more frequently in compulsive gaming and gambling [15,21]. Although boys show typically more geek behaviour, some studies have evidenced such behaviour is increasing among girls, too [22]. Furthermore, research stresses that being an adolescent is a well-established risk factor for internet gaming addiction [23].

Nomophobia is conceptualized as the intense fear of being disconnected from smartphone communication on a nonclinical level [24]. Messaging apps have increased nomophobia, especially in teenagers, the age group most affected by this problem. Currently, the accelerating development of smartphones devices and the fact that they are easy to carry around all the time, means that mobile phones are replacing the internet as a primary addictive source [25]. In relation to gender, studies suggest that girls use their smartphones more often than boys and that their use is, therefore, more problematic [16]. Hence, addictive behaviour in boys is associated with gaming apps, while in girls smartphone use is more related to online communication and social interaction [15]. With regard to age, adolescents have been targeted as a group at risk for smartphone addiction [26] and a recent meta-analysis suggests that one in four adolescents presents problematic smartphone use [27].

Social cognitive theory recognizes "personal factors" as potential predictors of internet addiction disorder and nomophobia, considering gender, age, ethnicity or beliefs as the most relevant. However, the contribution of other psychological factors such as self-esteem and personality traits should be considered in order to better understand of such risk behaviour online.

1.2. Body Self-Esteem and the Relationship with Problematic Social Media and Internet Use

Body self-esteem is another psychological variables that has been positively associated with the social networking addiction [28]. The internet is the virtual mirror of the current body image by exposing a manipulated online version with an increased physical and erotic appeal [29]. Frequent social media users have greater body dissatisfaction because they have internalized the model of a thin and long-legged body [30]. On the one hand, physical attractiveness -the emotional dimension of body self-esteem- may correlate positively with social networking and internet addiction [31], while body satisfaction -the cognitive dimension of body self-esteem- may correlate negatively. Recent studies reveal that adolescents who are dissatisfied with their bodies tend to selectively disclose their very positive and attractive features (e.g., their most attractive photographs). Such behaviour may work as a self-preservation strategy that helps to restore their self-confidence and receive positive feedback from their peers, which in turn generates positive feelings about themselves [32]. These idealized self-preservation motivations are positively associated with social networking addiction [33].

On the other hand, body satisfaction seems to be related to social desirability, thus, peers play a decisive role in the matter of public online appearance. Peris et al. [34] reported that an increasing number of adolescents edit their pictures before posting them on social networking sites to improve their image according to how they wish to see themselves and be seen by others. Following this line of research, studies indicate a negative association between adolescent body satisfaction and the use of social networking sites. According to Liu et al. [35], adolescents who are unsatisfied with their body, experience low self-esteem in real life and, therefore, more often seek positive responses through social media. In turn, positive responses from online social interactions may reinforce young people's use of smartphones, enhancing the risk of developing smartphone addiction. If adolescents would receive more positive responses and support in real life, they might refrain from frequent online interactions, even if they have low body satisfaction.

1.3. Personality Traits and the Association with Social Networking and Internet Addiction

In addition to self-esteem, personality has an important bearing on explaining individual differences in problematic internet use [33,36]. One of the most studied personality dimension related to internet addiction is extraversion, which refers to people's varying tendency to be friendly, sociable, and talk active. Studies suggest that there is a positive relationship between being extroverted and an increased risk of internet addiction [37]. Regarding the internet and social media use, they are both strongly and positively related to extraversion [38]. Extraverted individuals use social networking apps to engage in social interaction and they may attain more social resources online. In relation to the geek behaviour, Dieris-Hirche et al. [39] reported that gamers with a problematic internet use scored lower on extraversion and higher on neuroticism than those with non-problematic behaviour. Furthermore, as smartphones allow for the permanent connection demanded by extroverted people, extraversion has been identified as a predictor variable of problematic smartphone use [40]. However, some recent studies did not confirm such relationship [41]. These discrepancies may suggest the importance of a differential conceptualization of social networking and internet addiction in relation to different kinds of risk behaviour.

Another main personality factor is neuroticism, which refers to individual differences in emotional stability and psychological adjustment. Neuroticism has been shown a positive association with problematic internet use [42]. Studies suggest that introverted adolescents show increased online activity when compared with emotionally stable users, which may enhance the risk for internet addiction [27]. Regarding the social aspect of internet use, neurotic adolescents tend to show a more passive behaviour on social networking sites such as posting comments and Likes on other people's profiles, which may represent a way to seek social relationships [33,36,43]. This association can be interpreted in terms of their difficulties to pursue social relationships in real life and poor coping skills with emotional situations [42]. Research on the relationship between neuroticism and geek behaviour is limited. Furthermore, neurotics tend to show an increased or problematic use of online games [39]. With regard to nomophobia, neuroticism appears to predict obsessive smartphone use [44].

Besides extraversion and neuroticism, disinhibition is an other relevant personality domain, which is associated with increased internet addiction symptoms [45]. Research indicates that problematic internet and social media use is linked to different aspects of disinhibited behaviour such as poor self-control, impulsivity and sensation seeking. Disinhibited personality is common among teenagers, but usually decreases with age from adolescence to adulthood, supporting the hypothesis of normative psychological maturation. Apparently, for disinhibited adolescents the internet may represent a useful tool for satisfying their elevated urge of sensation seeking [45] by allowing relief from unpleasant feelings through quick online behaviour without thinking much of the consequences (e.g., by quickly engaging in anonymous virtual relationships) [31]. In view of the relevance of this personality domain related to addictive internet behaviour, it appears negligent that only few studies have included disinhibition as a potential predictor of social networking and internet addiction.

Moreover, narcissism has been related to problematic internet use in adolescence [46]. Narcissism may be understood as a dynamic self-regulation system related to grandiosity, self-absorption and a constant need for external approval. There is evidence to suggest that higher levels of narcissism are linked to an increased risk of social networking and internet addiction [28,37]. Social media use appear to fulfil two basic social needs that may explain the problematic internet use in young people with narcissistic personality: the need for self-presentations through social online profiles featured by pictures, status updates, personal notes, etc.; and the need to belong to a social group seeking repeated approval from their peers [47]. Furthermore, narcissistic teenagers tend to use social media such as Facebook more often because such platforms enable users to create visible profiles, that allow idealized self-promotion in a virtual environment [37]. In addition to problematic social media use, narcissism is strongly associated with smartphone addiction [48].

1.4. Rationale for the Study

Drawing from previous studies, there are psychological risk factors of significant relevance during adolescence such as gender and age, body self-esteem and personality traits, including extraversion, neuroticism, disinhibition and narcissism that predict internet-related addictions [16,21,31,37]. Most studies use a very limited conceptualization of social networking and internet addiction rather than broadening their view and focus on specific addictive behaviours (e.g., [5,45]). To our knowledge, there are no studies that have investigated the combined effects of body self-esteem and personality traits in order to determine different types of risk behaviour regarding the social networking and internet addiction in nonclinical adolescents, involving internet addiction symptoms, social media use, geek behaviour, and nomophobia. Additionally, the role of demographic factors such as gender and age have been extensively studied in relation to adolescent problematic internet use, however with mixed findings [8–10,16]. In addition, most research on social networking and internet addiction has used methodology based on linear regression models (e.g., [1,49]). These models are focused on the individual prediction and do not consider the possibility of different pathways that would lead to the same result [50,51]. In contrast, fuzzy-set qualitative comparative analysis (fsQCA) is a methodology that allows a more in-depth analysis of how a set of causal conditions contribute to a hypothesised outcome [52,53]. QCA is based on the assumption that such outcome depends on a combination of different factors rather than on individual levels of those factors [54,55]. In the field of psychological research, limited studies have used this technique despite it's considerable potential [56].

1.5. Purpose of the Study

The present study aimed to estimate the combined contribution of body self-esteem (body satisfaction and physical attractiveness), personality traits (extraversion, neuroticism, disinhibition and narcissism), and demographic factors (gender and age) in the prediction of four types of adolescent's social networking and internet addiction (internet addiction symptoms, social media use, geek behaviour, and nomophobia). Based on the reviewed research, we hypothesized as follows (1) the primary risk factors that predict internet addiction symptoms will be gender and age (girls of older age), low body satisfaction, and high levels of physical attractiveness, neuroticism, extraversion, disinhibition and narcissism; (2) the potential risk factor of social media use will be gender and age (girls of younger age), low body satisfaction, but high levels of physical attractiveness, neuroticism, extraversion, disinhibition and narcissism; (3) the significant risk factors that predict geek behaviour will be gender (boys), low body satisfaction and extraversion, but high physical attractiveness neuroticism, disinhibition and narcissism; and (4) the main risk factors of nomophobia will be gender (girls), low body satisfaction, but high levels of physical attractiveness, neuroticism, extraversion, disinhibition and narcissism.

2. Materials and Methods

2.1. Participants

The present research involved 447 adolescents aged 13–16 years ($M = 14.90$, $SD = 0.81$). The gender distribution of the sample was equitable (women: $n = 251$; 56.2%). The participants were students from public ($n = 201$; 45%) and private ($n = 246$; 55%) high schools in the Northern Regions of Spain. The following inclusion criteria applied: (a) school board gave their permission to collaborate with the research group; (b) students were not older than 16 years; (c) parents or guardians were asked to sign a written consent to allow adolescents' participation in the research. The simple random probability sampling method has been employed.

2.2. Variables and Instruments

2.2.1. Demographic Variables

Personal data referring to the students' gender, age and high school were collected administrating an ad hoc questionnaire.

2.2.2. Social Networking and Internet Addiction

The social networking and internet addiction was assessed using the Scale of risk of addiction to social media and the internet for adolescents (ERA-RSI) [6]. The scale is composed of 29 items divided into four subscales: internet addiction symptoms, social media use, geek behaviour, and nomophobia. The internet addiction symptoms subscale assesses behaviours of addiction to non-toxic substances (e.g., "I have been losing sleep over social media and watching online shows"; 9 items). The social media use subscale assesses adolescent "online socialization" behaviours (e.g., "I check my friends' profiles"; 8 items). The geek behaviour subscale includes aspects such as joining special interest groups, playing online and role-playing games or having sexual encounters (e.g., "I spend time on social media and the Internet to play online and/or role-playing games"; 6 items). The nomophobia subscale is related to feelings of anxiety and control when using a mobile phone (e.g., "I have a smartphone and I start feeling anxious or distressed when people don't answer immediately to my messages"; 6 items). Participants were asked to rate the items on a 4-point Likert scale (1 = *never or hardly never*; 4 = *many times or almost always*). The reliability indexes were adequate in this sample for the global scale of addiction ($\alpha = 0.90$) and all subscales: internet addiction symptoms ($\alpha = 0.84$), social media use ($\alpha = 0.83$), geek behaviour ($\alpha = 0.69$), nomophobia ($\alpha = 0.80$).

2.2.3. Body Self-Esteem

The body self-esteem was measured with the Body Self-esteem Scale (BSS) [57]. This scale is composed of 26 items and participants evaluate the degree of satisfaction with each part of their body. The first 20 assess body satisfaction, the cognitive dimension of body self-esteem, and are grouped into four body areas: face (e.g., "Are you satisfied with your eyes/mouth?"), upper torso (e.g., "Are you satisfied with your breasts/pectorals?"), lower torso (e.g., "Are you satisfied with your butt?"); and anthropometry (e.g., "Are you satisfied with your height/size?"). The score ranges on a 10-point Likert scale (1 = *very dissatisfied*, 10 = *very satisfied*). Physical attractiveness, the emotional dimension of body self-esteem, is assessed through 6 items and six aspects are evaluated: physically interesting, socially charming, sexy, attractive, sensual and erotic (e.g., "To what extent do you consider yourself a physically attractive person?"). The score ranges on a 10-point Likert scale (1 = *not attractive at all*, 10 = *very attractive*). This scale shows good psychometric properties in the studied sample (Body satisfaction $\alpha = 0.93$; Physical attractiveness $\alpha = 0.93$; Body Self-esteem $\alpha = 0.95$).

2.2.4. Personality Factors

The extraversion and neuroticism dimensions were assessed using NEO Five Factory inventory (NEO-FFI) [58], the brief version of NEO-PI-R. The scale is composed of 60 items distributed equally on five factors (12 items in each factor). Two subscales have been used in the present study: neuroticism (e.g., "I am quite emotionally stable"–inversed item) and extraversion (e.g., "I like having a lot of people around me"). They were chosen for two reasons. First, they are the ones that accumulate the most research with regard to social media and internet use, and second, because it was necessary to reduce the application time due to the fatigue and tiredness. The scale uses a 5-point Likert scale (0 = *strongly disagree*; 4 = *strongly agree*). Both factors have shown suitable reliability in this sample (extraversion: α= 0.81; neuroticism: α= 0.79).

The disinhibition was evaluated using the Sensation Seeking Scale (SSS-Q) [59], Spanish adaptation [60]. The scale is composed of 40 items, which give rise to 4 subscales (10 items each). We selected the subscale of disinhibition for the aim of our study, given that its content is most relevant for risk behaviour and addiction. The participants provide information about their own disinhibition behaviours (e.g., "I like wild parties without limits"). The adolescents can answer the items *affirmatively* (1) or *negatively* (0). The scale has shown good reliability in this sample ($\alpha = 0.64$).

The narcissism was evaluated using the Narcissistic Personality Inventory (NPI) [61], Spanish adaptation (NP-15) [62]. The scale is composed of 15 items and the narcissism includes facets such as a need for recognition, an exalted and distorted image or a feeling of special status (e.g., "It's very important that others pay attention to me and admire what I do"). The scale uses a 6-point Likert scale (1 = *absolutely false*; 6 = *absolutely true*). The scale has shown adequate reliability in this sample ($\alpha = 0.81$).

2.3. Procedure

The present research was approved by the Ethics Commission for Research with Human Beings (CEISH) of the University of the Basque Country Euskal Herriko Unibersitatea (UPV/EHU), with the registration number CEISH/136/2012/PERIS HERNANDEZ. The declaration of the file was in the Basque Agency of Data Protection with the registration number 2080310015-INA0004. The study applied the ethical principles of the Declaration of Helsinki, the requirements established by the Ethics Commission for Human Research of the University of the Basque Country (UPV) and the State Government, as well as the deontological regulations of the Official College of Psychologists for experimentation on human beings.

First, the headmasters and psychologists of each school were contacted and information about the study and a copy of the research project were provided. Once they accepted to participate, the parents of the adolescents were contacted. An information letter and informed consent was sent to their homes. The informed consent was signed and returned to the school by the adolescent's parents or legal guardians in order to participate. The adolescents were informed about the questionnaires and confidentiality, and they signed the informed consent. The data collection took place in the school classrooms during the tutorial hour. The questionnaires were administered by evaluators with a Degree in Psychology and lasted approximately 40 min.

2.4. Data Analysis

Firstly, descriptive analysis followed by bivariate correlations and linear regression analyses were conducted using the statistical package IBM SPSS V.25 for Windows (IBM Corporation, Foster City, CA, USA). A three-stage hierarchical multiple regression analyses was performed to examine the predictive power of demographic variables (age and gender), body self-esteem (body satisfaction and physical attractiveness), and personality factors (neuroticism, extraversion, disinhibition and narcissism) on social networking and internet addiction. A total of four regression models were carried out, one for each dimension of addiction and predictors were entered in three stages: (a) Gender and age, (b) body self-esteem and (c) personality factors.

Secondly, we performed fuzzy-set qualitative comparative analysis (fsQCA) using the Fs/QCA 3.0 software (University of California, Irvine, CA, USA). Prior to conducting the analysis, we estimated the calibration scores, transformed raw data responses into fuzzy-set responses, and removed missing data. In order to obtain the constructs (variables) and increase the variability, the items of each scale were multiplied [63]. Following the multiplication of the items, the extraversion and neuroticism scales were divided by one hundred to avoid excessively large numbers that the program could not handle. The rest of the variables remained undivided. In addition, each variable was then recalibrated in three categories: percentile 10 (low levels or condition is absent), percentile 50 (intermediate level, condition is neither absent nor present) and percentile 90 (high levels or condition is present) [64]. All scores must range between 0 and 1. Thus, gender and age scores were calibrated manually. Age scores were coded according at four points equidistant between 0 and 1, gender scores were recoded with 0 for boys and 1 for girls. We conducted descriptive analyses with the transformed scores. Finally, necessary and sufficient conditions analyses estimated the combined influence of the demographic variables, body self-esteem and personality factors on high levels of internet addiction symptoms, social media use, geek behaviour and nomophobia. There is no theoretical number of combinations that produces the outcome.

3. Results

3.1. Descriptive Analysis and Relationships Between Variables Studied

Descriptive statistics (means and standard deviations) and correlations between the study variables are displayed in Table 1. Results indicated that age was significantly and in a negative way related to social media use, nomophobia, neuroticism and extraversion, while the associations with disinhibition and narcissism are positive. In general, personality was positively and significant related to social networking and internet addiction. Specifically, disinhibition was associated with the four dimensions of addiction; neuroticism and extraversion were related to internet addiction symptoms, social media use and nomophobia, but not with geek behaviour; and there were a positive association between narcissism and internet addiction symptoms, geek behaviour and nomophobia, but there are not with social media use. With regard to body self-esteem, physical attractiveness was significantly and positively correlated with the four dimensions of addiction to the internet and online social networking, whereas body satisfaction only correlated to internet addiction symptoms and the relationship was negative. As regards the relationship between personality and self-esteem, neuroticism was negatively and significant related to body satisfaction and attraction, whereas extraversion, disinhibition and narcissism were related positively, although there was not relationship between disinhibition and body satisfaction.

Table 1. Bivariate correlations between all variables studied.

	1	2	3	4	5	6	7	8	9	10	11
1. Age	–										
2. IAS	−0.07	–									
3. SMU	−0.17 ***	0.56 ***	–								
4. GB	−0.02	0.29 ***	0.30 ***	–							
5. NP	−0.16 **	0.61 ***	0.55 ***	0.32 ***	–						
6. NT	−0.12 *	0.36 ***	0.29 ***	0.03	0.25 ***	–					
7. EX	−0.11 *	0.12 ***	0.20 ***	0.04	0.09 ***	−0.19 ***	–				
8. DI	0.16 ***	0.40 ***	0.23 ***	0.16 ***	0.23 ***	0.12 **	0.20 ***	–			
9. NA	0.11 *	0.21 ***	0.02	0.18 ***	0.18 ***	0.06	−0.11 *	0.20 ***	–		
10. BS	−0.01	−0.15 ***	−0.02	0.06	−0.03	−0.32 ***	0.20 ***	−0.03	0.10 *	–	
11. PA	0.06	0.13 **	0.11 **	0.15 ***	0.13 **	−0.20 ***	0.25 ***	0.22 ***	0.26 ***	0.66 ***	–
M	14.90	19.05	21.17	9.31	12.62	32.89	46.27	14.73	40.45	6.58	6.18
SD	0.81	5.77	4.86	2.77	4.21	7.31	6.33	2.13	10.13	1.32	1.67

Note. IAS = Internet addiction symptoms. SMU = Social media use. GB = Geek behaviour. NP = Nomophobia. NT = Neuroticism. EX = Extraversion. DI = Disinhibition. NA = Narcissism. BS = Body satisfaction. PA = Physical attractiveness. *M* = mean. *SD* = standard deviation. * $p \leq 0.05$. ** $p \leq 0.01$. *** $p \leq 0.001$.

3.2. Demographic and Psychological Predictors of Social Networking and Internet Addiction

Predictive analysis of social networking and internet addiction was conducted with a three-step hierarchical multiple regressions (Table 2).

Regarding the first prediction model, three sets of variables were established, which explained 37% of the variance of adolescents' internet addiction symptoms. In the first step, which included demographic variables, specifically gender and age, explained 9% of the variance. The second next step, which included two dimensions of body self-esteem, accounted for an additional 10% of explained variance. The third and final step included the four dimensions of personality explained an additional 19% of variance. The final regression model indicates that gender, body satisfaction, physical attractiveness, neuroticism, extraversion, disinhibition and narcissism are significant predictors of emotional symptoms.

The second prediction model consisted of three sets of variables, which explained 35% of the variance of social media use. In the first step, gender and age were included and together they explained 23% of the variance. In the second step, the two dimensions of body self-esteem were entered and accounted for a significant increase of 5% of the variance. In the final step, the four personality dimensions were included, which accounted for an additional 10% of variance. In this final model, gender, age, physical attractiveness, neuroticism, extraversion, and disinhibition significantly predicted social media use.

With regard to the third prediction model, the overall regression model was significant but only explained 5% of geek behaviour. Following the same *modus operandi*, demographic variables were included in step 1 explaining 1% of the variance, followed by both dimensions of body self-esteem in step 2 increasing the explained variance by 3% and finally all personality four dimensions were entered in step accounting for an additional 3% of the variance. The resulting model suggested that only disinhibition and narcissism significantly predict geek behaviour.

The forth the prediction model was established in three steps and explained 21% of the variance of nomophobia. Firstly, gender and age were entered in the first step and explained 8% of the variance. Secondly, body satisfaction and physical attractiveness were included in the second step elevating the explained variance by 8%. Thirdly, neuroticism, extraversion, disinhibition and narcissism were added in the third step and together explained an additional 9 % of the variance. In this overall model, gender, age, physical attractiveness, neuroticism, disinhibition and narcissism significantly predicted nomophobia.

Table 2. Results of hierarchical multiple regression analyses.

Variable	Internet Addiction Symptoms				Social Media Use				Geek Behaviour				Nomophobia			
	ΔR^2	ΔF	β	*t*	ΔR^2	ΔF	β	*t*	ΔR^2	ΔF	β	*t*	ΔR^2	ΔF	β	*t*
Step 1	0.09	21.93 ***			0.23	64.84 ***			0.01	1.62			0.08	20.43 ***		
Gender			0.27	6.50 ***			0.45	10.67 ***			−0.06	−1.08			0.26	5.51 ***
Age			−0.06	−1.45			−0.08	−2.07 *			−0.07	−1.34			−0.14	−3.12 **
Step 2	0.10	28.34 ***			0.05	12.47 ***			0.03	5.77 **			0.05	13.08 ***		
Body satisfaction			−0.18	−3.34 ***			0.04	0.76			−0.04	−0.54			−0.05	−0.87
Physical attractiveness			0.21	3.89 ***			0.11	1.99 *			0.12	1.76			0.16	2.59 **
Step 3	0.19	33.64 ***			0.10	16.51 ***			0.03	3.52 **			0.09	12.66 ***		
Neuroticism			0.23	5.44 ***			0.18	4.16 ***			0.03	0.59			0.15	3.20 ***
Extraversion			0.09	2.11 *			0.15	3.44 ***			0.01	0.18			0.05	1.15
Disinhibition			0.30	7.06 ***			0.20	4.63 ***			0.10	2.02 *			0.17	3.56 ***
Narcissism			0.16	3.93 ***			0.04	1.00			0.13	2.57 *			0.17	3.68 ***
Durbin-Watson		1.27				1.26				1.61				1.33		
R^2		0.37 ***				0.35 ***				0.05 **				0.21 ***		

Note. ΔR^2 = change in R^2; ΔF = change in *F*; ß = regression coefficient; *t* = value of *t*-test statistic; * $p \leq 0.05$. ** $p \leq 0.01$. *** $p \leq 0.001$.

3.3. Combined Contribution of Body Self-Esteem, Personality Traits and Personal Predictors of Social Networking and Internet Addiction

The descriptive statistics of the variables under study and the calibration values were calculated first (Table 3). The dimensions of social networking and internet addiction based on personality, body self-esteem and personal factors were then examined by fuzzy-set QCA. The necessary and sufficient conditions were estimated. On the one hand, the analysis of necessary conditions or variables allows us to determine whether there are any variables that are always required to be present for the prediction of the outcome (in our study, high levels of addiction). For a condition/variable to be necessary, the consistency score must be above 0.90 [51]. The results showed that none of the variables studied have to be considered necessary condition because their consistencies scores were all below 0.90. On the other hand, the sufficient conditions refer to those variables that predict the outcome, but the prediction may be possible without them. All logically possible combinations of the causal conditions are captures in the truth table together with the result of each setting [50]. The analysis provides three types of solutions: a parsimonious one (less restrictive), an intermediate one and a complex one (most restrictive). The literature recommends, therefore, focusing on the intermediate solution [51], which is presented in this study.

Table 3. Descriptive statistics and calibration scores.

		IAS	SMU	GB	NP	NT	EX	DI	NA	BS	PA
Mean		5164.83	6055.23	32.66	181.55	13,973.16	210,162.93	69.35	746,999.25	2261.04	113609.19
Standard deviation		19,216.76	10,541.97	93.71	328.93	54,894.57	336,946.74	112.60	4,809,591.82	1705.54	156,534.70
Minimum		1.00	1.00	1.00	1.00	0.01	0.25	1.00	0.02	21.96	1.00
Maximum		262,144.00	65,536.00	972.00	2304.00	703,125.00	2,441,406.25	1024.00	84,375,000.00	10,000.00	1,000,000.00
Calibration scores											
	10	5.60	70.40	1.00	1.00	11.52	6635.52	4.00	6.48	439.84	1896.00
Percentile	50	384.00	1944.00	6.00	48.00	518.40	78,643.20	32.00	2764.80	1837.68	63,504.00
	90	10,368.00	15,552.00	72.00	524.80	22,413.31	562,500.00	128.00	813,957.12	4724.43	290,304.00

Note. IAS = Internet addiction symptoms. SMU = Social media use. GB = Geek behaviour. NP = Nomophobia. NT = Neuroticism. EX = Extraversion. DI = Disinhibition. NA = Narcissism. BS = Body satisfaction. PA = Physical attractiveness.

In the social networking and internet addiction, the combination of conditions resulting in high levels of internet addiction symptoms, social media use, geek behaviour and nomophobia were analysed (Table 4). The solutions were found to be adequate, considering that a fsQCA model is acceptable when consistency is above 0.70 [50]. The main three pathways for high levels of all four dimensions of social networking and internet addiction are shown in Table 4.

The solution showed eight pathways, which explained 46% of high levels of internet addiction symptoms. The first pathway was the result of the combined contribution of a high gender score (girl), high age, high disinhibition and high narcissism. The second pathway to predict high levels of internet addiction symptoms was the combined contribution of a high gender score (girl), high neuroticism, high extraversion and high disinhibition and the third was the result of the interaction of a high gender score (girl), high neuroticism, high extraversion, high disinhibition and high narcissism.

The solution showed 12 pathways, which explained 56% of high levels of social media use. The first pathway was the result of the combined contribution of high gender score (girl), high neuroticism and high physical attractiveness. The second pathway was the combined contribution of high gender score (girl), low neuroticism and high disinhibition. The third pathway was the result of the interaction of a high gender score (girl), high age, high extraversion and high physical attractiveness.

The solution showed eight pathways, which explained 33% of high levels of geek behaviour. The first combination was the result of the interaction of a high gender score (girl), high age, high neuroticism, high extraversion, high disinhibition and high narcissism. The second pathway was the result of the combined contribution of high gender score (girl), high age, high neuroticism, high extraversion, high narcissism, high body satisfaction and high physical attractiveness, and the third pathway was the result of the interaction of a low gender score (boy), high neuroticism, high narcissism and low physical attractiveness.

Table 4. Three main pathways for the high levels of social networking and internet addiction.

Frequency Cutoff: 1	High Internet Addiction Symptoms			High Social Media Use			High Geek Behaviour			High Nomophobia		
	Consistency Cutoff: 0.90			Consistency Cutoff: 0.90			Consistency Cutoff: 0.90			Consistency Cutoff: 0.90		
	1	2	3	1	2	3	1	2	3	1	2	3
Gender	•	•	•	•	•	•	•	•	○	•		
Age	•					•	•	•				○
Body satisfaction								•			○	○
Physical attractiveness				•		•		•	○		•	
Neuroticism		•	•	•	○		•	•	•		•	•
Extraversion		•				•	•	•		•	•	
Disinhibition	•	•	•		•		•				•	•
Narcissism	•		•				•	•	•	•	•	•
Raw Coverage	0.24	0.23	0.23	0.29	0.28	0.26	0.13	0.12	0.12	0.21	0.17	0.17
Unique Coverage	0.012	0.044	0.010	0.017	0.020	0.004	0.026	0.015	0.008	0.042	0.028	0.016
Consistency	0.88	0.90	0.91	0.91	0.88	0.91	0.89	0.89	0.89	0.85	0.94	0.90
Overall Solution Coverage			0.46			0.56			0.33			0.41
Overall Solution Consistency			0.86			0.83			0.85			0.85

Note. • = presence of condition/high levels, ○ = absence of condition/low levels. Gender: • = girls; ○ = boys. All sufficient conditions are adequate.

Finally, the solution showed 11 pathways, which explained 41% of high levels of nomophobia. The first pathway was the result of the interaction of a high gender score (girl), high extraversion and high narcissism. The second pathway was the result of the interaction of a high neuroticism, high extraversion, high disinhibition, high narcissism, low body satisfaction and high physical attractiveness. The third pathway predicting a high nomophobia was the result of the combined contribution of a low age, high neuroticism, high disinhibition, high narcissism and low body satisfaction.

The solution showed 12 pathways which explained 56% of high levels of social media use. The first pathway was the result of the combined contribution of high gender score (girl), high neuroticism and high physical attractiveness. The second pathway was the combined contribution of high gender score (girl), low neuroticism and high disinhibition. The third pathway was the result of the interaction of a high gender score (girl), high age, high extraversion and high physical attractiveness.

The solution showed eight pathways which explained 33% of high levels of geek behaviour. The first combination was the result of the interaction of a high gender score (girl), high age, high neuroticism, high extraversion, high disinhibition and high narcissism. The second pathway was the result of the combined contribution of high gender score (girl), high age, high neuroticism, high extraversion, high narcissism, high body satisfaction and high physical attractiveness, and the third pathway was the result of the interaction of a low gender score (boy), high neuroticism, high narcissism and low physical attractiveness.

Finally, the solution showed 11 pathways which explained 41% of high levels of nomophobia. The first pathway was the result of the interaction of a high gender score (girl), high extraversion and high narcissism. The second pathway was the result of the interaction of a high neuroticism, high extraversion, high disinhibition, high narcissism, low body satisfaction and high physical attractiveness. The third pathway predicting a high nomophobia was the result of the combined contribution of a low age, high neuroticism, high disinhibition, high narcissism and low body satisfaction.

4. Discussion

Although research on social networking and internet addiction is on the rise, a vast majority of investigations have conceptualized internet addictive as a one-dimensional construct, focusing on problematic or compulsive use rather than specific risk behaviour. We aimed to estimate the combined contribution of body self-esteem (body satisfaction and physical attractiveness), personality traits (extraversion, neuroticism, disinhibition and narcissism), and personal factors (gender and age) in the prediction of four types of adolescents' social networking and internet addiction, involving internet addiction symptoms, social media use, geek behaviour, and nomophobia. Several studies have investigated the relations between addiction to the internet and social networking and several of the variables included within this study [65]; however, the addiction literature using a novel methodological approach by comparing linear regression models with fsQCA models in the prediction of adolescents' internet addiction is sparse. Thus, this study makes an innovative contribution to the addiction literate by evaluating a more comprehensive model of addiction to internet and social media, examine different pathways of variables representing important biologically based personality traits, variables of body self-esteem, which are of special relevance during adolescence and taking into account gender and age differences. The findings extend our understanding of the psychological risk factors of social networking and internet addiction among nonclinical adolescents.

4.1. Risk Factors of Addiction Symptoms

The first hypothesis has been supported by the results from the hierarchical regression and fsQCA models, which showed a significant influence of gender, body self-esteem and personality traits on internet addiction symptoms. Both methodologies match the result that gender and disinhibition were the most relevant predictors of internet addiction symptoms. The regression model suggests that adolescents, more girls than boys, who are more disinhibited, neurotic, narcissistic and extraverted, and experience lower body satisfaction but higher physical attractiveness, present more internet addiction symptoms. The results from fsQCA suggested that the combination of gender (being a girl) and high levels of disinhibition were the most significant predictors of internet addiction symptoms. The two latter in combination with high levels of narcissism, neuroticism and extraversion also predicted an increase in internet addiction symptoms, but to a lesser extent. In contrast to the regression models, results from QCA suggests that body self-esteem does not seem to be an important predictor of internet addiction symptoms, since they have not been included in the three most significant pathways. These findings are in line with recent research, indicating that girls tend to present more addiction symptoms in comparison with boys [8,16]. However, these results are inconsistent with other studies that suggest that boys are more likely to show addictive internet behavior [10]. Evidentially, boys and girls may be at risk of potential internet addiction, but probably for different reasons. While girls are usually more interested in the social interactions over the internet, boys, contrarily, use the internet primarily for online gaming [3]. Furthermore, disinhibition has been the primary psychological risk factor for adolescents' internet addiction symptoms in our study, highlighting the role of impulsive behaviour as a manifestation of poor self-regulating skills in relation to problematic internet use during adolescence [45]. In general, personality factors appear to be more important than body self-esteem in the prediction of internet addiction, matching research finding that have pointed out the important of personality traits in explaining individual differences in symptoms of internet addiction [33,36].

4.2. Risk Factors of Social Media Use

Overall, the second hypothesis is also supported by the results obtained in our study, suggesting that the combination of demographic and psychological variables influence adolescents' problematic social media use. Both regression and QCA models indicate that gender combined with physical attractiveness seem to be the most relevant predictors of social media use. On the one hand, results from the hierarchical regression models revealed that girls more than boys, younger adolescents more than the older ones, with a physically attractive body image and a disinhibited, neurotic and extraverted personality, tend to use social media more often. On the other hand, fsQCA models suggest that there are different pathways to predict the problematic social media use. One of the main pathways shows that girls with higher physical attractiveness in combination with higher neuroticism report more excessive social media use. Another pathway suggests that girls with high levels of disinhibition in combination with low levels of neuroticism also show increased social internet use. A third pathway combines girls of older age with high physical attractiveness and high extraversion, in the prediction of addictive use of social networking sites. These differences may suggest that the combination of personality factors and body self-esteem may vary producing different patterns of risk behaviour. Stressing the role of neuroticism is critical, given that both high and low levels of emotional stability in combination with other psychological risk factors predict social media use. Our findings are relatively conforming with previous studies, which have provided hard evidence that girls use social media more often than boys [8,15]. In a recent study, Escario and Wilkinson (2020) [21] have provided evidence that men and women relate differently to the internet. For instance, while women tend to participate more in social activities, using chat rooms, sending e-mails and visiting social networking sites such as Facebook or Instagram, men spend more time on online games, online gambling, and visit pornographic sites more frequently. Our findings also corroborate previous studies that have reported a positive relationship between adolescents' body image and the use of social media [31]. In fact, teenage girls feel more pressured to present themselves with an idealized physical attractiveness by editing their self-portraits before sharing them in social media, which in turn generates positive feelings by receiving increased external approval, hence, higher risk of social networking addiction [2,7]. With regard to personality factors, our findings were consistent with previous studies, showing that disinhibition, neuroticism and extraversion were positively associated with addictive use of social networking sites [42,45].

4.3. Risk Factors of Geek Behaviour

Additionally, the third hypothesis has been supported by our results in terms of adolescents' geek behaviour, however the predictive capacity of demographic and psychological factors is lower than for the other types, irrespective of the methodology used. The personality factors narcissism and neuroticism appear to be the most relevant predictor of geek behaviour. The results from regression models indicate that high levels of narcissism and disinhibition predict geek behaviour, explaining only a small amount of variance. In fsQCA analysis, the model included more factors than in the other models in order to improve the prediction outcome, producing pathways with many different combinations of variables. For instance, the main pathway suggests that girls of older age with high levels of narcissism, neuroticism, extraversion and disinhibition tend to be geekier. Furthermore, the second pathway shows that girls of older age, who are neurotic, extraverted and narcissistic and experience higher levels of body satisfaction and physical attractiveness, also present more geek behaviour. Lastly, the third pathway indicates that boys with high levels of narcissism and neuroticism combined with an unattractive body image also predict greater geek behaviour. Both boys and girls, especially the older ones, appear to score high on geek behaviour, depending on the combination of several psychological risk factors. In fact, a single dimension of personality traits or body self-esteem is not sufficient in order to explain the individual difference in this type of internet addiction. These unexpected results are, however, compatible with some previous research, which suggests that both boys and girls may present geek behaviour [8]. Traditionally, geek behaviour related to excessive online gaming and gambling has been related to the male gender [15,21,65]. However, girls seem also to be at risk of such behaviour especially related to online eroticism, which is a novel insight regarding current literature. Another important finding of this study is that narcissism and neuroticism are the main predictors of geek behaviour. There is some evidence, that neurotic and introverted adolescents show increased online gaming behaviour [39]. Nevertheless, the role of narcissism in the prediction of geek behaviour has not been established in prior research, thus, our findings may give rise to a new line of research. The strong link between narcissism and geek behaviour may be explained by the "compensatory perspective", which suggest that online gaming might fulfil a compensatory purpose for narcissistic individuals with emotional dysregulation [66]. Regarding the role of body self-esteem, our results are less conclusive. Previous research have associated gamers' body dissatisfaction or negative body attitudes with the exposure to ideal video game bodies [67].

4.4. Risk Factors of Nomophobia

With regard to nomophobia, the results of this study confirm the forth hypothesis, demonstrating a significant impact of demographic factors, body self-esteem and personality traits. However, it should be highlighted that narcissism has been the most influential factor in the prediction of nomophobia in both methodologies. Based on the results of the hierarchical regressions, more girls than boys, younger adolescents more than the older ones, with an attractive body image and a primarily narcissistic, disinhibited and neurotic personality, report higher levels of nomophobia. Similarly, gender and age appear in the primary combinations of fsQCA that predict nomophobia. In addition, low levels of body satisfaction in combination of high levels of physical attractiveness also contribute to the prediction of nomophobia. Even though narcissism is the most significant predictor, high levels of neuroticism, extraversion and disinhibition also appear in two of the three main combinations. In line with previous studies, our findings provide further evidence that personality in general and narcissism specifically are significant predictors of adolescents' nomophobia [28,43]. Results from a recent studies indicate that the link between narcissism and smartphone distress may be explained by increased attention-seeking [48]. One of the characteristics of nomophobia is the constant checking for instant notifications, which may act as reward, but on the other hand, increases the level of anxiety and distress [68]. Hence, the need for affirmation by narcissistic adolescents may encourage greater dependence on social media that manifests itself in increased anxiety or signs of nomophobia. Furthermore, our findings match previous research on the relative effect of demographic variables on nomophobia. On the one hand, girls appear to feel more anxious and insecure regarding their smartphones, and on the other hand, nomophobia levels increase along with age [49]. Finally, our findings suggest different patterns for cognitive versus emotional components of body self-esteem in the prediction of nomophobia [69]. It can be argued that adolescents, who are less satisfied with their body and therefore promote a more attractive image of themselves in social media and internet, tend to feel more anxious and insecure when they are disconnected from their smartphones, similar as the experience of social rejection.

In conclusion, the results of both regression and QCA analyses suggest that demographic variables, body self-esteem and personality traits significantly influence adolescent social networking and internet addiction. On the one hand, girls with a disinhibited personality, in addition to high levels of neuroticism, narcissism and extraversion but to a lesser extend, tend to present more symptoms of internet addiction. Meanwhile, girls, who feel physically attractive and describe themselves as disinhibited and extraverted tend to use social media more often. Neuroticism also seems to be an influence with different patterns: low levels of neuroticism in combination with high levels of disinhibition, or high levels of neuroticism in combination with greater physical attractiveness, predict increased social media use. On the other hand, adolescents, both girls and boys, with high levels of narcissism and neuroticism show greater geek behaviour. The influence of body self-esteem is unclear. Finally, adolescents, mostly girls of younger age, who are less satisfied with their body but show greater physical attractiveness, with high levels of narcissism, and also high levels of neuroticism, extraversion and disinhibition but to a lesser extend, are more likely to present more signs of nomophobia.

4.5. Strengths, Limitations and Further Research

The strengths of this present research are both theoretical and methodological. On the one hand, our study focuses on internet addiction as multidimensional concept rather than a one-dimensional approach. The instrument of measurement of addiction to social media and the internet has been specifically designed for adolescents and contemplates the four different types of addictive online behaviour: (a) symptoms that excessive internet use entails; (b) the use of social networking sites for virtual interactions with peers, due to the fact that at this age peer relationship are more and more important; (c) geek behaviour, which is characterized by an intense level of interest in online gaming and online sexuality typical of adolescence; and (d) finally nomophobia defined as the problematic use of the mobile phone, which starts at early adolescence. On the other hand, the methodological approach is the second strengths of this study. If both methodologies are compared in the prediction of social networking and Internet addiction in adolescence, fsQCA models cover a greater number of factors than regression models. In addition, fsQCA allows for multiple pathways to be estimated by combining the predictors in different ways, depending on the relationships between variables. For these reasons, fsQCA methodology may be considered a complementary analytical approach to traditional regression models.

It is necessary to stress the limitations of the study. With regard to sampling method and size, the sample size was appropriate for conducting multiple regressions and fsQCA, however probability sampling does not guarantee the generalization of the results obtained. It would be recommendable in future research to carry out a cluster-stratified random sampling that includes adolescents from all over the country. With regard to data collection, we believe that self-reports completed by the adolescents were appropriate for the purpose of the study. Future research, however, may use mixed methods (qualitative and quantitative data), multiple reports from parents and peers, which would provide more in-depth information about adolescents' addiction. Finally, one of the main limitations of fsQCA is the limited number of predictor variables that can be included in the analyses. While in regression models an increase in sample size allows an increase in the number of predictors, in fsQCA the maximum number of conditions is invariable.

This research contributes to the study of potential risk factors of social networking and internet addiction in adolescence, regardless of the limitations that has been considered. Moreover, this study offers a more comprehensive conceptualization of internet addiction by identifying four different types of addictive online behaviour in young people and describing the different pathways of variables representing important biologically based personality traits, variables of body self-esteem, which are of special relevance during adolescence and taking into account gender and age differences. Our findings extend previous addiction literature, providing an in-depth analysis of several combinations of psychological and demographic variables that may increase the risk of potential behavioural addictions.

5. Conclusions

The implications of this study are both theoretical and practical. Overall, this study makes a unique contribution to the literature on addictive use of social networking sites and broadens the way for further research that may provide additional evidence towards other adolescent-relevant variables. Such research would provide health professionals with relevant information about individual differences in the four types of social networking and internet addiction, and therefore, enriching their professional experience when working with affected adolescents. Furthermore, from a practical point of view, the findings of our study may help psychotherapists to make decisions on who to prioritize a certain intervention and treatment approach based on the relevance of the reported risk factors such as personality and gender. Such stable characteristics may be detected quickly in adolescents in order to identify those who are most at risk and thus start a preventive intervention at an early stage. Thus, the study of youth internet addiction is essential in order to improve prevention and early intervention. Finally, this study has identified important risk factors that underlie the psychological mechanisms of social networking and internet addiction in adolescents. Stressing the benefits of identifying adolescent internet addiction symptoms, problematic social media use, geek behaviour and nomophobia, through a specific instrument that allows for a multidimensional assessment.

Author Contributions: All authors contributed equivalently to this research. K.S. and I.M.-C. developed and designed the research, supervising the writing of the manuscript; M.P. and U.d.l.B. were in charge of data collection and conducted the formal analysis; M.P. and U.d.l.B. wrote the first draft of the manuscript; and K.S. and I.M.-C. reviewed, edited, and modified the manuscript critically. All authors have read and agreed to the published version of the manuscript.

Funding: This research was supported by grants from the Department of Education, Universities and Research of the Basque Government (BFI-2012-40).

Acknowledgments: The authors appreciate the participation and support of the secondary schools.

Conflicts of Interest: The authors state that there is no conflict of interest.

References

1. Buctot, D.B.; Kim, N.; Kim, J.J. Factors associated with smartphone addiction prevalence and its predictive capacity for health-related quality of life among Filipino adolescents. *Child. Youth Serv. Rev.* **2020**, *110*, 104758. [CrossRef]
2. Kuss, D.J.; Griffiths, M.D. Social networking sites and addiction: Ten lessons learned. *Int. J. Environ. Res. Public Health* **2017**, *14*, 311. [CrossRef] [PubMed]
3. Laconi, S.; Kaliszewska-Czeremska, K.; Gnisci, A.; Sergi, I.; Barke, A.; Jeromin, F.; Groth, J.; Gamez-Guadix, M.; Ozcan, N.K.; Demetrovics, Z.; et al. Cross-cultural study of problematicinternet use in nine European countries. *Comput. Human Behav.* **2018**, *84*, 430–440. [CrossRef]
4. Machimbarrena, J.M.; González-Cabrera, J.; Ortega-Barón, J.; Beranuy-Fargues, M.; Álvarez-Bardón, A.; Tejero, B. Profiles of problematic internet use and its impact on adolescents' health-related quality of life. *Int. J. Environ. Res. Public Health* **2019**, *16*, 20. [CrossRef] [PubMed]
5. Fischer-Grote, L.; Kothgassner, O.D.; Felnhofer, A. Risk factors for problematic smartphone use in children and adolescents: A review of existing literature. *Neuropsychiatrie* **2019**, *33*, 179–190. [CrossRef]
6. Peris, M.; Maganto, C.; Garaigordobil, M. Escala de riesgo de adicción-adolescente a las redes sociales e internet: Fiabilidad y validez (ERA-RSI) [Scale of risk of addiction to medias and Internet for adolescents: Reliability and validity (ERA-RSI)]. *Rev. Psic Clin. con Niños y Adolesc* **2018**, *5052*, 30–36.

7. Aparicio-Martínez, P.; Ruiz-Rubio, M.; Perea-Moreno, A.J.; Martínez-Jiménez, M.P.; Pagliari, C.; Redel-Macías, M.D.; Vaquero-Abellán, M. Gender differences in the addiction to social networks in the Southern Spanish university students. *Telemat Inform.* **2020**, *46*, 101304. [CrossRef]
8. Lopez-Fernandez, O. Generalised versus specific internet use-related addiction problems: A mixed methods study on internet, gaming, and social networking behaviours. *Int. J. Environ. Res. Public Health* **2018**, *15*, 2913. [CrossRef]
9. Yudes-Gómez, C.; Baridon-Chauvie, D.; González-Cabrera, J.-M. Cyberbullying and problematic Internet use in Colombia, Uruguay and Spain: Cross-cultural study. *Comunicar* **2018**, *XXXVI*, 49–58. [CrossRef]
10. Munno, D.; Cappellin, F.; Saroldi, M.; Bechon, E.; Guglielmucci, F.; Passera, R.; Zullo, G. Internet addiction disorder: Personality characteristics and risk of pathological overuse in adolescents. *Psychiatry Res.* **2017**, *248*, 1–5. [CrossRef]
11. Gómez, P.; Rial, A.; Braña, T.; Golpe, S.; Varela, J. Screening of problematic internet use among Spanish adolescents: Prevalence and related variables. *Cyberpsychogy Behav. Soc. Netw.* **2017**, *20*, 259–267. [CrossRef] [PubMed]
12. Carbonell, X.; Chamarro, A.; Oberst, U.; Rodrigo, B.; Prades, M. Problematic use of the internet and smartphones in university students: 2006–2017. *Int. J. Environ. Res. Public Health* **2018**, *15*, 475. [CrossRef] [PubMed]
13. Durán, M.; Guerra, J.M. Spanish university students' uses of the social network Tuenti and their addiction levels: The protective role of the positive attitude toward mothers' presence as contact. *Anal. Psicol.* **2015**, *31*, 260–267.
14. Ostovar, S.; Allahyar, N.; Aminpoor, H.; Moafian, F.; Binti, M.; Griffiths, M.D. Internet addiction and its psychosocial risks (depression, anxiety, stress and loneliness) among Iranian adolescents and young adults: A structural equation model in a cross-sectional study. *Int. J. Ment. Health Addict.* **2016**, *14*, 157. [CrossRef]
15. Chen, B.; Liu, F.; Ding, S.; Ying, X.; Wang, L.; Wen, Y. Gender differences in factors associated with smartphone addiction: A cross-sectional study among medical college students. *BMC Psychiatry* **2017**, *17*, 341. [CrossRef] [PubMed]
16. Cimadevilla, R.; Jenaro, C. Impact on psychological health of internet and mobile phone abuse in a spanish sample of secondary students. *Rev. Argentina Clin. Psicol.* **2019**, *XXVIII*, 339–347.
17. Bischof-Kastner, C.; Kuntsche, E.; Wolstein, J. Identifying problematic internet users: Development and validation of the Internet Motive Questionnaire for Adolescents (IMQ-A). *J. Med. Internet Res.* **2016**, *16*, e230. [CrossRef]
18. Brenley, D.B.; Covey, J. Risky behavior via social media: The role of reasoned and social reactive pathways. *Comput. Hum. Behav.* **2018**, *78*, 183–191. [CrossRef]
19. Wiederhold, B.K. VR Online dating: The new safe sex. *Cyberpsychogy Behav. Soc. Netw.* **2016**, *19*, 297–298. [CrossRef]
20. Kokkinakis, A.V.; Lin, J.; Pavlas, D.; Wade, A.R. What's in a name? Ages and names predict the valence of social interactions in a massive online game. *Comput. Hum. Behav.* **2016**, *55*, 605–613. [CrossRef]
21. Escario, J.J.; Wilkinson, A.V. Exploring predictors of online gambling in a nationally representative sample of Spanish adolescents. *Comput. Hum. Behav.* **2020**, *102*, 287–292. [CrossRef]
22. Salonen, A.H.; Alho, H.; Castrén, S. Attitudes towards gambling, gambling participation, and gambling-related harm: Cross-sectional Finnish population studies in 2011 and 2015. *BMC Public Health* **2017**, *17*, 122. [CrossRef] [PubMed]
23. Wittek, C.T.; Finseras, T.R.; Pallesen, S.; Mentzoni, R.A.; Hanss, D.; Molde, H. Prevalence and predictors of video fame addiction: A study based on a national representative sample of gamers. *Int. J. Ment. Health Addict.* **2015**, *14*, 672–686. [CrossRef] [PubMed]
24. King, A.L.; Valença, A.; Nardi, A.E. Nomophobia: The mobile phone in panic disorder with agoraphobia: Reducing phobias or worsening of dependence? *Cogn. Behav. Neurol.* **2010**, *23*, 52–54. [CrossRef]
25. Barnes, S.J.; Pressey, A.D.; Scornavacca, E. Mobile ubiquity: Understanding the relationship between cognitive absorption, smartphone addiction and social network services. *Comput. Hum. Behav.* **2019**, *90*, 246–258. [CrossRef]
26. Cha, S.S.; Seo, B.K. Smartphone use and smartphone addiction in middle school students in Korea: Prevalence, social networking service, and game use. *Health Psychol. Open* **2018**, *5*, 1–15. [CrossRef]

27. Sohn, S.; Rees, P.; Wildridge, B.; Kalk, N.J.; Carter, B. Correction to: Prevalence of problematic smartphone usage and associated mental health outcomes amongst children and young people: A systematic review, meta-analysis and GRADE of the evidence. *BMC Psychiatry* **2019**, *19*, 1–11. [CrossRef]
28. Andreassen, C.S.; Pallesen, S.; Griffiths, M.D. The relationship between addictive use of social media, narcissism, and self-esteem: Findings from a large national survey. *Addict. Behav.* **2017**, *64*, 287–293. [CrossRef]
29. Maganto, C.; Peris, M. La corporalidad de los adolescentes en las redes sociales [The body image in adolescents in social network]. *Cuadernos de Psiquiatría y Psicoterapia del Niño y del Adolescente* **2013**, *55*, 53–62.
30. De Vries, D.A.; Peter, J.; de Graaf, H.; Nikken, P. Adolescents' social network site use, peer appearance-related feedback, and body dissatisfaction: Testing a mediation model. *J. Youth Adolesc.* **2016**, *45*, 211–224. [CrossRef]
31. Peris, M. *Adicción y erotización en las redes sociales e internet: Diseño y estandarización de la batería En-Red-A2 [Addiction and eroticization in social networks and the internet: Design and standardization of the En-Red-A2 battery]*; Universidad del País Vasco: País Vasco, Spain, 2017.
32. Wang, D. A study of the relationship between narcissism, extraversion, body-esteem, social comparison orientation and selfie-editing behavior on social networking sites. *Personal. Individ. Differ.* **2019**, *146*, 127–129. [CrossRef]
33. Kircaburun, K.; Alhabash, S.; Tosuntaş, Ş.B.; Griffiths, M.D. Uses and gratifications of problematic social media use among university students: A simultaneous examination of the Big Five of personality traits, social media platforms, and social media use motives. *Int. J. Ment. Health Addict.* **2018**, *18*, 525–547. [CrossRef]
34. Peris, M.; Maganto, C.; Arigita, A.; Barrientos, A.; León, A.; Sánchez-Cabrero, R. El derecho al olvido digital la imagen corporal virtual en adolescentes jóvenes. In *Intervención e Investigación en Contextos Clínicos de la Salud*; Pérez-Fuentes, M.C., Molero, M.M., Gázquez, J.J., Martos, A., Barragán, A.B., Simón, M.M., Oropesa, N.F., Pino, R.M., Eds.; ASUNIVEP: Murcia, Spain, 2019; pp. 135–143.
35. Liu, Q.; Sun, J.; Li, Q.; Zhou, Z. Body dissatisfaction and smartphone addiction among Chinese adolescents: A moderated mediation model. *Child. Youth Serv. Rev.* **2020**, *108*, 104613. [CrossRef]
36. Blachnio, A.; Przepiorka, A.; Senol-Durak, E.; Durak, M.; Sherstyuk, L. The role of personality traits in Facebook and internet addictions: A study on Polish, Turkish, and Ukrainian samples. *Comput. Hum. Behav.* **2017**, *68*, 269–275. [CrossRef]
37. Hussain, Z.; Pontes, H.M. Personality, internet addiction, and other technological addictions: A psychological examination of personality traits and technological addictions. In *Psychological, Social, and Cultural Aspects of Internet Addiction*; Bozoglan, B., Ed.; IGI Global: Hershey, PA, USA, 2018; pp. 45–71.
38. Cheng, C.; Wang, H.; Sigerson, L.; Chau, C.I. Do the socially rich get richer? A nuanced perspective on social network site use and online social capital accrual. *Psychol. Bull.* **2019**, *145*, 734–764. [CrossRef]
39. Dieris-Hirche, J.; Pape, M.; te Wildt, B.T.; Kehyayan, A.; Esch, M.; Aicha, S.; Herpertz, S.; Bottel, L. Problematic gaming behavior and the personality traits of video gamers: A cross-sectional survey. *Comput. Hum. Behav.* **2020**, *106*, 106272. [CrossRef]
40. Pivetta, E.; Harkin, L.; Billieux, J.; Kanjo, E.; Kuss, D. Problematic smartphone use: An empirically validated model. *Comput. Hum. Behav.* **2019**, *100*, 105–117. [CrossRef]
41. Mitchell, L.; Hussain, Z. Predictors of problematic smartphone use: An examination of the Integrative Pathways Model and the role of age, gender, impulsiveness, excessive reassurance seeking, extraversion, and depression. *Behav. Sci.* **2018**, *8*, 74. [CrossRef] [PubMed]
42. Marengo, D.; Poletti, I.; Settanni, M. The interplay between neuroticism, extraversion, and social media addiction in young adult Facebook users: Testing the mediating role of online activity using objective data. *Addict. Behav.* **2020**, *102*, 106150. [CrossRef] [PubMed]
43. Lachmann, B.; Sindermann, C.; Sariyska, R.Y.; Luo, R.; Melchers, M.C.; Becker, B.; Cooper, A.J.; Montag, C. The role of empathy and life satisfaction in internet and smartphone use disorder. *Front. Psychol.* **2018**, *9*, 398. [CrossRef]
44. Hussain, Z.; Griffiths, M.D.; Sheffield, D. An investigation into problematic smartphone use: The role of narcissism, anxiety, and personality factors. *J. Behav. Addict.* **2017**, *6*, 378–386. [CrossRef] [PubMed]
45. Gervasi, A.M.; La Marca, L.; Lombardo, E.M.C.; Mannino, G.; Iacolino, C.; Schimmenti, A. Maladaptive personality traits and internet addiction symptoms among young adults: A study based on the alternative DSM-5 model for personality disorders. *Clin. Neuropsychiatry* **2017**, *14*, 20–28.

46. Clancy, E.M.; Klettke, B.; Hallford, D.J. The dark side of sexting—Factors predicting the dissemination of sexts. *Comput. Hum. Behav.* **2019**, *92*, 266–272. [CrossRef]
47. Sarabia, I.; Estévez, A. Sexualized behaviors on Facebook. *Comput. Hum. Behav.* **2016**, *61*, 219–226. [CrossRef]
48. Hawk, S.T.; van den Eijnden, J.J.M.; van Lissa, C.J.; ter Bogt, T.F.M. Narcissistic adolescents' attention-seeking following social rejection: Links with social media disclosure, problematic social media use, and smartphone stress. *Comput. Hum. Behav.* **2019**, *92*, 65–75. [CrossRef]
49. Yildiz Durak, H. Investigation of nomophobia and smartphone addiction predictors among adolescents in Turkey: Demographic variables and academic performance. *Soc. Sci. J.* **2019**, *56*, 492–517. [CrossRef]
50. Eng, S.; Woodside, A.G. Configural analysis of the drinking man: Fuzzy-set qualitative comparative analyses. *Addict. Behav.* **2012**, *37*, 541–543. [CrossRef]
51. Ragin, C.C. *Redesigning Social Inquiry. Fuzzy Sets and Beyond*; University of Chicago Press: Chicago, IL, USA, 2008.
52. Calabuig, F.; Prado-Gascó, V.; Crespo-Hervás, J.; Núñez-Pomar, J.; Añó, V. Predicting future intentions of basketball spectators using SEM and fsQCA. *J. Bus. Res.* **2016**, *69*, 1396–1400. [CrossRef]
53. Legewie, N. An introduction to applied data analysis with Qualitative Comparative Analysis (QCA). *Forum Qual. Soc. Res.* **2013**, *14*, 1–45.
54. De la Barrera, U.; Schoeps, K.; Gil-Gómez, J.-A.; Montoya-Castilla, I. Predicting adolescent adjustment and well-being: The interplay between socio-emotional and personal factors. *Int. J. Environ. Res. Public Health* **2019**, *16*, 4650. [CrossRef]
55. Giménez-Espert M del, C.; Valero-Moreno, S.; Prado-Gascó, V.J. Evaluation of emotional skills in nursing using regression and QCA models: A transversal study. *Nurse Educ. Today* **2019**, *74*, 31–37. [CrossRef] [PubMed]
56. De la Barrera, U.; Villanueva, L.; Prado-Gascó, V. Emotional and personality predictors that influence the appearance of somatic complaints in children and adults. *Psicothema* **2019**, *31*, 407–413. [PubMed]
57. Peris, M.; Maganto, C.; Garaigordobil, M. Escala de Autoestima Corporal: Datos psicométricos de fiabilidad y validez. *Rev. Psicol. Clínica con Niños y Adolesc.* **2016**, *3*, 51–58.
58. Costa, P.T.; McCrae, R.R. *Inventario de Personalidad NEO Revisado (NEO PIR) e Inventario NEO Reducido de Cinco Factores (NEO-FFI)*; TEA Ediciones: Madrid, Spain, 2008.
59. Zuckerman, M.; Eysenck, S.B.G.; Eysenck, H.J. Sensation seeking in England and America: Cross-cultural, age, and sex comparisons. *J. Consult. Clin. Psychol.* **1978**, *46*, 139–149. [CrossRef]
60. Pérez, J.; Torrubia, R. Fiabilidad y validez de la versión española de la Escala de Búsqueda de Sensaciones (Forma V). *Rev. Latinoam Univ.* **1986**, *18*, 7–22.
61. Raskin, R.; Terry, H. Principal-components analysis of the narcissistic personality inventory and further evidence of its construct validity. *J. Pers. Soc. Psychol.* **1988**, *54*, 890–902. [CrossRef]
62. Trechera, J.L.; Millán, G.; Fernández-Morales, E. Estudio empírico del trastorno narcisista de la personalidad (TNP). *Acta Colomb. Psicol.* **2008**, *11*, 25–36.
63. Villanueva, L.; Montoya-Castilla, I.; Prado-Gascó, V. The importance of trait emotional intelligence and feelings in the prediction of perceived and biological stress in adolescents: Hierarchical regressions and fsQCA models. *Stress* **2017**, *20*, 355–362. [CrossRef]
64. Woodside, A.G. Moving beyond multiple regression analysis to algorithms: Calling for adoption of a paradigm shift from symmetric to asymmetric thinking in data analysis and crafting theory. *J. Bus. Res.* **2013**, *66*, 463–472. [CrossRef]
65. Su, W.; Han, X.; Jin, C.; Yan, Y.; Potenza, M.N. Are males more likely to be addicted to the internet than females? A meta-analysis involving 34 global jurisdictions. *Comput. Hum. Behav.* **2019**, *99*, 86–100. [CrossRef]
66. Di Blasi, M.; Giardina, A.; Lo Coco, G.; Giordano, C.; Billieux, J.; Schimmenti, A. A compensatory model to understand dysfunctional personality traits in problematic gaming: The role of vulnerable narcissism. *Personal. Individ. Differ.* **2020**, *160*, 109921. [CrossRef]
67. Matthews, N.L.; Lynch, T.; Martins, N. Real ideal: Investigating how ideal and hyper-ideal video game bodies affect men and women. *Comput. Hum. Behav.* **2016**, *59*, 155–164. [CrossRef]

68. Anshari, M.; Alas, Y.; Sulaiman, E. Smartphone addictions and nomophobia among youth. *Vulnerable Child. Youth Stud.* **2019**, *14*, 242–247. [CrossRef]
69. Peris, M.; Maganto, C.; Kortabarria, L. Body self-esteem, virtual image in social networks and sexuality in adolescent. *Eur. J. Investig. Health Psychol. Educ.* **2013**, *3*, 171–180. [CrossRef]

International Journal of
Environmental Research and Public Health

MDPI

Article

The Matthew Effect in Recovery from Smartphone Addiction in a 6-Month Longitudinal Study of Children and Adolescents †

Seung-Yup Lee [1], Hae Kook Lee [2], Jung-Seok Choi [3], Soo-young Bang [4], Min-Hyeon Park [1], Kyu-In Jung [1] and Yong-Sil Kweon [2,*]

1 Department of Psychiatry, Eunpyeong St. Mary's Hospital, College of Medicine, The Catholic University of Korea, Seoul 03312, Korea; seungyup@catholic.ac.kr (S.-Y.L.); neominnie@hanmail.net (M.-H.P.); cki@catholic.ac.kr (K.-I.J.)
2 Department of Psychiatry, Uijeongbu St. Mary's Hospital, College of Medicine, The Catholic University of Korea, Seoul 06591, Korea; nplhk@catholic.ac.kr
3 Department of Psychiatry, SMG-SNU Boramae Medical Center, Seoul 07061, Korea; psychoresi@hanmail.net
4 Department of Psychiatry, College of Medicine, Eulji University, Seoul 01830, Korea; dresme@daum.net
* Correspondence: yskwn@catholic.ac.kr; Tel.: +82-31-820-3032; Fax: +82-31-847-3630
† Running title: Predictors to persistent problematic smartphone use.

Received: 18 May 2020; Accepted: 30 June 2020; Published: 1 July 2020

Abstract: The clinical course of problematic smartphone use (PSU) remains largely unknown due to a lack of longitudinal studies. We recruited 193 subjects with smartphone addiction problems for the present study. After providing informed consent, the subjects completed surveys and underwent comprehensive interviews regarding smartphone usage. A total of 56 subjects among the 193 initially recruited subjects were followed up for six months. We compared baseline characteristics between persistent addicted users and recovered users at the end of the 6-month follow-up. Persistent problematic smartphone users displayed higher baseline smartphone addiction severity and were more prone to develop mental health problems at the follow-up. However, baseline depressive or anxiety status did not significantly influence the course of PSU. PSU behaved more like an addictive disorder rather than a secondary psychiatric disorder. Harm avoidance, impulsivity, higher Internet use, and less conversation time with mothers were identified as poor prognostic factors in PSU. Lower quality of life, low perceived happiness, and goal instability also contributed to persistent PSU, while recovery increased these scores as well as measures of self-esteem. These findings suggest that the Matthew effect is found in the recovery of PSU with better premorbid psychosocial adjustment leading to a more successful recovery. Greater clinical resources are required for interventions in vulnerable populations to modify the course of this increasingly prevalent problematic behavior worldwide.

Keywords: problematic phone use; internet; pain; dry eye; depression; anxiety; quality of life; recovery; prognosis; cohort

1. Introduction

Smartphone use has become an integral part of everyday life. While many conveniences are associated with widespread smartphone use, public health concerns regarding problematic smartphone use (PSU) have also increased over the past decade. In particular, lack of knowledge about the long-term consequences of PSU has raised further concerns in developmental studies of children and adolescents.

Even though connectivity is generally believed to be increased by smartphones, people may also snub others by touching their smartphones rather than engaging in person. This novel type of social exclusion phenomenon is called "phubbing" and was reported to be linked to social anxiety [1]. Thus, we can understand that novel problems may arise following novel technologies.

South Korea reported a smartphone penetration rate of 95% in 2018, the highest in the world [2]. In 2019, a nationally representative survey of South Koreans revealed PSU prevalence of 20.0% across all age groups [3], which corresponds to the global prevalence of 23.3% reported in a recent systemic review [4]. While many reports outline general difficulties in controlling excessive smartphone use, the groups at the highest risk in the 2019 Korean national survey were children and adolescents, with PSU prevalence of 22.9% and 30.2%, respectively [3].

There is contradictory evidence in the literature regarding whether smartphone addiction truly exists [5–7]. PSU is generally not regarded as a serious health issue like substance abuse. In a British survey, the majority of parents report that they do not regulate their children's smartphone time [8].

PSU has been associated with depression, anxiety, aggression, decreased sleep duration/quality, and other addictive behaviors [4,9–14]. It is also related to physical health risks such as dry eye symptoms, musculoskeletal pain, and accidents [15–20]. While such physical problems may arise as a direct consequence of uncontrolled excessive smartphone use, the relationship between psychosocial problems and PSU is yet to be established. For instance, PSU was associated with depression and lower subjective happiness; attention deficit hyperactivity disorder (ADHD) and impulsivity; anxiety, and lower quality of life in cross-sectional studies [10,21,22]. However, it is yet unclear whether such negative mental and behavioral variables act mainly as risk factors that contribute to PSU development or such were derived from PSU as adverse outcomes. To complicate matters, factors such as self-esteem, goal instability, or parent–child relationships may also exert influence over the course of PSU among the developing children and adolescents. To verify their temporal order and relative contributions, well designed longitudinal studies are required.

However, longitudinal studies are scarce, and all existing longitudinal studies recruited convenient samples rather than surveying the clinical population [23–26]. The pre-existing longitudinal studies are limited by drawing conclusions exclusively by conducting surveys on non-clinical convenient samples. Non-clinical samples are more likely to have lower PSU severity and fewer related mental or physical health problems compared to clinical samples [27]. Despite being categorized as smartphone addiction, the severity of the subjects recruited from the non-clinical community settings may have not been severe enough to capture the true clinical picture of the population with clinically significant impairment by PSU.

Together with the lack of any longitudinal studies that demonstrate the diagnostic stability of PSU, the lower severity of the pre-existing studies due to their reliance on non-clinical samples may have contributed to the prior argument that PSU lacks health consequences severe enough to make it a legitimate addictive disorder [5]. Therefore, there is a strong need to investigate the clinical course of PSU in clinical settings and its diagnostic stability in the long term.

The lack of evidence regarding the long-term consequences of PSU is especially alarming since PSU is highly prevalent among children and adolescents [3]. The mental and behavioral consequences of PSU may not only deteriorate current function but may linger through the loss of future opportunities and altered psychosocial development. Furthermore, previous studies indicate that excessive Internet use heightens the risk of substance use [28], and problematic gaming elevates the future risk of developing gambling problems. Addictive use of social media was reported to be related to reduced well-being [29]. These findings demand closer attention to whether PSU occurring during this critical developmental period leads to mental health problems later [30].

The information technologies were suggested to potentially reduce social inequalities by increasing the reachability of information and educational resources to a wider population. However, the "digital divide" created by the disparity in access to online information may further broaden the social health inequalities of our society at the same time [31]. With the growing smartphone penetration around the globe, however, the inequalities by the lack of accessibility will likely decrease. In contrast, other forms of inequalities may emerge due to excessive smartphone use and its long-term consequences.

For instance, children from higher socioeconomic status were more likely to use their digital devices safely while punishing parental attitude was associated with increased risk of PSU [32,33]. While education is generally considered as the "great equalizer", PSU was related to drug consumption and poor academic achievement [34]. The undermined educational opportunity in children and adolescents due to PSU may not only impact their lives in schools but lead to a lifelong socioeconomic disparity through employment and income across the generation by the missed opportunities in the crucial period of development.

"For to everyone who has, more will be given and he will grow rich; but from the one who has not, even what he has will be taken away." (Matthew 25:29) [35]. The Matthew effect, a term derived from the Gospel of Matthew by Merton [36], will also be useful to describe the cumulative disadvantage that may potentially occur when PSU is not properly mitigated in children with individual and environmental vulnerabilities.

Therefore, in this study, we explore the course of PSU and prognostic factors that may influence its course. To the best of our knowledge, this is the first longitudinal study of PSU conducted in a clinical sample. In addition to smartphone-related variables, we performed follow-up measurements of psychosocial variables to determine their contributions to PSU. The identification of risk and prognostic factors related to PSU will facilitate the development of preventive and interventional strategies against this worldwide public health risk among children and adolescents.

2. Materials and Methods

2.1. Study Procedure and Participants

This study was conducted as a part of the clinical Cohort for Understanding of Internet addiction Rescue factors in Early life (c-CURE) study. C-CURE is a multi-center study involving three university hospitals in the Seoul metropolitan area of South Korea that explores risk and protective factors regarding harmful digital media use in children and adolescents. The sample includes subjects who visited the outpatient departments of these hospitals for the treatment of excessive Internet, game, or smartphone use.

After obtaining written informed consent from both participants and their parents/guardians, a brief 15–20 min session of parental coaching was conducted with the additional provision of a 12-page pocket reference outlining parental guidance on digital media use at baseline. The content of the pocket guide includes a parental checklist for children's excessive use, explanation of risk factors (neglect, ADHD, depression, impulse control, and social phobia), explanation of game genres and motivations of children, parenting coaching tips, setting rules, fostering alternative recreational activities, and how to act when childrens relapse (checklist for warning signs, converse more with attentive listening, help children to find other activities that may interest them, provide emotional support, and seek expert's help).

Of the initial 188 participants, we included 85 subjects (64.7% males and 35.3% females) aged between 7 and 18 years old (mean age of 13.2 years old). The inclusion criterion was demonstrating scores equal to or greater than 31 on the smartphone addiction scale-short version (SAS-SV) on the baseline assessment. The subjects were evaluated at three and six months of follow-up.

This study was conducted following the Declaration of Helsinki and was reviewed and approved by the Institutional Review Boards of Uijeongbu St. Mary's Hospital, the Catholic University of Korea (UC15ONMI0072), Seoul Metropolitan Government, Seoul National University Boramae Medical Center (16 April 2016), and Eulji University Seoul Eulji Hospital (EMCS2015-05-020-001).

2.2. Measures

2.2.1. Smartphone Addiction Scale-Short Version (SAS-SV)

This ten-item self-measurement was used to evaluate PSU. It is a shortened version of the original Smartphone addiction scale with 33 items that assess six components of addiction model—excessive use, tolerance, withdrawal, disturbance in daily-life, over-valued online relationship, and positive anticipation [37]. While the first four components correspond to diagnostic criteria of substance use disorder by the 5th edition of Diagnostic and Statistical Manual of Mental Disorders, the last two are unique to smartphone addiction scale.

The SAS-SV is rated on a six-point Likert scale (1 = strongly disagree to 6 = strongly agree) with the total scores ranging from 10 to 60. It questions items such as "Missing planned work due to smartphone use." A higher score indicates greater severity of PSU, and a cutoff score of 31 was used in this study as previously suggested [38]. Cronbach's alpha was 0.63 in this study.

2.2.2. Internet Addiction Test (IAT)

We used the IAT to evaluate the severity of problematic Internet use. It questions items such as "How often do you choose to spend more time online over going out with others?" Rated on a 5-point Likert scale (1 = rarely to 5 = always), this 20 item self-measurement indicates more severe problematic Internet use with increasing scores. Cronbach's alpha was 0.88 in this study.

2.2.3. Survey on the Pattern of Digital-Media Use

Comprehensive surveys were conducted to evaluate digital media use. Participants answered questions about smartphone access such as starting age, average daily time spent on smartphones or the Internet, type of media content, and patterns of use such as bedtime smartphone usage. In addition, participants responded to questions about weight and height for calculating body mass index (BMI) and presence of musculoskeletal pain in the neck, shoulders, and hands/wrists/fingers to assess potential physical consequences of PSU.

2.2.4. Dry Eye Symptoms (DES) Checklist

Questions about dry eye symptoms described jointly by the Korean ministry of health and welfare, the Korean Academy of Medical Science, and the Korean Ophthalmological Society were posed separately to the participants (foreign object sensation in the eyes, heaviness of the eyelid, eye-stiffness, blurry eyes, ocular fatigue, ocular soreness, frequent eye redness, dryness, photosensitivity, ocular pain, and string-like ocular discharge) [39]. The total number of positive responses to the DES checklist was used to measure the severity of DES. Cronbach's alpha was 0.82 in this study.

2.2.5. Junior Temperament Character Inventory (JTCI)

We utilized a child and adolescent version of the Temperament and Character Inventory, which was developed based on the biopsychosocial personality model of Cloninger [40]. The JTCI has been previously validated in Korean [41]. JTCI inventory assesses the four temperament types (novelty seeking, harm avoidance, reward dependence, and persistence) with yes/no responses to the 82 items. As high levels of novelty seeking and harm avoidance are well-known risk factors for addictive disorders such as gaming disorder [42], individual scores were anticipated to shift right in our study subjects. To explore the contributions of these variables within our study sample, individual scores were standardized to T-scores for further analyses.

2.2.6. Depression Assessment

While considering participant age, the 27-item children's depression inventory (CDI) was used to screen for depressive symptoms two weeks prior to assessment in children aged 12 or less. Each item is rated on a 3-point Likert scale from zero to two with descriptive statements such as "I feel sad sometimes, most, or always". Participants with total scores of 22 or greater were classified as having depressive status, as previously suggested in a Korean validation study [43]. Cronbach's alpha was 0.86 in this study.

The Beck Depression Inventory-II (BDI-II) was used to screen for depression in adolescents between 13 and 18 years of age. It contains five subscales of negative mood, interpersonal problems, ineffectiveness, anhedonia, and negative self-esteem. Each item is rated on a 4-point Likert scale from zero to three, and answers refer to the two weeks prior to assessment. It uses descriptions such as "I am so sad and unhappy that I cannot stand it" for each scale. Scores of 10 or greater on this 21-item self-questionnaire were used to identify participants with depression according to Beck [44]. Higher scores indicate more severe depressive symptoms in both tests. Cronbach's alpha was 0.93 in this study.

2.2.7. Anxiety Assessment

The Korean version of the State Anxiety Inventory for Children (SAIC), a part of the State-Trait Anxiety Inventory for Children, was used to assess anxiety in children aged 12 or less. This 20-item inventory includes questions regarding how frequently respondents feel worried, bothered, or nervous on a 3-point Likert scale (1 = almost never to 3 = almost always). With total scores ranging from 20 to 60, higher scores indicate greater anxiety. For this study, we used a score of 41 or higher as a cut-off for defining an anxious state, as previously suggested in a Korean validation study [45]. Cronbach's alpha was 0.81 in this study.

The anxiety levels of adolescents between 13 and 18 years of age were examined with the State-Trait Anxiety Inventory-form X (STAI-X). Scores for this 20-item inventory range from 20 to 80 and the previously-defined cut-off point of 52 was used to classify subjects with high anxiety [46]. Cronbach's alpha was 0.90 in this study.

2.2.8. Barratt Impulsiveness Scale-11 (BIS-11)

The Korean version of the BIS-11, which includes three subcomponents of cognitive, motor, and non-planned impulsivity, was utilized to measure impulsivity [47]. This 23-item self-measurement is rated on a 4-point Likert scale (1 = rarely or never to 4 = almost always or always) with higher scores indicating greater impulsivity [48]. It questions items such as "I say things without thinking." Cronbach's alpha was 0.73 in this study.

2.2.9. Conners-Wells' Adolescent Self-Report Scale-Short Form (CASS-S)

We used the Korean version of the CASS-S, which is a 27-item instrument that assesses ADHD [49]. It is answered on a 4-point Likert scale (0 = not true at all to 3 = very often, very frequently). CASS-S has three subcomponents (conduct problems, cognitive problems, and hyperactivity), with higher scores suggesting more severe ADHD symptoms. CASS-S contains questions such as "I have difficulty sitting still." Cronbach's alpha was 0.83 in this study.

2.2.10. Goal Instability

This inventory was developed from concepts derived from the self-psychology of Heinz Kohut that measure goal-directedness, self-competencies, and career decidedness [50]. Participants responded to the Korean version of the goal instability tool, which includes 10 items rated on a 6-point Likert scale [51]. It questions items such as "I have confusion about who I am." Endorsement of each item results in higher scores, which indicate greater goal instability of the respondents. Cronbach's alpha was 0.88 in this study.

2.2.11. Adolescents Happiness Index (AHI)

We used the 30-item AHI to examine levels of positive emotion and well-being. This inventory is rated on a 4-point Likert scale (1 = strongly disagree to 3 = strongly agree) and has four sub-domains including self-concept, family relationship, leisure, and peer relationship [52]. Higher scores indicate higher levels of happiness. AHI asks questions such as "I am satisfied with myself." Cronbach's alpha was 0.92 in this study.

2.2.12. Rosenberg Self-Esteem Scale (RSES)

Self-esteem was measured using a 10-item scale with each item rated on a 5-point Likert scale (1 = strongly disagree to 5 = strongly agree) [53]. With total scores ranging from 10 to 50, the RSES was developed to measure feelings of self-acceptance, self-respect, and positive evaluation of self. A higher score suggests better self-esteem. RSES contains questions such as "I am able to do things as well as most other people." Cronbach's alpha was 0.84 in this study.

2.2.13. Pediatric Quality of Life (pedsQL)

The pedsQL^{TM} Generic Core 4.0 was used to evaluate the quality of life. This is a 23-item tool with responses on a 5-point Likert scale (conversions 0 = 100, 1 = 75, 2 = 50, 3 = 25, and 4 = 0 scores). The pedsQL has four subcomponents including physical, emotional, social, and academic functioning. Better quality of life is suggested by higher scores [54]. It questions items such as "hard to run" Cronbach's alpha was 0.93 in this study.

2.3. *Statistical Analyses*

Fifty-six participants remained in the study among the 85 subjects included at baseline, for a follow-up rate of 65.9%. We grouped participants into a persistent group and a recovered group according to SAS-SV scores at the endpoint of 6-month follow-up. Participants whose SAS-SV scores were equal to or greater than 31 were classified in the persistent group, while those with SAS-SV scores less than 31 were considered recovered at the endpoint.

To identify prognostic factors of PSU, baseline sociodemographic and clinical variables between the two groups were compared. In addition, PSU-related variables and psychological variables were compared at 3- and 6-month follow-ups.

Chi-square or Fisher's exact tests were performed for categorical variables. Mann–Whitney U-tests were conducted for continuous variables due to the limit of the final sample size at 6 months. To measure the effect size of PSU to variables at the follow-up, Cohen's d, and Cramer's V were calculated for continuous and categorical variables, respectively. All analyses were performed using SPSS Statistics for Windows, Version 24.0 (IBM Corp., Armonk, NY, USA) with two-tailed statistical significance set at $p = 0.05$.

3. Results

The baseline comparison between the PSU recovered and persistent PSU group was performed for the sociodemographic, smartphone-related, psychological, and physical measures to identify the prognostic factors. In addition, we compared the significant variables identified by the analyses at the baseline at the 3- and 6-month follow-ups to explore the clinical course.

There were no significant demographic differences between the recovered group and the persistent group including age, sex, or socioeconomic status (Table 1). However, the persistent group had fewer conversations with their mothers on weekdays ($p = 0.002$). The persistent group also showed significantly higher PSU at baseline as indicated by SAS-SV scores ($p = 0.043$), the average daily amount of time for mobile/smartphone use on both weekdays ($p = 0.026$) and weekends ($p = 0.037$), and Internet use time on weekdays ($p = 0.044$) (Table 1).

Table 1. Baseline comparison of sociodemographic and smartphone-related measures between recovered and persistent groups.

	Recovered Group (n = 21)	Persistent Group (n = 35)	p
Age (years)	12.2 ± 3.0	13.7 ± 2.5	0.083
Sex			0.822
Male	15 (71.4%)	24 (68.6%)	
Female	6 (28.6%)	11 (31.4%)	
Education status of mother			0.288
High school or lower	13 (61.9%)	22 (75.9%)	
College or higher	8 (38.1%)	7 (24.0%)	
Subjective economic status †			0.537
Low	7 (36.8%)	11 (35.5%)	
Middle	6 (31.6%)	6 (19.4%)	
Upper-middle or above	6 (31.6%)	14 (45.2%)	
Living without father	7 (33.3%)	12 (35.3%)	0.882
Living without mother	0 (0%)	5 (14.7%)	0.065
Conversation time with mother (min)			
Weekday (average)	150.7 ± 166.6	71.7 ± 92.7	0.002 **
Weekend (average)	215.5 ± 327.1	142.8 ± 264.0	0.092
Smartphone use start (age)	8.5 ± 1.5	9.2 ± 2.3	0.369
Ownership of smartphone (+) †	20 (95.2%)	34 (97.1%)	>0.999
Smartphone Addiction Scale-Short	37.1 ± 5.6	40.4 ± 6.0	0.043 *
Internet Addiction Test score	56.2 ± 14.9	60.22 ± 14.7	0.371
Mobile/smartphone use time (min)			
Weekday (average)	202.4 ± 111.8	279.43 ± 146.9	0.026 *
Weekend (average)	299.5 ± 137.5	426.00 ± 226.2	0.037 *
Internet use time (min)			
Weekday (average)	211.4 ± 84.0	282.9 ± 132.2	0.044 *
Weekend (average)	352.9 ± 158.7	444.0 ± 200.2	0.108
Internet content use			
News	9 (42.9%)	11 (31.4%)	0.388
Adult material	3 (14.3%)	12 (34.3%)	0.102
Online game	13 (61.9%)	26 (74.3%)	0.329
Blog	9 (42.9%)	6 (17.1%)	0.035
Social networking service	14 (66.7%)	20 (57.1%)	0.480
Bedtime phone use before sleep (min)	31.4 ± 38.2	57.86 ± 61.1	0.143

† Fisher's exact test; * $p < 0.05$, ** $p < 0.01$.

The online content consumed by the groups differed. The persistent group consumed more adult material, but the difference between groups was not statistically significant. The recovered group was more likely to be involved in blogging ($p = 0.035$).

In the comparison of psychological variables, the two groups did not differ significantly regarding the history of psychiatric disorders, current depression, or current anxiety. However, when stratified by age, younger children (less than 13 years of age) in the persistent group showed significantly higher SAIC scores ($p = 0.035$). Furthermore, the persistent group tended to display higher CDI, but the difference between groups was not significant ($p = 0.06$). While the persistent PSU group had significantly higher levels of harm avoidance ($p = 0.014$), BIS-11 ($p = 0.036$), and goal instability ($p = 0.008$), they had significantly lower pedsQL ($p < 0.001$) and AHI ($p = 0.006$) in the whole group comparison (Table 2).

Table 2. Baseline comparison of psychological and physical measures between recovered and persistent groups.

	Recovered Group (n = 21)	Persistent Group (n = 35)	p
Psychological Assessment			
Novelty seeking (T-score)	52.1 ± 8.3	52.0 ± 10.3	0.658
Harm avoidance (T-score)	46.8 ± 12.0	55.8 ± 10.2	0.014 *
History of Psychiatric diagnosis †			
Depression (+)	2 (9.5%)	8 (22.9%)	0.290
Attention deficit hyperactivity disorder (+)	4 (19.0%)	7 (20.0%)	>0.999
Depression (+) status	7 (33.3%)	20 (58.8%)	0.066
CDI (n = 14) ∮	10.6 ± 4.8	18.0 ± 6.9	0.060
BDI (n = 41) ∮	17.3 ± 17.9	18.7 ± 12.8	0.306
Anxiety (+) status †	3 (14.3%)	10 (28.6%)	0.330
SAIC (n = 14) ∮	27.0 ± 3.5	35.2 ± 2.9	0.002 **
STAI-X (n = 42) ∮	43.2 ± 14.9	47.4 ± 13.4	0.355
Barratt Impulsiveness Scale-II	56.1 ± 9.1	61.0 ± 6.9	0.036 *
Goal instability	29.9 ± 13.0	38.9 ± 10.8	0.008 **
CASS-Short form	25.9 ± 9.8	28.8 ± 10.6	0.436
Rosenberg Self-Esteem Scale	27.6 ± 6.4	24.4 ± 5.4	0.100
Quality of life	1911.9 ± 350.3	1472.1 ± 420.4	<0.001 **
Happiness scale	60.1 ± 16.1	49.1 ± 12.9	0.006 **
Physical Assessment			
Body mass index	22.6 ± 5.0	21.9 ± 5.9	0.264
Dry Eye Scale	1.6 ± 2.2	2.2 ± 2.1	0.153
Musculoskeletal pain (+)			
Neck	2 (9.5%)	15 (42.9%)	0.009 **
Shoulder †	1 (4.8%)	9 (25.7%)	0.072
Hand/Wrist/Finger †	2 (9.5%)	7 (20.0%)	0.459

† Fisher's exact test, ∮ Stratified by age as CDI and SAIC ≤ 12; BDI and STAI-X > 12; * $p < 0.05$, ** $p < 0.01$. CDI, Children's Depression Inventory; BDI, Beck Depression Inventory; SAIC, State Anxiety Inventory for Children; STAI-X, State-Trait Anxiety Inventory form X; CASS, Conners-Wells' Adolescent Self-Report Scale.

When examining the physical consequences of PSU, baseline variables such as BMI, DES, shoulder, and hand/wrist/finger pain did not differ between the two groups. However, the symptom presentation of neck pain was significantly higher in persistent problematic users ($p = 0.009$) (Table 2).

In follow-up comparisons of variables that showed significant differences at baseline, BIS-11 and weekday conversation time with the mother did not demonstrate significant differences at follow-up visits. However, the recovered group continued to demonstrate higher pedsQL at both 3- and 6-month follow-ups ($p < 0.001$ for both) and also showed greater AHI at 6-month follow-up ($p = 0.003$). In contrast, the persistent PSU group showed higher goal instability ($p = 0.011$) and neck pain ($p = 0.043$) at the endpoint (Table 3).

The two groups continued to display significant differences in SAS-SV and the daily average amount of time for mobile/smartphone use on both weekdays and weekends, with a steeper reduction in the recovered group. IAT scores, which did not differ between groups at baseline ($p = 0.371$), became significantly different at both 3- and 6-month follow-ups ($p = 0.006$, and $p < 0.001$, respectively) (Figure 1).

Table 3. Follow-up findings of the recovered and persistent groups.

		Recovered Group (n = 21)	Persistent Group (n = 35)	p	Cohens' d
Barratt Impulsiveness Scale-II	3 months	54.3 ± 11.1	57.0 ± 9.3	0.373	0.26
	6 months	53.0 ± 10.1	57.0 ± 8.5	0.1	0.43
Goal instability	3 months	28.2 ± 10.6	33.9 ± 11.5	0.073	0.52
	6 months	25.5 ± 12.1	34.7 ± 13.0	0.011 *	0.73
Quality of life	3 months	1994.1 ± 246.7	1591.7 ± 405.9	<0.001 **	1.2
	6 months	2078.6 ± 271.7	1612.1 ± 451.7	<0.001 **	1.25
Happiness scale	3 months	60.9 ± 16.8	52.0 ± 16.1	0.077	0.54
	6 months	60.7 ± 11.1	50.0 ± 14.2	0.003 **	0.84
					Cramer's V
Neck pain	3 months	3 (14.3%)	8 (24.2%)	0.691 †	0.17 (p = 0.46)
	6 months	2 (10.0%)	10 (35.7%)	0.043 *	0.29 (p = 0.04)

† Fisher's exact test; * $p < 0.05$, ** $p < 0.01$.

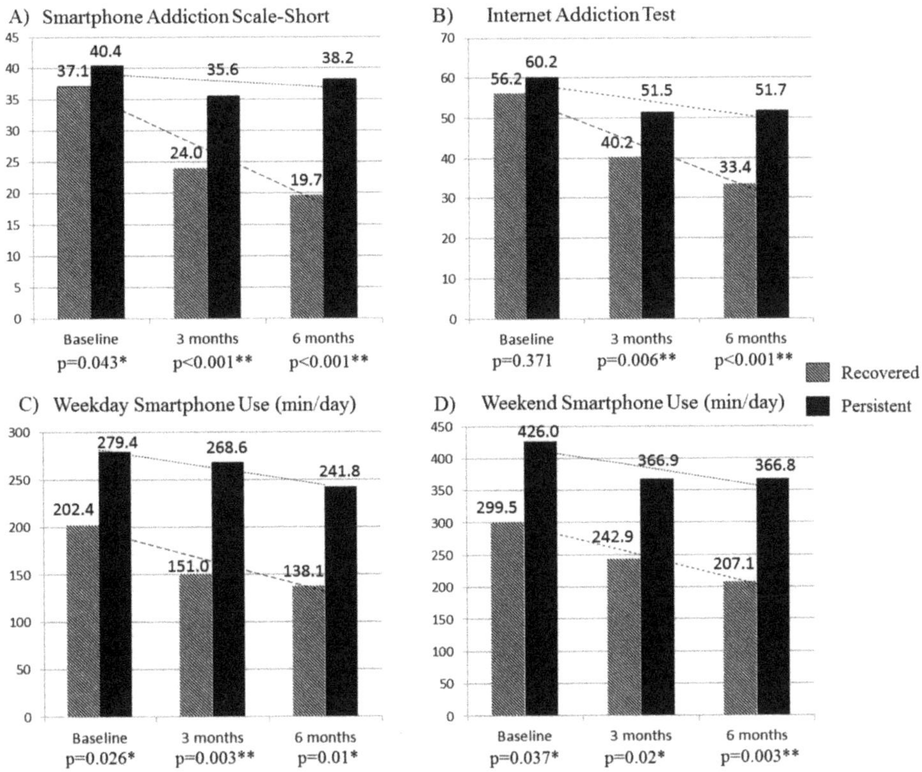

Figure 1. Longitudinal course of smartphone problems by recovery status. * $p < 0.05$, ** $p < 0.01$.

We also conducted analyses at follow-up points of variables that were not significantly different at the baseline. The duration of bedtime phone usage before sleep was significantly lower in the recovered group at both 3- and 6-month follow-ups (p = 0.036, and p = 0.019, respectively). Recovered problematic users were significantly less likely to suffer from DES at 6-month follow-up (p = 0.019). The persistent group, however, was more likely to exhibit depressive and anxious states at 3- and 6-month follow-ups, respectively (Figure 2). RSES, which was not significantly different either at baseline or 3-month follow-up, was significantly different between groups at 6-month follow-up ($p < 0.001$) (Figure 3).

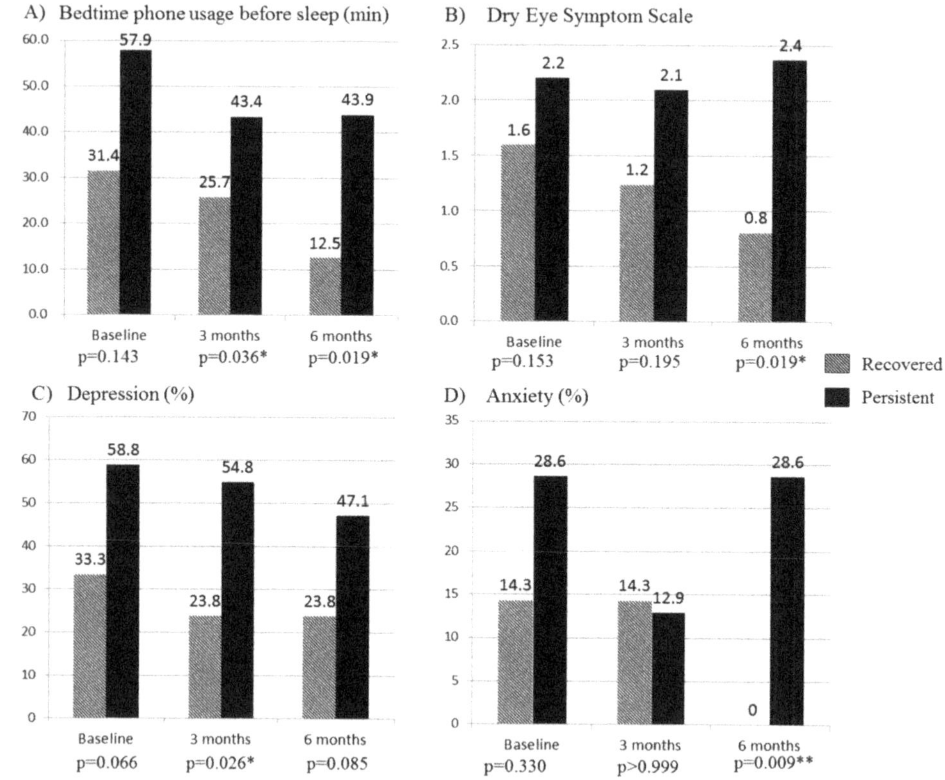

Figure 2. Physical and mental consequences of problematic smartphone use. * $p < 0.05$, ** $p < 0.01$.

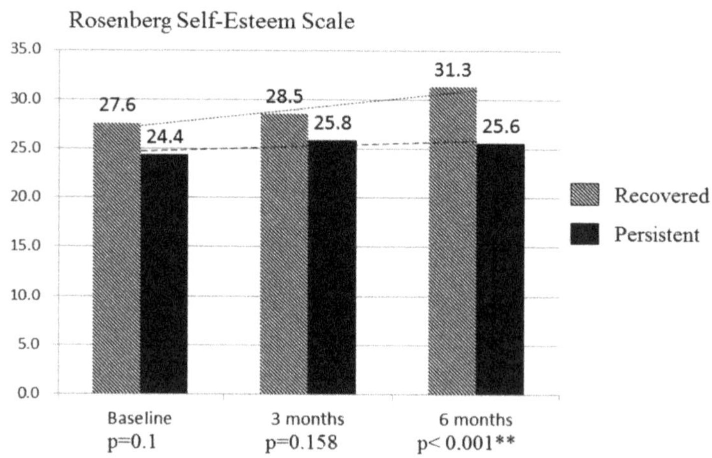

Figure 3. Follow-up comparison of self-esteem between the recovered and the persistent problematic smartphone users. ** $p < 0.01$.

4. Discussion

PSU is increasingly prevalent worldwide. Depression, anxiety, sleep disturbance, poor academic achievement, and physical hazards such as increased risk of accidents, musculoskeletal problems, and ophthalmic problems have been reported in numerous cross-sectional studies [11,12,14–20]. However, the long-term consequences of PSU remain largely unknown. The ever-increasing penetration of smartphones among children and adolescents is particularly alarming since the long-term consequences of PSU on developmental trajectories remain unknown. In addition to direct adverse outcomes on mental or physical health, severely affected children may suffer indirectly by loss of future opportunities through the impoverished acquisition of knowledge or skills during critical periods of development.

The objective of this study was to identify prognostic factors of PSU among children and adolescents. Currently, there are ongoing debates concerning excessive smartphone use, whether it should be considered "addiction" or merely "problematic use" and whether such behavior is "primary" or "secondary". Therefore, we also attempted to explore the temporal relationships between PSU and psychiatric problems, in addition to exploring the clinical course of PSU.

While there were no significant differences between the recovered group and the persistent PSU group in terms of psychiatric diagnosis, depression, or anxiety scores at baseline (Table 2), the persistent problematic smartphone users displayed a significantly higher rate of depression at 3-month follow-up and a significantly higher rate of anxiety at 6-month follow-up (Figure 2). Therefore, our findings support the designation of PSU as a primary disorder rather than a secondary phenomenon following negative emotional experiences in the clinical population. On the contrary, greater depressive or anxiety symptoms displayed by the PSU persistence suggest that adverse psychiatric consequences may develop unless PSU is adequately managed.

Although no significant differences were found in the whole group comparison, the anxiety level was significantly higher in children with persistent PSU aged 12 or less. Furthermore, although the difference was not statistically significant, a similar tendency was observed for depressive status in this younger group. These findings suggest that smartphones may serve as a compensatory measure against negative emotional experiences in younger children. Thus, PSU may also occur as a secondary disorder in young children. However, to confirm such a hypothesis, these findings should be replicated in a larger sample. When PSU is reported in children, considering both age and the context of use may help clinicians to understand the relationship between PSU and mental health.

The psychosocial vulnerabilities associated with persistent PSU at baseline were less conversation time with mothers, goal instability, low quality of life, lower level of perceived happiness, harm avoidance, and impulsivity (Table 2).

Parents exert significant influence over the development of their children. Parental perception of the social media use and parent–child interaction insecure attachment have been suggested as a major risk factors for addictive disorders [55], and punishing parental attitude was previously reported to be related to PSU [33]. Less conversation time with mothers may indicate weaker mother–child relationships. The lack of maternal protective factor was found to significantly increase the risk of continued problematic use of smartphones in this study. Despite not reaching a statistically significant level ($p = 0.065$), the persistent group was more likely to be living without their mothers, and it is in line with the aforementioned finding. Our finding is in line with the previous literature. In a large study of Chinese adolescents, mother–child relationships exerted more significant influence on problematic internet use than father–child relationships, and separated or "left-behind adolescents" were associated with subcomponents of problematic internet use [56].

The persistent PSU group showed lower quality of life and happiness but greater goal instability at baseline as well as at follow-up (Table 3). Further studies are indicated to investigate whether recovery from PSU by active therapeutic intervention also improves the quality of life, perceived happiness, and goal instability.

In short, the likelihood of recovery from PSU was higher among happy children with less goal instability, who interacted more with their mothers and enjoyed a better quality of life. The recovered users also spent significantly less time using smartphones at follow-up. Rosenberg Self-Esteem Scale scores, which demonstrated no significant differences at baseline, significantly increased at the endpoint among recovered participants (Figure 3). This indicates that recovered participants also exhibited increased self-esteem at 6-month follow-up.

The Matthew effect, which describes a psychosocial phenomenon of deepening inequality in which the rich become richer while the poor become even poorer [36], is thereby seen in the recovery process of PSU. The chance of recovery from PSU was lower in children with psychosocial disadvantages. PSU is likely to further widen the gap between children with psychological vulnerabilities or unfavorable family backgrounds and those without. The persistence of PSU may exert a significant influence over the developmental trajectory of self-esteem. While the self-esteem of the recovered children is boosted by the recognition of their success in achieving the recovery and better psychosocial outcomes such as higher quality of life and subjective happiness, the children with persistent PSU are more likely to fall behind in other important areas such as socialization and academic performance. Therefore, greater clinical attention and more intensive resources should be offered to PSU subjects with higher risk factor burdens.

Harm-avoidant temperament and impulsivity are well-known risk factors of PSU and other addictive disorders [12,42,57,58]. These risk factors may act as biological vulnerabilities to PSU. A previous structural imaging study demonstrated a significantly lower gray matter volume of the orbitofrontal cortex, an important structure involved in impulse control, in subjects with PSU compared to normal smartphone users [59]. These findings indicate shared mechanisms underlying PSU and addictive disorders. Further neurobiological studies are required to confirm the potential mechanism. Even though the impulsivity scores were higher in the persistent group throughout the follow-up, statistical significance was not sustained in the follow-up unlike the baseline. Therefore, further studies with a large sample size are required to determine whether impulsivity as a trait has a prognostic value. Another finding suggesting the addictive nature of PSU is the demonstrated prognostic value of PSU severity. Persistent problematic smartphone users demonstrated significantly higher initial PSU severity and spent greater amounts of time using smartphones, findings that are in line with those of a previous longitudinal study [24]. The fact that PSU severity per se at baseline, but not mental health problems, exert significant influence on the clinical course highlights the possibility that addiction plays a key role in PSU.

When participants with PSU were followed for 6 months, the majority (62.5%) continued to demonstrate problematic patterns of smartphone use. This finding suggests that the clinical course of PSU is stable over time and that more active management than parental education is needed to modify its course. The demonstrated diagnostic stability of PSU suggests that it may not be a benign behavioral problem lasting only temporarily.

Comparisons between the recovered group and the persistent group in this study demonstrated various physical consequences of PSU. Those with persistent smartphone problems were more likely to display neck pain not only at baseline but also at 6-month follow-up, which is in line with previous findings in the literature [16,19]. Considering that shoulder and hand/wrist/finger pain did not differ between the two groups, the assessment of neck pain seems to be the most clinically useful musculoskeletal symptom to screen for PSU and to monitor as a sign of PSU recovery.

Although no significant difference was observed at the baseline, the persistent group also showed significantly higher DES at the endpoint (Figure 2). In cross-sectional studies, longer smartphone usage was reported to be associated with DES [15,18]. This study is the first longitudinal study to demonstrate the development of DES as a consequence of persistent PSU.

Another longitudinal study previously showed that smartphone owners sleep significantly less during follow-up than subjects who do not own smartphones [60]. Exposure to blue light from smartphones at night was associated with disruptions of circadian rhythms and decreased performance at attentional tasks [61]. We found that the recovered group spent significantly less time using smartphones at bedtime at 3- and 6-month follow-ups. Therefore, bedtime smartphone usage should be evaluated in subjects with PSU to better manage potential sleep problems or difficulty concentrating in the daytime.

The two groups in this study were compared in the context of online content consumption. The consumption of adult material tended to be higher in the persistent group, but the difference was not statistically significant. However, the recovered group was more likely to engage in blogging. Still, it is too early to conclude that blogging is protective against PSU persistence. Maintaining a blog often involves sharing information such as the bloggers' point of view, and it usually requires more time to write a blog than to write on social networking services such as Instagram. Therefore, bloggers may display lower harm avoidance or impulsivity. Further studies are needed to verify whether recovery is facilitated by such individual traits or actually by the maintenance of blogs.

To summarize, the results of our study support the application of the addiction model to PSU. PSU severity but not psychiatric comorbidity increased the risk of persistent PSU. Moreover, children and adolescents were more likely to develop physical and mental health problems with persistent problematic smartphone use. PSU not only showed characteristics of a primary disorder but also demonstrated diagnostic stability. As PSU shares common risk factors with addictive disorders such as impulsivity and harm avoidance, it manifests characteristics of addiction. However, it was previously argued that PSU is not an addictive disorder [5].

Several factors may have contributed to the traditional view. First, the traditional view was drawn from studies that utilized convenient samples and cross-sectional designs. To the best of our knowledge, the present study is the first longitudinal study conducted in clinical settings. The severity of problematic smartphone use in the non-clinical population is likely to be lower than in the clinical population, with higher self-remission. Therefore, the benign nature of PSU in the non-clinical population may have played a role in the formulation of such a benign view. In addition to differences in severity, the wide use of smartphones in everyday life may also make researchers more reluctant to label PSU as addictive behavior, in a desire to avoid criticism due to pathologizing and medicalizing daily behaviors. Even though we are also against pathologizing the simple overuse of smartphones by referring to it as an addictive disorder, PSU may have a wide spectrum in terms of severity and longitudinal course. Heavy users at the end of the continuum may display clinical characteristics resembling those of addicts. For such users, broader management options should be available, and the allocation of greater therapeutic resources will be required to improve clinical outcomes.

The strength of our study lies in its comprehensive assessment of demographic, psychosocial, and digital media-related risk factors. However, this study also has some limitations. While the majority of participants were followed throughout the study period, a substantial number (34.1%) dropped out at the 6-month follow-up. We conducted secondary analyses comparing the completed group and dropouts. There were no significant differences in demographics between groups. However, the baseline SAS-SV scores of the dropouts were significantly lower compared to the participants that remained in the study. Subjects with lower PSU severity may have been less motivated to continue the study.

Although our sample size is relatively large compared to those of other longitudinal studies, we were unable to perform parametric analyses for continuous variables due to the sample size. Lack of such analyses generally leads to a loss in statistical power, and small potential differences that otherwise might have been detected in parametric analyses may not have been found in the non-parametric analyses. Another limitation is the lack of objective measurements to validate the responses of the participants. Although the participants were reassured regarding the confidentiality of assessments to parents or guardians unless immediate harm was anticipated, the reliance on self-reports may result in an under-estimation of problems or severity. Significant discordance was reported between self-reported and clinician-rated severity of problematic gaming in a previous study [62]. Social desirability is known to contribute to under-reporting of deviant or socially undesirable behaviors [63]. Therefore, the utilization of objective measures such as smartphone applications that analyze patterns of use is required to manage the potential for under-reported PSU in children and adolescents.

5. Conclusions

In this longitudinal study, diagnostic stability was observed for PSU, with the majority of participants displaying persistent PSU six months after the beginning of the study. Prognostic factors of PSU were individual traits such as harm avoidance and impulsivity; psychological vulnerabilities such as high SAS-SV, low quality of life, low perceived happiness, and goal instability; and social risk factors such as shorter weekday conversations with mothers. We suggest that the bio-psychosocial framework could be useful in the evaluation of PSU and in predicting its course. The exploration of causality between PSU and psychiatric problems showed that PSU increased the risk of developing psychiatric problems later, which supports the interpretation of PSU as a primary disorder. While the recovered group demonstrated fewer psychological risk factors at baseline than persistent problematic users, they also benefited from further gains in psychological well-being and were less afflicted by physical consequences. This Matthew effect in the recovery of PSU calls for additional attention to be paid to disadvantaged children or adolescents with PSU, to better facilitate their recovery and reduce the harmful long-term consequences of PSU as a secondary prevention strategy.

Author Contributions: S.-Y.L. analyzed and interpreted the data and drafted the manuscript; J.-S.C., S.-y.B. and Y.-S.K. participated in the recruiting and data collection; H.K.L., and Y.-S.K. developed concepts and supervised; M.-H.P., and K.-I.J. reviewed and edited the manuscript. All authors have read and agree to the published version of the manuscript.

Funding: This research was funded by the Korean Mental Health Technology R&D Project, Ministry of Health & Welfare, Republic of Korea (HM14C2603).

Acknowledgments: This study was supported by a grant of the Korean Mental Health Technology R&D Project, Ministry of Health & Welfare, Republic of Korea (HM14C2603).

Conflicts of Interest: The authors declare no conflict of interest. The funders had no role in the design of the study; in the collection, analyses, or interpretation of data; in the writing of the manuscript, or in the decision to publish the results.

References

1. Guazzini, A.; Duradoni, M.; Capelli, A.; Meringolo, P. An Explorative Model to Assess Individuals' Phubbing Risk. *Future Internet* **2019**, *11*, 21. [CrossRef]
2. O'Dea, S. Smartphone Penetration Rate by Country 2018. Available online: https://www.statista.com/statistics/539395/smartphone-penetration-worldwide-by-country/ (accessed on 4 April 2020).
3. Ministry of Science and ICT; National Information Society Agency. 2019 The Survey on Smartphone Overdependence. 2020. Available online: https://www.nia.or.kr/site/nia_kor/ex/bbs/View.do?cbIdx=65914&bcIdx=21939&parentSeq=21939 (accessed on 1 May 2020).
4. Sohn, S.; Rees, P.; Wildridge, B.; Kalk, N.J.; Carter, B. Prevalence of problematic smartphone usage and associated mental health outcomes amongst children and young people: A systematic review, meta-analysis and GRADE of the evidence. *BMC Psychiatry* **2019**, *19*, 356. [CrossRef]
5. Panova, T.; Carbonell, X. Is smartphone addiction really an addiction? *J. Behav. Addict.* **2018**, *7*, 252–259. [CrossRef]
6. Yu, S.; Sussman, S. Does Smartphone Addiction Fall on a Continuum of Addictive Behaviors? *Int. J. Environ. Res. Public Health* **2020**, *17*, 422. [CrossRef] [PubMed]
7. Elhai, J.D.; Levine, J.C.; Hall, B.J. The relationship between anxiety symptom severity and problematic smartphone use: A review of the literature and conceptual frameworks. *J. Anxiety Disord.* **2019**, *62*, 45–52. [CrossRef] [PubMed]
8. Quarter of Children Under Six Have A Smartphone, Study Finds. Agency reporter, 8 April 2018.
9. Beyens, I.; Nathanson, A.I. Electronic Media Use and Sleep Among Preschoolers: Evidence for Time-Shifted and Less Consolidated Sleep. *Health Commun.* **2019**, *34*, 537–544. [CrossRef]
10. Grant, J.E.; Lust, K.; Chamberlain, S.R. Problematic smartphone use associated with greater alcohol consumption, mental health issues, poorer academic performance, and impulsivity. *J. Behav. Addict.* **2019**, *8*, 335–342. [CrossRef]

11. Kim, Y.J.; Jang, H.M.; Lee, Y.; Lee, D.; Kim, D.J. Effects of Internet and Smartphone Addictions on Depression and Anxiety Based on Propensity Score Matching Analysis. *Int. J. Environ. Res. Public Health* **2018**, *15*, 859. [CrossRef]
12. Lee, S.Y.; Lee, D.; Nam, C.R.; Kim, D.Y.; Park, S.; Kwon, J.G.; Kweon, Y.S.; Lee, Y.; Kim, D.J.; Choi, J.S. Distinct patterns of Internet and smartphone-related problems among adolescents by gender: Latent class analysis. *J. Behav. Addict.* **2018**, *7*, 454–465. [CrossRef]
13. Yang, J.; Fu, X.; Liao, X.; Li, Y. Association of problematic smartphone use with poor sleep quality, depression, and anxiety: A systematic review and meta-analysis. *Psychiatry Res.* **2020**, *284*, 112686. [CrossRef]
14. Vahedi, Z.; Saiphoo, A. The association between smartphone use, stress, and anxiety: A meta-analytic review. *Stress Health J. Int. Soc. Investig. Stress* **2018**, *34*, 347–358. [CrossRef] [PubMed]
15. Moon, J.H.; Kim, K.W.; Moon, N.J. Smartphone use is a risk factor for pediatric dry eye disease according to region and age: A case control study. *Bmc Ophthalmol.* **2016**, *16*, 188. [CrossRef]
16. Lee, S.; Choi, Y.H.; Kim, J. Effects of the cervical flexion angle during smartphone use on muscle fatigue and pain in the cervical erector spinae and upper trapezius in normal adults in their 20s. *J. Phys. Ther. Sci.* **2017**, *29*, 921–923. [CrossRef]
17. Eitivipart, A.C.; Viriyarojanakul, S.; Redhead, L. Musculoskeletal disorder and pain associated with smartphone use: A systematic review of biomechanical evidence. *Hong Kong Physiother. J. Off. Publ. Hong Kong Physiother. Assoc. Ltd. Wu Li Chih Liao* **2018**, *38*, 77–90. [CrossRef] [PubMed]
18. Choi, J.H.; Li, Y.; Kim, S.H.; Jin, R.; Kim, Y.H.; Choi, W.; You, I.C.; Yoon, K.C. The influences of smartphone use on the status of the tear film and ocular surface. *PLoS ONE* **2018**, *13*, e0206541. [CrossRef] [PubMed]
19. Alsalameh, A.M.; Harisi, M.J.; Alduayji, M.A.; Almutham, A.A.; Mahmood, F.M. Evaluating the relationship between smartphone addiction/overuse and musculoskeletal pain among medical students at Qassim University. *J. Fam. Med. Prim. Care* **2019**, *8*, 2953–2959. [CrossRef]
20. Kim, H.J.; Min, J.Y.; Kim, H.J.; Min, K.B. Accident risk associated with smartphone addiction: A study on university students in Korea. *J. Behav. Addict.* **2017**, *6*, 699–707. [CrossRef]
21. Guo, N.; Luk, T.T.; Ho, S.Y.; Lee, J.J.; Shen, C.; Oliffe, J.; Chan, S.S.; Lam, T.H.; Wang, M.P. Problematic Smartphone Use and Mental Health in Chinese Adults: A Population-Based Study. *Int. J. Environ. Res. Public Health* **2020**, *17*, 844. [CrossRef] [PubMed]
22. Gao, Q.; Sun, R.; Fu, E.; Jia, G.; Xiang, Y. Parent-child relationship and smartphone use disorder among Chinese adolescents: The mediating role of quality of life and the moderating role of educational level. *Addict. Behav.* **2020**, *101*, 106065. [CrossRef] [PubMed]
23. Karsay, K.; Schmuck, D.; Matthes, J.; Stevic, A. Longitudinal Effects of Excessive Smartphone Use on Stress and Loneliness: The Moderating Role of Self-Disclosure. *Cyberpsychology Behav. Soc. Netw.* **2019**, *22*, 706–713. [CrossRef] [PubMed]
24. Lapierre, M.A.; Zhao, P.; Custer, B.E. Short-Term Longitudinal Relationships Between Smartphone Use/Dependency and Psychological Well-Being Among Late Adolescents. *J. Adolesc. Health Off. Publ. Soc. Adolesc. Med.* **2019**, *65*, 607–612. [CrossRef]
25. Lee, J.E.; Jang, S.-I.; Ju, Y.J.; Kim, W.; Lee, H.J.; Park, E.-C. Relationship between Mobile Phone Addiction and the Incidence of Poor and Short Sleep among Korean Adolescents: A Longitudinal Study of the Korean Children & Youth Panel Survey. *J. Korean Med. Sci.* **2017**, *32*, 1166–1172.
26. Tossell, C.; Kortum, P.; Shepard, C.; Rahmati, A.; Zhong, L. Exploring Smartphone Addiction: Insights from Long-Term Telemetric Behavioral Measures. *Int. J. Interact. Mob. Technol.* **2015**, *9*, 7. [CrossRef]
27. Feliciano, L.; Yochim, B.; Steers, M.E.; Jay, A.A.; Segal, D.L. *Understanding Research in Clinical and Counseling Psychology*, 2nd ed.; Jay, C., Hersen, T.M., Eds.; Routledge Taylor & Francis: New York, NY, USA, 2011.
28. Lee, B.H.; Lee, H.K. Longitudinal study shows that addictive Internet use during adolescence was associated with heavy drinking and smoking cigarettes in early adulthood. *Acta Paediatrica (Oslo Norway 1992)* **2017**, *106*, 497–502. [CrossRef]
29. Duradoni, M.; Innocenti, F.; Guazzini, A. Well-Being and Social Media: A Systematic Review of Bergen Addiction Scales. *Future Internet* **2020**, *12*, 24. [CrossRef]
30. Vadlin, S.; Aslund, C.; Nilsson, K.W. A longitudinal study of the individual—And group-level problematic gaming and associations with problem gambling among Swedish adolescents. *Brain Behav.* **2018**, *8*, e00949. [CrossRef]

31. Latulippe, K.; Hamel, C.; Giroux, D. Social Health Inequalities and eHealth: A Literature Review with Qualitative Synthesis of Theoretical and Empirical Studies. *J. Med. Internet Res.* **2017**, *19*, e136. [CrossRef]
32. Wang, X.; Xing, W. Exploring the Influence of Parental Involvement and Socioeconomic Status on Teen Digital Citizenship: A Path Modeling Approach. *J. Educ. Technol. Soc.* **2018**, *21*, 186–199.
33. Lee, H.; Kim, J.W.; Choi, T.Y. Risk Factors for Smartphone Addiction in Korean Adolescents: Smartphone Use Patterns. *J. Korean Med. Sci.* **2017**, *32*, 1674–1679. [CrossRef]
34. Muñoz-Miralles, R.; Ortega-González, R.; López-Morón, M.R.; Batalla-Martínez, C.; Manresa, J.M.; Montellà-Jordana, N.; Chamarro, A.; Carbonell, X.; Torán-Monserrat, P. The problematic use of Information and Communication Technologies (ICT) in adolescents by the cross sectional JOITIC study. *BMC Pediatr.* **2016**, *16*, 140. [CrossRef]
35. United States Conference of Catholic Bishops Matthew 25:29. Available online: http://www.usccb.org/bible/mt/25:29 (accessed on 5 May 2020).
36. Merton, R.K. The Matthew effect in science: The reward and communication systems of science are considered. *Science* **1968**, *159*, 56–63. [CrossRef] [PubMed]
37. Kwon, M.; Lee, J.-Y.; Won, W.-Y.; Park, J.-W.; Min, J.-A.; Hahn, C.; Gu, X.; Choi, J.-H.; Kim, D.-J. Development and Validation of a Smartphone Addiction Scale (SAS). *PLoS ONE* **2013**, *8*, e56936. [CrossRef] [PubMed]
38. Kwon, M.; Kim, D.J.; Cho, H.; Yang, S. The smartphone addiction scale: Development and validation of a short version for adolescents. *PLoS ONE* **2013**, *8*, e83558. [CrossRef] [PubMed]
39. Ministry of Health & Welfare; Korean Academy of Medical Science; Korean Ophthalmological Society. The Symptoms and Complications of Dry Eye Syndrome. Available online: http://health.cdc.go.kr/health/mobileweb/content/group_view.jsp?CID=F98951A125 (accessed on 1 March 2020).
40. Cloninger, C.R.; Svrakic, D.M.; Przybeck, T.R. A psychobiological model of temperament and character. *Arch. Gen. Psychiatry* **1993**, *50*, 975–990. [CrossRef]
41. Lyoo, I.K.; Han, C.H.; Lee, S.J.; Yune, S.K.; Ha, J.H.; Chung, S.J.; Choi, H.; Seo, C.S.; Hong, K.E. The reliability and validity of the junior temperament and character inventory. *Compr. Psychiatry* **2004**, *45*, 121–128. [CrossRef]
42. Jeong, H.; Lee, H.K.; Kwon, Y.-S.; Yim, H.W.; Lee, S.-Y. Gaming disorder and bidirectional relationships with aggression and impulsivity. *Curr. Opin. Behav. Sci.* **2020**, *31*, 69–75. [CrossRef]
43. Cho, S.C.; Lee, Y.S. Development of the Korean form of the Kovacs' Childeren's Depression Inventory. *J. Korean Neuropsychiatr. Assoc.* **1990**, *29*, 943–956.
44. Beck, A.; Steer, R.; Brown, G. Manual for the Beck depression inventory-II (BDI-II). *Behav. Res. Ther.* **1996**, *37*, 183–190.
45. Cho, S.C.; Choi, C.J. Development of the Korean form of the State-Trait anxiety Inventory for Children. *Seoul J. Psychiatry* **1989**, *14*, 150–157.
46. Kim, J.T. *Relationship between State-Train Anxiety and Socializing*; Korea University: Seoul, Korea, 1978.
47. Patton, J.H.; Stanford, M.S.; Barratt, E.S. Factor structure of the Barratt impulsiveness scale. *J. Clin. Psychol.* **1995**, *51*, 768–774. [CrossRef]
48. HS, L. *Manual for Impulsivity Test*; Guidance Korea: Seong Nam, Korea, 1992.
49. Bahn, G.H.; Shin, M.S.; Cho, S.C.; Hong, K.E. A Preliminary Study For The Development Of The Assessment Scale For Adhd In Adolescents:Reliability And Validity For Cass(S). *J. Korean Acad. Child Adolesc. Psychiatry* **2001**, *12*, 218–224.
50. Robbins, S.B.; Payne, E.C.; Chartrand, J.M. Goal instability and later life adjustment. *Psychol. Aging* **1990**, *5*, 447–450. [CrossRef]
51. Lee, E.H. The role of goal instability and resolving style of problems between nurses' job stressors and emotional exhaustion. *Korean J. Woman Psychol.* **2008**, *13*, 397–413. [CrossRef]
52. Kim, B.H.; Bang, S.Y.; Lee, H.K.; Cho, S.S.; Choi, S.; Ryu, C.S. *Developing and Study of happiness index of Korean Adolescents*; Happy Home Foundation: Seoul, Korea, 2007.
53. Rosenberg, M. *Society and the Adolescent Self-Image*; Princeton University Press: Princeton, NJ, USA, 1965.
54. Choi, E.S. Psychometric test of the PedsQLTM 4.0 Generic Core Scale in Korean adolescents. 2004. Available online: http://dcollection.yonsei.ac.kr/public_resource/pdf/000000125930_20200701160223.pdf (accessed on 3 May 2020).

55. Nakhoul, L.; Obeid, S.; Sacre, H.; Haddad, C.; Soufia, M.; Hallit, R.; Akel, M.; Salameh, P.; Hallit, S. Attachment style and addictions (alcohol, cigarette, waterpipe and internet) among Lebanese adolescents: A national study. *BMC Psychol.* **2020**, *8*, 33. [CrossRef] [PubMed]
56. Xu, J.; Shen, L.X.; Yan, C.H.; Hu, H.; Yang, F.; Wang, L.; Kotha, S.R.; Ouyang, F.; Zhang, L.N.; Liao, X.P.; et al. Parent-adolescent interaction and risk of adolescent internet addiction: A population-based study in Shanghai. *BMC Psychiatry* **2014**, *14*, 112. [CrossRef] [PubMed]
57. Hanafi, E.; Siste, K.; Wiguna, T.; Kusumadewi, I.; Nasrun, M.W. Temperament profile and its association with the vulnerability to smartphone addiction of medical students in Indonesia. *PLoS ONE* **2019**, *14*, e0212244. [CrossRef]
58. Alvarez, A.; Avila, J.J.; Palao, D.J.; Montejo, A.L. Influence of Personality Traits on the Severity of Alcohol Use Disorders. *J. Clin. Med.* **2018**, *7*, 127. [CrossRef]
59. Lee, D.; Namkoong, K.; Lee, J.; Lee, B.O.; Jung, Y.C. Lateral orbitofrontal gray matter abnormalities in subjects with problematic smartphone use. *J. Behav. Addict.* **2019**, *8*, 404–411. [CrossRef]
60. Schweizer, A.; Berchtold, A.; Barrense-Dias, Y.; Akre, C.; Suris, J.C. Adolescents with a smartphone sleep less than their peers. *Eur. J. Pediatr.* **2017**, *176*, 131–136. [CrossRef]
61. Heo, J.Y.; Kim, K.; Fava, M.; Mischoulon, D.; Papakostas, G.I.; Kim, M.J.; Kim, D.J.; Chang, K.J.; Oh, Y.; Yu, B.H.; et al. Effects of smartphone use with and without blue light at night in healthy adults: A randomized, double-blind, cross-over, placebo-controlled comparison. *J. Psychiatr. Res.* **2017**, *87*, 61–70. [CrossRef]
62. Jeong, H.; Yim, H.W.; Lee, S.Y.; Lee, H.K.; Potenza, M.N.; Kwon, J.H.; Koo, H.J.; Kweon, Y.S.; Bhang, S.Y.; Choi, J.S. Discordance between self-report and clinical diagnosis of Internet gaming disorder in adolescents. *Sci. Rep.* **2018**, *8*, 10084. [CrossRef] [PubMed]
63. Johnson, T.; Fendrich, M. Modeling sources of self-report bias in a survey of drug use epidemiology. *Ann. Epidemiol.* **2005**, *15*, 381–389. [CrossRef] [PubMed]

International Journal of
Environmental Research and Public Health

MDPI

Article

Relationship between Difficulty in Emotion Regulation and Internet Addiction in College Students: A One-Year Prospective Study

Jui-Kang Tsai [1,2,†], Wei-Hsin Lu [3,4,†], Ray C. Hsiao [5,6], Huei-Fan Hu [7,*] and Cheng-Fang Yen [8,9,*]

1 Department of Psychiatry, Kaohsiung Armed Forces General Hospital, Kaohsiung 80284, Taiwan; regon720@gmail.com
2 Department of Nursing, Meiho University, Pingtung 91202, Taiwan
3 Department of Psychiatry, Ditmanson Medical Foundation Chia-Yi Christian Hospital, Chia-Yi City 60002, Taiwan; wiiseen@gmail.com
4 Department of Senior Citizen Service Management, Chia Nan University of Pharmacy and Science, Tainan 71710, Taiwan
5 Department of Psychiatry and Behavioral Sciences, University of Washington School of Medicine, Seattle, WA 98105, USA; rhsiao@u.washington.edu
6 Department of Psychiatry, Children's Hospital and Regional Medical Center, Seattle, WA 98105, USA
7 Department of Psychiatry, Tainan Municipal Hospital (managed by Show Chwan Medical Care Corporation), Tainan 70173, Taiwan
8 Department of Psychiatry, School of Medicine and Graduate Institute of Medicine, College of Medicine, Kaohsiung Medical University, Kaohsiung 80708, Taiwan
9 Department of Psychiatry, Kaohsiung Medical University Hospital, Kaohsiung 80708, Taiwan
* Correspondence: cych07205@gmail.com (H.-F.H.); chfaye@cc.kmu.edu.tw (C.-F.Y.); Tel.: +886-6-2609926 (H.-F.H.); +886-7-3124941 (C.-F.Y.)
† J.-K.T. and W.-H.L. contributed equally to this study.

Received: 31 May 2020; Accepted: 30 June 2020; Published: 2 July 2020

Abstract: This prospective study evaluated the predictive effect of difficulty in emotion regulation on the occurrence and remission of Internet addiction (IA) and determined whether IA has a role in changing emotion regulation among college students during a follow-up period of 1 year. A total of 500 college students (262 women and 238 men) were recruited. In baseline and follow-up investigations, the levels of IA and difficulty in emotion regulation were evaluated using the Chen Internet Addiction Scale and the Difficulties in Emotion Regulation Scale (DERS), respectively. The results indicated that the subscale of impulse control difficulties on the DERS predicted the incidence of IA during the follow-up period of 1 year in male participants ($t = -2.875$, $p = 0.005$), whereas no subscale on the DERS predicted the remission of IA. IA did not predict the change in difficulties in emotion regulation. The subscale of impulse control difficulties on the DERS predicted the occurrence of IA in the college students and warrants early intervention.

Keywords: internet addiction; emotion regulation; college student

1. Introduction

The Internet has become a key resource in society and everyday life. At social and individual levels, the Internet facilitates interpersonal communication, provides entertainment, and helps individuals create new social networks [1,2]. However, inappropriate or excessive Internet use may result in Internet addiction (IA) and negative life outcomes [3]. People who have IA may have difficulties in controlling their Internet usage and develop problems in occupational or academic performance and in social relationships, as well as psychological and physical problems [4]. Although the Diagnostic and

Statistical Manual of Mental Disorders, Fifth Edition (DSM-5) does not list IA as a formal psychiatric diagnosis, it calls for research to examine Internet gaming disorder, a common type of IA, in terms of its course, symptoms, and underlying mechanisms [5]. Therefore, IA is an issue that warrants further study.

A multinational meta-analysis provided estimates of the global prevalence of IA, which varies across world regions, with the highest prevalence of 10.9% in the Middle East and the lowest prevalence of 2.6% in Northern and Western Europe [6]. The prevalence rate of IA among college students is 3.7% in Japan [7], 8–13% in Taiwan [8], and 13.6% in China [9]. College students may use the Internet for studying, gaming, social networking, gambling, chatting, shopping, and watching pornographic videos [10,11]. However, given that college students have free and unlimited access to the Internet, have flexible schedules, and are free from their parents' interference, they were identified as having a high risk for IA [12].

There have many studies examining the relationship between IA and personality characteristics such as neuroticism [13–15], conscientiousness [13–15], agreeableness [13–15], boredom proneness [16], borderline personality characteristics [17], reinforcement sensitivity [18], and frustration intolerance [18]. Emotion regulation is the attempt to alter emotional experiences via the initiation, maintenance, or modification of frequency, intensity, or duration of emotional experiences [19]. Difficulties with emotion regulation are believed to be risk factors for addiction [20,21]. Gratz and Roemer proposed that emotion regulation involves multiple conceptions, including (a) awareness and understanding of emotions, (b) acceptance of emotions, (c) ability to control impulsive behaviors and behave in accordance with desired goals when experiencing negative emotions, and (d) ability to use situationally appropriate emotion regulation-n strategies flexibly to modulate emotional responses as desired in order to meet individual goals and situational demands [22]. The relative absence of any or all of these abilities would indicate the presence of difficulties in emotion regulation [22].

Research has found that young people with IA had more difficulty in identifying and describing emotions, understanding emotional reactions, and controlling their impulsive behaviors in response to negative emotional experiences, and were less able to use effective emotion regulation strategies than adolescents without IA [23–27]. It is hypothesized that inadequate prefrontal cognitive control to suppress their negative emotions may result in impulsivity and increase the risk of IA [26]. A two-year prospective study also reported that emotional problems, such as depression and social phobia, predict the occurrence of IA in adolescents [28]. However, research also found that Internet use is a common strategy that people use to alleviate emotional distress [29–31]. Based on the principle of functional analysis in behavioral psychology [32], many forms of problematic behaviors and psychopathology can be conceptualized as efforts to escape and avoid emotions, thoughts, memories, and other private experiences [33]. Although IA is generally considered an unhealthy effort to cope with stress and negative emotions [29], whether IA deteriorates or improves the function of emotion regulation warrants survey.

No prospective study has examined the bidirectional relationship between IA and difficulties in emotion regulation. The prospective study design can provide the temporal relationships between IA and difficulties in emotion regulation and infer the causal relationship between them. The aims of this prospective study were to evaluate the predictive effect of difficulty in emotion regulation on the occurrence and remission of IA, and to determine whether IA can predict the change in difficulty in emotion regulation among college students during the follow-up period of 1 year. In particular, given that emotion regulation contains multiple dimensions of conceptions [22], this study aimed to examine whether the relationships between IA and emotion regulation vary according to various dimensions of emotion regulation. Based on the results of previous studies described above, we hypothesized that difficulty in emotion regulation may predict the occurrence of IA and the non-remission of IA during the period of one-year follow-up. Moreover, we also hypothesized that IA may worsen emotion regulation during the period of 1 year.

2. Materials and Methods

2.1. Participants

Participants were recruited using an advertisement posted for college students aged between 20 and 30 years. A total of 500 college students (262 women and 238 men) participated in this study. Regarding the sample size, a previous study found that 8–13% of college students in Taiwan had IA [8]. The sample of 500 participants was determined as adequate based on the estimation with 80% power, 95% confidence interval, and statistically significant level at 5% [34]. The mean age of the participants was 22.1 years (standard deviation (SD): 1.8 years). Informed consent was obtained from all the participants prior to assessment. The study was approved by the Institutional Review Board of Kaohsiung Medical University Hospital (KMUH-IRB-20120249).

2.2. Measures

2.2.1. Chen Internet Addiction Scale (CIAS)

We used the self-administered Chen IA Scale (CIAS) to assess the participants' severity of IA in the month preceding the study [35]. The CIAS contains 26 items rated on a 4-point Likert scale with the minimum and maximum scores of 26 and 104, respectively [35]. A higher total score indicates more severe IA. The internal reliability (Cronbach's α) of the CIAS in the present study was 0.93. According to diagnostic criteria for IA, a score of 68 out of the total CIAS score has the highest diagnostic accuracy and accepted sensitivity and specificity for IA [36]. Accordingly, those with CIAS scores of 68 or more were classified as those with IA.

2.2.2. Difficulties in Emotion Regulation Scale (DERS)

The DERS is a 36-item self-reported measure developed to assess clinically relevant difficulties in emotion regulation [22]. Items are scored on six scales: Nonacceptance of Emotional Responses (6 items), Difficulties Engaging in Goal-Directed Behavior (5 items), Impulse Control Difficulties (6 items), Lack of Emotional Awareness (6 items), Limited Access to Emotion Regulation Strategies (8 items), and Lack of Emotional Clarity (5 items). Participants are asked to indicate how often each of the 36 items applied to them on a 5-point Likert scale ranging from 1 (almost never) to 5 (almost always). Subscale scores are obtained by summating corresponding items. Cronbach's α for the Taiwanese version of the DERS ranged from 0.72 to 0.81. Higher scores on the DERS indicate greater difficulties in emotion regulation.

2.2.3. Demographic Characteristics

The demographic characteristics of participants collected in this study were sex (female or male) and age (years).

2.3. Study Process and Statistical Analysis

In the initial assessment (Stage 1), all the participants were invited to complete the CIAS and DERS. One year later (Stage 2), the participants were invited to complete the CIAS and DERS again. According to the scores of the CIAS at Stage 1 and Stage 2, the participants were classified into one of four groups (Figure 1). The participants deemed not to be addicted at Stage 1 were stratified into the Persistent no IA (PNIA) and New IA (NIA) groups based on subsequent non-addiction and addiction statuses at Stage 2, respectively. The remaining participants, who were initially deemed to be addicted to the Internet at Stage 1, were stratified into the groups of Remitted IA (RIA) and Persistent IA (PIA) based on remission and continuation of the behavior at Stage 2, respectively.

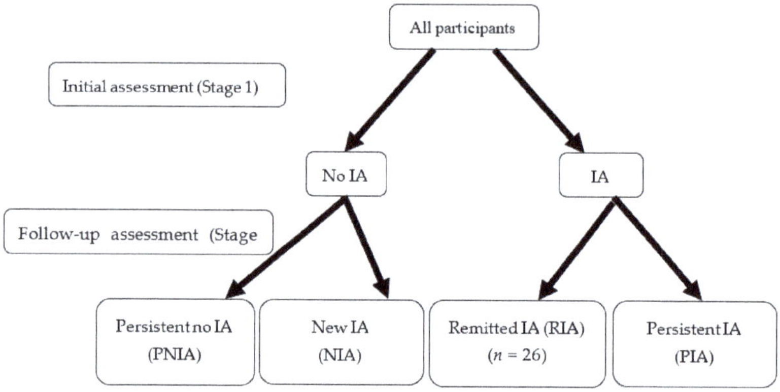

Figure 1. Study participants were assigned to one of four addiction groups. IA: Internet addiction

The incidence rate of IA at Stage 2 and its relationship with difficulty in emotion regulation at Stage 1 were examined in the PNIA and NIA groups. The remission rate of IA at Stage 2 and its relationship with difficulty in emotion regulation at Stage 1 were examined in the RIA and PIA groups. The chi-square test was utilized to evaluate category variables, and Student's t-test was used for continuous variables.

We evaluated whether changes in the difficulty of emotion regulation during the one-year period were different between the college students with and without IA at Stage 1. A repeated-measures two-way analysis of variance (ANOVA) with investigation stages (Stage 1 vs. Stage 2) as a within-subjects factor and the groups (with or without IA at Stage 1) as a between-subjects factor was performed for DERS scores. All statistical analyses were performed using SPSS version 20.0 (SPSS Inc., Chicago, IL, USA). Because of the existence of multiple comparisons, p-values of <0.008 (0.05/6) were considered statistically significant.

3. Results

A total of 324 college students (65.8%, 169 women and 155 men) underwent the follow-up assessment 1 year later. Of the 176 participants (93 women and 83 men) who did not participate in follow-up assessment, 96 (54.5%) could not be contacted, 48 (27.3%) refused to participate in follow-up assessment, and 32 (18.2%) had the desire to participate but were unable to do so due to work or army service. The results of comparing demographic data and the levels of IA on the CIAS and difficulties in emotion regulation on the DERS at initial assessment between participants who participated in and who did not participate in follow-up assessment are shown in Table 1. No differences were found in gender, age, and the levels of IA and difficulties in emotion regulation between these two groups (all $p > 0.008$).

Of the 268 participants in the PNIA and NIA groups who had no IA at Stage 1, 20 were deemed to have IA at Stage 2 (the NIA group), resulting in an incidence rate of 7.5% during the one-year period. Of the 56 individuals in the RIA and PIA groups who had IA at Stage 1, 26 were classified as being without IA at Stage 2 (the RIA group), indicating a one-year remission rate of 46.4%.

Comparisons of demographic characteristics and difficulties in emotion regulation between the PNIA and NIA groups are presented in Table 2. The results revealed that compared with the PNIA group, the NIA group had more severe impulse control difficulties on the DERS at Stage 1, revealing that impulse control difficulties at Stage 1 predicted the incidence of IA at Stage 2 during the follow-up period of 1 year. We further stratified the participants by gender and found that the prediction of impulse control difficulties for the incidence of IA existed in only male ($t = -2.875$, $p = 0.005$) but not in female participants ($t = -1.270$, $p = 0.206$).

Table 1. Comparisons of demographic data and difficulties in emotion regulation between participants who participated in and who did not participate in follow-up assessment.

	No Follow-Up (n =176) n (%) or mean (SD)	Follow-Up (n = 324) n (%) or mean (SD)	χ^2 or t
Gender, n (%)			
Female	93 (52.8)	169 (52.2)	0.884
Male	83 (47.2)	155 (47.8)	
Age (years), mean (SD)	21.9 (1.5)	22.3 (1.9)	0.047
CIAS, mean (SD)	56.6 (13.2)	55.0 (14.2)	1.189
DERS, mean (SD)			
Non-acceptance of emotional responses	12.7 (4.8)	12.6 (3.9)	0.273
Difficulties engaging in goal-directed behavior	12.2 (3.9)	12.0 (3.6)	0.539
Impulse control difficulties	11.2 (4.0)	11.5 (4.2)	−0.802
Lack of emotional awareness	15.3 (4.5)	14.7 (3.9)	1.364
Limited access to emotion regulation strategies	17.1 (5.7)	16.7 (5.3)	0.819
Lack of emotional clarity	9.9 (3.4)	10.0 (2.9)	−0.350

CIAS: Chen Internet Addiction Scale; DERS: Difficulties in Emotion Regulation Scale; SD: Standard deviation.

Table 2. Comparisons of demographic data and difficulties in emotion regulation between the PNIA and NIA groups and the RIA and PIA groups.

	PNIA Group (n = 248) n (%) or mean (SD)	NIA Group (n = 20) n (%) or mean (SD)	χ^2 or t	RIA Group (n = 26) n (%) or mean (SD)	PIA Group (n = 30) n (%) or mean (SD)	χ^2 or t
Gender, n (%)						
Female	129 (50.0)	11 (55)	0.066	15 (57.7)	14 (46.7)	0.678
Male	119 (48.0)	9 (45)		11 (42.3)	16 (53.3)	
Age (years), mean (SD)	22.3 (2.0)	22.4 (1.9)	−0.347	21.6 (1.2)	22.6 (1.9)	−2.421
DERS, mean (SD)						
Non-acceptance of emotional responses	12.2 (3.9)	13.6 (3.2)	−1.532	13.7 (3.6)	14.3 (4.7)	−0.601
Difficulties engaging in goal-directed behavior	11.4 (3.3)	11.9 (3.1)	−0.605	14.8 (3.4)	14.8 (4.3)	−0.029
Impulse control difficulties	10.8 (3.8)	13.4 (4.1)	−2.902 *	14.0 (4.7)	14.2 (5.0)	−0.179
Lack of emotional awareness	14.5 (4.1)	14.0 (2.8)	0.514	16.3 (3.4)	16.0 (3.1)	0.312
Limited access to emotion regulation strategies	15.7 (4.8)	18.5 (4.2)	−2.481	20.1 (6.0)	20.4 (5.6)	−0.229
Lack of emotional clarity	9.7 (2.9)	10.5 (2.9)	−1.168	11.0 (2.1)	11.0 (2.7)	−0.052

* $p < 0.008$. DERS: Difficulties in Emotion Regulation Scale; NIA: No Internet addiction; PIA: Persistent Internet addiction; PNIA: Persistent no Internet addiction; RIA: Remitted Internet addiction; SD: Standard deviation.

Comparisons of demographic characteristics and difficulties in emotion regulation at Stage 1 between the RIA and PIA groups are presented in Table 1. No dimension of difficulty in emotion regulation at Stage 1 was significantly associated with the remission of IA at Stage 2.

The results of the repeated-measures ANOVA for the effect of IA at Stage 1 on changes in DERS scores from Stage 1 to Stage 2 are shown in Table 3. Regarding the results of within-subject analysis, scores for the subscales of lack of emotional awareness (p = 0.017) and limited access to emotion regulation strategies (p = 0.022) tended to decrease more from Stage 1 to Stage 2 in the IA group than in the non-IA group; however, the results did not reach a statistically significant level. Regarding the results of between-subject analysis, there were significant differences in non-acceptance of emotional responses (p = 0.006), difficulties engaging in goal-directed behavior ($p < 0.001$), impulse control difficulties ($p < 0.001$), limited access to emotion regulation strategies ($p < 0.001$), and lack of emotional clarity ($p < 0.001$) between the participants with and without IA at Stage 1.

Table 3. Interactions between Internet addiction and changes in difficulty in emotion regulation analyzed by repeated-measures ANOVA.

	Within-Subject Analysis					Between-Subject Analysis			
	df	Mean Square	F	p		df	Mean Square	F	p
Non-acceptance of emotional responses									
Cohort	1	0.28	0.03	0.859	Intercept	1	64,390.83	2769.62	<0.001
Cohort × Internet addiction	1	10.75	1.22	0.271	Internet addiction	1	178.02	7.66	0.006
	322	8.84			Error	322	23.25		
Difficulties engaging in goal-directed behavior									
Cohort	1	0.61	0.10	0.755	Intercept	1	63,252.18	3477.03	<0.001
Cohort × Internet addiction	1	12.34	1.96	0.162	Internet addiction	1	831.19	45.69	<0.001
Error	322	6.29			Error	322	18.19		
Impulse control difficulties									
Cohort	1	45.45	5.81	0.016	Intercept	1	55,047.18	2220.87	<0.001
Cohort × Internet addiction	1	24.46	3.13	0.078	Internet addiction	1	655.32	26.44	<0.001
Error	322	7.82			Error	322	24.79		
Lack of emotional awareness									
Cohort	1	331.13	46.49	<0.001	Intercept	1	76,173.83	3796.15	<0.001
Cohort × Internet addiction	1	40.72	5.72	0.017	Internet addiction	1	96.71	4.82	0.029
Error	322	7.12			Error	322	20.07		
Limited access to emotion regulation strategies									
Cohort	1	44.38	3.87	0.050	Intercept	1	116,832.71	3059.68	<0.001
Cohort × Internet addiction	1	60.46	5.28	0.022	Internet addiction	1	1150.13	30.12	<0.001
Error	322	11.46			Error	322	38.19		
Lack of emotional clarity									
Cohort	1	41.63	10.91	0.001	Intercept	1	37,474.75	3280.47	<0.001
Cohort × Internet addiction	1	0.07	0.02	0.892	Internet addiction	1	151.88	13.30	<0.001
Error	322	3.82			Error	322	11.42		

To explore the effects of IA on the changes in difficulties in emotional regulation, the scores on the DERS were further compared between the Stage 1 and Stage 2 investigations using a paired t-test in the participants with and without IA at Stage 1 (Table 4). The results of the paired t-test demonstrated that in the IA group, the scores on the subscale of lack of emotional awareness decreased significantly from Stage 1 to Stage 2. In the non-IA group, the scores on the subscales of lack of emotional awareness and lack of emotional clarity decreased significantly from Stage 1 to Stage 2.

Table 4. Comparisons of difficulties in emotion regulation between the initial (Stage 1) and follow-up (Stage 2) investigations among the participants with and without Internet addiction at Stage 1.

	Stage 1 Mean (SD)	Stage 2 Mean (SD)	Paired *t*	*p*
Internet addiction group (n = 56)				
Non-acceptance of emotional responses	14.0 (4.2)	13.7 (4.2)	0.582	0.563
Difficulties engaging in goal-directed behavior	14.8 (3.9)	14.3 (3.8)	0.787	0.435
Impulse control difficulties	14.1 (4.8)	12.9 (4.3)	1.931	0.059
Lack of emotional awareness	16.1 (3.2)	13.6 (3.4)	5.500	<0.001
Limited access to emotion regulation strategies	20.3 (5.7)	18.8 (5.3)	2.084	0.042
Lack of emotional clarity	11.0 (2.4)	10.4 (3.1)	1.520	0.134
Non-Internet addiction group (n = 268)				
Non-acceptance of emotional responses	12.3 (3.8)	12.7 (4.1)	−1.504	0.134
Difficulties engaging in goal-directed behavior	11.4 (3.3)	11.7 (3.6)	−1.372	0.171
Impulse control difficulties	11.0 (3.9)	10.8 (3.9)	0.808	0.420
Lack of emotional awareness	14.4 (4.0)	13.2 (3.5)	5.243	<0.001
Limited access to emotion regulation strategies	16.0 (4.8)	16.1 (4.9)	−0.407	0.684
Lack of emotional clarity	9.8 (2.9)	9.1 (2.6)	4.275	<0.001

4. Discussion

This study revealed that more severe impulse control difficulties predicted a higher incidence of IA during the follow-up period of 1 year in male participants. Related cross-sectional studies have found a significant association between impulsivity and Internet gaming disorder in adolescents and young adults [37,38]. A longitudinal study also indicated that impulsivity is a risk factor for Internet gaming disorders among adolescents [39]. Loss of control of Internet use is an essential criterion for IA [36]; therefore, impulse control difficulties could make individuals more susceptible to the rewarding effects of Internet use and contribute to vulnerability to IA. Impulse control difficulties may also negatively affect young adults' relationships with others and thus increase their desire to seek out friendship and joy from the Internet. Moreover, a Go/NoGo study reported that college students with IA had lower electrophysiological activation in the conflict detection stage than did those without IA [40], indicating that they had lower efficiency in information processing than did their peers without IA. Low efficiency in information processing may limit individuals' ability to access effective emotion regulation strategies. Cognitive-behavioral therapy for IA aims to increase the clients' ability of impulse control by training clients' skills to monitor their inner feelings and control impulse behaviors [41–43]. Research also found that electro-acupuncture had an advantage over psychological intervention in terms of impulsivity control in adolescents with IA [44]. The present study found a gender difference in the prediction of impulse control difficulties for the incidence of IA. Previous studies have also found gender differences in the psychopathology of IA [45,46]. However, the numbers of participants with the incidence of IA during the one-year follow-up period were small. Further study with more participants is warranted to replicate this result.

The one-year remission rate of our study was 46.4%. A previous study on young adolescents in Taiwan found the one-year remission rate of IA was 49.5% [28]. A study on Hong Kong Chinese secondary school students found the one-year remission rate of IA was 59.29/100 person-years [47]. A study on German adolescents found that the one-year remission rate of IA was 71.6% [48]. The results of the present and previous studies indicated that incidence of remission was high without noticeable interventions during adolescence and adulthood. Although the present study did not find a predictive effect of difficulties in emotion regulation on the remission of IA, a previous study found that a lower level of maladaptive emotion regulation strategies significantly predicted for remission one year later [48], indicating that programs enhancing emotion regulation are important for helping people to reduce the severity of IA.

The results of ANOVA examining the changes in difficulties in emotion regulation from Stage 1 to Stage 2 did not show statistically significant differences between the IA and non-IA groups. The results did not support our hypothesis that IA may worsen emotion regulation. The score on the subscale of lack of emotional awareness decreased significantly from Stage 1 to Stage 2 in both IA and non-IA groups, whereas the score on the subscale of lack of emotional clarity decreased significantly in only the non-IA group. However, it is noteworthy that the IA group had higher scores of all subscales of the DERS than the non-IA group at the initial investigation (p-values ranging from <0.001 to 0.003). Research has found that people with negative emotional states, such as depression, anxiety, and feelings of loneliness, were more likely to use the Internet to manage their emotional problems [29–31]. Internet use might be a way for people with negative emotions to experience pleasure and avoid unpleasant experiences [49]. Research found that maladaptive emotion regulation strategies could predict the maintenance of problematic Internet use, whereas a targeted positive development of emotion regulation in childhood and adolescence could promote spontaneous remission of problematic Internet use [48]. Therefore, mental health professionals need to provide intervention programs for people with IA to improve their ability of emotion regulation.

The present study has several limitations that should be addressed. First, the data were exclusively self-reported, and we did not obtain additional information regarding mental health diagnoses or treatment history. Second, although participants recruited from the community are more representative compared with those recruited from clinical units, the volunteers may have had various motivations for participating in the study. Third, we did not assess the content of the Internet activity nor psychiatric diagnoses. Fourth, we did not survey factors that may relate to change in IA in college students, for example, personal traits, socialization process, and enthusiasm devoted to study. Lastly, the four subgroups (PNIA, NIA, RIA, and PIA) were of unequal size; two were comparably small (n = 20 in NIA; n = 30 in PIA). The small sample size might limit the power of statistical analysis.

5. Conclusions

The results of this study indicated that impulse control difficulties predicted the incidence of IA during the follow-up period of 1 year among male college students, whereas IA did not significantly predict the change in difficulties in emotion regulation. Mental health professionals should help college students develop effective strategies for emotion regulation.

Author Contributions: J.-K.T. and C.-F.Y. wrote the report; R.C.H., W.-H.L., H.-F.H., and C.-F.Y. conceived of and performed the study; C.-F.Y. analyzed the data. All authors have read and agreed to the published version of the manuscript.

Funding: This research was funded by Kaohsiung Medical University Hospital, grant number KMUH103-3M38.

Conflicts of Interest: The authors declare no conflict of interest.

References

1. Rice, R.E. Influences, usage, and outcomes of Internet health information searching: Multivariate results from the Pew surveys. *Int. J. Med. Inform.* **2006**, *75*, 8–28. [CrossRef] [PubMed]
2. Rodgers, S.; Chen, Q. Internet community group participation: Psychosocial benefits for women with breast cancer. *J. Comput. Mediat. Commun.* **2005**, *10*, 21–26. [CrossRef]
3. Jelenchick, L.A.; Eickhoff, J.; Christakis, D.A.; Brown, R.L.; Zhang, C.; Benson, M.; Moreno, M.A. The Problematic and Risky Internet Use Screening Scale (PRIUSS) for adolescents and young adults: Scale development and refinement. *Comput. Hum. Behav.* **2014**, *35*, 171–178. [CrossRef]

4. Tao, R.; Huang, X.; Wang, J.; Zhang, H.; Zhang, Y.; Li, M. Proposed diagnostic criteria for internet addiction. *Addiction* **2010**, *105*, 556–564. [CrossRef] [PubMed]
5. American Psychiatric Association. *Diagnostic and Statistical Manual of Mental Disorders*, 5th ed.; Arlington American Psychiatric Association: Washington, DC, USA, 2013.
6. Cheng, C.; Li, A.Y. Internet addiction prevalence and quality of (real) life: A meta-analysis of 31 nations across seven world regions. *Cyberpsychol. Behav. Soc. Netw.* **2014**, *17*, 755–760. [CrossRef]
7. Tateno, M.; Teo, A.R.; Shirasaka, T.; Tayama, M.; Watabe, M.; Kato, T.A. Internet addiction and self-evaluated attention-deficit hyperactivity disorder traits among Japanese college students. *Psychiatry Clin. Neurosci.* **2016**, *70*, 567–572. [CrossRef] [PubMed]
8. Ko, C.H.; Hsiao, S.; Liu, G.C.; Yen, J.Y.; Yang, M.J.; Yen, C.F. The characteristics of decision making, potential to take risks, and personality of college students with Internet addiction. *Psychiatry Res.* **2010**, *175*, 121–125. [CrossRef]
9. Yang, T.; Yu, L.; Oliffe, J.L.; Jiang, S.; Si, Q. Regional contextual determinants of internet addiction among college students: A representative nationwide study of China. *Eur. J. Public Health* **2017**, *27*, 1032–1037. [CrossRef]
10. Anderson, K.J. Internet use among college students: An exploratory study. *J. Am. Coll. Health* **2001**, *50*, 21–26. [CrossRef]
11. Balhara, P.S.Y.; Doric, A.; Stevanovic, D.; Knez, R.; Swarndeep ingh, S.; Chowdhury, M.R.R.; Kafali, H.Y.; Sharma, P.; Vally, Z.; Vu, T.V.; et al. Correlates of problematic Internet use among college and university students in eight countries: An international cross-sectional study. *Asian J. Psychiatry* **2019**, *45*, 113–120. [CrossRef]
12. Kuss, D.J.; Griffiths, M.D.; Binder, J.F. Internet addiction in students: Prevalence and risk factors. *Comput. Hum. Behav.* **2013**, *29*, 959–966. [CrossRef]
13. Chang, Y.H.; Lee, Y.T.; Hsieh, S. Internet interpersonal connection mediates the association between personality and Internet addiction. *Int. J. Environ. Res. Public Health* **2019**, *16*, 3537. [CrossRef]
14. Chwaszcz, J.; Lelonek-Kuleta, B.; Wiechetek, M.; Niewiadomska, I.; Palacz-Chrisidis, A. Personality traits, strategies for coping with stress and the level of Internet addiction-A study of Polish secondary-school students. *Int. J. Environ. Res. Public Health* **2018**, *15*, 987. [CrossRef] [PubMed]
15. Shi, M.; Du, T.J. Associations of personality traits with internet addiction in Chinese medical students: The mediating role of attention-deficit/hyperactivity disorder symptoms. *BMC Psychiatry* **2019**, *19*, 183. [CrossRef] [PubMed]
16. Chou, W.J.; Chang, Y.P.; Yen, C.F. Boredom proneness and its correlation with Internet addiction and Internet activities in adolescents with attention-deficit/hyperactivity disorder. *Kaohsiung J. Med. Sci.* **2018**, *34*, 467–474. [CrossRef]
17. Lu, W.H.; Lee, K.H.; Ko, C.H.; Hsiao, R.C.; Hu, H.F.; Yen, C.F. Relationship between borderline personality symptoms and Internet addiction: The mediating effects of mental health problems. *J. Behav. Addict.* **2017**, *6*, 434–441. [CrossRef]
18. Lu, W.H.; Chou, W.J.; Hsiao, R.C.; Hu, H.F.; Yen, C.F. Correlations of Internet addiction severity with reinforcement sensitivity and frustration intolerance in adolescents with attention-deficit/hyperactivity disorder: The moderating effect of medications. *Front. Psychiatry* **2019**, *10*, 268. [CrossRef] [PubMed]
19. Hormes, J.M.; Kearns, B.; Timko, C.A. Craving Facebook? Behavioral addiction to online social networking and its association with emotion regulation deficits. *Addiction* **2014**, *109*, 2079–2088. [CrossRef] [PubMed]
20. Aldao, A.; Nolen-Hoeksema, S.; Schweizer, S. Emotion regulation strategies across psychopathology: A meta-analytic review. *Clin. Psychol. Rev.* **2010**, *30*, 217–237. [CrossRef]
21. Hayes, S.C.; Strosahl, K.D.; Wilson, K.G.; Bissett, R.T.; Pistorello, J.; Toarmino, D.; Polusny, M.A.; Dykstra, T.A.; Batten, S.V.; Bergan, L.; et al. Measuring experiential avoidance: A preliminary test of a working model. *Psychol. Rec.* **2004**, *54*, 553–578. [CrossRef]
22. Gratz, K.L.; Roemer, L. Multidimensional assessment of emotion regulation and dysregulation: Development, factor structure, and initial validation of the difficulties in emotion regulation scale. *J. Psychopathol. Behav. Assess.* **2004**, *26*, 41–54. [CrossRef]
23. Ko, C.H.; Yen, J.Y.; Yen, C.F.; Chen, C.S.; Chen, C.C. The association between Internet addiction and psychiatric disorder: A review of the literature. *Eur. Psychiatry* **2012**, *27*, 1–8. [CrossRef] [PubMed]

24. Estévez, A.; Jáuregui, P.; Sánchez-Marcos, I.; López-González, H.; Griffiths, M.D. Attachment and emotion regulation in substance addictions and behavioral addictions. *J. Behav. Addict.* **2017**, *6*, 534–544. [CrossRef] [PubMed]
25. Karaer, Y.; Akdemir, D. Parenting styles, perceived social support and emotion regulation in adolescents with internet addiction. *Compr. Psychiatry* **2019**, *92*, 22–27. [CrossRef] [PubMed]
26. Lee, J.; Lee, S.; Chun, J.W.; Cho, H.; Kim, D.J.; Jung, Y.C. Compromised prefrontal cognitive control over emotional interference in adolescents with Internet gaming disorder. *Cyberpsychol. Behav. Soc. Netw.* **2015**, *18*, 661–668. [CrossRef] [PubMed]
27. Wang, W.; Li, D.; Li, X.; Wang, Y.; Sun, W.; Zhao, L.; Qiu, L. Parent-adolescent relationship and adolescent internet addiction: A moderated mediation model. *Addict. Behav.* **2018**, *84*, 171–177. [CrossRef]
28. Ko, C.H.; Yen, J.Y.; Chen, C.S.; Yeh, Y.C.; Yen, C.F. Predictive values of psychiatric symptoms for internet addiction in adolescents: A 2-year prospective study. *Arch. Pediatr. Adolesc. Med.* **2009**, *163*, 937–943. [CrossRef]
29. Whang, L.S.; Lee, S.; Chang, G. Internet over-users' psychological profiles: A behavior sampling analysis on internet addiction. *Cyberpsychol. Behav.* **2003**, *6*, 143–150. [CrossRef]
30. Bakken, I.J.; Wenzel, H.G.; Götestam, K.G. Internet addiction among Norwegian adults: A stratified probability sample study. *Scand. J. Psychol.* **2009**, *50*, 121–127. [CrossRef]
31. Van den Eijnden, R.; Vermulst, A.; van Rooij, A.J.; Scholte, R.; van de Mheen, D. The bidirectional relationships between online victimization and psychosocial problems in adolescents: A comparison with real-life victimization. *J. Youth Adolesc.* **2014**, *43*, 790–802. [CrossRef] [PubMed]
32. Kanter, J.W.; Cautilli, J.D.; Busch, A.M.; Baruch, D.E. Toward a comprehensive functional analysis of depressive behavior: Five environmental factors and a possible sixth and seventh. *Behav. Anal. Today* **2005**, *6*, 65–81. [CrossRef]
33. Hayes, S.C.; Wilson, K.G.; Gifford, E.V.; Follette, V.M.; Strosahl, K. Experimental avoidance and behavioral disorders: A functional dimensional approach to diagnosis and treatment. *J. Consult. Clin. Psychol.* **1996**, *64*, 1152–1168. [CrossRef] [PubMed]
34. Hsieh, F.Y. Sample size tables for logistic regression. *Stat. Med.* **1989**, *8*, 795–802. [CrossRef] [PubMed]
35. Chen, S.H.; Weng, L.J.; Su, Y.J.; Wu, H.M.; Yang, P.F. Development of Chinese Internet Addiction Scale and its psychometric study. *Chin. J. Psychol.* **2003**, *45*, 279–294.
36. Ko, C.H.; Yen, J.Y.; Chen, S.H.; Yang, M.J.; Yen, C.F. Proposed diagnostic criteria and the screening and diagnosing tool of Internet addiction in college student. *Compr. Psychiatry* **2009**, *50*, 378–384. [CrossRef]
37. Ding, W.N.; Sun, J.H.; Sun, Y.W.; Chen, X.; Zhou, Y.; Zhuang, Z.G.; Xu, J.R.; Du, Y.S. Trait impulsivity and impaired prefrontal impulse inhibition function in adolescents with Internet gaming addiction revealed by a go/no-go fMRI study. *Behav. Brain. Funct.* **2014**, *10*, 20. [CrossRef]
38. Ko, C.H.; Hsieh, T.J.; Wang, P.W.; Lin, W.C.; Yen, C.F.; Chen, C.S.; Yen, J.Y. Altered gray matter density and disrupted functional connectivity of the amygdala in adults with Internet gaming disorder. *Prog. Neuropsychopharmacol. Biol. Psychiatry* **2015**, *57*, 185–192. [CrossRef] [PubMed]
39. Gentile, D.A.; Choo, H.; Liau, A.; Sim, T.; Li, D.; Fung, D.; Khoo, A. Pathological video game use among youths: A two-year longitudinal study. *Pediatrics* **2011**, *127*, e319–e329. [CrossRef]
40. Dong, G.; Zhou, H.; Zhao, X. Impulse inhibition in people with Internet addiction disorder: Electrophysiological evidence from a Go/NoGo study. *Neurosci. Lett.* **2010**, *485*, 138–142. [CrossRef]
41. Du, Y.S.; Jiang, W.; Vance, A. Longer term effect of randomized, controlled group cognitive behavioural therapy for Internet addiction in adolescent students in Shanghai. *Aust. N. Z. J. Psychiatry* **2010**, *44*, 129–134. [CrossRef]
42. Park, S.Y.; Kim, S.M.; Roh, S.; Soh, M.A.; Lee, S.H.; Kim, H.; Han, D.H. The effects of a virtual reality treatment program for online gaming addiction. *Comput. Methods Programs Biomed.* **2016**, *129*, 99–108. [CrossRef] [PubMed]
43. Torres-Rodríguez, A.; Griffiths, M.D.; Carbonell, X.; Oberst, U. Treatment efficacy of a specialized psychotherapy program for Internet Gaming Disorder. *J. Behav. Addict.* **2018**, *7*, 939–952. [CrossRef] [PubMed]
44. Yang, Y.; Li, H.; Chen, X.X.; Zhang, L.M.; Huang, B.J.; Zhu, T.M. Electro-acupuncture treatment for internet addiction: Evidence of normalization of impulse control disorder in adolescents. *Chin. J. Integr. Med.* **2017**, *23*, 837–844. [CrossRef] [PubMed]

45. Bonnaire, C.; Baptista, D. Internet gaming disorder in male and female young adults: The role of alexithymia, depression, anxiety and gaming type. *Psychiatry Res.* **2019**, *272*, 521–530. [CrossRef] [PubMed]
46. Stockdale, L.; Coyne, S.M. Video game addiction in emerging adulthood: Cross-sectional evidence of pathology in video game addicts as compared to matched healthy controls. *J. Affect. Disord.* **2018**, *225*, 265–272. [CrossRef] [PubMed]
47. Lau, J.T.F.; Wu, A.M.S.; Gross, D.L.; Cheng, K.M.; Lau, M.M.C. Is Internet addiction transitory or persistent? Incidence and prospective predictors of remission of Internet addiction among Chinese secondary school students. *Addict. Behav.* **2017**, *74*, 55–62. [CrossRef]
48. Wartberg, L.; Lindenberg, K. Predictors of spontaneous remission of problematic Internet use in adolescence: A one-year follow-up study. *Int. J. Environ. Res. Public Health* **2020**, *17*, 448. [CrossRef]
49. Chou, W.P.; Lee, K.H.; Ko, C.H.; Liu, T.L.; Hsiao, R.C.; Lin, H.F.; Yen, C.F. Relationship between psychological inflexibility and experiential avoidance and internet addiction: Mediating effects of mental health problems. *Psychiatry Res.* **2017**, *257*, 40–44. [CrossRef]

International Journal of
Environmental Research and Public Health

MDPI

Article

Mind over Matter: Testing the Efficacy of an Online Randomized Controlled Trial to Reduce Distraction from Smartphone Use

Melina A. Throuvala [1,*], Mark D. Griffiths [1], Mike Rennoldson [2] and Daria J. Kuss [1]

[1] International Gaming Research Unit, Psychology Department, Nottingham Trent University, Nottingham NG1 4FQ, UK; mark.griffiths@ntu.ac.uk (M.D.G.); daria.kuss@ntu.ac.uk (D.J.K.)

[2] Psychology Department, Nottingham Trent University, Nottingham NG1 4FQ, UK; mike.rennoldson@ntu.ac.uk

* Correspondence: melina.throuvala@ntu.ac.uk

Received: 31 May 2020; Accepted: 30 June 2020; Published: 5 July 2020

Abstract: Evidence suggests a growing call for the prevention of excessive smartphone and social media use and the ensuing distraction that arises affecting academic achievement and productivity. A ten-day online randomized controlled trial with the use of smartphone apps, engaging participants in mindfulness exercises, self-monitoring and mood tracking, was implemented amongst UK university students ($n = 143$). Participants were asked to complete online pre- and post-intervention assessments. Results indicated high effect sizes in reduction of smartphone distraction and improvement scores on a number of self-reported secondary psychological outcomes. The intervention was not effective in reducing habitual behaviours, nomophobia, or time spent on social media. Mediation analyses demonstrated that: (i) emotional self-awareness but not mindful attention mediated the relationship between intervention effects and smartphone distraction, and (ii) online vigilance mediated the relationship between smartphone distraction and problematic social media use. The present study provides preliminary evidence of the efficacy of an intervention for decreased smartphone distraction and highlights psychological processes involved in this emergent phenomenon in the smartphone literature. Online interventions may serve as complementary strategies to reduce distraction levels and promote insight into online engagement. More research is required to elucidate the mechanisms of digital distraction and assess its implications in problematic use.

Keywords: distraction; smartphones; social media; intervention; randomized controlled trial; social media addiction

1. Introduction

Attentional focus is one of the most fundamental resources and a key to successful and high-order work [1]. In the attention economy [2], multiple online and offline activities compete for an alternative share of attention [3]. This trend is expected to grow in the face of increasing communication complexity and information overload [4], which is becoming even more prevalent partially due to the vast online accessibility, immediacy and convenience of smartphones, acting as a major motivational pull for engagement [5] and prompting constant multitasking and frequent attentional loss [6]. There are currently more than 3.5 billion smartphone users [7] and smartphone use is an emergent area of research [8–10]. Emerging evidence on cognitive function has shown that smartphone availability and daily interruptions compete with higher-level cognitive processes creating a cognitive interference effect [11–15], associated with poorer cognitive functioning [16–19], performance impairments in daily life [20] and potential supplanting of analytical thinking skills by "offloading thinking to the device" [21] (p. 473). In spite of such initial evidence, there are cognitive correlates within the smartphone context, such as distraction, which have been less explored in the literature. Studies report that students use their smartphones for more than 25% of effective class duration, and smartphone distractions occur every 3–4 min, for over a minute in duration [22]. Student focus on any single task is reported to last 3–5 min [23] with excessive smartphone use hindering academic performance as a result of allowing goal-irrelevant information to compete with goal-relevant tasks [24,25]. Therefore, examining the processes involved in the occurrence of distraction as well as protective strategies for its containment is timely. The present study evaluates the efficacy of evidence-based mediating strategies in reducing distraction employed in an online randomized controlled trial.

Distraction is an emotion regulation coping strategy used to deflect attention from the task at hand in order to relieve emotional distress, reflected as difficulty in concentrating and maintaining goal-focused behaviour, with an adaptive function in negative affect situations [26–30]. Smartphone distraction constitutes an emergent concern, operationally defined as the disruption in attention due to: (i) external cues received (i.e., notifications), (ii) cognitive salience (i.e., internal cues) of the smartphone and social media, or (iii) cognitive avoidance (i.e., coping mechanism) for emotion regulation [17,31–33]. Checking behaviours, frequently engaged in during smartphone use, are associated with repeated external or internal interruptions, leading to attentional micro-disengagements and distraction [20,31,34]. According to the control model of social media engagement [5], this may occur as need to control online content, relationships and self-presentation produces an attentional conflict (offline vs. online or platform/activity switch), arousal and distraction, leading either to facilitation (by the presence of online others) [35,36] and heightened engagement or shallow processing, when engaged in parallel cognitively demanding tasks. Therefore, constant disruptions may cause a rise in attention problems and hyperactivity levels [37] as a result of allowing goal-irrelevant information to compete with goal-relevant tasks [24,25] with impacts on wellbeing, productivity and academic achievement, particularly amongst young people [22,38–41]. A large contributor to this effect is excessive social media use, which has been suggested as a vulnerability factor for problematic smartphone use [42–44]. To date, the effects of smartphone use on student outcomes may still be small [45].

1.1. Distraction and Its Relation to Other Psychological Constructs in the Smartphone Literature

Online vigilance. Distraction by smartphone use appears to be activated by internal thoughts or external cues interfering with other tasks, which may be driven by online vigilance—a constant preoccupation with online content, leading to salience, monitoring and prompting urges to check [46], resulting in strong habitual behaviour [47,48]. Salience of online content has been found to be negatively associated with affective wellbeing and life satisfaction, particularly when thoughts are negative [49].

Attention impulsiveness and habitual smartphone use. Attention impulsiveness has also been implicated in smartphone distraction, reinforced by rewarding, habitual checking behaviours [47], and has a significant relationship with problematic smartphone use [50]. Recent evidence also suggests symptom severity of problematic social media use to be mainly associated with attention impulsiveness and difficulties with inhibitory control or executive control functions [51], task performance [52] and chronic media multitasking [25]. This is intensified in a low interest academic context, reducing lecture comprehension [53], level of motivation, and fluid intelligence [54].

Fear of Missing Out (FoMO) and Nomophobia (NoMO). FoMO—the fear of being excluded from rewarding social experiences – and NoMO – the fear of no access to a mobile device—have both been evidenced in the smartphone literature as triggering a need to be in constant contact and reinforcing use [55–62]. Therefore, FoMO could be a main driver of distraction due to the propensity to be present in the positive experiences others are having, depicted in online content. FoMO has been associated with depression, smartphone addiction, anxiety, mindfulness and wellbeing [63], negative affectivity, problematic smartphone use, and levels of online social engagement [60].

Stress, anxiety, emotion regulation and problematic use. Socio-emotional correlates of FoMO have included negative affect, rejection sensitivity, and high stress levels [64], and reviews have suggested a small-to-medium association between smartphone use and stress and anxiety [65]. Therefore, negative emotional states may be a precursor to smartphone distraction and its use may be motivated by emotion regulation. Relief of negative emotions and psychological states along with emotional gains from smartphone use have been found to be significantly higher for Generation Z (individuals born between 1995 and 2015) [66] and could be an outcome of difficulties with emotion regulation, creating a vicious cycle sustaining overreliance for coping [67] and dysfunctional metacognitive beliefs among problematic users [68]. Smartphone unavailability and intolerance of uncertainty have been evidenced in problematic smartphone use [69,70], and affect perceived stress and mental wellbeing [71]. Concerns for the emotional and behavioural consequences of excessive smartphone and social media use have been addressed [9,72–75]. However, what constitutes problematic online behaviour needs constant conceptual and methodological re-evaluation [76] as engagement with new products/platforms emerges.

Mindfulness, self-monitoring and mood tracking. Self-monitoring of social media activity, self-exclusion from specific platforms, and the practice of mindfulness are considered successful wellbeing practices [77,78]. Mindfulness, defined as the purposeful, non-judgemental awareness of the presenting experience [79], facilitates the sustaining of on-task behaviours [80], affecting attention, affect regulation, body awareness, and self-perception [81–83], and has been used in gambling harm-reduction and substance use disorders, with intervention effects reducing cravings, post-traumatic symptoms, and negative affect [84–90]. Mindfulness has been negatively associated with distraction, suggesting that one's awareness of own thought wandering (meta-awareness) may decrease the frequency of distraction [17] and aid academic attainment [91]. Self-monitoring of mood (also defined as mood tracking) has been found to boost overall emotional self-awareness [92], which can in turn lead to improvements in emotional self-regulation [93]. Therefore, these strategies could be trialled to help diminish attentional bias occurring within the context of social media and smartphone use [94,95].

1.2. Smartphone Mental Health Apps (MHapps) and Online Randomized Controlled Trials

Digital wellbeing apps or MHapps (apps that track an individual's behaviour, i.e., time spent online, or that aid cognitive, emotional and/or behavioural wellbeing) [96] have been suggested as supporting self-awareness and self-regulation [97] and utilized in mental healthcare given their functionality, accessibility, higher adherence rates, real-time assessment, low-cost and for their intervention potential [98,99]. The literature suggests that evidence-based apps may be efficacious in raising self-awareness, mental health literacy and wellbeing, self-efficacy, and ability to cope [96,100–102]. Online psychological interventions are becoming more prominent in the digital age [103], rendering numerous positive health outcomes [102,104–108], complementing service provision and recognized by governmental health institutions (e.g., National Institute for Health and Care Excellence (NICE) in the UK) [109]. However, more research is required to determine the comparative effectiveness of these therapies and their components [110] in improving mental health and wellbeing and rigorous objective evaluation beyond their developers is required.

To date, there have been a small number of internet-based interventions associated with device use in university settings. Distraction is not considered a dysfunctional construct by itself, but has been implicated in emotion regulation, ADHD, and other disorders [111–113], and has been minimally examined in the context of the digital environment with no evidence to date as to strategies that could ameliorate its occurrence [114]. Therefore, the aim of the present study was to test the preliminary efficacy of an online intervention based on cognitive behavioural principles (i.e., self-monitoring, mood tracking, and mindfulness) to reduce distraction and related psychological outcomes (i.e., stress) among university students. Given: (i) young adults are keen users of smartphone apps, with increased vulnerability to self-regulation and technology use [74], (ii) the high stakes for academic achievement, and (iii) the similarity in processes observed between gambling addiction and social media overuse [115], the strategies of *mindfulness, activity monitoring*, and *mood tracking* utilized in gambling harm-reduction [86,116,117] are employed in the present study. These strategies were delivered and facilitated through the use of smartphone MHapps and were tested for their efficacy in reducing levels of distraction and related psychological outcomes and their role in inducing changes in wellbeing [118–120]. The following hypotheses were formulated:

Hypothesis 1 (H1). *Compared to the control condition at follow-up, students receiving the intervention would report: (i) lower rates of smartphone distraction, smartphone and social media use duration, impulsivity, stress, problematic social media use, FoMO and NoMO and (ii) higher levels of mindful attention, emotional self-awareness, and self-efficacy.*

Hypothesis 2 (H2). *At follow-up, high distractors (HDs) compared to low distractors (LDs) (based on a median-split analysis) would show a greater reduction in distraction and significant improvement in outcomes.*

Hypothesis 3 (H3). *The intervention will mediate the relationship between (i) mindful attention and smartphone distraction, and (ii) emotional awareness and smartphone distraction. Additionally, online vigilance will mediate the relationship between smartphone distraction and problematic social media use.*

To the authors' knowledge and given the novelty of the construct of smartphone distraction, this is the first study to examine a preliminary online randomized controlled trial via MHapps for the reduction of smartphone distraction. The present study fills a gap in the smartphone literature by assessing the efficacy of engaging with behaviour change strategies (i.e., mindfulness, self-monitoring, and mood-tracking) used successfully in gambling harm prevention for the reduction of distraction.

2. Materials and Methods

2.1. Design

The present study tested the efficacy of a ten-day online app-delivered randomized controlled trial (RCT) based on cognitive-behavioural principles to reduce distraction (primary outcome) and a number of secondary psychological outcomes: self-awareness, mindful attention, FoMO, anxiety, and depression among university students. RCTs are considered the gold standard in intervention effectiveness despite limitations addressed by scholars [121,122], primarily for the lack of external validity or methodological choices [123]. A pragmatic psychosocial intervention with an RCT design was chosen [124]. The duration of the intervention was set given a pragmatic consideration of the free use period of one of the apps (*Headspace*) and, secondly, due to the preliminary nature of this investigation. Consolidated Standards of Reporting Trials (CONSORT) guidelines were followed in the protocol and the procedures and reporting of the intervention [125].

The intervention involved the active engagement for the period of ten consecutive days with three smartphone apps serving three different functions: to assess smartphone and social media use, conduct mindfulness sessions with an emphasis on eliminating distraction, and track mood and assess its impact on distraction, stress, self-regulation, and other measures. Interaction with apps was encouraged to: (i) raise emotional awareness of common mood states, such as feeling down, worried, or stressed through mindfulness, (ii) guide basic smartphone monitoring, focusing skills, and awareness, and (iii) provide insight through mood tracking (Table 1). To further support active engagement with these intervention components, eligible participants were asked to keep a daily online activity log for the duration of the intervention (i.e., the number of screen-unlocks and the time of day and number of minutes for which the smartphone was used, usefulness of apps, etc.), to aid time perception of daily activities, raise awareness levels, and help increase the accuracy of self-reporting and adherence to the intervention [126,127]. Promoting self-awareness of media use and understanding of own behaviour was a key target of the intervention in order to curb distraction. The study was reviewed and approved (No. 2018/226) by the research team's university ethics committee.

Table 1. The three components of the intervention.

Intervention Components	Smartphone App Used	Evidence-Based Benefits	Psychological Evidence for Benefits
		Mindfulness	
Brief mindfulness sessions	Headspace app	• Mindfulness practice and mood tracking offer benefits in emotion regulation, attention, stress and low mood levels & meta-awareness • Evidence for replenishing students' focused engagement in mental tasks (i.e., homework)	[128] [80] [82,83] [129]
		Self-monitoring and Self-exclusion	
Social media and smartphone use Abstinence option	Anti-Social app	• Self-monitoring & exclusion (minutes on social media, times of unlocking smart phone each day, favourite and most time consuming and accessed apps) aid emotion regulation • Reflection on dependence on smartphone, extent of use, lost attention, checking frequency • Performance feedback & meta-awareness	[130] [131] [77] [132] [133] [134] [101]
		Mood-tracking	
	Pacifica app	• Mood tracking can boost overall emotional self-awareness which can in turn lead to improvements in emotional self-regulation	[93] [135] [136]

Daily reminders and messages via blogging were sent as a reminder to maintain routine and reflect on levels of activity [126,137].

2.2. Participants

Participants were recruited using convenience and snowball sampling techniques. After gaining institutional ethical approval, the study was advertised to students through the research credit scheme, in university lectures and labs, and to the public through social media as an online intervention to assess the reduction of smartphone distraction. This experimental intervention demanded a significant time involvement and offering incentives increased the chances of participation and completion of the full ten-day intervention. In return for participation, students were offered either research credits or entry in a prize draw (£50 gift cards). Participants were included in the study based on two screening criteria: regular smartphone and social media usage. Only those affirming both and granting consent were able to continue with participation. Following the completion of the survey, participants were allocated to one of the two conditions (intervention [IG] or control [CG]) and further instructions for participation in the intervention were provided depending on the allocation condition. After initially providing age and gender demographics, participants responded to survey items regarding habitual smartphone and social media behaviour (estimates of duration of use), smartphone distraction severity, trait self-regulation, trait mindfulness and other psychological constructs (detailed in "Materials"). The survey took approximately 25 min to complete.

A total of 261 participants were recruited who participated in the baseline assessment. Of these, 155 were undergraduate Psychology students in the UK (59.3%). The sample comprised 47 males (18%) and 214 females (82%), with an age range of 18 to 32 years (M = 20.72, SD = 3.12). Figure 1 depicts the flow of participants through the study procedures. After the baseline assessment, during the intervention period two individuals of the intervention group withdrew from the study and were not considered in the analysis. From the 259 remaining participants, seven were removed due to providing 90% incomplete data. The final sample considered at baseline was 252 participants (intention to treat (ITT) group) and included 123 participants in the intervention group and 129 in the control group. Participants who completed both assessments were considered in the per-protocol analysis (PP) (*n* = 143, 56% of the original sample), with 72 participants comprising the IG and 71 participants the CG.

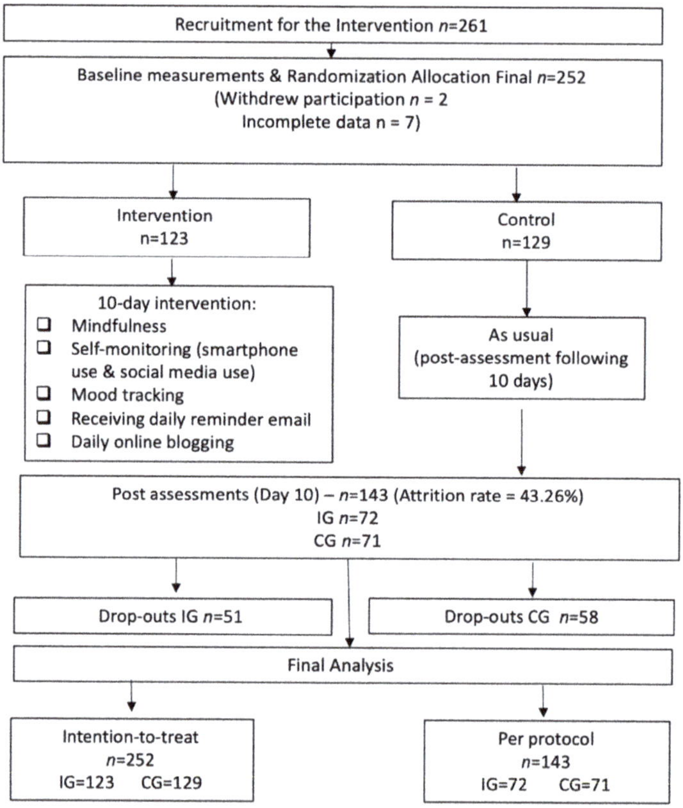

Figure 1. Participant flow in the intervention.

2.3. Materials

The survey consisted of sociodemographic and usage data (questions related specifically to smartphone and social media use [hours per day]). The demographic questions and user-related questions had open responses (i.e., "How many hours per day do you use social media?"). The following scales were used for the psychological measures of the study:

The Smartphone Distraction Scale [138] is a newly developed scale comprising of 16 Likert-type items. The scale comprises four factors: attention impulsiveness, online vigilance, emotion regulation, and multitasking. Scores range from 1 (almost never) to 5 (almost always) with higher scores representing a greater degree of distraction. Individual items on the test were summed to give composite scores. Sample items included in the scale are the following: "I get distracted by my phone notifications", and "I constantly check my phone to see who liked my recent post while doing important tasks". The scale has demonstrated good psychometric properties [138] and excellent reliability in the present study with a Cronbach's alpha of 0.90 for Time 1 (T1) and 0.88 for Time 2 (T2).

The Mindful Attention Awareness Scale (MAAS) [139] is a 15-item assessment tool that assesses the dispositional tendency of participants to be mindful in everyday life and has been validated among young people, university students and community samples [139,140]. Item statements reflect experience of mindfulness, mindlessness in general and specific daily situations and are distributed across a range of cognitive, emotional, physical, interpersonal, and general domains. Response options are based on a six-point Likert scale from 1 (almost always) to 6 (almost never). Scores were averaged across the 15 items to obtain an overall mindfulness score with higher scores reflecting higher levels of dispositional mindfulness. Sample items include "I could be experiencing some emotion and not be aware of it until sometime later" and "I find it difficult to stay focused on what's happening in the present" and exhibited a high degree of internal consistency in the present study with a Cronbach's alpha of 0.92 for T1 and 0.93 for T2.

The Emotional Self-Awareness Scale (ESAS) [92] was used to assess ESA and comprises five variables: recognition, identification, communication, contextualization, and decision making. The scale consists of 32 items (e.g., "I usually know why I feel the way I do") rated from 0 (strongly disagree) to 4 (strongly agree). The total ESA score ranged from 0 to 128, and sub-scale items are combined to produce a composite score with higher scores indicating higher ESA. The ESAS has presented reasonable internal consistency (Cronbach's alpha = 0.72, 0.69, and 0.76 for pre-test, post-test and six-week follow-up) [92]. The scale has demonstrated good validity in prior studies [92,101] and adequate internal consistency in the present study (Cronbach's alpha of 0.87 for T1 and 0.86 for T2).

The Perceived Stress Scale (PSS) [141] is one of the most widely used scales to assess perceived stress and the degree of unpredictability, uncontrollability, and burden in various situations. The scale used was the 10-item version rated from 0 (never) to 4 (very often) with sample items such as "In the last month, how often have you felt that you were unable to control the important things in your life?", and "In the last month, how often have you felt that you were on top of things?" Scores are obtained by summing the items, with the higher score indicating more perceived stress. The scale possesses good psychometric properties [142] and its internal consistency in the present study was 0.86 for T1 and 0.83 for T2.

The seven-item Generalized Anxiety Disorder Scale (GAD-7) [143] is a brief clinical measure that assesses for the presence and severity of Generalized Anxiety Disorder (GAD). The self-report scale asks how often during the last two weeks individuals experienced symptoms of GAD. Total scores range from 0–21 with cut-off scores of 5, 10, and 15 being indicative of mild, moderate, and severe anxiety, respectively. Increasing scores on the GAD-7 are strongly associated with greater functional impairment in real-world settings. Sample items are rated from 0 (not at all) to 3 (nearly every day) and sample items include: "Feeling nervous, anxious or on edge" and "Trouble relaxing". The scale has been widely used and considered a valid and reliable screening tool in previous research, presenting good reliability, factorial and concurrent validity [144,145], and demonstrated excellent internal consistency in the present study ($\alpha = 0.93$ T1 and $\alpha = 0.90$ for T2).

The Self-Report Behavioural Automaticity Index (SRBAI) [146] was used to assess habitual strength. The four-item scale was used to assess the degree of automaticity and contained items such as: "Using social media on my smartphone is something ... I do automatically" and "I start doing before I realize I'm doing it". Participants indicate their agreement with each item on a Likert scale ranging from 1 (does not apply at all) to 7 (fully applies). Scores were averaged across items to obtain an overall habit score, with higher scores indicating stronger habitual smartphone use behaviour. The scale has been reported as psychometrically sound in previous studies with good reliability, convergent and predictive validity [146,147] and demonstrated good internal consistency in the present study with a Cronbach's alpha of 0.87 (T1) and 0.89 (T2).

The Generalized Self-Efficacy Scale (GSE) [148] is a widely used psychometric instrument comprising ten items that assess perceived self-efficacy ("I can always manage to solve difficult problems if I try hard enough."). Items are rated on a four-point scale ranging from 1 (not at all true) to 4 (exactly true). The GSE has demonstrated satisfactory internal consistency and validity across studies [149,150]. Cronbach's alpha in the present study was 0.90 (T1) and 0.88 (T2).

The Online Vigilance Scale (OVS) [46] is a 12-item Likert scale which assesses a relatively new construct in the internet-related literature, referring to individuals' cognitive orientation towards online content, expressed as cognitive salience, reactivity to online cues and active monitoring of online activity. Sample items include "My thoughts often drift to online content" and "I constantly monitor what is happening online". Scale items are rated on a four-point Likert scale from 1 (does not apply at all) to 4 (fully applies). Higher mean scores indicate a higher degree of online vigilance. The scale has evidenced sound construct and nomological validity and high internal consistency [46,49,78]. The Cronbach's alpha in the present study was 0.89 (T1) and 0.87 (T2).

The eight-item Barratt Impulsiveness Scale-Alternative Version (BIS-8) [151] is a psychometrically improved abbreviated version of the 11-item BIS scale [151] presenting good construct and concurrent validity in young populations [152,153]. The scale assesses impulsive behaviour and poor self-inhibition and uses a four-point Likert scale from 1 (do not agree) to 4 (agree very much). Sample items include: "I do things without thinking" and "I act on the spur of the moment". Cronbach's alpha coefficient in the present study was 0.85 (T1) and 0.86 (T2).

The Deficient Self-Regulation Measure [154] is a seven-item scale assessing deficient self-regulation in videogame playing adapted for unregulated internet use [155]. The scale is rated on a seven-point Likert scale from 1 (almost never) to 7 (almost always) and has demonstrated sound psychometric properties [154]. The scale was adapted for smartphone use with sample items such as "I would go out of my way to satisfy my urges to use social media" and "I have to keep using social media more and more to get my thrill". The original scale and its adaptation has presented satisfactory psychometric properties [154,155]. The Cronbach's alpha coefficient in the present study was 0.89 (T1) and 0.87 (T2).

The Bergen Social Media Addiction Scale (BSMAS) [115,156–158] is a six-item self-report scale for assessing social media addiction severity based on the framework of the components model of addiction (salience, mood modification, tolerance, withdrawal, conflict, and relapse) [159]. Each item examines the experience of using social media over the past year and is rated on a five-point Likert scale from 1 (very rarely) to 5 (very often), producing a composite score ranging from 6 to 30. Higher BSMAS scores indicate greater risk of social media addiction severity. A sample question from the BSMAS is "How often during the last year have you used social media so much that it has had a negative impact on your job/studies?" A cut-off score over 19 indicates problematic social media use [160]. The BSMAS has presented sound psychometric properties [115,156–158] with high internal consistency (α = 0.82) [161]. The Cronbach's alpha in the present study was 0.91 (T1) and 0.87 (T2).

The Fear of Missing Out Scale (FoMOS) [162] includes ten items and asks participants to evaluate the extent to which they experience symptoms of FoMO. The scale is rated on a seven-point Likert scale from 1 (not at all true) to 5 (extremely true of me). The statements include: "I fear others have more rewarding experiences than me... I get anxious when I don't know what my friends are up to...It bothers me when I miss an opportunity to meet up with friends...". A total score was calculated by averaging the scores, with higher mean scores indicating a greater level of FoMO. This instrument has demonstrated good construct validity [162,163], and good internal consistency with Cronbach's alphas of α = 0.93 [164] and 0.87 [64] with α = 0.87 in the present study.

The Nomophobia Questionnaire (NMP-Q) [165] comprises 20 items rated using a seven-point Likert scale from 1 (strongly disagree) to 7 (strongly agree). Total scores are calculated by summing up responses to each item, resulting in a nomophobia score ranging from 20 to 140, with higher scores corresponding to greater nomophobia severity. NMP-Q scores are interpreted in the following way: 20 = absence of nomophobia; 21–59 = mild level of nomophobia; 60–99 = moderate level of nomophobia; and 100+ = severe nomophobia. The scale has demonstrated good psychometric properties [165,166] with Cronbach's alphas of 0.94 [165] and 0.95 [167]. In the present study, internal consistency was: 0.89 for (T1) and 0.88 for (T2) respectively.

2.4. The Intervention

The intervention initially involved the search and identification of appropriate mobile apps (in both the Apple iTunes store and the Android Google Play store) for daily self-monitoring of social media activity for mindfulness practices and mood tracking. The apps needed to be freely available in order to be accessible by the participants. Due to time limitations, the development of an app that would encompass all three features (mindfulness of distraction, self-monitoring, and mood-tracking) was deemed adequate for the study given the ample availability of well-designed products offering these services. The following three freely available smartphone lifestyle apps were utilized: (i) Antisocial (screen time): to self-monitor screen time/social media use and for voluntary self-exclusion (block app after time limit is reached), (ii) Headspace (mindfulness): brief mindfulness sessions, (iii) Pacifica (mood tracking): the app encouraged monitoring and tracking an individual's emotional state at various times during the day to enhance awareness.

At the outset of the study, participants were directed to an information statement followed by the digital provision of informed consent before responding to the questions. At the end of the survey, they were automatically assigned through the automatic randomization procedure used by the online survey platform *Qualtrics* to either an intervention or a control group. Therefore, the intervention was double-blind (to participants and investigators). Participants assigned to the *IG* were asked to download the apps onto their smartphones and to actively engage with all three apps daily for 10 days, which was the maximum free period offered by one of these apps. Participants were encouraged to engage with mindfulness/focusing exercises to track their emotional state during the day and monitor patterns in their wellbeing as well as report daily on smartphone usage rates. Thereafter, participants received daily notifications via email for the duration of the intervention to remind them to provide online reports about their own social media usage rates, apps accessed, checking frequency, potential self-restriction from use, and satisfaction with the intervention. This process was used to motivate engagement with the apps and accountability. Efficacy was evaluated by having a CG condition where participants did not engage in any app use and only completed assessments on the first and tenth day. The target of the intervention was to induce a more mindful state, raise awareness of media and smartphone use, enhance self-regulation and therefore reduce distractions and time spent on smartphones and indirectly on social media by using these apps.

2.5. Data Analysis

2.5.1. Sample Size Estimation

The sample size for the RCT was determined a priori using G*Power v.3 software for the expected increased effectiveness of the intervention compared to control on the primary outcome distraction at post-assessment (T2). Empirical reviews [168] have suggested a median standardised target effect size of 0.30 (interquartile range: 0.20–0.38), with the median standardised observed effect size 0.11 (IQR 0.05–0.29). The present study was a low-threshold intervention for a non-clinical population, so a mean effect of $d = 0.30$ was expected. With a power of $1\text{-}\beta = 0.8$, and a significance level of $\alpha = 0.05$, the sample size was calculated to be $n = 95$ participants per group to find between- and within-group effects. To account for attrition rates in online interventions and control for both Type I and II error rates, $n = 125$ participants per group were targeted for recruitment [169].

2.5.2. Data Cleaning, Assumption Testing and Descriptive Analysis

All data were analysed through SPSS v.25 (Chicago, IL, USA). Preliminary data analyses included examining the data for data entry errors, normality testing, outliers, and missing data. Seven cases were treated with listwise deletion due to a very high percentage of incomplete data at baseline, resulting in a final sample size of 252. For the rest of the dataset, Little's Missing Completely at Random (MCAR) test showed that data were missing completely at random ($p = 0.449$). Multiple imputation was used to complete the dataset for the baseline analysis and for the non-completers from post-intervention assessment based on patterns of missingness. The data were also checked to ensure that all assumptions for the outlined statistical analyses were satisfied. The Kolmogorov-Smirnov test was used to evaluate the normal distribution of the variables, and skewness and kurtosis values were examined. For both assessments, all self-report data were normally distributed. Assumptions of *t*-tests included normality, homogeneity of variance, and independence of observations. Violations of the assumption of homogeneity of variance were tested using Levene's test of equality of variances [170]. Descriptive statistics were conducted to summarize the demographic characteristics of the sample as well as scores for the self-reported and performance-based measures of interest (i.e., stress). Pearson's correlations examined bivariate relationships between smartphone distraction and psychological variables, and frequency of smartphone and social media use (presented in Table 3).

2.5.3. Randomization and Risk of Bias

While allocation randomisation aimed to reduce any differences between the groups at baseline, a series of independent sample *t*-tests for the continuous variables and chi-square tests for the categorical variables (gender, ethnicity and education and relationship status) were conducted to analyse group mean differences and compare the baseline and post-intervention outcomes for the control and intervention groups. These were also applied at post-intervention outcomes for both the control and the intervention group. A decrease from the baseline to the post-intervention assessment was hypothesised for the primary outcomes of smartphone distraction, stress, anxiety, deficient self-regulation, FoMO and NoMO and an increase was hypothesized for mindful attention, self-awareness and self-efficacy.

Following the descriptive analysis, data from the baseline and post-intervention assessments were analysed to test each of the hypotheses provided to inform the assessment of the intervention efficacy. Two approaches to analysis were adopted. First, to isolate any effect of the intervention, a per-protocol (PP) analysis was conducted to maintain the baseline equivalence of the intervention group produced by random allocation [171]. However, given the limitations to this first analysis approach and to minimise biases resulting from noncompliance, non-adherence, attrition or withdrawal [172,173], analysis was performed also on an intention-to-treat (ITT) basis [172]. However, these results were not reported in the present study.

2.5.4. Analysis of Intervention Effects and Testing of Hypothesized Mechanisms

The effects of the intervention were assessed with an analysis of covariance (ANCOVA), with a minimum significance level at $p < 0.05$. ANCOVA was chosen given that it is quite robust with regard to violations of normality, with minimal effects on significance or power [174,175] with any differences between the groups at baseline, for the various assessments being used as covariates in the model and considered artefacts of the randomisation [176]. Co-varying for baseline scores supported the analysis in two ways. First, while randomisation aimed to reduce any pre-intervention differences between the groups, residual random differences may have occurred. Accounting for such differences isolated the effect of the intervention. Partial eta-squared were used as measures of strength of association [177]. To better understand the effect size of the intervention, it has been recommended to use the differences in adjusted means (standardized mean difference effect sizes) between the two groups, as standardising can easily distort judgements of the magnitude of an effect (due to changes to the sample SD but not the population SD, which may bias the estimate of the effect size measure, such as Cohen's *d*) [178]. As Cohen's *d* has been reported in other RCT and pre-post intervention studies, Cohen's *d* was estimated [179]. Finally, because the sample sizes of the two groups were unequal, Type III Sums of Squares were used for the ANCOVA.

To test the third hypothesis and the hypothesized psychological mechanisms underlying the intervention results, three different mediation analyses were performed across the chosen psychological constructs using SPSS Statistics (version 25) and PROCESS (Model 4; [180–183]), using a non-parametric resampling method bootstrap with 5000 bootstrapped samples and bias-corrected 95% confidence intervals, to probe conditional indirect effects for the variables examined. These analyses were performed on the ITT sample in post-intervention results.

3. Results

3.1. Baseline Equivalence Evaluation

The *t*-test results for the pre-test scores found no significant differences between the groups, indicating independence. The post-test scores were significantly lower in the intervention group. For the smartphone distraction scale, the mean pre-test score was 58.06 (SD = 7.69) for the intervention group and 59.72 (SD = 8.08) for the control group. The mean post-test score was 39.70 (SD = 17.67) for the intervention and 58.78 (SD = 17.47) for the control group, respectively. The pre-test score mean was not significantly different between groups (t = −0.70, *ns*), but the post-test score mean was significantly lower for the intervention group than for the comparison group (t = −6.69, $p < 0.001$). The pattern was similar in the results for the other variables except for NoMO, habitual behaviour, and social media use per day. Table 2 provides a summary of the baseline *t*-test and chi-square outcomes and internal consistency for each scale at each measurement period. All scales demonstrated good internal consistency for the sample considered.

Table 2. Per protocol baseline sociodemographic, usage data, psychological variables and pre-post intervention scale reliabilities.

	Intervention (n = 72)		Control (n = 71)		Chi Square/ *t*-Tests		
Socio/demographics	*n*	%	*n*	%		-	-
Gender (female)	60	83.33	62	87.32	1.83, *ns* [a]		
Education (under graduates %)	67	93.05	65	91.54	1.03, *ns*		
Relationship status (% not in relation)	40	55.55	38	53.52	1.35, *ns*		
Ethnicity (White %)	49	68.05	42	59.15	1.63, *ns*		
	M (SD)		*M (SD)*		*t* Tests	Cronbach's α T1	Cronbach's α T2
Age	20.69 (3.27)		20.82 (3.70)		−0.20, *ns*	-	-
Smart hours/day	4.55 (2.28)		5.23 (1.89)		−0.28, *ns*	-	-
SM hours/day	2.17 (1.430		2.47 (1.28)		−1.36, *ns*	-	-
Smart. distraction	59.52 (7.69)		57.55 (8.08)		−0.70, *ns*	0.90	0.88
Self-awareness	74.71(8.20)		75.00 (9.38)		−0.20, *ns*	0.87	0.86
Mindful Attention	3.28 (0.52)		3.40 (0.56)		−1.32, *ns*	0.92	0.93
Stress	24.44 (4.72)		28.78 (6.05)		−0.33, *ns*	0.86	0.83
Anxiety	15.93 (5.94)		16.63 (4.94)		−0.77, *ns*	0.93	0.90
Online vigilance	2.43 (0.48)		2.38 (0.52)		0.63, *ns*	0.89	0.87
Efficacy	28.04 (4.35)		28.96 (4.55)		−2.51, *ns*	0.90	0.88
FoMO	3.48 (1.36)		3.54 (1.34)		−0.32, *ns*	0.89	0.90
NoMO	77.17 (22.40)		86.32 (23.68)		−0.49., *ns*	0.89	0.88
Def. self-regulation	14.15 (5.32)		15.35 (5.39)		−1.50, *ns*	0.89	0.87
Impulsivity	14.74 (3.39)		16.27 (3.52)		−0.264, *ns*	0.85	0.86
Prob. SM use	17.15 (4.95)		17.18 (5.42)		−0.035, *ns*	0.91	0.89
Automaticity	5.14 (1.33)		5.11 (1.20)		−0.88, *ns*	0.87	0.89

a ns = non-significant. FoMO = Fear of Missing Out; NoMO = Nomophobia; Def. self-regulation = Deficient self-regulation; Prob. SM use = Problematic social media use.

A series of Bivariate Pearson's r correlation analyses was conducted to examine the results obtained amongst SDs and the secondary outcomes (Table 3). Smartphone distraction correlated significantly with problematic social media use ($r(252) = 0.63$, $p < 0.01$), anxiety (r (252) = 0.46, $p < 0.01$), online vigilance (r (252) = 0.51, $p < 0.01$), automaticity (r (252) = 0.57, $p < 0.01$), impulsivity ($r(252) = 0.45$, $p < 0.01$), deficient self-regulation ($r(252) = 0.33$, $p < 0.01$), smartphone use/day ($r(252) = 0.31$, $p < 0.01$), $p < 0.01$), FoMO ($r(252) = 0.28$, $p < 0.01$) and NoMO ($r(252) = 0.51$, $p < 0.01$). However, smartphone distraction correlated negatively with two variables: mindful attention ($r(252) = -0.52$, $p < 0.01$) and self-awareness ($r(252) = -0.34$, $p < 0.01$).

Table 3. Bivariate Pearson's *r* correlation analyses.

	1	2	3	4	5	6	7	8	9	10	11	12	13	14
1. Distraction	1													
2. Stress	0.199 **	1												
3. Pr. SM use	0.631 **	0.173 **	1											
4. Mind.Att.	−0.523 **	−0.145 *	−0.455 **	1										
5. Self-Aware	−0.340 **	0.057	−0.318 **	−0.209 **	1									
6. Anxiety	0.460 **	0.380 **	0.435 **	0.450 **	0.242 **	1								
7. Onl. Vigil.	0.507 **	0.280 **	0.620 **	0.380 **	0.223 **	0.283 **	1							
8. Efficacy	−0.107	−0.343 **	−0.149 *	−0.101	0.148 *	−0.399 **	−0.056	1						
9. Automat	0.575 **	0.286 **	0.466 *	0.324 **	0.194 **	0.304 **	0.348 **	−0.179 **	1					
10. Impuls.	0.455 **	0.006	−0.053	−0.037	−0.522	−0.026	0.035	0.086	0.037	1				
11. Def. Self-reg.	0.333 **	0.048	0.017	0.048	−0.068	0.007	0.074	0.025	0.049	0.859 **	1			
12. Smart/day	0.314 **	−0.280	0.013	−0.128	−0.025	−0.161	0.082	0.021	−0.145	−0.008	−0.004	1		
13. SM/day	0.116	0.004	−0.025	−0.008	−0.109	0.024	−0.035	−0.111	0.061	0.154	0.168 *	0.423 **	1	
14. FoMO	0.281 **	0.323 **	0.382 **	0.103	0.310 **	0.369 **	−0.032	−0.164 **	0.235 **	0.026	0.035	0.183 **	0.180 **	1
15. NoMO	0.513 **	0.375 **	0.421 **	0.007	0.142 *	0.312 **	0.136 *	−0.209 **	0.392 **	−0.084	−0.084	0.189 **	0.096	0.341 **

* $p < 0.05$; ** $p < 0.01$; *** $p < 0.001$. Pr. SM use: Problematic social media use; Mind. Att: Mindful attention; Onl. Vigil.: Online vigilance; FoMO: Fear of Missing Out; NoMO: Nomophobia; Def. self-regulation: Deficient self-regulation; SM/day; Social Media use/day.

3.2. Intervention Efficacy Evaluation

To test H1 and assess the effect of the intervention on smartphone distraction, two separate ANCOVAs were conducted. First, to isolate any effect of the intervention, a per-protocol analysis was conducted. As depicted in Table 4, distraction outcomes decreased significantly for the intervention group from the baseline (intervention: $M = 58.06$, $SD = 7.69$; control: $M = 59.72$, $SD = 8.08$) to the post-intervention assessment (intervention: $M = 39.70$, $SD = 17.67$; control: $M = 58.78$, $SD = 17.47$), with a non-significant difference for the control group. As confirmed by Levene's test, the outcome variances were homogenous. Confirming the homogeneity of the regression slopes, the interaction between the baseline scores and the experimental group was significant. There was a main effect of the intervention group on post-intervention distraction scores after controlling for baseline outcomes ($F(1, 140) = 46.59$, $p < 0.001$, $\eta p^2 = 0.250$). The baseline scores were not a significant predictor of post-intervention values ($F(1, 140) = 18.62$, $p = 0.117$). Post-hoc tests indicated there was a statistically significant adjusted mean difference ($M = -18.95$, $SD = 2.77$, ($p < 0.001$) in reduction between IG compared to CG (Figure 2). For the ITT analysis, a main effect on the intervention group on post-intervention SDS outcomes after controlling for the baseline values was found ($F(1, 250) = 96.88$, $p < 0.001$, $\eta p^2 = 0.28$). As indicated in Figure 2, post-hoc tests indicated there was a significant difference between IG and CG ($p < 0.001$). Comparing the estimated marginal means showed that there was an adjusted mean difference in reduction between IG ($M = 39.56$) compared to CG ($M = 58.93$). Consequently, across both analyses, this hypothesis was supported.

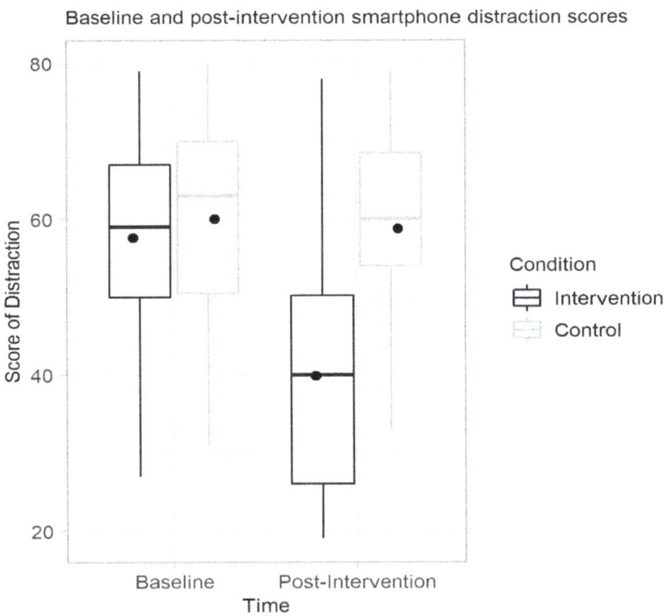

Figure 2. Per protocol smartphone distraction outcomes before and after the intervention.

Table 4. Per protocol sample (n = 143) primary and secondary measures, means, SDs, effect sizes and F-values for between-group comparisons.

Measure	Experimental (n = 72)		Control (n = 71)		Effect	Effect Size	Cohen's d
	Pre	Post	Pre	Post			
	M(SD)	*M(SD)*	*M(SD)*	*M(SD)*	*F*	$\eta p2$	*d*
Smart.Distraction	58.06 (7.69)	39.70 (17.67)	59.72 (8.08)	58.78 (17.47)	46.59 ***	0.25	1.11
Self-awareness	74.71 (8.20)	83.30 (9.89)	75.00 (9.38)	76.25 (10.25)	18.19 ***	0.12	0.69
Mind.Attention	3.28 (0.52)	3.97 (0.69)	3.40 (0.56)	3.37 (0.76)	16.24 ***	0.22	0.82
Stress	24.44 (4.72)	24.10 (4.63)	28.78 (6.05)	27.94 (5.24)	23.11 ***	0.14	0.77
Anxiety	15.93 (5.94)	14.75 (4.43)	16.63 (4.95)	17.44 (4.42)	12.42 ***	0.08	0.60
Vigilance	2.43 (0.49)	1.98 (0.63)	2.38 (0.52)	2.39 (0.52)	18.66 ***	0.12	0.70
Self-efficacy	28.04 (4.36)	32.32 (5.08)	28.96 (4.55)	29.99 (5.05)	9.40 ***	0.06	0.46
FoMO	3.48 (1.36)	2.86 (1.16)	3.54 (1.34)	3.32 (1.22)	5.49 ***	0.04	0.39
NoMO	77.17 (2.40)	78.03 (2.72)	86.32 (23.6)	79.50 (2.74)	7.71	-	-
Def. self-reg.	17.16 (6.70)	14.00 (5.32)	17.61 (6.91)	15.32 (5.39)	6.60 ***	0.04	0.25
Impulsivity	17.32 (3.79)	14.74 (3.41)	17.65 (3.92)	16.27 (3.51)	15.91 ***	0.10	0.44
Probl. SM use	17.15 (4.95)	15.12 (4.40)	17.18 (5.42)	17.24 (5.11)	6.96 ***	0.05	0.44
Automaticity	5.14 (1.33)	4.77 (1.30)	5.11 (1.20)	4.98 (1.59)	0.78	-	-
SM. use/day	2.92 (1.75)	2.17 (1.44)	2.89 (1.52)	2.47 (1.28)	3.70	-	-
Smart. use/day	4.51 (2.28)	3.51 (1.88)	4.45 (1.89)	4.11 (1.68)	4.43 ***	0.03	0.34

* $p < 0.05$; ** $p < 0.01$; *** $p < 0.001$.

ANCOVA analyses for the secondary outcomes were also tested across both PP and ITT samples. Specifically, for the PP sample, main effects of the experimental group on post-intervention outcomes after controlling for baseline scores were found for self-awareness ($F(1, 140) = 18.19$, $p < 0.001$, $\eta p^2 = 0.115$), mindful attention ($F(1, 140) = 16.24$, $p < 0.001$, $\eta p^2 = 0.22$), anxiety ($F(1, 140) = 12.42$, $p < 0.001$, $\eta p^2 = 0.08$), stress ($F(1, 140) = 23.11$, $p < 0.001$, $\eta p^2 = 0.14$), online vigilance ($F(1, 140) = 18.66$, $p < 0.001$, $\eta p^{2=}$ 0.118), FoMO ($F(1, 140) = 5.49$, $p < 0.001$, $\eta p^2 = 0.04$), deficient self-regulation ($F(1, 140) = 6.60$, $p < 0.001$, $\eta p^2 = 0.045$), self-efficacy ($F(1,140) = 9.40$, $p < 0.001$, $\eta p^2 = 0.063$), impulsivity ($F(1, 140) = 15.91$, $p < 0.001$, $\eta p^2 = 0.10$), problematic social media use ($F(1, 140) = 6.96$, $p < 0.001$, $\eta p^2 = 0.05$), and smartphone use/day ($F(1, 140) = 4.43$, $p < 0.001$, $\eta p^2 = 0.03$). No intervention effects were found for the intervention group for the variables of social media use/day ($F(1, 140) = 3.697$, $p = 0.06$), habit strength ($F(1, 140) = 0.78$, $p = 0.78$), and NoMO ($F(1, 140) = 7.714$, $p = 0.91$). ITT analyses demonstrated similar patterns to the PP samples' outcomes.

3.3. *Intervention Effects Based on Distraction Severity*

In order to evaluate the effects of the intervention in the intervention group based on level of distraction and to assess whether the effects were consistent in the intervention group independent of degree of distraction, participants were classed into two categories of high distractors vs. low distractors depending on perceived distraction level. A median-split analysis with high vs. low distractor levels was determined by scores above vs. below the median and these were separately analysed inside the intervention group. Therefore, a two-way mixed ANOVA with time (pre-test and post-test) as within-factor and distraction severity (high and low distraction) as between-factor was performed to investigate the impact of the intervention (time) and degree of distraction (high vs. low) as assessed at baseline on distraction levels at post-intervention. This analysis was conducted only for the dependent variable for which the interactions were found to be significant.

Results indicated there was a significant main effect of the intervention $F(1,70) = 77.17$, $p < 0.001$. There was a significant main effect of distraction $F(1,70) = 21.48$, $p < 0.001$ with high distractors ($M = 48.67$) benefiting more than the low distractors ($M = 33.54$). Additionally, there was a significant interaction between the distraction status (high vs. low) and the degree of distraction $F(1,70) = 20.10$, $p < 0.001$. No significant interactions were found for self-awareness ($F(1,70) = 1.07$, $p = 0.32$); stress ($F(1,70) = 0.17$, $p = 0.68$); online vigilance ($F(1,70) = 0.98$, $p = 0.32$), deficient self-regulation ($F(1,70) = 0.22$, $p = 0.64$), self-efficacy ($F(1,70) = 0.22$, $p = 0.64$), anxiety ($F(1,70) = 1.73$, $p = 0.19$), and social media use ($F(1,70) = 19.28$, $p = 0.30$). However, significant main effects were also found for self-awareness

($F(1,70) = 30.05$, $p < 0.001$), deficient self-regulation $F(1,70) = 20.10$, $p < 0.001$, stress ($F(1,70) = 47.95$, $p < 0.001$), online vigilance $F(1,70) = 42.07$, $p < 0.001$, problematic social media use $F(1,70) = 9.94$, $p < 0.05$; FoMO ($F(1,70) = 10.33$, $p < 0.001$) and smartphone use/day ($F(1,70) = 53.12$, $p < 0.001$).

3.4. Mediation Analyses

More specifically for mediation 1, the intervention group was the proposed independent variable in these analyses, mindfulness was the proposed mediator, and smartphone distraction was the outcome variable. For mediation 2, stress was the proposed independent variable in these analyses, online vigilance was the proposed mediator, and smartphone distraction was the outcome variable. For mediation 3, smartphone distraction was the predictor, social media addiction was the outcome and online vigilance was the mediator. Analysed variables included the T1 scores on the constructs examined as covariates to account for pre-intervention performance.

For mediation 1, it was hypothesized that mindful attention would mediate the relationship between the intervention and smartphone distraction (Table 5). No mediation effect was found for mindful attention on the variables. However, a main effect of the intervention on smartphone distraction (path a: $b = -0.67$, $t = -8.23$, $p < 0.001$) was found, but no main effect of mindful attention on smartphone distraction (path b; $b = 1.16$, $t = 0.67$, *ns*).

Table 5. Mediation effects of mindful attention and emotional self-awareness on intervention effects and smartphone distraction and of online vigilance on smartphone distraction and social media addiction ($n = 252$).

Predictor	Outcome	Mediator	*ab (B)*	*a*	*b*	*c*	*c′*
Intervention	Smart.Distract.	Mindful Att.	−0.79 [−3.10, −1.59]	−0.67 [−0.84, −0.51]	1.16 [−2.25, 4.58]	20.75 [16.35, 25.16]	21.55 [16.62,26.48]
Intervention	Smart.Distract.	Self-aware	−2.02 [−3.97, −0.35]	−6.78 [−9.15, −4.40]	0.30 [0.07, 0.52]	20.91 [16.59, 25.22]	22.93 [18.38, 27.48]
Smart. distract.	Probl. SM use	On.vigilance	0.02 [0.01, 0.03]	0.01 [0.010, 0.015]	1.66 [0.78, 2.54]	0.11 [0.08, 0.13]	0.089 [0.06, 0.12]

For mediation 2, it was hypothesized that self-awareness would mediate the relationship between the intervention and smartphone distraction (Table 5). An indirect effect was found on self-awareness on the variables (a × b: $b = -2.02$, BCa CI = [−3.10, −1.59]), indicating mediation. The intervention significantly predicted self-awareness (path a; $b = -6.78$, $t = -4.32$, $p < 0.001$) and self-awareness significantly predicted lower levels of smartphone distraction (path b; $b = 0.30$, $t = 4.02$, $p < 0.001$).

For mediation 3, it was hypothesized that online vigilance would mediate the relationship between distraction and social media addiction (Table 5). An indirect effect was found on self-awareness on the variables (a × b: $b = 0.02$, BCa CI = [0.01, 0.03]), indicating mediation. The intervention significantly predicted self-awareness (path a; $b = -0.01$, $t = -3.32$, $p < 0.001$) and self-awareness significantly predicted lower levels of smartphone distraction (path b; $b = 1.66$, $t = 4.02$, $p < 0.001$).

4. Discussion

The present study tested the efficacy of an online intervention employing an integrative set of strategies—consisting of mindfulness, self-monitoring and mood tracking—in assisting young adults to decrease levels of smartphone distraction and improve on a variety of secondary psychological outcomes, such as mindful attention, emotional awareness, stress and anxiety, and perceived self-efficacy, as well as to reduce stress, anxiety, deficient self-regulation, problematic social media use and smartphone-related psychological outcomes (i.e., online vigilance, FoMO and NoMO). Results of the present study provided support for the online intervention effectiveness in impacting these outcomes. Findings suggested that students receiving the intervention reported a significant reduction in the primary outcome of smartphone distraction, unlike students in the control group who reported a non-significant reduction in smartphone distraction. In terms of the secondary outcomes, participants in the intervention condition experienced a significant increase in self-awareness, mindful attention, and self-efficacy, and a significant decrease in smartphone use/day, impulsivity, stress, anxiety, deficient self-regulation, FoMO, and problematic use. No significant results were found for social media use per day, habitual/automated use and NoMO.

According to the findings of the present intervention, it appears likely that practising mindfulness and monitoring mood and smartphone activity could lead to a desired behavioural change towards less distraction and less perceived stress with carry-over effects in self-awareness and self-efficacy, similar to interventions for other mental health problems [83,85,87,91,93,184,185]. These findings are consistent with the growing body of research indicating that mindfulness and self-monitoring are effective strategies to increase self-awareness and reduce stress [84–90,186]. Mindful attention could enhance awareness of individual media behaviour by: (i) raising understanding and awareness of disruptive media multitasking activities (i.e., predictors, patterns and effects), and (ii) raising awareness of different strategies for coping with digital distraction and of which strategies are most effective. Second, self-monitoring could help in developing an understanding of media habits and time spent on smartphone and social media activities and could curb perceived excess smartphone interaction, consistent with other study findings [92,101,187,188]. Therefore, strategies employing increased mindfulness practice and self-monitoring could aid attentional capacity and self-awareness, which is considered a necessary condition in the behaviour change process of risky behaviours [189,190].

Third, mood tracking could enhance awareness of triggers of negative mood and ensuing negative emotional states acting as drivers for distraction. It appears that the same technologies which may impact negatively on young people may be used to leverage smartphone use [100] and deflect psychological distress if evidence-based behaviour change strategies are applied. Intervention strategies such as mindfulness and self-monitoring may encourage increased self-awareness and thus help reduce distraction levels and increase mindful attention.

The intervention was also successful in reducing secondary outcomes, such as stress levels and FoMO, and it had a positive effect on emotion regulation and loss of control levels. Distraction appears to be associated with higher access to social media content and is mediated by online vigilance. Salience of smartphone-mediated social interactions (i.e., the salience dimension of online vigilance) has been found to be negatively related to affective wellbeing [49]. It has been reported that emotional dysregulation mediates the relationship between psychological distress and problematic smartphone use [191]. Higher self-regulation online has been identified as a moderator between need to belong and problematic social media use in young people [192] and emotion dysregulation as a mediator between insecure attachment and addiction [193]. Although distraction is an emotion regulation strategy with a protective function against emotionally distressing states [111] and dysphoric mood [194], or is used for adaptive coping [195,196], deficits in attentional control, such as distraction, may also be implicated in stress, anxiety or other affective disorders [197] and in generalized anxiety disorder with core cognitive symptoms related to excessive thoughts and deficits associated with increased perseverative worry [198]. Therefore, higher mindful attention and monitoring of mood may have influenced the reduction of distraction and the enhancement of emotional control.

Mediation analyses were also performed to understand the relationships between intervention effects on smartphone distraction via two mediators, mindful attention and self-awareness, and of online vigilance on the relationship between distraction and social media addiction. Mediation effects were significant for the relationship among intervention effects and distraction via self-awareness, and for distraction and problematic social media use via online vigilance, indicating that self-awareness could be a potential behaviour strategy to mitigate distraction levels. However, the relationship among intervention effects and distraction was not significant via mindful attention as a mediator. Therefore, in the present study it appeared that despite its statistically significant increase, mindful attention was not a mediating factor for distraction in the intervention. Mindful attention could potentially be the vehicle to increasing emotional self-awareness [93,184,199], prompting more controlled smartphone interactions. On the contrary, online vigilance was found to be a mechanism associated with smartphone distraction and problematic social media use, given the strong preoccupation with the content prompted even by the mere presence of smartphones, confirming previous findings [200].

Therefore, despite its protective function, distraction may concurrently serve as a gateway to increased smartphone engagement and time spent on devices. Time spent alone is not a defining factor and it has been argued instead that the interaction of content, context and time spent, as well as the meaning attached to these interactions, may determine the level of problematic media use [5,201]. Within smartphone use, distraction is a salient behaviour with evidence that distraction and mind-wandering are associated with online vigilance, which via reduced mindfulness may be associated with decreased wellbeing [78]. Furthermore, inattention symptoms have been implicated in risk for smartphone addiction and problematic smartphone use [202]. Therefore, handling distraction, which has neural correlates [203], may be the means to resisting cue reactivity, implicated in smartphone addiction, in reduced cognitive performance [113] or in obsessive-compulsive symptoms [204]. Further research is required to assess these cognitive and emotive dimensions of smartphone distraction and its effects on engagement in line with current trends [205]. However, it has been proposed that the construct of distraction extends beyond the debate on smartphone addiction by considering the role of the smartphone in coping with negative emotions and addressing preference for online vs. offline communications [206].

Research is still conflicted in relation to the cognitive function of distraction. Experimental smartphone research has provided initial evidence that social apps compared to non-social apps on smartphones do not capture attention despite their perceived high reward value [207,208], but other studies support a high interference effect [209]. Therefore, more research is required to elucidate the mechanisms of digital distraction and delineate how digital technologies, individual choices, and contexts affect individuals' attention spans and attentional loss, as well as mental health conditions, such as ADHD and anxiety and overall psychological wellbeing [210]. The present RCT assessed the effectiveness of the impact of the use of mindfulness, self-monitoring, and mood tracking delivered through interaction with smartphone apps in reducing distraction arising from recreational smartphone use and social media use. The findings suggest that engaging with the aforementioned practices was effective in reducing distraction levels, stress, anxiety, deficient self-regulation, impulsivity and smartphone-related psychological outcomes, and improving mindful attention and emotional self-awareness and self-efficacy.

Limitations, Implications, and Recommendations

Some limitations need to be taken into consideration. First, a convenience sample of university students was used, which hinders the generalizability of the findings to other groups (i.e., older adults or children). However, this population was considered of primary interest for the study because university students are digital natives liable to experience negative academic consequences due to vulnerability to problematic smartphone use [211].

The effect sizes found in this RCT were medium to large for the variables examined, exceeding the expected range for low-intensity, non-clinical interventions [212]. However, as a result of the main recruitment protocol, the intervention may have attracted participants who had an interest in the outcomes and a potential self-assessed vulnerability. Therefore, the voluntary, self-selected nature of participation could have introduced a significant degree of participant response and confirmation bias [213], resulting in the medium to high effect sizes. Additionally, the high drop-out rates, consistent with other online RCTs [214], could have significantly affected the strength of the findings [215], and the use of a passive control group might have led to an overestimation of the effects [216]. Due to the use of market-available apps, actual adherence and engagement with the intervention was not accounted for, nor were reasons for dropout [217]. Therefore, the findings should be treated with caution and replicated in future designs. Future studies should systematically address response bias and include methods in the RCT to improve the accuracy of self-reported data [218,219]. Combining self-report with behavioural data [220], ecological momentary sampling [221], psycho-informatics and digital phenotyping, the provision of a digital footprint for prognostic, diagnostic and intervention purposes [222], could enhance the ecological validity of the study. Equally, incorporating the measurement of brain activity using magnetic resonance imaging (MRI) in interventions could greatly enhance accuracy of assessment of prevention efforts and understanding of the role of neurobiology in behaviour [223,224].

The impact of the intervention on gender was not examined because this university student sample consisted mainly of female participants. Considering the gender differences reported in smartphone use [48,225] and in attention processes [226], future studies should explore its effect, which could have significant implications for the intervention and prevention of attention failures and poor student outcomes [227]. Additionally, the study design did not manage to provide a longer intervention period due to the lack of freely available apps for participants to use and did not include a second follow-up period to track maintenance of long-term effects, as is customary in RCTs, or the use of qualitative process evaluation for a critical understanding of impact of the intervention components [228]. Finally, social, economic and family conditions as well as other issues, which are critical to young people's psycho-emotional states and sense of identity, were not accounted for in the present study [229,230].

Despite these limitations, the study provides initial evidence for efficacy of strategies in curbing smartphone distraction and adds to the limited body of knowledge of cognitive-emotive processes in smartphone and social media use [205]. It also contributed to the still limited knowledge on interventions in smartphone distraction and constitutes a simple, first-step, low key intervention programme, which may be practised by individuals seeking support for attentional difficulties on a self-help basis or within a stepped-care clinical framework for prevention purposes [96]. Experiencing distraction from smartphones and social media content, interferes with high-level cognitive processes and has productivity and emotional implications (i.e., stress) in various contexts and situations [51,231–234], being further compromised by digital triggers and the structural design of smartphones prompting salience and reactivity [235].

These results have clinical implications as low-intensity interventions may prevent small scale emotional problems from developing into clinical disorders and can reduce incidences of mental health problems [236,237]. Practitioners may also find value in using mindfulness and monitoring practices as an adjunct to therapy for problematic use of smartphones. It may be of high value for academic institutions to build specific university-based programmes on maintaining balanced technology use, tackling unregulated and promoting positive smartphone use, or guiding students towards suitable methods to address attention problems more effectively [238,239]. Apps may also be utilized by schools for students that are faced with attentional/excessive use difficulties and in assisting young people to become aware of their emotions in preparation for learning more adaptive coping strategies. Distraction is an emergent phenomenon in the digital era considering that the boundaries between work and recreation are increasingly blurred with both domains arguably dependent on the use of digital media [240]. More research on attentional processes within smartphone use could aid the understanding of these processes and impacts experienced across different age groups.

5. Conclusions

Psychological low-cost interventions may be effective in addressing precursors of problematic behaviours and enhancing wellbeing dimensions. The aim of the present study was to assess the efficacy of an RCT combining evidence-based cognitive-behavioural strategies to reduce distraction from smartphone use, increase mindful attention, emotional self-awareness and self-efficacy and reduce stress, anxiety, deficient self-regulation and smartphone related psychological outcomes (i.e., online vigilance, FoMO and NoMO). Second, it tested the mediating effect of mindful attention and self-awareness of the intervention on distraction, and of online vigilance on the relationship between distraction and social media addiction.

Findings suggested that students receiving the intervention reported a significant reduction in the primary outcome of smartphone distraction, whereas students in the control group reported a non-significant reduction in smartphone distraction. In terms of the secondary outcomes, participants in the intervention condition experienced a significant increase in self-awareness, mindful attention and self-efficacy and a significant decrease in smartphone use/day, impulsivity, stress and anxiety levels, FoMO, deficient self-regulation and problematic social media use. No significant results were found for duration of social media use/day, habitual use and NoMO. Mediation effects of the intervention were also observed on distraction and problematic social media use via the mediators of emotional self-awareness and online vigilance in mitigating distraction levels. Mindful attention was not found to be a mediating process for reducing distraction in the intervention.

Research on digital distraction is still scarce, yet there is increasing interest in cognitive impacts within digital environments. More evidence is required to assess the nature of attention failures and difficulties occurring both in normative and excessive online use. This evidence would allow an understanding of the prevalence and the nature of these difficulties, as well as their integration in intervention media literacy and risk prevention programmes, enhancing wellbeing, productivity and academic performance.

Author Contributions: Conceptualization, M.A.T.; methodology, M.A.T., M.D.G., M.R., D.J.K.; formal analysis, M.A.T.; investigation, M.A.T.; data curation, M.A.T.; supervision, M.D.G., M.R., D.J.K.; writing—original draft preparation, M.A.T.; review and editing, M.A.T., M.D.G., M.R., D.J.K.; All authors have read and agreed to the published version of the manuscript.

Funding: No funding was received for the present study.

Conflicts of Interest: The authors declare no conflict of interest.

References

1. Ashinoff, B.K.; Abu-Akel, A. Hyperfocus: The forgotten frontier of attention. *Psychol. Res.* **2019**. [CrossRef] [PubMed]
2. Atchley, P.; Lane, S. Cognition in the attention economy. In *Psychology of Learning and Motivation*; Elsevier: Amsterdam, The Netherlands, 2014; pp. 133–177; Available online: https://linkinghub.elsevier.com/retrieve/pii/B9780128002834000046 (accessed on 11 June 2019).
3. Falkinger, J. Attention economies. *J. Econ. Theory* **2007**, *133*, 266–294. [CrossRef]
4. Lee, A.R.; Son, S.-M.; Kim, K.K. Information and communication technology overload and social networking service fatigue: A stress perspective. *Comput. Hum. Behav.* **2016**, *55*, 51–61. [CrossRef]
5. Throuvala, M.A.; Griffiths, M.D.; Rennoldson, M.; Kuss, D.J. A 'control model' of social media engagement in adolescence: A grounded theory analysis. *Int.J. Environ. Res. Public Health* **2019**, *16*, 4696. [CrossRef]
6. Rosen, C. *The Myth of Multitasking*; Center for the Study of Technology and Society: Washington, DC, USA, 2008; pp. 105–110; Available online: https://www.jstor.org/stable/43152412 (accessed on 5 April 2020).
7. Statista. Number of Smartphone Users Worldwide from 2016 to 2021. 2020. Available online: https://www.statista.com/statistics/330695/number-of-smartphone-users-worldwide/ (accessed on 20 March 2020).
8. Busch, P.A.; McCarthy, S. Antecedents and consequences of problematic smartphone use: A systematic literature review of an emerging research area. *Comput. Hum. Behav.* **2020**, 106414. [CrossRef]
9. Kuss, D.; Harkin, L.; Kanjo, E.; Billieux, J. Problematic smartphone use: Investigating contemporary experiences using a convergent design. *Int. J. Environ. Res. Public Health* **2018**, *15*, 142. [CrossRef]
10. Montag, C.; Wegmann, E.; Sariyska, R.; Demetrovics, Z.; Brand, M. How to overcome taxonomical problems in the study of Internet use disorders and what to do with "smartphone addiction"? *J. Behav. Addict.* **2019**, 1–7. [CrossRef] [PubMed]
11. Beuckels, E.; Kazakova, S.; Cauberghe, V.; Hudders, L.; De Pelsmacker, P. Freedom makes you lose control: Executive control deficits for heavy versus light media multitaskers and the implications for advertising effectiveness. *Eur. J. Mark.* **2019**, *53*, 848–870. [CrossRef]
12. Fitz, N.; Kushlev, K.; Jagannathan, R.; Lewis, T.; Paliwal, D.; Ariely, D. Batching smartphone notifications can improve well-being. *Comput. Hum. Behav.* **2019**, *101*, 84–94. [CrossRef]
13. Marsh, E.J.; Rajaram, S. The digital expansion of the mind: Implications of Internet usage for memory and cognition. *J. Appl. Res. Mem. Cogn.* **2019**, *8*, 1–14. [CrossRef]

14. Ward, A.F.; Duke, K.; Gneezy, A.; Bos, M.W. Brain drain: The mere presence of one's own smartphone reduces available cognitive capacity. *J. Assoc. Consum. Res.* **2017**, *2*, 140–154. [CrossRef]
15. Thornton, B.; Faires, A.; Robbins, M.; Rollins, E. The mere presence of a cell phone may be distracting: Implications for attention and task performance. *Soc. Psychol.* **2014**, *45*, 479–488. [CrossRef]
16. Hartanto, A.; Yang, H. Is the smartphone a smart choice? The effect of smartphone separation on executive functions. *Comput. Hum. Behav.* **2016**, *64*, 329–336. [CrossRef]
17. Wilmer, H.H.; Sherman, L.E.; Chein, J.M. Smartphones and cognition: A review of research exploring the links between mobile technology habits and cognitive functioning. *Front. Psychol.* **2017**, *8*. [CrossRef] [PubMed]
18. Canale, N.; Vieno, A.; Doro, M.; Rosa Mineo, E.; Marino, C.; Billieux, J. Emotion-related impulsivity moderates the cognitive interference effect of smartphone availability on working memory. *Sci. Rep.* **2019**, *9*, 18519. [CrossRef]
19. Hadar, A.; Hadas, I.; Lazarovits, A.; Alyagon, U.; Eliraz, D.; Zangen, A. Answering the missed call: Initial exploration of cognitive and electrophysiological changes associated with smartphone use and abuse. *PLoS ONE* **2017**, *12*, e0180094. [CrossRef] [PubMed]
20. Duke, É.; Montag, C. Smartphone addiction, daily interruptions and self-reported productivity. *Addict. Behav. Rep.* **2017**, *6*, 90–95. [CrossRef]
21. Barr, N.; Pennycook, G.; Stolz, J.A.; Fugelsang, J.A. The brain in your pocket: Evidence that smartphones are used to supplant thinking. *Comput. Hum. Behav.* **2015**, *48*, 473–480. [CrossRef]
22. Kim, I.; Kim, R.; Kim, H.; Kim, D.; Han, K.; Lee, P.H.; Mark, D.; Lee, U. Understanding smartphone usage in college classrooms: A long-term measurement study. *Comput. Educ.* **2019**, *141*, 103611. [CrossRef]
23. Rostain, A.L. Book review of the "The Distracted Mind: Ancient Brains in a High-Tech World". *J. Am. Acad. Child. Adolesc. Psychiatry* **2019**, *58*, 829–831. Available online: https://linkinghub.elsevier.com/retrieve/pii/S0890856719304125 (accessed on 3 August 2019). [CrossRef]
24. Mendoza, J.S.; Pody, B.C.; Lee, S.; Kim, M.; McDonough, I.M. The effect of cellphones on attention and learning: The influences of time, distraction, and nomophobia. *Comput. Hum. Behav.* **2018**, *86*, 52–60. [CrossRef]
25. Uncapher, M.R.K.; Thieu, M.; Wagner, A.D. Media multitasking and memory: Differences in working memory and long-term memory. *Psychon Bull. Rev.* **2016**, *23*, 483–490. [CrossRef]
26. McRae, K.; Hughes, B.; Chopra, S.; Gabrieli, J.D.E.; Gross, J.J.; Ochsner, K.N. The neural bases of distraction and reappraisal. *J. Cogn. Neurosci.* **2010**, *22*, 248–262. [CrossRef] [PubMed]
27. Nolen-Hoeksema, S.; Wisco, B.E.; Lyubomirsky, S. Rethinking rumination. *Perspect. Psychol. Sci.* **2008**, *3*, 400–424. [CrossRef] [PubMed]
28. Sheppes, G.; Scheibe, S.; Suri, G.; Gross, J.J. Emotion-regulation choice. *Psychol. Sci.* **2011**, *22*, 1391–1396. [CrossRef] [PubMed]
29. Moyal, N. Cognitive strategies to regulate emotions—current evidence and future directions. *Front. Psychol.* **2014**, *4*, 1019. [CrossRef]
30. Gross, J.J. The emerging field of emotion regulation: An integrative review. *Rev. Gen. Psychol.* **1998**. [CrossRef]
31. Gazzaley, A.; Rosen, L.D. *The Distracted Mind: Ancient Brains in a High-Tech. World*; MIT Press: Cambridge, MA, USA, 2016.
32. Rosen, L.D.; Lim, A.F.; Carrier, L.M.; Cheever, N.A. An empirical examination of the educational Impact of text message-induced task switching in the classroom: Educational implications and strategies to enhance learning. *Rev. Psicol. Educ.* **2011**, *17*, 163–177.
33. Stothart, C.; Mitchum, A.; Yehnert, C. The attentional cost of receiving a cell phone notification. *J. Exp. Psychol. Hum. Percept. Perform.* **2015**, *41*, 893–897. [CrossRef]
34. Peña-Sarrionandia, A.; Mikolajczak, M.; Gross, J.J. Integrating emotion regulation and emotional intelligence traditions: A meta-analysis. *Front. Psychol.* **2015**, *6*, 160. [CrossRef]
35. Baron, R.S. Distraction-conflict theory: Progress and problems. *Adv Exp. Soc. Psychol.* **1986**, *19*, 1–40.
36. Baron, R.S.; Moore, D.; Sanders, G.S. Distraction as a source of drive in social facilitation research. *J. Pers. Soc. Psychol.* **1978**, *36*, 816–824. [CrossRef]

37. Montagni, I.; Guichard, E.; Kurth, T. Association of screen time with self-perceived attention problems and hyperactivity levels in French students: A cross-sectional study. *BMJ Open* **2016**, *6*, e009089. [CrossRef] [PubMed]
38. Baumgartner, S.E.; van der Schuur, W.A.; Lemmens, J.S.; Te Poel, F. The relationship between media multitasking and attention problems in adolescents: Results of two longitudinal studies: Media multitasking and attention problems. *Hum. Commun. Res.* **2017**. [CrossRef]
39. Kuznekoff, J.H.; Titsworth, S. The impact of mobile phone usage on student learning. *Commun. Educ.* **2013**, *62*, 233–252. [CrossRef]
40. Giunchiglia, F.; Zeni, M.; Gobbi, E.; Bignotti, E.; Bison, I. Mobile social media usage and academic performance. *Comput. Hum. Behav.* **2018**, *82*, 177–185. [CrossRef]
41. Felisoni, D.D.; Godoi, A.S. Cell phone usage and academic performance: An experiment. *Comput. Educ.* **2018**, *117*, 175–187. [CrossRef]
42. Csibi, S.; Griffiths, M.D.; Demetrovics, Z.; Szabo, A. Analysis of problematic smartphone use across different age groups within the 'components model of addiction'. *Int. J. Ment. Health Addict.* **2019**. [CrossRef]
43. Lopez-Fernandez, O.; Kuss, D.J.; Romo, L.; Morvan, Y.; Kern, L.; Graziani, P.; Rousseau, A.; Rumpf, H.-J.; Bischof, A.; Gässler, A. Self-reported dependence on mobile phones in young adults: A European cross-cultural empirical survey. *J. Behav. Addict.* **2017**, *6*, 168–177. [CrossRef]
44. Kuss, D.J. Mobile technology and social media: The "extensions of man" in the 21st century. *Hum. Dev.* **2017**, *60*, 141–143. [CrossRef]
45. Kates, A.W.; Wu, H.; Coryn, C.L.S. The effects of mobile phone use on academic performance: A meta-analysis. *Comput. Educ.* **2018**, *127*, 107–112. [CrossRef]
46. Reinecke, L.; Klimmt, C.; Meier, A.; Reich, S.; Hefner, D.; Knop-Huelss, K.; Rieger, D.; Vorderer, P. Permanently online and permanently connected: Development and validation of the Online Vigilance Scale. *PLoS ONE* **2018**, *13*, e0205384. [CrossRef] [PubMed]
47. Oulasvirta, A.; Rattenbury, T.; Ma, L.; Raita, E. Habits make smartphone use more pervasive. *Pers. Ubiquitous Comput.* **2012**, *16*, 105–114. [CrossRef]
48. Van Deursen, A.J.A.M.; Bolle, C.L.; Hegner, S.M.; Kommers, P.A.M. Modeling habitual and addictive smartphone behavior. *Comput. Hum. Behav.* **2015**, *45*, 411–420. [CrossRef]
49. Johannes, N.; Meier, A.; Reinecke, L.; Ehlert, S.; Setiawan, D.N.; Walasek, N.; Dienlin, T.; Buijzen, M.; Veling, H. The relationship between online vigilance and affective well-being in everyday life: Combining smartphone logging with experience sampling. *PsyArXiv* **2019**. Available online: https://osf.io/t3wc2 (accessed on 4 January 2020).
50. Billieux, J.; Van der Linden, M.; d'Acremont, M.; Ceschi, G.; Zermatten, A. Does impulsivity relate to perceived dependence on and actual use of the mobile phone? *Appl. Cogn. Psychol* **2007**, *21*, 527–537. [CrossRef]
51. Wegmann, E.; Müller, S.M.; Turel, O.; Brand, M. Interactions of impulsivity, general executive functions, and specific inhibitory control explain symptoms of social-networks-use disorder: An experimental study. *Sci. Rep.* **2020**, *10*, 3866. [CrossRef]
52. Wu, E.X.W.; Liaw, G.J.; Goh, R.Z.; Chia, T.T.Y.; Chee, A.M.J.; Obana, T.; Rosenberg, M.D.; Yeo, B.T.; Asplund, C.L. Overlapping attentional networks yield divergent behavioral predictions across tasks: Neuromarkers for diffuse and focused attention? *NeuroImage* **2020**, *20*, 116535. [CrossRef]
53. Gupta, N.; Irwin, J.D. In-class distractions: The role of Facebook and the primary learning task. *Comput. Hum. Behav.* **2016**, *55*, 1165–1178. [CrossRef]
54. Unsworth, N.; McMillan, B.D. Attentional disengagements in educational contexts: A diary investigation of everyday mind-wandering and distraction. *Cogn. Res. Princ. Implic.* **2017**, *2*, 32. [CrossRef]
55. Buglass, S.L.; Binder, J.F.; Betts, L.R.; Underwood, J.D.M. Motivators of online vulnerability: The impact of social network site use and FOMO. *Comput. Hum. Behav.* **2017**, *66*, 248–255. [CrossRef]
56. Cheever, N.A.; Rosen, L.D.; Carrier, L.M.; Chavez, A. Out of sight is not out of mind: The impact of restricting wireless mobile device use on anxiety levels among low, moderate and high users. *Comput. Hum. Behav.* **2014**, *37*, 290–297. [CrossRef]
57. Clayton, R.B.; Leshner, G.; Almond, A. The extended iSelf: The impact of iPhone separation on cognition, emotion, and physiology. *J. Comput. Mediat. Commun.* **2015**, *20*, 119–135. [CrossRef]

58. Eide, T.A.; Aarestad, S.H.; Andreassen, C.S.; Bilder, R.M.; Pallesen, S. Smartphone restriction and its effect on subjective withdrawal related scores. *Front. Psychol* **2018**, *9*. [CrossRef] [PubMed]
59. Elhai, J.D.; Levine, J.C.; Dvorak, R.D.; Hall, B.J. Fear of missing out, need for touch, anxiety and depression are related to problematic smartphone use. *Comput. Hum. Behav.* **2016**, *63*, 509–516. [CrossRef]
60. Elhai, J.D.; Yang, H.; Montag, C. Fear of missing out (FOMO): Overview, theoretical underpinnings, and literature review on relations with severity of negative affectivity and problematic technology use. *Braz. J. Psychiatr.* **2020**. [CrossRef]
61. Franchina, V.; Vanden Abeele, M.; van Rooij, A.; Lo Coco, G.; De Marez, L. Fear of missing out as a predictor of problematic social media use and phubbing behavior among Flemish adolescents. *Int. J. Environ. Res. Public Health* **2018**, *15*, 2319. [CrossRef]
62. Lai, C.; Altavilla, D.; Ronconi, A.; Aceto, P. Fear of missing out (FOMO) is associated with activation of the right middle temporal gyrus during inclusion social cue. *Comput. Hum. Behav.* **2016**, *61*, 516–521. [CrossRef]
63. O'Connell, C. How FOMO (Fear of Missing Out), the smartphone, and social media may be affecting university students in the Middle East. *N. Am. J. Psychol.* **2020**, *22*, 83–102.
64. Browne, B.; Aruguete, M.S.; McCutcheon, L.E.; Medina, A.M. Social and emotional correlates of the fear of missing out. *N. Am. J. Psychol* **2018**, *20*, 341–354.
65. Vahedi, Z.; Saiphoo, A. The association between smartphone use, stress, and anxiety: A meta-analytic review. *Stress Health* **2018**, *34*, 347–358. [CrossRef] [PubMed]
66. Zhitomirsky-Geffet, M.; Blau, M. Cross-generational analysis of predictive factors of addictive behavior in smartphone usage. *Comput. Hum. Behav.* **2016**, *64*, 682–693. [CrossRef]
67. Marino, C.; Gini, G.; Angelini, F.; Vieno, A.; Spada, M.M. Social norms and e-motions in problematic social media use among adolescents. *Addict. Behav. Rep.* **2020**, *11*, 100250. [CrossRef] [PubMed]
68. Balıkçı, K.; Aydın, O.; Sönmez, İ.; Kalo, B.; Ünal-Aydın, P. The relationship between dysfunctional metacognitive beliefs and problematic social networking sites use. *Scand. J. Psychol.* **2020**. [CrossRef]
69. Kneidinger-Müller, B. When the smartphone goes offline: A factorial survey of smartphone users' experiences of mobile unavailability. *Comput. Hum. Behav.* **2019**, *98*, 1–10. [CrossRef]
70. Rozgonjuk, D.; Elhai, J.D.; Täht, K.; Vassil, K.; Levine, J.C.; Asmundson, G.J.G. Non-social smartphone use mediates the relationship between intolerance of uncertainty and problematic smartphone use: Evidence from a repeated-measures study. *Comput. Hum. Behav.* **2019**, *96*, 56–62. [CrossRef]
71. Rasmussen, E.E.; Punyanunt-Carter, N.; LaFreniere, J.R.; Norman, M.S.; Kimball, T.G. The serially mediated relationship between emerging adults' social media use and mental well-being. *Comput. Hum. Behav.* **2020**, *10*, 206–213. [CrossRef]
72. Griffiths, M.D.; Lopez-Fernandez, O.; Throuvala, M.A.; Pontes, H.; Kuss, D.J. Excessive and Problematic Use of Social Media in Adolescence: A Brief Overview. Report Submitted to the UK Parliament Science and Technology Committee (Impact of Social Media and Screen-Use on Young People's Health Inquiry). 2018. Available online: http://data.parliament.uk/writtenevidence/committeeevidence.svc/evidencedocument/science-and-technology-committee/social-media-and-mental-health/written/81105.pdf (accessed on 30 May 2018).
73. Griffiths, M.D.; Kuss, D.J.; Billieux, J.; Pontes, H.M. The evolution of Internet addiction: A global perspective. *Addict. Behav.* **2016**, *53*, 193–195. [CrossRef]
74. Kuss, D.J.; Griffiths, M.D.; Binder, J.F. Internet addiction in students: Prevalence and risk factors. *Comput. Hum. Behav.* **2013**, *29*, 959–966. [CrossRef]
75. Kelly, Y.; Zilanawala, A.; Booker, C.; Sacker, A. Social media use and adolescent mental health: Findings from the UK Millennium cohort study. *EClinicalMedicine* **2019**, *6*, 59–68. [CrossRef]
76. Ellis, D.A. Are smartphones really that bad? Improving the psychological measurement of technology-related behaviors. *Comput. Hum. Behav.* **2019**, *97*, 60–66. [CrossRef]
77. Bentley, F.; Tollmar, K. The power of mobile notifications to increase wellbeing logging behavior. In Proceedings of the SIGCHI Conference on Human Factors in Computing Systems-CHI '13; ACM Press: Paris, France, 2013; p. 1095.
78. Johannes, N.; Veling, H.; Dora, J.; Meier, A.; Reinecke, L.; Buijzen, M. Mind-wandering and mindfulness as mediators of the relation between online vigilance and well-being. *Cyberpsychol. Behav. Soc. Netw.* **2018**. [CrossRef] [PubMed]

79. Kabat-Zinn, J. Mindfulness-based interventions in context: Past, present, and future. *Clin. Psychol. Sci. Pract.* **2003**, *10*, 144–156. [CrossRef]
80. Kasson, E.M.; Wilson, A.N. Preliminary evidence on the efficacy of mindfulness combined with traditional classroom management strategies. *Behav. Anal. Pract.* **2017**, *10*, 242–251. [CrossRef]
81. Begun, A.L.; Murray, M.M. Mindfulness practices in addictive behavior prevention, treatment, and recovery. In *The Routledge Handbook of Social Work and Addictive Behaviors*; Routledge: Melbourne, Australia, 2020.
82. Broderick, P.C.; Jennings, P.A. Mindfulness for adolescents: A promising approach to supporting emotion regulation and preventing risky behavior. *New Dir. Youth Dev.* **2012**, *2012*, 111–126. [CrossRef]
83. Hölzel, B.K.; Lazar, S.W.; Gard, T.; Schuman-Olivier, Z.; Vago, D.R.; Ott, U. How does mindfulness meditation work? Proposing mechanisms of action from a conceptual and neural perspective. *Perspect. Psychol. Sci.* **2011**, *6*, 537–559. [CrossRef] [PubMed]
84. Auer, M.M.; Griffiths, M.D. The use of personalized behavioral feedback for online gamblers: An empirical study. *Front. Psychol.* **2015**, *6*, 1406. [CrossRef]
85. Bennike, I.H.; Wieghorst, A.; Kirk, U. Online-based mindfulness training reduces behavioral markers of mind wandering. *J. Cogn. Enhanc.* **2017**, *1*, 172–181. [CrossRef]
86. Bonello, M.; Griffiths, M.D. Behavioural tracking, responsible gambling tools and online voluntary self-exclusion: Implications for problem gamblers. *Casino Gambl. Int.* **2019**, *38*, 41–45.
87. Cavanagh, K.; Strauss, C.; Cicconi, F.; Griffiths, N.; Wyper, A.; Jones, F. A randomised controlled trial of a brief online mindfulness-based intervention. *Behav. Res. Ther.* **2013**, *51*, 573–578. [CrossRef]
88. Gainsbury, S.M. Review of self-exclusion from gambling venues as an intervention for problem gambling. *J. Gambl. Stud.* **2014**, *30*, 229–251. [CrossRef] [PubMed]
89. Glück, T.M.; Maercker, A. A randomized controlled pilot study of a brief web-based mindfulness training. *BMC Psychiatr.* **2011**, *11*. [CrossRef] [PubMed]
90. Ophir, E.; Nass, C.; Wagner, A.D. Cognitive control in media multitaskers. *Proc. Natl. Acad. Sci. USA* **2009**, *106*, 15583–15587. [CrossRef] [PubMed]
91. Bakosh, L.S.; Tobias Mortlock, J.M.; Querstret, D.; Morison, L. Audio-guided mindfulness training in schools and its effect on academic attainment: Contributing to theory and practice. *Learn. Instr.* **2018**, *58*, 34–41. [CrossRef]
92. Kauer, S.D.; Reid, S.C.; Crooke, A.H.D.; Khor, A.; Hearps, S.J.C.; Jorm, A.F.; Sanci, L.; Patton, G. Self-monitoring using mobile phones in the early stages of adolescent depression: Randomized controlled trial. *J. Med. Internet Res.* **2012**, *14*, e67. [CrossRef]
93. Hill, C.L.M.; Updegraff, J.A. Mindfulness and its relationship to emotional regulation. *Emotion* **2012**, *12*, 81–90. [CrossRef]
94. Gibb, B.E.; McGeary, J.E.; Beevers, C.G. Attentional biases to emotional stimuli: Key components of the RDoC constructs of sustained threat and loss. *Am. J. Med. Genet. B Neuropsychiatr. Genet.* **2016**, *171*, 65–80. [CrossRef]
95. Zhang, M.; Heng, S.; Song, G.; Fung, D.S.; Smith, H.E. Co-designing a mobile gamified attention bias modification intervention for substance use disorders: Participatory research study. *JMIR MHealth UHealth* **2019**, *7*, e15871. [CrossRef]
96. Bakker, D.; Kazantzis, N.; Rickwood, D.; Rickard, N. Mental health smartphone apps: Review and evidence-based recommendations for future developments. *JMIR Ment. Health* **2016**, *3*, e7. [CrossRef]
97. Király, O.; Potenza, M.N.; Stein, D.J.; King, D.L.; Hodgins, D.C.; Saunders, J.B.; Griffiths, M.D.; Gjoneska, B.; Billieux, J.; Brand, M.; et al. Preventing problematic internet use during the COVID-19 pandemic: Consensus guidance. *Compr. Psychiatr.* **2020**, *100*, 152180. [CrossRef]
98. Gupta, T.; Mittal, V.A. Advances in clinical staging, early intervention, and the prevention of psychosis. *F1000Research* **2019**, *8*, 2027. [CrossRef] [PubMed]
99. Linardon, J.; Cuijpers, P.; Carlbring, P.; Messer, M.; Fuller-Tyszkiewicz, M. The efficacy of app-supported smartphone interventions for mental health problems: A meta-analysis of randomized controlled trials. *World Psychiatr.* **2019**, *18*, 325–336. [CrossRef] [PubMed]
100. Bakker, D.; Kazantzis, N.; Rickwood, D.; Rickard, N. A randomized controlled trial of three smartphone apps for enhancing public mental health. *Behav. Res. Ther.* **2018**, *109*, 75–83. [CrossRef] [PubMed]
101. Bakker, D.; Rickard, N. Engagement in mobile phone app for self-monitoring of emotional wellbeing predicts changes in mental health: *MoodPrism*. *J. Affect. Disord.* **2018**, *227*, 432–442. [CrossRef] [PubMed]

102. Howells, A.; Ivtzan, I.; Eiroa-Orosa, F.J. Putting the 'app' in happiness: A randomised controlled trial of a smartphone-based mindfulness intervention to enhance wellbeing. *J. Happiness Stud.* **2016**, *17*, 163–185. [CrossRef]
103. Kummervold, P.; Chronaki, C.; Lausen, B.; Prokosch, H.-U.; Rasmussen, J.; Santana, S.; Staniszewski, A.; Wangberg, S. eHealth tends in Europe 2005-2007: A population-based survey. *J. Med. Internet Res.* **2008**, *10*, e42. [CrossRef]
104. Howarth, A.; Smith, J.G.; Perkins-Porras, L.; Ussher, M. Effects of brief mindfulness-based interventions on health-related outcomes: A systematic review. *Mindfulness* **2019**, *10*, 1957–1968. [CrossRef]
105. Melia, R.; Francis, K.; Hickey, E.; Bogue, J.; Duggan, J.; O'Sullivan, M.; Young, K. Mobile health technology interventions for suicide prevention: Systematic review. *JMIR MHealth UHealth* **2020**, *8*, e12516. [CrossRef]
106. Neary, M.; Schueller, S.M. State of the field of mental health apps. *Cogn. Behav. Pract.* **2018**, *25*, 531–537. [CrossRef]
107. Plaza, I.; Demarzo, M.M.P.; Herrera-Mercadal, P.; García-Campayo, J. Mindfulness-based mobile applications: Literature review and analysis of current features. *JMIR Mhealth Uhealth* **2013**, *1*, e24. [CrossRef]
108. Torous, J.; Powell, A.C. Current research and trends in the use of smartphone applications for mood disorders. *Internet Interv.* **2015**, *2*, 169–173. [CrossRef]
109. Sanderson, C.; Kouzoupi, N.; Hall, C.L. Technology matters: The human touch in a digital age—a blended approach in mental healthcare delivery with children and young people. *Child. Adolesc. Ment. Health* **2020**. [CrossRef] [PubMed]
110. Gainsbury, S.; Blaszczynski, A. A systematic review of Internet-based therapy for the treatment of addictions. *Clin. Psychol. Rev.* **2011**, *31*, 490–498. [CrossRef]
111. Borkovec, T.D.; Roemer, L. Perceived functions of worry among generalized anxiety disorder subjects: Distraction from more emotionally distressing topics? *J. Behav. Ther. Exp. Psychiatr.* **1995**, *26*, 25–30. [CrossRef]
112. Cho, S.; Lee, E. Distraction by smartphone use during clinical practice and opinions about smartphone restriction policies: A cross-sectional descriptive study of nursing students. *Nurse Educ. Today* **2016**, *40*, 128–133. [CrossRef]
113. Dolcos, F.; Iordan, A.D.; Kragel, J.; Stokes, J.; Campbell, R.; McCarthy, G.; Cabeza, R. Neural correlates of opposing effects of emotional distraction on working memory and episodic memory: An event-related fMRI investigation. *Front. Psychol.* **2013**, *4*, 180–195. [CrossRef] [PubMed]
114. Aagaard, J. Media multitasking, attention, and distraction: A critical discussion. *Phenomenol. Cogn. Sci.* **2015**, *14*, 885–896. [CrossRef]
115. Andreassen, C.S.; Billieux, J.; Griffiths, M.D.; Kuss, D.J.; Demetrovics, Z.; Mazzoni, E.; Pallesen, S. The relationship between addictive use of social media and video games and symptoms of psychiatric disorders: A large-scale cross-sectional study. *Psychol. Addict. Behav.* **2016**, *30*, 252–262. [CrossRef]
116. Calado, F.; Alexandre, J.; Griffiths, M.D. Gambling among adolescents and emerging adults: A cross-cultural study between Portuguese and English youth. *Int. J. Ment. Health Addict.* **2018**. [CrossRef]
117. Canale, N.; Vieno, A.; Griffiths, M.D. The extent and distribution of gambling-related harms and the prevention paradox in a British population survey. *J. Behav. Addict.* **2016**, *5*, 204–212. [CrossRef]
118. Patton, G.C.; Sawyer, S.M.; Santelli, J.S.; Ross, D.A.; Afifi, R.; Allen, N.B.; Arora, M.; Azzopardi, P.; Baldwin, W.; Bonell, C.; et al. Our future: A Lancet commission on adolescent health and wellbeing. *Lancet* **2016**, *387*, 2423–2478. [CrossRef]
119. Sohn, S.; Rees, P.; Wildridge, B.; Kalk, N.J.; Carter, B. Prevalence of problematic smartphone usage and associated mental health outcomes amongst children and young people: A systematic review, meta-analysis and GRADE of the evidence. *BMC Psychiatr.* **2019**, *19*, 356.
120. Spijkerman, M.P.J.; Pots, W.T.M.; Bohlmeijer, E.T. Effectiveness of online mindfulness-based interventions in improving mental health: A review and meta-analysis of randomised controlled trials. *Clin. Psychol. Rev.* **2016**, *45*, 102–114. [CrossRef] [PubMed]
121. Cartwright, N. Are RCTs the gold standard? *BioSocieties* **2007**, *2*, 11–20. [CrossRef]
122. Sullivan, G.M. Getting off the "gold standard": Randomized controlled trials and education research. *J. Grad. Med. Educ.* **2011**, *3*, 285–289. [CrossRef] [PubMed]
123. Deaton, A.; Cartwright, N. Understanding and misunderstanding randomized controlled trials. *Soc. Sci. Med.* **2018**, *210*, 2–21. [CrossRef]

124. Ruggeri, M.; Lasalvia, A.; Bonetto, C. A new generation of pragmatic trials of psychosocial interventions is needed. *Epidemiol. Psychiatr. Sci.* **2013**, *22*, 111–117. [CrossRef] [PubMed]
125. Campbell, M.K.; Elbourne, D.R.; Altman, D.G. CONSORT statement: Extension to cluster randomised trials. *BMJ* **2004**, *328*, 702–708. [CrossRef]
126. Schell, B.; Gillen, G. *Willard and Spackman's Cccupational Therapy*, 13th ed.; Lippincott Williams & Wilkins: Philadephia, PA, USA, 2018; 2828p.
127. Oussedik, E.; Foy, C.G.; Masicampo, E.J.; Kammrath, L.K.; Anderson, R.E.; Feldman, S.R. Accountability: A missing construct in models of adherence behavior and in clinical practice. *Patient Prefer. Adherence* **2017**, *11*, 1285–1294. [CrossRef]
128. Lanier, J.; Noyes, E.; Biederman, J. Mind wandering (internal distractibility) in ADHD: A literature review. *J. Atten. Disord.* **2019**. [CrossRef]
129. Franklin, M.S.; Mrazek, M.D.; Anderson, C.L.; Johnston, C.; Smallwood, J.; Kingstone, A.; Schooler, J. Tracking distraction: The relationship between mind-wandering, meta-awareness, and ADHD symptomatology. *J. Atten. Disord.* **2017**, *21*, 475–486. [CrossRef] [PubMed]
130. Bandura, A. Social cognitive theory of self-regulation. *Organ. Behav. Hum. Decis Process.* **1991**, *50*, 248–287. [CrossRef]
131. Tseng, V.W.-S.; Lee, M.L.; Denoue, L.; Avrahami, D. Overcoming distractions during transitions from break to work using a conversational website-blocking system. In Proceedings of the CHI 2019—2019 CHI Conference on Human Factors in Computing Systems, Glasgow, UK, 4–9 May 2019. [CrossRef]
132. Evans, C.; King, D.L.; Delfabbro, P.H. Effect of brief gaming abstinence on withdrawal in adolescent at-risk daily gamers: A randomized controlled study. *Comput. Hum. Behav.* **2018**, *88*, 70–77. [CrossRef]
133. King, D.L.; Kaptsis, D.; Delfabbro, P.H.; Gradisar, M. Effectiveness of brief abstinence for modifying problematic Internet gaming cognitions and behaviors. *J. Clin. Psychol.* **2017**, *73*, 1573–1585. [CrossRef]
134. Turel, O. Short abstinence from online social networking sites reduces perceived stress, especially in excessive users. *Psychiatr. Res.* **2018**, *270*, 947–953. [CrossRef]
135. Caldeira, C.; Chen, Y.; Chan, L.; Pham, V.; Chen, Y.; Zheng, K. Mobile apps for mood tracking: An analysis of features and user reviews. *AMIA Annu. Symp. Proc.* **2017**, *2017*, 495–504.
136. Bakker, D.; Kazantzis, N.; Rickwood, D.; Rickard, N. Development and pilot evaluation of smartphone-delivered cognitive behavior therapy strategies for mood- and anxiety-related problems: MoodMission. *Cogn. Behav. Pract.* **2018**, *25*, 496–514. [CrossRef]
137. Haug, S.; Meyer, C.; Dymalski, A.; Lippke, S.; John, U. Efficacy of a text messaging (SMS) based smoking cessation intervention for adolescents and young adults: Study protocol of a cluster randomised controlled trial. *BMC Public Health.* **2012**, *12*, 51. [CrossRef]
138. Throuvala, M.A.; Pontes, H.; Griffiths, M.D.; Rennoldson, M.; Kuss, D.J. Development and validation of the Smartphone Distraction Scale (SDS): A cognitive and emotion regulation construct. *Manuscr. Submitt. Publ.* **2020**.
139. Brown, K.W.; Ryan, R.M. The benefits of being present: Mindfulness and its role in psychological well-being. *J. Pers. Soc. Psychol.* **2003**, *84*, 822–848. [CrossRef]
140. Black, D.S.; Sussman, S.; Johnson, C.A.; Milam, J. Psychometric assessment of the Mindful Attention Awareness Scale (MAAS) among Chinese adolescents. *Assessment* **2012**, *19*, 42–52. [CrossRef]
141. Cohen, S.; Kamarck, T.; Mermelstein, R. A global measure of perceived stress. *J. Health Soc. Behav.* **1983**, *24*, 385–396. [CrossRef] [PubMed]
142. Lee, E.-H. Review of the psychometric evidence of the Perceived Stress Scale. *Asian Nurs. Res.* **2012**, *6*, 121–127. [CrossRef] [PubMed]
143. Spitzer, R.L.; Kroenke, K.; Williams, J.B.W.; Löwe, B. A brief measure for assessing generalized anxiety disorder: The GAD-7. *Arch. Intern. Med.* **2006**, *166*, 1092–1097. [CrossRef] [PubMed]
144. Donker, T.; van Straten, A.; Marks, I.; Cuijpers, P. Quick and easy self-rating of Generalized Anxiety Disorder: Validity of the Dutch web-based GAD-7, GAD-2 and GAD-SI. *Psychiatr. Res.* **2011**, *188*, 58–64. [CrossRef]
145. Sousa, T.V.; Viveiros, V.; Chai, M.V.; Vicente, F.L.; Jesus, G.; Carnot, M.J.; Gordo, A.C.; Ferreira, P. Reliability and validity of the Portuguese version of the Generalized Anxiety Disorder (GAD-7) scale. *Health Qual. Life Outcomes* **2015**, *13*, 50. [CrossRef]

146. Gardner, B.; Abraham, C.; Lally, P.; de Bruijn, G.-J. Towards parsimony in habit measurement: Testing the convergent and predictive validity of an automaticity subscale of the Self-Report Habit Index. *Int. J. Behav. Nutr. Phys. Act.* **2012**, *9*, 102. [CrossRef]
147. Marchant, G.; Chevance, G.; Boiché, J. Intention and automaticity toward physical and sedentary screen-based leisure activities in adolescents: A profile perspective. *J. Sport Health Sci.* **2018**, *7*, 481–488. [CrossRef]
148. Schwarzer, R.; Jerusalem, M. Generalized Self-Efficacy scale. In *Measures in Health Psychology: A User's Portfolio. Causal and Control Beliefs*; NFER-NELSON: Windsor, UK, 1995; pp. 35–37.
149. Schwarzer, R.; Mueller, J.; Greenglass, E. Assessment of perceived general self-efficacy on the Internet: Data collection in cyberspace. *Anxiety Stress Coping Int. J.* **1999**, *12*, 145–161. [CrossRef]
150. De las Cuevas, C.; Peñate, W. Validation of the General Self-Efficacy Scale in psychiatric outpatient care. *Psicothema* **2015**, *27*, 410–415.
151. Morean, M.E.; DeMartini, K.S.; Leeman, R.F.; Pearlson, G.D.; Anticevic, A.; Krishnan-Sarin, S.; Krystal, J.H.; O'Malley, S. Psychometrically improved, abbreviated versions of three classic measures of impulsivity and self-control. *Psychol. Assess.* **2014**, *26*, 1003–1020. [CrossRef] [PubMed]
152. Mathias, C.W.; Stanford, M.S.; Liang, Y.; Goros, M.; Charles, N.E.; Sheftall, A.H.; Mullen, J.; Hill-Kapturczak, N.; Acheson, A.; Olvera, R.L.; et al. A test of the psychometric characteristics of the BIS-Brief among three groups of youth. *Psychol. Assess.* **2018**, *30*, 847–856. [CrossRef] [PubMed]
153. Steinberg, L.; Sharp, C.; Stanford, M.S.; Tharp, A.T. New tricks for an old measure: The development of the Barratt Impulsiveness Scale–Brief (BIS-Brief). *Psychol. Assess.* **2013**, *25*, 216–226. [CrossRef]
154. Davies, J.J.; Hemingway, T.J. Guitar hero or zero?: Fantasy, self-esteem, and deficient self-regulation in rhythm-based music video games. *J. Media Psychol.* **2014**, *26*, 189–201. [CrossRef]
155. LaRose, R.; Lin, C.A.; Eastin, M.S. Unregulated Internet usage: Addiction, habit, or deficient self-regulation? *Media Psychol.* **2009**, *5*, 225–253. [CrossRef]
156. Lin, C.-Y.; Broström, A.; Nilsen, P.; Griffiths, M.D.; Pakpour, A.H. Psychometric validation of the Persian Bergen Social Media Addiction Scale using classic test theory and Rasch models. *J. Behav. Addict.* **2017**, *6*, 620–629. [CrossRef]
157. Monacis, L.; de Palo, V.; Griffiths, M.D.; Sinatra, M. Social networking addiction, attachment style, and validation of the Italian version of the Bergen Social Media Addiction Scale. *J. Behav. Addict.* **2017**, *6*, 178–186. [CrossRef]
158. Pontes, H.M.; Andreassen, C.S.; Griffiths, M.D. Portuguese validation of the Bergen Facebook Addiction Scale: An empirical study. *Int. J. Ment. Health Addict.* **2016**, *14*, 1062–1073. [CrossRef]
159. Griffiths, M.D. A 'components' model of addiction within a biopsychosocial framework. *J. Subst Use* **2005**, *10*, 191–197. [CrossRef]
160. Bányai, F.; Zsila, Á.; Király, O.; Maraz, A.; Elekes, Z.; Griffiths, M.D.; Andreassen, C.S.; Demetrovics, Z. Problematic social media use: Results from a large-scale nationally representative adolescent sample. *PLoS ONE* **2017**, *12*, e0169839. [CrossRef]
161. Yam, C.-W.; Pakpour, A.H.; Griffiths, M.D.; Yau, W.-Y.; Lo, C.-L.M.; Ng, J.M.T.; Lin, C.-Y.; Leung, H. Psychometric testing of three Chinese online-related addictive behavior instruments among Hong Kong University students. *Psychiatr. Q.* **2019**, *90*, 117–128. [CrossRef]
162. Przybylski, A.K.; Murayama, K.; DeHaan, C.R.; Gladwell, V. Motivational, emotional, and behavioral correlates of fear of missing out. *Comput. Hum. Behav.* **2013**, *29*, 1841–1848. [CrossRef]
163. Can, G.; Satici, S.A. Adaptation of fear of missing out scale (FoMOs): Turkish version validity and reliability study. *Psicol. Reflex E Crítica* **2019**, *32*, 3. [CrossRef] [PubMed]
164. Perrone, M.A. #FoMO: Establishing Validity of the Fear of Missing Out Scale with an Adolescent Population 2016. Available online: https://aura.alfred.edu/handle/10829/7399 (accessed on 30 May 2020).
165. Yildirim, C.; Correia, A.-P. Exploring the dimensions of nomophobia: Development and validation of a self-reported questionnaire. *Comput. Hum. Behav.* **2015**, *49*, 130–137. [CrossRef]
166. Lee, S.; Kim, M.; Mendoza, J.S.; McDonough, I.M. Addicted to cellphones: Exploring the psychometric properties between the nomophobia questionnaire and obsessiveness in college students. *Heliyon* **2018**, *4*, e00895. [CrossRef] [PubMed]
167. Tams, S.; Legoux, R.; Leger, P.-M. Smartphone withdrawal creates stress: A moderated mediation model of nomophobia, social threat, and phone withdrawal context. *Comput. Hum. Behav.* **2018**, *81*, 1–9. [CrossRef]

168. Rothwell, J.C.; Julious, S.A.; Cooper, C.L. A study of target effect sizes in randomised controlled trials. *Trials* **2018**, *19*, 544. [CrossRef]
169. Bhide, A.; Shah, P.S.; Acharya, G. A simplified guide to randomized controlled trials. *Acta Obstet. Gynecol. Scand.* **2018**, *97*, 380–387. [CrossRef]
170. Grimm, L.G.; Yarnold, P.R. *Reading and Understanding Multivariate Statistics*; Reading and Understanding Multivariate Statistics; American Psychological Association: Washington, DC, USA, 1995; 373p.
171. Gupta, S.K. Intention-to-treat concept: A review. *Perspect. Clin. Res.* **2011**, *2*, 109–112. [CrossRef]
172. Montori, V.M.; Guyatt, G.H. Intention-to-treat principle. *CMAJ Can. Med. Assoc. J.* **2001**, *165*, 1339–1341.
173. Altman, D.G.; Doré, C.J. Randomisation and baseline comparisons in clinical trials. *Lancet* **1990**, *335*, 149–153. [CrossRef]
174. Anderson, S.R.; Auquier, A.; Hauck, W.W.; Oakes, D.; Vandaele, W.; Weisberg, H.I. *Statistical Methods for Comparative Studies: Techniques for Bias Reduction*; John Wiley & Sons: Hoboken, NJ, USA, 2009; 310p.
175. Olejnik, S.F.; Algina, J. Parametric ANCOVA and the rank transform ANCOVA when the data are conditionally non-normal and heteroscedastic. *J. Educ. Stat.* **1984**, *9*, 129–149. [CrossRef]
176. Keppel, G.; Wickens, T.D. *Design and Analysis: A Researcher's Handbook*, 4th ed.; Prentice Hall: Upper Saddle River, NJ, USA, 2004.
177. Pierce, C.A.; Block, R.A.; Aguinis, H. Cautionary note on reporting eta-squared values from multifactor ANOVA designs. *Educ. Psychol. Meas* **2004**, *64*, 916–924. [CrossRef]
178. Olejnik, S.; Algina, J. Measures of effect size for comparative studies: Applications, interpretations, and limitations. *Contemp. Educ. Psychol* **2000**, *25*, 241–286. [CrossRef]
179. Cohen, J. A power primer. *Psychol. Bull.* **1992**, *112*, 155–159. [CrossRef] [PubMed]
180. Hayes, A.F.; Montoya, A.K.; Rockwood, N.J. The analysis of mechanisms and their contingencies: PROCESS versus structural equation modeling. *Australas Mark. J. AMJ* **2017**, *25*, 76–81. [CrossRef]
181. Hayes, A.F.; Rockwood, N.J. Regression-based statistical mediation and moderation analysis in clinical research: Observations, recommendations, and implementation. *Behav. Res. Ther.* **2017**, *98*, 39–57. [CrossRef]
182. Hayes, A.F.; Preacher, K.J. Statistical mediation analysis with a multicategorical independent variable. *Br. J. Math. Stat. Psychol.* **2014**, *67*, 451–470. [CrossRef]
183. Hayes, A.F. PROCESS: A Versatile Computational Tool for Observed Variable Mediation, Moderation, and Conditional Process Modelling; White Paper; 2012. Available online: http://www.afhayes.com/public/process2012.pdf (accessed on 12 June 2018).
184. Gámez-Guadix, M.; Calvete, E. Assessing the relationship between mindful awareness and problematic internet use among adolescents. *Mindfulness* **2016**, *7*, 1281–1288. [CrossRef]
185. Mrazek, M.D.; Franklin, M.S.; Phillips, D.T.; Baird, B.; Schooler, J.W. Mindfulness training improves working memory capacity and GRE performance while reducing mind wandering. *Psychol. Sci* **2013**, *24*, 776–781. [CrossRef]
186. Calvete, E.; Gámez-Guadix, M.; Cortazar, N. Mindfulness facets and problematic Internet use: A six-month longitudinal study. *Addict. Behav* **2017**, *72*, 57–63. [CrossRef] [PubMed]
187. Biddle, S.J.; Edwardson, C.L.; Wilmot, E.G.; Yates, T.; Gorely, T.; Bodicoat, D.H.; Ashra, N.; Khunti, K.; Nimmo, M.A.; Davies, M. A randomised controlled trial to reduce sedentary time in young adults at risk of Type 2 Diabetes Mellitus: Project STAND (sedentary time and diabetes). *PLoS ONE* **2015**, *10*, e0143398. [CrossRef] [PubMed]
188. Kazdin, A.E. Reactive self-monitoring: The effects of response desirability, goal setting, and feedback. *J. Consult. Clin. Psychol* **1974**, *42*, 704–716. [CrossRef] [PubMed]
189. Prochaska, J.O.; Velicer, W.F. The transtheoretical model of health behavior change. *Am. J. Health Promot. AJHP* **1997**, *12*, 38–48. [CrossRef] [PubMed]
190. Prochaska, J.O.; DiClemente, C.C.; Norcross, J.C. In search of how people change. Applications to addictive behaviors. *Am. Psychol.* **1992**, *47*, 1102–1114. [CrossRef] [PubMed]
191. Squires, L.R.; Hollett, K.B.; Hesson, J.; Harris, N. Psychological distress, emotion dysregulation, and coping behaviour: A theoretical perspective of problematic smartphone use. Advance online publication. *Int. J. Ment. Health Addict.* **2020**. [CrossRef]
192. Ostendorf, S.; Wegmann, E.; Brand, M. Problematic social networks use in German children and adolescents—The interaction of need to belong, online self-regulative competences, and age. *Int. J. Environ. Res. Public Health* **2020**, *17*, 2518. [CrossRef]

193. Liese, B.S.; Kim, H.S.; Hodgins, D.C. Insecure attachment and addiction: Testing the mediating role of emotion dysregulation in four potentially addictive behaviors. *Addict. Behav.* **2020**, 106432. [CrossRef]
194. Kuehner, C.; Huffziger, S.; Liebsch, K. Rumination, distraction and mindful self-focus: Effects on mood, dysfunctional attitudes and cortisol stress response. *Psychol. Med.* **2009**, *39*, 219–228. [CrossRef]
195. Colder Carras, M.; Kalbarczyk, A.; Wells, K.; Banks, J.; Kowert, R.; Gillespie, C.; Latkin, C. Connection, meaning, and distraction: A qualitative study of video game play and mental health recovery in veterans treated for mental and/or behavioral health problems. *Soc. Sci. Med.* **2018**, *216*, 124–132. [CrossRef]
196. Huffziger, S.; Reinhard, I.; Kuehner, C. A longitudinal study of rumination and distraction in formerly depressed inpatients and community controls. *J. Abnorm. Psychol.* **2009**, *118*, 746–756. [CrossRef] [PubMed]
197. Denkova, E.; Wong, G.; Dolcos, S.; Sung, K.; Wang, L.; Coupland, N.; Dolcos, F. The impact of anxiety-inducing distraction on cognitive performance: A combined brain imaging and personality investigation. *PLoS ONE* **2010**, *5*, e14150. [CrossRef]
198. Armstrong, T.; Zald, D.H.; Olatunji, B.O. Attentional control in OCD and GAD: Specificity and associations with core cognitive symptoms. *Behav. Res. Ther.* **2011**, *49*, 756–762. [CrossRef] [PubMed]
199. Mrazek, M.D.; Smallwood, J.; Schooler, J.W. Mindfulness and mind-wandering: Finding convergence through opposing constructs. *Emotion* **2012**, *12*, 442–448. [CrossRef]
200. Johannes, N.; Veling, H.; Verwijmeren, T.; Buijzen, M. Hard to resist? The effect of smartphone visibility and notifications on response inhibition. *J. Media Psychol. Theor Methods Appl.* **2019**, *31*, 214–225. [CrossRef]
201. Griffiths, M.D.; Szabo, A. Is excessive online usage a function of medium or activity? *Empirical Pilot Study. J. Behav. Addict.* **2014**, *3*, 74–77. [CrossRef]
202. Panagiotidi, M.; Overton, P. Attention deficit hyperactivity symptoms predict problematic mobile phone use. *Curr. Psychol.* **2020**. [CrossRef]
203. Schmitgen, M.M.; Horvath, J.; Mundinger, C.; Wolf, N.D.; Sambataro, F.; Hirjak, D.; Kubera, K.; Koenig, J.; Christian Wolf, R. Neural correlates of cue reactivity in individuals with smartphone addiction. *Addict. Behav.* **2020**, *108*, 106422. [CrossRef]
204. Kempf, C.A.; Ehrhard, K.A.; Stoner, S.C. Evaluation of obsessive-compulsive symptoms in relation to smartphone use. *Ment. Health Clin.* **2020**, *10*, 44–48. [CrossRef]
205. Wegmann, E.; Brand, M. Cognitive correlates in gaming disorder and social networks use disorder: A comparison. *Curr. Addict. Rep.* **2020**. [CrossRef]
206. Pancani, L.; Preti, E.; Riva, P. The psychology of smartphone: The development of the Smartphone Impact Scale (SIS). *Assessment* **2019**. [CrossRef]
207. Johannes, N.; Dora, J.; Rusz, D. Social smartphone apps do not capture attention despite their perceived high reward value. *Collabra Psychol.* **2019**, *5*, 14. [CrossRef]
208. Levine, L.E.; Waite, B.M.; Bowman, L.L. Use of instant messaging predicts self-report but not performance measures of inattention, impulsiveness, and distractibility. *Cyberpsychology Behav. Soc. Netw* **2013**, *16*, 898–903. [CrossRef] [PubMed]
209. Boer, M.; Stevens, G.; Finkenauer, C.; Eijnden, R. Attention deficit hyperactivity disorder-symptoms, social media use intensity, and social media use problems in adolescents: Investigating directionality. *Child. Dev.* **2019**. [CrossRef] [PubMed]
210. Kushlev, K.; Proulx, J.; Dunn, E.W. 'Silence Your Phones': Smartphone notifications increase inattention and hyperactivity symptoms. In Proceedings of the 2016 CHI Conference on Human Factors in Computing Systems-CHI '16; ACM Press: Santa Clara, CA, USA, 2016; pp. 1011–1020. Available online: http://dl.acm.org/citation.cfm?doi:d=2858036.2858359 (accessed on 17 July 2019).
211. Rozgonjuk, D.; Elhai, J.D.; Ryan, T.; Scott, G.G. Fear of missing out is associated with disrupted activities from receiving smartphone notifications and surface learning in college students. *Comput. Educ.* **2019**, *140*, 103590. [CrossRef]
212. Richards, D.; Farrand, P.; Christensen, H.; Griffiths, K.; Kavanagh, D.; Klein, B.; Lau, M.A.; Proudfoot, J. *Oxford Guide to Low Intensity CBT Interventions*, 1st ed.; Bennett-Levy, J., Ed.; Oxford University Press: Oxford, UK, 2010; 628p.
213. Althubaiti, A. Information bias in health research: Definition, pitfalls, and adjustment methods. *J. Multidiscip. Healthc.* **2016**, *9*, 211–217. [CrossRef]
214. Melville, K.M.; Casey, L.; Kavanagh, D.J. Dropout from Internet-based treatment for psychological disorders. *Br. J. Clin. Psychol.* **2009**, *49*, 455–471. [CrossRef]

215. Dumville, J.C.; Torgerson, D.J.; Hewitt, C.E. Reporting attrition in randomised controlled trials. *BMJ* **2006**, *332*, 969–971. [CrossRef]
216. Furukawa, T.A.; Noma, H.; Caldwell, D.M.; Honyashiki, M.; Shinohara, K.; Imai, H.; Chen, P.; Hunot, V.; Churchill, R. Waiting list may be a nocebo condition in psychotherapy trials: A contribution from network meta-analysis. *Acta Psychiatr. Scand.* **2014**, *130*, 181–192. [CrossRef]
217. Christensen, H.; Griffiths, K.M.; Farrer, L. Adherence in Internet interventions for anxiety and depression: Systematic review. *J. Med. Internet Res.* **2009**, *11*, e13. [CrossRef]
218. Andrews, S.; Ellis, D.A.; Shaw, H.; Piwek, L. Beyond self-report: Tools to compare estimated and real-world smartphone use. Pietschnig, J.; editor. *PLoS ONE* **2015**, *10*, e0139004. [CrossRef]
219. Lin, Y.-H.; Lin, Y.-C.; Lee, Y.-H.; Lin, P.-H.; Lin, S.-H.; Chang, L.-R.; Tseng, H.-W.; Yen, L.-Y.; Yang, C.C.H.; Kuo, T.B.J. Time distortion associated with smartphone addiction: Identifying smartphone addiction via a mobile application (App). *J. Psychiatr. Res.* **2015**, *65*, 139–145. [CrossRef] [PubMed]
220. Bauman, A.E.; Petersen, C.B.; Blond, K.; Rangul, V.; Hardy, L.L. The descriptive epidemiology of sedentary behaviour. In *Sedentary Behaviour Epidemiology*; Springer Series on Epidemiology and Public Health; Leitzmann, M.F., Jochem, C., Schmid, D., Eds.; Springer International: New York, NY, USA, 2018; pp. 73–106.
221. Davis, K.; Charmaraman, L.; Weinstein, E. Introduction to Special Issue: Adolescent and Emerging Adult Development in an Age of Social Media. *J. Adolesc. Res.* **2019**. [CrossRef]
222. Baumeister, H.; Montag, C. Digital phenotyping and mobile sensing: Rapidly evolving interdisciplinary research endeavor. In *Digital Phenotyping and Mobile Sensing: New Developments in Psychoinformatics*; Springer: Cham, Switzerland, 2019.
223. Garrison, K.A.; Potenza, M.N. Neuroimaging and biomarkers in addiction treatment. *Curr. Psychiatr. Rep.* **2014**, *16*, 513. [CrossRef]
224. Suckling, J.; Nestor, L.J. The neurobiology of addiction: The perspective from magnetic resonance imaging present and future. *Addict. Abingdon Engl.* **2017**, *112*, 360–369. [CrossRef] [PubMed]
225. Andone, I.; Błaszkiewicz, K.; Eibes, M.; Trendafilov, B.; Montag, C.; Markowetz, A. How age and gender affect smartphone usage. In *Proceedings of the 2016 ACM International JoInt. Conference on Pervasive and Ubiquitous Computing*; ACM: Heidelberg, Germany, 2016; pp. 9–12. [CrossRef]
226. Feng, Q.; Zheng, Y.; Zhang, X.; Song, Y.; Luo, Y.-J.; Li, Y.; Talhelm, T. Gender differences in visual reflexive attention shifting: Evidence from an ERP study. *Brain Res.* **2011**, *1401*, 59–65. [CrossRef]
227. Wu, J.-Y.; Cheng, T. Who is better adapted in learning online within the personal learning environment? Relating gender differences in cognitive attention networks to digital distraction. *Comput. Educ.* **2018**, *128*, 312–329. [CrossRef]
228. Albright, K.; Gechter, K.; Kempe, A. Importance of mixed methods in pragmatic trials and dissemination and implementation research. *Acad. Pediatr.* **2013**, *13*, 400–407. [CrossRef]
229. Alegría, M.; NeMoyer, A.; Falgas, I.; Wang, Y.; Alvarez, K. Social determinants of mental health: Where we are and where we need to go. *Curr. Psychiatr. Rep.* **2018**, *20*, 95. [CrossRef]
230. Macintyre, A.; Ferris, D.; Gonçalves, B.; Quinn, N. What has economics got to do with it? The impact of socioeconomic factors on mental health and the case for collective action. *Palgrave Commun.* **2018**, *4*, 10. [CrossRef]
231. Posner, M.I. Orienting of attention. *Q. J. Exp. Psychol.* **1980**, *32*, 3–25. [CrossRef] [PubMed]
232. Risko, E.F. Examining the implications of Internet usage for memory and cognition: Prospects and promise. *J. Appl. Res. Mem. Cogn* **2019**, *8*, 36–39. [CrossRef]
233. Russell, D.M. Commentary on the paper: "The digital expansion of the mind: Implications of Internet usage for memory and cognition". *J. Appl. Res. Mem. Cogn.* **2019**, *8*, 33–35. [CrossRef]
234. Garland, E.L.; Farb, N.A.; Goldin, P.R.; Fredrickson, B.L. Mindfulness broadens awareness and builds eudaimonic meaning: A process model of mindful positive emotion regulation. *Psychol. Inq.* **2015**, *26*, 293–314. [CrossRef] [PubMed]
235. Wegmann, E.; Oberst, U.; Stodt, B.; Brand, M. Online-specific fear of missing out and Internet-use expectancies contribute to symptoms of Internet-communication disorder. *Addict. Behav. Rep.* **2017**, *5*, 33–42. [CrossRef] [PubMed]
236. Boehm, J.K.; Vie, L.L.; Kubzansky, L.D. The promise of well-being interventions for improving health risk behaviors. *Curr. Cardiovasc. Risk Rep.* **2012**, *6*, 511–519. [CrossRef]

237. Trudel-Fitzgerald, C.; Millstein, R.A.; von Hippel, C.; Howe, C.J.; Tomasso, L.P.; Wagner, G.R.; VanderWeele, T. Psychological well-being as part of the public health debate? Insight into dimensions, interventions, and policy. *BMC Public Health* **2019**, *19*, 1712. [CrossRef]
238. Stellefson, M.; Hanik, B.; Chaney, B.; Chaney, D.; Tennant, B.; Chavarria, E.A. eHealth literacy among college students: A systematicreview with implications for eHealth education. *J. Med. Internet Res.* **2011**, *13*, e102. [CrossRef]
239. Van der Heide, I.; Snoeijs, S.; Boerma, W.G.W.; Schellevis, F.G.; Rijken, M.P. How to Strengthen Patient-Centredness in Caring for People with Multimorbidity in Europe? Policy Brief 22 ICARE4EU Consortium. 2017; p. 32. Available online: https://www.ncbi.nlm.nih.gov/books/NBK464537/pdf/Bookshelf_NBK464537.pdf (accessed on 26 April 2020).
240. Chen, L.; Nath, R.; Tang, Z. Understanding the determinants of digital distraction: An automatic thinking behavior perspective. *Comput. Hum. Behav.* **2020**, *104*, 106195. [CrossRef]

International Journal of
Environmental Research and Public Health

MDPI

Article

Autonomy Need Dissatisfaction in Daily Life and Problematic Mobile Phone Use: The Mediating Roles of Boredom Proneness and Mobile Phone Gaming

Wei Hong [1], Ru-De Liu [1,*], Yi Ding [2], Rui Zhen [3], Ronghuan Jiang [1] and Xinchen Fu [1]

1 Beijing Key Laboratory of Applied Experimental Psychology, National Demonstration Center for Experimental Psychology Education (Beijing Normal University), Faculty of Psychology, Beijing Normal University, Beijing 100875, China; psyhongwei@163.com (W.H.); jrh_psy@163.com (R.J.); fxc_psy@163.com (X.F.)

2 Graduate School of Education, Fordham University, New York, NY 10023, USA; yding4@fordham.edu

3 Institute of Psychological Sciences, Hangzhou Normal University, Hangzhou 311121, China; zhenrui1206@126.com

* Correspondence: rdliu@bnu.edu.cn; Tel.: +86-10-58806324

Received: 28 May 2020; Accepted: 22 July 2020; Published: 23 July 2020

Abstract: Psychological needs dissatisfaction has been identified as hindering adaptive development, in which autonomy need dissatisfaction, as one core component, may be associated with adolescents' maladaptive online behaviors. Sporadic research has examined the association between autonomy need dissatisfaction and problematic mobile phone use (PMPU). Boredom proneness and mobile phone gaming were suggested to be linked to this association. This study aimed to examine the mediating effects of boredom proneness and mobile phone gaming in the association between autonomy need dissatisfaction and PMPU. A total of 358 secondary school students completed questionnaires at three waves; autonomy need dissatisfaction was measured in time 1 (T1); boredom proneness and mobile phone gaming were measured one year later (time 2, T2); PMPU was measured two years later (time 3, T3). The structural equation model results showed that T1 autonomy need dissatisfaction not only directly predicted T3 PMPU, but also exerted effects via the mediating role of T2 boredom proneness and the chain mediating role of T2 boredom proneness and T2 mobile phone gaming. These findings reveal the unique role of specific psychological need in engaging PMPU, which provides support to targeted interventions, such that promoting autonomy need satisfaction may be an instrumental procedure to prevent adolescents from addiction-like online behaviors.

Keywords: autonomy need dissatisfaction; problematic mobile phone use; boredom proneness; mobile phone gaming; multiple mediation

1. Introduction

Mobile phones, as the most accessible device to connect to the internet, have penetrated every aspect of daily lives, such that they help people obtain information, maintain social connectedness, and entertain themselves [1,2]. According to a national survey in China, there were 897 million mobile phone users as of March 2020, accounting for 99.3% of the Internet users [3]. Some of the users invested an excessive amount of time and resources into their mobile phones. This behavior can be described as problematic mobile phone use (PMPU), which refers to a constellation of emerging addiction symptoms, including cravings, withdrawal, and loss of control [4]. Numerous negative consequences occur after engaging in PMPU. For instance, PMPU has been identified to lead to sleep problems [5,6], poor academic performance and school adjustment [7,8], cognitive failures [6,9], and physical and mental health problems [10–12]. Furthermore, it was found that 10% of British adolescents were problematic users and 20.5% of them were potential problematic users [13]. A recent research showed that the prevalence of PMPU was 29% in young adults from the United Arab Emirates [1]. Such potential hazards and the high prevalence of PMPU stimulate public concerns and gain increasing scholarly attention.

People are active and purposive when engaging on the internet via mobile phones because it can satisfy specific psychological needs, as postulated by uses and gratification theory [14]. This perspective implies that people with unsatisfied needs in daily life tend to use mobile phones as a compensator to cope with the negative life situation. Combined with the model of compensatory internet use [15], this compensatory use of the internet via mobile phones is more likely to result in problematic use and addiction tendencies. A substantial body of literature has revealed that psychological needs dissatisfaction in daily life contributes to pathological Internet use (PIU) [16] and PMPU [17].

Based on self-determination theory, humans have three inherent psychological needs, including the need for autonomy, competence, and relatedness. Deci and Ryan [18] proposed that psychological needs satisfaction is an essential nutriment for psychological growth and wellness; its dissatisfaction hinders self-integrity and lead to problematic outcomes. Given that different needs play different roles in behavior patterns and social functioning [19], an increasing number of studies attempted to differentiate the unique effect of each type of need. One of the very few studies found that only autonomy (not relatedness and competence) need dissatisfaction significantly predicted problematic online behaviors [20]. It is known that autonomy need is described as the need to self-regulate their experiences and actions [21]. That is, when behaviors are volitional and self-endorsed, individuals would experience high levels of autonomy need satisfaction. Stated differently, autonomy need dissatisfaction suggests what individuals do is not congruent with their intrinsic motivation and authentic interests. As a result, ameliorating behaviors, such as engaging on the internet through mobile phones, are activated to compensate for the lack of fulfillment of this kind of need, which increases the probability of problematic use [15]. In short, it seems that autonomy need dissatisfaction is positively associated with PMPU.

Furthermore, autonomy need dissatisfaction as the perceptions of the external environment can be considered to be a distal factor in explaining the etiology of addictive symptoms of PIU; these distal factors exert effects on maladaptive online behaviors via the mediating effects of proximal factors, as postulated in the cognitive-behavioral model of PIU [22]. For instance, boredom proneness results from the external environment without autonomy [23], and servers as a contributor to PMPU [24,25]. Similarly, online gaming, as a specific behavioral response to cope with autonomy need dissatisfaction, is an important predictor of PMPU [26]. These relations indicate that boredom proneness and mobile phone gaming may be potential mediators in the process. However, there has been a lack of empirical research to support this relation. To address this issue, this study aimed to examine the mediating roles of boredom proneness and mobile phone gaming in the association between autonomy need dissatisfaction in daily life and PMPU.

1.1. Boredom Proneness as a Mediator

Boredom proneness may be a potential mediator between autonomy need dissatisfaction and PMPU. Specifically, boredom refers to a general tendency to experience boredom in situations with deficits in interest, meaning, excitement, and challenge [27]. People with autonomy need dissatisfaction have relatively few opportunities to make their own decisions, and have to engage in activities incongruent with their authentic interests [21]. Thus, non-interest-orientated activities may lead to low levels of psychological arousal and high levels of boredom proneness [28]. This notion has been supported by the various findings that psychological needs (including autonomy need) satisfaction/dissatisfaction significantly predicts boredom in sports activities [29], in academic settings [30], and in work domains [31]. In a 2009 study, adolescent soccer athletes who perceived less autonomy reported more boredom experience [23]. Thus, autonomy need dissatisfaction in daily life seems to be positively associated with boredom proneness.

Regarding the second stage of the mediation process, boredom proneness has been identified as a high-risk factor for PIU [32,33] and PMPU [24,34]. Adolescents with high levels of boredom proneness tend to experience low levels of internal motivation and external stimulation [27]. One approach to cope with boredom is to engage in online activities as they may help to increase the feelings of excitement and sensation [32,35]. Chronic and habitual use of this approach would increase the risk of engaging in PIU [32,33]. Similarly, previous research has found that boredom proneness positively predicts PMPU among adolescents [24,34]. Altogether, it appears that autonomy need dissatisfaction is positively associated with boredom proneness, which in turn is positively associated with PMPU.

1.2. Mobile Phone Gaming as a Mediator

Another potential mediator may be mobile phone gaming, because distal causes and proximal factors jointly facilitate an excessive use of specific internet functions (e.g., online gaming), which further leads to behavioral symptoms of PIU [22]. As stated earlier, individuals with autonomy need dissatisfaction may activate ameliorating behaviors to compensate for the deficits in this kind of need. As reviewed by Ryan and Deci [21], a key characteristic of games is providing opportunities for actions. For instance, players are free to choose the types of games and activities that they want to engage in, to decide the avatars and roles, and to fulfill the game missions. Experimental evidence has indicated that the autonomy character of a game would facilitate immersion-related experiences, further increasing enjoyment and decreasing boredom in the game world [36]. In this sense, people who experience autonomy need dissatisfaction might have the motivation to engage in gaming as a way to compensate, thus exhibiting longer game-playtime on a weekly basis [37]. More important, empirical research has found that psychological needs dissatisfaction [38] and autonomy need dissatisfaction [20] in the real world positively predict the excessive use of video games. Thus, autonomy need dissatisfaction appears to be positively associated with frequent mobile phone gaming.

Moreover, when adolescents have a history of mobile phone use for gaming, desirable game experiences (e.g., flow experience in the game world) may encourage them to repeatedly engage in this activity [39]. In the long run, they are more likely to frequently act on mobile phones and become addicted to using mobile phones [40]. In support of this notion, frequent online gaming has been shown to positively predict PIU in cross-sectional research [41] and predict PIU one year later in the longitudinal research [42]. Similarly, Jeong, Kim, Yum and Hwang [26] and Lee, Kim and Choi [40] found that frequent mobile gaming contributed to PMPU. Altogether, it appears that autonomy need dissatisfaction is positively associated with mobile phone gaming, which in turn is positively associated with PMPU.

1.3. A Multiple Mediation Model

The mediating roles of boredom proneness and mobile phone gaming have been advanced to describe the relation between autonomy need dissatisfaction and PMPU wherein boredom proneness was argued to positively associate with frequent mobile phone use [43]. For instance, Chou, et al. [44] found that adolescents with high boredom proneness are more easily to perceive low levels of external stimulation and are more likely to engage in online gaming for self-entertainment. Likewise, Biolcati, Mancini and Trombini [25] supported this finding and found that adolescents with higher boredom proneness reported higher levels of participation in mobile phone gaming in comparison to adolescents with lower boredom proneness.

Taken together, individuals with autonomy need dissatisfaction in the real world cannot voluntarily make choices and engage in activities congruent with their authentic interests [18]. Thus, they are prone to having low intrinsic motivation and exhibit low psychological involvement, which may increase the tendency to experience boredom [21,45,46]. Furthermore, bored individuals are more likely to play mobile games as a compensator of boredom [15]. In this regard, frequent mobile phone gaming increases the risk of problematic use and addictive symptoms [26,40]. Accordingly, it is possible that autonomy need dissatisfaction is indirectly associated with PMPU via the multiple mediating role of boredom proneness and mobile phone gaming.

1.4. The Present Study

According to the above findings, autonomy need dissatisfaction in daily life has been argued to be a contributing factor to PMPU. Autonomy need dissatisfaction as the perceptions of the external environment may exert effects on psychological symptoms via individual characteristics. Based on self-determination theory [18], the model of compensatory internet use [15], and the cognitive-behavioral model of PIU [22], boredom proneness as a dispositional factor can be partially attributed to the lack of autonomy from the external environment; and gaming involved a specific mobile phone usage can be considered as coping strategies. That is, boredom proneness and mobile phone gaming can be postulated as potential mediators to elucidate how autonomy need dissatisfaction was associated with PMPU. Nevertheless, there has been a lack of empirical research, especially using cross-temporal designs, to examine whether autonomy need dissatisfaction is associated with PMPU via the mediating roles of boredom proneness and mobile phone gaming. To this end, we attempted to assess the independent variable in Time 1 (T1), the mediating variables in Time 2 (T2), and the dependent variable in Time 3 (T3). As shown in Figure 1, this study was guided by the following hypotheses:

H1: *T1 autonomy need dissatisfaction in daily life is positively associated with T3 PMPU.*

H2: *T2 boredom proneness mediates the association between T1 autonomy need dissatisfaction and T3 PMPU.*

H3: *T2 mobile phone gaming mediates the association between T1 autonomy need dissatisfaction and T3 PMPU.*

H4: *T1 autonomy need dissatisfaction in daily life is indirectly associated with T3 PMPU through the multiple mediating role of T2 boredom proneness and T2 mobile phone gaming.*

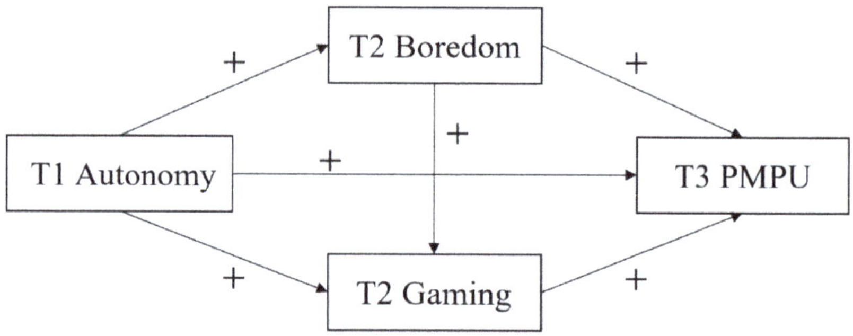

Figure 1. The conceptual model between autonomy need dissatisfaction and PMPU. Note. T1 Autonomy = Autonomy need dissatisfaction in Time 1; T2 Boredom = Boredom proneness in Time 2; T2 Gaming = Mobile phone gaming in Time 2; T3 PMPU = Problematic mobile phone use in Time 3.

2. Materials and Methods

2.1. Participants and Procedures

A sample of 1060 students from a regular secondary school in Beijing, China, was recruited to participate in the first data collection (T1). Due to graduation, 819 students participated in the second data collection (T2) after one year and 358 students participated in the third data collection (T3) after two years. This study focused on the participants who completed the questionnaires at three waves. The sample comprised 154 (43.0%) boys and 204 girls (57.0%). They had an average age of 13.19

years (standard deviation (*SD*) = 1.44), ranging from 12 to 16 years. Each participant reported having a constant Internet-accessible mobile phone.

This research was approved by the Academic Ethics Committee of the Faculty of Psychology at Beijing Normal University. Before the formal investigation, participants and their parents or legal guardians were provided with written consent forms, which informed them that personal information would be kept confidential and their responses would be used only for research purposes. Additionally, students were informed that they had the right to opt out of the research at any time. The research assistants distributed and collected the self-report questionnaires in the regular classrooms. Data collection took approximately 15 min, and participants were compensated with a small gift (e.g., a pen).

2.2. Measure

2.2.1. Autonomy Need Dissatisfaction

The level of autonomy need dissatisfaction was measured in T1 by the Basic Need Satisfaction in General Scale, which consists of the domains of autonomy, competence, and relatedness needs satisfaction [47]. This scale has been tested and used in the Chinese context [48]. The autonomy subscale contains three negative items (e.g., There is not much opportunity for me to decide for myself how to do things in my daily life), which has been used to assess autonomy need dissatisfaction [49]. Participants rated the items on a 5-point Likert scale (1 = *not at all true*, 5 = *very true*), with higher scores indicating higher levels of autonomy need dissatisfaction. The internal consistency of this scale showed acceptable reliability (Cronbach $\alpha = 0.60$).

2.2.2. Boredom Proneness

The level of boredom proneness was measured in T2 by the short version of the Boredom Proneness Scale [27]. This scale contains eight items (e.g., I often find myself at "loose ends," not knowing what to do) with one-dimensional structure. Participants rated the items on a 7-point Likert scale (1 = *strongly disagree*, 7 = *strongly agree*), with higher scores indicating higher levels of boredom proneness. The internal consistency of this scale showed satisfactory reliability (Cronbach $\alpha = 0.94$).

2.2.3. Mobile Phone Gaming

The measure of mobile phone gaming in T2 was adapted from the Chinese Internet Usage Questionnaire [50] and the Mobile Phone Use Patterns Questionnaire [43]. In total, there were 10 items regarding Internet use and 17 items regarding mobile phone use. After instructing "according to your daily routine, … ", only one item (i.e., I play mobile games on my phone) was used to assess the mobile phone gaming frequency on a daily basis. Participants rated the items on a 5-point Likert scale (1 = *never*, 5 = *always*), with higher scores indicating more frequent use for mobile games.

2.2.4. Problematic Mobile Phone Use (PMPU)

Participants' severity of PMPU was assessed in T3 by the short version of the Mobile Phone Problem Use Scale [51], which has been validated in the Chinese context and showed good validity and reliability [17,52]. This scale contains 10 items (e.g., I find it difficult to switch off my mobile phone) with five aspects, including craving, withdrawal, peer dependence, loss of control, and negative life consequences. Participants rated the items on a 5-point Likert scale (1 = *strongly disagree*, 5 = *strongly agree*), with higher scores indicating more severe PMPU. The internal consistency of this scale showed satisfactory reliability (Cronbach's $\alpha = 0.87$).

2.3. Data Analyses

Means, standard deviations, and Pearson correlations were calculated using SPSS 19.0. The hypothesized multiple mediation model was tested by structural equation modeling (SEM) using Mplus 7.1 [53]. The model was evaluated by following model fit indices: the chi-square values ($\chi2$), the comparative fit index (CFI), the Tucker–Lewis fit index (TLI), the root mean square error of approximation (RMSEA), and the standardized root mean square residual (SRMR). The CFI and TLI at 0.90 or above, and the RMSEA and SRMR at 0.08 or lower, indicate that the model is acceptable [54].

3. Results

3.1. Descriptive Statistics and Correlations

Means, standard deviations, and Pearson correlations are presented in Table 1. As shown, autonomy need dissatisfaction was significantly and positively correlated with boredom proneness and PMPU, but it was not correlated with mobile phone gaming. Furthermore, each two elements of boredom proneness, mobile phone gaming, and PMPU had a positive association.

Table 1. Means, standard deviations, and correlations among the main variables.

Variables	M	SD	1	2	3	4
1 T1 Autonomy	2.91	0.84	-			
2 T2 Boredom	3.82	1.33	0.32 ***	-		
3 T2 Gaming	3.24	1.17	0.06	0.19 ***	-	
4 T3 PMPU	2.70	0.76	0.25 ***	0.27 ***	0.21 ***	-

Note. T1 Autonomy = Autonomy need dissatisfaction in Time 1; T2 Boredom = Boredom proneness in Time 2; T2 Gaming = Mobile phone gaming in Time 2; T3 PMPU = Problematic mobile phone use in Time 3; *** $p < 0.001$.

3.2. Examinations of the Measurement Model

Before testing the hypothesized model by SEM, it was necessary to examine the measurement model. According to the recommendation from Wu and Wen [55], autonomy need dissatisfaction could be loaded by the three observed items; boredom proneness that has eight items with one-dimensional structure could be parceled into three indicators; mobile phone gaming with only one item could be loaded by the one item; PMPU with five aspects could be loaded by the five substructures. Altogether, the CFA results of the measurement model showed a good model fit: $\chi^2/df = 3.17$, CFI = 0.95, TLI = 0.93, RMSEA = 0.08, SRMR = 0.05, in that all the loadings on latent variables were significant ($p < 0.001$).

3.3. Examinations of the Structural Model

As hypothesized, a multiple model with T1 autonomy need dissatisfaction as the independent variable, T2 boredom proneness and mobile phone gaming as the mediators, and T3 PMPU as the dependent variable was established. The SEM results showed a good model fit: $\chi^2/df = 3.15$, CFI = 0.94, TLI = 0.92, RMSEA = 0.05, SRMR = 0.05. As shown in Figure 2, T1 autonomy need dissatisfaction significantly predicted T3 PMPU. Similarly, T1 autonomy need dissatisfaction positively predicted T2 boredom proneness, which in turn positively predicted T3 PMPU. However, T1 autonomy need dissatisfaction did not predict T2 mobile phone gaming, although T2 mobile phone gaming positively predicted T3 PMPU.

To further examine the significance of the indirect effects, bias-corrected bootstrap tests derived with 1000 samples were used. That the 95% confidence interval did not contain zero indicated statistical significance [56]. As shown in Table 2, T1 autonomy need dissatisfaction positively predicted T3 PMPU, supporting H1. Furthermore, T2 boredom proneness significantly mediated the association between T1 autonomy need dissatisfaction and T3 PMPU, supporting H2. Whereas, mobile phone gaming did not mediate the association between T1 autonomy need dissatisfaction and T3 PMPU, rejecting H3. Additionally, the chain of T2 boredom proneness and T2 mobile phone gaming significantly mediated the association between T1 autonomy need dissatisfaction and T3 PMPU, supporting H4.

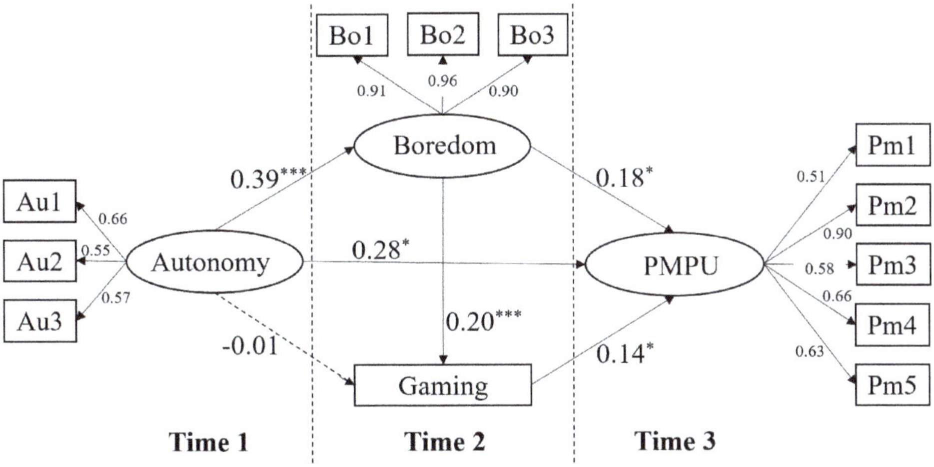

Figure 2. The mediation model of the association between T1 autonomy need dissatisfaction and T3 PMPU. Note. All the loadings on latent variables were significant ($p < 0.001$). Autonomy = Autonomy need dissatisfaction, Boredom = Boredom proneness, Gaming = Mobile phone gaming, PMPU = Problematic mobile phone use. * $p < 0.05$, *** $p < 0.001$.

Table 2. Bias-corrected bootstrap tests on the direct and indirect effects.

Paths	Standardized(β)	95% Confidence Interval		Hypotheses
		Low	High	
T1 Autonomy → T3 PMPU	0.277	0.112	0.443	Supporting H1
T1 Autonomy → T2 Boredom → T3 PMPU	0.069	0.007	0.013	Supporting H2
T1 Autonomy → T2 Gaming → T3 PMPU	−0.001	−0.024	0.022	Rejecting H3
T1 Autonomy → T2 Boredom → T2 Gaming → T3 PMPU	0.011	0.001	0.025	Supporting H4

Note. T1 Autonomy = Autonomy need dissatisfaction in Time 1; T3 PMPU = Problematic mobile phone use in Time 3; T2 Boredom = Boredom proneness in Time 2; T2 Gaming = Mobile phone gaming in Time 2.

4. Discussion

This study focused on autonomy need dissatisfaction and examined its potential effect on PMPU. Boredom proneness and mobile phone gaming were suggested to be incorporated into this association to elucidate the underlying mechanism. Based on three-wave data, the SEM model results showed that T1 autonomy need dissatisfaction not only directly predicted T3 PMPU, but also exerted effects on T3 PMPU via the mediating role of T2 boredom proneness and via the chain mediating role of T2 boredom proneness and T2 mobile phone gaming. Altogether, the findings provide empirical evidence to support the relation between specific psychological need and PMPU, which lends further insight into targeted prevention and interventions of problematic online behaviors.

4.1. Autonomy Need Dissatisfaction, Boredom Proneness, and PMPU

This study demonstrated that T1 autonomy need dissatisfaction directly predicted T3 PMPU, it also indirectly predicted T3 PMPU through the mediating role of T2 boredom proneness. According to self-determination theory [18], adolescents with autonomy need dissatisfaction have few opportunities to volitionally make choices and self-organize actions in daily life. Thus, they may have to participate in activities with little intrinsic motivation, which increases a tendency to experience boredom [29,45]. For instance, they may execute what others compel them to do, such as participating in extracurricular courses that are arranged by their parents. In this sense, they are more likely to experience boredom.

This finding was consistent with the previous studies that the higher levels of autonomy need dissatisfaction that adolescents perceive, the more likely they would experience boredom [31].

Furthermore, bored adolescents are more likely to seek external stimulation to cope with boredom [27], and thus they may spend much time and resources on the internet (or via mobile phones), which further increases the risk of problematic behaviors, including PIU [32,33] and PMPU [24]. Altogether, adolescents with autonomy need dissatisfaction cannot freely make decisions and volitionally engage in what they are interested in, which chronically contributes to boredom proneness. These bored adolescents are more likely to frequently act on mobile phones, leading to problematic use. Thus, it seems that autonomy need dissatisfaction in daily life gives rise to boredom proneness, which in turn increases the risk of subsequent PMPU.

4.2. Autonomy Need Dissatisfaction, Mobile Phone Gaming, and PMPU

This study showed that T1 autonomy need dissatisfaction did not predict T2 mobile phone gaming although T2 gaming positively predicted T3 PMPU. This finding weakly supported the mediating role of mobile phone gaming in the mediation process because the first stage was not significant. One possibility may be that psychological need dissatisfaction plays a double-edged role in determining online gaming [37]. As mentioned earlier, games that provide adolescents with opportunities for actions can assist in compensating for unsatisfied autonomy need in the real world [21,57]. Therefore, adolescents with autonomy need dissatisfaction may resort to the internet (or via mobile phones) to compensate for this dissatisfaction [15]. For instance, when individuals feel psychologically pressured and constrained, they would use mobile phones for gaming to alleviate these undesirable feelings as a way to compensate because they are free to do whatever they want in the game world. This perspective implies that the higher levels of autonomy need dissatisfaction that adolescents perceive, the more frequent mobile phone gaming they would engage in [38].

Nevertheless, adolescents with autonomy need dissatisfaction have few opportunities to decide for themselves even though they may think that gaming is tempting [39,47]. Specifically, adolescents with autonomy need dissatisfaction may be under the restrictions of their parents, particularly when engaging in mobile phone gaming. This perspective suggests that the higher levels of autonomy need dissatisfaction that adolescents perceive, the fewer opportunities they might have to play mobile games. Taken these two perspectives together, the former compensatory effect (i.e., autonomy need dissatisfaction motivates mobile gaming as a compensator) may neutralize the later restriction effect (i.e., autonomy need dissatisfaction indicates few opportunities for mobile gaming). Thus, it is not surprising that autonomy need dissatisfaction in daily life was weakly associated with mobile phone gaming. Future studies are warranted to further examine the complicated association between autonomy need dissatisfaction and online gaming.

4.3. A Multiple Mediation Model

One intriguing finding was that the chain of T2 boredom proneness and T2 mobile phone gaming significantly mediated T1 autonomy need dissatisfaction and T3 PMPU. Consistent with the etiology of addictive symptoms of PIU [22], psychological needs dissatisfaction as a distal factor exerts effects on addiction tendencies through the mediating variables (i.e., boredom proneness and mobile phone gaming). Specifically, adolescents with autonomy need dissatisfaction in real life tend to have relatively fewer opportunities to make decisions; thus, they may have to engage what is not congruent with their authentic interests [47]. For instance, the parents of Chinese students may arrange for them to engage in repetitive and monotonous academic activities. In a long run, they may possess low levels of intrinsic motivation and exhibit high levels of boredom proneness. Concurring with the earlier findings [25,44], bored adolescents may engage in online gaming to alleviate boredom as they can obtain external stimulation and gain flow experiences when fully involving in gaming [39]. For instance, participants reported that they played a kind of multiplayer online battle arena game named *Arena of Valor* on mobile phones because they felt that doing so can swipe away boring time. Additionally, mobile phone gaming has been identified as a high-risk factor for PMPU [26,40]. That is, adolescents who frequently engage in mobile phone gaming are at risk in developing problematic use and nurturing addiction tendencies. Taken together, it seems that autonomy need dissatisfaction in daily life positively predicts boredom proneness that contributes to frequent mobile phone gaming, which in turn leads to subsequent PMPU.

4.4. Limitations, Future Directions, and Implications

There are several limitations of this study. First, self-reported data may produce response bias although there was no serious common method bias using Harman's single factor test [58]. Future studies could record time of generalized use and gaming on mobile phones, which may provide objective data and enhance reliability and validity. Second, the reliability of the measure of autonomy need dissatisfaction appeared to be somewhat low although it has been used in several studies [47–49]. Thus, this scale should be further improved in future research. Third, this study recruited secondary school students only from a regular secondary school; therefore, generalization of the conclusions to other groups should be made with caution. Future studies could focus on adults and/or clinical groups, which may contribute to a broader application of these findings. Fourth, the mediating effects seemed to be relatively small, however small effects can assist in developing theories when the findings support the theoretical hypotheses [59]. In addition, small effects should not be disregarded because they might be accumulated to generate large effects with the changing conditions [60]. In this digital age in particular, the use of mobile phones has exponentially grown and corresponding problems (e.g., addiction tendencies) have been increasing and appear severe, thus possibly leading to large effects on PMPU in future studies.

Despite the limitations, notable implications are twofold. From a theoretical perspective, this study was the first of its kind to use a cross-temporal design and to exclusively examine the effect of autonomy need dissatisfaction on subsequent PMPU. On the one hand, this study focused on the role of specific need (i.e., autonomy need), instead of psychological needs as a single entity, in explaining maladaptive online behaviors. On the other hand, boredom proneness as an individual characteristic and gaming as possible coping strategies helped to elucidate the potential etiology of addiction-like symptoms associated with mobile phone use in the framework of self-determination. These findings based on the three-wave data revealed that autonomy need dissatisfaction not only directly predicted subsequent PMPU, but also exerted indirect effects via the mediating roles of boredom proneness and mobile phone gaming. These findings help to develop a better understanding of the formation process of PMPU, which provides support for prevention and intervention programs. For instance, excessive parental restrictions on children's online behaviors (e.g., limiting use time, monitoring online content) may backfire because these children perceive autonomy need dissatisfaction and may increase addiction-like tendencies [61,62]. In contrast, we recommend that families and schools provide adolescents with a certain degree of autonomy and encourage adolescents to self-organize their behaviors, which can reduce their tendencies to experience boredom. Accordingly, these less-bored adolescents are less likely to engage in high-frequency game play, thus decreasing the risk of engaging in PMPU. Additionally, families and school personnel could try to purposefully increase diverse activities to avoid boredom from adolescents, as well as to guide adolescents to increase appropriate mobile phone use and decrease excessive mobile gaming, which may be instrumental to prevent from addiction-like online behaviors.

5. Conclusions

This is one of very few studies to focus on the association between autonomy need dissatisfaction in daily life and PMPU. With boredom proneness and mobile phone gaming introduced, the mediation model may contribute to explaining the potential mechanism of this association. Based on three-wave data, the results showed that T1 autonomy need dissatisfaction not only directly predicted T3 PMPU, but also exerted effects on T3 PMPU via the mediating role of T2 boredom proneness and via the chain mediating role of T2 boredom proneness and T2 mobile phone gaming. Altogether, these findings reveal the unique role of specific psychological need satisfaction in PMPU, which suggests that promoting autonomy need satisfaction may prevent adolescents from mobile phone addiction.

Author Contributions: Conceptualization, W.H. and R.-D.L.; Formal analysis, W.H.; Funding acquisition, R.-D.L.; Investigation, R.Z., R.J. and X.F.; Project administration, R.-D.L.; Supervision, R.-D.L.; Validation, R.-D.L. and Y.D.; Writing—original draft, W.H.; Writing—review and editing, W.H. and Y.D. All authors have read and agreed to the published version of the manuscript.

Funding: This research was funded by the National Social Science Fund of China, grant number 17BSH102.

Acknowledgments: We appreciated the support from the participating schools' students and teachers.

Conflicts of Interest: The authors declare no conflict of interest.

References

1. Vally, Z.; El Hichami, F. An examination of problematic mobile phone use in the United Arab Emirates: Prevalence, correlates, and predictors in a college-aged sample of young adults. *Addict. Behav. Rep.* **2019**, *9*, 100185. [CrossRef] [PubMed]
2. Elhai, J.D.; Dvorak, R.D.; Levine, J.C.; Hall, B.J. Problematic smartphone use: A conceptual overview and systematic review of relations with anxiety and depression psychopathology. *J. Affect. Disord.* **2017**, *207*, 251–259. [CrossRef] [PubMed]
3. CNNIC. The 45th China Statistical Report on Internet Development. 2020. Available online: http://www.cnnic.net.cn/hlwfzyj/hlwxzbg/hlwtjbg/202004/P020200428596599037028.pdf (accessed on 28 April 2020).
4. Bianchi, A.; Phillips, J.G. Psychological predictors of problem mobile phone use. *Cyberpsychol. Behav.* **2005**, *8*, 39–51. [CrossRef] [PubMed]
5. Liu, Q.Q.; Zhou, Z.K.; Yang, X.J.; Kong, F.C.; Niu, G.F.; Fan, C.Y. Mobile phone addiction and sleep quality among Chinese adolescents: A moderated mediation model. *Comput. Hum. Behav.* **2017**, *72*, 108–114. [CrossRef]
6. Hong, W.; Liu, R.-D.; Ding, Y.; Sheng, X.; Zhen, R. Mobile phone addiction and cognitive failures in daily life: The mediating roles of sleep duration and quality and the moderating role of trait self-regulation. *Addict. Behav.* **2020**, *107*, 106383. [CrossRef] [PubMed]
7. Seo, D.G.; Park, Y.; Kim, M.K.; Park, J. Mobile phone dependency and its impacts on adolescents' social and academic behaviors. *Comput. Hum. Behav.* **2016**, *63*, 282–292. [CrossRef]
8. Jun, S. Longitudinal influences of depressive moods on problematic mobile phone use and negative school outcomes among Korean adolescents. *Sch. Psychol. Int.* **2019**, *40*, 294–308. [CrossRef]
9. Hadlington, L.J. Cognitive failures in daily life: Exploring the link with internet addiction and problematic mobile phone use. *Comput. Hum. Behav.* **2015**, *51*, 75–81. [CrossRef]
10. Chen, L.; Yan, Z.; Tang, W.; Yang, F.; Xie, X.; He, J. Mobile phone addiction levels and negative emotions among Chinese young adults: The mediating role of interpersonal problems. *Comput. Hum. Behav.* **2016**, *55*, 856–866. [CrossRef]
11. Kim, E.; Cho, I.; Kim, E.J. Structural equation model of smartphone addiction based on adult attachment theory: Mediating effects of loneliness and depression. *Asian Nurs. Res.* **2017**, *11*, 92–97. [CrossRef]
12. Asante, K.O.; Nyako, J. The physical and behavioural consequences of Facebook use among university students. *Mediterr. J. Soc. Sci.* **2014**, *5*, 774–781. [CrossRef]
13. Lopez-Fernandez, O.; Honrubia-Serrano, L.; Freixa-Blanxart, M.; Gibson, W. Prevalence of problematic mobile phone use in British adolescents. *Cyberpsychol. Behav. Soc. Netw.* **2014**, *17*, 91–98. [CrossRef] [PubMed]
14. Katz, E.; Blumler, J.; Gurevitch, M. Utilization of mass communication by the individual. In *The Uses of Mass Communications: Current Perspectives on Gratifications Research*; Katz, E., Blumler, J., Eds.; Sage: Beverly Hills, CA, USA, 1974; pp. 19–32.
15. Kardefelt-Winther, D. A conceptual and methodological critique of internet addiction research: Towards a model of compensatory internet use. *Comput. Hum. Behav.* **2014**, *31*, 351–354. [CrossRef]
16. Liu, Q.X.; Fang, X.Y.; Wan, J.J.; Zhou, Z.K. Need satisfaction and adolescent pathological internet use: Comparison of satisfaction perceived online and offline. *Comput. Hum. Behav.* **2016**, *55*, 695–700. [CrossRef]

17. Hong, W.; Liu, R.-D.; Oei, T.-P.; Zhen, R.; Jiang, S.; Sheng, X. The mediating and moderating roles of social anxiety and relatedness need satisfaction on the relationship between shyness and problematic mobile phone use among adolescents. *Comput. Hum. Behav.* **2019**, *93*, 301–308. [CrossRef]
18. Deci, E.L.; Ryan, R.M. The "what" and "why" of goal pursuits: Human needs and the self-determination of behavior. *Psychol. Inq.* **2000**, *11*, 227–268. [CrossRef]
19. Chen, B.; Vansteenkiste, M.; Beyers, W.; Boone, L.; Deci, E.L.; Van der Kaap-Deeder, J.; Duriez, B.; Lens, W.; Matos, L.; Mouratidis, A.; et al. Basic psychological need satisfaction, need frustration, and need strength across four cultures. *Motiv. Emot.* **2014**, *39*, 216–236. [CrossRef]
20. Mills, D.J.; Milyavskaya, M.; Heath, N.L.; Derevensky, J.L. Gaming motivation and problematic video gaming: The role of needs frustration. *Eur. J. Soc. Psychol.* **2018**, *48*, 551–559. [CrossRef]
21. Ryan, R.M.; Deci, E.L. *Self-Determination Theory: Basic Psychological Needs in Motivation, Development, and Wellness*; Guilford Publications: New York, NY, USA, 2017.
22. Davis, R.A. A cognitive-behavioral model of pathological internet use. *Comput. Hum. Behav.* **2001**, *17*, 187–195. [CrossRef]
23. Alvarez, M.S.; Balaguer, I.; Castillo, I.; Duda, J.L. Coach autonomy support and quality of sport engagement in young soccer players. *Span. J. Psychol.* **2009**, *12*, 138–148. [CrossRef]
24. Ksinan, A.J.; Mališ, J.; Vazsonyi, A.T. Swiping away the moments that make up a dull day: Narcissism, boredom, and compulsive smartphone use. *Curr. Psychol.* **2019**. Published online. [CrossRef]
25. Biolcati, R.; Mancini, G.; Trombini, E. Proneness to boredom and risk behaviors during adolescents' free time. *Psychol. Rep.* **2018**, *121*, 303–323. [CrossRef] [PubMed]
26. Jeong, S.H.; Kim, H.; Yum, J.Y.; Hwang, Y. What type of content are smartphone users addicted to?: SNS vs. games. *Comput. Hum. Behav.* **2016**, *54*, 10–17. [CrossRef]
27. Struk, A.A.; Carriere, J.S.; Cheyne, J.A.; Danckert, J. A short boredom proneness scale: Development and psychometric properties. *Assessment* **2015**, *24*, 346–359. [CrossRef] [PubMed]
28. Farmer, R.; Sundberg, N.D. Boredom proneness: The development and correlates of a new scale. *J. Pers. Assess.* **1986**, *50*, 4–17. [CrossRef] [PubMed]
29. González, L.; Castillo, I.; Balaguer, I. Exploring the role of resilience and basic psychological needs as antecedents of enjoyment and boredom in female sports. *Rev. Psicodidáctica* **2019**, *24*, 131–137.
30. Sulea, C.; van Beek, I.; Sarbescu, P.; Virga, D.; Schaufeli, W.B. Engagement, boredom, and burnout among students: Basic need satisfaction matters more than personality traits. *Learn. Individ. Differ.* **2015**, *42*, 132–138. [CrossRef]
31. Van Hooff, M.L.M.; Van Hooft, E.A.J. Boredom at work: Towards a dynamic spillover model of need satisfaction, work motivation, and work-related boredom. *Eur. J. Work Organ. Psychol.* **2016**, *26*, 133–148. [CrossRef]
32. Skues, J.; Williams, B.; Oldmeadow, J.; Wise, L. The effects of boredom, loneliness, and distress tolerance on problem internet use among university students. *Int. J. Ment. Health Addict.* **2015**, *14*, 167–180. [CrossRef]
33. Nichols, L.A.; Nicki, R. Development of a psychometrically sound internet addiction scale: A preliminary step. *Psychol. Addict. Behav.* **2004**, *18*, 381–384. [CrossRef]
34. Leung, L. Linking psychological attributes to addiction and improper use of the mobile phone among adolescents in Hong Kong. *J. Child. Media* **2008**, *2*, 93–113. [CrossRef]
35. Lin, C.H.; Yu, S.F. Adolescent internet usage in Taiwan: Exploring gender differences. *Adolescence* **2008**, *43*, 317–331. [PubMed]
36. Kim, K.; Schmierbach, M.G.; Bellur, S.; Chung, M.Y.; Fraustino, J.D.; Dardis, F.; Ahern, L. Is it a sense of autonomy, control, or attachment? Exploring the effects of in-game customization on game enjoyment. *Comput. Hum. Behav.* **2015**, *48*, 695–705. [CrossRef]
37. Mills, D.J.; Allen, J.J. Self-determination theory, internet gaming disorder, and the mediating role of self-control. *Comput. Hum. Behav.* **2020**, *105*, 106209. [CrossRef]
38. Scerri, M.; Anderson, A.; Stavropoulos, V.; Hu, E. Need fulfilment and internet gaming disorder: A preliminary integrative model. *Addict. Behav. Rep.* **2019**, *9*, 100144. [CrossRef]
39. Leung, L. Exploring the relationship between smartphone activities, flow experience, and boredom in free time. *Comput. Hum. Behav.* **2020**, *103*, 130–139. [CrossRef]
40. Lee, H.; Kim, J.W.; Choi, T.Y. Risk factors for smartphone addiction in Korean adolescents: Smartphone use patterns. *J. Korean Med. Sci.* **2017**, *32*, 1674–1679. [CrossRef]

41. Škařupová, K.; Ólafsson, K.; Blinka, L. The effect of smartphone use on trends in European adolescents' excessive internet use. *Behav. Inf. Technol.* **2015**, *35*, 68–74. [CrossRef]
42. van Rooij, A.J.; Schoenmakers, T.M.; van de Eijnden, R.J.; Van de Mheen, D. Compulsive internet use: The role of online gaming and other internet applications. *J. Adolesc. Health* **2010**, *47*, 51–57. [CrossRef]
43. Hao, Z.; Jin, L.; Li, Y.; Akram, H.R.; Saeed, M.F.; Ma, J.; Ma, H.; Huang, J. Alexithymia and mobile phone addiction in Chinese undergraduate students: The roles of mobile phone use patterns. *Comput. Hum. Behav.* **2019**, *97*, 51–59. [CrossRef]
44. Chou, W.J.; Chang, Y.P.; Yen, C.F. Boredom proneness and its correlation with internet addiction and internet activities in adolescents with attention-deficit/hyperactivity disorder. *Kaohsiung J. Med. Sci.* **2018**, *34*, 467–474. [CrossRef] [PubMed]
45. Mcleod, C.R.; Vodanovich, S.J. The relationship between self-actualization and boredom proneness. *J. Soc. Behav. Pers.* **1991**, *6*, 137–146.
46. Pekrun, R.; Goetz, T.; Daniels, L.M.; Stupnisky, R.H.; Perry, R.P. Boredom in achievement settings: Exploring control–value antecedents and performance outcomes of a neglected emotion. *J. Educ. Psychol.* **2010**, *102*, 531–549. [CrossRef]
47. Johnston, M.M.; Finney, S.J. Measuring basic needs satisfaction: Evaluating previous research and conducting new psychometric evaluations of the basic needs satisfaction in general scale. *Contemp. Educ. Psychol.* **2010**, *35*, 280–296. [CrossRef]
48. Shen, C.X.; Liu, R.D.; Wang, D. Why are children attracted to the internet? The role of need satisfaction perceived online and perceived in daily real life. *Comput. Hum. Behav.* **2013**, *29*, 185–192. [CrossRef]
49. Wang, C.; Hsu, H.C.K.; Bonem, E.M.; Moss, J.D.; Yu, S.; Nelson, D.B.; Levesque-Bristol, C. Need satisfaction and need dissatisfaction: A comparative study of online and face-to-face learning contexts. *Comput. Hum. Behav.* **2019**, *95*, 114–125. [CrossRef]
50. Shen, C.X.; Liu, R.D.; Wang, D. The relationship between internet use and children's loneliness: A moderating effect of personality. *J. Psychol. Sci.* **2013**, *36*, 1140–1145. (In Chinese)
51. Foerster, M.; Roser, K.; Schoeni, A.; Röösli, M. Problematic mobile phone use in adolescents: Derivation of a short scale MPPUS-10. *Int. J. Public Health* **2015**, *60*, 277–286. [CrossRef]
52. Liu, R.D.; Hong, W.; Ding, Y.; Oei, T.P.; Zhen, R.; Jiang, S.; Liu, J. Psychological distress and problematic mobile phone use among adolescents: The mediating role of maladaptive cognitions and the moderating role of effortful control. *Front. Psychol.* **2019**, *10*, 1589. [CrossRef]
53. Cheung, M.W.L. Comparison of approaches to constructing confidence intervals for mediating effects using structural equation models. *Struct. Equ. Modeling* **2007**, *14*, 227–246. [CrossRef]
54. Wen, Z.L.; Hau, K.T.; Marsh, H.W. Structural equation model testing: Cutoff criteria for goodness of fit indices and chi-square test. *Acta Psychol. Sin.* **2004**, *36*, 186–194. (In Chinese)
55. Wu, Y.; Wen, Z. Item parceling strategies in structural equation modeling. *Adv. Psychol. Sci.* **2011**, *19*, 1859–1867. (In Chinese)
56. MacKinnon, D.P.; Lockwood, C.M.; Williams, J. Confidence limits for the indirect effect: Distribution of the product and resampling methods. *Multivar. Behav. Res.* **2004**, *39*, 99–128. [CrossRef] [PubMed]
57. Allen, J.J.; Anderson, C.A. Satisfaction and frustration of basic psychological needs in the real world and in video games predict internet gaming disorder scores and well-being. *Comput. Hum. Behav.* **2018**, *84*, 220–229. [CrossRef]
58. Zhou, H.; Long, L. Statistical remedies for common method biases. *Adv. Psychol. Sci.* **2004**, *12*, 942–950. (In Chinese)
59. Gall, M.D.; Gall, J.P.; Borg, W.R. *Educational Research: An Introduction*, 8th ed.; Pearson: Boston, MA, USA, 2007.
60. Ellis, P.D. *The Essential Guide to Effect Sizes—Statistical Power, Meta-Analysis, and the Interpretation of Research Results*; Cambridge University Press: New York, NY, USA, 2010.

61. Hefner, D.; Knop, K.; Schmitt, S.; Vorderer, P. Rules? Role model? Relationship? The impact of parents on their children's problematic mobile phone involvement. *Media Psychol.* **2018**, *22*, 82–108. [CrossRef]
62. Fu, X.; Liu, J.; Liu, R.D.; Ding, Y.; Wang, J.; Zhen, R.; Jin, F. Parental monitoring and adolescent problematic mobile phone use: The mediating role of escape motivation and the moderating role of shyness. *Int. J. Environ. Res. Public Health* **2020**, *17*, 1487. [CrossRef]

International Journal of
Environmental Research and Public Health

MDPI

Article

Problematic Relationships with Smartphones of Spanish and Colombian University Students

Verónica Marín-Díaz *, Juan Manuel Muñoz-González and Begoña-Esther Sampedro-Requena

Education department, University of Cordoba, 14071 Córdoba, Spain; juan.manuel@uco.es (J.M.M.-G.); bsampedro@uco.es (B.-E.S.-R.)
* Correspondence: vmarin@uco.es; Tel.: +34-957212617

Received: 17 June 2020; Accepted: 22 July 2020; Published: 25 July 2020

Abstract: The presence of smartphones in the lives of the population in general, and of youth in particular, is evident, and is derived from elements such as the diversity of prices as well as the ease of access of all the resources that can be reached through the internet. With the use of a descriptive approach using a quantitative poll, the objective of the present study was to discover the opinions of university students in Spain and Colombia about smartphone use, as well as the consequences of its use, and if this use could derive into so-called problematic smartphone use (PSU). For gathering the information, the *Mobile Phone Problematic Use Scale* (MPPUSA) was utilized, with a sample size $n = 4009$. The main result reached was that the model obtained is structured around six factors that determined the elements in light of PSU. The initial conclusion found was that the model applied can be utilized with Colombian students, with young Spanish women and students in the macro area of Social Sciences, the ones who had problematic behavior with the devices, as compared to the Health Sciences students who did not have it.

Keywords: problematic smartphone use; university student; internet; MPPUSA

1. Introduction

It is a known fact that the present society is digital and that the so-called Information and Communication Technologies (from here on ICT) coexist with us systematically, allowing for the interactions with fellow humans to occur at anytime and anywhere [1]. All of this is thanks to the universalization of the internet in general, and the tools which we can use through it, as these tools provide abundant means for socializing with others, perform consultations of any kind, search for knowledge, entertain ourselves, etc. [1–3].

The IWS [4] points out that in 2019, there were more than 4 billion people connected in the world, and in the specific case of Spain [5] this figure was the non-negligible amount of 42.4 million users and in Colombia more than 31 million people from a population of more than 49 million [4,6]. In both countries, 96% and 63% of users accessed the internet through their smartphones, respectively.

Nowadays, independently of the country where we live, the smartphone is one of the key technologies for the population at large. The diversity of the possible functions of these devices, apart from the use for which they were originally conceived [7], has made them become a key feature in the processes of socialization and learning of children, adolescents and young adults. As indicated by many studies [8–12], these devices allow accessing a great variety of content presented in different formats, messaging services and social networks, allowing for the development of different skills and competencies beyond the digital, allowing access to other technological resources, etc., which makes them even more attractive.

Related to this, diverse studies [13,14] have shown that Spanish university students, most of whom are younger than 35 years old, use smartphones from a social, academic and personal point of view. Likewise, for Colombian university students [15], smartphone use has different aims, such as navigating on the internet, or checking e-mail. It is unquestionable that the use of this device has increased around the world in the last few years, especially in the population within this age range (university students younger than 35). This has extended its use to the area of education, where some studies [16] have shown the student's habits within and outside university spaces, where in the first scenario it is commonly used in transportation and outdoors; while in the second scenario the classrooms, the hallways and the libraries are the most exploited places. It is precisely the attractiveness they possess that has resulted on their use increasing exponentially, with some authors [3,17–19] already speaking about the problematic use of smartphones.

Before continuing, it should be mentioned that within the scientific community, there is no clear definition of this behavioral disorder, as we have found authors who discuss the problematic use of the device [3,17–19], while others prefer to talk about addiction [1,2,20–22]. However, the purpose of this study is not to delve into this debate, as the starting position of this work is that only a problematic relationship exists [17,18], detailed on the fifth version of the *Diagnostic and Statistical Manual of Mental Disorders* [23] which points to this relationship not having an addiction factor, given that this publication only mentions addiction in cases of pathological gambling and a behavioral disorders. Thus, it is understood that only a problematic use of the device exists, indicating that a medical condition recognized by the entire community does not exist [11].

On the other hand, some authors [17,24–27] point to factors such as the lack of sleep, hyperactivity, insecurity when driving, anxiety, depression, personality disorders, etc., associated to harmful behaviors such as in the use drugs including heroin, alcohol, cannabis, speed, etc., that although not linked to technologies, have been observed in the population who have a problematic use of mobile phones. Nevertheless, some research studies [28] suggest that there is a common factor among the addictive behaviors in the use of substances and the internet, the impulsivity which is manifested, among other aspects, by low academic performance or low levels of concentration, however, this cannot be extrapolated to an addictive use of the smartphone.

At the same time, it should be pointed out that as [11] indicate, the population, who along with adolescents, are the main population at risk of having a problematic use of the devices, are the university students, given that they consider the device as part of them [11,24]. On the other hand, the fear of being left out of relationships and developing the so-called FOMO syndrome [29], makes this population group more susceptible to developing a behavioral disorder, which is translated into the problematic use of the phone.

Mobile phones, as previously indicated [1–3,30], provide their users with numerous gratifications which begs the question of what type of relationship we as individuals have developed towards them or with them. Some research studies have determined that a variable that determines the presence of problematic use is the amount of time spent utilizing the devices [11,30,31], while others indicate that the variable gender is the one that defines this relationship, with women showing a greater relationship with them. Thus, works such as [32] point out that younger people aged 18–29 years old have a high dependency on their mobile phones, which can lead to the individuals developing an emotional support relationship [18]. These aspects, in the education sphere for example, have resulted in some education centers prohibiting their presence in the centers [8].

As a result and given the high number of connections produced with smartphones in Spain and in Colombia [4–6], it is necessary to clarify the attitude of university students towards these devices. In this sense, the main objective of the present research, which will be described below, is to discover the opinions of the university students in Spain and Colombia on the use of smartphones, as well as the consequences of this use. Studies conducted among Spanish and Colombian university students [8,12,17,18] have brought to light that this scale fits the smartphone user profile, as well as the needs of the context.

The main conclusions found are: on the one hand, that younger students show a problematic use of smartphones [18,32], and in general, women more frequently display this behavioral disorder [11,30,31], without differences also found as a function of the country of origin.

The main objective of the research was to analyze the behavior of the six MPPUSA factors according to sex, age, country, macro area and self-perception in the problematic use of the smartphones by university students from Spain and Colombia. This general objective, at the same time, is concretized in the following specific objectives:

- To discover the existence of a problematic use of the mobile phone in Spanish and Colombian university students.
- To analyze if there are differences between the students objects of the study with respect to the problematic use of the mobile phone as a function of their age and sex, as well as their country of origin and macro area.
- To verify if there are relationships between the different dimensions of the questionnaire
- To explain which factors have an influence on the perceptions of the students of the problematic use of mobile phones, as well as its consideration as a function of the country, of their age and sex.

2. Materials and Methods

The design of the study utilized a descriptive approach with a cross-sectional, quantitative survey, due to the numerical and reliable nature of the data collected, and the use of a deductive and structured research strategy.

2.1. Sample

To select the sample, non-probabilistic, convenience sampling [33] was utilized, given that the questionnaire was provided to the students whom the study researchers had access to during 2017–2018 academic year.

The sample was comprised by a total of 4009 students, of which 60.1% were female and 39.9% male, distributed in the following manner: 2965 were enrolled in the National Open and Distance University (UNAD), Colombia (74.0%) and 1044 from the University of Cordoba, Spain (26%). As for the student profiles, the macro area of enrollment was considered, addressing the five university macro areas established in both countries, which were: Social and Judicial Sciences (S and JS, $n = 1911$, 47.7%), Health Sciences (HS, $n = 726$, 18.1%), Arts and Humanities (A and H, $n = 164$, 4.1%), Experimental Sciences (ES, $n = 168$, 4.2%), and Engineering and Architecture (E and A, $n = 1039$, 25.9%), as well as age, as shown in Table 1.

Table 1. Profile of the students as a function of the macro area, age and country.

				Macro Area						
				SJS	HS	AH	ES	EA	TOTAL	TOTAL Country
Country	Colombia	Age	From 18 to 20 y.o.	3.5%	4.5%	1.5%	1.0%	6.5%	16.9%	74.0%
			From 21 to 23 y.o.	3.2%	3.1%	0.8%	0.5%	5.0%	12.6%	
			From 24 to 26 y.o.	3.6%	2.4%	0.4%	0.7%	4.4%	11.5%	
			Older than 26	11.4%	8.1%	1.4%	2.0%	10.1%	33.0%	
	Spain	Age	From 18 to 20 y.o.	13.9%	0.0%	0.0%	0.0%	0.0%	13.9%	26.0%
			From 21 to 23 y.o.	8.9%	0.0%	0.0%	0.0%	0.0%	8.9%	
			From 24 to 26 y.o.	2.1%	0.0%	0.0%	0.0%	0.0%	2.1%	
			Older than 26	1.1%	0.0%	0.0%	0.0%	0.0%	1.1%	
			TOTAL	47.7%	18.1%	4.1%	4.2%	25.9%	100%	

Notes: SJS = Social and Judicial Sciences; HS = Health Sciences; AH = Arts and Humanities; ES = Experimental Sciences; EA = Engineering and Architecture.

2.2. Instrument

The data collection instrument utilized in this research was based on the Mobile Phone Problematic Use Scale for Adolescent (MPPUSA) [34], it is contextualized to the Spanish-speaking population.

This questionnaire is anonymous, with closed-ended questions and polythematic, with a Likert-type response scale, ranging from complete disagreement (1) to complete agreement (5). The factorial structure was determined through an Exploratory Factorial Analysis (EFA), using the principal components method for extracting the factors, with the following indices of adjustment: KMO = 0.98, Bartlett's sphericity test with $p < 0.001$, and a total explained variance of 56.83%. The classification of the MPPUSA items by [35] was used as a reference providing the following structure with six dimensions as a result:

- Tolerance (Cronbach's alpha = 0.904): this dimension alludes to aspects related with the usage time of the smartphone and the impossibility of decreasing it. It encompasses a total of three items.
- Escape route (Cronbach's alpha = 0.907): comprised by three items, it alludes to the device as a resource for evading specific problems. It encompasses elements such as the increase in the feeling of well-being, its use when feeling lonely, as well as its prioritization against urgent matters or tasks.
- Disconnection (Cronbach's alpha = 0.912): this dimension covers aspects such as the appearance of worry if the subject cannot answer a call, the inability of disconnecting the mobile phone or the appearance of the feeling of disorientation in the case of not having the device available.
- Anxiety (Cronbach's alpha = 0.919): composed by three items that allude to the inability of the subjects to reduce the smartphone usage time, the appearance of nervousness in the case of not being able to read messages or answer calls, as well as the anger when they have the obligation of turning the device off.
- Negative consequences (Cronbach's alpha = 0.935): it constitutes the broadest dimension of the questionnaire, as it encompasses a total of 12 items that refer to the prioritization of the use of the smartphone against other tasks, the reduction of hours of sleep of the subjects, the making of economic investments beyond their means, the appearance of tardiness, the appearance of inconveniences from its use or the decrease in academic and/or professional performance, among other aspects.
- Social motivations (Cronbach's alpha = 0.901): this dimension alludes to aspects such as the impossibility of maintaining communication with one's peers in the case of not having the device, as well as the perception of anger by them in the case that the terminal is not connected; it encompasses two items.

As for reliability, it was measured through the internal consistency approach, obtaining a Cronbach's alpha value of > 0.90, for the dimensions as a set, as well as individually, demonstrating the high reliability of the instrument.

Lastly, an item with a dichotomous answer (yes/no) was included about the self-perception they considered they had about the problematic use of Smartphones (explaining it according to [19], as a persistent deterioration or anguish), formulated in the following terms: «I believe I have a problematic use of the Smartphone», as well as a series of independent variables from the academic area (macro area) and sociodemographic areas (age, gender and country).

Lastly, the instrument had a total of 26 items written as statements and structured into the six dimensions previously mentioned,

2.3. Procedure

The filling of the questionnaire was done in person and anonymously, with the study researchers present in the classroom to help answer any possible doubts of the students in the understanding of the items in the instrument. In the case of the Colombian students, a videoconference system was established, which allowed for the resolution of problems in real time during the completion of the instrument.

2.4. Analysis Conducted

In this study, the following analysis were utilized to reach the objectives set:

First, a descriptive analysis of the 26 variables of the questionnaire was performed, through the measurements of central tendency (mean) and dispersion (standard deviation). And the variable of the self-perception of the problematic use of the Smartphone.

In second place, a descriptive analysis was performed of the six dimensions of the questionnaire, which was similar to the analysis described above. In third place, an analysis of variance was conducted to verify the existence or not of differences in each of the dimensions of the questionnaire as a function of the independent variables (country, gender, age and macro area), through Student's t-test and ANOVA (both with n.s = 0.05). Afterwards, the relationship between the dimensions that comprised the questionnaire were verified through bivariate correlations.

Lastly, binary logistic regressions were utilized with a forward selection Wald's test and considering the goodness-of-fit of Hosmer-Lemeshow and the Confidence Interval (CI) for exp(B), to explain the influence of the six factors studied on the self-perception of the problematic use of the smartphone and the influence of the country.

All of these analyses were performed with the SPSS statistical package, version 23 (IBM Corp., Armonk, NY).

3. Results

3.1. Descriptive Study

In first place, the descriptive results (mean and standard deviation) of the 26 items that compose the *MPPUSA Questionnaire* used in this research study are listed (see Table 2).

Table 2. Frequency distribution of the items from the MPPUSA Questionnaire.

MPPUSA Items	n	M	SD
I can never spend enough time on my mobile phone	4009	2.61	1.279
I have used my mobile phone to make myself feel better when I was feeling down	4009	2.83	1.423
I find myself occupied on my mobile phone when I should be doing other things, and it causes problems	4009	2.82	1.413
I have tried to hide from others how much time I spend on my mobile phone	4009	1.77	1.209
I lose sleep due to the time I spend on my mobile phone	4009	3.06	1.457
I have spent with the mobile phone more than I should have	4009	2.36	1.421
When out of range for some time, I become worried about the thought of missing a call	4009	2.81	1.442
Sometimes, when I am on my mobile phone and I am doing other things, I get carried away with the conversation and I don't pay attention to what I am doing	4009	2.78	1.368
The time I spend on my mobile phone has increased over the last 12 months	4009	3.54	1.269
I have used my mobile phone to talk to others when I was feeling isolated	4009	3.76	1.301
I have attempted to spend less time on my mobile phone but am unable to	4009	2.23	1.343
I find it difficult to switch off/switch to silent my mobile phone	4009	2.67	1.518
I feel anxious if I have not checked for messages or switched on my mobile phone for some time	4009	2.37	1.391
I have frequent dreams about my mobile phone	4009	1.39	.973
My friends and family complain about my use of the mobile phone	4009	2.45	1.406
If I don't have a mobile phone, my friends would find it hard to get in touch with me	4009	3.62	1.287
My academic performance has decreased as a direct result of the time I spend on my mobile phone	4009	2.15	1.333
I have aches and pains that are associated with my mobile phone use	4009	2.08	1.336
I find myself using on my mobile phone for longer periods of time than intended	4009	2.86	1.406
There are times when I would rather use my mobile phone than deal with other more urgent matters	4009	2.11	1.364
I am often late for appointments because I'm talking on my mobile phone when I shouldn't be	4009	1.80	1.245
I become irritable if I have to switch off/to silent my mobile phone for classes, meals, or at the cinema	4009	1.56	1.086
I have been told that I spend too much time on my mobile phone	4009	2.69	1.445
More than once I have been in trouble because my mobile phone has gone off during a class, at the cinema, or in a theatre	4009	2.37	1.438
My friends don't like it when my mobile phone is switched off/to silent	4009	2.84	1.410
I feel lost without my mobile phone	4009	2.43	1.420

Notes: M = mean; SD= standard deviation. Source: Author created.

It can be observed that the items 5, 9, 10 and 16 point to being more in agreement with the statements as compared with items 4, 14, 21 and 22, which point to being more in disagreement.

Likewise, the results obtained in the self-perception on the problematic use of the Smartphone, point that 79.9% (n = 3205) considered that they did not have a problematic use, as compared to 20.1% (n = 804) who did.

3.1.1. Tolerance

The results obtained highlight that the students were indifferent as for the tolerance in the use of the smartphone (M = 3.00, SD = 0.87).

With respect to the comparison of means related with country and gender, the Student's t-test for independent samples showed statistically significant differences in the first case, as $t = 6.167$, $p < 0.050$, with the Colombian students being the ones who obtained a greater score in this dimension (M = 3.04 vs. M = 2.87).

Lastly, for age and the macro area, the analysis of variance (ANOVA) only pointed to the existence of statistically significant differences in the second case, as [F (4, 4003) = 5.373; $p < 0.050$]. The post hoc multiple comparisons using the Games-Howell test showed within which macro area the differences in means were found. These macro areas were Experimental Sciences along with Engineering and Architecture, with respect to Social and Judicial Sciences, which had the greatest score (μ = 3.13 and y μ = 3.05 vs. μ = 2.93).

3.1.2. Escape Route

Once the analysis was performed, the results pointed out that the students were partially in disagreement as for the use of the smartphone as an escape route (M = 2.90, SD = 0.94).

As for the comparison of means, the Student's t-test for independent samples provided evidence of the existence of statistically significant differences for the country, exclusively ($t = 2.262$, $p < 0.050$), with the mean for the Colombian subjects being higher than the Spanish ones (2.91 vs. 2.85).

Lastly, for age and macro area, the analysis of variance (ANOVA), showed statistically significant differences in the first case [F (3, 4005) = 20.038; $p < 0.050$], with the difference of means found in students older than 26 with respect to those aged 18 to 20, 21 to 23, 24 to 26, respectively (μ = 2.74 vs. μ = 3.01, μ = 2.97 and μ = 2.89).

3.1.3. Disconnection

The results obtained showed that the students were partially in disagreement with respect to the disconnection of the mobile phone (M = 2.63, SD = 1.09).

The Student's t-test for independent samples, performed to detect possible differences with respect to the country and gender, highlighted statistically significant differences in both cases. In the first ($t = -3.577$, $p < 0.050$), the Spanish students obtained a greater score with respect to the Colombian students (M = 2.73 vs. M = 2.60). As for gender ($t = -3.456$, $p = 0.001$), women were the ones who obtained a greater score with respect to the men (M = 2.68 vs. M = 2.56).

On the other hand, the analysis of variance (ANOVA) utilized to find statistically significant differences between age and the macro area, showed statistically significant differences in both cases, with the students older than 26 obtaining the lowest score [F (3, 4005) = 6.778; $p = 0.000$], as compared to those aged 18–20, 21 to 23 and 24 to 26, respectively (μ = 2.54 vs. μ = 2.71, μ = 2.70 and μ = 2.58). On the other hand, as for the macro area [F (4, 4003) = 2.634; $p < 0.050$], differences were found in the area of Social Sciences with respect to Health Sciences, with this last obtaining the lowest score (μ = 2.54 vs. μ = 2.68).

3.1.4. Anxiety

The results obtained highlight that the students had an opinion that was partially in disagreement for the anxiety dimension of the use of the smartphone (M = 2.19, SD = 1.04).

With respect to the comparison of means related with country and gender, Student's *t*-test for independent samples showed statistically significant differences only in the second case, as $t = -2.225$, $p = 0.020$, with the women receiving a higher score in this dimension (M = 2.22 vs. M = 2.14).

Lastly, for age and the macro area, the analysis of variance (ANOVA) only pointed to the existence of statistically significant differences in the first case (age), as [F (3, 4005) = 7.180; $p = 0.000$]. The multiple post-hoc comparisons with the Games-Howell test allowed us to find exactly in what age range were these differences in means found, with those older than 26 being the ones who had a lower score with respect to students with ages between 18 and 20, and 21 and 23, respectively $\mu = 2.09$ vs. $\mu = 2.28$ and $\mu = 2.21$)

3.1.5. Negative Consequences

Once the analysis was performed, the results showed that the students were partially in disagreement about the negative consequences derived from the use of the smartphone (M = 2.31, SD = 0.84).

For the comparison of means, Student's *t*-test for independent samples showed the existence of statistically significant differences for country, exclusively ($t = 8.891$, $p = 0.000$), with the means for the Colombian students being higher than for the Spanish ones (2.36 vs. 2.14).

Lastly, for the age and the macro area, the analysis of variance (ANOVA) indicated statistically significant differences in both cases. In the first case [F (3, 4005) = 13.181; $p = 0.000$], finding the differences in means of the students aged between 18 and 20, with respect to the rest ($\mu = 2.42$ vs. $\mu = 2.30$, $\mu = 2.29$ and $\mu = 2.21$), and in the second [F (4, 4003) = 9.024; $p = 0.000$], the differences in means were found in the students belonging to the macro area of Social and Judicial Sciences with respect to those in Arts and Humanities and Engineering and Architecture, respectively ($\mu = 2.23$ vs. $\mu = 2.48$, $\mu = 2.39$).

3.1.6. Social Motivations

The results obtained showed that the students were partially in disagreement with respect to the social motivations derived from the use of the mobile phone (M = 3.22, SD = 1.03).

Student's *t*-test for independent samples, performed to detect possible differences with respect to country and gender, highlighted statistically significant differences only in the first case, $t = 5.079$, $p = 0.000$, with the Colombian students the ones who had the highest scores as compared to the Spanish ones (M = 3.27 vs. M = 3.09).

Also, the ANOVA performed to verify statistically significant differences between age and macro area, showed that there were no statistically significant differences in any of the cases.

3.2. Correlational Analysis

This section addresses the correlational study between the six dimensions of the questionnaire. The data obtained, after the use of a Spearman's correlation test to observe the relationship between the six dimensions of the scale, can be observed below (see Table 3):

Table 3. Results of the bivariate correlations of the items from the 6 dimensions of the questionnaire.

		T	E	D	A	C	M
T	r	1.000	0.443 **	0.466 **	0.444 **	0.592 **	0.307 **
	p		0.000	0.000	0.000	0.000	0.000
E	r		1.000	0.473 **	0.495 **	0.598 **	0.328 **
	p			0.000	0.000	0.000	0.000
D	r			1.000	0.754 **	0.589 **	0.354 **
	p				0.000	0.000	0.000
A	r				1.000	0.678 **	0.314 **
	p					0.000	0.000
C	r					1.000	0.334 **
	p						0.000
M	r						1.000
	p						

Note: T = Tolerance; E = Escape; D = Disconnection; A = Anxiety; C = Consequences; M = Motivations.
** The correlation is significant at 0.01 (2-tailed).

As a function of the data obtained, it is observed that all the dimensions were correlated amongst themselves, with these correlations being high and moderate [36], using as a reference the value of their respective coefficients. In this sense, the following had a high correlation: the dimension Disconnection with the dimension Anxiety, as Rho = 0.754 and $p = 0.000$; as well as this last with the dimension Negative consequences (Rho = 0.678 and $p = 0.000$). As for moderate correlations (as defined by the authors cited previously), these were observed in the rest of the dimensions, as they obtained values between 0.300 and 0.600.

3.3. Binary Logistic Regression Analysis

The best fitting model with respect to the self-perception shown by the university students from Spain and Colombia on the problematic use they display, was produced through a binary logistic regression. As this independent dependent variable is dichotomous in nature, with Yes/No values, the forward selection Wald test was utilized [37]. This model had a high specificity (98.8%) and high sensitivity (95%), with an overall percentage of 98% for the cut-off value of 50%, which indicates that the model classifies equally well those who consider themselves to have a self-perception on the problematic use of the Smartphone as well as those who do not.

The results shown from the analysis point out, for a sample of 4009 subjects, where 3205 answered No (79.9%) and 804 Yes (20.1%), that the goodness-of-fit of the best fitting model shows: a -2log of likelihood = 354.509, a Cox and Snell $R^2 = 0.599$; and an R^2 of Nagelkerke = 94.6% (0.946); aside from the value of the Hosmer and Lemeshow test which shows a good fit, as its significance is 0.988 (≥ 0.05 according to [38]). Likewise, the values obtained in the Chi-square test for step ($\chi^2 = 3663.81$, gl = 6 and $p < 0.05$); block ($\chi^2 = 3663.81$, gl = 6 and $p < 0.05$); and model ($\chi^2 = 3663.81$, gl = 6 and $p < 0.05$), shows that the fitted model is significantly differentiated from the initial or base model.

As shown in Table 4, it can be observed how all the factors that appear in the MPPUSA intervene in the explanation in a significant manner. From them and tending to their relevance according to the value of B, these would be ordered as Consequences; Tolerance; Disconnection; Escape; Motivations and Anxiety.

Thus, the logistic equation would be:

$$\begin{aligned} \text{Y (self-perception in the problematic use)} = -89.07 + 1.13\text{Tolerance} + 1.11\text{Escape} + \\ 1.12\text{Disconnection} + 0.54\text{Anxiety} + 1.18\text{Consequences} + 0.99\text{Motivations} \end{aligned} \quad (1)$$

Thus, it can be observed that all the factors resulting from the MPPUSA positively and significantly influence the self-perception of considering that a problem with the use of the smartphone exists (reference category).

Table 4. Binary logistic regression of the self-perception on the problematic smartphone use [a] of Spanish and Colombian university students with respect to the MPPUSA.

	B	E.T.	Wald	gl	Sig.	Exp (B)	C.I. 95% for EXP(B)	
							Lower	Upper
Tolerance	1.129	0.122	86.236	1	0.000	3.093	2.437	3.926
Escape	1.105	0.107	106.641	1	0.000	3.020	2.448	3.724
Disconnection	1.115	0.106	110.914	1	0.000	3.051	2.479	3.755
Anxiety	0.540	0.077	49.069	1	0.000	1.715	1.475	1.995
Consequences	1.179	0.091	166.350	1	0.000	3.252	2.718	3.890
Motivations	0.987	0.112	77.352	1	0.000	2.683	2.153	3.343
Constant	−89.073	6.691	177.220	1	0.000	0.000		

Note. [a] = Dependent variable: Self-perception of having a problematic use of the smartphone Yes and No.

When analyzing the explanatory model of the self-perception of the problematic Smartphone use, having in mind the country of origin as the selection variable, it is observed, for 2965 Colombian students, of which 2266 answered No (76.4%) and 699 Yes (23.6%), that the goodness-of-fit had the following values: a -2log of likelihood = 290.826; a Cox and Snell $R^2 = 0.630$; and a, R^2 from Nagelkerke = 94.8% (0.948); with the Hosmer and Lemeshow value = 0.999, which demonstrates a good fit, as its significance is ≥0.05 [38]. Also, it has a high specificity (98.5) and a high sensitivity (95%), with an overall percentage of 97.7% for the cut-off value of 50%, which indicates that the model classifies those who consider themselves to have a self-perception on the problematic use of the Smartphone as well as those who do not equally well.

Table 5 shows how all the factors obtained from the MPPUSA significantly intervene in the explanation, as it occurred in the general model, in the same order according to their relevance (B value).

Table 5. Binary logistic regression of the self-perception of the problematic smartphone use [a] of Colombian university students with respect to the MPPUSA.

	B	E.T.	Wald	gl	Sig.	Exp (B)	C.I. 95% for EXP(B)	
							Lower	Lower
Tolerance	1.124	0.131	73.716	1	0.000	3.077	2.381	3.977
Escape	1.062	0.113	87.622	1	0.000	2.892	2.315	3.612
Disconnection	1.089	0.114	91.520	1	0.000	2.970	2.376	3.712
Anxiety	0.495	0.080	37.976	1	0.000	1.640	1.401	1.919
Consequences	1.191	0.103	133.351	1	0.000	3.291	2.688	4.028
Motivations	0.952	0.121	61.350	1	0.000	2.590	2.041	3.286
Constant	−88.133	7.340	144.163	1	0.000	0.000		

Note. [a] = Dependent variable: Self-perception of having a problematic use of the Smartphone Yes and No.

In this occasion, the logistical equation is the following:

$$\begin{gathered}\text{Y (self-perception on the problematic use of Colombian students)} = -88.13 + \\ 1.12\text{Tolerance} + 1.06\text{Escape} + 1.09\text{Disconnection} + 0.50\text{Anxiety} + 1.19\text{Consequences} + \\ 0.95\text{Motivations}\end{gathered} \quad (2)$$

As a result, the self-perception of the Colombian students when considering if they have a problematic smartphone use (reference category), all the factors from the MPPUSA have a positive and significant influence.

Lastly, Table 6 shows the results of the explanatory model of the self-perception on the problematic use of the smartphone of Spanish university students (selection variable). Of the 1044 students, 939 who answered No (89.9%) and 105 Yes (10.1%). The goodness-of-fit of the best fitting model obtained the following values: a -2log of likelihood= 54.446; a Cox and Snell $R^2 = 0.451$; and a Nagelkerke $R^2 = 94.2\%$ (0.942); with the Hosmer and Lemeshow test obtaining a value of 1.0, which demonstrates a good fit, as its significance is ≥0.05 [38]. The specificity is high (99.4%), as well as its sensitivity (94.3%), with an overall percentage of 98.9%) for the cut-off value of 50%), which indicates that the model classifies those who consider themselves to have a self-perception on the problematic use of the Smartphone as well as those who do not equally well.

Table 6. Binary logistic regression of the self-perception on the problematic smartphone use [a] of Spanish university students with respect to the MPPUSA.

	B	E.T.	Wald	gl	Sig.	Exp (B)	C.I. 95% for EXP(B)	
							Lower	Lower
Tolerance	1.461	0.399	13.394	1	0.000	4.309	1.971	9.422
Escape	1.475	0.354	17.312	1	0.000	4.369	2.181	8.751
Disconnection	1.084	0.306	12.516	1	0.000	2.957	1.622	5.392
Anxiety	1.247	0.347	12.923	1	0.000	3.478	1.763	6.863
Consequences	1.266	0.232	29.758	1	0.000	3.548	2.251	5.593
Motivations	1.461	0.367	15.821	1	0.000	4.312	2.099	8.859
Constant	−108.940	19.958	29.795	1	0.000	0.000		

Note. [a] = Dependent variable: Self-perception of having a problematic use of the smartphone Yes and No.

Thus, the equation is:

$$\text{Y (self-perception of the problematic use of the smartphone of Spanish university students)} = -108.94 + 1.46\text{Tolerance} + 1.48\text{Escape} + 1.08\text{Disconnection} + 1.25\text{Anxiety} + 1.27\text{Consequences} + 1.46\text{Motivations}. \quad (3)$$

On Table 6, it is observed that the factors that appear in the MPPUSA significantly intervene in the explanation; however, when taking into account their relevance (B value), it is observed that the order is different from the general model and the order found for the Colombian students, being: Escape; Motivations; Tolerance; Consequences; Anxiety; Disconnection.

In summary, for the self-perception of the Spanish students who consider if they have a problem with the use of smartphones (reference category), all the factors found from the MPPUSA have a positive and significant influence.

When comparing the Spanish students with the Colombian ones, it is observed that although all the factors contribute to the best fitting models, in Spain the factor Escape had more weight than in Colombia, with Consequences being more significant for the latter.

The analysis of the explanatory model of the self-perception on the problematic Smartphone use, with sex as the selection variable, shows that for 1601 male students, of which 1270 answered No (79.3%) and 331 Yes (20.7%), the goodness-of-fit had the following values: a -2log of likelihood = 134.565; a Cox and Snell $R^2 = 0.607$; and a R^2 from Nagelkerke = 95% (0.950); with a Hosmer and Lemeshow value = 0.994. This demonstrates a good fit, as its significance is ≥0.05 [38]. Also, it had a high specificity (98.8%) and a high sensitivity (94.3%), with an overall percentage of 97.9% for the cut-off value of 50%, indicating that the model classifies both of those who consider themselves to have a self-perception on the problematic use of the smartphone as well as those who do not, equally well.

Table 7 shows that all the factors obtained from the MPPUSA significantly intervened in the explanation, as in the general model, in the same order according to their relevance (B value).

Table 7. Binary logistic regression of the self-perception of the problematic Smartphone use [a] of male university students with respect to the MPPUSA.

	B	E.T.	Wald	gl	Sig.	Exp (B)	C.I. 95% for EXP(B)	
							Inferior	Superior
Tolerance	1.258	0.210	35.838	1	0.000	3.517	2.330	5.308
Escape	1.223	0.175	48.813	1	0.000	3.397	2.411	4.787
Disconnection	1.164	0.179	42.275	1	0.000	3.204	2.256	4.552
Anxiety	0.598	0.121	24.422	1	0.000	1.819	1.435	2.307
Consequences	1.216	0.153	63.448	1	0.000	3.373	2.501	4.550
Motivations	0.899	0.174	26.843	1	0.000	2.457	1.749	3.453
Constant	−93.506	11.329	68.122	1	0.000	0.000		

Note. [a] = Dependent variable: Self-perception of having a problematic use of the smartphone Yes and No.

On this occasion, the logistical equation is the following:

$$\text{Y (self-perception on the problematic use of male students)} = -93.51 + 1.26\text{Tolerance} + 1.22\text{Escape} + 1.16\text{Disconnection} + 0.60\text{Anxiety} + 1.22\text{Consequences} + 0.90\text{Motivations} \quad (4)$$

As a result, all the factors from the MPPUSA have a positive and significant influence on the self-perception of the male students when considering if they have a problematic use of the smartphone (reference category).

Table 8 shows the results of the explanatory model for female university students (selection variable). Of these 2408 students, 1935 answered No (80.4%) and 473 Yes (19.6%). The goodness-of-fit of the best fitting model had the following values: a -2log of likelihood = 215.28; a Cox and Snell $R^2 = 0.594$; and a Nagelkerke $R^2 = 94.5\%$ (0.945); with the Hosmer and Lemeshow test obtaining a value of 0.897, which demonstrates a good fit, as its significance was ≥0.05 [38]. The specificity and the sensitivity were high (98.7% and 94.3%, respectively), with an overall percentage of 97.8% for the cut-off value of 50%, indicating that the model classifies both of those who consider themselves to have a self-perception on the problematic use of the Smartphone as well as those who do not, equally well.

Table 8. Binary logistic regression of the self-perception on the problematic smartphone use [a] of female university students with respect to the MPPUSA.

	B	E.T.	Wald	gl	Sig.	Exp (B)	C.I. 95% for EXP(B)	
							Lower	Lower
Tolerance	1.086	0.154	49.572	1	0.000	2.963	2.190	4.010
Escape	1.029	0.140	54.110	1	0.000	2.799	2.127	3.682
Disconnection	1.113	0.138	64.931	1	0.000	3.045	2.323	3.992
Anxiety	0.483	0.103	21.964	1	0.000	1.621	1.324	1.983
Consequences	1.194	0.120	98.472	1	0.000	3.301	2.607	4.179
Motivations	1.081	0.152	50.779	1	0.000	2.947	2.189	3.967
Constant	−88.400	8.595	105.794	1	0.000	0.000		

Note. [a] = Dependent variable: Self-perception of having a problematic use of the smartphone Yes and No.

Thus, the equation is

$$\text{Y (self-perception of the problematic use of the smartphone of female university students)} = -88.4 + 1.09\text{Tolerance} + 1.03\text{Escape} + 1.11\text{Disconnection} + 0.48\text{Anxiety} + 1.19\text{Consequences} + 1.08\text{Motivations}. \quad (5)$$

In Table 8, it is observed that the factors from the MPPUSA significantly intervened in the explanation; however, when taking into account their relevance (B value), it was observed that the order was different from both the general model and the one for the male students. Thus, for females, the order was: Consequences; Disconnection; Tolerance; Motivations; Escape; Anxiety.

Having in mind age as the selection variable, the explanatory model for 1237 university students aged from 18 to 20 years old, of whom 979 answered No (79.1%) and 258 Yes (20.9%), had the following goodness-of-fit values: a -2log of likelihood = 146.052; a Cox and Snell $R^2 = 0.596$; and a R^2 from Nagelkerke = 93% (0.930); with the Hosmer and Lemeshow value = 0.993, which demonstrates a good fit, as its significance was ≥0.05 [38]. Also, it had a high specificity (98.1%) and a high sensitivity (93.8%), with an overall percentage of 97.2% for the cut-off value of 50%, which indicates that the model classifies those who considered themselves to have a self-perception on the problematic use of the Smartphone as well as those who do not, equally well.

Table 9 shows how all the factors obtained from the MPPUSA significantly intervene in the explanation, as it occurred in the general model, in the same order according to their relevance (B value).

Table 9. Binary logistic regression of the self-perception of the problematic smartphones use [a] of university students aged between 18 and 20 years old with respect to the MPPUSA.

	B	E.T.	Wald	gl	Sig.	Exp (B)	C.I. 95% for EXP(B)	
							Inferior	Superior
Tolerance	0.875	0.179	23.795	1	0.000	2.399	1.688	3.410
Escape	0.951	0.159	36.032	1	0.000	2.590	1.898	3.533
Disconnection	0.988	0.153	41.600	1	0.000	2.686	1.989	3.627
Anxiety	0.490	0.117	17.473	1	0.000	1.632	1.297	2.054
Consequences	1.034	0.129	64.056	1	0.000	2.813	2.184	3.625
Motivations	0.829	0.168	24.417	1	0.000	2.292	1.649	3.185
Constant	−76.999	9.391	67.220	1	0.000	0.000		

Note. [a] = Dependent variable: Self-perception of having a problematic use of the smartphone Yes and No.

On this occasion, the logistical equation is the following:

$$\text{Y (self-perception on the problematic use of smartphone of students between 18 and 20 years)} = -76.99 + 0.88\text{Tolerance} + 0.95\text{Escape} + 0.99\text{Disconnection} + 0.49\text{Anxiety} + 1.03\text{Consequences} + 0.83\text{Motivations} \quad (6)$$

As a result, all the factors of the MPPUSA have a positive and significant influence on the self-perception of university students aged between 18 and 20 years old when considering whether they have a problematic use of smartphones (reference category).

Table 10 shows the results of the explanatory model of students aged from 21 to 23 years old (age selection variable). Of these 860 students, 696 answered No (80.9%) and 164 Yes (19.1%). The goodness-of-fit of the best fitting model obtained the following values: a -2log of likelihood= 45.820; a Cox and Snell $R^2 = 0.602$; and a Nagelkerke $R^2 = 96.7\%$ (0.967); with the Hosmer and Lemeshow test obtaining a value of 0.995, which demonstrates a good fit, as its significance is ≥0.05 [38]. Both the specificity and sensitivity were high (99.1% and 97%, respectively), with an overall percentage of 98.7% for the cut-off value of 50%, which indicates that the model classifies those who consider themselves to have a self-perception on the problematic use of the smartphone as well as those who do not equally well.

Table 10. Binary logistic regression of the self-perception on the problematic smartphone use [a] of university students between 21 to 23 years of age with respect to the MPPUSA.

	B	E.T.	Wald	gl	Sig.	Exp (B)	C.I. 95% for EXP(B)	
							Lower	Lower
Tolerance	1.323	0.382	12.004	1	0.001	3.753	1.776	7.931
Escape	1.339	0.338	15.669	1	0.000	3.816	1.966	7.406
Disconnection	1.594	0.362	19.439	1	0.000	4.924	2.424	10.003
Anxiety	0.770	0.226	11.617	1	0.001	2.159	1.387	3.361
Consequences	1.952	0.446	19.174	1	0.000	7.044	2.940	16.876
Motivations	1.939	0.560	12.001	1	0.001	6.952	2.321	20.821
Constant	−134.749	29.206	21.286	1	0.000	0.000		

Note. [a] = Dependent variable: Self-perception of having a problematic use of the smartphone Yes and No.

Thus, the equation is:

$$\text{Y (self-perception on the problematic use of smartphone of students between 21 and 23 years)} = -134.75 + 1.32\text{Tolerance} + 1.34\text{Escape} + 1.59\text{Disconnection} + 0.77\text{Anxiety} + 1.95\text{Consequences} + 1.94\text{Motivations}. \quad (7)$$

For university students between the ages of 24 and 26, the results of the explanatory self-perception model had the following aspects (see Table 11): of the 546 students, 442 answered No (81%) and 104 Yes (19%). The goodness-of-fit of the best fitting model obtained the following values: a -2log of likelihood = 56.168; a Cox and Snell $R^2 = 0.581$; and a Nagelkerke $R^2 = 93.4\%$ (0.934); with the Hosmer and Lemeshow test obtaining a value of 0.876, which demonstrates a good fit, as its significance is ≥0.05 [38]. The specificity and sensitivity were high (99.1% and 94.2%, respectively), with an overall percentage of 98.2% for the cut-off value of 50%, indicating that the model classifies those who consider themselves to have a self-perception on the problematic use of the smartphone as well as those who do not, equally well.

Table 11. Binary logistic regression of the self-perception on the problematic smartphone use [a] of university students between 24 to 26 years of age with respect to the MPPUSA.

	B	E.T.	Wald	gl	Sig.	Exp (B)	C.I. 95% for EXP(B)	
							Lower	Lower
Tolerance	1.058	0.257	16.939	1	0.000	2.882	1.741	4.771
Escape	0.922	0.247	13.943	1	0.000	2.514	1.550	4.078
Disconnection	0.790	0.246	10.282	1	0.001	2.203	1.359	3.569
Anxiety	0.472	0.191	6.071	1	0.014	1.603	1.101	2.332
Consequences	1.060	0.203	27.378	1	0.000	2.885	1.940	4.291
Motivations	0.829	0.295	7.875	1	0.005	2.291	1.284	4.088
Constant	−77.268	14.469	28.517	1	0.000	0.000		

Note. [a] = Dependent variable: Self-perception of having a problematic use of the smartphone Yes and No.

Thus, the equation is:

$$\text{Y (self-perception on the problematic use of smartphone of students between 24 and 26 years)} = -77.27 + 1.06\text{Tolerance} + 0.92\text{Escape} + 0.79\text{Disconnection} + 0.47\text{Anxiety} + 1.06\text{Consequences} + 0.83\text{Motivations}. \quad (8)$$

Table 12 shows the results of the explanatory model of students older than 26 (age selection variable). Of the 1366 students, 1088 answered No (79.6%) and 278 Yes (20.4%). The goodness-of-fit of the best fitting model obtained the following values: a -2log of likelihood = 86.276; a Cox and Snell $R^2 = 0.612$; and a Nagelkerke $R^2 = 96.3\%$ (0.963); with the Hosmer and Lemeshow test obtaining a value of 0.987, which demonstrates a good fit, as its significance is ≥0.05 [38]. Both the specificity and its sensitivity were high (99.2% and 96.4%, respectively), with an overall percentage of 98.6% for the cut-off value of 50%, which indicates that the model classifies those who consider themselves to have a self-perception on the problematic use of the Smartphone as well as those who do not, equally well.

Table 12. Binary logistic regression of the self-perception on the problematic smartphone use [a] of students older than 26 years old with respect to the MPPUSA.

	B	E.T.	Wald	gl	Sig.	Exp (B)	C.I. 95% for EXP(B)	
							Lower	Lower
Tolerance	1.721	0.346	24.689	1	0.000	5.591	2.836	11.024
Escape	1.562	0.285	30.009	1	0.000	4.768	2.727	8.337
Disconnection	1.486	0.275	29.122	1	0.000	4.418	2.576	7.579
Anxiety	0.747	0.193	14.941	1	0.000	2.111	1.445	3.083
Consequences	1.476	0.241	37.435	1	0.000	4.375	2.727	7.019
Motivations	1.222	0.229	28.590	1	0.000	3.394	2.169	5.312
Constant	−117.614	18.738	39.397	1	0.000	0.000		

Note. [a] = Dependent variable: Self-perception of having a problematic use of the smartphone Yes and No.

Thus, the equation is:

$$\begin{aligned} &\text{Y (self-perception on the problematic use of smartphone of students older than 26 years)} \\ &= -117.61 + 1.72\text{Tolerance} + 1.56\text{Escape} + 1.49\text{Disconnection} + 0.75\text{Anxiety} + \\ &1.48\text{Consequences} + 1.22\text{Motivations} \end{aligned} \quad (9)$$

On Table 12, it is observed that the factors that appeared in the MPPUSA significantly intervened on the explanation; However, having in mind its relevance (B value), it was observed that the order was different from the general model and the order found for different ages, being Tolerance; Escape; Disconnection; Consequences; Motivations; Anxiety for those over 26 years old.

4. Discussion

We are in agreement with [39] in that the technologies themselves are not elements that can provoke a problematic behavior in its use, but it is the individuals who develop this behavior, which may or may not be problematic [9].

In this sense, after utilizing the MPPUSA, the results obtained provided us with a profile of a young Spanish and Colombian university student who has a problematic behavior with the mobile device as a function of 6 factors relative to the tolerance towards the time spent using the device, its regard as a means of escape or the disconnection to the world that surrounds them, linked to the anxiety that could be caused if they do not use it, just as the results reached by [17], who showed that the university students with whom the same instrument was utilized, felt better when they utilized the smartphone to evade or avoid situations they found themselves in. The results also showed that the participating students felt that not having the device available gave them anxiety and stress, the negative consequences its use could have due to their using it most of their time in different ways, and lastly, the social motivations related to its use of not, in line with data from [12,17,18,34]. It is meaningful that these results coincide with those from [40], who reached similar results after the use of an instrument called Questionnaire of Experiences Related to the Mobile Phone (Spanish acronym: CERM) with university students, although its short length (10 items) did not analyze in detail aspects such as feelings of avoidance or FOMO.

Taking into account the objectives established, it was verified that the students self-perceived not having a problematic use of the device. This result should be pondered, given that the study by [13] pointed that the university students have control over the addiction to the Internet (85.17%), and in our case, the perception of not having a problematic use was 79.9%. This shows that the university students younger than 35 years old do not consider a problematic or addictive use of the Smartphone or the Internet.

As opposed to results reached by [17,18,21], after the use of the same instrument along the same line, and also as opposed from results reached by [41], who reported an increasing trend of cell phone use after the use of MMPUS. Nevertheless, it is significant that in general, the Colombian students were more aware of the negative consequences that its use could imply, pointing out that this is linked to the time they spend connected or using it in a generalized manner or having to hide this usage time from family and friends, data that coincides with those from [42], who after the application of the HUTL, pointed out the same aspects for this study population. They are also more aware that their state of rest and their academic performance will the affected due to the excessive use of the Smartphone [18,43], with this aspect being common to all the research studies performed with other instruments such as MMPUS, CERM or HUTL [17,40–43]. Nevertheless, these behaviors that are detrimental, in one way or another, to daily life, are also observed in the use of substances o addiction to the Internet, where the study by [44] points that the altered behaviors related to the Internet, taking into account the sex of the students, could be due to the use or contents consumed. Therefore, we should ask ourselves if the devices are understood as mere devices or associated to the consumption of contents they enjoy.

They also indicated that when they are bored, sad or alone, they use the mobile phone. On other hand, they also commented that if they did not have a device such as this, it would be difficult for their friends to locate them and that they did not like it when the turned their phones off, and this is where the importance a smartphone has in their social relations is derived from just as the results found after the use of the MMPUS [42]. On the other hand, the Spanish students have features of the FOMO syndrome, manifesting that they are worried about not having the device with them, and they would miss calls or messages, and that it is difficult for them to turn the phone off [12,18,29,43].

If we center our attention to the age of the participants, it is verified, as in the work by [12], that the younger students from both countries had a PSU as compared to those who were older, as opposed to the results from [45], given that significant differences were found in the factors found that referred to the use of the mobile phone as a means of escape, to disconnect, anxiety and the negative consequences due to its use.

Gender could be pointed out as an element that determines the attitude towards the mobile phone [32,43] an aspect that coincides with other tests applied [40–42]. As opposed to the work by [12,41–43], it was the women, and more specifically Spanish women, the ones who had a problematic attitude with the device, such as the shown in the results from works by [18,20,46–49], as the participating sample expressed feeling anxiety and disconnection with their surroundings if they did not have a working phone [47–50].

On the other hand, if attention is placed on the macro area of study, as referred to the awareness of the consequences, the Arts and Humanities students, as well as the Engineering and Architecture ones, as compared to the other macro areas, such in the case of Social Sciences, are ones who had features of the so-called FOMO [10,29,45]. As for the elements that comprise the factor Tolerance, it was verified that the students from the macro areas of Experimental Sciences, Engineering and Architecture, were more aware of having less tolerance (being in a bad mood, using the device due to boredom or for feeling alone, and to make superfluous calls) [46,50]. It should be highlighted that the students from the area of Health Sciences did not have any prevalence of problematic behavior.

With respect to the factors reached in this study, it can be confirmed that the previous works [30,34,49,50] have shown the existence of 6 factors, and that the grouping of the items, aside from the correlations between themselves, determine the elements that define a problematic behavior towards smartphones, and about which work models should be developed to be able to perform an intervention.

Lastly, it was verified that the model of 6 factors created in line with the one from [35], follows the trail of the data from [17,30,49,50], and points to the Colombian students being closer to this model as compared to the Spanish students, as it describes the factor Consequences as being more significant in the model, as compared to Escape from the Spanish group. On the other hand, age and sex do not exactly follow the general model.

The section should not end without indicating the limitations of this international study, where the Latin American context, as well as the geography of the terrain, could define uses, trends, and views. One of these limitations could language used. Although in this case the language used was Spanish, it should be point out that just as with other languages, it has local variations depending on the country we find ourselves in. Therefore, the initial instrument developed in the Spanish utilized in Spain had to be adapted to Colombian Spanish, resulting in a delay of the research study.

On the other hand, although the sample was large, the Spanish representation should be increased, not only for the generalization of the data, but also for the performing a comparison that is not catalogued as cross-sectional, and that contributes a PSU analysis model for the entire Latin American context, which entails the intention of widening the study to other countries in the American continent.

5. Conclusions

Based on the above, it can be concluded that on the one hand, women in general, and in particular the Spanish women who studied within the macro area of Social Sciences, were the ones who had a problematic use of the mobile phone as compared to the men in both countries, showing the same factors as a pharmacological addiction. On the other hand, the students who resided in Colombia utilized the smartphone as a means of escape from boredom. These same students are aware of the consequences of the excessive use of the device on their personal and academic life. Also, the younger Spanish students had symptomatic signs of the FOMO syndrome. Along this line, in both countries, the students who showed these signs were enrolled in the area of Social Sciences. With respect to the Experimental Sciences and Engineering and Architecture students, these recognized spending too many hours being preoccupied with the phone. It should be pointed out that in the macro areas of Arts and Humanities, and Engineering and Architecture, their students were aware of the negative consequences of its use on their personal and academic lives.

Ultimately, it should be indicated, specifically for the differences found according to age and the country of origin, that the younger students, independently of the country, had a problematic use of smartphones and that Colombian students were closer to the general model, while for the Spanish ones, the order of relevance of each of the factors varied, while the general model does not fit the age or sex of the students.

Author Contributions: Conceptualización, V.M.-D.; curación de datos, J.M.M.-G. and B.-E.S.-R.; investigación, J.M.M.-G. and B.-E.S.-R.; metodología, J.M.M.-G. and B.-E.S.-R.; administración de proyecto, V.M.-D.; supervisión, V.M.-D. and B.-E.S.-R.; redacción-borrador original, V.M.-D. All authors have read and agreed to the published version of the manuscript.

Funding: This research received no external funding.

Conflicts of Interest: The authors declare no conflict of interest.

References

1. Babakr, Z.H.; Majeed, K.; Soran, P.M.; Kakamad, K. Internet addiction in kurdistan university students: Prevalence and association with self-control. *Euro. J. Edu. Res.* **2019**, *8*, 867–873. [CrossRef]
2. Longstreet, P.; Brooks, S.; Gonzalez, E.S. Internet addiction: When the positive emotions are not so positive. *Tech. Soc.* **2019**, *57*, 76–85. [CrossRef]
3. Mendes, I.; Silvia, I. Uso problemático da internet em adultos: Que relação com sintomas clínicos? *Revista De Estudios E Investigación En Psicología Y Educación* **2017**, *13*, 129–133. [CrossRef]
4. Internet World Stast. Usage and Population Statistics. 2020. Available online: https://www.internetworldstats.com/top20.htm (accessed on 10 April 2020).
5. DITRENDIA. Informe Ditrendia: Mobile en España y en el mundo 2019. 2019. Available online: https://ditrendia.es/informe-mobile-espana-mundo-2019/ (accessed on 15 April 2020).
6. DANE. Indicadores básicos de TIC en Hogares. 2019. Available online: https://www.dane.gov.co/index.php/estadisticas-por-tema/tecnologia-e-innovacion/tecnologias-de-la-informacion-y-las-comunicaciones-tic/indicadores-basicos-de-tic-en-hogares (accessed on 10 April 2020).

7. Fernández-Castillo, E.; Choliz, M.; Martínez, A.C.; Rodríguez, Y.C.; Monteagudo, P. Evaluación de las propiedades psicométricas del Test de dependencia al Teléfono Móvil en estudiantes universitarios cubanos. *Katharsis Rev. de Cienc. Soc.* **2019**, *27*, 73–85.
8. Besolí, G.; Palomas, N.; Chamorro, A. Uso del móvil en padres, niños y adolescentes: Creencias a cerca de sus riesgos y beneficios. *Aloma* **2018**, *36*, 29–39.
9. Fombona, J.; Roza Martin, P. Uso de los dispositivos móviles en educación infantil. *Edmetic* **2016**, *5*, 158–181. [CrossRef]
10. Kumcagiz, H.; Gunduz, Y. Relationship between psychological well-being and smartphone addiction of university students. *Int. J. High. Edu.* **2016**, *5*, 144–156. [CrossRef]
11. Panova, T.; Carbonell, J. Smartphone addiction really addiction? *J. Behav. Addic.* **2018**, *7*, 250–259. [CrossRef]
12. Romero, J.M.; Aznar, I. Análisis de la adicción al smartphone en estudiantes universitarios. *Revista de Educación a Distancia* **2019**, 19. [CrossRef]
13. Sánchez Vega, E.; Cantero-Téllez, R.; Rodríguez-Martínez, M.C. Adicción a internet en futuros profesores. Un estudio de caso con alumnado universitario. *Int. J. Technol. Educ. Innov.* **2016**, *2*, 46–53. [CrossRef]
14. Cabrera, Y. Hábitos de uso y consumo de la generación millennial en sus teléfonos móviles. adComunica. *Sci. J. Stra. Tren. Inn. Comm.* **2019**, *18*, 21–40. [CrossRef]
15. Rodríguez, H.; Restrepo, L.; García, G. Habilidades digitales y uso de teléfonos inteligentes (smartphones) en el aprendizaje en la educación superior. *Virt. J. Uni. Cat. Nor.* **2017**, *50*, 126–142.
16. Vázquez-Cano, E.; Sevillano García, L. El smartphone en la educación superior. Un estudio comparativo del uso educativo, social y ubicuo en universidades españolas e hispanoamericanas. *Signo y Pensamiento* **2015**, *34*, 132–149. [CrossRef]
17. Marín, V.; Vega, E.; Sampedro, E. Uso problemático del smartphone en estudiantes universitarios. *Rev. Revista Española De Drogodependencias* **2018**, *43*, 62–76.
18. Olivencia-Carrión, M.A.; Ramírez-Uctes, I.; Holgado-Tello, P.; López-Torrecillas, F. Validation of a Spanish questionnaire on mobile phone abuse. *Fron. Psych.* **2018**, *9*, 824. [CrossRef]
19. Elhai, J.D.; Dvorak, R.D.; Levine, J.C.; Hall, B.J. Problematic smartphone use: A conceptual overview and systematic review of relations with anxiety and depression psychopathology. *J. Affec. Disor.* **2017**, *207*, 251–259. [CrossRef]
20. Roberts, J.A.; Yaya, L.H.; Manolis, C. The invisible addiction: Cell-phone activities and addiction among male and female college students. *J. Behav. Addic.* **2014**, *3*, 254–265. [CrossRef] [PubMed]
21. Roberts, J.A.; Pulling, C.; Manolis, C. I need my smartphone a hierarchical model of personality and cell-phone addiction. *Perso. Indi. Diff.* **2015**, *79*, 13–19. [CrossRef]
22. Sigerson, L.; Li, A.Y.-L.; Cheung, M.W.-L.; Cheng, C. Examining common information technology addictions and their relationships with non-technology-related addictions. *Comp. Hum. Behav.* **2017**, *75*, 520–526. [CrossRef]
23. American Psychiatric Association (APA). *Diagnostic and Statistical Manual of Mental Disorders;* American Psychiatric Association: Arlington, VA, USA, 2013.
24. Biglu, M.H.; Ghavami, M. Factors influencing dependence on mobile phone. *J. Anal. Res. Clin. Med.* **2016**, *4*, 158–162. [CrossRef]
25. Bunyalug, M.; Kanchanakhan, N. Effect of using smartphone before bed on sleep quality among undergraduate students at Chulalongkorn University. *Thai. J. Health Res.* **2017**, *31* (Suppl. 2), S225–S231. [CrossRef]
26. Echeburúa, E.; del Corral, P. Adicción a las nuevas tecnologías y a las redes sociales en jóvenes: Un nuevo reto. *Adicciones* **2010**, *22*, 91–96. [CrossRef]
27. Rozgonjuk, D.; Levine, J.C.; Hall, B.J.; Elhai, J.D. The association between problematic smartphone use, depression and anxiety symptom severity, and objectively measured smartphone use over one week. *Comp. Hum. Behav.* **2018**, *87*, 10–17. [CrossRef]
28. Zsidó, A.N.; Darnai, G.; Inhóf, O.; Perlaki, G.; Orsi, G.; Nagy, S.; Lábadi, B.; Lénárd, K.; Kovács, N.; Dóczi, T.; et al. Differentiation between young adult Internet addicts, smokers, and healthy controls by the interaction between impulsivity and temporal lobe thickness. *J. Behav. Addict.* **2019**, *8*, 35–47. [CrossRef] [PubMed]
29. Przybylski, A.K.; Murayama, K.; DeHaan, C.R.; Gladwell, V. Motivational, emotional, and behavioral correlates of fear of missing out. *Comp. Hum. Behav.* **2013**, *29*, 1841–1848. [CrossRef]

30. Ruíz, C. Perfil de uso del teléfono móvil e internet en una muestra de universitarios españoles: ¿usan o abusan? *Bordón* **2016**, *68*, 131–145. [CrossRef]
31. Polo, M.I.; Mendo, S.; León, B.; Felipe, E. Abuso del móvil en estudiantes universitarios y perfiles de victimización y agresión. *Adicciones* **2017**, *29*, 245–255. [CrossRef] [PubMed]
32. Smith, A. US smartphone use in 2015. 2015. Available online: https://www.pewresearch.org/internet/2015/04/01/us-smartphone-use-in-2015/ (accessed on 18 April 2020).
33. Otzen, T.; Manterola, C. Técnicas de muestreo sobre una población a estudio. *Mor. Int. J.* **2017**, *35*, 227–232. [CrossRef]
34. López-Fernández, O.; Honrubia-Serrano, L.; Freixa-Blanxart, M. Adaptación española del "mobile phone problem use scale" para población adolescente. *Adiciones* **2012**, *24*, 123–130. [CrossRef]
35. López-Fernández, O.; Honrubia-Serrano, L.; Freixa-Blanxart, M.; Gibson, W. Prevalence of problematic mobile phone use in British Adolescents. *Cyberpsych. Behav. Soc. Netw.* **2014**, *17*, 1–8. [CrossRef]
36. Pérez, R.; García, J.L.; Gil, J.A.; Galán, A. *Estadística Aplicada A La Educación*; Pearson Prentice Hall: Madrid, Spain, 2009.
37. Berlanga-Silvente, V.; Vilà-Baños, R. Cómo obtener un modelo de regresión logística binaria con SPSS. *REIRE. Rev.* **2014**, *7*, 105–118. [CrossRef]
38. López-Roldán, P.; Fachelli, S. *Metodología De La Investigación Social Cuantitativa*; UAB: Barcelona, Spain, 2015.
39. García-Oliva, C.; Piqueras, J.A.; Marzo, J.C. Uso problemático de internet, el móvil y los videojuegos en una muestra de adolescentes alicantinos. *Heal. Addic.* **2017**, *17*, 189–200. [CrossRef]
40. Beranuy, M.; Chamarro, A.; Graner, C.; Carbonell, X. Validación de dos escalas breves para evaluar la adicción a Internet y el abuso de móvil. *Psicothema* **2009**, *21*, 480–485.
41. Medrano, J.L.; Rosales, F.L.; Loving, R.D. Conducta adictiva a las redes sociales y su relación con el uso problemático del móvil. *Acta De Investigación Psicológica* **2017**, *7*, 2832–2838. [CrossRef]
42. Olivencia-Carrión, M.A.; Pérez-Marfil, M.N.; Ramos-Revelles, M.B.; López-Torrecillas, F. Personalidad y su relación con el uso versus abuso del teléfono móvil. *Acción Psicológica* **2016**, *13*, 109–118. [CrossRef]
43. Ruiz-Palmero, J.; Sánchez-Rivas, E.; Gómez-García, M.; Sánchez, E. Future Teachers' Smartphone Uses and Dependence. *Educ. Sci.* **2019**, *9*, 194. [CrossRef]
44. Inhóf, O.; Zsidó, A.N.; Perlaki, G.; Orsi, G.; Lábadi, B.; Kovács, N.; Szente, A.; Dóczi, T.; Janszky, J.; Darnai, G. Internet addiction associated with right pars opercularis in females. *J. Behav. Addict.* **2019**, *8*, 162–168. [CrossRef]
45. González, N.; Estévez, A. El apoyo social percibido moderador entre el uso problemático de Internet y la sintomatología depresiva en jóvenes adultos. *Heal. Addic.* **2017**, *17*, 53–62. [CrossRef]
46. Carbonell, X.; Chamarro, A.; Griffiths, M.; Oberst, U.; Cladellas, R.; Talarn, A. Problematic Internet and cell phone use in Spanish teenagers and young students. *Anales De psicología* **2012**, *28*, 789–796. [CrossRef]
47. Loconi, C.; Alvitres, V.; Ortega, B.; Zapirain, I. Uso del Smartphone en estudiantes de 2° año escuela administración de empresas de universidad particular de chiclayo. *Manglar* **2017**, *12*, 63–70. [CrossRef]
48. Demirhan, E.; Randler, C.; Horzum, M.B. Is problematic mobile phone use explained by chronotype and personality? *Chronob. Int.* **2016**, *33*, 821–831. [CrossRef] [PubMed]
49. de la Villa, M.; Suárez, C. Factores de riesgo en el uso problemático de Internet y del teléfono móvil en adolescentes españoles. *Revista Iberoamericana De Psicología Y Salud* **2016**, *7*, 69–78. [CrossRef]
50. Aljomaa, S.S.; Al.Qudah, M.F.; Albursan, I.S.; Bakhiet, S.F.; Abduljabbar, S.F. Smartphone addiction among university students in the light of some variables. *Comp. H. Behav.* **2016**, *61*, 155–164. [CrossRef]

International Journal of
Environmental Research and Public Health

MDPI

Article

Assessing the Psychometric Properties of the Internet Addiction Test in Peruvian University Students

Arnold Alejandro Tafur-Mendoza [1], Julio César Acosta-Prado [2,3,*], Rodrigo Arturo Zárate-Torres [4] and Duván Emilio Ramírez-Ospina [3]

1 Research Center (CIUP), Universidad del Pacífico, Lima 15072, Peru; aa.tafurm@up.edu.pe
2 School of Business Science, Universidad del Pacífico, Lima 15072, Peru
3 School of Accounting, Economic and Business Sciences, Universidad de Manizales, Manizales 170001, Colombia; merca2@umanizales.edu.co
4 Colegio de Estudios Superiores de Administración, Bogota 111071, Colombia; rodrigo.zarate@cesa.edu.co
* Correspondence: jc.acostap@up.edu.pe; Tel.: +511-219-0100 (ext. 2660)

Received: 30 May 2020; Accepted: 24 July 2020; Published: 10 August 2020

Abstract: The use of the Internet has been gradually and unstoppably gaining ground in all areas of life, from recreational activities to how social relations are established. However, the existence of clinical cases indicates that the addictive use of the Internet is a problem that seriously affects some people. Among the instruments that measure this construct, the Internet Addiction Test (IAT) stands out. However, instrumental studies of this test are scarce in Latin America. The present study sought to analyze the psychometric properties of the IAT in a sample of 227 Peruvian undergraduate university students. Confirmatory factor analysis was used to provide validity evidence based on the internal structure, and evidence based on the relationship with other variables was also provided. Reliability was estimated through the ordinal alpha coefficient. The results indicated that the IAT adequately fits a bifactor model (with two specific factors, time/control and stress/compensate), obtaining good levels of reliability. Additionally, the IAT scores correlate significantly with the average number of hours per day on the internet and social skills. The results lead to the conclusion that the scores in the IAT have evidence of validity and reliability for its use.

Keywords: internet addiction test; university students; Peruvian sample; psychometric properties

---

1. Introduction

In the framework of a society in which both communication and the free flow of information are closely related to the development of the network, it is necessary to know to what extent reality and the virtual sphere intermingle. From the lowering of costs, the multiplicity of ways of accessing the network, and the advent of social networks, the Internet grows exponentially every day [1]. As studies in Sweden [2] and Spain [3] seem to indicate, the fierce proliferation of the network as a means of communication has brought negative consequences (cyberbullying, problematic Internet use, sexting, nomophobia, etc.) that have a stronger impact on the young population, having identified a series of problems among which stand out the addiction to this environment and that affect above all the social sphere of the individual.

However, the use of the Internet has expanded gradually and unstoppably in all areas of life [4]. The existence of clinical cases indicates that maladaptive use of the Internet is an existing problem that seriously affects some people, mainly those with special emotional needs and young people and adolescents [5]. The use or abuse of the Internet arises from disciplines, such as psychology or psychiatry [6]. In this sense, it is not surprising that the term Internet addiction was used for the first time by the psychiatrist Ivan Golberg in 1995 [7]. The literature on problematic use of the Internet shows a very varied terminology to describe the problems derived from the use of the Internet, among which are: Computer addiction, excessive use of the Internet, pathological use of the Internet, Internet dependence, compulsive use of the Internet, disorder by impulsive use, and compulsive use of the Internet or Internet addiction [8,9].

Kimberly Young [10], after reviewing a series of investigations, which indicated that some online users became addicted to the Internet in the same way that others became addicted to drugs, alcohol, or gambling, suggests the need to empirically investigate the concept of addictive use of the Internet. In addition to the review, the study sought to identify whether excessive use of the Internet can be considered addictive and to know the magnitude of the problems created by these abuses. However, Young's approaches to Internet addiction sparked a controversial debate among doctors and academics at the time [11].

On the other hand, there are proposals [12] that suggested that Internet addiction is a disorder that should have been considered in the fifth edition of the Diagnostic and Statistical Manual of Mental Disorders (DSM-V) since it is an impulsive-compulsive spectrum disorder [13], which consists of the use of computers. However, these recommendations or proposals were not reflected in the DSM-V, since the term Internet addiction does not appear in the manual [14]. Among the diagnoses referenced in the DSM-V, Internet gaming disorder (IGD) is strongly related to the pathological nature of Internet use. The excessive use of the Internet not only involves the reproduction of online games (for example, it also implies the excessive use of social networks, such as Facebook or Twitter). Therefore, IGD is within the Internet addiction. Thus, research on other excessive uses of the Internet should follow guidelines analogous to those suggested with IGD [14].

Among the theoretical models proposed to understand Internet addiction, Davis' cognitive-behavioral model [15] is the one that has had the most development and has been empirically tested in different contexts [16]. Two parts are distinguished in this model, one specific and the other general. The first involves excessive use of an aspect of the internet, while generalized use encompasses multiple excessive uses of the Internet [15]. Generalized use is associated with social support and social service uses of the Internet, where maladaptive cognitions are a strong predictor of this component and to a lesser extent of specific use [17]. Under this model, symptoms of Internet addiction are primarily affective or behavioral and are usually preceded or caused by cognitive symptoms, mediated by specific and general pathological Internet use [15,16]. The value of the cognitive-behavioral model lies in that it contemplates a continuum of severity regarding Internet use, allowing a better understanding by mental health specialists of the way and degree to which the excessive use of the internet can affect the lives of people [18].

To measure Internet addiction, various tests have been developed, which, mainly, are based on the diagnosis of this disorder, including the Problematic Internet Use Questionnaire (PIUQ) [19], the Compulsive Internet Use Scale (CIUS) [20], Internet-Related Problem Scale [21], and the Internet-Related Experiences Questionnaire (IREQ) [22]. The PIUQ, made up of 18 items, was built for the general population and measures problems related to Internet use through three subscales: Obsession, neglect, and control disorder. The CUIS is made up of 14 items that measure the severity of compulsive Internet use. The Internet-Related Problem Scale has 20 items that address DSM-IV symptoms for substance abuse: Tolerance, escape from other problems, reduced activities, loss of control, negative effects, withdrawal, craving, and introversion. The IREQ was developed in middle-school students and is made up of 10 items that measure possible Internet addiction based on intrapersonal and interpersonal experiences. However, the test that has had the most development and that has been adapted to various contexts is the Internet Addiction Test (IAT) [23]. Thus, the objective of this study was to analyze the psychometric properties of the IAT in a sample of Peruvian undergraduate university students.

2. Literature Review

For this study, in a search conducted in Scopus and Web of Science (WoS) on studies of the psychometric properties of the IAT, 46 studies were found (April 2020). Table 1 presents a summary of 39 of the 46 studies found, which were those that analyzed the internal structure of the IAT, an aspect that does not have a consensus in all the studies carried out. Most of the studies were carried out in Europe and Asia; in North America, only one study was found in the United States and another in Canada, while, in Latin America, the IAT has been studied in Colombia and Chile. Regarding the populations where these studies have been carried out, the majority were in university students, probably due to the accessibility of the sample and that it is the age group that spends the most time on the Internet (young people between 18 and 25 years old).

Table 1. Previous studies on psychometric properties of the Internet Addiction Test (IAT).

Study	Country	Sample Size	Factors	Data Analysis	Items	Reliability (α)
1. Widyanto et al. (2004) [24]	-	86	6	PCA	20	>0.50 [c]
2. Ngai (2007) [25]	Hong Kong	988	4	PCA	20	>0.60 [c]
3. Chang et al. (2008) [26]	Hong Kong	410	3	PCA, CFA [a]	20	>0.80 [d]
4. Khazaal et al. (2008) [27]	France	246	1	EFA, CFA	20	0.93
5. Widyanto et al. (2011) [28]	-	225	3	PCA	20	-
6. Panayides et al. (2012) [29]	Cyprus	604	1	PCA	20	0.89
7. Jelenchick et al. (2012) [30]	United States	215	2	EFA	20	>0.80 [c]
8. Barke et al. (2012) [31]	Germany	841	2	PCA, CFA	20	0.89
9. Puerta-Cortés et al. (2012) [32]	Colombia	1117	3	PCA	20	0.89
10. Faraci et al. (2013) [33]	Italy	485	1, 2	EFA, CFA	20, 18	>0.70 [c]
11. Watters et al. (2013) [34]	Canada	1948	2	CFA [b]	20	0.93
12. Pawlikowski et al. (2013) [35]	Germany	1049	2	PCA, CFA	11	>0.80 [c]
13. Lee et al. (2013) [36]	Korea	279	4	PCA	20	0.91
14. Hawi (2013) [37]	Lebanon	817	1	PCA, CFA	20	0.92
15. Lai et al. (2013) [38]	Hong Kong	844	3	CFA [a]	20	0.93
16. Pontes et al. (2014) [39]	Portugal	593	1	CFA	20	0.90
17. Karim et al. (2014) [40]	Bangladesh	177	4	PCA	18	0.89
18. Tsimtsiou et al. (2014) [41]	Greece	151	3	EFA	20	0.91
19. Chong et al. (2015) [42]	Malaysia	162	5	PCA	20	0.91
20. Fernández-Villa et al. (2015) [43]	Spain	963	2	EFA, CFA	19	0.91
21. Lu et al. (2015) [44]	Malaysia	104	6	EFA	20	0.93 [e]
22. Dhir et al. (2015) [45]	India	1914	1	EFA, CFA	20	0.88
23. Lai et al. (2015) [46]	Hong Kong, Japan and Malaysia	2535	3	CFA	20	-
24. Fioravanti et al. (2015) [47]	Italy	840	2	EFA, CFA	20	>0.80 [c]
25. Hawi (2015) [48]	Poland	1245	2	PCA, CFA	20	0.90
26. Kaya (2016) [49]	Turkey	990	4	EFA, CFA	20	0.92
27. Servidio (2017) [50]	Italy	659	2	PCA, CFA	18	0.89
28. Boysan et al. (2017) [51]	Turkey	455	1	PCA, CFA	20	0.93
29. Samaha et al. (2018) [52]	Lebanon	256	4	EFA, CFA	19	0.91
30. Waqas et al. (2018) [53]	Pakistan	522	1	EFA, CFA	20	0.90
31. Neelapaijit et al. (2018) [54]	Thailand	324	3	EFA, CFA	20	0.89
32. Tsermentseli et al. (2018) [55]	Greece	725	3	EFA, CFA [b]	19	>0.70 [f]
33. Hernández et al. (2018) [56]	Chile	425	2	CFA	10	0.85
34. Černja et al. (2019) [57]	Croatia	352	3	PCA, CFA	20	0.91
35. Tudorel et al. (2019) [58]	Romania	421	2	EFA, CFA	20	0.86
36. Ndasauka et al. (2019) [59]	Pakistan	506	4	EFA	20	0.88
37. Yaffe et al. (2019) [60]	Israel	180	2	PCA, CFA	18	>0.70 [c]
38. Talwar et al. (2019) [61]	Malaysia	307	3	PCA, CFA	19	>0.70 [c]
39. Lu et al. (2019) [62]	Malaysia	1120	4	EFA, CFA	17	-

Note. [a] Hierarchical model; [b] Bifactor model; [c] α coefficients of the factors; [d] Construct reliability; [e] Rasch model (person reliability); [f] ω coefficients of the factors; PCA = Principal Component Analysis; EFA = Exploratory Factor Analysis; CFA = Confirmatory Factor Analysis.

The reliability of the IAT has usually been estimated through the alpha coefficient, obtaining, in most cases, acceptable values above 0.70, and even exceeding 0.90. However, most of these studies have not considered the assumptions that this coefficient has, such as the one-dimensionality of the data, that the measurement model is tau-equivalent, and that there are no correlated errors [63]. Violation of any of these assumptions would distort the values reported in these studies. Additionally, there is no consideration of the ordinality of the items, which must be considered when working with Likert-type scales [64]. A meta-analysis of the internal consistency of the IAT (reliability generalization) indicates a combined alpha of 0.90 for samples of university students [65].

Regarding the internal structure of the IAT, the literature shows factorial models with one to six factors. Although, the diversity of models increases, because two studies may have the same number of factors, but they differ in the number of items used (not all studies end their analyses with the original 20 items but eliminate some of them because they do not fit their proposed model), in the addition of correlated errors (in many cases to improve the adjustment indices in the confirmatory factor analysis, without taking into account the substantive justification of the correlated errors [66]) or in the distribution of the items by the factor (two models can coincide in the number of factors and items, but the grouping of the items is different). This large number of models does not allow for a uniform IAT structure to be available, and for studies seeking to study this aspect in a particular sample, they have to resort to exploratory techniques or to test a large number of existing models.

An aspect linked to the above is the analysis technique used to find the factor structure in these studies. Almost half of the studies reported in Table 1 indicate that the principal component analysis (PCA) has been performed as the first or only technique, which is not suitable for psychological constructs, since it is a variable reduction technique; therefore, it seeks to explain the greater proportion of explained variance, considering both the variance shared by the items and the error variance. On the contrary, the factor analysis only considers the extraction of the factors the variance that the items share among themselves [67]. The use of the PCA could lead to obtaining structures that do not fit the data, especially since the measurement nature of the data is not considered.

In Peru, only one study of the psychometric properties of IAT [68] in middle-school students between the ages of 13 and 19 years was found. Item 1 was removed from the scale because it presented a low discriminatory capacity; therefore, this version had 19 items. Reliability was estimated through the alpha coefficient, obtaining a value of 0.870. On the other hand, the internal structure was evaluated by PCA, finding a factorial solution of four components. This structure was corroborated in the same sample through confirmatory factor analysis (CFA). It is important to note that the use of middle-school students and university students is different. University students use the Internet primarily for information search; also, they spend more time online and have a wider range of Internet uses than middle-school students [69].

Another source of validity evidence that has been explored in these background studies is evidence based on the relationship with other variables, specifically convergent evidence, where the hours of daily Internet use was used as a measure, in addition to the IAT [38,41,45,50]. In these studies, correlation coefficients between 0.29 and 0.48 were found. Additionally, Internet use has an impact on a person's social life, and studies have shown that Internet addiction is negatively related to social skills [70–72].

Considering all the above, the IAT is one of the most used instruments for measuring Internet addiction and that, in Peru, it has only worked with middle-school students in the adolescence stage but not in other samples, such as university students, which is precisely in this group where most adaptations have been made in other countries. Furthermore, the statistical procedures used in previous studies may be questionable and, in some cases, inadequate. Thus, the IAT does not have a version for Peruvian university students, so its psychometric properties are not known in this population and the extrapolation of results in other contexts would not be appropriate, since validity and reliability are not inherent in a test but correspond to the specific interpretations and uses of the scores obtained in a test [73]. Therefore, the objective of this study was to analyze the psychometric properties of the IAT in a sample of Peruvian undergraduate university students.

3. Materials and Methods

3.1. Design

According to the classification system of research designs in psychology [74], the objective of the study corresponds to an instrumental investigation, since the psychometric properties of a psychological instrument are analyzed in a specific sample. For the development of the study, various standards, guidelines, good practices, and recommendations were followed in instrumental studies in the behavioral and health sciences [73,75].

3.2. Participants

The selection of the participants was carried out through an intentional non-probability sampling [76], where the individuals in the sample were directly and selectively chosen, seeking to obtain similar proportions in each of the categories of the variables sex, year of study, and academic discipline. Regarding the sample size, a priori statistical power analysis was performed to determine the minimum recommended sample size. Statistical power is the ability of a statistical test (for example, U Mann–Whitney or H Kruskal–Wallis) to reject a null hypothesis when it is false; in other words, it is the probability of not committing the type II error [77]. The input parameters for this analysis were based on a simple two-tailed correlational model (similar to the one that will be carried out in the collection of validity evidence based on the relationship with other variables), the significance level (α) being 0.05, the expected effect size equal to 0.20 (recommended minimum value representing practical significance in social science data [78]), and an expected statistical power of 0.80 (recommended minimum in behavioral science [79]). The recommended minimum sample size was 191 (Figure 1).

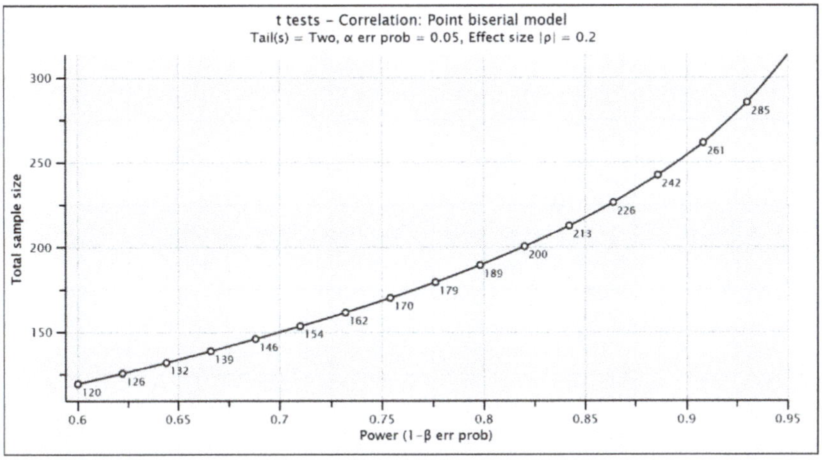

Figure 1. A priori statistical power analysis to determine the minimum recommended sample size.

The final sample was made up of 227 Peruvian undergraduate university students from a public university located in Metropolitan Lima (Peru). The ages of the participants were between 18 and 40 years (M = 20.81, SD = 2.92). The highest proportion of students were females (57.30%), in their second year of studies (30.00%), and belonged to professional careers in the engineering area (19.80%). Table 2 presents a more detailed description of the sample characteristics.

Table 2. Sociodemographic characteristics of participants (n = 227).

Characteristic	n	%
Gender		
Male	97	42.70
Female	130	57.30
Year of study		
First	24	10.60
Second	68	30.00
Third	63	27.80
Fourth	42	18.50
Fifth	30	13.20
Academic discipline		
Health Sciences	41	18.10
Humanities	30	13.20
Social Sciences	27	11.90
Basic sciences	43	18.90
Engineering	45	19.80
Economic-Business	41	18.10

3.3. Instruments

3.3.1. IAT—Internet Addiction Test

The Spanish version of the IAT was used (Appendix A), adapted to a sample of Colombian university students [32]. The IAT is made up of 20 items, all in a positive sense, on a five-point Likert-type scale (1 = rarely; 2 = occasionally; 3 = frequently; 4 = usually; 5 = always), the minimum score being 20 and the maximum score 100, where the student was asked to choose the alternative that best suits their reality. The IAT items were reviewed by the authors of this study, pointing out that it was not necessary to adjust their wording. To corroborate this assumption, a pilot study was carried out with 10 university students, who indicated that they had no problems in understanding the items.

In the adaptation study, the total scale had a good level of reliability (α = 0.89). Likewise, through an analysis of the main components, they found an internal structure of three factors (consequences for the use of the Internet, cognitive-emotional dimension, and time control), which together had an explained variance of 47.80%, and correlated positively with the number of hours of daily Internet access, although the magnitudes were low [32].

For the application of this study, in addition to the IAT, a sociodemographic and Internet usage file was added to collect information on the career that the participants studied, as well as the year of studies they were studying, age, gender, and the average daily hours spent on the Internet.

3.3.2. EHS—Social Skills Scale

The Social Skills Scale, developed by Gismero [80], is made up of 33 items. Five items are written in the direct sense, while 28 items are in an inverse sense, seeking to detect the lack of assertion or deficit in social skills. The original instrument is composed of six factors: (1) Self-expression in social situations, eight items; (2) defense of one's rights as a consumer, five items; (3) expression of anger or disagreement, four items; (4) assertiveness, six items; (5) making requests, five items; and (6) starting interactions with the opposite sex, five items. Each item is answered using a four-point response format (A = I do not identify myself at all; B = Rather it does not have to do with me, even if it ever happens to me; C = It describes me approximately, although I do not always act or feel like this; and D = Strongly agree, and I would feel or act like this in most cases). A higher global score indicates that the person has a higher level of social skills and better insertion in various contexts or situations.

This instrument has been studied in various countries, presenting adequate psychometric properties, good levels of reliability, adequate adjustment to the six-factor structure, and discriminant evidence [81]. In the present study, the scores on the items showed good internal consistency at the global level (ω = 0.890). Likewise, in the factors, the omega coefficient (ω) varied from 0.500 to 0.748.

3.4. Procedure

The data collection process began with the request of permits, both to the directors of the schools of each academic discipline and the teachers in charge of some courses for the application of the instrument in classrooms. Before the application of the IAT, the students were given an informed consent form that contained the objective of the evaluation and where they were guaranteed the confidentiality and anonymity of their answers, as well as the possibility of withdrawing from their participation at any time of the evaluation without consequence. Only those students who voluntarily signed the informed consent participated in the study.

The application was collective, with an average duration of 15 min. Data collection lasted approximately five weeks. At the end of each evaluation, the examiners reviewed the application protocols to verify that there are no unanswered items or items with more than one marked answer option. If one or both situations described occurred, the evaluated person was asked to correct and provide their definitive answer. After data collection, each evaluation protocol was coded for the elaboration of the database in a Microsoft Excel 2016 spreadsheet.

During the evaluation process, the ethical guidelines for working with humans outlined in the code of ethics of the American Psychological Association (APA) were followed. Additionally, the ethical principles of the Declaration of Helsinki were respected, including its recent updates and regulations for research on human beings.

3.5. Data Analysis

The statistical analysis was divided into four stages: (1) Item analysis; (2) validity evidence based on the internal structure of the test through CFA; (3) validity evidence based on the relationship with other variables from the convergent evidence; and (4) reliability analysis using the internal consistency method.

The descriptive analysis of the items was performed using the mean and standard deviation, as measures of central tendency and dispersion, respectively. As descriptive measures of the distribution of the items, the skewness and kurtosis coefficients were used, where values between −2.00 and 2.00 indicate that the items follow a normal distribution [82]. Likewise, item discrimination was estimated through the item-rest polyserial correlation, which considers the ordinality of the items, considering values above 0.30 as adequate [83]. Finally, possible floor and ceiling effects in the items were analyzed, that is, to identify the proportion of participants who chose the lowest alternative (floor effect) and the highest (ceiling effect), taking as acceptable effects those that were between 1% and 15% [84].

The CFA was carried out using the 23 models presented in Table 1, considering only those models made up of the 20 original items and whose factors had at least three items. The estimation method used was the weighted least squares with mean and variance adjusted (WLSMV), appropriate for observable ordinal variables and that performs well against slightly non-normal underlying distributions [85]. The WLSMV involves the use of the diagonal weighted least squares (DWLS) estimator with robust standard errors and a statistical test with adjusted mean and variance (using a scale-shifted approach). To assess the level of adequacy of the models, the following fit indices were used [86,87]: $S\chi^2/df < 2.00$, root mean square error of approximation (RMSEA) < 0.08, comparative fit index (CFI) > 0.90, Tucker–Lewis index (TLI) > 0.90, standardized root mean square residual (SRMR) < 0.08, and weighted root mean square residual (WRMR) < 1.00.

The convergent evidence involved correlating the scores obtained in the IAT with another measure that seeks to evaluate the same construct or other construct with which it is expected is correlated; in this case, these measures were the average of daily hours that the study participants spent on the Internet and a social skills scale. Before selecting the correlation statistic to be used, the presence of outliers in these two variables with which the IAT scores were correlated was examined. Univariate analysis of outliers was visually inspected through boxplots for each variable (including the six dimensions of the social skills scale). A few outliers were found in two factors of the social skills scale and in the average of daily hours that the study participants spent on the Internet. Therefore, a robust statistic was chosen for the correlation between variables, the skipped correlation coefficient. This coefficient is robust to slight changes in any distribution and has the advantage of treating outliers in a way that considers the general structure of a data set [88]. As a measure of the effect size, the coefficient of determination was used, considering its interpretation values of 0.04, 0.25, and 0.64 as the minimum, moderate, and strong effect, respectively [78].

The internal consistency of the items was estimated using the ordinal alpha coefficient [89] since it works with the inter-item polychoric correlation matrix. Values above 0.70 were considered acceptable [90].

A priori statistical power analysis to determine the minimum sample size required for the study was performed using the G*Power 3.1.9.7 software [91]. The other analyses were performed in the R software version 4.0.2 (R Foundation for Statistical Computing, Vienna, Austria, 2016) [92], using the following packages: pacman 0.5.1 [93], readxl 1.3.1 [94], tidyverse 1.3.0 [95], psych 1.9.12.31 [96], lavaan 0.6–6 [97], BifactorIndicesCalculator 0.2.0 [98], semTools 0.5–2.925 [99], and WRS 0.36 [100].

4. Results

4.1. Item Analysis

Table 3 presents the results of the item analysis. The means of the items were between 1.352 (item 20) and 3.084 (items 1 and 7), indicating that the participants' responses were concentrated on the lowest alternatives (rarely, occasionally, and frequently). Likewise, the standard deviation fluctuated between 0.658 (item 20) and 1.211 (item 7), showing low variability in the data. Regarding the skewness and kurtosis measures, most of the items presented values between −2.00 and 2.00. However, items 9, 15, 19, and 20 showed an excess of kurtosis, and these last two also indicated an excess of skewness. Therefore, these four items showed a deviation from a normal distribution.

Table 3. Item analysis for the Internet Addiction Test (IAT).

Item	M	SD	Sk	Ku	Item-Rest Correlation	Floor (%)	Ceiling (%)
Item 1	3.084	1.200	0.115	−0.989	0.378	8	16
Item 2	2.066	0.902	0.734	0.375	0.570	28	1
Item 3	1.885	1.146	1.118	0.203	0.381	53	4
Item 4	1.670	0.826	1.329	1.882	0.381	51	1
Item 5	1.925	1.021	1.092	0.775	0.560	42	3
Item 6	1.828	0.908	1.013	0.437	0.532	43	0
Item 7	3.084	1.211	0.168	−1.016	0.371	7	17
Item 8	1.797	0.889	0.968	0.163	0.591	45	0
Item 9	1.656	0.860	1.506	2.404	0.471	53	1
Item 10	1.753	0.913	1.024	0.245	0.465	51	0
Item 11	1.855	0.898	0.798	−0.072	0.496	43	0
Item 12	1.626	0.900	1.489	1.868	0.451	59	1
Item 13	1.771	0.960	1.275	1.200	0.527	50	2
Item 14	1.877	1.006	1.053	0.442	0.515	45	2
Item 15	1.502	0.772	1.571	2.251	0.684	64	0
Item 16	2.286	1.094	0.833	0.149	0.443	24	6
Item 17	2.093	1.066	0.906	0.138	0.558	33	3
Item 18	1.736	1.000	1.415	1.409	0.538	54	2
Item 19	1.414	0.796	2.275	5.499	0.626	72	1
Item 20	1.352	0.658	2.176	5.640	0.512	73	0

Note. M = Mean; SD = Standard Deviation; Sk = Skewness; Ku = Kurtosis.

Concerning the item-rest polyserial correlation, all the items were above the threshold of 0.30, ranging from 0.371 (item 7) to 0.684 (item 15), indicating good discrimination by the IAT items. On the other hand, in the analysis of the response options, a soil effect was found in all the items, except for items 1 and 7. However, regarding the ceiling effect, most of the items had acceptable values (between 1% and 15%). These effects reflect the tendency of the participants to choose the lowest response options, being the most prominent in items 19 and 20, where more than 70% of the sample chose the response option "rarely". On the contrary, the "always" alternative was selected in a low percentage, even reaching 0% in some items (items 6, 8, 10, 11, 15, and 20).

4.2. Validity Evidence Based on the Internal Structure

The models tested were taken from Table 1. However, of the 39 studies presented, the studies selected were those where the factor structure was configured by all the items of the IAT (20 items) and that the proposed factors have at least three items in its composition, which is recommended to achieve an adequate representation of a factor [86]. Twenty-four studies met the two requirements and their factor models were tested. Four studies had the same unifactorial structure [29,45,51,53]. Whereas, five models [26,32,38,46,59] presented problems in the analysis: Covariance matrix of latent variables was not positive definite [59], some estimated latent variable variances were negative [26,38,46], or the model did not converge [32].

Table 4 presents the fit indices of the models tested. The bifactor model of Watters et al. (2013) [34] was the one that obtained the best results ($SS\chi^2/df$ < 2.00, RMSEA < 0.08, CFI > 0.90, TLI > 0.90, SRMR < 0.08, and WRMR < 1.00). Figure 2 shows the factor structure of the bifactor model, made up of a general factor that measures Internet addiction through 20 items and two specific factors, time/control and stress/compensate, the first factor consisting of five items (1, 2, 7, 16, and 17) and the second factor of 11 items (3, 4, 9, 10, 11, 12, 13, 15, 18, 19, and 20). Figure 2 also presents the factor loadings of the items in the general factor and the corresponding specific factors.

Table 4. Confirmatory factor analysis for the IAT.

Model	$SS\chi^2$	*df*	$SS\chi^2/df$	RMSEA (90% CI)	CFI	TLI	SRMR	WRMR
1. One-factor [29,45,51,53]	387.285	161	2.405	0.079 (0.069; 0.089)	0.890	0.871	0.092	1.168
2. Khazaal et al. (2008) [27]	449.357	169	2.659	0.086 (0.076; 0.095)	0.864	0.847	0.101	1.288
3. Widyanto et al. (2011) [28]	423.779	167	2.358	0.082 (0.073; 0.092)	0.876	0.859	0.095	1.244
4. Jelenchick et al. (2012) [30]	393.896	169	2.331	0.077 (0.067; 0.087)	0.891	0.878	0.092	1.189
5. Barke et al. (2012) [31]	382.215	167	2.289	0.076 (0.066; 0.086)	0.896	0.881	0.091	1.163
6. Watters et al. (2013) [34]	285.741	154	1.855	0.062 (0.050; 0.073)	0.936	0.921	0.073	0.939
7. Hawi (2013) [37]	462.205	166	2.784	0.089 (0.079; 0.099)	0.857	0.836	0.102	1.311
8. Lee et al. (2013) [36]	340.237	164	2.075	0.069 (0.059; 0.079)	0.915	0.901	0.084	1.086
9. Pontes et al. (2014) [39]	437.216	168	2.602	0.084 (0.075; 0.094)	0.870	0.853	0.099	1.267
10. Tsimtsiou et al. (2014) [41]	375.281	167	2.247	0.074 (0.064; 0.084)	0.899	0.885	0.090	1.156
11. Fioravanti et al. (2015) [47]	390.065	165	2.364	0.078 (0.068; 0.088)	0.891	0.875	0.093	1.182
12. Hawi (2015) [48]	342.871	169	2.029	0.067 (0.057; 0.078)	0.916	0.905	0.085	1.097
13. Dhir et al. (2015) [45]	437.235	166	2.634	0.085 (0.075; 0.095)	0.869	0.850	0.099	1.266
14. Waqas et al. (2018) [53]	475.709	170	2.798	0.089 (0.080; 0.099)	0.852	0.835	0.104	1.336
15. Tudorel et al. (2019) [58]	388.449	167	2.326	0.077 (0.067; 0.087)	0.893	0.878	0.092	1.176
16. Černja et al. (2019) [57]	399.818	167	2.394	0.079 (0.069; 0.088)	0.887	0.872	0.093	1.200

Note. RMSEA = Root Mean Square Error of Approximation; CI = Confidence Interval; CFI = Comparative Fit Index; TLI = Tucker–Lewis Index; SRMR = Standardized Root Mean Square Residual; WRMR = Weighted Root Mean Square Residual.

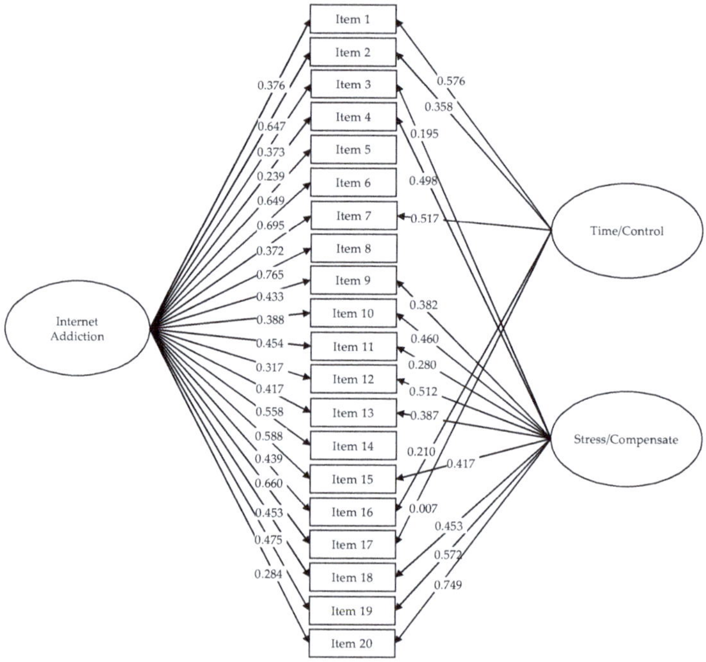

Figure 2. Factorial structure of the bifactor model.

Additionally, complementary statistical indices were calculated that allowed assessment of the robustness of the general factor, as well as the contribution of specific factors [101]. The omega hierarchical was used for the general factor and the specific factors, expecting values above 0.70 (ω_H) for the first and greater than 0.30 (ω_{HS}) for the seconds. The results indicate a value of 0.704 for the general factor, 0.234 for time/control, and 0.478 for stress/compensate. The H coefficient was also calculated, considering values greater than 0.70 to be adequate. The general factor obtained a coefficient of 0.888: For the time/control factor it was 0.785 while, for the stress/compensate factor, the H coefficient was equal to 0.513.

The explained common variance (ECV) for the general factor was 0.612, while the factors (ECV of a specific factor concerning itself) showed an ECV of 0.368 for time/control and 0.562 for stress/compensate. On the other hand, the proportion of uncontaminated correlations (PUCs) was equal to 0.658. At the item level, the ECV index (I-ECV) was between 0.126 and 1.000. Thus, the results allow the use of this bifactor model [102].

4.3. Validity Evidence Based on Relations to Other Variables

The convergent evidence was collected from the correlation between the two specific factors of IAT and the general factor with the average hours per day that study participants spent on the Internet and social skills. The minimum time of hours per day was less than one hour and the maximum 12 h (M = 3.368, SD = 2.277). The skipped correlation coefficients between the average daily hours on the Internet with Internet addiction (and time/control factor) were statistically significant (Table 5). Likewise, the effect size (skipped correlation coefficient squared) in the time/control factor and total Internet addiction was greater than 0.04, indicating a recommended minimum effect size that represents practical significance in social science data [78]. Regarding social skills, the correlations between Internet addiction (and its two factors) with social skills (and the defense of rights factor) were statistically significant (Table 5). Similar results were obtained for correlations between Internet addiction (and stress/compensate factor) with self-expression, disagreement, and assertiveness (Table 5). The size of the effect between Internet addiction and social skills presented practical significance. These findings provide convergent evidence to the IAT.

Table 5. Interfactorial matrix, convergent evidence, and reliability of the IAT.

Variable	Time/Control	Stress/Compensate	Internet Addiction
1. Time/Control	-		
2. Stress/Compensate	0.400 ***	-	
3. Internet Addiction	0.804 ***	0.861 ***	-
4. Self-Expression	−0.117	−0.197 **	−0.331 ***
5. Defense of rights	−0.185 **	−0.189 **	−0.350 ***
6. Disagreement	−0.055	−0.205 **	−0.311 ***
7. Assertiveness	−0.054	−0.189 **	−0.196 **
8. Making requests	0.019	−0.029	−0.118
9. Starting interactions	−0.127	0.058	−0.118
10. Social Skills	−0.137 *	−0.186 **	−0.257 ***
11. Hours on the Internet	0.397 ***	0.112	0.337 ***
12. Ordinal Alpha	0.727	0.856	0.888

Note. * $p < 0.05$; ** $p < 0.01$; *** $p < 0.001$.

On the other hand, the correlation between the specific factors of the IAT presented a minimum effect size (skipped correlation coefficient squared > 0.04). For the correlation between time/control and stress/compensate with Internet addiction, the effect size was strong (skipped correlation coefficient squared > 0.64). In all three cases, the correlations were statistically significant (Table 5).

4.4. Reliability

Reliability was evaluated by the internal consistency method, using for this purpose the ordinal alpha coefficient, which considers the ordinal nature of the items for its calculation. The specific factors and the general factor obtained satisfactory values, above 0.70 (Table 5). Likewise, the coefficients obtained were not remarkably high (0.90 or higher), indicating that the IAT, in the sample studied, does not include redundant items [103].

5. Discussion

The present study analyzed the psychometric properties of the IAT in a sample of Peruvian university students. The 20 items that make up the instrument presented adequate levels of discrimination, although the analysis of the response options indicated the presence of a floor effect in 18 items. Likewise, the levels of skewness and kurtosis were acceptable in most of the items. Regarding the validity evidence based on the internal structure, different models were tested through CFA, being the bifactor model with two specific factors (time/control and stress/compensate), the one that presented the best indexes of adjustment. Another source of validity evidence that was used was that based on the relationship with other variables, specifically the convergent evidence, finding statistically significant correlations between the two specific factors and the general factor with the average number of hours per day on the Internet and social skills. Finally, the reliability, estimated through the ordinal alpha coefficient, was acceptable for the general factor and the specific factors.

The items showed discrimination indexes above 0.30, which implies that each item is related to the other items taken together, which would also justify the presence of an underlying general factor. These results agree with a previous study [43], where all the items were higher than the cut-off point used in this study. However, other studies showed problems only with item 7 ("How often do you check your email before something else that you need to do?"), finding values of 0.170 [42], 0.250 [41], −0.098 [104], and 0.195 [32], while all the other items were greater than 0.30. In this study, item 7 had the lowest item-rest correlation (0.371).

Item 7 is problematic in the literature because it would be mainly relevant for university students or people whose jobs involve communication by this means. In university students, virtual communication occurs mainly with their professors by email, for sending papers, receiving corrected papers, or notifying activities on virtual platforms, as well as with their peers to exchange information (e.g., articles or books) or share working documents. It is important to highlight that, in this study, the item-rest correlation was estimated using the polyserial correlation, unlike previous studies that worked with the Pearson correlation coefficient. The polyserial correlation coefficient considers the ordinality of the items, being more precise in estimating the degree of item discrimination.

The presence of a floor effect in the items, together with low averages in these, shows the low level of Internet addiction of the participants. This result may be due to the characteristics of the study participants, who belong entirely to a public sector university, where most of the population belongs to a medium or medium-low socioeconomic level, having some limitations regarding the Internet accessibility as it involves spending on devices (cell phones, tablets, laptops, etc.) and mobile data for connectivity. On the other hand, Internet addiction is a clinical construct, so its presence in a non-clinical population (university students) should be low under normal conditions.

Regarding the internal structure of the IAT, this study recollected a large part of the models reported in previous studies to test them and find out how the IAT is structured in the Peruvian sample. The bifactor model (one general factor and two specific factors) [34] presented the best fit. In the reviewed literature, the other study that reported a bifactor model found a general factor and three specific factors [56]. Comparing the results obtained in this study with those reported by Watters et al. [34], many points of agreement were observed, both at the level of fit indices and in factor loadings. In both studies, the factor loadings had higher values in the general factor than in the specific factors. In both studies, several items presented factor loadings below 0.30 or 0.40, which are the usual cut-off points in this type of study. However, the results of the present bifactor model indicate that those items that had low factor loadings in the general factor, had higher factor loadings in the specific factors, and vice versa.

The characteristic described above is typical of bifactor models, which allow the simultaneous evaluation of the influence of the general factor and specific factors on the variability of each item. The evaluation of the bifactor model through the omega hierarchical (ω_H and ω_{HS}) and H coefficients (also known as construct reliability), ECV, I-ECV, and PUC, provided evidence regarding the relevance of the model. The remaining 15 models reviewed and tested presented convergence or adjustment problems. The difference in fit between the previous studies and the present study is probably due to the different estimation methods used. In this study, the WLSMV estimator was used that considers the categorical nature of the items. Furthermore, many of the previous studies had problems in choosing the appropriate statistical technique or methods.

The use of the PCA for work with psychological variables is not appropriate since in its conception it considers formative models, useful in other disciplines (economics, marketing, among others), where it seeks to group indicators or reduce the number of variables. By contrast, factor analysis works with reflective models, where an underlying variable (factor) causes certain behaviors (indicators or items). For the factor analysis, the exploratory or non-restrictive version involves making a series of decisions during the analysis, the most critical being the determination of the number of factors. Additionally, the choice of the rotation method should be justified by how the factors are related. In the reviewed antecedents, the use, in most of the studies, of the "Little Jiffy" was observed, which supposes a routine of analysis in the three mentioned aspects: PCA, eigenvalues greater than one (method to choose the number of factors) and Varimax rotation (consider that the factors are not correlated). The use of Little Jiffy has been heavily criticized and its use is not recommended, as it may lead to the acceptance of erroneous factor models [67].

From a theoretical point of view, the time/control factor is related to behavioral symptoms (e.g., neglect household chores to spend more time online), while the stress/compensate factor is made up of cognitive and affective symptoms (e.g., block out disturbing thoughts about your life with soothing thoughts of the Internet or snap, yell, or act annoyed if someone bothers you while you are online). This bifactor model is framed within the cognitive-behavioral perspective, explaining the symptoms of Internet addiction (of varied nature) from specific and generalized uses, which simultaneously influence the symptoms [15]. The model obtained in this study theoretically differs from the models presented in the literature review, because a component (behavioral, cognitive, or affective) is not emphasized, but rather, the components are worked on simultaneously.

Convergent evidence from the IAT was also provided, finding statistically significant correlations with the hours of daily Internet use, and time/control and total Internet addiction presented a recommended minimum effect size that represents practical significance in social science data. These results were like those obtained by other researchers [38,41,45,50]. Additionally, the IAT negatively correlates with social skills measures (total score and self-expression, disagreement, and assertiveness factors). Previous studies also report these relationships with similar degrees of correlation. Internet addiction is associated with greater difficulties in social skills, probably since the emotional burden produced by being connected to the Internet interferes with social aspects. In this way, the development of social skills is left aside due to the few social interactions that the subject experiences, since most of his time is online [2,71].

Regarding the reliability of the scores on the IAT, the ordinal alpha coefficient showed acceptable levels for the general factor and the specific factors. On this point, most of previous studies coincide, including meta-analytical studies [65].

Regarding the limitations of the study, the main one focuses on the size and variety of the sample. Regarding the first aspect, although globally, the sample size is justified in an a priori statistical power analysis, the number of participants within the groups of sociodemographic variables is small, which limits the possibility of carrying out additional analyses in the items. For example, knowing the differential functioning of items or knowing the factorial invariance of the IAT. Regarding the variety of the sample, the students belonged to a public university; therefore, they share various characteristics that make it a homogeneous group, and therefore, the variability in the responses to the items was low.

To know and deepen other characteristics of the IAT, future studies should focus their objectives on analyses that provide evidence of its clinical utility. In this way, the appropriate cut-off points should be determined to be able to classify people addicted to the Internet and, in turn, evaluate the intensity of this addiction. Likewise, it is necessary to previously know how the instrument works in people with a presumption of Internet addiction, as well as to what extent it is related to other tests that measure clinical constructs (e.g., depression or anxiety), being relevant measures to obtain validity evidence based on relations to other variables. Therefore, working with clinical samples is a necessity, since the IAT could have different uses in the diagnosis and treatment of Internet addiction.

Likewise, the study of the IAT in non-academic populations must be accompanied by a review of the content of the items, since some of them may only be valid for the population of university students, for example, item 6 "How often do your grades or school work suffer because of the amount of time you spend online? ", being, in this case, the rewriting of the item or its exclusion from the test. Additionally, given the complexity of the IAT structure, other multivariate techniques could be tried to corroborate what was found here or to propose more stable structures. Techniques, such as network analysis, exploratory structural equation modeling (ESEM), or Bayesian approaches to factor analysis, would help in this regard. Regarding reliability, it is relevant to obtain evidence on temporal stability (test-retest reliability), particularly useful in the IAT, due to the high variability that scores in this type of test can have.

6. Conclusions

This study represents a contribution to the study of the IAT in Latin America, where it has been little studied, unlike other contexts, such as Europe or Asia. The findings indicate that the IAT, in the sample of Peruvian university students, is made up of a general factor and two specific factors (time/control and stress/compensate), or a bifactor model. Likewise, added to the validity evidence based on the internal structure, the IAT showed evidence based on the relationship with other variables (average hours per day on the Internet). On the other hand, the reliability of the scores was acceptable.

The results lead to the conclusion that the scores in the IAT have evidence of validity and reliability for its use. This has implications for both researchers and those who are primarily involved in the patient practice. For researchers, the IAT constitutes an instrument that would allow studies on Internet addiction to be carried out, for example, knowing its prevalence in certain groups, identifying the factors associated with its genesis and evolution in people, or knowing the degree of sensitivity and specificity with which one can diagnose a person with Internet addiction. For professionals, the IAT is a tool that would help diagnose Internet addiction, and it would also allow evaluation of the effects produced by a treatment or therapy that seeks to decrease the level of Internet addiction.

Author Contributions: Conceptualization, A.A.T.-M., J.C.A.-P., R.A.Z.-T. and D.E.R.-O.; methodology, A.A.T.-M., J.C.A.-P., R.A.Z.-T. and D.E.R.-O.; software, A.A.T.-M., J.C.A.-P., R.A.Z.-T. and D.E.R.-O.; validation, A.A.T.-M., J.C.A.-P., R.A.Z.-T. and D.E.R.-O.; formal analysis, A.A.T.-M., J.C.A.-P., R.A.Z.-T. and D.E.R.-O.; investigation, A.A.T.-M., J.C.A.-P., R.A.Z.-T. and D.E.R.-O.; resources, A.A.T.-M., J.C.A.-P., R.A.Z.-T. and D.E.R.-O.; data curation, A.A.T.-M., J.C.A.-P., R.A.Z.-T. and D.E.R.-O.; writing—original draft preparation, A.A.T.-M., J.C.A.-P., R.A.Z.-T. and D.E.R.-O.; writing—review and editing, A.A.T.-M., J.C.A.-P., R.A.Z.-T. and D.E.R.-O.; visualization, A.A.T.-M., J.C.A.-P., R.A.Z.-T. and D.E.R.-O.; supervision, A.A.T.-M., J.C.A.-P., R.A.Z.-T. and D.E.R.-O.; project administration, A.A.T.-M., J.C.A.-P., R.A.Z.-T. and D.E.R.-O. All authors have read and agreed to the published version of the manuscript.

Funding: This research received no external funding.

Conflicts of Interest: The authors declare no conflict of interest.

Appendix A

English and Spanish version of the Internet Addiction Test (IAT)

1. How often do you find that you stay online longer than you intended? ¿Con qué frecuencia se conecta a internet más de lo previsto?
2. How often do you neglect household chores to spend more time online? ¿Con qué frecuencia descuida las actividades de la casa para estar más tiempo conectado?
3. How often do you prefer the excitement of the internet to intimacy with your partner? ¿Con qué frecuencia prefiere más la emoción que le produce estar conectado a la intimidad con su pareja o la relación directa con sus amigos?
4. How often do you form new relationships with fellow online users? ¿Con qué frecuencia forma nuevas relaciones con usuarios de Internet?
5. How often do others in your life complain to you about the amount of time you spend online? ¿Con qué frecuencia las personas cercanas a usted se quejan por la cantidad de tiempo que permanece conectado?
6. How often do your grades or school work suffer because of the amount of time you spend online? ¿Con qué frecuencia sus calificaciones o actividades académicas se afectan negativamente por la cantidad de tiempo que permanece en Internet?
7. How often do you check your email before something else that you need to do? ¿Con qué frecuencia revisa su correo electrónico antes de realizar otra tarea que necesita hacer?
8. How often does your job performance or productivity suffer because of the internet? ¿Con qué frecuencia el tiempo que pasa en Internet afecta negativamente su desempeño o productividad en el trabajo?
9. How often do you become defensive or secretive when anyone asks you what you do online? ¿Con qué frecuencia está a la defensiva o se muestra reservado cuando alguien le pregunta qué hace en Internet?
10. How often do you block out disturbing thoughts about your life with soothing thoughts of the internet? ¿Con qué frecuencia bloquea los pensamientos desagradables de su vida con pensamientos agradables relacionados con Internet?

11. How often do you find yourself anticipating when you will go online again? ¿Con qué frecuencia anticipa cuando estará conectado de nuevo?
12. How often do you fear that life without the internet would be boring, empty, and joyless? ¿Con qué frecuencia teme que la vida sin Internet sería aburrida, vacía o triste?
13. How often do you snap, yell, or act annoyed if someone bothers you while you are online? ¿Con qué frecuencia se enoja si alguien lo molesta mientras está conectado?
14. How often do you lose sleep due to being online? ¿Con qué frecuencia se queda sin dormir por conectarse durante la noche?
15. How often do you feel preoccupied with the internet when off-line, or fantasize about being online? ¿Con qué frecuencia se siente preocupado por no estar conectado o imagina estarlo?
16. How often do you find yourself saying "just a few more minutes" when online? ¿Con qué frecuencia dice: "unos minutos más", cuando está conectado?
17. How often do you try to cut down the amount of time you spend online and fail? ¿Con qué frecuencia trata de disminuir el tiempo que pasa en Internet y no lo logra?
18. How often do you try to hide how long you've been online? ¿Con qué frecuencia intenta ocultar el tiempo que permanece conectado?
19. How often do you choose to spend more time online over going out with others? ¿Con qué frecuencia prefiere pasar más tiempo en Internet que salir con otras personas?
20. How often do you feel depressed, moody, or nervous when you are off-line, which goes away once you are back online? ¿Con qué frecuencia se siente deprimido, malhumorado o nervioso cuando no está conectado, pero se siente mejor cuando se conecta de nuevo?

References

1. Freeman, C.G. The implications of online connectivity for world heritage in a digital platform society. *Hist. Environ.* **2018**, *30*, 84–95.
2. Engelberg, E.; Sjöberg, L. Internet use, social skills, and adjustment. *Cyberpsychol. Behav.* **2004**, *7*, 41–47. [CrossRef] [PubMed]
3. Gracia Blanco, M.D.; Vigo Anglada, M.; Fernández Perez, M.J.; Marcó Arbones, M. Problemas conductuales relacionados con el uso de Internet: Un estudio exploratorio. *Psicologia* **2002**, *18*, 273–292.
4. Mayer, M.A. The use of the Internet among adolescents; Risks and benefits. *Aten. Primaria* **2011**, *43*, 287–288. [CrossRef] [PubMed]
5. Sánchez-Carbonell, X.; Beranuy, M.; Castellana, M.; Chamarro, A.; Oberst, U. La adicción a Internet y al móvil: ¿moda o trastorno? *Adicciones* **2008**, *20*, 149–160. [CrossRef] [PubMed]
6. Garcia del Castillo, J.A.; Terol, M.D.C.; Nieto, M.; Lledó, A.; Sánchez, S.; Martín-Aragón, M.; Sitges, E. Uso y abuso de Internet en jóvenes universitarios. *Adicciones* **2008**, *20*, 131–142. [CrossRef] [PubMed]
7. Domínguez, C.; Geijo, S.; Sánchez, I.; Imaz, C.; Cabús, G. Revision of the mechanisms involved in the problematic use of the Internet. *Rev. Asoc. Esp. Neuropsiquiatr.* **2012**, *32*, 691–705.
8. Kuss, D.J.; Lopez-Fernandez, O. Internet addiction and problematic Internet use: A systematic review of clinical research. *World J. Psychiatry* **2016**, *6*, 143. [CrossRef]
9. Starcevic, V.; Aboujaoude, E. Internet addiction: Reappraisal of an increasingly inadequate concept. *CNS Spectr.* **2017**, *22*, 7–13. [CrossRef]
10. Young, K.S. Internet addiction: The emergence of a new clinical disorder. *Cyberpsychol. Behav.* **1998**, *1*, 237–244. [CrossRef]
11. Young, K.S. Internet addiction: A new clinical phenomenon and its consequences. *Am. Behav. Sci.* **2004**, *48*, 402–415. [CrossRef]
12. Block, J.J. Issues for DSM-V: Internet addiction. *Am. J. Psychiatry* **2008**, *165*, 306–307. [CrossRef] [PubMed]
13. Young, K.S.; Nabuco de Abreu, C. *Internet Addiction: A Handbook and Guide to Evaluation and Treatment*; John Wiley & Sons, Ltd.: Hoboken, NJ, USA, 2010; ISBN 978-0-470-55116-5.
14. American Psychiatric Association. *Diagnostic and Statistical Manual of Mental Disorders*, 5th ed.; American Psychiatric Association: Washington, DC, USA, 2013; ISBN 978-0-89042-554-1.

15. Davis, R.A. A cognitive-behavioral model of pathological Internet use. *Comput. Hum. Behav.* **2001**, *17*, 187–195. [CrossRef]
16. Teo, C.P.A.; Lim, K.; Oei, T.P. Problematic Internet Use: Variants of cognitive-behavioural model under path analysis. *Int. J. Psychol. Behav. Anal.* **2017**, *3*, 130. [CrossRef] [PubMed]
17. Díaz-Aguado, M.J.; Martín-Babarro, J.; Falcón, L. Problematic Internet use, maladaptive future time perspective and school context. *Psicothema* **2018**, *30*, 195–200.
18. Fernandes, B.; Rodrigues, B.; Pontes, H.M. Internet addiction or problematic internet use ? Which term should be used ? *Psicol. USP* **2019**, *30*, e190020. [CrossRef]
19. Demetrovics, Z.; Szeredi, B.; Rózsa, S. The three-factor model of Internet addiction: The development of the Problematic Internet Use Questionnaire. *Behav. Res. Methods* **2008**, *40*, 563–574. [CrossRef]
20. Meerkerk, G.-J.; Van Den Eijnden, R.J.J.M.; Vermulst, A.A.; Garretsen, H.F.L. The Compulsive Internet Use Scale (CIUS): Some psychometric properties. *Cyberpsychol. Behav.* **2008**, *12*, 1–6. [CrossRef]
21. Armstrong, L.; Phillips, J.G.; Saling, L.L. Potential determinants of heavier Internet usage. *Int. J. Hum. Comput. Stud.* **2000**, *53*, 537–550. [CrossRef]
22. Casas, J.A.; Ruiz-Olivares, R.; Ortega-Ruiz, R. Validation of the internet and social networking experiences questionnaire in Spanish adolescents. *Int. J. Clin. Health Psychol.* **2013**, *13*, 40–48. [CrossRef]
23. Young, K.S. *Caught in the Net: How to Recognize the Signs of Internet Addiction—And a Winning Strategy for Recovery*; John Wiley & Sons, Ltd.: New York, NY, USA, 1998; ISBN 978-0-471-19159-9.
24. Widyanto, L.; McMurran, M. The psychometric properties of the Internet Addiction Test. *Cyberpsychol. Behav.* **2004**, *7*, 443–450. [CrossRef]
25. Ngai, S.S.Y. Exploring the validity of the Internet Addiction Test for students in grades 5–9 in Hong Kong. *Int. J. Adolesc. Youth* **2007**, *13*, 221–237. [CrossRef]
26. Chang, M.K.; Man, S.P. Factor structure for Young's Internet Addiction Test: A confirmatory study. *Comput. Hum. Behav.* **2008**, *24*, 2597–2619. [CrossRef]
27. Khazaal, Y.; Billieux, J.; Thorens, G.; Khan, R.; Louati, Y.; Scarlatti, E.; Theintz, F.; Lederrey, J.; Van Der Linden, M.; Zullino, D. French validation of the Internet Addiction Test. *Cyberpsychol. Behav.* **2008**, *11*, 703–706. [CrossRef] [PubMed]
28. Widyanto, L.; Griffiths, M.D.; Brunsden, V. A psychometric comparison of the Internet Addiction Test, the Internet-Related Problem Scale, and self-diagnosis. *Cyberpsychol. Behav. Soc. Netw.* **2011**, *14*, 141–149. [CrossRef] [PubMed]
29. Panayides, P.; Walker, M.J. Evaluation of the psychometric properties of the Internet Addiction Test (IAT) in a sample of Cypriot high school students: The Rasch measurement perspective. *Eur. J. Psychol.* **2012**, *8*, 327–351. [CrossRef]
30. Jelenchick, L.A.; Becker, T.; Moreno, M.A. Assessing the psychometric properties of the Internet Addiction Test (IAT) in US college students. *Psychiatry Res.* **2012**, *196*, 296–301. [CrossRef]
31. Barke, A.; Nyenhuis, N.; Kröner-Herwig, B. The German version of the Internet Addiction Test: A validation study. *Cyberpsychol. Behav. Soc. Netw.* **2012**, *15*, 534–542. [CrossRef]
32. Puerta-Cortés, D.X.; Carbonell, X.; Chamarro, A. Análisis de las propiedades psicométricas de la versión en español del Internet Addiction Test. *Trastor. Adict.* **2012**, *14*, 99–104. [CrossRef]
33. Faraci, P.; Craparo, G.; Messina, R.; Severino, S. Internet Addiction Test (IAT): Which is the best factorial solution? *J. Med. Internet Res.* **2013**, *15*, e225. [CrossRef]
34. Watters, C.A.; Keefer, K.V.; Kloosterman, P.H.; Summerfeldt, L.J.; Parker, J.D.A. Examining the structure of the Internet Addiction Test in adolescents: A bifactor approach. *Comput. Hum. Behav.* **2013**, *29*, 2294–2302. [CrossRef]
35. Pawlikowski, M.; Altstötter-Gleich, C.; Brand, M. Validation and psychometric properties of a short version of Young's Internet Addiction Test. *Comput. Hum. Behav.* **2013**, *29*, 1212–1223. [CrossRef]
36. Lee, K.; Gyeong, H.; Yu, B.; Song, Y.M.; Lee, H.K.; Kim, D. Reliability and validity of the Korean version of the Internet Addiction Test among college students. *J. Korean Med. Sci.* **2013**, *28*, 763–768. [CrossRef] [PubMed]
37. Hawi, N.S. Arabic validation of the Internet Addiction Test. *Cyberpsychol. Behav. Soc. Netw.* **2013**, *16*, 200–204. [CrossRef] [PubMed]
38. Lai, C.-M.; Mak, K.-K.; Watanabe, H.; Ang, R.P.; Pang, J.S.; Ho, R.C.M. Psychometric properties of the Internet Addiction Test in Chinese adolescents. *J. Pediatr. Psychol.* **2013**, *38*, 794–807. [CrossRef]

39. Pontes, H.M.; Patrão, I.M.; Griffiths, M.D. Portuguese validation of the Internet Addiction Test: An empirical study. *J. Behav. Addict.* **2014**, *3*, 107–114. [CrossRef]
40. Karim, A.K.M.R.; Nigar, N. The Internet Addiction Test: Assessing its psychometric properties in bangladeshi culture. *Asian J. Psychiatr.* **2014**, *10*, 75–83. [CrossRef]
41. Tsimtsiou, Z.; Haidich, A.B.; Kokkali, S.; Dardavesis, T.; Young, K.S.; Arvanitidou, M. Greek version of the Internet Addiction Test: A validation study. *Psychiatr. Q.* **2014**, *85*, 187–195. [CrossRef]
42. Chong, N.; Isa, S.M.; Hashim, A.H.; Kumar, S.; Harbajan, M.K. Validity of the Malay version of the internet addiction test: A study on a group of medical students in Malaysia. *Asia-Pac. J. Public Heath* **2015**, *27*, NP2210–NP2219. [CrossRef]
43. Fernández-Villa, T.; Molina, A.J.; García-Martín, M.; Llorca, J.; Delgado-Rodríguez, M.; Martín, V. Validation and psychometric analysis of the Internet Addiction Test in Spanish among college students. *BMC Public Health* **2015**, *15*, 953. [CrossRef]
44. Lu, X.; Yeo, K.J. Psychometric properties of the Internet Addiction Test in a sample of Malaysian undergraduate students. *Psicol. Educ.* **2015**, *21*, 17–25. [CrossRef]
45. Dhir, A.; Chen, S.; Nieminen, M. Psychometric validation of Internet Addiction Test with Indian adolescents. *J. Educ. Comput. Res.* **2015**, *53*, 15–31. [CrossRef]
46. Lai, C.M.; Mak, K.K.; Cheng, C.; Watanabe, H.; Nomachi, S.; Bahar, N.; Young, K.S.; Ko, H.C.; Kim, D.; Griffiths, M.D. Measurement invariance of the Internet Addiction Test among Hong Kong, Japanese, and Malaysian adolescents. *Cyberpsychol. Behav. Soc. Netw.* **2015**, *18*, 609–617. [CrossRef] [PubMed]
47. Fioravanti, G.; Casale, S. Evaluation of the psychometric properties of the Italian internet addiction test. *Cyberpsychol. Behav. Soc. Netw.* **2015**, *18*, 120–128. [CrossRef] [PubMed]
48. Hawi, N.S.; Blachnio, A.; Przepiorka, A. Polish validation of the Internet Addiction Test. *Comput. Hum. Behav.* **2015**, *48*, 548–553. [CrossRef]
49. Kaya, F.; Delen, E.; Young, K.S. Psychometric properties of the Internet Addiction Test in Turkish. *J. Behav. Addict.* **2016**, *5*, 130–134. [CrossRef]
50. Servidio, R. Assessing the psychometric properties of the Internet Addiction Test: A study on a sample of Italian university students. *Comput. Hum. Behav.* **2017**, *68*, 17–29. [CrossRef]
51. Boysan, M.; Kuss, D.J.; Barut, Y.; Ayköse, N.; Güleç, M.; Özdemir, O. Psychometric properties of the Turkish version of the Internet Addiction Test (IAT). *Addict. Behav.* **2017**, *64*, 247–252. [CrossRef]
52. Samaha, A.A.; Fawaz, M.; El Yahfoufi, N.; Gebbawi, M.; Abdallah, H.; Baydoun, S.A.; Ghaddar, A.; Eid, A.H. Assessing the psychometric properties of the Internet Addiction Test (IAT) among Lebanese college students. *Front. Public Health* **2018**, *6*, 365. [CrossRef]
53. Waqas, A.; Farooq, F.; Raza, M.; Javed, S.T.; Khan, S.; Ghumman, M.E.; Naveed, S.; Haddad, M. Validation of the Internet Addiction Test in students at a Pakistani medical and dental school. *Psychiatr. Q.* **2018**, *89*, 235–247. [CrossRef]
54. Neelapaijit, A.; Pinyopornpanish, M.; Simcharoen, S.; Kuntawong, P.; Wongpakaran, N.; Wongpakaran, T. Psychometric properties of a Thai version Internet Addiction Test. *BMC Res. Notes* **2018**, *11*, 69. [CrossRef] [PubMed]
55. Tsermentseli, S.; Karipidis, N.; Samaras, P.; Thompson, T. Assessing the factorial structure of the Internet Addiction Test in a sample of Greek adolescents. *Hell. J. Psychol.* **2018**, *15*, 274–288.
56. Hernández, C.; Rivera, D. A Chilean transcultural adaptation and assessment of the Internet Addiction Test's factor structures: Development of a brief version. *Rev. Iberoam. Diag. Eval.* **2018**, *49*, 143–155.
57. Černja, I.; Vejmelka, L.; Rajter, M. Internet Addiction Test: Croatian preliminary study. *BMC Psychiatry* **2019**, *19*, 388. [CrossRef]
58. Tudorel, O.I.; Vintilă, M.; Vlaicu, L.; Bălăuță, D.; Goian, C.; Rusu, A. Romanian version of the Internet Addiction Test: Psychometric properties and cross-gender invariance. *Int. J. Ment. Health Addict.* **2019**, *17*, 234–246. [CrossRef]
59. Ndasauka, Y.; Pitafi, A.; Kayange, G.M. Psychometric properties of Young's Internet Addiction Test (IAT) in Urdu language. *Asian J. Psychiatr.* **2019**, *40*, 39–44. [CrossRef]
60. Yaffe, Y.; Seroussi, D.-E. Further evidence for the psychometric properties of young's Internet Addiction Test (IAT): A study on a sample of Israeli-Arab male adolescents. *Am. J. Health Behav.* **2019**, *43*, 1030–1039. [CrossRef]

61. Talwar, P.; Mohamad, F.S.; Mustaffa, M.S.; Othman, M.K.; Bee, O.Y.; Hian, C.C. Internet addiction among university students: Psychometric properties of three factor model Internet Addiction Test. *Online J. Health Allied Sci.* **2019**, *18*, 1–7.
62. Lu, X.; Yeo, K.J.; Guo, F.; Zhao, Z. Factor structure and a multiple indicators multiple cause model of Internet Addiction Test: The effect of socio-demographic and internet use variables. *Curr. Psychol.* **2019**. [CrossRef]
63. Raykov, T.; Marcoulides, G.A. Thanks coefficient alpha, we still need you! *Educ. Psychol. Meas.* **2019**, *79*, 200–210. [CrossRef]
64. Viladrich, C.; Angulo-Brunet, A.; Doval, E. A journey around alpha and omega to estimate internal consistency reliability. *Psicoogia* **2017**, *33*, 755–782.
65. Moon, S.J.; Hwang, J.S.; Kim, J.Y.; Shin, A.L.; Bae, S.M.; Kim, J.W. Psychometric properties of the Internet Addiction Test: A systematic review and meta-analysis. *Cyberpsychol. Behav. Soc. Netw.* **2018**, *21*, 473–484. [CrossRef] [PubMed]
66. Marcus, C.; Peter, H. Questionable research practices when using confirmatory factor analysis. *J. Manag. Psychol.* **2019**, *34*, 18–30.
67. Sellbom, M.; Tellegen, A. Factor analysis in psychological assessment research: Common pitfalls. and recommendations. *Psychol. Assess.* **2019**, *31*, 1428–1441. [CrossRef] [PubMed]
68. Matalinares, M.; Raymundo, O.; Baca, D. Psychometric adaptation of the Test of Internet Addiction. *Rev. Peru. Psicol. Trab. Soc.* **2014**, *3*, 45–66.
69. Wang, L.; Luo, J.; Gao, W.; Kong, J. The effect of Internet use on adolescents' lifestyles: A national survey. *Comput. Hum. Behav.* **2012**, *28*, 2007–2013. [CrossRef]
70. Zegarra, C.O.; Cuba, M.S. Frequency of Internet addiction and development of social skills in adolescents in an urban area of Lima. *Medwave* **2017**, *17*, e6857. [CrossRef]
71. Torrente, E.; Piqueras, J.A.; Orgilés, M.; Espada, J.P. Association of Internet addiction with social anxiety and lack of social skills in spanish adolescents. *Ter. Psicol.* **2014**, *32*, 175–184. [CrossRef]
72. Chou, W.; Huang, M.; Chang, Y.; Chen, Y.; Hu, H.; Yen, C. Social skills deficits and their association with Internet addiction and activities in adolescents with attention-deficit/hyperactivity disorder. *J. Behav. Addict.* **2017**, *6*, 42–50. [CrossRef]
73. American Educational Research Association; American Psychological Association; National Council on Measurement in Education. *Standards for Educational and Psychological Testing*; American Educational Research Association: Washington, DC, USA, 2014; ISBN 978-0935-302-356.
74. Ato, M.; López, J.J.; Benavente, A. A classification system for research designs in psychology. *Psicologia* **2013**, *29*, 1038–1059.
75. Zickar, M.J. Measurement development and evaluation. *Annu. Rev. Organ. Psychol. Organ. Behav.* **2020**, *7*, 213–232. [CrossRef]
76. Kerlinger, F.N.; Lee, H.B. *Foundations of Behavioral Research*, 4th ed.; Harcourt College Publishers: Fort Worth, TX, USA, 2000; ISBN 9780155078970.
77. Turner, D.P.; Houle, T.T. The importance of statistical power calculations. *Headache J. Head Face Pain* **2018**, *58*, 1187–1191. [CrossRef] [PubMed]
78. Ferguson, C.J. An effect size primer: A guide for clinicians and researchers. *Prof. Psychol. Res. Pr.* **2009**, *40*, 532–538. [CrossRef]
79. Cohen, J. *Statistical Power Analysis for the Behavioral Sciences*, 2nd ed.; Lawrence Erlbaum Associates: Hillsdale, NJ, USA, 1988; ISBN 0-8058-0283-5.
80. Gismero, E. *EHS: Escala de Habilidades Sociales*; TEA Ediciones: Madrid, Spain, 2000; ISBN 84-7174-609-3.
81. Miranda-Zapata, E.; Riquelme-Mella, E.; Cifuentes-Cid, H.; Riquelme-Bravo, P. Análisis factorial confrmatorio de la Escala de habilidades sociales en universitarios chilenos. *Rev. Lat. Psicol.* **2014**, *46*, 73–82.
82. Tabachnick, B.G.; Fidell, L.S. *Using Multivariate Statistics*, 7th ed.; Pearson: New York, NY, USA, 2019; ISBN 9780134792866.
83. Meyer, J.P. *Applied Measurement with Jmetrik*; Routledge: New York, NY, USA, 2014; ISBN 978-0-203-11519-0.
84. McHorney, C.A.; Tarlov, A.R. Individual-patient monitoring in clinical practice: Are available health status surveys adequate? *Qual. Life Res.* **1995**, 293–307. [CrossRef] [PubMed]
85. Li, C.-H. Confirmatory factor analysis with ordinal data: Comparing robust maximum likelihood and diagonally weighted least squares. *Behav. Res. Methods* **2016**, *48*, 936–949. [CrossRef]

86. Keith, T.Z. *Multiple Regression and Beyond: An Introduction to Multiple Regression and Structural Equation Modeling*, 3rd ed.; Routledge: New York, NY, USA, 2019; ISBN 9781351667937.
87. DiStefano, C.; Liu, J.; Jiang, N.; Shi, D. Examination of the Weighted Root Mean Square Residual: Evidence for trustworthiness? *Struct. Equ. Model. A Multidiscip. J.* **2018**, *25*, 453–466. [CrossRef]
88. Wilcox, R.R.; Rousselet, G.A.; Pernet, C.R. Improved methods for making inferences about multiple skipped correlations. *J. Stat. Comput. Simul.* **2018**, *88*, 3116–3131. [CrossRef]
89. Zumbo, B.D.; Gadermann, A.M.; Zeisser, C. Ordinal versions of coefficients alpha and theta for Likert rating scales. *J. Mod. Appl. Stat. Method* **2007**, *6*, 21–29. [CrossRef]
90. Nunnally, J.; Bernstein, I. *Psychometric Theory*, 3rd ed.; McGraw-Hill: New York, NY, USA, 1994; ISBN 9780070478497.
91. Erdfelder, E.; Faul, F.; Buchner, A.; Lang, A.-G. Statistical power analyses using G*Power 3.1: Tests for correlation and regression analyses. *Behav. Res. Methods* **2009**, *41*, 1149–1160.
92. R Core Team. *R: A Language and Environment for Statistical Computing*; R Foundation for Statistical Computing: Vienna, Austria, 2020; Available online: https://www.r-project.org (accessed on 9 June 2020).
93. Rinker, T.W.; Kurkiewicz, D. Pacman: Package Management for R. 2018. Available online: http://github.com/trinker/pacman (accessed on 9 June 2020).
94. Hadley, W.; Bryan, J. Readxl: Read Excel Files. 2019. Available online: https://cran.r-project.org/package=readxl (accessed on 9 June 2020).
95. Wickham, H.; Averick, M.; Bryan, J.; Chang, W.; Mcgowan, L.D.A.; François, R.; Grolemund, G.; Hayes, A.; Henry, L.; Hester, J.; et al. Welcome to the tidyverse. *J. Open Source Softw.* **2019**, *4*, 1686. [CrossRef]
96. Revelle, W. Psych: Procedures for Psychological, Psychometric, and Personality Research. 2019. Available online: https://cran.r-project.org/package=psych (accessed on 9 June 2020).
97. Rosseel, Y. Lavaan: An R package for structural equation modeling. *J. Stat. Softw.* **2012**, *48*, 1–36. [CrossRef]
98. Dueber, D. BifactorIndicesCalculator: Bifactor Indices Calculator. 2020. Available online: https://cran.r-project.org/package=BifactorIndicesCalculator (accessed on 9 June 2020).
99. Jorgensen, T.D.; Pornprasertmanit, S.; Schoemann, A.M.; Rosseel, Y. SemTools: Useful tools for structural equation modeling. 2020. Available online: https://cran.r-project.org/package=semTools (accessed on 9 June 2020).
100. Wilcox, R.R.; Schönbrodt, F.D. The WRS Package for Robust Statistics in R. 2019. Available online: https://github.com/nicebread/WRS (accessed on 9 June 2020).
101. Dominguez-Lara, S.; Rodriguez, A. Statistical indices from bifactor models. *Interacciones* **2017**, *3*, 59–65. [CrossRef]
102. Rodriguez, A.; Reise, S.P.; Haviland, M.G. Evaluating bifactor models: Calculating and interpreting statistical indices. *Psychol. Methods* **2016**, *21*, 137–150. [CrossRef]
103. Tavakol, M.; Dennick, R. Making sense of Cronbach's alpha. *Int. J. Med. Educ.* **2011**, *2*, 53–55. [CrossRef]
104. Conti, M.A.; Pelegrino, A.; Hearst, N.; Athanássios, T.; Tavares, H.; Nabuco de Abreu, C. Evaluation of semantic equivalence and internal consistency of a Portuguese version of the Internet Addiction Test (IAT). *Rev. Psiquiatr. Clin.* **2012**, *39*, 106–110. [CrossRef]

International Journal of
Environmental Research and Public Health

MDPI

Article

Empirical Relationships between Problematic Alcohol Use and a Problematic Use of Video Games, Social Media and the Internet and Their Associations to Mental Health in Adolescence

Lutz Wartberg [1] and Rudolf Kammerl [2,*]

[1] Department Psychology, Faculty of Life Sciences, MSH Medical School Hamburg, 20457 Hamburg, Germany; lutz.wartberg@medicalschool-hamburg.de

[2] Department of Education, Chair for Pedagogy with a Focus on Media Education, Friedrich-Alexander-University Erlangen-Nuremberg, 90478 Nuremberg, Germany

* Correspondence: rudolf.kammerl@fau.de; Tel.: +49-911-5302-520

Received: 2 August 2020; Accepted: 18 August 2020; Published: 21 August 2020

Abstract: Adolescents frequently show risky behavior, and these problematic behavior patterns often do not occur in isolation, but together. Problematic alcohol use is widespread among youth, as is problematic use of the Internet and of specific online applications (video games or social media). However, there is still a lack of findings for minors regarding the relations between these behavioral patterns (particularly between problematic alcohol use and problematic gaming or problematic social media use). Standardized instruments were used to survey problematic alcohol use, problematic gaming, problematic social media use, problematic Internet use and mental health among 633 adolescents (mean age: 15.79 years). Bivariate correlation and multivariable linear regression analyses were conducted. The correlation analyses showed statistically significant positive bivariate relationships between all four behavioral patterns each. Antisocial behavior was related to all problematic behavioral patterns. Whereas, emotional distress, self-esteem problems and hyperactivity/inattention were associated with substance-unrelated problematic behavior patterns only. Anger control problems were related to problematic alcohol use and problematic gaming. In adolescence, the findings revealed small effect sizes between substance-related and substance-unrelated problematic behavior patterns, but moderate to large effect sizes within substance-unrelated behavioral patterns. Similarities and differences were found in the relations between the behavioral patterns and mental health.

Keywords: Internet addiction; pathological Internet use; Internet gaming disorder; gaming disorder; social networking site addiction; social media addiction; Facebook addiction; problem drinking; alcohol; adolescent

1. Introduction

In adolescence, new developmental tasks can pose great challenges and psychological stress for minors [1]. In addition to effective problem-solving strategies, dysfunctional stress management strategies (e.g., use of psychotropic substances) are often used to cope with these developmental tasks [1], which can also serve the striving for independence (e.g., from parents) or the search for identity [2]. According to the Problem-Behavior Theory of Jessor [3], these dysfunctional stress management strategies or problematic behavior patterns often do not occur in isolation, but together. Problematic alcohol use (often referred to as problem drinking) is explicitly mentioned by Jessor [3] (p. 602). The extensive technological changes in recent years and decades call for the concept of Problem-Behavior Theory to be expanded in content. Some authors (e.g., De Leo & Wulfert [4]) now also include a problematic use of the Internet or its applications in this context. Considering the widespread occurrence of substance-related problematic behavior patterns and problematic use of digital media in adolescence, it seems quite relevant to examine whether there are empirical relationships. Initial studies (e.g., [5]) have conducted a combined investigation of substance-related problem behavior (e.g., problematic alcohol use) and problem behavior unrelated to substance use (e.g., problematic Internet use or problematic use of specific online applications). Another important approach specifically to explain Internet-use disorders is the Interaction of Person-Affect-Cognition-Execution (I-PACE) model by Brand et al. [6]. In the I-PACE model specific Internet-use disorders "... are considered to be the consequence of interactions between predisposing factors" (e.g., biopsychological constitution, psychopathology, personality, social cognitions and specific motives for using), "... coping styles and Internet-related cognitive biases... " and "... affective and cognitive responses to situational triggers in combination with reduced executive functioning" (p. 252).

The increasing relevance of problem behavior unrelated to substance use was demonstrated by the American Psychiatric Association (APA) with the inclusion of "Internet Gaming Disorder" in the appendix of the DSM-5 [7] (as a condition for which further studies are required), and by the World Health Organization (WHO) with the inclusion of "Gaming Disorder" in the ICD-11 [8]. In addition to use of video games (in the following referred to as gaming), the use of social media is often mentioned as another Internet application with potential for problematic use by adolescents (e.g., [9]). The collective term "social media" includes the use of websites of social networks and messengers as well as blogs, etc. [10]. In addition, an increasing number of studies is examining problematic use of the Internet on a more general level [11]. According to available empirical findings, it seems appropriate to differentiate between general problematic Internet use and the use of specific applications (such as online gaming, in cross-section: (e.g., [12]), in longitudinal section: [13]).

As an important substance-related behavior pattern, problematic consumption of alcohol and its consequences in adolescence have been a relevant aspect of public health efforts and scientific research for a long time (in Germany, for example, an increase in the age of first alcohol use from 14.1 years in 2004 to 15.0 years in 2018 has been observed [14]). Indeed, alcohol use among adolescents remains widespread in many countries. Following the European School Survey Project on Alcohol and Other Drugs (ESPAD) risky patterns of alcohol consumption are still highly prevalent in youth. In 1995, in the first ESPAD study, 36% of the examined European students reported heavy episodic drinking. In 2015, in the last ESPAD investigation, the observed prevalence for heavy episodic drinking was 35% [15]. Besides, a substantial percentage of youth already shows a problematic alcohol use. In Germany (where the present study was conducted), the one-year prevalence estimation for problematic alcohol use in a representative sample of adolescents was 5.0% [16].

In addition to substance-related problem behavior (especially alcohol consumption), the relevance of problem behavior unrelated to substance use (such as problematic Internet use, problematic gaming, and problematic social media use) has often been discussed and empirically investigated especially in adolescence. Surveys of representative samples of youth show high prevalence values for problematic Internet use, problematic gaming, and problematic social media use. Based on latent profile and latent class analyses in representative samples, prevalence estimates between 3.2% and 4.7% were observed for problematic Internet use in German minors [17–19]. For problematic gaming, in representative samples prevalence values of 2.6% among Slovenian students and 3.5% in German adolescents were reported [20,21]. Bányai et al. [22], investigated a representative sample of Hungarian students, conducted a latent profile analysis and obtained a prevalence estimate of 4.5% for problematic social media use. At present, it is still largely unclear how often problematic alcohol use and problematic use of the Internet use or specific online applications occur combined in adolescence.

According to the Problem-Behavior Theory of Jessor [3], associations between problematic alcohol and problematic use of the Internet or specific online applications seem not unlikely, but relatively few empirical findings on these relations have been published. Relationships between problematic Internet use and general substance use in adolescents have been investigated more frequently (see for example the systematic review of Lanthier-Labonté et al. [23]). However, it is important to distinguish between general alcohol consumption (which in Germany occurs among a majority of minors) and problematic alcohol use in adolescence (which only affects a minority, but which clearly can have serious consequences, see for instance the review of McCambridge et al. [24]).

In the few available studies, relations between problematic alcohol use and problematic Internet use in adolescence were usually examined in cross-sectional surveys. Pallanti et al. [25] observed a statistically significant positive correlation (0.38) between problematic alcohol use and problematic Internet use among Italian youth (average age: 16.67 years). In the study by Ha et al. [26], problematic alcohol use in South Korean adolescents (mean age: 15.8 years) did not occur more frequently in youth with problematic Internet use than among those without problematic Internet use. In contrast, Ko et al. [5] reported that problematic alcohol use occurred more often in Taiwanese adolescents with problematic Internet use than without problematic Internet use (average age of the sample was 16.26 years). Golpe et al. [27] found a statistically significant positive correlation (0.36) between problematic alcohol use and problematic Internet use among Spanish youth (mean age: 14.52 years). Rial et al. [28] obtained more problematic alcohol use among Spanish minors (average age: 14.41 years) with problematic Internet use than among adolescents without problematic Internet use. In addition, in a longitudinal study Gámez-Guadix et al. [29] showed that negative consequences of problematic Internet use predicted an increase in problematic alcohol use in Spanish minors (mean age: 14.92 years) six months later.

For relations between problematic alcohol use and problematic gaming there are so far only findings from studies with adults. Na et al. [30] observed relatively often (in 165 of the 1819 surveyed 20- to 49-year olds) a comorbid occurrence of problematic alcohol use and problematic gaming in South Korea and this comorbidity group showed more severe psychopathological impairments compared to the adults affected by only one of the two problem behavioral patterns. In contrast, the survey by Erevik et al. [31] showed a negative association between "high-level gaming" and problematic alcohol use among Norwegian university students (average age: 25.8 years). Problematic gaming was a protective factor against problematic alcohol use in this investigation. For relations between problematic alcohol use and problematic social media use only one result (also in adults) has been published so far [32]. Lyvers et al. [32] obtained a statistically significant positive correlation (0.42) between problematic alcohol use and problematic social media use in Australian adults (mean age: 26.1 years).

Furthermore, in both the I-PACE model ("psychopathology") [6] and the Problem-Behavior Theory (e.g., "low self-esteem") [3] associations to mental health were also mentioned. Problematic alcohol use (e.g., [33]), problematic gaming (e.g., [9]), problematic social media use (e.g., [9]) and problematic Internet use (e.g., [12]) are each associated with a higher psychopathological burden, but there are very few studies that have examined several of these patterns and their links to mental health in one sample (e.g., [33]).

To sum up, empirical evidence predominantly suggests relations between problematic alcohol use and problematic Internet use in adolescence [5,25,27–29]. Whether such associations exist between problematic alcohol use and problematic gaming or problematic social media use is currently unclear for youth. Accordingly, the present study is the first to investigate empirically associations between problematic alcohol use and problematic gaming and problematic social media use in adolescence. Furthermore, we wanted to explore and compare associations between problematic alcohol use, problematic gaming, problematic social media use, problematic Internet use and different aspects of mental health.

2. Materials and Methods

2.1. Research Questions

In the present study the following research questions (RQs) were examined:

RQ1 What is the relationship between problematic alcohol use and problematic gaming in adolescents?
RQ2 What is the relationship between problematic alcohol use and problematic social media use in adolescents?
RQ3 What is the relationship between problematic alcohol use and problematic Internet use in adolescents?
RQ4 What are the relationships between problematic alcohol use, problematic gaming, problematic social media use, problematic Internet use and different aspects of mental health?

2.2. Procedure

Data collection was conducted in Germany within the framework of the VEIF project in accordance with the Declaration of Helsinki. The Ethics Committee of the German Educational Research Association (Deutsche Gesellschaft für Erziehungswissenschaft, DGfE) approved the proceedings (approval number: 01/2018/DGfE). Informed consent was obtained both from the surveyed adolescents and from one of their parents. The findings reported below are based on nation-wide data collected in the first quarter of 2019 by a market research institute. The VEIF project is a longitudinal study in which data had previously been collected at annual intervals in the first quarter 2016, in the first quarter 2017 and in the first quarter 2018. At the time of the first data collection (2016), youth were aged between 12 and 14, and in 2019 between 14 and 17. As formulated in the research questions, relations between problematic alcohol use and problematic use of the Internet and specific online applications should be investigated. For the first time in the 2019 data collection, due to the increased age of the sample, a substantial and sufficient number of adolescents (n = 57) reported problematic alcohol use. The data collection was carried out in 2019 by 134 interviewers directly at the families' homes. Using computer-assisted personal interviewing (CAPI), one adolescent and one parent (dyad) were surveyed each. For the dyads, the parent was examined first and afterwards the adolescent. In the VEIF project, a sample with an increased risk of problematic use of digital media compared to the general population is examined. For this purpose, an oversampling of youth with an increased risk of problematic use of digital media was carried out before the first data collection in 2016 (a more detailed description of the study design and the recruitment process conducted at the beginning of the VEIF project can be found at Wartberg et al. [34]).

2.3. Measures

Problematic alcohol use in the last 12 months was measured with the Alcohol Use Disorders Identification Test-Consumption (AUDIT-C, Bush et al. [35]). We used the German version by Lampert and Kuntz [36]. The screening questionnaire comprises three items. In the first AUDIT-C question, the adolescents were surveyed how often they have a drink containing alcohol (five-step response format: 0 = "never", 1 = "monthly or less", 2 = "2–4 times a month", 3 = "2–3 times a week", 4 = "4 or more times a week"). In the second AUDIT-C question, the youth should rate how many drinks containing alcohol they consume on a typical day when they are drinking (five-step response format: 0 = "1–2", 1 = "3–4", 2 = "5–6", 3 = "7–9", 4 = "10 or more"). In the third AUDIT-C question, the adolescents were asked how often they consume six or more drinks on one occasion (five-step response format: 0 = "never", 1 = "less than monthly", 2 = "monthly", 3 = "weekly", 4 = "daily or almost daily"). An AUDIT-C sum value is formed by adding up the answers to all three questions and a higher sum value indicates a more pronounced problematic alcohol use. The reliability of the AUDIT-C was 0.72 in the investigated sample.

Problematic gaming in the past 12 months was assessed with the Internet Gaming Disorder Scale (IGDS, Lemmens et al. [37], we used the German version by Wartberg et al. [38]). The IGDS consists of nine questions with a binary answer format (0 = "no", 1 = "yes"). An IGDS total value is calculated from the nine items. A higher total figure indicates a more pronounced problematic gaming. The reliability of the IGDS was 0.82 in the examined sample.

Problematic social media use in the last 12 months was measured with the Social Media Disorder Scale (SMDS, van den Eijnden et al. [10]). The SMDS also comprises nine items with a binary response format (0 = "no", 1 = "yes"). An SMDS sum value can be calculated from the answers to the nine questions and a higher value indicates a more pronounced problematic social media use. The reliability of the SMDS was 0.83 in the surveyed sample.

To assess problematic Internet use, we utilized the Young Diagnostic Questionnaire (YDQ, Young [39], validation of the German version: Wartberg et al. [40]). The YDQ consists of eight items with a binary answer format (0 = "no", 1 = "yes") from which a sum value can be formed. A higher figure indicates a more pronounced problematic Internet use. The reliability of the YDQ was 0.70 in the investigated sample.

To assess adolescent mental health within the last six months, we applied the German adaption of the Reynolds Adolescent Adjustment Screening Inventory [41]: Screening psychischer Störungen im Jugendalter-II (SPS-J-II) [42]. The SPS-J-II consists of 32 questions (3-level response format: 0 = "never or almost never", 1 = "sometimes", 2 = "nearly all the time"). The questionnaire is divided into four scales assessing the frequency of adolescent antisocial behavior, anger control problems, emotional distress (combined measure of anxiety and depressiveness), and self-esteem problems. In each of the four scales, a higher sum value indicates a greater degree of adolescent psychopathological burden. We observed the following reliability coefficients (Cronbach's α): antisocial behavior: $\alpha = 0.77$, anger control problems: $\alpha = 0.79$, emotional distress: $\alpha = 0.88$ and self-esteem problems: $\alpha = 0.68$.

To collect a parental rating of adolescent hyperactivity/inattention over the last six months, we utilized the scale hyperactivity/inattention of the well-established Strengths and Difficulties Questionnaire (SDQ) [43]. This SDQ subscale comprises five questions with a 3-level response format (0 = "not true", 1 = "somewhat true", 2 = "certainly true"). A higher total value in this scale indicates a higher level of adolescent hyperactivity/inattention. In our sample, the reliability coefficient of the SDQ scale was $\alpha = 0.75$. Furthermore, socio-demographic characteristics (e.g., gender of the adolescent and age at the time of the survey) were collected.

2.4. Statistical Analyses

Overall, 319 adolescents had stated in the first question of the AUDIT-C [35] that they had never consumed alcohol in the last year. All these 319 cases did not have to answer the following two questions of the AUDIT-C and their AUDIT-C sum value was set to "0". Similarly, 40 minors had reported that they had never played video games in the last twelve months. These 40 cases did not have to answer the IGDS questions [37] and their IGDS total value was set to "0". Frequencies, mean values, standard deviations, reliability coefficients, correlations as well as multivariable linear regression analyses were calculated with SPSS version 25 (IBM, 2017, New York, NY, USA). First, bivariate correlation analyses were calculated. Subsequently, multivariable linear regression analyses were conducted (whereby all explanatory variables were simultaneously included in the regression models). In the four multivariable linear regression analyses problematic alcohol use, problematic gaming, problematic social media use, problematic Internet use were the dependent variable in one regression model each. In addition to gender, the different aspects of mental health (antisocial behavior, anger control problems, emotional distress, self-esteem problems, and hyperactivity/inattention) were utilized as explanatory variables.

3. Results

3.1. Descriptive Statistics

The sociodemographic characteristics of the total sample ($n = 633$) are presented in Table 1. In line with expectations (given the age of the sample and compulsory schooling in Germany), most of the adolescents interviewed were still attending school (93.5% of the total sample or 592 cases). For those youth who were still attending school, parents were asked to provide a prognosis for their future school leaving certificate (forecast). A total of 40 minors had already finished school (eight adolescents had achieved a graduation on a medium education level and another 32 adolescents on a high education level) and one girl did not go to school anymore (at the time of the survey she had no school leaving certificate).

Table 1. Sociodemographic characteristics of the sample.

Variable	Total Sample (n = 633) % or M (SD)
Gender	
Female	47.2%
Male	52.8%
Age [a]	15.79 (0.96)
Achieved or prospective level of graduation [b]	
Low-educational level	9.3%
Medium educational level	46.1%
High-educational level	44.5%

Note. [a] In years. [b] Prospective level: Forecast for all adolescents still attending school.

3.2. Bivariate Correlation Analyses

Answers to RQ1, RQ2 and RQ3: In the bivariate correlation analyses, we observed statistically significant positive associations between problematic alcohol use and problematic gaming behavior, problematic social media use and problematic Internet use (see Table 2). Furthermore, statistically significant positive relations between problematic gaming behavior, problematic social media use and problematic Internet use were found (see Table 2).

Table 2. Bivariate correlation analyses regarding the relationships of problematic alcohol use, problematic gaming behavior, problematic social media use and problematic Internet use in adolescents.

Variable	1	2	3	4
(1) Problematic alcohol use [a]	–			
(2) Internet gaming disorder [b]	0.12 **	–		
(3) Problematic social media use [c]	0.14 ***	0.45 ***	–	
(4) Problematic Internet use [d]	0.13 **	0.57 ***	0.63 ***	–

Note. [a] AUDIT-C sum value. [b] IGDS sum value. [c] SMDS sum value. [d] YDQ sum value; ** $p < 0.01$; *** $p < 0.001$.

3.3. Multiple Linear Regression Models

Answer to RQ4: In the multiple linear regression models problematic alcohol was associated with stronger antisocial behavior and lower anger control problems (see Table 3). Problematic gaming behavior was related to male gender and increased burden in all mental health aspects (antisocial behavior, anger control problems, emotional distress, self-esteem problems and hyperactivity/inattention) investigated. Problematic social media use was associated with female gender and higher levels of antisocial behavior, emotional distress, self-esteem problems and hyperactivity/inattention. Problematic Internet use was related to stronger antisocial behavior, higher emotional distress and more pronounced hyperactivity/inattention.

Table 3. Multiple linear regression models regarding the associations between mental health aspects and problematic alcohol use, problematic gaming behavior, problematic social media use and problematic Internet use in adolescents.

Mental Health Aspects	Problematic Alcohol Use	Internet Gaming Disorder	Problematic Social Media Use	Problematic Internet Use
	Standardized Beta Coefficients (95% CI)	Standardized Beta Coefficients (95% CI)	Standardized Beta Coefficients (95% CI)	Standardized Beta Coefficients (95% CI)
Gender [a]	−0.01 (−0.08; 0.07)	−0.33 *** (−0.39; −0.27)	0.18 *** (0.12; 0.24)	−0.02 (−0.08; 0.05)
Antisocial behavior	0.52 *** (0.42; 0.62)	0.18 *** (0.11; 0.26)	0.26 *** (0.17; 0.34)	0.19 *** (0.10; 0.27)
Anger control problems	−0.15 ** (−0.25; −0.04)	0.13 *** (0.04; 0.21)	0.08 (−0.01; 0.17)	0.02 (−0.08; 0.11)
Emotional distress	−0.06 (−0.15; 0.03)	0.09 * (0.01; 0.16)	0.13 ** (0.05; 0.20)	0.28 *** (0.20; 0.36)
Self-esteem problems	0.02 (−0.06; 0.10)	0.10 ** (0.03; 0.16)	0.12 ** (0.05; 0.18)	0.06 (−0.01; 0.13)
Hyperactivity/inattention	0.01 (−0.09; 0.10)	0.22 *** (0.15; 0.29)	0.26 *** (0.19; 0.34)	0.26 *** (0.19; 0.33)

Note. [a] Coding: 0 = male, 1 = female. * $p < 0.05$; ** $p < 0.01$; *** $p < 0.001$.

4. Discussion

In the present study, we investigated associations between problematic alcohol use and problematic use of the Internet and its specific applications (video games and social media) in adolescents. The first published findings already suggested relations between problematic alcohol use and problematic Internet use in adolescence [5,25,27–29]. We also found relationships between problematic alcohol use (operationalized via the established screening instrument AUDIT-C, which is recommended for use in adolescents, for example by Rumpf et al. [44]) and problematic Internet use (operationalized via the internationally frequently used YDQ [39]) in our sample of adolescents. Accordingly, our findings complement and confirm the international state of research on problematic alcohol use and problematic Internet use in adolescence.

Going beyond the current state of research, this study was the first to show empirically associations between problematic alcohol use and problematic gaming or problematic social media use in adolescence (a few findings have been reported from samples of adults). According to published empirical results (cross-sectional: e.g., Rosenkranz et al. [12], longitudinal: Wartberg et al. [13]), it does not seem advisable to transfer findings for a general problematic Internet use to a problematic use of specific online applications without further examination. In the context of the present study, it also seems helpful to distinguish between general problematic Internet use (in the cognitive-behavioral model of Davis [45] that would correspond to a "generalized pathological Internet use") and the problematic use of specific applications (such as video games or social media, in Davis' model [45] that could be a "specific pathological Internet use" each).

Regarding the associations between problematic alcohol use and problematic gaming, there have so far been only two results in adults [30,31]. However, these studies revealed heterogeneous findings. While the study by Na et al. [30] showed a positive relation between problematic alcohol use and problematic gaming, Erevik et al. [31] observed a negative association. The present survey could not confirm Erevik et al.'s finding [31] that problematic gaming in adults functioned as a protective factor against problematic alcohol use. This result of the present investigation indicates a positive relationship between problematic alcohol use and problematic gaming in adolescence, but naturally requires verification in further empirical surveys. Concerning the relation between problematic alcohol use and problematic social media use, only one study in adults [32] is available. Lyvers et al. [32] reported a positive correlation between both problematic behavioral patterns, which we observed in the present study also for adolescents. Again, this finding needs to be verified in future studies.

According to Jessors Problem-Behavior Theory [3], problematic behavior in adolescence often does not occur in isolation but in combination (this applies not only to problematic alcohol use but also to the examined three problematic patterns of Internet use or its applications). However, a combined survey of substance-unrelated problematic behavioral patterns has so far only been carried out very rarely. Estévez et al. [46] investigated the importance of emotion regulation in a mixed sample of adolescents and young adults in Spain for various problematic behaviors related and unrelated to substance use. In the correlation table of all constructs studied, positive statistically significant correlations were found for alcohol abuse with both problematic Internet use (0.28) and video game addiction (0.13) and between problematic Internet use and video game addiction (0.35) [46]. In our sample of adolescents, we also observed statistically significant correlations between problematic alcohol use and problematic gaming ($r = 0.12$), problematic social media use ($r = 0.14$) and problematic Internet use ($r = 0.13$, see Table 2). However, it should be noted that, according to the classification of Cohen [47], these correlations were small, while the correlation between problematic gaming and problematic social media use was moderate ($r = 0.45$). There were large correlations between problematic Internet use and problematic gaming ($r = 0.57$) as well as between problematic Internet use and problematic social media use ($r = 0.63$). Correspondingly, the associations between problem behaviors without substance relations were much more pronounced than to the substance-related problem behavior. Large correlation coefficients between problematic Internet use and problematic gaming in adolescents have been reported before by Király et al. [48] with 0.59 (p. 752) and by van Rooij et al. [49] with 0.63 (p. 509).

With regard to the associations between the problematic behaviors and mental health aspects (as described previously in the Problem-Behavior Theory [3] and the I-PACE model [6]), we observed both similarities and differences. More pronounced antisocial behavior was related to problematic alcohol use, problematic gaming behavior, problematic social media use and problematic Internet use and therefore seems to be consistently relevant for all four problematic behavioral patterns. Whereas, emotional distress, self-esteem problems and hyperactivity/inattention were associated with substance-unrelated problematic behavior patterns, but not with problematic alcohol use. This finding is a further contribution to the question of whether problematic alcohol consumption is related to depressiveness, where the published empirical findings are still heterogeneous (e.g., [50]). More anger control problems were associated with problematic gaming, but surprisingly less pronounced anger control problems with problematic alcohol use. The finding that problematic alcohol use is associated with fewer anger control problems and more pronounced antisocial behavior is novel and needs further investigation. Whereas the relationship between a problematic alcohol use and more externalizing problems (e.g., conduct problems or antisocial behavior) is considered empirically well established.

The present survey has various limitations. The VEIF project does not examine a representative sample, because a higher percentage of youth with a higher risk of problematic use of digital media was included than in the general population (oversampling). This may limit the transferability of the results to the general population as well as to other age groups (e.g., adults). Findings of a cross-sectional analysis were reported. Therefore, no cause-effect relationships or conclusions that one behavior leads to another can be deduced between the characteristics examined (this would require a longitudinal design). In addition to the investigated behavior patterns, further problematic behavior patterns unrelated to substance use are conceivable (e.g., problematic gambling), which could also be relevant for problematic alcohol use in adolescence, but were not considered in the study design of the VEIF project. Therefore, we cannot provide a complete model to explain adolescent problematic alcohol use (too many relevant variables, such as substance use of the peers, were not collected).

5. Conclusions

Despite the limitations mentioned above, the present investigation revealed some interesting new findings. The empirical associations between problematic alcohol use and a problematic gaming and a problematic social media use among adolescents extended existing research. With regard to a common etiology and for the development of comprehensive prevention programs (e.g., life skills trainings), it seems quite promising to examine several problematic behavior patterns and their associations to mental health together. Therefore, the findings of the present survey can be used to develop or revise preventive measures. Furthermore, if there are indications of problem behaviors in youth related or unrelated to substance use, they should be screened for further substance-related or non-substance-related problematic behavior patterns as part of a comprehensive diagnostic procedure (e.g., before the start of an intervention measure in an outpatient or inpatient setting) in order to exclude or take them into account as comorbidities in treatment.

Author Contributions: Conceptualization, R.K. and L.W.; methodology, L.W.; software, L.W.; validation, R.K. and L.W.; formal analysis, L.W.; investigation, R.K. and L.W.; resources, R.K. and L.W.; data curation, L.W.; writing—original draft preparation, L.W.; writing—review and editing, R.K.; visualization, L.W.; supervision, R.K.; project administration, R.K.; funding acquisition, R.K. and L.W. All authors have read and agreed to the published version of the manuscript.

Funding: This research was funded by the German Research Foundation (DFG, grant number KA 1611/6-2) and the APC was funded by the Open Access Publication Fund of the Friedrich-Alexander-University Erlangen-Nuremberg.

Acknowledgments: We thank all adolescents and parents who participated in the study.

Conflicts of Interest: The authors declare no conflict of interest.

References

1. Faltermaier, T. *Gesundheitspsychologie [Health Psychology]*; Kohlhammer: Stuttgart, Germany, 2017.
2. Scheithauer, H.; Hayer, T.; Niebank, K. *Problemverhalten und Gewalt im Jugendalter: Erscheinungsformen, Entstehungsbedingungen, Prävention und Intervention*; Kohlhammer: Stuttgart, Germany, 2008.
3. Jessor, R. Risk behavior in adolescence: A psychosocial framework for understanding and action. *J. Adolesc. Health* **1991**, *12*, 597–605. [CrossRef]
4. De Leo, J.A.; Wulfert, E. Problematic Internet use and other risky behaviors in college students: An application of problem-behavior theory. *Psychol. Addict. Behav.* **2013**, *27*, 133–141. [CrossRef]
5. Ko, C.H.; Yen, J.Y.; Yen, C.F.; Chen, C.S.; Weng, C.C.; Chen, C.C. The Association between Internet Addiction and Problematic Alcohol Use in Adolescents: The Problem Behavior Model. *Cyberpsychol. Behav.* **2008**, *11*, 571–576. [CrossRef] [PubMed]
6. Brand, M.; Young, K.S.; Laier, C.; Wölfling, K.; Potenza, M.N. Integrating psychological and neurobiological considerations regarding the development and maintenance of specific Internet-use disorders: An Interaction of Person-Affect-Cognition-Execution (I-PACE) model. *Neurosci. Biobehav. Rev.* **2016**, *71*, 252–266. [CrossRef] [PubMed]
7. American Psychiatric Association. *Diagnostic and Statistical Manual of Mental Disorders, DSM-5*, 5th ed.; American Psychiatric Publishing: Washington, DC, USA, 2013; ISBN 978-0890425558.
8. WHO. Gaming Disorder. 2018. Available online: who.int/news-room/q-a-detail/gaming-disorder (accessed on 25 May 2020).
9. Mérelle, S.M.Y.; Kleiboer, A.M.; Schotanus, M.; Cluitmans, T.L.M.; Waardenburg, C.M.; Kramer, D.; van de Mheen, D.; van Rooij, T. Which health-related problems are associated with problematic video-gaming or social media use in adolescents? A large-scale cross-sectional public health study. *Clin. Neuropsychiatry* **2017**, *14*, 11–19.
10. Van den Eijnden, R.J.; Lemmens, J.S.; Valkenburg, P.M. The social media disorder scale. *Comput. Hum. Behav.* **2016**, *61*, 478–487. [CrossRef]
11. Lopez-Fernandez, O. Generalised Versus Specific Internet Use-Related Addiction Problems: A Mixed Methods Study on Internet, Gaming, and Social Networking Behaviours. *Int. J. Environ. Res. Public Health* **2018**, *15*, 2913. [CrossRef]
12. Rosenkranz, T.; Müller, K.W.; Dreier, M.; Beutel, M.E.; Wölfling, K. Addictive potential of internet applications and differential correlates of problematic use in internet gamers versus generalized internet users in a representative sample of adolescents. *Eur. Addict. Res.* **2017**, *23*, 148–156. [CrossRef]
13. Wartberg, L.; Zieglmeier, M.; Kammerl, R. An Empirical Exploration of Longitudinal Predictors for Problematic Internet Use and Problematic Gaming Behavior. *Psychol. Rep.* **2020**. [CrossRef]
14. Orth, B.; Merkel, C. *Der Alkoholkonsum Jugendlicher und Junger Erwachsener in Deutschland. Ergebnisse des Alkoholsurveys 2018 und Trends*; BZgA-Forschungsbericht: Köln, Germany, 2019. [CrossRef]
15. EMCDDA and ESPAD. *ESPAD Report 2015—Results from the European School Survey Project on Alcohol and Other Drugs*; Publications Office of the European Union: Luxembourg, 2016.
16. Wartberg, L.; Kriston, L.; Thomasius, R. Prevalence of problem drinking and associated factors in a representative German sample of adolescents and young adults. *J. Public Health* **2019**, *41*, 543–549. [CrossRef]

17. Wartberg, L.; Kriston, L.; Kammerl, R.; Petersen, K.-U.; Thomasius, R. Prevalence of pathological Internet use in a representative German sample of adolescents: Results of a latent profile analysis. *Psychopathology* **2015**, *48*, 25–30. [CrossRef] [PubMed]
18. Rumpf, H.-J.; Vermulst, A.A.; Bischof, A.; Kastirke, N.; Gürtler, D.; Bischof, G.; Meerkerk, G.-J.; John, U.; Meyer, C. Occurence of internet addiction in a general population sample: A latent class analysis. *Eur. Addict. Res.* **2014**, *20*, 159–166. [CrossRef]
19. Wartberg, L.; Kriston, L.; Bröning, S.; Kegel, K.; Thomasius, R. Adolescent problematic Internet use: Is a parental rating suitable to estimate prevalence and identify familial correlates? *Comput. Hum. Behav.* **2017**, *67*, 233–239. [CrossRef]
20. Pontes, H.M.; Macur, M.; Griffiths, M.D. Internet Gaming Disorder among Slovenian primary schoolchildren: Findings from a nationally representative sample of adolescents. *J. Behav. Addict.* **2016**, *5*, 304–310. [CrossRef] [PubMed]
21. Wartberg, L.; Kriston, L.; Thomasius, R. Internet gaming disorder and problematic social media use in a representative sample of German adolescents: Prevalence estimates, comorbid depressive symptoms and related psychosocial aspects. *Comput. Hum. Behav.* **2020**, *103*, 31–36. [CrossRef]
22. Bányai, F.; Zsila, Á.; Király, O.; Maraz, A.; Elekes, Z.; Griffiths, M.D.; Andreassen, C.S.; Demetrovics, Z. Problematic social media use: Results from a large-scale nationally representative adolescent sample. *PLoS ONE* **2017**, *12*, e0169839. [CrossRef]
23. Lanthier-Labonté, S.; Dufour, M.; Milot, D.M.; Loslier, J. Is problematic Internet use associated with alcohol and cannabis use among youth? A systematic review. *Addict. Behav.* **2020**, *106*, 106331. [CrossRef]
24. McCambridge, J.; McAlaney, J.; Rowe, R. Adult consequences of late adolescent alcohol consumption: A systematic review of cohort studies. *PLoS Med.* **2011**, *8*, e1000413. [CrossRef]
25. Pallanti, S.; Bernardi, S.; Quercioli, L. The Shorter PROMIS Questionnaire and the Internet Addiction Scale in the assessment of multiple addictions in a high-school population: Prevalence and related disability. *CNS Spectr.* **2006**, *11*, 966–974. [CrossRef]
26. Ha, J.H.; Kim, S.Y.; Bae, S.C.; Bae, S.; Kim, H.; Sim, M.; Lyoo, I.K.; Cho, S.C. Depression and Internet addiction in adolescents. *Psychopathology* **2007**, *40*, 424–430. [CrossRef]
27. Golpe, S.; Gómez, P.; Braña, T.; Varela, J.; Rial, A. The relationship between consumption of alcohol and other drugs and problematic Internet use among adolescents. *Adicciones* **2017**, *29*, 268–277. [CrossRef] [PubMed]
28. Rial, A.; Golpe, S.; Isorna, M.; Braña, T.; Gómez, P. Minors and problematic Internet use: Evidence for better prevention. *Comput. Hum. Behav.* **2018**, *87*, 140–145. [CrossRef]
29. Gámez-Guadix, M.; Calvete, E.; Orue, I.; Las-Hayas, C. Problematic Internet use and problematic alcohol use from the cognitive–behavioral model: A longitudinal study among adolescents. *Addict. Behav.* **2015**, *40*, 109–114. [CrossRef] [PubMed]
30. Na, E.; Lee, H.; Choi, I.; Kim, D.J. Comorbidity of Internet gaming disorder and alcohol use disorder: A focus on clinical characteristics and gaming patterns. *Am. J. Addict.* **2017**, *26*, 326–334. [CrossRef]
31. Erevik, E.K.; Torsheim, T.; Andreassen, C.S.; Krossbakken, E.; Vedaa, Ø.; Pallesen, S. The associations between low-level gaming, high-level gaming and problematic alcohol use. *Addict. Behav. Rep.* **2019**, *10*, 100186. [CrossRef]
32. Lyvers, M.; Narayanan, S.S.; Thorberg, F.A. Disordered social media use and risky drinking in young adults: Differential associations with addiction-linked traits. *Aust. J. Psychol.* **2019**, *71*, 223–231. [CrossRef]
33. Wartberg, L.; Brunner, R.; Kriston, L.; Durkee, T.; Parzer, P.; Fischer-Waldschmidt, G.; Resch, F.; Sarchiapone, M.; Wasserman, C.; Hoven, C.W.; et al. Psychopathological factors associated with problematic alcohol and problematic Internet use in a sample of adolescents in Germany. *Psychiatry Res.* **2016**, *240*, 272–277. [CrossRef]
34. Wartberg, L.; Kriston, L.; Kammerl, R. Associations of social support, friends only known through the Internet, and health-related quality of life with Internet Gaming Disorder in adolescence. *Cyberpsychol. Behav. Soc. Netw.* **2017**, *20*, 436–441. [CrossRef]
35. Bush, K.; Kivlahan, D.R.; McDonell, M.B.; Fihn, S.D.; Bradley, K.A. The AUDIT alcohol consumption questions (AUDIT-C): An effective brief screening test for problem drinking. *Arch. Intern. Med.* **1998**, *158*, 1789–1795. [CrossRef]

36. Lampert, T.; Kuntz, B. Tabak- und Alkoholkonsum bei 11- bis 17-jährigen Jugendlichen. Ergebnisse der KiGGS-Studie—Erste Folgebefragung (KiGGS Welle 1). *Bundesgesundheitsblatt Gesundh. Gesundh.* **2014**, *57*, 830–839. [CrossRef]
37. Lemmens, J.S.; Valkenburg, P.M.; Gentile, D.A. The Internet Gaming Disorder Scale. *Psychol. Assess.* **2015**, *27*, 567–582. [CrossRef] [PubMed]
38. Wartberg, L.; Kriston, L.; Thomasius, R. The Prevalence and Psychosocial Correlates of Internet Gaming Disorder. *Dtsch. Arztebl. Int.* **2017**, *114*, 419–424. [CrossRef] [PubMed]
39. Young, K.S. Internet Addiction: The Emergence of a New Clinical Disorder. *Cyberpsychol. Behav.* **1998**, *1*, 237–244. [CrossRef]
40. Wartberg, L.; Durkee, T.; Kriston, L.; Parzer, P.; Fischer-Waldschmidt, G.; Resch, F.; Sarchiapone, M.; Wasserman, C.; Hoven, C.W.; Carli, V.; et al. Psychometric Properties of a German Version of the Young Diagnostic Questionnaire (YDQ) in two Independent Samples of Adolescents. *Int. J. Ment. Health Addict.* **2017**, *15*, 182–190. [CrossRef]
41. Reynolds, W.M. *Reynolds Adolescent Adjustment Screening Inventory™ (RAASI™): Professional Manual*; Psychological Assessment Resources: Lutz, Germany, 2001.
42. Hampel, P.; Petermann, F. Screening psychischer Störungen im Jugendalter-II (SPS-J-II). In *Deutschsprachige Adaptation des Reynolds Adolescent Adjustment Screening Inventory (RAASI) von William M. Reynolds (2. erweiterte Auflage)*, 2nd ed.; Huber: Bern, Switzerland, 2012.
43. Goodman, R. The Strengths and Difficulties Questionnaire: A research note. *J. Child Psychol. Psychiatry* **1997**, *38*, 581–586. [CrossRef]
44. Rumpf, H.J.; Wohlert, T.; Freyer-Adam, J.; Grothues, J.; Bischof, G. Screening Questionnaires for Problem Drinking in Adolescents: Performance of AUDIT, AUDIT-C, CRAFFT and POSIT. *Eur. Addict. Res.* **2013**, *19*, 121–127. [CrossRef]
45. Davis, R.A. A cognitive-behavioral model of pathological Internet use. *Comput. Hum. Behav.* **2001**, *17*, 187–195. [CrossRef]
46. Estévez, A.; Jauregui, P.; Sanchez-Marcos, I.; Lopez-Gonzalez, H.; Griffiths, M.D. Attachment and emotion regulation in substance addictions and behavioral addictions. *J. Behav. Addict.* **2017**, *6*, 534–544. [CrossRef]
47. Cohen, J. *Statistical Power Analysis for the Behavioral Sciences*; Lawrence Erlbaum Associates: Hillsdale, MI, USA, 1988.
48. Király, O.; Griffiths, M.D.; Urbán, R.; Farkas, J.; Kökönyei, G.; Elekes, Z.; Tamás, D.; Demetrovics, Z. Problematic Internet use and problematic online gaming are not the same: Findings from a large nationally representative adolescent sample. *Cyberpsychol. Behav. Soc. Netw.* **2014**, *17*, 749–754. [CrossRef]
49. van Rooij, A.J.; Schoenmakers, T.M.; Van den Eijnden, R.J.J.M.; Vermulst, A.A.; van de Mheen, D. Video game addiction test: Validity and psychometric characteristics. *Cyberpsychol. Behav. Soc. Netw.* **2012**, *15*, 507–511. [CrossRef]
50. Pesola, F.; Shelton, K.H.; Heron, J.; Munafò, M.; Hickman, M.; van den Bree, M.B. The developmental relationship between depressive symptoms in adolescence and harmful drinking in emerging adulthood: The role of peers and parents. *J. Youth Adolesc.* **2015**, *44*, 1752–1766. [CrossRef] [PubMed]

MDPI
St. Alban-Anlage 66
4052 Basel
Switzerland
Tel. +41 61 683 77 34
Fax +41 61 302 89 18
www.mdpi.com

International Journal of Environmental Research and Public Health Editorial Office
E-mail: ijerph@mdpi.com
www.mdpi.com/journal/ijerph

www.ingramcontent.com/pod-product-compliance
Lightning Source LLC
Chambersburg PA
CBHW061451220926
51850CB00004B/66
9783036512747